Probability Theory and
Mathematical Statistics

Probability Theory and Mathematical Statistics

MAREK FISZ

Late Professor of Mathematics
New York University

THIRD EDITION

John Wiley & Sons, Inc.
New York • London • Sydney

Published in colloboration with PWN, Polish Scientific Publishers

AUTHORIZED TRANSLATION FROM THE POLISH.

TRANSLATED BY R. BARTOSZYNSKI

THE FIRST AND SECOND EDITIONS OF THIS BOOK
WERE PUBLISHED AND COPYRIGHTED IN POLAND UNDER THE TITLE

RACHUNEK PRAWDOPODOBIENSTWA I STATYSTYKA
MATEMATYCZNA

THIRD PRINTING, JANUARY, 1967

LIBRARY OF CONGRESS CATALOG CARD NUMBER: 63-7554

PRINTED IN THE UNITED STATES OF AMERICA

To Olga and Aleksander

Preface to the English Edition

The opening sentence of this book is "Probability theory is a part of mathematics which is useful in discovering and investigating the regular features of random events." However, even in the not very remote past this sentence would not have been found acceptable (or by any means evident) either by mathematicians or by researchers applying probability theory. It was not until the twenties and thirties of our century that the very nature of probability theory as a branch of mathematics and the relation between the concept of "probability" and that of "frequency" of random events was thoroughly clarified. The reader will find in Section 1.3 an extensive (although, certainly, not exhaustive) list of names of researchers whose contributions to the field of basic concepts of probability theory are important. However, the foremost importance of Laplace's *Théorie Analytique des Probabilités*, von Mises' *Wahrscheinlichkeit, Statistik und Wahrheit*, and Kolmogorov's *Grundbegriffe der Wahrscheinlichkeitsrechnung* should be stressed. With each of these three works, a new period in the history of probability theory was begun. In addition, the work of Steinhaus' school of *independent functions* contributed greatly to the clarification of fundamental concepts of probability theory.

The progress in foundations of probability theory, along with the introduction of the theory of characteristic functions, stimulated the exceedingly fast development of modern probability theory. In the field of limit theorems for sums of random variables, a fairly general theory was developed (Khintchin, Lévy, Kolmogorov, Feller, Gnedenko, and others) for independent random variables, whereas for dependent random variables some important particular results were obtained (Bernstein, Markov, Doeblin, Kolmogorov, and others). Furthermore, the theory of stochastic processes became a mathematically rigorous branch of probability theory (Markov, Khintchin, Lévy, Wiener, Kolmogorov, Feller, Cramér, Doob, and others).

Some ideas on application of probability theory that are now used in mathematical statistics are due to Bayes (estimation theory), Laplace (quality control of drugs), and to Gauss (theory of errors). However, it was not until this century that mathematical statistics grew into a self-contained scientific subject. In order to restrict myself to the names of just a few of the principal persons responsible for this growth, I mention only K. Pearson, R. A. Fisher, J. Neyman, and A. Wald, whose ideas and systematic research contributed so much to the high status of modern mathematical statistics.

At present, the development of the theory of probability and mathematical statistics is going on with extreme intensity. On the one hand, problems in classical probability theory unsolved, as of now, are attracting much attention, whereas, on the other hand, much work is being done in an attempt to obtain highly advanced generalizations of old concepts, particularly by considering probability theory in spaces more general than the finite dimensional Euclidean spaces usually treated. That probability theory is now closely connected with other parts of mathematics is evidenced by the fact that almost immediately after the formulation of the distribution theories by L. Schwartz and Mikusinski their probabilistic counterparts were thoroughly discussed (Gelfand, Itô, Urbanik). Moreover, probability theory and mathematical statistics are no longer simply "customers" of other parts of mathematics. On the contrary, mutual influence and interchange of ideas between probability theory and mathematical statistics and the other areas of mathematics are constantly going on. One example is the relation between analytic number theory and probability theory (Borel, Khintchin, Linnik, Erdös, Kac, Rényi, and others), and another example is the use of the theory of games in mathematical statistics and its stimulating effect on the development of the theory of games itself (v. Neumann, Morgenstern, Wald, Blackwell, Karlin, and others).

Areas of application of probability theory and mathematical statistics are increasing more and more. Statistical methods are now widely used in physics, biology, medicine, economics, industry, agriculture, fisheries, meteorology, and in communications. Statistical methodology has become an important component of scientific reasoning, as well as an integral part of well-organized business and social work.

The development of probability theory and mathematical statistics and their applications is marked by a constantly increasing flood of scientific papers that are published in many journals in many countries.

* * *

Having described briefly the position of modern probability theory and mathematical statistics, I now state the main purpose of this book:

1. To give a systematic introduction to modern probability theory and mathematical statistics.

2. To present an outline of many of the possible applications of these theories, accompanied by descriptive concrete examples.

3. To provide extensive (however, not exhaustive) references of other books and papers, mostly with brief indications as to their contents, thereby giving the reader the opportunity to complete his knowledge of the subjects considered.

Although great care has been taken to make this book mathematically rigorous, the intuitive approach as well as the applicability of the concepts and theorems presented are heavily stressed.

For the most part, the theorems are given with complete proofs. Some proofs, which are either too lengthy or require mathematical knowledge far beyond the scope of this book, were omitted.

The entire text of the book may be read by students with some background in calculus and algebra. However, no advanced knowledge in these fields or a knowledge in measure and integration theory is required. Some necessary advanced concepts (for instance, that of the Stieltjes integral) are presented in the text. Furthermore, this book is provided with a Supplement, in which some basic concepts and theorems of modern measure and integration theory are presented.

Every chapter is followed by "Problems and Complements." A large part of these problems are relatively easy and are to be solved by the reader, with the remaining ones given for information and stimulation.

This book may be used for systematic one-year courses either in probability theory or in mathematical statistics, either for senior undergraduate or graduate students. I have presented parts of the material, covered by this book, in courses at the University of Warsaw (Poland) for nine academic years, from 1951/1952 to 1959/1960, at the Peking University (China) in the Spring term of 1957, and in this country at the University of Washington and at Stanford, Columbia and New York Universities for the last several years.

This book is also suitable for nonmathematicians, as far as concepts, theorems, and methods of application are concerned.

* * *

I started to write this book at the end of 1950. Its first edition (374 pages) was published in Polish in 1954. All copies were sold within a few months. I then prepared the second, revised and extended, Polish edition, which was published in 1958, simultaneously with its German translation. Indications about changes and extensions introduced into the second Polish edition are given in its preface. In comparison with the second Polish edition, the present English one contains many extensions and changes, the most important of which are the following:

Entirely new are the Problems and Complements, the Supplement, and the Sections: 2.7,C, 2.8,C, 3.2,C, 3.6,G, 4.6,B, 6.4,B, 6.4,C, 6.12,D, 6.12,F, 6.15, 8.11, 9.4,B, 9.6,E, 9.9,B, 10.10,B, 10.11,E, 12.6,B, 13.5,E, 13.7,D, 14.2,E, 14.4,D, 15.1,C, 15.3,C, 16.3,D, 16.6, and 17.10,A. Section 8.10 (Stationary processes) is almost entirely new.

Considerably changed or complemented are Sections: 2.5,C, 3.5, 3.6,C, 4.1, 4.2, 5.6,B, 5.7, 5.13,B, 6.2, 6.4,A, 6.5, 6.12,E, 6.12,G, 7.5,B, 8.4,D, 8.8,B, 8.12, 9.1, 9.7, 10.12, 10.13, 12,4,C, 12.4,D, 13.3, 16.2,C.

These changes and extensions have all been made to fulfill more completely the main purpose of this book, as stated previously.

* * *

J. Lukaszewicz, A. M. Rusiecki, and W. Sadowski read the manuscript of the first edition and suggested many improvements. The criticism of E. Marczewski and the reviews by Z. W. Birnbaum and S. Zubrzycki of the first edition were useful in preparing the second edition; also useful were valuable remarks of K. Urbanik, who read the manuscript of the second edition. Numerous remarks and corrections were suggested by J. Wojtyniak, and R. Zasępa (first edition), and by L. Kubik, R. Sulanke, and J. Wloka (second edition). R. Bartoszynski, with the substantial collaboration of Mrs H. Infeld, translated the book from the Polish. J. Karush made valuable comments about the language. Miss D. Garbose did the editorial work. B. Eisenberg assisted me in the reading of the proofs. My sincere thanks go to all these people.

New York   MAREK FISZ
October, 1962

Contents

PART 2   MATHEMATICAL STATISTICS

PART 1

Probability
Theory

CHAPTER 1

Random Events

1.1 PRELIMINARY REMARKS

A   Probability theory is a part of mathematics which is useful in discovering and investigating the regular features of random events. The following examples show what is ordinarily understood by the term random event.

Example 1.1.1. Let us toss a symmetric coin. The result may be either a head or a tail. For any one throw, we cannot predict the result, although it is obvious that it is determined by definite causes. Among them are the initial velocity of the coin, the initial angle of throw, and the smoothness of the table on which the coin falls. However, since we cannot control all these parameters, we cannot predetermine the result of any particular toss. Thus the result of a coin tossing, head or tail, is a random event.

Example 1.1.2. Suppose that we observe the average monthly temperature at a definite place and for a definite month, for instance, for January in Warsaw.[1] This average depends on many causes such as the humidity and the direction and strength of the wind. The effect of these causes changes year by year. Hence Warsaw's average temperature in January is not always the same. Here we can determine the causes for a given average temperature, but often we cannot determine the reasons for the causes themselves. As a result, we are not able to predict with a sufficient degree of accuracy what the average temperature for a certain January will be. Thus we refer to it as a random event.

B   It might seem that there is no regularity in the examples given. But if the number of observations is large, that is, if we deal with a mass phenomenon, some regularity appears.

Let us return to example 1.1.1. We cannot predict the result of any particular toss, but if we perform a long series of tossings, we notice that the number of times heads occur is approximately equal to the number of times tails appear. Let n denote the number of all our tosses and m the number of times heads appear. The fraction m/n is called the

[1] See example 12.5.1.

3

frequency of appearance of heads. The frequency of appearance of tails is given by the fraction $(n - m)/n$. Experience shows that if n is sufficiently large, thus if the tossings may be considered as a mass phenomenon, the fractions m/n and $(n - m)/n$ differ little; hence each of them is approximately $\frac{1}{2}$. This regularity has been noticed by many investigators who have performed a long series of coin tossings. Buffon

TABLE 1.1.1

FREQUENCY OF BIRTHS OF BOYS AND GIRLS

Year of Birth	Number of Births		Total Number of Births	Frequency of Births	
	Boys m	Girls f	$m + f$	Boys p_1	Girls p_2
1927	496,544	462,189	958,733	0.518	0.482
1928	513,654	477,339	990,993	0.518	0.482
1929	514,765	479,336	994,101	0.518	0.482
1930	528,072	494,739	1,022,811	0.516	0.484
1931	496,986	467,587	964,573	0.515	0.485
1932	482,431	452,232	934,663	0.516	0.484
Total	3,032,452	2,833,422	5,865,874	0.517	0.483

tossed a coin 4040 times, and obtained heads 2048 times; hence the ratio of heads was $m/n = 0.50693$. In 24,000 tosses, K. Pearson obtained a frequency of heads equal to 0.5005. We can see quite clearly that the observed frequencies oscillate about the number 0.5.

As a result of long observation, we can also notice certain regularities in example 1.1.2. We investigate this more closely in example 12.5.1.

Example 1.1.3. We cannot predict the sex of a newborn baby in any particular case. We treat this phenomenon as a random event. But if we observe a large number of births, that is, if we deal with a mass phenomenon, we are able to predict with considerable accuracy what will be the percentages of boys and girls among all newborn babies. Let us consider the number of births of boys and girls in Poland in the years 1927 to 1932. The data are presented in Table 1.1.1.

In this table m and f denote respectively the number of births of boys and girls in particular years. Denote the frequencies of births by p_1 and p_2, respectively; then

$$p_1 = \frac{m}{m + f}, \qquad p_2 = \frac{f}{m + f}$$

One can see that the values of p_1 oscillate about the number 0.517, and the values of p_2 oscillate about the number 0.483.

Example 1.1.4. We throw a die. As a result of a throw one of the faces $1, \ldots, 6$ appears. The appearance of any particular face is a random event. If, however, we perform a long series of throws, observing all those which give face one as a result, we will notice that the frequency of this event will oscillate about the number $\frac{1}{6}$. The same is true for any other face of the die.

This observed regularity, that the frequency of appearance of any random event oscillates about some fixed number when the number of experiments is large, is the basis of the notion of probability.

Concluding these preliminary remarks, let us stress the fact that the theory of probability is applicable only to events whose frequency of appearance can (under certain conditions) be either directly or indirectly observed or deduced by logical analysis.

1.2 RANDOM EVENTS AND OPERATIONS PERFORMED ON THEM

A   We now construct the mathematical definition of a random event, the colloquial meaning of which was discussed in the preceding section.

The primitive notion of the axiomatic theory of probability is that of the *set of elementary events*. This set is denoted by E.

For every particular problem we must decide what is called the elementary event; this determines the set E.

Example 1.2.1. Suppose that when throwing a die we observe the frequency of the event, an even face. Then, the appearance of any particular face i, where $i = 1, \ldots, 6$, is an elementary event, and is denoted by e_i. Thus the whole set of elementary events contains 6 elements.

In our example we are investigating the random event A that an even face will appear, that is, the event consisting of the elementary events, face 2, face 4, and face 6. We denote such an event by the symbol (e_2, e_4, e_6). The random event (e_2, e_4, e_6) occurs if and only if the result of a throw is either face 2 or face 4 or face 6.

If we wish to observe the appearance of an arbitrary face which is not face 1, we will have a random event consisting of five elements $(e_2, e_3, e_4, e_5, e_6)$.

Let us form the set Z of random events which in this example is the set of all subsets of E.

We include in Z all the single elements of E: $(e_1), (e_2), (e_3), (e_4), (e_5), (e_6)$, where for instance, the random event (e_4) is simply the appearance of the elementary event, face 4.

Besides the 6 one-element random events $(e_1), \ldots, (e_6)$, there also belong to the set Z 15 two-element subsets $(e_1, e_2), \ldots, (e_5, e_6)$, 20 three-element subsets $(e_1, e_2, e_3), \ldots, (e_4, e_5, e_6)$, 15 four-element subsets $(e_1, e_2, e_3, e_4), \ldots, (e_3, e_4, e_5, e_6)$, and 6 five-element subsets $(e_1, e_2, e_3, e_4, e_5), \ldots, (e_2, e_3, e_4, e_5, e_6)$. But these are not all.

Now consider the whole set E as an event. It is obvious that as a result of a throw we shall certainly obtain one of the faces $1, \ldots, 6$, that is, we are sure that one of the elementary events of the set E will occur. Usually, if the occurrence of an event is sure, we do not consider it a random event; nevertheless we shall consider a sure event as a random event and include it in the set Z of random events.

Finally, in throwing a die, consider the event of a face with more than 6 dots appearing. This event includes no element of E; hence as a subset of E, it is an empty set. Such an event is, of course, impossible, and usually is not considered as a random event. However, we shall consider it as a random event and we shall include it in the set Z of random events, denoting it by the symbol (0).

Including the impossible and sure events, the set Z of random events in our example has 64 elements.

Generally, if the set E contains n elements, then the set Z of random events contains 2^n elements, namely,

1 impossible event   (empty set),

$\binom{n}{1}$ one-element events,

$\binom{n}{2}$ two-element events,

. . .

$\binom{n}{n-1}$ $(n-1)$-element events,

1 sure event (the whole set of elementary events E).

B   In example 1.2.1., the set E of elementary events was finite; in the theory of probability we also consider situations where the set E is denumerable or is of power continuum. In the latter case the set Z of random events does not contain all events, that is, it does not contain all subsets of the set E. We shall restrict our considerations to a set Z which is a Borel field of subsets of E. The definition of such a set Z is given at the end of this section since this book is to be available to the readers who do not know the set operations which are involved in the definition of a Borel field.

We now give the definition of a random event. The notion of the set Z appears in this definition. But since it has not been given precisely, we return to the notion of a random event once more (see definition 1.2.10).

Definition 1.2.1.   Every element of the Borel field Z of subsets of the set E of elementary events is called a *random event*.

Definition 1.2.2.   The event containing all the elements of the set E of elementary events is called the *sure event*.

Definition 1.2.3.   The event which contains no elements of the set E of elementary events is called the *impossible event*.

The impossible event is denoted by (0).

Definition 1.2.4.   We say that *event A is contained in event B* if every elementary event belonging to A belongs to B.

We write

$$A \subset B$$

and read: A is contained in B.

We illustrate this notion by Fig. 1.2.1, where square E represents the set of elementary events and circles A and B denote subsets of E. We see that A is contained in B.

Definition 1.2.4'. Two events A and B are called *equal* if A is contained in B and B is contained in A.

We write

$$A = B.$$

We now postulate the following properties of Z.

Property 1.2.1. *The set Z of random events contains as an element the whole set E.*

Property 1.2.2. *The set Z of random events contains as an element the empty set* (0).

These two properties state that the set Z of random events contains as elements the sure and the impossible events.

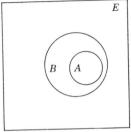

Fig. 1.2.1

Definition 1.2.5. We say that two events A and B are *exclusive* if they do not have any common element of the set E.

Example 1.2.2. Consider the random event A that two persons from the group of n persons born in Warsaw in 1950 will still be alive in the year 2000 and the event B that two or more persons from the group considered will still be alive in the year 2000. Events A and B are not exclusive.

If, however, we consider the event B' that only one person will still be alive in the year 2000, events A and B' will be exclusive.

Let us analyze this example more closely. In the group of n elements being considered it may happen that 1, or 2, or 3 ... up to n persons will still be alive in the year 2000, and it may happen that none of them will be alive at that time. Then the set E consists of $n + 1$ elementary events $e_0, e_1, \ldots, e_n$, where the indices $0, 1, \ldots, n$ denote the number of persons from the group being considered who will still be alive in the year 2000. The random event A in this example contains only one element, namely, the elementary event e_2. The random event B contains $n - 1$ elementary events, namely, $e_2, e_3, \ldots, e_n$. The common element of the two events A and B is the elementary event e_2, and hence these two events are not exclusive. However, event B' contains only one element, namely, the elementary event e_1. Thus events A and B' have no common element, and are exclusive.

C   We now come to a discussion of operations on events. Let $A_1, A_2, \ldots$ be a finite or denumerable sequence of random events.

Definition 1.2.6. The event A which contains those and only those elementary events which belong to at least one of the events $A_1, A_2, \ldots$ is called the *alternative (or sum or union)* of the events $A_1, A_2, \ldots$.

We write

$$A = A_1 \cup A_2 \cup \ldots \qquad \text{or} \quad A = A_1 + A_2 + \ldots, \qquad \text{or} \quad A = \sum_i A_i$$

and read: A_1 or A_2 or $\ldots$.

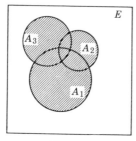

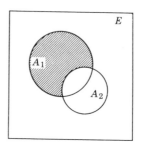

Fig. 1.2.2   Fig. 1.2.3

Let us illustrate the alternative of events by Fig. 1.2.2.

On this figure, square E represents the set of elementary events and circles A_1, A_2, A_3 denote three events; the shaded area represents the alternative $A_1 + A_2 + A_3$.

In our definition the alternative of random events corresponds to the set-theoretical sum of the subsets $A_1, A_2, \ldots$, of the set of elementary events.

The alternative of the events $A_1, A_2, \ldots$ occurs if and only if at least one of these events occurs.

The essential question which arises here is whether the alternative of an arbitrary (finite or denumerable) number of random events belongs to Z and hence is a random event. A positive answer to this question results from the following postulated property of the set Z of random events.

Property 1.2.3. *If a finite or denumerable number of events $A_1, A_2, \ldots$ belong to Z, then their alternative also belongs to Z.*

It is easy to verify that for every event A the following equalities are true:

$$A \cup A = A, \quad A \cup E = E, \quad A \cup (0) = A.$$

For example, we prove that

$$A \cup A = A.$$

In fact, every elementary event belonging to $A \cup A$ belongs to A; hence $(A \cup A) \subset A$. Similarly, $A \subset (A \cup A)$; thus $A \cup A = A$.

Definition 1.2.7. The random event A containing those and only those elementary events which belong to A_1 but do not belong to A_2 is called the *difference* of the events A_1 and A_2.

We write

$$A = A_1 - A_2.$$

The difference of events is illustrated by Fig. 1.2.3, where square E represents the set of all elementary events and circles A_1 and A_2 represent two events; the shaded area represents the difference $A_1 - A_2$.

The difference $A_1 - A_2$ occurs if and only if event A_1 but not event A_2 occurs.

If events A_1 and A_2 are exclusive, the difference $A_1 - A_2$ coincides with the event A_1.

As before, we postulate the following property of the set Z of random events.

Property 1.2.4. *If events A_1 and A_2 belong to Z, then their difference also belongs to Z.*

Example 1.2.3. Suppose that we investigate the number of children in a group of families. Consider the event A that a family chosen at random[1] has only one child and the event B that the family has at least one child. The alternative $A + B$ is the event that the family has at least one child.

If it is known that in the group under investigation there are no families having more than n children, the set of elementary events consists of $n + 1$ elements which, as in example 1.2.2. is denoted by $e_0, e_1, \ldots, e_n$. Event A contains only one elementary event e_1, and event B contains n elementary events $e_1, \ldots e_n$. The difference $A - B$ is, of course, an impossible event since there is no elementary event which belongs to A and not to B. However, the difference $B - A$ contains the elements $e_2, e_3, \ldots, e_n$ and is the event that the family has more than one child.

Definition 1.2.8. The event A which contains those and only those elements which belong to all the events $A_1, A_2, \ldots$ is called the *product* (*or intersection*) of these events.

We write

$$A = A_1 \cap A_2 \cap \ldots, \quad \text{or} \quad A = A_1 A_2 \ldots, \quad \text{or} \quad A = \prod_i A_i$$

and read: A_1 and A_2 and $\ldots$.

The product of events is illustrated by Fig. 1.2.4, where square E represents the set of elementary events, and circles A_1, A_2, A_3 represent three events; the shaded area represents the product $A_1 A_2 A_3$.

In our definition the product of events $A_1, A_2, \ldots$, corresponds to the set-theoretical product of subsets $A_1, A_2, \ldots$, of the set of elementary events. A product of events occurs if and only if all these events occur.

We postulate the following property of Z.

Property 1.2.5. *If a finite or denumerable number of events $A_1, A_2, \ldots$ belong to Z, then their product also belongs to Z.*

It is easy to verify that for an arbitrary event A the following equalities are true:

$$A \cap A = A, \quad A \cap E = A, \quad A \cap (0) = (0).$$

Example 1.2.4. Consider the random event A that a farm chosen at random has at least one horse and at least one plow, with the additional condition that the maximum number of plows as well as the maximum number of horses are

---

[1] We shall discuss later the methods of making such a choice.

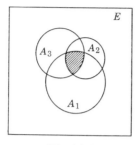
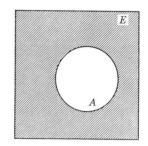

Fig. 1.2.4   Fig. 1.2.5

two. Consider also the event B that on the farm there is exactly one horse and at most one plow. We find the product of events A and B.

In this example the set of elementary events has 9 elements which are denoted by the symbols

$$e_{00}, e_{01}, e_{02}, e_{10}, e_{11}, e_{12}, e_{20}, e_{21}, e_{22},$$

the first index denoting the number of horses, and the second the number of plows.

The random event A contains four elementary events, $e_{11}, e_{12}, e_{21}, e_{22}$ and the random event B contains two elementary events, e_{10} and e_{11}. The product $A \cap B$ contains one elementary event e_{11}, and hence the event $A \cap B$ occurs if and only if on the chosen farm there is exactly one horse and exactly one plow.

Definition 1.2.9.   The difference of events $E - A$ is called the *complement* of the event A and is denoted by $\bar{A}$.

The complement of an event is illustrated by Fig. 1.2.5, where square E represents the set of elementary events, and circle A denotes some event; the shaded area represents the complement $\bar{A}$ of A.

This definition may also be formulated in the following way: Event $\bar{A}$ occurs if and only if event A does not occur.

According to properties 1.2.1 and 1.2.4 of the set Z of random events, the complement $\bar{A}$ of A is a random event.

Example 1.2.5.   Suppose we have a number of electric light bulbs. We are interested in the time t that they glow. We fix a certain value t_0 such that if the bulb burns out in a time shorter than t_0, we consider it to be defective. We select a bulb at random. Consider the random event A that we select a defective bulb. Then the random event that we select a good one, that is, a bulb that glows for a time no shorter than t_0, is the event $\bar{A}$, the complement of the event A.

We now give the definition (see Supplement) of the Borel field of events which was mentioned earlier.

Definition 1.2.10.   A set Z of subsets of the set E of elementary events with properties 1.2.1 to 1.2.5 is called *a Borel field of events*, and its elements are called *random events*.

In the sequel we consider only random events, and often instead of writing "random event" we simply write "event."

D   The following definitions will facilitate some of the formulations and proofs given in the subsequent parts of this book.

Definition 1.2.11.   The sequence $\{A_n\}(n = 1, 2, \ldots)$ of events is called *nonincreasing* if for every n we have

$$A_n \supset A_{n+1}.$$

The product of a nonincreasing sequence of events $\{A_n\}$ is called the *limit* of this sequence. We write

$$A = \prod_{n \geqslant 1} A_n = \lim_{n \to \infty} A_n.$$

Definition 1.2.12.   The sequence $\{A_n\}(n = 1, 2, \ldots)$ of events is called *nondecreasing* if for every n we have

$$A_{n+1} \supset A_n.$$

The sum of a nondecreasing sequence $\{A_n\}$ is called the *limit* of this sequence.
We write

$$A = \sum_{n \geqslant 1} A_n = \lim_{n \to \infty} A_n.$$

1.3 THE SYSTEM OF AXIOMS OF THE THEORY OF PROBABILITY

A   In everyday language the notion of probability is used without a precise definition of its meaning. However, probability theory, as a mathematical discipline, must make this notion precise. This is done by constructing a system of axioms which formalize some basic properties of probability, or in brief, by the axiomatization of the theory of probability.[1] The additional properties of probability can be obtained as consequences of these axioms.

In mathematics, the notion of random event defined in the preceding section corresponds to what is called a random event in everyday use. The system of axioms which is about to be formulated makes precise the notion of the probability of a random event. It is the mathematical formalization of certain regularities in the frequencies of occurrence of

---

[1] Many works have been devoted to the axiomatization of the theory of probability. We mention here the papers of Bernstein [1], Łomnicki [1], Rényi [1], Steinhaus [1], and the book by Mazurkiewicz [1]. The system of axioms given in this section was constructed by Kolmogorov [7]. (The numbers in brackets refer to the number of the paper quoted in the references at the end of the book.) The basic notions of probability theory are also discussed in the distinguished work of Laplace [1], and by Hausdorff [1], Mises [1, 2], Jeffreys [1], and Barankin [2].

random events (this last to be understood in the intuitive sense) observed during a long series of trials performed under constant conditions.

Suppose we are given the set of elementary events E and a Borel field Z of its subsets. As has already been mentioned (Section 1.1), it has been observed that the frequencies of occurrence of random events oscillate about some fixed number when the number of experiments is large. This observed regularity of the frequency of random events and the fact that the frequency is a non-negative fraction less or equal to one have led us to accept the following axiom.

Axiom I.   *To every random event A there corresponds a certain number $P(A)$, called the probability of A, which satisfies the inequality*

$$0 \leqslant P(A) \leqslant 1.$$

The following simple example leads to the formulation of axiom II.

Example 1.3.1.   Suppose there are only black balls in an urn. Let the random experiment consist in drawing a ball from the urn. Let m/n denote, as before, the frequency of appearance of the black ball. It is obvious that in this example we shall always have $m/n = 1$. Here, drawing the black ball out of the urn is a sure event and we see that its frequency equals one.

Taking into account this property of the sure event, we formulate the following axiom.

Axiom II.   *The probability of the sure event equals one.*
We write

$$P(E) = 1.$$

We shall see in Section 2.3 that the converse of axiom II is not true: if the probability of a random event A equals one, or $P(A) = 1$, the set A may not include all the elementary events of the set E.

We have already seen that the frequency of appearance of face 6 in throwing a die oscillates about the number $\frac{1}{6}$. The same is true for face 2. We notice that these two events are exclusive and that the frequency of occurrence of either face 6 or face 2 (that is, the frequency of the alternative of these events), which equals the sum of their frequencies, oscillates about the number $\frac{1}{6} + \frac{1}{6} = \frac{1}{3}$.

Experience shows that if a card is selected from a deck of 52 cards (4 suits of 13 cards each) many times over, the frequency of appearance of any one of the four aces equals about $\frac{4}{52}$, and the frequency of appearance of any spade equals about $\frac{13}{52}$. Nevertheless, the frequency of appearance of the alternative, ace or spade, oscillates not about the number $\frac{4}{52} + \frac{13}{52} = \frac{17}{52}$ but about the number $\frac{16}{52}$. This phenomenon is explained by the fact that ace and spade are not exclusive random events (we could select the ace of spades). Therefore the frequency of the alternative, ace or spade, is not equal to the sum of the frequencies of ace and spade. Taking into account this property of the frequency of the alternative of events, we formulate the last axiom.

Axiom III. *The probability of the alternative of a finite or denumerable number of pairwise exclusive events equals the sum of the probabilities of these events.*

Thus, if we have a finite or countable sequence of pairwise exclusive events $\{A_k\}$, $k = 1, 2, \ldots$, then, according to axiom III, the following formula holds:

$$(1.3.1) \qquad P\left(\sum_k A_k\right) = \sum_k P(A_k).$$

In particular, if a random event contains a finite or countable number of elementary events e_k and $(e_k) \in Z$ $(k = 1, 2, \ldots)$,

$$P(e_1, e_2, \ldots) = P(e_1) + P(e_2) + \cdots$$

The property expressed by axiom III is called the *countable* (or *complete*) *additivity* of probability.[1]

Axiom III concerns only the sums of pairwise exclusive events. Now let A and B be two arbitrary random events, exclusive or not. We shall find the probability of their alternative.

We can write

$$A \cup B = A \cup (B - AB),$$
$$B = AB \cup (B - AB).$$

The right sides of these expressions are alternatives of exclusive events. Therefore, according to axiom III, we have

$$P(A \cup B) = P(A) + P(B - AB),$$
$$P(B) = P(AB) + P(B - AB).$$

From these two equations we obtain the probability of the alternative of two events

$$(1.3.2) \qquad P(A \cup B) = P(A) + P(B) - P(AB).$$

Let $A_1, A_2, \ldots, A_n$, where $n \geqslant 3$, be arbitrary random events. It is easy to deduce the formula (due to Poincaré [1])

$$(1.3.2') \quad P\left(\sum_{k=1}^{n} A_k\right) = \sum_{k=1}^{n} P(A_k) - \sum_{\substack{k_1, k_2 = 1 \\ k_1 < k_2}}^{n} P(A_{k_1} A_{k_2})$$

$$+ \sum_{\substack{k_1, k_2, k_3 = 1 \\ k_1 < k_2 < k_3}}^{n} P(A_{k_1} A_{k_2} A_{k_3}) + \cdots + (-1)^{n+1} P(A_1 \ldots A_n).$$

---

[1] We could have said that the probability $P(A)$, satisfying axioms I to III, is a *normed, non-negative, and countably additive measure* on the Borel field Z of subsets of E.

B   Consider a finite or countable number of random events A_k, where $k = 1, 2, \ldots$. If every elementary event of the set E belongs to at least one of the random events $A_1, A_2, \ldots$, we say that these events *exhaust the set of elementary events E*. The alternative $\sum_k A_k$ contains all the elementary events of the set E and therefore is the sure event. By axiom II we obtain

Theorem 1.3.1.   *If the events* $A_1, A_2, \ldots$ *exhaust the set of elementary events E,*

$$(1.3.3) \qquad\qquad P\left(\sum_k A_k\right) = 1.$$

Example 1.3.2.   Let the set of all non-negative integers form the set of elementary events. Let (e_n) be the event of obtaining the number n, where $n = 0, 1, 2, \ldots$. Suppose that

$$P(e_n) = \frac{c}{n!},$$

where c is some constant. From theorem 1.3.1 and axiom III it follows that

$$P\left(\sum_{n=0}^{\infty} e_n\right) = c \sum_{n=0}^{\infty} \frac{1}{n!}.$$

But $P\left(\sum_{n=0}^{\infty} e_n\right) = 1$ and $\sum_{n=0}^{\infty} 1/n! = e$, where e is the base of natural logarithms. We then have $1 = ce$; hence

$$c = e^{-1}.$$

In the following chapters it turns out that in this example we have considered a particular case of the *Poisson distribution* which appears very often in practice.

We now prove the following theorem.

Theorem 1.3.2.   *The sum of the probabilities of any event A and its complement $\bar{A}$ is one.*

Proof.   From the definition of $\bar{A}$ it follows that the alternative $A \cup \bar{A}$ of A and $\bar{A}$ is the sure event; therefore, according to axiom II we have

$$P(A \cup \bar{A}) = 1.$$

But since events A and $\bar{A}$ are exclusive, we have, by axiom III,

$$P(A \cup \bar{A}) = P(A) + P(\bar{A})$$

and finally

$$(1.3.4) \qquad\qquad P(A) + P(\bar{A}) = 1.$$

Let A be the impossible event. We prove the next theorem.

Theorem 1.3.3. *The probability of the impossible event is zero.*

Proof. For every random event A we have the equality

$$A \cup E = E.$$

If A is the impossible event (does not contain any of the elementary events), A and E are exclusive because they have no common element. Applying axiom III, we obtain

$$P(A) + P(E) = P(E).$$

It follows immediately that

$$P(A) = 0.$$

We shall see in Section 2.3 that the converse is not true; from the fact that the probability of some event equals zero it does not follow that this event is impossible.

C   The following two theorems have numerous applications.

Theorem 1.3.4. *Let $\{A_n\}$, $n = 1, 2, \ldots$, be a nonincreasing sequence of events and let A be their product. Then*

$$(1.3.5) \qquad P(A) = \lim_{n \to \infty} P(A_n).$$

Proof. If the sequence $\{A_n\}$ is nonincreasing, then for every n we have

$$A_n = \sum_{k=n}^{\infty} A_k \bar{A}_{k+1} + A.$$

It follows from formula (1.3.2) that

$$(1.3.6) \qquad P(A_n) = P\left(\sum_{k=n}^{\infty} A_k \bar{A}_{k+1} \right) + P(A) - P\left(A \sum_{k=n}^{\infty} A_k \bar{A}_{k+1} \right).$$

We note that

$$A \sum_{k=n}^{\infty} A_k \bar{A}_{k+1} = \sum_{k=n}^{\infty} A A_k \bar{A}_{k+1}.$$

For every k, the event $A A_k \bar{A}_{k+1}$ is the impossible event; therefore $P(A A_k \bar{A}_{k+1}) = 0$. By axiom III, we obtain

$$P\left(\sum_{k=n}^{\infty} A A_k \bar{A}_{k+1} \right) = 0.$$

Since the events under the summation sign on the right-hand side of formula (1.3.6) are exclusive, we have

$$(1.3.7) \qquad P(A_n) = \sum_{k=n}^{\infty} P(A_k \bar{A}_{k+1}) + P(A).$$

However, the series

$$\sum_{k=1}^{\infty} P(A_k \bar{A}_{k+1})$$

is convergent, being a sum of non-negative terms whose partial sums are bounded by one. It follows that as $n \to \infty$ the sum in (1.3.7) tends to zero. Thus, finally,

$$\lim_{n \to \infty} P(A_n) = P(A).$$

Theorem 1.3.5. Let $\{A_n\}$, $n = 1, 2, \ldots$, be a nondecreasing sequence of events and let A be their alternative. Then we have

(1.3.8)
$$P(A) = \lim_{n \to \infty} P(A_n).$$

Proof. Consider the sequence of events $\{\bar{A}_n\}$ which are the complements of the events A_n. From the assumption that $\{A_n\}$ is a nondecreasing sequence it follows that $\{\bar{A}_n\}$ is a nonincreasing sequence. Let $\bar{A}$ be the product of events $\bar{A}_n$. From theorem 1.3.4 it follows that

$$P(\bar{A}) = \lim_{n \to \infty} P(\bar{A}_n).$$

Hence

$$P(A) = 1 - P(\bar{A}) = 1 - \lim_{n \to \infty} P(\bar{A}_n) = 1 - \lim_{n \to \infty} \left[1 - P(A_n)\right] = \lim_{n \to \infty} P(A_n)$$

and the theorem is proved.

We give one more simple theorem.

Theorem 1.3.6. If events A and B satisfy the condition

$$A \subset B,$$

then

$$P(A) \leqslant P(B).$$

Proof. Let us write

$$B = A + (B - A).$$

Events A and $B - A$ are exclusive; hence, according to axiom III,

$$P(B) = P(A) + P(B - A).$$

Since $P(B - A) \geqslant 0$, we have $P(B) \geqslant P(A)$.

1.4 APPLICATION OF COMBINATORIAL FORMULAS FOR COMPUTING PROBABILITIES

In some problems we can compute probabilities by applying combinatorial formulas. We illustrate this by some examples.

Example 1.4.1. Suppose we have 5 balls of different colors in an urn. Assume that the probability of drawing any particular ball is the same for any ball and equals p.

Here E consists of 5 elements and by hypothesis each has the same probability. Hence by theorem 1.3.1, we have $5p = 1$, or $p = \frac{1}{5}$.

Example 1.4.2. Suppose we have in the urn 9 slips of paper with the numbers 1 to 9 written on them, and suppose there are no two slips marked with the same number. Then E has 9 elementary events. Denote by A the event that on the slip of paper selected at random an even number will appear. What is the probability of this event?

As before, we suppose that the probability of selecting any particular slip is the same for any slip, and hence equals $\frac{1}{9}$. We shall obtain a slip with an even number if we draw one of the slips marked with 2, 4, 6 or 8. According to axiom III, the required probability equals

$$P(A) = \tfrac{1}{9} + \tfrac{1}{9} + \tfrac{1}{9} + \tfrac{1}{9} = \tfrac{4}{9}.$$

If in the example considered we wish to compute the probability of selecting a slip with an odd number, we may notice that this random event is the complement of A (we denote it by $\bar{A}$) and, by theorem 1.3.2, we have

$$P(\bar{A}) = 1 - P(A) = \tfrac{5}{9}.$$

Example 1.4.3. Let us toss a coin three times. What is the probability that heads appear twice?

The number of all possible combinations which may occur as a result of three successive tosses equals $2^3 = 8$. Denote the appearance of heads by H and the appearance of tails by T. We have the following possible combinations:

$$HHH,\ HHT,\ HTH,\ THH,\ HTT,\ THT,\ TTH,\ TTT.$$

Consider each of these combinations as an elementary event and the whole collection of them as the set E. Suppose that the occurrence of each of them has the same probability. We then have that the probability of each particular combination equals $1/2^3$. From the table we see that heads appear twice in three elementary events (HHT, HTH, THH); hence by axiom III the required probability is $\frac{3}{8}$.

If in the example just considered we had n tosses instead of 3 and looked for the probability of obtaining heads m times, our reasoning would have been as follows.

The number of all possible combinations with n tosses equals 2^n. The number of combinations in which heads appear m times equals the number of combinations of m elements from n elements given by

$$\binom{n}{m} = \frac{n!}{m!\,(n-m)!}.$$

If every possible result of n successive tosses of a coin is equally likely, the required probability is

(1.4.1)
$$\frac{n!}{2^n m!\,(n-m)!}.$$

Example 1.4.4. Compute the probability that heads appear at least twice in three successive tosses of a coin.

The random event under consideration will occur if in three tosses heads appear two or three times. According to formula (1.4.1), the probability that heads appear three times equals

$$\frac{3!}{2^3 \, 3! \, 0!} = \frac{1}{8},$$

and the probability that heads appear twice equals $\frac{3}{8}$, as we already know. Hence, according to axiom III, the required probability is

$$\tfrac{1}{8} + \tfrac{3}{8} = \tfrac{1}{2}.$$

In examples 1.4.1 to 1.4.4 the equiprobability of all elementary events was assumed. This assumption was obviously satisfied in our examples, but it is not always acceptable.

1.5 CONDITIONAL PROBABILITY

A   Let us first consider some examples.

Example 1.5.1. A. Markov [4] has investigated the probability of the appearance of these pairs of letters in Russian:

Vowel after vowel,
Vowel after consonant.

To compute these probabilities he counted the corresponding pairs of letters in Pushkin's poem *Eugene Onegin* on the basis of a text of 20,000 letters, and he accepted the observed frequencies as probabilities.[1] The experiment yielded the following results: there were 8638 vowels, and the pair "vowel after vowel" appeared 1104 times.

Let us analyze this example. Denote a vowel by a and a consonant by b. As elementary events we shall consider the pairs aa, ba, ab, bb, the set of elementary events is then (aa, ab, ba, bb).

Consider event B that a pair of letters will appear in which a vowel is in second place. Event B may be written as (aa, ba). It is known that a vowel appears 8638 times. These vowels follow either another vowel (in the pairs aa) or a consonant (in the pairs ba). Because no vowel appears at the beginning of the text considered

"Мой дядя самых честных правил...,"[2]

event B occurs 8638 times. Thus

$$P(B) = \frac{8638}{20,000} = 0.432.$$

Consider now event A that the pair of letters occurs with a vowel in first place. Event A may be written as (aa, ab).

[1] The methods of verification of such hypotheses are given in Part 2 of this book.
[2] It means, "My uncle's shown his good intentions."

The question "What is the frequency of a vowel followed by a vowel?" might now be formulated as follows.

What is the probability of event A in cases when event B has already occurred? We are not interested here in the probability of event A in the whole set E of elementary events but in the conditional probability which would correspond to the conditional frequency of event A provided event B has occurred; in other words, the probability of event A in the set (aa, ba) considered as the whole set of elementary events.

In our example we are interested in the probability of the event (aa). The experiment showed that this event appeared 1104 times, and, since event B appeared 8638 times, the probability we are looking for equals

$$\frac{1104}{8638} = 0.128.$$

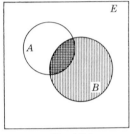

Fig. 1.5.1

B   In general, let B be an event in the set of elementary events E. The set B is then an element of the Borel field Z of subsets of the set E of all elementary events. Suppose $P(B) > 0$. Let us consider B as a new set of elementary events and denote by Z' the Borel field of all subsets of B which belong to the field Z.

Consider an arbitrary event A from the field Z. It may happen in particular cases that the event A belongs to the field Z', namely, when A is a subset of B. If, however, A contains any element of E which does not belong to B, A is not an element of Z'; yet some part of A may be a random event in Z', namely, when A and B have common elements, that is, when the product AB is not empty.

Now let B denote a fixed element of the field Z, where $P(B) > 0$, while A runs over all possible elements of Z; then all elements of Z' are products of the form AB. To stress the fact that the product AB is now being considered as an element of Z' (and not of Z) we denote it by the symbol $A \mid B$ and read: "A provided that B" or "A provided that B has occurred."

If A contains B, $A \mid B$ is the sure event (in the field Z').

Event $A \mid B$ is illustrated by Fig. 1.5.1. Here square E represents the set of all elementary events, and circles A and B denote some random events. The shaded area represents the random event B, and the doubly shaded area represents the random event $A \mid B$, that is, "event A provided that B has occurred."

The probability of the event $A \mid B$ in the field Z' will be denoted by $P(A \mid B)$ and read: The conditional probability of A provided B has occurred.

As will be shown shortly this probability can be defined by using the probability in the field Z; hence there is no need to postulate separately the existence of the probability $P(A \mid B)$ and its properties.

C   To facilitate the understanding of the definition of $P(A \mid B)$, let us consider the following.

Suppose we have performed n random experiments and have obtained the event B m times. Moreover, in $k(k \leqslant m)$ of these experiments we also obtained the random event A. The frequency of AB equals k/n, and the frequency of B equals m/n; the frequency of the random event A, provided the random event B has occurred, equals k/m.

Applying the equality

$$(1.5.1) \qquad\qquad \frac{k}{m} = \frac{k/n}{m/n},$$

to the probabilities instead of the frequencies, we accept the following definition.

Definition 1.5.1.   Let the probability of an event B be positive. *The conditional probability of the event A provided B has occurred* equals the probability of AB divided by the probability of B.

Thus

$$(1.5.2) \qquad\qquad P(A \mid B) = \frac{P(AB)}{P(B)}, \qquad \text{where} \quad P(B) > 0.$$

Similarly,

$$(1.5.3) \qquad\qquad P(B \mid A) = \frac{P(AB)}{P(A)}, \qquad \text{where} \quad P(A) > 0.$$

From (1.5.2) and (1.5.3) we obtain

$$(1.5.4) \qquad\qquad P(AB) = P(B)P(A \mid B) = P(A)P(B \mid A).$$

This formula is to be read: *The probability of the product AB of two events equals the product of the probability of B times the conditional probability of A provided B has occurred or, what amounts to the same thing, to the probability of A times the probability of B provided A has occurred.*

Let A_1, A_2, A_3 denote three events from the same field Z. Consider the expression $P(A_3 \mid A_1 A_2)$, or the probability of A_3 provided the product $A_1 A_2$ has occurred. According to (1.5.2) this probability, assuming that $P(A_1 A_2) > 0$, equals

$$(1.5.5) \qquad\qquad P(A_3 \mid A_1 A_2) = \frac{P(A_1 A_2 A_3)}{P(A_1 A_2)}.$$

From (1.5.5) and (1.5.3) we obtain for the probability of the product of three events the relations

$$(1.5.6) \qquad P(A_1 A_2 A_3) = P(A_1 A_2)P(A_3 \mid A_1 A_2)$$
$$= P(A_1)P(A_2 \mid A_1)P(A_3 \mid A_1 A_2).$$

This formula is to be read: *The probability of the product of three events equals the probability of the first event times the conditional probability of the second event provided the first event has occurred times the probability of the third event provided the product of the first two events has occurred.*

Now let $A_1, A_2, \ldots, A_n$ be random events. We could consider the conditional probabilities $P(A_{k_1} A_{k_2} \ldots A_{k_r} \mid A_{k_{r+1}} \ldots A_{k_n})$ of the product of some subgroup consisting of r events $(1 \leqslant r \leqslant n - 1)$ provided the product of the remaining $n - r$ events has occurred. By a reasoning similar to that stated we obtain

(1.5.7) $\quad P(A_1 A_2 \ldots A_n)$

$$= P(A_1)P(A_2 \mid A_1)P(A_3 \mid A_1 A_2) \ldots P(A_n \mid A_1 \ldots A_{n-1}).$$

D We shall show that the conditional probability satisfies axioms I to III.

We notice that the following inequality is true:

(1.5.8) $\qquad\qquad P(AB) \leqslant P(B).$

In fact, event B may occur either when event A occurs, or when event A does not occur; hence

$$B = AB \cup \bar{A}B,$$

where $\bar{A}$ is the complement of A. Thus $AB \subset B$, and from theorem 1.3.6, we obtain (1.5.8).

Since $P(AB) \geqslant 0$ and $P(B) > 0$ we obtain, from formula (1.5.8),

$$0 \leqslant P(A \mid B) \leqslant 1,$$

which is the property expressed by axiom I.

Now let $A \mid B$ be the sure event in field Z', that is, let $AB = B$. Then

$$P(AB) = P(B),$$

and hence

$$P(A \mid B) = 1.$$

This is the property expressed by axiom II.

Consider now the alternative $\sum_i (A_i \mid B)$ of pairwise exclusive events. We can write

$$\sum_i (A_i \mid B) = \left(\sum_i A_i \right) \bigg| B,$$

and hence

$$P\left[\sum_i (A_i \mid B) \right] = P\left[\left(\sum_i A_i \right) \bigg| B \right].$$

According to (1.5.2) and axiom III we have

$$P\left[\left(\sum_i A_i\right)\Big| B\right] = \frac{P\left[\left(\sum_i A_i\right)B\right]}{P(B)} = \frac{P\left(\sum_i A_iB\right)}{P(B)}$$

$$= \sum_i \frac{P(A_iB)}{P(B)} = \sum_i P(A_i \mid B).$$

This formula expresses the countable additivity of conditional probability. Since all the axioms are satisfied for the conditional probabilities, the theorems derived from these axioms hold for the conditional probabilities.

1.6 BAYES THEOREM

A   Before we start the general consideration let us consider an example.

Example 1.6.1. We have two urns. There are 3 white and 2 black balls in the first urn and 1 white and 4 black balls in the second. From an urn chosen at random we select one ball at random. What is the probability of obtaining a white ball if the probability of selecting each of the urns equals 0.5?

Denote by A_1 and A_2 respectively, the events of selecting the first or second urn, and by B the event of selecting a white ball. Event B may happen either together with event A_1 or together with event A_2; hence we have

$$B = A_1B + A_2B,$$

and since events A_1B and A_2B are exclusive, we have

$$P(B) = P(A_1B) + P(A_2B).$$

Applying formula (1.5.4) we obtain

(1.6.1) $P(B) = P(A_1)P(B \mid A_1) + P(A_2)P(B \mid A_2).$

In this example we have $P(A_1) = P(A_2) = 0.5$, $P(B \mid A_1) = 0.6$, and $P(B \mid A_2) = 0.2$. Placing these values into (1.6.1) we obtain $P(B) = 0.4$.

Formula (1.6.1) obtained in this example is a special case of the *theorem of absolute probability*, which is now given.

Theorem 1.6.1. *If the random events $A_1, A_2, \ldots$ are pairwise exclusive and exhaust the set E of elementary events, and if $P(A_i) > 0$ for $i = 1, 2, \ldots$, then for any random event B we have*

(1.6.2) $P(B) = P(A_1)P(B \mid A_1) + P(A_2)P(B \mid A_2) + \ldots$

In fact, from the assumptions it follows that B may happen together with one and only one of the events A_i. We then have

$$B = A_1B + A_2B + \ldots$$

and

(1.6.3) $P(B) = P(A_1B) + P(A_2B) + \ldots$

According to (1.5.4) we obtain for every i,

(1.6.4) $P(A_iB) = P(A_i)P(B \mid A_i)$.

Substituting values (1.6.4) into (1.6.3) we get (1.6.2).

B   Again let the events A_i satisfy the assumptions of theorem 1.6.1. Suppose that the event B has occurred. Now what is the probability of A_i? This question is answered by the following theorem due to Bayes.

Theorem 1.6.2.   *If the events $A_1, A_2, \ldots$ satisfy the assumptions of the theorem of absolute probability and $P(B) > 0$, then for $i = 1, 2, \ldots$ we have*

$$(1.6.5) \quad P(A_i \mid B) = \frac{P(A_i)P(B \mid A_i)}{P(A_1)P(B \mid A_1) + P(A_2)P(B \mid A_2) + \ldots}.$$

In fact, substituting A_i for A in formula (1.5.4), we obtain

$$P(A_i \mid B) = \frac{P(A_i)P(B \mid A_i)}{P(B)},$$

and introducing in the denominator expression (1.6.2) for $P(B)$, we obtain (1.6.5).

Formula (1.6.5) is called *Bayes formula* or the *formula for a posteriori probability*. The latter name is explained by the fact that this formula gives us the probability of A_i after B has occurred. On the other hand, the probabilities $P(A_i)$ in this formula are called the *a priori probabilities*.

Bayes formula plays an important role in applications.

Example 1.6.2.   Guns 1 and 2 are shooting at the same target. It has been found that gun 1 shoots on the average nine shots during the same time gun 2 shoots ten shots. The precision of these two guns is not the same; on the average, out of ten shots from gun 1 eight hit the target, and from gun 2, only seven.

During the shooting the target has been hit by a bullet, but it is not known which gun shot this bullet. What is the probability that the target was hit by gun 2?

Denote by A_1 and A_2 the events that a bullet is shot by gun 1 and gun 2, respectively. Taking into consideration the ratio of the average number of shots made by gun 1 to the average number of shots made by gun 2, we can put $P(A_1) = 0.9P(A_2)$.[1] Denote by B the event that the target is hit by the bullet. According to the data about the precision of the guns we have $P(B \mid A_1) = 0.8$ and $P(B \mid A_2) = 0.7$. According to Bayes formula

$$P(A_2 \mid B) = \frac{P(A_2)P(B \mid A_2)}{P(A_1)P(B \mid A_1) + P(A_2)P(B \mid A_2)}$$

$$= \frac{0.7\,P(A_2)}{0.9P(A_2) \cdot 0.8 + 0.7P(A_2)} = 0.493.$$

[1] The methods of verifying such hypotheses will be given in Part 2.

1.7 INDEPENDENT EVENTS

A   In general, the conditional probability $P(A \mid B)$ differs from $P(A)$. However, the case when we have the equality

(1.7.1) $P(A \mid B) = P(A)$

is of special importance. Then the fact that B has occurred does not have any influence on the probability of A, or we could say, the probability of A is independent of the occurrence of B.

We notice that if (1.7.1) is satisfied, formula (1.5.4) gives

(1.7.2) $P(AB) = P(A)P(B).$

Formula (1.7.2) is also satisfied if

(1.7.3) $P(B \mid A) = P(B)$

Formula (1.7.2) was derived from formula (1.5.4) where it was assumed that $P(A) > 0$ and $P(B) > 0$; nevertheless this formula is also valid when one of these probabilities equals zero.

We now define the notion of *independence* of two random events.

Definition 1.7.1.   Two events A and B are called *independent* if their probabilities satisfy (1.7.2), that is, if the probability of the product AB is equal to the product of the probabilities of A and B.

It follows from this definition that the notion of independence of two events is symmetric with respect to these events.

As we have already established, formula (1.7.2) can be obtained from either of the formulas (1.7.1) and (1.7.3). We also notice that formulas (1.5.4) and (1.7.2) with $P(A) > 0$ and $P(B) > 0$ give formulas (1.7.1) and (1.7.3). We thus deduce that *each of the last formulas is a necessary and sufficient condition for the independence of two events with positive probabilities.*

B   The notion of independence of two events can be generalized to the case of any finite number of events.

Definition 1.7.2.   Events $A_1, A_2, \ldots, A_n$ are *independent* if for all integer indices $k_1, k_2, \ldots k_s$ satisfying the conditions

$$1 \leqslant k_1 < k_2 < \ldots < k_s \leqslant n$$

we have

(1.7.4) $P(A_{k_1} A_{k_2} \ldots A_{k_s}) = P(A_{k_1}) P(A_{k_2}) \ldots P(A_{k_s}),$

that is, if the probability of the product of every combination $A_{k_1}, A_{k_2}, \ldots, A_{k_s}$ of events equals the product of the probabilities of these events.

It is possible that $A_1, A_2, \ldots, A_n$ are pairwise independent, that is, each two events among $A_1, A_2, \ldots, A_n$ are independent, that each three of them are independent, and so on, and yet $A_1, A_2, \ldots, A_n$ are not independent. This is illustrated by the example given by Bernstein.

Example 1.7.1. There are four slips of paper of identical size in an urn. Each slip is marked with one of the numbers $110, 101, 011, 000$ and there are no two slips marked with the same number. Consider event A_1 that on the slip selected the number 1 appears in the first place, event A_2 that 1 appears in the second place, and A_3 that 1 appears in the third place. The number of slips of each category is 2, the number of all slips is 4; hence if we assume that each slip has the same probability of being selected, we have

$$P(A_1) = P(A_2) = P(A_3) = \tfrac{1}{2}.$$

Let A denote the product $A_1 A_2 A_3$. $P(A) = 0$ since event A is impossible (there is no slip marked with 111) and since

$$P(A_1)P(A_2)P(A_3) = \tfrac{1}{8} \neq 0 = P(A),$$

events A_1, A_2, and A_3 are not independent. We shall show, however, that these events are pairwise independent. In fact, for the pair A_1, A_2 we have

$$P(A_2 \mid A_1) = \tfrac{1}{2} = P(A_2),$$

since there are only two slips having 1 in the first place, and only one among them with 1 in the second place. In a similar way we could show that the remaining pairs are independent.

Independence of a countable number of events is defined in the following way.

Definition 1.7.3. Events $A_1, A_2, \ldots$ are *independent* if for every $n = 2, 3, \ldots$ events $A_1, A_2, \ldots, A_n$ are independent.

From definitions 1.7.3 and 1.7.2 it follows that if the random events $A_1, A_2, \ldots$ are independent, for $n = 2, 3, \ldots$ and for arbitrary indices $k_1, k_2, \ldots, k_n$, events $A_{k_1}, A_{k_2}, \ldots, A_{k_n}$ are independent.

To stress the fact that we consider the independence in the sense of definitions 1.7.2 and 1.7.3, and not the independence of pairs, triples and so on, the term *independence en bloc*, or *mutual independence*, is used in probability theory. We shall avoid these terms, and independence will always be understood in the sense of the given definitions.

Problems and Complements

1.1. Prove that the operations of addition and multiplication of random events are commutative and satisfy the following associative and distributive laws:

$$A_1 + A_2 = A_2 + A_1,$$
$$A_1 A_2 = A_2 A_1,$$
$$A_1 + (A_2 + A_3) = (A_1 + A_2) + A_3,$$
$$A_1(A_2 A_3) = (A_1 A_2)A_3,$$
$$A_1(A_2 + A_3) = A_1 A_2 + A_1 A_3.$$

1.2. Prove the relations:

$$A_1 A_2 = A_1 - (A_1 - A_2),$$
$$A_1 + A_2 = A_1 + (A_2 - A_1),$$
$$A_1 - A_2 = A_1 - A_1 A_2,$$
$$A_1(A_2 - A_3) = A_1 A_2 - A_3.$$

1.3. Prove that $A_1 \subset A_2$ implies $A_1 + A_2 = A_2$ and $A_1 A_2 = A_1$.

1.4. (a) Prove the following two identities, called *de Morgan's laws:*

$$\overline{A_1 + A_2} = \bar{A_1} \bar{A_2}, \qquad \overline{A_1 A_2} = \bar{A_1} + \bar{A_2}.$$

(b) Generalize these identities to the case of n ($n > 2$) random events.

1.5. (a) Prove that for $n = 2, 3, \ldots$

$$A_1 + \ldots + A_n = A_1 + (A_2 - A_1 A_2) + \ldots + (A_n - A_1 A_n \ldots - A_{n-1} A_n).$$

Note that the terms on the right-hand side are pairwise exclusive.

(b) Show that

$$\bigcup_{i=1}^{\infty} A_i = A_1 + (A_2 - A_1 A_2) + (A_3 - A_1 A_3 - A_2 A_3) + \ldots.$$

1.6. Prove that properties 1.2.2 and 1.2.5 follow from properties 1.2.1, 1.2.3, and 1.2.4.

1.7. Let $\{A_n\}$ ($n = 1, 2, \ldots$) be an arbitrary sequence of random events. The random event A^* which contains all the elementary events which belong to an infinite number of the events A_n will be called the *upper limit* of the sequence $\{A_n\}$,

$$A^* = \limsup_{n \to \infty} A_n.$$

The random event A_* which contains all the elementary events which belong to all but a finite number of the events A_n will be called the *lower limit* of the sequence $\{A_n\}$,

$$A_* = \liminf_{n \to \infty} A_n.$$

(a) Prove that

$$A^* = \prod_{n=1}^{\infty} \sum_{k=n}^{\infty} A_k,$$

$$A_* = \sum_{n=1}^{\infty} \prod_{k=n}^{\infty} A_k.$$

(b) Show that

$$A_* \subset A^*.$$

1.8. (Notation as in the preceding problem) If $A^* = A_*$, then $A = A^* = A_*$ is called the *limit* of the sequence $\{A_n\}$. Then we write

$$A = \lim_{n \to \infty} A_n.$$

(a) Prove that if $\{A_n\}$ is a nondecreasing sequence,

$$A = A^* = A_* = \sum_{n=1}^{\infty} A_n.$$

(b) Prove that if $\{A_n\}$ is a nonincreasing sequence, then

$$A = A^* = A_* = \prod_{n=1}^{\infty} A_n.$$

1.9. (a) Prove that for an arbitrary sequence of random events $\{A_n\}$

$$P\left(\limsup_{n\to\infty} A_n\right) \geq \limsup_{n\to\infty} P(A_n),$$

and

$$P\left(\liminf_{n\to\infty} A_n\right) \leq \liminf_{n\to\infty} P(A_n).$$

Hint: Use theorems 1.3.4 and 1.3.5.

(b) Deduce that if $\lim_{n\to\infty} A_n$ exists, then

$$P\left(\lim_{n\to\infty} A_n\right) = \lim_{n\to\infty} P(A_n).$$

1.10. Using Problems 1.5(a) and (b), prove the inequalities

$$P\left(\bigcup_{i=1}^{n} A_i\right) \leq \sum_{i=1}^{n} P(A_i),$$

$$P\left(\bigcup_{i=1}^{\infty} A_i\right) \leq \sum_{i=1}^{\infty} P(A_i).$$

In combinatorial Problems 1.11 to 1.16 assume that all the possible outcomes have the same probability.

1.11. A deck of cards contains 52 cards. Player G has been dealt 13 of them. Compute the probability that player G has

(a) exactly 3 aces,

(b) at least 3 aces,

(c) any 3 face cards of the same face value,

(d) any 3 cards of the same face value from the five highest denominations,

(e) any 3 cards of the same face value from the eight lowest denominations,

(f) any 3 cards of the same face value,

(g) three successive spades,

(h) at least three successive spades,

(i) three successive cards of any suit,

(j) at least three successive cards of any suit.

1.12. Three dice are thrown once. Compute the probability of obtaining

(a) face 2 on one die,

(b) face 3 on at least one die,

(c) an even sum,

(d) a sum divisible by 3,

(e) a sum exceeding 7,

(f) a sum smaller than 12,

(g) a sum which is a prime number.

1.13. (*Chevalier de Méré's problem*) Find which of the following two events is more likely to occur: (1) to obtain at least one ace in a simultaneous throw of four dice; (2) at least one double ace in a series of 24 throws of a pair of dice.

1.14. (*Banach's problem*) A mathematician carries two boxes of matches, each of which originally contained n matches. Each time he lights a cigarette he selects one box at random. Compute the probability that when he eventually selects an empty box, the other will contain r matches, where $r = 0, 1, \ldots, n$.

1.15. An urn contains m white and $n - m$ black balls. Two players successively draw balls at random, putting the drawn ball back into the urn before the next drawing. The player who first succeeds in drawing a white ball wins. Compute the probability of winning by the player who starts the game.

1.16. There are 28 slips of paper; on each of them one letter is written. The letters and their frequencies are presented in the following table:

Letter	a	c	e	h	i	j	l	m	n	o	s	t	y	ż
Number of slips	3	1	3	1	1	2	2	1	2	2	2	4	2	2

The slips are then arranged in random order. What is the probability of obtaining the sentence:[1] "Sto lat sto lat niech żyje żyje nam"?

1.17. The famous poet Goethe once gave his guest, the famous chemist Runge, a box of coffee beans. Runge used this gift—at that time very valuable—for scientific experiments, and for the first time obtained pure caffeine. Is it possible to compute the probability of this event? If so, is the answer unique? What are the factors which determine the precise formulation of the random event whose probability we compute?

1.18. (*Bertrand's paradox*) A circle is drawn around an equilateral triangle with side a. Then a random chord is drawn in this circle. The event A occurs if and only if the length l of this chord satisfies the relation $l > a$. State the conditions under which (a) $P(A) = 0.5$, (b) $P(A) = 1/3$, (c) $P(A) = 1/4$. Should these results be considered as paradoxical?

1.19. (*Buffon's problem*) A needle of length $2l$ is thrown at random on a plane on which parallel lines are drawn at a distance $2a$ apart ($a > l$). What is the probability of the needle intersecting any of these lines?

1.20. The probability that both of a pair of twins are boys equals 0.32 and the probability that both of them are girls equals 0.28. Find the conditional probability that

(a) the second twin is a boy, provided the first is a boy,

(b) the second twin is a girl, provided the first is a girl.

Hint: Use example 1.1.3.

1.21. (a) What should n be in order that the probability of obtaining the face 6 at least once in a series of n independent throws of a die will exceed $3/4$?

(b) The events A_1, $A_2, \ldots$ are independent and $P(A_j) = p$ ($j = 1, 2, \ldots$). Find the least n such that

$$P\left(\bigcup_{k=1}^{n} A_k \right) \geq p_0,$$

where p_0 is a given number.

1.22. (a) The events $A_1, A_2, \ldots$ are independent and $P(A_k) = p_k$. What is the probability that none of the events A_k occurs?

(b) Answer the question of Problem 1.22(a) without the assumption of independence.

1.23. Prove that if the events A and B are independent, the same is true for the events $\bar{A}$ and $\bar{B}$.

---

[1] The beginning of a Polish birthday song. It means something similar to, "May he live a hundred years, a hundred years."

CHAPTER 2

Random Variables

2.1 THE CONCEPT OF A RANDOM VARIABLE

We can assign a number to every elementary event from a set E of elementary events. In the coin-tossing example we assigned the number 1 to the appearance of heads and the number 0 to the appearance of tails. Then the probability of obtaining the number 1 as a result of an experiment will be the same as the probability of obtaining a head, and the probability of obtaining the number 0 will be the same as the probability of obtaining a tail.

Similarly, in the example of throwing a die we can assign to every result of a throw one of the numbers i ($i = 1, \ldots, 6$) corresponding to the number of dots appearing on the face resulting from our throw. In general, let e denote an elementary event of a set E of elementary events. On the set E we define a single-valued real function $X(e)$ such that, roughly speaking, the probability that this function will assume certain values is defined. To formulate precisely the conditions which are to be satisfied by this function let us introduce the notion of an inverse image.

Definition 2.1.1. Let $X(e)$ be a single-valued real function defined on the set E of elementary events. The set A of all elementary events to which the function $X(e)$ assigns values in a given set S of real numbers is called the *inverse image of the set S.*

It is clear that the inverse image of the set R of all real numbers is the whole set E.

Definition 2.1.2. A single-valued real function $X(e)$ defined on the set E of elementary events is called a *random variable*[1] if the inverse image of every interval I on the real axis of the form $(-\infty, x)$ is a random event.

We shall set the probability $P^{(x)}(I)$ that the random variable $X(e)$ takes on a value in the interval I equal to the probability $P(A)$ of the inverse image A of I.

[1] The notion of a random variable corresponds in the theory of real functions to the notion of a *function measurable with respect to the field of sets* being considered.

From now on, we shall usually write X instead of $X(e)$ and remember that the function X is defined on the set E of elementary events e. Random variables are usually denoted by capital letters X, Y, ..., and their values by the corresponding small letters x, y, ...

From definition 2.1.2 it follows in particular that if a random event A is the inverse image of a point x, the probability that the random variable X takes on the value x equals the probability of the event A,

$$P^{(x)}(X = x) = P(A).$$

Since any interval I of the form $[a, b)$, where $a < b$, is the difference of the intervals $(-\infty, b)$ and $(-\infty, a)$, the existence of the probability $P^{(x)}(I)$ follows from definition 2.1.2.

We can show that the inverse image of the sum, difference, and product of arbitrary intervals is, respectively, the sum, difference, and product of the inverse images of these intervals.

Every Borel set of real numbers may be considered to be the result of a finite or denumerable number of operations of addition, subtraction, and multiplication performed on intervals (see also Supplement, example S2). Thus from the properties of the set Z of random events, from the properties of probability discussed in the previous chapter, and from the definition of a random variable, it follows that if X is a random variable, for every Borel set on the real axis, the probability $P^{(x)}(S)$ that X takes on a value in the set S is defined. In other words, we have defined the *probability distribution* of the random variable X. (See Supplement, Extension Theorem.) Instead of "the probability distribution" of the random variable X we shall often say in brief "the *distribution*" of the random variable X.

Definition 2.1.3.   The function $P^{(x)}(S)$ giving the probability that a random variable X takes on a value belonging to S, where S is an arbitrary Borel set on the real axis, is called the *probability function* of X.

We write

$$P^{(x)}(S) = P^{(x)}(X \in S).$$

In applications we shall usually deal with sets of numbers of simple structure like intervals or finite sums of intervals.

We notice that a random variable is a function and not a variable in the sense which is usually understood in mathematical analysis. This warning is very important and should be remembered constantly. In particular one has to be very careful when applying the formal rules for operations on variables to operations on random variables.

In this section, in Section 2.5A, and in the Supplement, we use the symbol $P^{(x)}$ to denote probabilities for random variables; hence for events which are subsets of a new set of elementary events $X(E)$, whereas the symbol P was used for random events which are subsets of the initial set E. In the remainder of the book there is no need to distinguish between them, and hence the symbol P is used throughout.

2.2 THE DISTRIBUTION FUNCTION

It is convenient to characterize the probability distribution by means of the distribution function which is now defined. Let us start with an example

Example 2.2.1. Consider tossings of a die. To every elementary event, that is, to every result of a throw, we can assign one of the numbers $1, 2, \ldots, 6$, the number of dots which appear on the resultant face. Here the random variable X may take on six values $x_i = i$ $(i = 1, \ldots, 6)$ with the same probability $P(X = x_i) = \frac{1}{6}$. The probability that X is less than 1, of course, equals zero,

$$P(X < 1) = 0.$$

If x is a number satisfying the inequality $1 < x \leqslant 2$,

$$P(X < x) = P(X = 1) = \tfrac{1}{6}.$$

If $2 < x \leqslant 3$,

$$P(X < x) = P(X = 1) + P(X = 2) = \tfrac{1}{3}.$$

Similarly for $5 < x \leqslant 6$, we have

$$P(X < x) = \sum_{i=1}^{5} P(X = i) = \tfrac{5}{6},$$

and finally for $x > 6$ we have

$$P(X < x) = P(X \leqslant 6) = \sum_{i=1}^{6} P(X = i) = 1.$$

If we graphed the function $P(X < x)$ in this example as a function of x, we would obtain the step function shown in Fig. 2.2.1.

We see that as x increases, the value of $P(X < x)$ does not decrease, and at the points x_i $(i = 1, \ldots, 6)$ this value increases by a constant number $P(X = x_i)$.

Definition 2.2.1. The function $F(x)$ defined as

(2.2.1) $$F(x) = P(X < x)$$

is called the *distribution function* of the random variable X.

It follows from the considerations of the preceding section that the probability distribution of a random variable X is determined by its distribution function.

The equalities

(2.2.2) $$F(-\infty) = 0, \quad F(+\infty) = 1$$

are obviously satisfied. We now show that the distribution function $F(x)$ is a nondecreasing function. In fact, let x_1 and x_2, where $x_1 < x_2$, be two

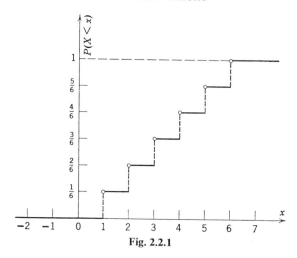

Fig. 2.2.1

points on the real axis. Since the interval $(-\infty, x_2)$ contains the interval $(-\infty, x_1)$, we have $P(X < x_2) \geqslant P(X < x_1)$ and hence $F(x_2) \geqslant F(x_1)$.

Every distribution function is continuous at least from the left. Indeed, let $x_1 < x_2 < \ldots < x$ be an arbitrary increasing sequence convergent to x. Denote by A_k the event that the random variable X takes on a value from the half-open interval $[x_k, x)$. If $k_1 < k_2$, from the occurrence of event A_{k_2} follows the occurrence of event A_{k_1}; hence $\{A_k\}$ is a nonincreasing sequence of events. The limit of the sequence $\{x_k\}$, that is, the point x, does not belong to any of the intervals being considered and therefore the product $A = \prod_{k=1}^{\infty} A_k$ is the impossible event and thus $P(A) = 0$. Thus, from theorem 1.3.4 we obtain

$$\lim_{k \to \infty} P(A_k) = \lim_{k \to \infty} P(x_k \leqslant X < x) = \lim_{k \to \infty} [F(x) - F(x_k)]$$
$$= F(x) - \lim_{k \to \infty} F(x_k) = 0,$$

or

$$\lim_{k \to \infty} F(x_k) = F(x).$$

Since $\{x_k\}$ is an arbitrary increasing sequence, convergent from the left to x, it follows that the distribution function $F(x)$ is continuous from the left.

We now show that every real function $F(x)$ which is nondecreasing, continuous from the left, and satisfying equalities (2.2.2) determines the probability distribution of some random variable. Let $F(x)$ satisfy these conditions. Take the interval $[0, 1]$ as the set of elementary events and the field of all Borel subsets of this interval as the field of random events. Take as a probability measure the Lebesgue measure (for definition, see Supplement); then the probability of a Borel set from the interval [0, 1] is equal to its Lebesgue measure; in particular, the probability of the interval [0, e], where $0 < e \leqslant 1$, equals the length e of this interval. We now define the random variable $X(e)$ in the following way:

$$X(e) = \inf_{F(y) = e} y \qquad (0 \leqslant e \leqslant 1).$$

Thus, for a given value e, the random variable $X(e)$ equals the least upper bound of the set of all y such that $F(y) = e$. Then the distribution function of $X(e)$ is the function F, since

$$P[X(e) < x] = P(\inf_{F(y) = e} y < x) = F(x) \qquad (-\infty < x < \infty).$$

Thus we have proved

Theorem 2.2.1. *The single-valued function $F(x)$ is a distribution function if and only if it is nondecreasing, continuous at least from the left and satisfies conditions (2.2.2).*

From this result it follows immediately that the values of the distribution function at the points of continuity determine this function everywhere.

We show that the set of points of discontinuity is at most countable. Denote by H_n the set of points at which the distribution function $F(x)$ has a jump not smaller than $1/n$. Then we have

$$H = H_1 + H_2 + \cdots$$

For every n the set H_n is finite; hence the set H is at most countable.

2.3 RANDOM VARIABLES OF THE DISCRETE TYPE AND THE CONTINUOUS TYPE

A   We shall deal mainly with random variables of two types, namely of the discrete type and of the continuous type. (For other cases, see Supplement.)

Definition 2.3.1. A random variable is said to be *of the discrete type* if it takes on, with probability 1, values belonging to a set S which is at most countable, and every value in the set S has positive probability. We call these values *jump points*, and their probabilities, *jumps*.

Example 2.3.1. A stock of goods contains good and defective items. We have here two elementary events—that we draw a good or a defective item. Denote the probability of drawing at random a good item by p and suppose that $0 < p < 1$. Assign to the drawing of a good item the number 1, and to the drawing of a defective item the number 0. Then we obtain a random variable of the discrete type, which can take on only two values with positive probability, 1 and 0, with probabilities p and $1 - p$ respectively.

Now let a random variable X of the discrete type take on the values x_i ($i = 1, 2, \ldots$) with probabilities p_i. According to definition 2.3.1 we have

$$(2.3.1) \qquad \qquad \sum_{i=1}^{n} p_i = 1,$$

if the number of jump points x_i is finite, and

$$(2.3.1') \qquad \qquad \sum_{i=1}^{\infty} p_i = 1,$$

if the number of jump points x_i is countable.

We can formulate definition 2.1.3 as follows.

Definition 2.3.2. Let x_i ($i = 1, 2, \ldots$) be an arbitrary jump point of a random variable X of the discrete type. The probability that the random variable X takes on the value x_i is called the *probability function* of the discrete-type random variable X and we write

$$(2.3.2) \qquad \qquad P(X = x_i) = p_i,$$

where the numbers p_i ($i = 1, 2, \ldots$) satisfy either (2.3.1) or (2.3.1').

The set of jumps of a probability function can be an arbitrary finite or countable set of positive numbers satisfying either condition (2.3.1) or (2.3.1').

The distribution function $F(x)$ has the form

$$(2.3.3) \qquad \qquad F(x) = \sum_{x_i < x} p_i,$$

where the summation is extended over all points x_i for which $x_i < x$.

Consider now a random variable X which has no jump points. The distribution function of such a random variable is a continuous function. We shall mainly consider a special class of such random variables called random variables of the continuous type.

Definition 2.3.3. A random variable X is said to be *of the continuous type* if there exists a non-negative function $f(x)$ such that for every real number x the following relation holds:

$$(2.3.4) \qquad \qquad F(x) = \int_{-\infty}^{x} f(x)\, dx,$$

where $F(x)$ is the distribution function of X. The function $f(x)$ is called the *probability density* of the random variable X.

Instead of "probability density" we often use the term "*density function*," or simply "*density*."

Every density function $f(x)$ satisfies the relation

(2.3.5) $$F(+\infty) = \int_{-\infty}^{\infty} f(x)\, dx = 1.$$

Moreover, for every real a and b, where $a < b$, we have

$$P(a \leqslant X \leqslant b) = F(b) - F(a) = \int_{a}^{b} f(x)\, dx.$$

It can be shown that for every Borel set S we have

(2.3.6) $$P(S) = \int_{S} f(x)\, dx.$$

If the function $f(x)$ is continuous at some point x,

(2.3.7) $$F'(x) = f(x).$$

Thus at the continuity points of the function $f(x)$ we have ·

(2.3.8) $$f(x) = \lim_{\Delta x \to 0} \frac{F(x + \Delta x) - F(x)}{\Delta x} = \lim_{\Delta x \to 0} \frac{P(x \leqslant X < x + \Delta x)}{\Delta x}.$$

Every real function $f(x)$ which is non-negative, integrable over the whole real axis, and satisfies (2.3.5) is the probability density of a random variable X of the continuous type.

Indeed, it can easily be shown that the function $F(x)$ defined by the formula

$$F(x) = \int_{-\infty}^{x} f(x)\, dx$$

has all the properties of a distribution function.

Example 2.3.2. On the set of all real numbers, define the density function $f(x)$ in the following way:

$$f(x) = \begin{cases} 0 & \text{for} \quad x < 0, \\ \tfrac{1}{2}x & \text{for} \quad 0 \leqslant x \leqslant 2, \\ 0 & \text{for} \quad x > 2. \end{cases}$$

The distribution function of a random variable X with this density has the form

$$F(x) = \begin{cases} 0 & \text{for} \quad x < 0, \\ \tfrac{1}{4}x^2 & \text{for} \quad 0 \leqslant x \leqslant 2, \\ 1 & \text{for} \quad x > 2. \end{cases}$$

B We now show (see Section 1.3) that if the probability of a random event equals zero, it does not follow that this event is impossible. Similarly we prove that if the probability of a random event equals one, it does not follow that this event is sure.

Consider a random variable X of the continuous type with the density function $f(x)$ and let x_0 be some fixed point. What is the probability that $X = x_0$?

From (2.3.6) we obtain

$$P(X = x_0) = \int_{x_0}^{x_0} f(x)\, dx = 0.$$

Nevertheless the point x_0 belongs to the set of elementary events which here coincides with the real axis.

Denote by R' the x-axis without point x_0. Then we have the relation

$$P(R') = \int_{R'} f(x)\, dx = 1.$$

Nevertheless, the event that X takes on a value belonging to set R' is not the sure event. Thus, if for a random variable X of the continuous type, the probability of a certain event equals zero, this event should not be considered to be the impossible event in the sense of definition 1.2.3, but it should be considered only as an event which is very unlikely to occur. Similarly, if for a random variable X of the continuous type the probability of an event equals one, this event should be considered as an event which is very likely to occur but not as the sure event in the sense of definition 1.2.2.

2.4   FUNCTIONS OF RANDOM VARIABLES

A.   Let us consider an example.

Example 2.4.1.   Suppose that the random variable X may take on two values $x_1 = 5$ and $x_2 = 10$ with probabilities $P(X = 5) = \frac{1}{3}$, and $P(X = 10) = \frac{2}{3}$. Let us transform the random variable X, taking $Y = 2X$.

The random variable Y can also take on two values, $y_1 = 10$ and $y_2 = 20$, where

$$P(Y = 10) = P(2X = 10) = P(X = 5) = \tfrac{1}{3},$$
$$P(Y = 20) = P(2X = 20) = P(X = 10) = \tfrac{2}{3}.$$

Thus Y may take on values $y_i = 2x_i$ $(i = 1, 2)$ with the same probabilities as X takes on the values x_i. The distribution function of X is, by (2.3.3),

$$F(x) = \begin{cases} 0 & \text{for} \quad x \leqslant 5, \\ \frac{1}{3} & \text{for} \quad 5 < x \leqslant 10, \\ 1 & \text{for} \quad x > 10. \end{cases}$$

Denote the distribution function of Y by $F_1(y)$. Then we have

$$F_1(y) = P(Y < y) = P(2X < y) = P(X < \tfrac{1}{2}y) = F(\tfrac{1}{2}y).$$

Hence

$$F_1(y) = \begin{cases} 0 & \text{for} \quad y \leqslant 10, \\ \frac{1}{3} & \text{for} \quad 10 < y \leqslant 20, \\ 1 & \text{for} \quad y > 20. \end{cases}$$

Thus, it is possible to obtain the distribution of Y from the distribution of X.

In general, if $Y = g(X)$ is a single-valued and continuous[1] transformation of a random variable X, Y is also a random variable, whose distribution function may be obtained from the distribution function of X.

We consider some simple transformations.

Let $F(x)$ be the distribution function of a random variable X. Consider the transformation

$$Y = -X.$$

Denote the distribution function of Y by $F_1(y)$. We have

$$(2.4.1) \qquad F_1(y) = P(Y < y) = P(-X < y)$$
$$= P(X > -y) = 1 - P(X \leqslant -y).$$

If the random variable X is of the continuous type,

$$(2.4.2) \qquad F_1(y) = 1 - F(-y).$$

Let us write (2.4.2) in the form

$$F_1(y) = 1 - \int_{-\infty}^{-y} f(x)\, dx.$$

Denoting the density of Y by $f_1(y)$ we obtain

$$(2.4.3) \qquad F_1'(y) = f_1(y) = f(-y).$$

If X is of the discrete type and $-y$ is its jump point,

$$(2.4.4) \qquad F_1(y) = 1 - F(-y) - P(X = -y).$$

Consider the general linear transformation

$$Y = aX + b.$$

---

[1] The assumption of continuity of the function $g(X)$ is not necessary. It is sufficient to assume that $g(X)$ is a Baire function, which means that $g(X)$ is single-valued and that the function $h(Y)$ inverse to the function $g(X)$ transforms every interval of the form $(-\infty, y)$ into a Borel set, which will be here denoted by $h(-\infty, y)$. Then the distribution function of the random variable Y is determined. Indeed

$$P(Y < y) = P[g(X) < y] = P[X \in h(-\infty, y)].$$

In the sequel, we shall always suppose, without mentioning it, that the functions under consideration have this property.

We consider two cases.

I. $a > 0$. In the preceding notations we have

(2.4.5) $\quad F_1(y) = P(aX + b < y) = P\left(X < \dfrac{y - b}{a}\right) = F\left(\dfrac{y - b}{a}\right)$.

Equation (2.4.5) is valid for the discrete as well as for the continuous case. Moreover in the latter case we have

(2.4.6) $\qquad\qquad F_1'(y) = f_1(y) = \dfrac{1}{a} f\left(\dfrac{y - b}{a}\right)$.

II. $a < 0$. We now have

(2.4.7) $\qquad\qquad F_1(y) = P(aX + b < y) = P\left(X > \dfrac{y - b}{a}\right)$

$$= 1 - P\left(X \leqslant \dfrac{y - b}{a}\right).$$

If the random variable X is of the continuous type, we obtain from (2.4.7)

(2.4.8) $\qquad\qquad F_1(y) = 1 - F\left(\dfrac{y - b}{a}\right)$

and

(2.4.9) $\qquad\qquad F_1'(y) = f_1(y) = -\dfrac{1}{a} f\left(\dfrac{y - b}{a}\right)$.

We notice that formulas (2.4.6) and (2.4.9) may be written as one formula, namely

(2.4.10) $\qquad\qquad F_1'(y) = \dfrac{1}{|a|} f\left(\dfrac{y - b}{a}\right)$.

Formula (2.4.3) is a particular case of formula (2.4.10).

If the random variable X is of the discrete type, by (2.4.7) we obtain

(2.4.11) $\qquad F_1(y) = 1 - F\left(\dfrac{y - b}{a}\right) - P\left(X = \dfrac{y - b}{a}\right)$,

if the point $(y - b)/a$ is a jump point of X. At the remaining points $P[X = (y - b)/a] = 0$ and formulas (2.4.11) and (2.4.8) coincide.

B   Now let X be a random variable with the distribution function $F(x)$. Consider the transformation $Y = X^2$. The random variable Y does not take on negative values. Let $F_1(y)$ be the distribution function of Y. Then we have

(2.4.12)

$$F_1(y) = \begin{cases} 0 & \text{for } y \leqslant 0, \\ P(Y < y) = P(X^2 < y) = P(-\sqrt{y} < X < \sqrt{y}) & \text{for } y > 0. \end{cases}$$

If the random variable X is of the continuous type, from (2.4.12) we obtain

$$(2.4.13) \qquad F_1(y) = \begin{cases} 0 & \text{for} \quad y \leqslant 0, \\ F(\sqrt{y}) - F(-\sqrt{y}) & \text{for} \quad y > 0. \end{cases}$$

If the random variable X has the density $f(x)$, by formula (2.4.13) the density of Y is

$$(2.4.14) \quad f_1(y) = \begin{cases} 0 & \text{for} \quad y \leqslant 0, \\ \dfrac{F'(\sqrt{y}) + F'(-\sqrt{y})}{2\sqrt{y}} = \dfrac{f(\sqrt{y}) + f(-\sqrt{y})}{2\sqrt{y}} & \text{for} \quad y > 0. \end{cases}$$

If, however, the random variable X is of the discrete type, from (2.4.12) we obtain

$$(2.4.15) \qquad F_1(y) = \begin{cases} 0 & \text{for} \quad y \leqslant 0, \\ F(\sqrt{y}) - F(-\sqrt{y}) - P(X = -\sqrt{y}) & \text{for} \quad y > 0. \end{cases}$$

If the point $-\sqrt{y}$ is not a jump point of the random variable X, then $P(X = -\sqrt{y}) = 0$ and formula (2.4.15) becomes identical with (2.4.13).

Let $x_1, x_2, \ldots$ be the jump points of the random variable X, and $y_1, y_2, \ldots$ be the points corresponding to them according to the relation $y_i = x_i^2$. We have

$$(2.4.16) \quad P(Y = y_i) = P(X^2 = y_i) = P(X = -\sqrt{y_i}) + P(X = \sqrt{y_i}).$$

Example 2.4.2. Suppose that the random variable X may take on only two values, $x_1 = -1$ and $x_2 = 1$, where $P(X = -1) = P(X = 1) = \frac{1}{2}$. Let $Y = X^2$. The random variable Y may take on only one value, $y = 1$, and we obviously have

$$P(Y = 1) = P(X = -1) + P(X = 1) = \frac{1}{2} + \frac{1}{2} = 1.$$

C Let X be a random variable of the continuous type with density $f(x)$. Consider a one-to-one transformation defined by a function $y = g(x)$ which has an everywhere continuous derivative $g'(x)$.

Let $[x_1, x_2)$ be an interval such that $g'(x) \neq 0$ for $x_1 \leqslant x < x_2$. Let $y_1 = g(x_1)$ and $y_2 = g(x_2)$. Denote by $x = h(y)$ the function inverse to the function $g(x)$. By assumptions, $h(y)$ is single-valued and its derivative $h'(y)$ is finite and continuous in the interval (y_1, y_2). Thus, we have

$$(2.4.17) \qquad P(x_1 \leqslant X < x_2) = \int_{x_1}^{x_2} f(x)\, dx = \int_{y_1}^{y_2} f[h(y)]h'(y)\, dy.$$

If the derivative $h'(y) > 0$, then $y_1 < y_2$ and the integral on the right-hand side of (2.4.17) expresses the probability $P(y_1 \leqslant Y < y_2)$, where $Y = g(X)$; if $h'(y) < 0$, then $y_2 < y_1$ and from (2.4.17) we obtain

$$(2.4.18) \quad P(x_1 \leqslant X < x_2) = -\int_{y_2}^{y_1} f[h(y)]h'(y)\, dy$$

$$= \int_{y_2}^{y_1} f[h(y)]\, |h'(y)|\, dy = P(y_2 \leqslant Y < y_1).$$

From (2.4.17) and (2.4.18) it follows that the random variable $Y = g(X)$ has the density

(2.4.19) $f[h(y)] |h'(y)|$.

Formula (2.4.10) is a special case of (2.4.19).

2.5 MULTIDIMENSIONAL RANDOM VARIABLES

A   The following example illustrates the notion of a multidimensional random variable.

Example 2.5.1.   Table 2.5.1 contains the data concerning the distribution of the population of Poland according to sex and age from the census of 1931.

TABLE 2.5.1

POPULATION OF POLAND BY SEX AND AGE
(Data from the census of 1931)

Age Group	Men	Women
	(in thousands)	
0–4	2020	1962
5–9	2005	1962
10–14	1405	1372
15–19	1474	1562
20–29	2931	3213
30–39	1999	2255
40–49	1391	1596
50–59	1052	1201
60–69	753	875
70 or more	386	474
Total	15,416	16,472

The element of investigation is an inhabitant of Poland in the year 1931. Every inhabitant of Poland is characterized in this table by two characteristics, sex and age. We can assign numerical values to these characteristics.

To analyze the results of a census, IBM cards are prepared for every person included in the census. To every characteristic under consideration a number is assigned on this card. For instance, to every man the number 1 is usually assigned and to every woman the number 0. Similarly, to every age group a certain number is assigned.

Consider the random event that a card chosen at random corresponds to a person belonging to a given sex and age group. This will be an elementary event. On the other hand, this event is equivalent to the event that a card chosen at random has been perforated to correspond to this sex and age group. Thus to every elementary event there corresponds a pair of numbers.

Let us consider some set E of elementary events. We have defined the probability $P(A)$ on a Borel field Z of subsets A of the set E of all elementary events. We define on the set E a collection of n real, single-valued functions, assigning to every elementary event an n-tuple $\{x_1, x_2, \ldots x_n\}$. In other words we assign to every elementary event of E a point in n-dimensional Euclidean space.

Definition 2.5.1. The collection of n real single-valued functions $X = (X_1, X_2, \ldots X_n)$ defined on E is called an *n-dimensional random variable* if the inverse image A of every generalized n-dimensional[1] interval I of the form $(-\infty, -\infty, \ldots, -\infty, a_1, a_2, \ldots, a_n)$ is a random event.

For every such interval we define the probability[2] $P^{(x)}(I)$ that the n-dimensional random variable takes on a value in the interval I, namely, $P^{(x)}(I) = P(A)$. As in Section 2.1 it is possible to show that for every Borel set S in the space $(x_1, x_2, \ldots, x_n)$ the last condition defines the probability $P^{(x)}(S)$ that the random variable $(X_1, X_2, \ldots, X_n)$ will take on a value in S, or defines the probability distribution of the random variable $(X_1, X_2, \ldots, X_n)$. As before, the function $P^{(x)}(S)$ is called the *probability function*. In place of "multidimensional random variable" the terms *joint random variable* or *random vector* are sometimes used.

B In further considerations we usually restrict ourselves to two-dimensional random variables.

Definition 2.5.2. The function $F(x, y)$ defined by

(2.5.1) $$F(x, y) = P(X < x, Y < y)$$

is called the *distribution function* of the random variable (X, Y).

As in Section 2.2 it is possible to show tnat the function $F(x, y)$ is non-decreasing and continuous at least from the left with respect to every variable. Moreover, it satisfies the equalities

(2.5.2) $$F(-\infty, y) = F(x, -\infty) = 0, \quad F(+\infty, +\infty) = 1.$$

We notice that we also have the equality

(2.5.3) $$\begin{aligned}
P(x_1 &\leqslant X < x_2, y_1 \leqslant Y < y_2) \\
&= P(X < x_2, Y < y_2) - P(X < x_2, Y < y_1) \\
&\quad - P(X < x_1, Y < y_2) + P(X < x_1, Y < y_1) \\
&= F(x_2, y_2) - F(x_2, y_1) - F(x_1, y_2) + F(x_1, y_1)
\end{aligned}$$

Figure 2.5.1 illustrates formula (2.5.3).

[1] *A generalized n-dimensional interval I* in n-dimensional Euclidean space is the set of points such that each coordinate is contained in some interval in the usual sense. The interval $(-\infty, -\infty, \ldots, -\infty, a_1, a_2, \ldots, a_n)$ is then the set of points in n-dimensional Euclidean space for which the inequalities $-\infty < x_i < a_i$ $(i = 1, 2, \ldots, n)$ are satisfied.

[2] See the last paragraph of Section 2.1.

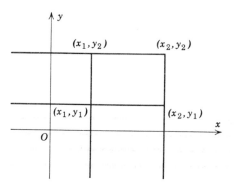

Fig. 2.5.1

In contrast to the one-dimensional distribution functions, in order that the function $F(x, y)$ be the distribution function of a two-dimensional random variable, it is not sufficient that this function be continuous from the left, nondecreasing with respect to each of the variables, and satisfy conditions (2.5.2). Indeed, if $F(x, y)$ is a distribution function, then as may be seen from (2.5.3), for all values x_2, x_1 $(x_2 > x_1)$ and y_2, y_1 $(y_2 > y_1)$ the relation

$$F(x_2, y_2) - F(x_2, y_1) - F(x_1, y_2) + F(x_1, y_1) \geqslant 0$$

must be satisfied.

From the following example it follows that the preceding conditions alone do not imply the last inequality. Let

$$F(x, y) = \begin{cases} 0, & \text{if } x + y \leqslant 0, \\ 1, & \text{if } x + y > 0. \end{cases}$$

Then the function $F(x, y)$ takes on the value 0 for the points on and below the line $y = -x$, and the value 1 for the points above this line. This function is nondecreasing, continuous from the left with respect to x and y, and satisfies equalities (2.5.2). Nevertheless, this function $F(x, y)$ does not satisfy the last inequality. For, consider any rectangle in the plane (x, y) with three vertices above the line $y = -x$ and the fourth below this line. For example, it may be the rectangle with vertices $(3, 3)$, $(3, -1)$, $(-1, 3)$, $(-1, -1)$. Applying formula (2.5.3) we obtain

$$P(-1 \leqslant X < 3, -1 \leqslant Y < 3)$$
$$= F(3, 3) - F(3, -1) - F(-1, 3) + F(-1, -1) = -1.$$

The following theorem is true.

__Theorem 2.5.1.__ A real single-valued function $F(x, y)$ is a distribution function of a certain two-dimensional random variable if and only if $F(x, y)$

is nondecreasing and continuous at least from the left with respect to both arguments x and y, satisfies equalities (2.5.2), and the inequality

(2.5.4) $\qquad F(x_2, y_2) - F(x_2, y_1) - F(x_1, y_2) + F(x_1, y_1) \geqslant 0$

holds for every (x_1, y_1), (x_2, y_2), *where* $x_1 < x_2$ *and* $y_1 < y_2$.

C   We shall mainly consider multidimensional random variables of the continuous or of the discrete type.

Definition 2.5.3.   The two-dimensional random variable (X, Y) is said to be of the *discrete type* if, with probability 1, it takes on pairs of values belonging to a set S of pairs that is at most countable, and every pair (x_i, y_k) is taken with positive probability p_{ik}. We call these pairs of values *jump points*, and their probabilities, *jumps*.

From definition 2.5.3 it follows that

$$\sum_i \sum_k p_{ik} = 1.$$

The distribution function $F(x, y)$ has the form

(2.5.5) $\qquad F(x, y) = \sum_{\substack{x_i < x \\ y_k < y}} p_{ik},$

where the summation is extended over all points (x_i, y_k) for which the inequalities $x_i < x$ and $y_k < y$ are satisfied.

The definition of the probability function may be given in the following form.

Definition 2.5.4.   Let (x_i, y_k), where $i = 1, 2, \ldots$ and $k = 1, 2, \ldots$, be an arbitrary jump point of the random variable (X, Y) of the discrete type. The probability that the random variable (X, Y) will take on the pair of values (x_i, y_k) is called the *probability function* of (X, Y). We write

(2.5.6) $\qquad P(X = x_i, Y = y_k) = p_{ik}.$

We now define the notion of a two-dimensional random variable of the continuous type.

Definition 2.5.5.   The two-dimensional random variable (X, Y) is called of the *continuous type*, if there exists a non-negative function $f(x, y)$ such that for every pair (x, y) of real numbers the following relation is satisfied:

(2.5.7) $\qquad F(x, y) = \int_{-\infty}^{x} \left[\int_{-\infty}^{y} f(x, y) \, dy \right] dx,$

where $F(x, y)$ is the distribution function of (X, Y). The function $f(x, y)$ is called the *density function*.

The density function $f(x, y)$ satisfies the equality

(2.5.8) $F(+\infty, +\infty) = \int_{-\infty}^{\infty} \int_{-\infty}^{\infty} f(x, y)\, dx\, dy = 1.$

If the density function $f(x, y)$ is continuous at the point (x, y),

(2.5.9) $\dfrac{\partial^2 F(x, y)}{\partial x\, \partial y} = f(x, y).$

Thus at the continuity points of (x, y) we have

(2.5.10) $f(x, y) = \lim\limits_{\substack{\Delta x \to 0 \\ \Delta y \to 0}} \dfrac{P(x \leqslant X < x + \Delta x,\, y \leqslant Y < y + \Delta y)}{\Delta x\, \Delta y}.$

All the notions derived in this paragraph for two-dimensional random variables can easily be generalized to random variables with more than two dimensions. In particular, the distribution function of the n-dimensional random variable $(X_1, X_2, \ldots, X_n)$ is the function defined as

(2.5.11) $F(x_1, x_2, \ldots, x_n) = P(X_1 < x_1, X_2 < x_2, \ldots, X_n < x_n).$

The generalization of (2.5.3) to the case of n-dimensional random vectors $(X_1, X_2, \ldots, X_n)$ is the following formula:

(2.5.12)

$P(x_1 \leqslant X_1 < x_1 + h_1, \ldots, x_n \leqslant X_n < x_n + h_n)$

$= F(x_1 + h_1, \ldots, x_n + h_n)$

$- \sum\limits_{i=1}^{n} F(x_1 + h_1, \ldots, x_{i-1} + h_{i-1}, x_i, x_{i+1} + h_{i+1}, \ldots, x_n + h_n)$

$+ \sum\limits_{\substack{i,j=1 \\ i<j}}^{n} F(x_1 + h_1, \ldots, x_i, \ldots, x_j, \ldots, x_n + h_n) + \ldots$

$+ (-1)^n F(x_1, \ldots, x_n),$

where $h_i > 0$ $(i = 1, 2, \ldots, n)$.

Theorem 2.5.1 can also be generalized to n-dimensional random variables. Instead of inequality (2.5.4) we would then have the inequality which states that the right-hand side of (2.5.12) is non-negative.

Since the distribution function $F(x_1, x_2, \ldots, x_n)$ is continuous from the left and nondecreasing with respect to every argument, it follows that the values of $F(x_1, \ldots, x_n)$ at its continuity points determine this function completely.

We notice one more essential difference between the one-dimensional and multidimensional distribution functions, namely, if the one-dimensional random variable X does not have jump points, its distribution function

$F(x)$ is everywhere continuous. Nevertheless the distribution function $F(x_1, \ldots, x_n)$ can have discontinuity points even if the random variable $(X_1, \ldots, X_n)$ does not have jump points. This will happen when there exists a hyperplane determined by the equalities

$$X_{j_1} = a_1, \ldots, X_{j_r} = a_r,$$

where $1 \leqslant r < n$ and $a_1, \ldots, a_r$ are constants such that

$$P(X_{j_1} = a_1, \ldots X_{j_r} = a_r) > 0.$$

Example 2.5.2. Let us consider the random vector (X, Y) with distribution function $F(x, y)$ of the form

$$F(x, y) = \begin{cases} 0 & \text{in the domains } (-\infty < x \leqslant 0, \quad -\infty < y < +\infty) \\ & \text{and } (0 < x < +\infty, \quad -\infty < y \leqslant 0), \\ xy & \text{in the domain } (0 \leqslant x \leqslant 1, \quad 0 \leqslant y \leqslant \tfrac{1}{2}), \\ x/2 & \text{in the domain } (0 \leqslant x \leqslant 1, \quad \tfrac{1}{2} \leqslant y < \infty), \\ y & \text{in the domain } (1 < x < \infty, \quad 0 \leqslant y \leqslant 1), \\ 1 & \text{in the domain } (x > 1, \quad y > 1). \end{cases}$$

The random vector (X, Y) takes on, with probability $\tfrac{1}{2}$, a point in $(x = 1, \tfrac{1}{2} \leqslant y \leqslant 1)$. Clearly, every point with coordinates $(1, y)$, where $\tfrac{1}{2} < y \leqslant \infty$, is a discontinuity point of the distribution function $F(x, y)$, although the random vector (X, Y) has no jump points.

If all the "vertices" of the generalized interval, given by the inequalities

$$x_1 \leqslant X_1 < x_1 + h_1, \qquad x_2 \leqslant X_2 < x_2 + h_2, \ldots, x_n \leqslant X_n < x_n + h_n,$$

are continuity points of the distribution function $F(x_1, \ldots, x_n)$, for the "surface" S of the interval we have

$$(2.5.13) \qquad P[(X_1, X_2, \ldots, X_n) \in S] = 0.$$

Definition 2.5.6. An interval, generalized or in the usual sense, for which relation (2.5.13) holds, is called a *continuity interval*.

We notice that an interval may be a continuity interval in spite of the fact that the distribution function may be discontinuous at some "vertices" of this interval. Thus in example 2.5.2 the rectangle with vertices $(1, 2)$, $(1, 3)$, $(2, 2)$ and $(2, 3)$ is a continuity interval in the sense of definition 2.5.6, and, nevertheless, the vertices $(1, 2)$ and $(1, 3)$ are discontinuity points of the distribution function $F(x, y)$.

We notice also that the generalized interval of the form $(-\infty, -\infty, \ldots, -\infty, a_1, a_2, \ldots, a_n)$ is a continuity interval if and only if the point $(a_1, a_2, \ldots, a_n)$ is a continuity point of the distribution function.

2.6   MARGINAL DISTRIBUTIONS

Let (X, Y) be a two-dimensional random variable of the discrete type which can take on the values (x_i, y_k). Then we have

$$P(X = x_i, Y = y_k) = p_{ik}.$$

Define

(2.6.1) $$p_{.k} = \sum_i p_{ik},$$

(2.6.2) $$p_{i.} = \sum_k p_{ik}.$$

We have

$$p_{.k} = \sum_i p_{ik}$$

$$= P(X = x_1, Y = y_k) + P(X = x_2, Y = y_k) + P(X = x_3, Y = y_k) + \ldots.$$

Hence $p_{.k}$ equals the probability that $Y = y_k$ when X takes on any of the possible values. We can write

$$p_{.k} = P(Y = y_k).$$

Furthermore, it is obvious that

$$\sum_k p_{.k} = \sum_i \sum_k p_{ik} = 1.$$

The collection of numbers $p_{.k}$ is then a set of jumps of a probability function. The distribution determined by these jumps is called the *marginal distribution* of the random variable Y. Similarly, the collection of numbers $p_{i.}$ is the set of jumps of the *marginal distribution* of X.

Example 2.6.1.   Suppose that we have 21 slips of paper in an urn. On each slip one of the numbers $1, 2, \ldots, 21$ is written, and there are no two slips marked with the same number. We are interested in two characteristics of these numbers, their divisibility by 2 and by 3. We choose one slip of paper at random. Let us assign to the appearance of an even number the number 1, and to the appearance of an odd number the number 0, and denote the random variable thus defined by X. Then the random variable X takes on two values $x_1 = 1$ and $x_2 = 0$. Similarly, let the random variable Y take on the value $y_1 = 1$ when a number divisible by 3 is chosen, and the value $y_2 = 0$ otherwise.

Among the 21 numbers under consideration we have the following types of numbers:

divisible by 2 and by 3	3,
divisible by 2 and not divisible by 3	7,
divisible by 3 and not divisible by 2	4,
divisible neither by 2 nor by 3	7.

If the probability of drawing each slip is the same, we obtain

$$p_{11} = P(X = 1, Y = 1) = \tfrac{3}{21}, \qquad p_{21} = P(X = 0, Y = 1) = \tfrac{4}{21},$$

$$p_{12} = P(X = 1, Y = 0) = \tfrac{7}{21}, \qquad p_{22} = P(X = 0, Y = 0) = \tfrac{7}{21}.$$

We can obtain two marginal distributions; one if we classify the numbers only according to their divisibility by 2, and the second if we classify the numbers only according to their divisibility by 3.

From equality (2.6.1) we obtain

$$p_{.1} = p_{11} + p_{21} = \tfrac{3}{21} + \tfrac{4}{21} = \tfrac{7}{21},$$

$$p_{.2} = p_{12} + p_{22} = \tfrac{7}{21} + \tfrac{7}{21} = \tfrac{14}{21}.$$

This is the marginal distribution of divisibility by 3. Similarly, from relation (2.6.2) we obtain

$$p_{1.} = p_{11} + p_{12} = \tfrac{3}{21} + \tfrac{7}{21} = \tfrac{10}{21},$$

$$p_{2.} = p_{21} + p_{22} = \tfrac{4}{21} + \tfrac{7}{21} = \tfrac{11}{21}.$$

This is the marginal distribution of divisibility by 2.

The name "marginal distribution" becomes clear by presenting this example in the form of Table 2.6.1.

TABLE 2.6.1

	Numbers Divisible by 2	Numbers Indivisible by 2	Total
Numbers divisible by 3	3	4	7
Numbers indivisible by 3	7	7	14
Total	10	11	21

As we see, the marginal distribution of divisibility by 2 is given in the marginal row at the bottom of the table and the marginal distribution of divisibility by 3 is given in the marginal column on the right side of Table 2.6.1.

In general, let $F(x, y)$ be the distribution function of a two-dimensional random variable (X, Y). The distribution function of the marginal distribution of X has the form

$$(2.6.3) \qquad F(x, \infty) = P(X < x, Y < \infty).$$

If (X, Y) is a random variable of the discrete type, (2.6.3) takes the form

$$(2.6.4) \qquad F(x, \infty) = \sum_{x_i < x} \sum_{k} p_{ik},$$

where the summation is extended over all the values of k, and those values of i for which $x_i < x$.

If (X, Y) is a random variable of the continuous type,

$$(2.6.5) \qquad F(x, \infty) = \int_{-\infty}^{x} \left[\int_{-\infty}^{+\infty} f(x, y) \, dy \right] dx.$$

The density of the marginal distribution of the random variable X has the form

$$(2.6.6) \qquad f_1(x) = \int_{-\infty}^{+\infty} f(x, y) \, dy.$$

Analogous formulas may be derived for the marginal distribution of the random variable Y.

Let $F(x_1, x_2, \ldots, x_n)$ be the distribution function of the random vector $(X_1, X_2, \ldots, X_n)$, where $n > 2$. Then we can obtain $\binom{n}{k}$ k-dimensional marginal distributions for $k = 1, 2, \ldots, n - 1$. Thus, for example, the distribution function of the random variable (X_1, X_2) has the form

$$F(x_1, x_2, \infty, \ldots, \infty) = P(X_1 < x_1, X_2 < x_2, X_3 < \infty, \ldots, X_n < \infty).$$

2.7 CONDITIONAL DISTRIBUTIONS

A   In Section 1.5 we defined the conditional probability $P(A \mid B)$, that is, the probability of the event A under condition B. We now investigate *conditional distributions*.

Let (X, Y) be a two-dimensional random variable of the discrete type, where X can take on the values x_i $(i = 1, 2, \ldots)$ and Y can take on the values y_k $(k = 1, 2, \ldots)$. Let

$$P(X = x_i, Y = y_k) = p_{ik}.$$

Then the marginal distributions are

$$P(X = x_i) = p_{i.} = \sum_k p_{ik},$$

$$P(Y = y_k) = p_{.k} = \sum_i p_{ik}.$$

Let us define for every i and k the probabilities

$$(2.7.1) \qquad P(X = x_i \mid Y = y_k) = \frac{p_{ik}}{p_{.k}}$$

and

$$(2.7.2) \qquad P(Y = y_k \mid X = x_i) = \frac{p_{ik}}{p_{i.}}.$$

When y_k is fixed and x_i varies over all possible jump points, expression (2.7.1) is the probability function of the random variable X of the discrete type under the condition $Y = y_k$. When x_i is fixed and y_k varies over all possible jump points, expression (2.7.2) is the probability function of the random variable Y of the discrete type under the condition $X = x_i$. In fact, the expressions on the right-hand sides of (2.7.1) and (2.7.2) are non-negative numbers bounded by one. Moreover, it is easy to verify that

$$\sum_i P(X = x_i \mid Y = y_k) = \frac{\sum_i p_{ik}}{p_{.k}} = \frac{p_{.k}}{p_{.k}} = 1,$$

$$\sum_k P(Y = y_k \mid X = x_i) = \frac{\sum_k p_{ik}}{p_{i.}} = \frac{p_{i.}}{p_{i.}} = 1.$$

B   Now let $f(x, y)$ be the density of a two-dimensional random variable (X, Y) of the continuous type. Consider the interval $[x, x + h)$ and the event $x \leqslant X < x + h$. Suppose that

$$P(x \leqslant X < x + h) > 0.$$

For every value of y and every interval $[x, x + h)$ we define the conditional probability

$$(2.7.3) \quad P(Y < y \mid x \leqslant X < x + h) = \frac{P(Y < y, x \leqslant X < x + h)}{P(x \leqslant X < x + h)}$$

$$= \frac{\int_x^{x+h} \left[\int_{-\infty}^y f(x, y) \, dy \right] dx}{\int_x^{x+h} \left[\int_{-\infty}^{+\infty} f(x, y) \, dy \right] dx}.$$

Obviously expression (2.7.3) is the conditional distribution function of the random variable Y, under the condition that the random variable X satisfies the inequality $x \leqslant X < x + h$, where it is assumed that $P(x \leqslant X < x + h) > 0$, since for fixed x and h the expression on the right-hand side of (2.7.3) is a nondecreasing function of y, continuous from the left, and takes on the value 0 when $y = -\infty$ and 1 when $y = +\infty$. Difficulties arise when we try to obtain the probability $P(Y < y \mid X = x)$, for then $P(X = x) = 0$. We shall here restrict ourselves to the case where the limit of expression (2.7.3) as $h \to 0$ exists. (See Section 3.6,C and Supplement.)

Suppose that the density function $f(x, y)$ is everywhere continuous and that the marginal density

$$(2.7.4) \qquad f_1(x) = \int_{-\infty}^{\infty} f(x, y) \, dy$$

is a continuous function of x. Suppose that at the point x being considered the function $f(x)$ is positive. Dividing the numerator and the denominator of the expression on the right-hand side of (2.7.3) by h and passing to the limit when $h \to 0$ we obtain

$$(2.7.5) \quad F(y \mid x) = \lim_{h \to 0} P(Y < y \mid x \leqslant X < x + h) = \frac{\displaystyle\int_{-\infty}^{y} f(x, y)\, dy}{f_1(x)}.$$

For a fixed value of x, expression (2.7.5) is the conditional distribution function of the random variable Y. From our assumptions it follows that it is a distribution function of a random variable of the continuous type. Its density function, which we shall denote by $g(y \mid x)$, is given by the formula

$$(2.7.6) \qquad g(y \mid x) = \frac{f(x, y)}{f_1(x)}.$$

Similarly we can express the density and the distribution function of the conditional distribution of the random variable X under the condition that $Y = y$.

Let us write formula (2.7.5) in the form

$$f_1(x) F(y \mid x) = \int_{-\infty}^{y} f(x, y)\, dy.$$

It follows that

$$(2.7.7) \qquad \int_{-\infty}^{\infty} f_1(x) F(y \mid x)\, dx = F_2(y).$$

Formula (2.7.7) is a generalization of theorem 1.6.1 to the case of random variables of the continuous type.

When we consider random variables with more than two dimensions, we can consider conditional distributions under the condition that a certain group of random variables takes on certain constant values. For example, let us consider the three-dimensional random variable (X_1, X_2, X_3) of the continuous type with the density $f(x_1, x_2, x_3)$ which is everywhere continuous and with all the marginal densities continuous. We can obtain here, for instance, the following conditional distribution functions:

$$(2.7.8) \qquad F(x_3 \mid x_1, x_2) = \frac{\displaystyle\int_{-\infty}^{x_3} f(x_1, x_2, x_3)\, dx_3}{\displaystyle\int_{-\infty}^{\infty} f(x_1, x_2, x_3)\, dx_3},$$

$$(2.7.9) \quad F(x_3, x_2 \mid x_1) = \frac{\displaystyle\int_{-\infty}^{x_2} \left(\int_{-\infty}^{x_3} f(x_1, x_2, x_3)\, dx_3 \right) dx_2}{\displaystyle\int_{-\infty}^{\infty} \int_{-\infty}^{\infty} f(x_1, x_2, x_3)\, dx_3\, dx_2}.$$

It is assumed, of course, that the denominators on the right-hand sides of (2.7.8) and (2.7.9) are positive. In an analogous way we can obtain the conditional densities.

Example 2.7.1. Let us return to Table 2.6.1. We want to determine the distribution of even numbers according to their divisibility by 3. In other words, we ask: What is the probability that a number chosen at random will be divisible by 3 given that it is even? With this formulation of the problem we are interested not in the distribution of divisibility by 3 in the whole set of 21 numbers, but only in some part of it, namely, in the set of 10 even numbers.

In the notation of example 2.6.1 we have, according to (2.7.2),

$$P(Y = 1 \mid X = 1) = \frac{P(Y = 1, X = 1)}{P(X = 1)} = \frac{p_{11}}{p_{1.}} = \frac{3}{10},$$

$$P(Y = 0 \mid X = 1) = \frac{P(Y = 0, X = 1)}{P(X = 1)} = \frac{p_{12}}{p_{1.}} = \frac{7}{10}.$$

Similarly, we can obtain the conditional distribution of odd numbers into those divisible and not divisible by 3,

$$P(Y = 0 \mid X = 0) = \frac{P(Y = 0, X = 0)}{P(X = 0)} = \frac{p_{22}}{p_{2.}} = \frac{7}{11},$$

$$P(Y = 1 \mid X = 0) = \frac{P(Y = 1, X = 0)}{P(X = 0)} = \frac{p_{21}}{p_{2.}} = \frac{4}{11}.$$

We can also obtain the conditional distributions of the numbers divisible by 3 and numbers not divisible by 3 into even and odd numbers.

C   The truncated distribution of a random variable X is a particular case of a conditional distribution of X.

Definition 2.7.1. Let X be a random variable and S a Borel set on the x-axis such that $0 < P(X \in S) < 1$. The conditional distribution defined for any real x by the expression $P(X < x \mid X \in S)$ is called the *truncated distribution* of X.

If X is of the discrete type with jump points x_i and jumps p_i, the probability function of the truncated distribution of X is of the form

$$P(X = x_i \mid X \in S) = \frac{P(X = x_i, X \in S)}{P(X \in S)} = \begin{cases} \dfrac{p_i}{\sum\limits_{x_j \in S} p_j} & \text{if } x_i \in S, \\ 0 & \text{if } x_i \notin S. \end{cases}$$

If X is of the continuous type with density $f(x)$, then

$$P(X < x \mid X \in S) = \frac{P(X < x, X \in S)}{P(X \in S)} = \frac{\int_{(-\infty,x) \cap S} f(x) \, dx}{\int_S f(x) \, dx}.$$

The density $g(x)$ of this distribution takes the form

$$g(x) = \begin{cases} \dfrac{f(x)}{\displaystyle\int_S f(x)\,dx} & \text{if } x \in S, \\[2ex] 0 & \text{if } x \notin S. \end{cases}$$

Example 2.7.2. Consider the random variable X with density

$$f(x) = \begin{cases} 1 & (0 \leqslant x \leqslant 1), \\ 0 & (x < 0, x > 1). \end{cases}$$

Let us take the interval $[0, \tfrac{1}{2})$ as the set S. We then have

$$P(X \in S) = \int_0^{1/2} dx = \tfrac{1}{2}.$$

The density $g(x)$ of the truncated distribution of X is of the form

$$g(x) = \begin{cases} 2 & (0 \leqslant x \leqslant \tfrac{1}{2}), \\ 0 & (x < 0, x > \tfrac{1}{2}). \end{cases}$$

2.8 INDEPENDENT RANDOM VARIABLES

A   In Section 1.7 we introduced the notion of independence of random events, which we shall now apply to random variables.

Let $F(x, y)$, $F_1(x)$ and $F_2(y)$ denote, respectively, the two-dimensional distribution function of the random variable (X, Y) and the marginal distribution functions of the random variables X and Y.

Definition 2.8.1. The random variables X and Y are said to be _independent_ if for every pair (x, y) of real numbers the equality

$$(2.8.1) \qquad F(x, y) = F_1(x)F_2(y)$$

is satisfied.

Now let (a, b) and (c, d), where $a < c$ and $b < d$, be two arbitrary points in the plane (x, y). We have

$$P(a \leqslant X < c) = F_1(c) - F_1(a),$$
$$P(b \leqslant Y < d) = F_2(d) - F_2(b).$$

Multiplying the right-hand sides and the left-hand sides of these relations and applying formula (2.8.1) gives

$$P(a \leqslant X < c)P(b \leqslant Y < d) = F(c, d) - F(c, b) - F(a, d) + F(a, b).$$

From (2.5.3) we obtain

$$(2.8.2) \qquad P(a \leqslant X < c, b \leqslant Y < d) = P(a \leqslant X < c)P(b \leqslant Y < d).$$

Let us consider a two-dimensional random variable (X, Y) of the discrete type with jump points (x_i, y_k) and jumps p_{ik}. Suppose that X and Y are independent. Then in the special case of (2.8.2) for which the rectangles $(a \leqslant X < c, b \leqslant Y < d)$ are reduced to the points $(X = a, Y = b)$ (that is, when $a \leftarrow c$ and $b \leftarrow d$) we obtain, by theorem 1.3.4, the equality

$$(2.8.3) \quad p_{ik} = P(X = x_i, Y = y_k) = P(X = x_i)P(Y = y_k) = p_{i.}\, p_{.k}$$

for every pair (x_i, y_k). It is easy to show that if equality (2.8.3) is satisfied for every pair (x_i, y_k), relation (2.8.1) is also satisfied. We have thus shown that *if (X, Y) is a random variable of the discrete type, equality* (2.8.3) *holding for every pair (x_i, y_k) is a necessary and sufficient condition for the independence of the random variables X and Y.*

From (2.7.1) and (2.7.2), for every pair of numbers (i, k), we obtain

$$(2.8.4) \quad \begin{aligned} P(X = x_i \mid Y = y_k) &= P(X = x_i), \\ P(Y = y_k \mid X = x_i) &= P(Y = y_k). \end{aligned}$$

From (2.8.4) it follows that if the random variables X and Y are independent, the distribution of X is the same for all values of the random variable Y. Thus no value obtained for the variable Y gives any information about the distribution of the variable X, and conversely, the conditional distribution of the random variable Y is identical for all values of X.

If the random variable (X, Y) is of the continuous type, differentiation of the expression (2.8.1) with respect to x and y, with the possible exception of the set of points at which the density function $f(x, y)$ is not continuous, gives

$$(2.8.5) \quad \frac{\partial^2 F(x, y)}{\partial x\, \partial y} = f(x, y) = F_1{}'(x)F_2{}'(y) = f_1(x)f_2(y).$$

Conversely, let equation (2.8.5) be satisfied. Then

$$\begin{aligned} F(x, y) &= \int_{-\infty}^{x} \left[\int_{-\infty}^{y} f(x, y)\, dy \right] dx = \int_{-\infty}^{x} \left[\int_{-\infty}^{y} f_1(x)f_2(y)\, dy \right] dx \\ &= \int_{-\infty}^{x} f_1(x)\, dx \int_{-\infty}^{y} f_2(y)\, dy = F_1(x)F_2(y). \end{aligned}$$

Hence we have shown that *if (X, Y) is a random variable of the continuous type whose density function $f(x, y)$ is everywhere continuous, the validity of* (2.8.5) *for arbitrary points (x, y) is a necessary and sufficient condition for the independence of the random variables X and Y.*

Let S_1 and S_2 be two arbitrary Borel sets on the x-axis and y-axis, respectively. It can be shown that if the random variables X and Y are independent,

$$(2.8.6) \quad P(X \in S_1, Y \in S_2) = P(X \in S_1)P(Y \in S_2).$$

From relations (2.7.5) and (2.8.5) we obtain, for every value of x

(2.8.7) $$F(y \mid x) = F_2(y)$$

and similarly, for every value of y

(2.8.8) $$F(x \mid y) = F_1(x).$$

Thus, if the random variables X and Y are independent, the conditional distribution function of Y, given $X = x$, does not depend on x. The same is true of the conditional distribution function of the random variable X, given $Y = y$. Equalities (2.8.7) and (2.8.8), and also equality (2.8.4), give the most intuitive sense of the notion of independence.

Example 2.8.1. Consider two consecutive tosses of a coin. The random variable X takes on the value 0 or 1 according to whether heads or tails appear as a result of the first toss. The random variable Y takes on the value 0 or 1 according to whether heads or tails appear as a result of the second toss.

As a result of two tosses, the two-dimensional random variable (X, Y) may take on one of the values, $(1, 1)$, $(1, 0)$, $(0, 1)$, $(0, 0)$.

The probability of each of these events is the same, and hence equals $\frac{1}{4}$. Both X and Y take on the values 0 and 1 with probability $\frac{1}{2}$. Thus we have

$$P(X = 1,\ Y = 1) = \tfrac{1}{4} = \tfrac{1}{2} \cdot \tfrac{1}{2} = P(X = 1)P(Y = 1),$$

$$P(X = 1,\ Y = 0) = \tfrac{1}{4} = P(X = 1)P(Y = 0),$$

$$P(X = 0,\ Y = 1) = \tfrac{1}{4} = P(X = 0)P(Y = 1),$$

$$P(X = 0,\ Y = 0) = \tfrac{1}{4} = P(X = 0)P(Y = 0).$$

Therefore X and Y are independent.

Let X_1 and X_2 be two independent random variables. Consider two single-valued functions $Y_1 = g_1(X_1)$ and $Y_2 = g_2(X_2)$. Y_1 and Y_2 are also random variables (see note on page 37). We shall show that Y_1 and Y_2 are independent. Denote by $h_1(-\infty, y_1)$ and $h_2(-\infty, y_2)$, respectively, the Borel sets into which the functions inverse to g_1 and g_2 map the intervals $(-\infty, y_1)$ and $(-\infty, y_2)$. We have

$$
\begin{aligned}
(2.8.9)\quad P(Y_1 < y_1,\ Y_2 < y_2) &= P[g_1(X_1) < y_1, g_2(X_2) < y_2] \\
&= P[X_1 \in h_1(-\infty, y_1), X_2 \in h_2(-\infty, y_2)] \\
&= P[X_1 \in h_1(-\infty, y_1)]P[X_2 \in h_2(-\infty, y_2)] \\
&= P(Y_1 < y_1)P(Y_2 < y_2).
\end{aligned}
$$

B   The notion of independence can be generalized to an arbitrary finite or countable number of random variables.

Definition 2.8.2. The random variables $X_1, X_2, \ldots, X_n$ are called *independent* if for n arbitrary real numbers $(x_1, \ldots, x_n)$ the following relation is satisfied:

$$(2.8.10) \qquad F(x_1, x_2, \ldots, x_n) = F_1(x_1)F_2(x_2) \ldots F_n(x_n),$$

where $F(x_1, \ldots, x_n)$ is the distribution function of the random variable $(X_1, \ldots, X_n)$ and $F_1(x_1), \ldots, F_n(x_n)$ are the marginal distribution functions of $X_1, \ldots, X_n$.

We notice that if the random variables $X_1, \ldots, X_n$ are independent, then for every $s \leqslant n$ the random variables $X_{k_1}, X_{k_2}, \ldots X_{k_s}$, where $1 \leqslant k_1 < k_2 \ldots < k_s \leqslant n$, are also independent. For simplicity in notation assume that $k_1 = 1$, $k_2 = 2, \ldots, k_s = s$. It follows from (2.8.10) that

$$F(x_1, x_2, \ldots, x_s, +\infty, \ldots, +\infty)$$

$$= \lim_{x_{s+1} \to \infty, \ldots, x_n \to \infty} F(x_1, x_2, \ldots, x_s, x_{s+1}, \ldots, x_n)$$

$$= \lim_{x_{s+1} \to \infty, \ldots, x_n \to \infty} [F_1(x_1)F_2(x_2) \ldots F_s(x_s)F_{s+1}(x_{s+1}) \ldots F_n(x_n)]$$

$$= F_1(x_1)F_2(x_2) \ldots F_s(x_s)F_{s+1}(+\infty) \ldots F_n(+\infty)$$

$$= F_1(x_1)F_2(x_2) \ldots F_s(x_s).$$

We now give the definition of independence of a countable number of random variables.

Definition 2.8.3. The random variables $X_1, X_2, \ldots, X_n, \ldots$ are called *independent* if for every $n = 2, 3, \ldots$ the random variables $X_1, X_2, \ldots, X_n$ are independent.

C The notion of independence can be extended to random vectors as follows.

Definition 2.8.4. The random vectors

$$X = (X_1, \ldots, X_j) \quad \text{and} \quad Y = (Y_1, \ldots, Y_r)$$

are independent if for every $j + r$ real numbers $x_1, \ldots, x_j, y_1, \ldots, y_r$ we have

$$F(x_1, \ldots, x_j, y_1, \ldots, y_r) = G(x_1, \ldots, x_j)H(y_1, \ldots, y_r),$$

where F, G, and H are the distribution functions of the random vectors $(X_1, \ldots, X_j, Y_1, \ldots, Y_r)$, $(X_1, \ldots, X_j)$, and $(Y_1, \ldots, Y_r)$, respectively.

We stress the fact that from the independence of the random vectors $X = (X_1, \ldots, X_j)$ and $Y = (Y_1, \ldots, Y_r)$ it does not follow that the components $X_1, \ldots, X_j$ of X or the components $Y_1, \ldots, Y_r$ of Y are independent.

The notion of independence can be generalized to the case of an arbitrary finite or countable number of random vectors. The definitions are analogous to definitions 2.8.2 and 2.8.3.

2.9 FUNCTIONS OF MULTIDIMENSIONAL RANDOM VARIABLES

A   The considerations of Section 2.4 can be applied without difficulty to functions of multidimensional random variables. Here we give the formula for the two-dimensional probability density of a function of a random variable (X, Y) of the continuous type, corresponding to formula (2.4.19) for a one-dimensional random variable. Let

(2.9.1) $U_1 = g_1(X, Y), \qquad U_2 = g_2(X, Y)$

be a continuous one-to-one mapping of the random variable (X, Y) with density $f(x, y)$. Suppose that the functions g_1 and g_2 have continuous partial derivatives with respect to x and y, and let $(a \leqslant X < b, c \leqslant Y < d)$ be a rectangle on which the Jacobian of the transformation (2.9.1) is different from zero.

Denote by $x = h_1(u_1, u_2)$ and $y = h_2(u_1, u_2)$ the inverse transformation. By our assumptions, the functions h_1 and h_2 are also one-to-one and have continuous partial derivatives with respect to u_1 and u_2. Denote by J the Jacobian of the inverse transformation

$$J = \begin{vmatrix} \dfrac{\partial x}{\partial u_1} & \dfrac{\partial x}{\partial u_2} \\[2ex] \dfrac{\partial y}{\partial u_1} & \dfrac{\partial y}{\partial u_2} \end{vmatrix}$$

By our assumptions, this Jacobian is finite and continuous in the domain S of the plane (u_1, u_2), where S is the image of the rectangle $(a \leqslant X < b, c \leqslant Y < c)$ given by transformation (2.9.1). We have

$$(2.9.2) \quad P(a \leqslant X < b, c \leqslant Y < d) = \int_a^b \left[\int_c^d f(x, y) \, dy \right] dx$$

$$= \iint\limits_{(S)} f[h_1(u_1, u_2), h_2(u_1, u_2)] \, |J| \, du_1 \, du_2.$$

It follows from (2.9.2) that the two-dimensional density of the random variable (U_1, U_2) has the form

$$(2.9.3) \qquad r(u_1, u_2) = f[h_1(u_1, u_2), h_2(u_1, u_2)] \, |J|.$$

B We now investigate the distribution of the sum, difference, product, and ratio of two random variables. They are given here as examples of continuous functions of multidimensional random variables, but at the same time the formulas involved are very important in probability theory and its applications.

Example 2.9.1. Consider again two consecutive throws of a die. Let the random variable X correspond to the result of the first throw and Y to the result of the second throw. The random variables X and Y are independent. Both X and Y take on the values $1, \ldots, 6$, each with probability $\frac{1}{6}$. Hence the two-dimensional random variable (X, Y) can take on the pairs of values (i, k), where i and k run over all integers from 1 to 6. Let us form the value of the sum $i + k$ for every possible pair (i, k). All possible values of the sum $i + k$ form the set of possible values of a new random variable which will be called the *sum* of the random variables X and Y. This sum is again a one-dimensional random variable and takes on the following values: $2, 3, 4, \ldots, 11, 12$.

Let $Z = X + Y$. We shall compute the probability function of Z. Because of the independence of X and Y we have

$$P(Z = 2) = P(X = 1)P(Y = 1) = \tfrac{1}{36},$$
$$P(Z = 3) = P(X = 1, Y = 2) + P(X = 2, Y = 1) = \tfrac{1}{18},$$
$$P(Z = 4) = P(X = 1, Y = 3) + P(X = 2, Y = 2) + P(X = 3, Y = 1) = \tfrac{1}{12},$$
$$\dotfill$$
$$P(Z = 12) = P(X = 6, Y = 6) = \tfrac{1}{36}.$$

The reader can verify that

$$P(Z = 2) + P(Z = 3) + \ldots + P(Z = 12) = 1.$$

In the example of the double throw of a die we could also have considered the random variable $V = X - Y$. The set of all possible values of this random variable consists of all possible values of the difference of the numbers i and k. The random variable V then takes on the values

$$-5, -4, \ldots, 0, 1, 2, \ldots, 5.$$

For example

$$P(V = 0) = P(X = 1, Y = 1) + P(X = 2, Y = 2) + P(X = 3, Y = 3)$$
$$+ P(X = 4, Y = 4) + P(X = 5, Y = 5) + P(X = 6, Y = 6) = \tfrac{1}{6},$$
$$P(V = -5) = P(X = 1, Y = 6) = \tfrac{1}{36}.$$

The reader can verify that

$$\sum_{k=-5}^{5} P(V = k) = 1.$$

Similarly, we could have considered the probability function of the random variable $T = XY$ and of $S = Y/X$.

We can see from this example what is understood by the sum, difference, product, and ratio of two random variables. Thus the random variable $X + Y$ is a function of the two-dimensional random variable (X, Y). The set of possible values of $X + Y$ is formed from all possible values of the sum $x + y$, where x is a possible value of X and y is a possible value of Y. Similarly, the sum of any finite number of random variables can be defined.

The set of possible values of the random variable $X - Y$ consists of all values $x - y$, where x is a possible value of X and y is a possible value of Y. The product and ratio of two random variables are defined in an analogous way.

C   Suppose we are given the distribution of the two-dimensional random variable (X, Y). We shall find the distributions of the random variables obtained as a result of four arithmetic operations performed on X and Y.

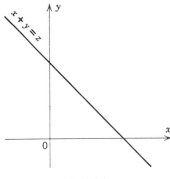

Fig. 2.9.1

We shall find the distribution function of the sum

$$(2.9.4) \qquad Z = X + Y$$

of two random variables or, in other words, the function

$$F(z) = P(X + Y < z).$$

If for a given value of z we draw on the plane (x, y) the line $x + y = z$ (see Fig. 2.9.1), then $F(z)$ will be the probability that the point with coordinates x, y lies below this line.

If (X, Y) is a random variable of the discrete type and takes on values (x_i, y_k),

$$(2.9.5) \qquad F(z) = \sum_{x_i + y_k < z} P(X = x_i, Y = y_k),$$

where the summation is extended over all the values (x_i, y_k) for which the inequality under the summation sign is satisfied.

Let (X, Y) be a random variable of the continuous type and let $f(x, y)$, $f_1(x)$, $f_2(y)$ denote respectively the densities of the random variables (X, Y), X and Y. Let us write equality (2.9.4) in the form

$$(2.9.6) \qquad X = X, Z = X + Y \qquad \text{or} \qquad X = X, Y = Z - X,$$

where the identity $X = X$ is added in order to reduce this problem to a special case of transformation (2.9.1) considered at the beginning of this section. We have

$$J = \begin{vmatrix} 1 & 0 \\ -1 & 1 \end{vmatrix} = 1.$$

From (2.9.3) it follows that the density of (X, Z) is

(2.9.7) $$f(x, z - x).$$

The density $\psi(z)$ of Z is obtained as a marginal density of the two-dimensional random variable (X, Z) by integrating (2.9.7) with respect to x from $-\infty$ to $+\infty$,

(2.9.8) $$\psi(z) = \int_{-\infty}^{+\infty} f(x, z - x)\, dx.$$

Finally we obtain

(2.9.9) $$F(z) = \int_{-\infty}^{z} \left[\int_{-\infty}^{+\infty} f(x, z - x)\, dx \right] dz.$$

If the random variables X and Y are independent, according to (2.8.5), we have

$$f(x, y) = f_1(x) f_2(y).$$

Hence

(2.9.8′) $$\psi(z) = \int_{-\infty}^{+\infty} f_1(x) f_2(z - x)\, dx$$

and

(2.9.9′) $$F(z) = \int_{-\infty}^{z} \left[\int_{-\infty}^{+\infty} f_1(x) f_2(z - x)\, dx \right] dz.$$

Because of the symmetry of the sum, we can replace in formulas (2.9.8), (2.9.8′), (2.9.9), and (2.9.9′), x by $z - y$ and $z - x$ by y.

Example 2.9.2. Consider the random variable X with the density

(2.9.10) $$f(x) = \frac{1}{\sqrt{2\pi}}\, e^{-x^2/2}.$$

Instead of the expression $e^{-x^2/2}$ we often write $\exp(-x^2/2)$. The reader will find in Franklin's book [1] the proof that

$$\frac{1}{\sqrt{2\pi}} \int_{-\infty}^{+\infty} e^{-x^2/2}\, dx = 1.$$

Expression (2.9.10) is the density of the *Gauss distribution*, which is also called the *normal distribution*. This distribution will be treated more extensively later.

Let the random variable Y have the density

$$f(y) = \frac{1}{\sqrt{2\pi}}\, e^{-y^2/2}.$$

Suppose that X and Y are independent. Thus the density of the joint random variable (X, Y) is

(2.9.11) $$f(x, y) = \frac{1}{\sqrt{2\pi}}\, e^{-x^2/2} \cdot \frac{1}{\sqrt{2\pi}}\, e^{-y^2/2} = \frac{1}{2\pi}\, e^{-(x^2+y^2)/2}.$$

Consider the random variable $Z = X + Y$. By (2.9.8′) we have for the density of Z

$$\psi(z) = \int_{-\infty}^{+\infty} \frac{1}{\sqrt{2\pi}} \exp\left(-\frac{x^2}{2}\right) \frac{1}{\sqrt{2\pi}} \exp\left(-\frac{(z-x)^2}{2}\right) dx$$

$$= \frac{1}{2\pi} \int_{-\infty}^{+\infty} \exp\left(-\frac{2x^2 - 2zx + z^2}{2}\right) dx.$$

Since

$$2x^2 - 2zx + z^2 = (x\sqrt{2})^2 - 2(x\sqrt{2})\frac{z}{\sqrt{2}} + \left(\frac{z}{\sqrt{2}}\right)^2 + \frac{z^2}{2} = \left(x\sqrt{2} - \frac{z}{\sqrt{2}}\right)^2 + \frac{z^2}{2},$$

we have

$$-\tfrac{1}{2}(2x^2 - 2zx + z^2) = -\tfrac{1}{2}\left(x\sqrt{2} - \frac{z}{\sqrt{2}}\right)^2 - \frac{z^2}{4},$$

and hence

$$\psi(z) = \frac{1}{2\pi} \int_{-\infty}^{+\infty} \exp\left[-\tfrac{1}{2}\left(x\sqrt{2} - \frac{z}{\sqrt{2}}\right)^2\right] \exp\left(-\frac{z^2}{4}\right) dx$$

$$= \frac{1}{2\sqrt{\pi}} \exp\left(-\frac{z^2}{4}\right) \frac{1}{\sqrt{\pi}} \int_{-\infty}^{+\infty} \exp\left[-\tfrac{1}{2}\left(x\sqrt{2} - \frac{z}{\sqrt{2}}\right)^2\right] dx.$$

Introducing the substitution $u = x\sqrt{2} - z/\sqrt{2}$ into the last integral we obtain

$$\frac{1}{\sqrt{\pi}} \int_{-\infty}^{+\infty} \exp\left[-\tfrac{1}{2}\left(x\sqrt{2} - \frac{z}{\sqrt{2}}\right)^2\right] dx = \frac{1}{\sqrt{2\pi}} \int_{-\infty}^{+\infty} \exp\left(-\frac{u^2}{2}\right) du = 1,$$

and finally

$$\psi(z) = \frac{1}{2\sqrt{\pi}} \exp\left(-\frac{z^2}{4}\right).$$

Thus the distribution function $F(z)$ is given by the formula

$$F(z) = \frac{1}{2\sqrt{\pi}} \int_{-\infty}^{z} e^{-z^2/4}\, dz.$$

We leave it to the reader as an exercise to derive the formulas for the distribution function and the density of the difference of two random variables.

Let us now consider the product of two random variables X and Y with the two-dimensional density $f(x, y)$ and the marginal densities $f_1(x)$ and $f_2(y)$, respectively. Let

(2.9.12) $Z = XY.$

This equality may be written as a system of equalities

$$X = X, \qquad Y = \frac{Z}{X}.$$

We have

$$J = \begin{vmatrix} 0 & 1 \\ \dfrac{1}{x} & \dfrac{-z}{x^2} \end{vmatrix} = -\frac{1}{x},$$

and hence $|J| = 1/|x|$. It follows from (2.9.3) that the two-dimensional density of the random variable (Z, X) has the form

$$(2.9.13) \qquad f\left(x, \frac{z}{x}\right)\frac{1}{|x|}.$$

The density $\psi(z)$ of Z can be obtained by integrating expression (2.9.13) from $-\infty$ to $+\infty$. We have

$$(2.9.14) \qquad \psi(z) = \int_{-\infty}^{+\infty} f\left(x, \frac{z}{x}\right)\frac{1}{|x|}\, dx.$$

The distribution function of Z has the form

$$(2.9.15) \qquad F(z) = \int_{-\infty}^{z}\left[\int_{-\infty}^{+\infty} f\left(x, \frac{z}{x}\right)\frac{1}{|x|}\, dx\right] dz.$$

If the random variables X and Y are independent, equalities (2.9.14) and (2.9.15) can be written as

$$(2.9.14') \qquad \psi(z) = \int_{-\infty}^{+\infty} \frac{1}{|x|} f_1(x) f_2\left(\frac{z}{x}\right) dx$$

and

$$(2.9.15') \qquad F(z) = \int_{-\infty}^{z}\left[\int_{-\infty}^{+\infty} \frac{1}{|x|} f_1(x) f_2\left(\frac{z}{x}\right) dx\right] dz.$$

We can replace in (2.9.14), (2.9.14'), (2.9.15), and (2.9.15'), x by z/y, z/x by y, and $|x|$ by $|y|$.

For the ratio of two random variables,

$$Z = \frac{X}{Y},$$

we obtain the following formulas:

$$(2.9.16) \qquad \psi(z) = \int_{-\infty}^{+\infty} f(yz, y)\, |y|\, dy$$

and

$$(2.9.17) \qquad F(z) = \int_{-\infty}^{z}\left[\int_{-\infty}^{+\infty} f(yz, y)\, |y|\, dy\right] dz.$$

If the random variables X and Y are independent,

$$(2.9.16')\qquad \psi(z) = \int_{-\infty}^{+\infty} f_1(yz)f_2(y)\,|y|\,dy,$$

$$(2.9.17')\qquad F(z) = \int_{-\infty}^{z} \left[\int_{-\infty}^{+\infty} f_1(yz)f_2(y)\,|y|\,dy\right]dz.$$

The formulas for the distribution of the sum, difference, product, and ratio of two independent random variables X and Y, without the assumption that the random vector (X, Y) is of the continuous type, are given in Section 6.5. (For the proofs of those formulas, see Supplement.)

2.10 ADDITIONAL REMARKS

In this chapter we have investigated random vectors, hence random variables which take on values from finite-dimensional Euclidean spaces. Recently the theory of much more general random variables has developed, namely, the theory of random variables which take on values from an infinite dimensional space, for instance, a functional space. Some remarks on this subject are found in Section 8.1. The basic information on this theory can be found in the papers of Kolmogorov [8], Mourier [1], Prohorov [3], Getoor [2], and in the monographs of Doob [5] and Loève [4].

Problems and Complements

2.1. X is a random variable. Is $|X|$ a random variable too?

2.2. The probability function of the random variable X is of the form

$$P(X = r) = e^{-\lambda}\frac{\lambda^r}{r!}\qquad (r = 0, 1, \ldots).$$

Find the probability functions of the random variables (a) $Y = -X$, (b) $Y = aX + b$, (c) $Y = X^2$, (d) $Y = \sqrt{X}$, (e) $Y = X^l$ (l an integer).

2.3. $F(x)$ and $f(x)$ are the distribution function and the density, respectively, of a random variable X. Find the distribution function and the density of the random variables (a) $Y = 1/X$, (b) $Y = e^X$, (c) $Y = \sin X$, (d) $Y = |X|$, (e) $Y = F(X)$.

2.4. The distribution of the random vector (X, Y) is given by the formulas

$$P(X = 1,\ Y = 1) = P(X = 1,\ Y = 2) = P(X = 2,\ Y = 2) = \tfrac{1}{3}.$$

(a) Find the distribution functions $F(x, y)$, $F_1(x)$, and $F_2(y)$.

(b) Check whether the points $(1, \tfrac{1}{2})$, $(1, 3)$, $(2, \tfrac{1}{2})$, and $(2, 3)$ are discontinuity points of $F(x, y)$.

2.5. The probability function of the random vector (X, Y) is of the form

$$P(X = 0, 1 \leqslant Y \leqslant 2) = P(1 \leqslant X \leqslant 2, Y = 0) = \tfrac{1}{2}.$$

(a) Find the discontinuity points of the distribution function $F(x, y)$.

(*b*) Check whether the rectangle with vertices $(0, 0)$, $(2, 0)$, $(0, \frac{1}{2})$ and $(2, \frac{1}{2})$ is a continuity rectangle.

2.6. Prove formula (2.5.12) (*a*) for $n = 3$, (*b*) for an arbitrary n.

2.7. Give an example of three dependent random variables X, Y, Z which are pairwise independent.

Hint: Modify example 1.7.1. For other examples, see Geisser and Mantel [1].

2.8. $F(x, y)$ and $f(x, y)$ are the distribution function and the density, respectively, of a random vector (X, Y). Find the distribution function and the density of the random vector (Z, U) defined as

$$Z = aX + bY + c,$$
$$U = dX + eY + f.$$

2.9. The random variables $X_1, X_2, \ldots, X_n$ are independent. Prove that (*a*) if at least one X_j is of the continuous type, the sum $X_1 + X_2 + \ldots + X_n$ is also of the continuous type, (*b*) if at least one X_j has a continuous distribution function the sum $X_1 + X_2 + \ldots + X_n$ also has a continuous distribution function.

Hint: Use formula (6.5.6).

2.10. Prove that if in the last problem we omit the assumption of independence, theorems (*a*) and (*b*) are no longer valid.

Hint: Give an example.

CHAPTER 3

Parameters of the Distribution
of a Random Variable

3.1 EXPECTED VALUES

With every distribution of a random variable there are associated certain numbers called the *parameters of the distribution*, which play an important role in mathematical statistics. The parameters of a distribution are the moments and functions of them and also the order parameters. We shall introduce these notions successively.

Let X be a random variable. Consider the single-valued function $g(X)$ of X.

Definition 3.1.1. Let X be a random variable of the discrete type with jump points x_k and jumps p_k. The series

$$(3.1.1) \qquad E[g(X)] = \sum_k p_k g(x_k)$$

is called the *expected value* of the random variable $g(X)$ if the following inequality is satisfied:

$$(3.1.1') \qquad \sum_k p_k |g(x_k)| < \infty.$$

Definition 3.1.2. Let X be a random variable of the continuous type with density function $f(x)$. Let $g(x)$ be Riemann integrable. The integral

$$(3.1.2) \qquad E[g(X)] = \int_{-\infty}^{\infty} g(x) f(x)\, dx$$

is called the *expected value* of the random variable $g(X)$ if the following inequality is satisfied:

$$(3.1.2') \qquad \int_{-\infty}^{\infty} |g(x)| f(x)\, dx < \infty.$$

64

When the right side of formula (3.1.1) exists but formula (3.1.1') is not satisfied, we say that the expected value of the random variable $g(X)$ does not exist.

Similarly, $E[g(X)]$ fails to exist when the right-hand side of (3.1.2) exists but the inequality (3.1.2') is not satisfied.

Let $g(X) = X$. The number $E(X)$ defined by formula (3.1.1) or (3.1.2) is called the *expected value* of the random variable X. (For the definition of $E(X)$, see also Supplement.) The expected value $E(X)$ is sometimes called the *mathematical expectation*, or, in brief, the *expectation* or the *mean value* of the random variable X.

We shall show that the definition of the mathematical expectation of the function $g(X)$ determines this value uniquely, that is, if $Y = g(X)$, the expected value $E[g(X)]$ equals the expected value $E(Y)$ computed directly from the distribution of the random variable Y. We shall restrict ourselves to the proof for the case when X is a random variable of the discrete type. Let X have jump joints x_k and jumps p_k, and let Y have jump joints y_j and jumps q_j. Notice that q_j ($j = 1, 2, \ldots$) equals the sum of the probabilities p_k for those k for which the equality $g(x_k) = y_j$ holds. Since, by the assumption about the existence of the expectation $E[g(X)]$, the series $\sum_R p_k g(x_k)$ is absolutely convergent, we can group its terms arbitrarily. Hence we obtain

$$E(Y) = \sum_j q_j y_j = \sum_k p_k g(x_k) = E[g(X)].$$

The following are some examples of computing expected values.

Example 3.1.1.   Suppose that the random variable X can take on two values $x_1 = -1$ with probability $p_1 = 0.1$ and $x_2 = +1$ with probability $p_2 = 0.9$.
The expected value of X equals

$$E(X) = 0.1 \cdot (-1) + 0.9 \cdot (+1) = 0.8.$$

Example 3.1.2.   Let the random variable X take on the values $k = 0, 1, 2, \ldots$ and let

(3.1.3) $$P(X = k) = \frac{\lambda^k}{k!} e^{-\lambda},$$

where $\lambda > 0$ is a certain positive constant. Formula (3.1.3) represents the *Poisson distribution*.

We shall compute the expected value of X,

(3.1.4) $$E(X) = \sum_{k=0}^{\infty} k \frac{\lambda^k}{k!} e^{-\lambda} = \lambda e^{-\lambda} \sum_{k=1}^{\infty} \frac{\lambda^{k-1}}{(k-1)!} = \lambda e^{-\lambda} \sum_{r=0}^{\infty} \frac{\lambda^r}{r!} = \lambda e^{-\lambda} e^{\lambda} = \lambda.$$

Example 3.1.3.   The random variable X takes on the values $r = 0, 1, 2, \ldots, n$ with

(3.1.5) $$P(X = r) = \frac{n!}{r! \, (n-r)!} p^r (1-p)^{n-r},$$

where p is a certain constant satisfying the double inequality $0 < p < 1$. This distribution is called the *binomial distribution*, since on the right-hand side of (3.1.5) we have the terms of the expansion of $[p + (1 - p)]^n$ according to Newton's formula.

Suppose, for instance, that in a stock of goods we have good and defective items and suppose that the probability that an item chosen at random is a good one equals p. We assume that when choosing n items at random, the results of the considered n random experiments are independent events (this can be achieved, for instance, by putting back each item before choosing the next one). The probability that among n items chosen at random there will be r good ones is given by (3.1.5). It is very easy to derive this formula. Denote a good and a defective item respectively by g and d. The probability of obtaining one of the possible combinations of the letters g and d such that g appears r times and d appears $n - r$ times equals, because of the independence of these events,

$$p^r(1 - p)^{n-r}.$$

Since the number of all such combinations of r among n elements equals

$$\binom{n}{r} = \frac{n!}{r!\,(n - r)!},$$

we obtain formula (3.1.5).

The expected value of X equals

$$(3.1.6) \quad E(X) = \sum_{r=0}^{n} r \frac{n!}{r!\,(n - r)!} p^r(1 - p)^{n-r}$$

$$= \sum_{r=1}^{n} r \frac{n!}{r!\,(n - r)!} p^r(1 - p)^{n-r} = np \sum_{r=1}^{n} \frac{(n - 1)!}{(r - 1)!\,(n - r)!} p^{r-1}(1 - p)^{n-r}$$

$$= np \sum_{k=0}^{n-1} \frac{(n - 1)!}{k!\,(n - 1 - k)!} p^k(1 - p)^{n-1-k} = np[p + (1 - p)]^{n-1} = np.$$

Example 3.1.4. A random variable X is of the continuous type with the density function

$$f(x) = \frac{1}{\sqrt{2\pi}} e^{-x^2/2}.$$

The expected value of this random variable is

$$E(X) = \frac{1}{\sqrt{2\pi}} \int_{-\infty}^{+\infty} x e^{-x^2/2}\, dx = -\frac{1}{\sqrt{2\pi}} \int_{-\infty}^{+\infty} (-x) e^{-x^2/2}\, dx$$

$$= -\frac{1}{\sqrt{2\pi}} \left[e^{-x^2/2} \right]_{-\infty}^{+\infty} = 0.$$

We now give two examples of random variables which do not have an expected value.

Example 3.1.5.   Let the random variable X take on the values $x_k = (-1)^k 2^k/k$ ($k = 1, 2, \ldots$) with probabilities $p_k = \dfrac{1}{2^k}$. We have

$$\sum_{k=1}^{\infty} p_k x_k = \sum_{k=1}^{\infty} (-1)^k/k = -\log 2.$$

Nevertheless, the expected value $E(X)$ does not exist in this example, since

(3.1.7)
$$\sum_{k=1}^{\infty} p_k |x_k| = \sum_{k=1}^{\infty} 1/k = \infty.$$

Example 3.1.6.   Consider the random variable Y defined by the formula $Y = |X|$, where X has the distribution given in the preceding example. From (3.1.7) it follows that the expected value of Y does not exist.

In Chapter 5 we give an example of a random variable of the continuous type whose expected value does not exist.

3.2   MOMENTS

A   The expected value $E(X)$, discussed in Section 3.1, is a particular case of the class of parameters called moments.

Definition 3.2.1.   The expected value of the function $g(X) = X^k$, that is

$$m_k = E(X^k),$$

is called the _moment of order k_ of the random variable X.

Thus, if the random variable X is of the discrete type with jump points x_l and jumps p_l, by (3.1.1), the moment of order k is given by the formula

(3.2.1)
$$m_k = E(X^k) = \sum_l x_l^k p_l.$$

If the random variable X is of the continuous type with the density function $f(x)$, by (3.1.2), the moment of order k is given by the formula

(3.2.2)
$$m_k = E(X^k) = \int_{-\infty}^{\infty} x^k f(x)\, dx.$$

We remind the reader that for the moment m_k to exist, it is necessary that the series (3.2.1) or the integral (3.2.2) be absolutely convergent. It follows that if the moment of order k exists, all the moments of order smaller than k also exist.

We notice that if the moment of order k of a random variable X exists,

$$\lim_{a \to \infty} a^k P(|X| > a) = 0,$$

where $a > 0$. Indeed, taking into consideration the existence of the moment of order k, we have [the proof is given only for random variables of the continuous type with the density $f(x)$]

$$\lim_{a \to \infty} a^k P(|X| > a) \leqslant \lim_{a \to \infty} \int_{|x| > a} |x|^k f(x) \, dx = 0.$$

This relation may be written in a more intuitive way.

$$P(|X| > a) = o\left(\frac{1}{a^k}\right).$$

We see that the existence of moments is connected with the probability that the random variable takes on large absolute values. If the set of all possible values of a random variable is bounded from both sides, moments of arbitrary order exist.

Now let $g_1(X)$ and $g_2(X)$ be two single-valued functions of a random variable X and let the expected values $E[g_1(X)]$ and $E[g_2(X)]$ exist. We shall show that

(3.2.3) $$E[g_1(X) + g_2(X)] = E[g_1(X)] + E[g_2(X)].$$

We restrict ourselves to the case when X is of the continuous type. Then we have

(3.2.4) $$E[g_1(X) + g_2(X)] = \int_{-\infty}^{\infty} [g_1(x) + g_2(x)] f(x) \, dx.$$

Since by assumption the inequalities

$$\int_{-\infty}^{\infty} |g_1(x)| f(x) \, dx < \infty, \qquad \int_{-\infty}^{\infty} |g_2(x)| f(x) \, dx < \infty,$$

hold, the expected value of the sum $g_1(X) + g_2(X)$ exists because

$$\int_{-\infty}^{\infty} |g_1(x) + g_2(x)| f(x) \, dx \leqslant \int_{-\infty}^{\infty} |g_1(x)| f(x) \, dx + \int_{-\infty}^{\infty} |g_2(x)| f(x) \, dx.$$

From (3.2.4) we obtain

$$E[g_1(X) + g_2(X)] = \int_{-\infty}^{\infty} g_1(x) f(x) \, dx + \int_{-\infty}^{\infty} g_2(x) f(x) \, dx$$

$$= E[g_1(X)] + E[g_2(X)]$$

and (3.2.3) is proved. This formula can easily be generalized to an arbitrary finite number of terms.

The reader can easily verify the formula

(3.2.5) $$E[(aX)^k] = a^k E(X^k),$$

where a is a constant. Taking into account the fact that for an arbitrary constant b we have

$$E(b) = b,$$

we obtain

(3.2.6) $E(aX + b) = aE(X) + b.$

Example 3.2.1. The random variable X can take on two values 2 and 4, where $P(X = 2) = 0.2$ and $P(X = 4) = 0.8$. We shall find $E(X^2)$.
From (3.2.1) we have

$$E(X^2) = 4 \cdot 0.2 + 16 \cdot 0.8 = 13.6.$$

Example 3.2.2. The random variable X has the normal distribution with density $f(x) = (1/\sqrt{2\pi})e^{-x^2/2}$. We shall find $E(X^2)$.
Integrating by parts, we obtain

$$E(X^2) = \frac{1}{\sqrt{2\pi}} \int_{-\infty}^{+\infty} x^2 e^{-x^2/2}\, dx = \frac{1}{\sqrt{2\pi}} \int_{-\infty}^{+\infty} (-x)(-x e^{-x^2/2})\, dx$$

$$= \frac{1}{\sqrt{2\pi}} \left[-x e^{-x^2/2} \right]_{-\infty}^{+\infty} + \frac{1}{\sqrt{2\pi}} \int_{-\infty}^{+\infty} e^{-x^2/2}\, dx = 1.$$

We can also compute expressions of the type

(3.2.7) $E[(X - c)]^k,$

where c is an arbitrary constant.

Definition 3.2.2. Expression (3.2.7) is called the *moment of order k with respect to the point c.*
Moments with respect to the expected value, that is, with respect to the point $c = m_1 = E(X)$, are of great importance.

Definition 3.2.3. Moments with respect to the expected value are called *central moments.*
We denote them by μ_k. We then have

(3.2.8) $\mu_k = E[X - E(X)]^k.$

Moments with respect to the point $c = 0$ are called *ordinary moments*. Transforming expression (3.2.8) we can, by taking into consideration equation (3.2.3), express the central moments in terms of ordinary ones. For the first three moments we have

$$\mu_1 = E[X - E(X)] = E(X - m_1) = E(X) - m_1 = m_1 - m_1 = 0,$$

$$\mu_2 = E[X - E(X)]^2 = E[(X - m_1)^2]$$

(3.2.9) $= E(X^2) - 2m_1 E(X) + m_1^2 = m_2 - m_1^2,$

$$\mu_3 = E[X - E(X)]^3 = E[(X - m_1)^3]$$

$$= E(X^3) - 3m_1 E(X^2) + 3m_1^2 E(X) - m_1^3$$

$$= m_3 - 3m_1 m_2 + 3m_1^3 - m_1^3 = m_3 - 3m_1 m_2 + 2m_1^3.$$

Let us consider more closely the case $k = 2$. The moment

$$E[(X - c)^2]$$

is also called the *mean quadratic deviation* of the random variable X from the point c.

Definition 3.2.4. The central moment of the second order, that is, the mean deviation of the random variable X from its expected value m_1 is called *the variance*.

Usually the variance is denoted by $D^2(X)$ or σ^2. From (3.2.9) we obtain

$$(3.2.10) \qquad D^2(X) = \sigma^2 = m_2 - m_1^2.$$

The parameter σ^2 is a measure of dispersion of the random variable around its expected value. The more concentrated is the distribution, the smaller is the value of σ^2.

Definition 3.2.5. The square root of the variance is called the *standard deviation*.

Example 3.2.5. Let us compute the variance of the binomial distribution (see example 3.1.3), where

$$P(X = r) = \binom{n}{r} p^r (1 - p)^{n-r} \qquad (r = 0, 1, \ldots, n).$$

Proceeding as in the computation of the expected value $E(X)$ in example 3.1.3, we obtain

$$E(X^2) = \sum_{r=0}^{n} r^2 \binom{n}{r} p^r (1 - p)^{n-r} = pn(q + pn),$$

where $q = 1 - p$. Since $m_1 = E(X) = np$, we obtain by (3.2.10)

$$(3.2.11) \qquad D^2(X) = pn(q + pn) - p^2 n^2 = pqn.$$

Suppose, for example, that $n = 2$. Then the random variable X can take on one of the values 0, 1, 2. Suppose that $p = \frac{1}{2}$. Then $E(X) = \frac{1}{2} \times 2 = 1$.

The distribution with the values $n = 2$ and $p = 0.5$ is represented by Fig. 3.2.1, where the values of the random variable are given on the horizontal line and the probabilities of these values are given on the vertical line.

According to equality (3.2.11) in this distribution we have $D^2(X) = \frac{1}{2}$.

Let $n = 3$ and $p = \frac{1}{3}$. Then the random variable X can take the values 0, 1, 2, 3. We have $E(X) = \frac{1}{3} \cdot 3 = 1$. The expected value is the same but the variance is bigger, since we have $D^2(X) = \frac{1}{3} \cdot \frac{2}{3} \cdot 3 = \frac{2}{3}$.

The corresponding distribution is represented by Fig. 3.2.2.

Comparing Figs. 3.2.1 and 3.2.2 we see that in the second case, the dispersion around the expected value of the random variable X (which equals 1) is larger than in the first case. These examples illustrate the fact that the variance may serve as a measure of dispersion.

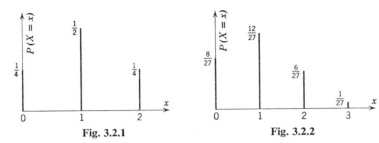

Fig. 3.2.1   Fig. 3.2.2

The variance has the following important property: *For every* $c \neq m_1$

(3.2.12) $D^2(X) < E[(X - c)^2]$.

In fact, from (3.2.3), (3.2.5), and (3.2.9) we have

$$E[(X - c)^2] = E[(X - m_1 + m_1 - c)^2]$$
$$= E[(X - m_1)^2] + 2(m_1 - c)E(X - m_1) + (m_1 - c)^2$$
$$= D^2(X) + (m_1 - c)^2.$$

From our assumption that $(m_1 - c)^2 \neq 0$, we obtain inequality (3.2.12).

We shall now find the variance of a linear function of the random variable X. Suppose that $Y = X + c$. We have

(3.2.13) $D^2(Y) = E[Y - E(Y)]^2 = E[X + c - E(X) - E(c)]^2$
$$= E[X - E(X)]^2 = D^2(X).$$

Thus, the variance is invariant under translation since the translation does not change its value.

Next, suppose that

$$Y = aX + b.$$

As a result of relations (3.2.3), (3.2.5), and (3.2.10) we obtain

(3.2.14) $D^2(Y) = E[(aX + b)^2] - [E(aX + b)]^2 = a^2E(X^2) - a^2[E(X)]^2$
$$= a^2\{E(X^2) - [E(X)]^2\} = a^2D^2(X).$$

Definition 3.2.6. A random variable X for which

$$E(X) = 0, \qquad D^2(X) = 1,$$

is called a *standardized random variable.*

If X is a random variable with expected value m_1 and standard deviation σ, the random variable Y defined as

$$Y = \frac{X - m_1}{\sigma}$$

is a standardized random variable. From formulas (3.2.6) and (3.2.14), we obtain $E(Y) = 0$ and $D^2(Y) = 1$.

Besides the standard deviation, the *mean deviation* can serve as a measure of dispersion. It is defined by the formula

(3.2.15)
$$\sum_{-\infty < x_i < \infty} p_i \,|x_i - m_1| \,,$$

if the variable X is of the discrete type with jump points x_i. It is defined by the formula

(3.2.15′)
$$\int_{-\infty}^{+\infty} |x - m_1| f(x)\,dx,$$

if X is a random variable of the continuous type. In Part 2 we shall give criteria by which we can decide which of these two measures of dispersion is better.

Sometimes it is convenient to characterize the dispersion not directly by the standard deviation, but by the ratio of the standard deviation to the expected value. This ratio is also a parameter of the distribution.

Definition 3.2.7. The ratio of the standard deviation to the expected value is called the *coefficient of variation*.

This ratio is denoted by v. We then have

(3.2.16)
$$v = \frac{\sigma}{m_1} \,.$$

The coefficient of variation v is equal to the standard deviation when the expected value equals one. In other words, the coefficient of variation is a measure of dispersion if the expected value is the unit of measurement.

Example 3.2.4. We have previously established that in the binomial distribution the expected value equals np (see example 3.1.3), and the standard deviation is npq (see example 3.2.3). Thus the coefficient of variation in the binomial distribution is given by the formula

$$v = \frac{\sqrt{npq}}{np} = \sqrt{q/np}.$$

B   We now introduce the notion of a symmetric distribution.

Definition 3.2.8. We say that the random variable X has a *symmetric distribution* if there exists a point a such that for every x the distribution function $F(x)$ of X satisfies the equation

$$F(a - x) = 1 - F(a + x) - P(X = a + x).$$

The point a is called the *center of symmetry*. In particular, if $a = 0$, for every x,

$$F(-x) = 1 - F(x) - P(X = x).$$

If a random variable with symmetric distribution is of the continuous type, its density function $f(x)$ satisfies the equation [excluding the discontinuity points, if any, of the function $f(x)$]

$$f(a - x) = f(a + x).$$

If X is of the discrete type, its jump points and their probabilities are placed symmetrically with respect to the center of symmetry.

As an immediate consequence we obtain: *If a random variable has a symmetric distribution and its expected value exists, this expected value equals the center of symmetry.*

It follows that for a symmetric distribution the central moments of odd orders (if they exist) are equal to zero.

Sometimes it is necessary to establish the extent by which the distribution departs from symmetry. Since in a symmetric distribution all the central moments of odd order equal zero, the value of every moment of odd order may serve as a measure of asymmetry of the distribution. It is convenient to express this value in the scale in which the unit of measurement is the standard deviation. Thus for the measure of asymmetry we usually employ the expression

(3.2.17) $$\gamma = \frac{\mu_3}{\sigma^3}.$$

This expression is called the *coefficient of skewness*. This coefficient may be positive or negative, and accordingly we shall speak about positive or negative asymmetry.

C   We now make some comments on the determination of the distribution function $F(x)$ of a random variable X when all the moments $m_k = E(X^k)$ $(k = 1, 2, \ldots)$ are given. We shall here state the following theorem without proof. (See Problem 4.10).

Theorem 3.2.1. *Suppose that the moments m_k $(k = 1, 2, \ldots)$ of a random variable X exist and the series*

$$\sum_{k=1}^{\infty} \frac{m_k}{k!} r^k$$

is absolutely convergent for some $r > 0$. Then the set of moments $\{m_k\}$ uniquely determines the distribution function $F(x)$ of X.

In particular, it follows from this theorem that if for some constant M,

$$|m_k| \leqslant M^k \qquad (k = 1, 2, \ldots),$$

the distribution function $F(x)$ is uniquely determined. Thus, if the set of all possible values of a random variable X is bounded from both sides, the set $\{m_k\}$ determines $F(x)$ uniquely.

If not all the moments exist, then those moments that do exist fail to determine the distribution function $F(x)$, as can be seen from the following example.

Example 3.2.5. Let the random variable X take on the values $x_k = 2^k/k^2$ ($k = 1, 2, \ldots$) with probabilities $p_k = 1/2^k$. We have

$$E(X) = \sum_{k=1}^{\infty} x_k p_k = \sum_{k=1}^{\infty} \frac{1}{k^2} < \infty,$$

$$\sum_{k=1}^{\infty} x_k^2 p_k = \sum_{k=1}^{\infty} \frac{2^k}{k^4} = \infty.$$

Let the random variable Y take on zero with probability $\frac{1}{2}$ and the values $y_k = 2^{k+1}/k^2$ with probabilities $p_k' = 1/2^{k+1}$. Then we have

$$E(Y) = \sum_{k=1}^{\infty} y_k p_k' = \sum_{k=1}^{\infty} \frac{1}{k^2} = E(X),$$

$$\sum_{k=1}^{\infty} y_k^2 p_k' = \sum_{k=1}^{\infty} \frac{2^{k+1}}{k^4} = \infty.$$

The random variables X and Y have the same moments of the first order and do not have moments of any order greater than one. The distributions of these random variables are obviously different.

Many mathematical papers have been devoted to the question of determining the distribution by the moments. We mention here the papers of Stieltjes [1] and Hamburger [1].

A condition under which the two-dimensional distribution of a random variable is determined by its moments has been given by Cramér and Wold [1].

3.3 THE CHEBYSHEV INEQUALITY

The role of the standard deviation as a parameter measuring the dispersion is particularly clear when we look at the famous *Chebyshev inequality*

$$(3.3.1) \qquad P(|X - m_1| \geqslant k\sigma) \leqslant \frac{1}{k^2},$$

which is valid for every positive k. We obtain the Chebyshev inequality as a consequence of the following theorem.

Theorem 3.3.1. *If a random variable Y can take on only non-negative values and has expected value $E(Y)$, then for an arbitrary positive number K*

$$(3.3.2) \qquad P(Y \geqslant K) \leqslant \frac{E(Y)}{K}.$$

Proof. We prove this theorem for a random variable of the continuous type (the proof for the discrete type is analogous). We have

$$E(Y) = \int_0^\infty yf(y)\,dy \geqslant \int_K^\infty yf(y)\,dy \geqslant K\int_K^\infty f(y)\,dy = KP(Y \geqslant K).$$

Inequality (3.3.2) follows immediately.

Suppose now that the random variable X has the expected value $E(X) = m_1$ and the standard deviation σ. Consider the random variable

$$(3.3.3) \qquad\qquad Y = (X - m_1)^2.$$

This random variable satisfies the assumptions of theorem 3.3.1, since $Y \geqslant 0$ and $E(Y)$ exists, where

$$E(Y) = E[(X - m_1)^2] = \sigma^2.$$

For the constant $K = k^2\sigma^2$, we obtain from (3.3.2)

$$(3.3.4) \qquad\qquad P[(X - m_1)^2 \geqslant k^2\sigma^2] \leqslant \frac{1}{k^2}.$$

Inequality (3.3.4) is equivalent to the Chebyshev inequality (3.3.1).

Inequality (3.3.1) is valid for arbitrary random variables whose variances exist.

Putting $k = 3$ in (3.3.1), we obtain

$$P[\,|X - m_1| \geqslant 3\sigma] \leqslant \tfrac{1}{9}.$$

From the following example it follows that in the class of random variables whose second order moment exists one cannot obtain a better inequality than the Chebyshev inequality.

Example 3.3.1. The random variable X has the probability function

$$P(X = -k) = P(X = k) = \frac{1}{2k^2}, \qquad P(X = 0) = 1 - \frac{1}{k^2},$$

where k is some positive constant. We have here $m_1 = 0$, $\sigma = 1$; hence for the k being considered,

$$P(\,|X - m_1| \geqslant k\sigma) = P(|X| \geqslant k)$$
$$= P(X = -k) + P(X = k) = \frac{1}{k^2}.$$

We notice that in smaller classes of random variables, for instance, for those random variables whose moment of order $2n$ ($n > 1$) exists, one can (see Cramér [2]) obtain an inequality better than (3.3.1).

An inequality, analogous to the Chebyshev inequality, for two-dimensional random variables has been given by Berge [1] (see Problem 3.9), and for random vectors with an arbitrary finite number of dimensions by Olkin and Pratt [1]. In Section 6.12,B we shall give the Kolmogorov inequality, which is a deep generalization of the Chebyshev inequality for sums of independent random variables.

3.4 ABSOLUTE MOMENTS

Definition 3.4.1. The expression $E(|X|^k)$ is called the *absolute moment of order k*. The absolute moments are denoted by β_k.

Thus, if X is a random variable of the discrete type with jump points x_i,

$$(3.4.1) \qquad \beta_k = E(|X|^k) = \sum_i |x_i|^k p_i.$$

If X is a random variable of the continuous type with density $f(x)$,

$$(3.4.1') \qquad \beta_k = E(|X|^k) = \int_{-\infty}^{+\infty} |x|^k f(x)\, dx.$$

The absolute moment of an even order equals the moment of the same order.

As we know (Section 3.2), from the existence of the moment of order k, the existence of the absolute moment of order k follows, and obviously the existence of all the absolute moments of orders smaller than k. We now prove the following theorem concerning absolute moments, which is called the *Lapunov inequality* [1].

Theorem 3.4.1. *If for a random variable X the absolute moment of order n exists, for arbitrary $k(k = 1, 2, \ldots, n - 1)$ the following inequality is then true:*

$$(3.4.2) \qquad \beta_k^{1/k} \leqslant \beta_{k+1}^{1/(k+1)}.$$

Proof. Suppose that the random variable X is of the continuous type. Let u and v be two arbitrary real numbers. Consider the non-negative expression

$$(3.4.3) \quad \int_{-\infty}^{+\infty} [u\,|x|^{(k-1)/2} + v(|x|^{(k+1)/2}]^2 f(x)\, dx$$

$$= \int_{-\infty}^{\infty} [u^2\,|x|^{k-1} + 2uv\,|x|^k + v^2\,|x|^{k+1}] f(x)\, dx$$

$$= \beta_{k-1} u^2 + 2\beta_k uv + \beta_{k+1} v^2.$$

Since the quadratic form on the right-hand side of (3.4.3) does not change sign, the inequality

$$(3.4.4) \qquad \beta_k^2 \leqslant \beta_{k-1}\beta_{k+1}$$

holds. Raising both sides of (3.4.4) to power k, we obtain

(3.4.5) $$\beta_k^{2k} \leqslant \beta_{k-1}^k \beta_{k+1}^k.$$

Putting $k = 1, 2, \ldots, n - 1$ in inequality (3.4.5), we obtain

$$\beta_1^2 \leqslant \beta_0 \beta_2,$$

$$\beta_2^4 \leqslant \beta_1^2 \beta_3^2,$$

$$\beta_3^6 \leqslant \beta_2^3 \beta_4^3,$$

$$\ldots$$

where

$$\beta_{n-1}^{2(n-1)} \leqslant \beta_{n-2}^{n-1} \beta_n^{n-1},$$

$$\beta_0 = \int_{-\infty}^{+\infty} |x|^0 f(x) \, dx = 1.$$

Multiplying the k successive inequalities, and taking into account the fact that $\beta_0 = 1$, for every $k = 1, 2, \ldots, n - 1$, we obtain the inequality

(3.4.6) $$\beta_k^{k+1} \leqslant \beta_{k+1}^k.$$

Raising both sides of (3.4.6) to power $1/[k(k + 1)]$, we obtain inequality (3.4.2).

The proof of the Lapunov inequality for a random variable of the discrete type is analogous.

3.5 ORDER PARAMETERS

Definition 3.5.1. The value x satisfying the inequalities

(3.5.1) $$P(X \leqslant x) \geqslant \tfrac{1}{2}, \qquad P(X \geqslant x) \geqslant \tfrac{1}{2}$$

is called the *median*, and is denoted by $x_{1/2}$. The pair of inequalities (3.5.1) is equivalent to the double inequality

(3.5.1') $$\tfrac{1}{2} - P(X = x) \leqslant F(x) \leqslant \tfrac{1}{2}.$$

If $P(X = x_{1/2}) = 0$, hence in particular, if the random variable X is of the continuous type, the median is the number x satisfying the equality

(3.5.2) $$F(x) = \tfrac{1}{2}.$$

It may happen that many values of x satisfy inequalities (3.5.1) or equation (3.5.2). Then each of them is called the *median*.

Example 3.5.1. Suppose that the random variable X can take on the values 0 and 1, where

$$P(X = 0) = \tfrac{1}{5}, \qquad P(X = 1) = \tfrac{4}{5}.$$

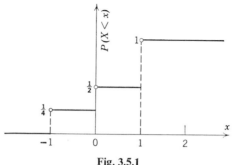

Fig. 3.5.1

The median is the point $x = 1$, since

$$P(X \leqslant 1)$$
$$= P(X = 0) + P(X = 1) = 1 > \tfrac{1}{2},$$
$$P(X \geqslant 1) = P(X = 1) = \tfrac{4}{5} > \tfrac{1}{2}.$$

Example 3.5.2. The random variable X is of the continuous type with the density defined as

$$f(x) = \begin{cases} 0 & \text{for} \quad x < 0, \\ \cos x & \text{for} \quad 0 \leqslant x \leqslant \tfrac{1}{2}\pi, \\ 0 & \text{for} \quad x > \tfrac{1}{2}\pi. \end{cases}$$

We find the median of X from equation

(3.5.2) $\quad \displaystyle\int_{-\infty}^{x_{1/2}} f(x)\, dx = \int_{0}^{x_{1/2}} \cos x\, dx = [\sin x]_0^{x_{1/2}} = \sin x_{1/2} = 0.5.$

Then

$$x_{1/2} = \tfrac{1}{6}\pi.$$

Example 3.5.3. Suppose that the random variable X can take on three values, $x_1 = -1$, $x_2 = 0$, $x_3 = +1$, with the probabilities $P(X = -1) = P(X = 0) = \tfrac{1}{4}$, $P(X = +1) = \tfrac{1}{2}$. Figure 3.5.1 represents the distribution function of this random variable.

Here each value x from the interval $(0, 1)$ is the median since for each such x the inequalities

$$P(X \leqslant x) \geqslant \tfrac{1}{2}, \qquad P(X \geqslant x) \geqslant \tfrac{1}{2}$$

are satisfied.

The median is a special case of the class of parameters called *quantiles*.

Definition 3.5.2. The value x satisfying the inequalities

(3.5.3) $\quad P(X \leqslant x) \geqslant p, \qquad P(X \geqslant x) \geqslant 1 - p \qquad (0 < p < 1)$

is called *the quantile of order p* and is denoted by x_p.

The pair of inequalities (3.5.3) is equivalent to the double inequality

(3.5.3') $\qquad\qquad p - P(X = x) \leqslant F(x) \leqslant p.$

If $P(X = x_p) = 0$, hence in particular, if the random variable X is of the continuous type, the quantile of order p is the number satisfying the equation

$$(3.5.4) \qquad\qquad F(x) = p.$$

It may happen that many numbers x satisfy inequalities (3.5.3) or equation (3.5.4). Each of them is then called the *quantile of order p*.

Quantiles and functions of them are called *order parameters*.

Example 3.5.4. Suppose that the random variable X has the normal distribution with density

$$f(x) = \frac{1}{\sqrt{2\pi}} e^{-x^2/2}.$$

We shall find the point x for which $F(x) = 0.1$. We use Table III of the normal distribution at the end of the book where we find that $x_{1/10} \cong -1.28$.

Some simple functions of the quantiles may also serve as measures of dispersion. One of them is the *semi-interquartile range* defined as

$$(3.5.5) \qquad\qquad \tfrac{1}{2}(x_{3/4} - x_{1/4}).$$

If the set of all possible values of a random variable is bounded from both sides, there exist finite upper and lower bounds of the values taken by this random variable. For a discrete random variable with a finite set of jump points these upper and lower bounds coincide with the maximal and minimal values which can be taken on by this random variable.

An important order parameter which also serves as a measure of dispersion is the range. If a and b are, respectively, the lower and upper bounds of the values taken on by the random variable, the *range* is defined by the formula

$$(3.5.6) \qquad\qquad d = b - a.$$

In example 3.5.2 the range equals $\tfrac{1}{2}\pi$ and in example 3.5.3 it equals 2.

3.6 MOMENTS OF RANDOM VECTORS

A   Let (X, Y) be a two-dimensional random variable. Consider the single-valued function $g(X, Y)$ of (X, Y).

Definition 3.6.1. Let (X, Y) be a two-dimensional random variable of the discrete type with jump points (x_i, y_k) and jumps p_{ik}. The series

$$(3.6.1) \qquad\qquad E[g(X, Y)] = \sum_{i,k} p_{ik} g(x_i, y_k)$$

is called the *expected value* of $g(X, Y)$ if the following inequality is satisfied:

$$(3.6.1') \qquad\qquad \sum_{i,k} p_{ik} |g(x_i, y_k)| < \infty.$$

Definition 3.6.2. Let (X, Y) be a random variable of the continuous type with density $f(x, y)$. Let $g(x, y)$ be Riemann integrable. The integral

$$(3.6.2) \qquad E[g(X, Y)] = \int_{-\infty}^{\infty} \int_{-\infty}^{\infty} g(x, y)f(x, y) \, dx \, dy$$

is called the *expected value* of $g(X, Y)$ if the following inequality is satisfied:

$$(3.6.2') \qquad \int_{-\infty}^{\infty} \int_{-\infty}^{\infty} |g(x, y)| f(x, y) \, dx \, dy < \infty.$$

Definition 3.6.3. The expected value of the function $g(X, Y) = X^l Y^n$, that is,

$$m_{ln} = E(X^l Y^n),$$

where l and n are non-negative integers, is called the *moment of order $l + n$* of the random variable (X, Y).

Thus, if (X, Y) is a random variable of the discrete type with jump points (x_i, y_k) and jumps p_{ik}, according to formula (3.6.1) we have

$$(3.6.3) \qquad m_{ln} = \sum_{i,k} p_{ik} x_i^{\,l} y_k^{\,n}.$$

If (X, Y) is a random variable of the continuous type with the density function $f(x, y)$, then according to formula (3.6.2) we have

$$(3.6.4) \qquad m_{ln} = \int_{-\infty}^{\infty} \int_{-\infty}^{\infty} x^l y^n f(x, y) \, dx \, dy.$$

We remind the reader that for the moment to exist it is necessary that the series (3.6.3) or the integral (3.6.4) be absolutely convergent.

Two moments of the first order can exist, m_{10} and m_{01}. We notice immediately that these moments are the expected values of the marginal distributions of the random variables X and Y; thus

$$(3.6.5) \qquad m_{10} = E(X) \qquad \text{and} \quad m_{01} = E(Y).$$

There can be three moments of the second order, m_{20}, m_{11}, and m_{02}, and we have

$$(3.6.6) \qquad m_{20} = E(X^2) \qquad \text{and} \quad m_{02} = E(Y^2).$$

The central moments are defined similarly to those for one-dimensional distributions. Denoting them by μ_{ln} we have

$$(3.6.7) \qquad \mu_{ln} = E[(X - m_{10})^l (Y - m_{01})^n].$$

In particular,

(3.6.8)
$$\mu_{10} = E(X - m_{10}) = 0,$$
$$\mu_{01} = E(Y - m_{01}) = 0,$$

(3.6.9)
$$\mu_{20} = E[(X - m_{10})^2] = \sigma_1^2,$$
$$\mu_{02} = E[(Y - m_{01})^2] = \sigma_2^2,$$

where σ_1 and σ_2 are the standard deviations of the random variables X and Y, respectively. The central moment μ_{11} is called the *covariance* and is denoted by cov (X, Y). Between the ordinary and central moments there are relations similar to those for one-dimensional distributions,

(3.6.10)
$$\mu_{20} = m_{20} - m_{10}^2,$$
$$\mu_{02} = m_{02} - m_{01}^2.$$

For the covariance we have

(3.6.11) $\mu_{11} = E[(X - m_{10})(Y - m_{01})]$
$$= E(XY) - m_{10}E(Y) - m_{01}E(X) + m_{10}m_{01}$$
$$= m_{11} - m_{10}m_{01} - m_{01}m_{10} + m_{10}m_{01} = m_{11} - m_{10}m_{01}.$$

B   We now investigate the expected value of the sum and the product of two random variables. Let X and Y be two random variables, dependent or not. Suppose that the expected values $E(X)$ and $E(Y)$ exist. We shall consider the random variable $Z = X + Y$ and its expected value $E(Z)$.

Let (X, Y) be a random variable of the discrete type with jump points (x_i, y_k) and jumps p_{ik}. Substituting $Z = X + Y = g(X, Y)$ into formula (3.6.1) and taking into consideration that $E(X)$ and $E(Y)$ exist, we obtain

(3.6.12) $E(Z) = \sum_{i,k} p_{ik}(x_i + y_k) = \sum_i x_i \sum_k p_{ik} + \sum_k y_k \sum_i p_{ik}$
$$= \sum_i x_i p_{i.} + \sum_k y_k p_{.k} = E(X) + E(Y).$$

Suppose that the random variable (X, Y) is of the continuous type with the density function $f(x, y)$. Then from (3.6.2) we obtain

(3.6.13) $E(Z) = \int_{-\infty}^{\infty} \int_{-\infty}^{\infty} (x + y)f(x, y)\, dx\, dy$
$$= \int_{-\infty}^{\infty} x \int_{-\infty}^{\infty} f(x, y)\, dy\, dx + \int_{-\infty}^{\infty} y \int_{-\infty}^{\infty} f(x, y)\, dx\, dy$$
$$= \int_{-\infty}^{\infty} x f_1(x)\, dx + \int_{-\infty}^{\infty} y f_2(y)\, dy = E(X) + E(Y).$$

We can generalize (3.6.12) and (3.6.13) by induction to an arbitrary finite number of random variables. Thus we obtain the following theorem.

Theorem 3.6.1. *The expected value of the sum of an arbitrary finite number of random variables, whose expected values exist, equals the sum of their expected values.*

For the product of random variables we have:

Theorem 3.6.2. *The expected value of the product of an arbitrary finite number of independent random variables, whose expected values exist, equals the product of the expected values of these variables.*

Proof. We restrict the proof to the product of two random variables, since it is easy to extend the proof by induction to an arbitrary finite number of random variables.

Suppose that (X, Y) is a random variable of the discrete type, that X and Y are independent, and that the expected values $E(X)$ and $E(Y)$ exist. From (3.6.1) we obtain, using the independence of X and Y,

$$(3.6.14) \qquad E(XY) = \sum_{i,k} p_{ik} x_i y_k = \sum_{i,k} p_{i.} p_{.k} x_i y_k$$

$$= \sum_i p_{i.} x_i \sum_k p_{.k} x_k = E(X)E(Y).$$

Similarly, if (X, Y) is of the continuous type, we have

$$E(XY) = \int_{-\infty}^{\infty} \int_{-\infty}^{\infty} xy f(x, y)\, dx\, dy = \int_{-\infty}^{\infty} \int_{-\infty}^{\infty} xy f_1(x) f_2(y)\, dx\, dy$$

$$(3.6.15)$$

$$= \int_{-\infty}^{\infty} x f_1(x)\, dx \int_{-\infty}^{\infty} y f_2(y)\, dy = E(X)E(Y).$$

Theorem 3.6.2 is proved (for the converse of this theorem, see Supplement).

We notice that from formula (3.6.11) and from theorem 3.6.2 we obtain that the covariance of two independent random variables equals 0. In fact, here we have

$$\mu_{11} = E(XY) - E(X)E(Y) = 0.$$

However, from the fact that the covariance of two random variables is zero it does not follow that they are independent.

Suppose that X and Y are two random variables with variances $D^2(X)$ and $D^2(Y)$. Consider the random variable

$$Z = X + Y.$$

The variance of Z is

$$D^2(Z) = D^2(X + Y) = E[(X + Y)^2] - [E(X + Y)]^2$$

$$= E(X^2) + 2E(XY) + E(Y^2) - [E(X)]^2 - [E(Y)]^2 - 2E(X)E(Y)$$

$$= E(X^2) - [E(X)]^2 + E(Y^2) - [E(Y)]^2 + 2E(XY) - 2E(X)E(Y).$$

Finally, we obtain

$$(3.6.16) \quad D^2(X + Y) = D^2(X) + D^2(Y) + 2E(XY) - 2E(X)E(Y).$$

If X and Y are independent we obtain, using formula (3.6.14),

$$(3.6.17) \qquad\qquad D^2(Z) = D^2(X) + D^2(Y).$$

This result can be extended to an arbitrary finite number of independent random variables.

Theorem 3.6.3. *The variance of the sum of an arbitrary finite number of independent random variables, whose variances exist, equals the sum of their variances.*

C   We now give the formulas for the moments of the conditional distribution of one of the random variables X, Y under the condition that the second takes on a certain given value. Let (X, Y) be a random variable of the discrete type with jump points (x_i, y_k) and jumps p_{ik}. Denoting by $E(Y^l \mid X = x_i)$ the conditional expected value of the random variable Y^l ($l = 1, 2, \ldots$) under the condition that $X = x_i$, we obtain from (2.7.2) and (3.2.1)

$$(3.6.18) \qquad\qquad E(Y^l \mid X = x_i) = \sum_k y_k^{\ l} \frac{p_{ik}}{p_{i.}}.$$

Similarly, for the conditional expected value of the random variable X^l ($l = 1, 2, \ldots$) when $Y = y_k$, we obtain the formula

$$(3.6.19) \qquad\qquad E(X^l \mid Y = y_k) = \sum_i x_i^{\ l} \frac{p_{ik}}{p_{.k}}.$$

Let (X, Y) be a two-dimensional random variable of the continuous type with the density function $f(x, y)$ and suppose that both conditional densities exist (see Section 2.7,B). From (2.7.6) and (3.2.2) we obtain

$$(3.6.20) \qquad\qquad E(Y^l \mid X = x) = \int_{-\infty}^{\infty} y^l \frac{f(x, y)\, dy}{f_1(x)}.$$

Similarly,

$$(3.6.21) \qquad\qquad E(X^l \mid Y = y) = \int_{-\infty}^{\infty} x^l \frac{f(x, y)}{f_2(y)}\, dx.$$

Formulas (3.6.18) to (3.6.21) are, of course, meaningful only if the moments being considered exist. We notice that the conditional moments of one random variable are constant for a fixed value of the second random variable. Thus, for example, for a constant x_i the moment $E(Y^l \mid X = x_i)$ is a fixed number. If, however, we remember that x_i is a value of the random variable X, we notice that we can consider the random variable $E(Y^l \mid x)$ which takes on the value $E(Y^l \mid X = x_i)$ with probability p_i. $(i = 1, 2, \ldots)$. By (3.6.18), for every subset S of the set of jump points x_i of X, we have the equation (see Section 2.7,C)

$$(3.6.22) \qquad E(Y^l \mid X \in S) = \sum_k y_k^l P(Y = y_k \mid X \in S)$$

$$= \sum_k y_k^l \left(\sum_{x \in S} p_{ik} \middle/ \sum_{x_i \in S} p_{i.} \right)$$

$$= \sum_{x_i \in S} p_{i.} \sum_k \frac{y_k^l p_{ik}}{p_{i.}} \middle/ \sum_{x_i \in S} p_{i.}$$

$$= \sum_{x_i \in S} \left(p_{i.} \middle/ \sum_{x_i \in S} p_{i.} \right) E(Y^l \mid X = x_i)$$

$$= E[E(Y^l \mid X = x_i) \mid X \in S].$$

In particular, if we take the set of all jump points x_i of the random variable X as the set S, we obtain

$$(3.6.22') \qquad E(Y^l) = \sum_i \sum_k y_k^l p_{ik} = \sum_i p_{i.} E(Y^l \mid X = x_i)$$

$$= E[E(Y^l \mid x)].$$

Similarly, for a random variable (X, Y) of the continuous type, for every Borel set S on the real axis for which $P(X \in S) > 0$, we obtain from (3.6.20) the equation

$$(3.6.23) \quad E(Y^l \mid X \in S) = \int_{-\infty}^{\infty} y^l \int_S f(x, y) \, dx \, dy / P(X \in S)$$

$$= \int_S f_1(x) \int_{-\infty}^{\infty} y^l \frac{f(x, y)}{f_1(x)} \, dy \, dx / P(X \in S)$$

$$= \int_S \frac{f_1(x)}{P(X \in S)} E(Y^l \mid X = x) \, dx = E[E(Y^l \mid X = x) \mid X \in S].$$

Putting $S = (-\infty, +\infty)$ we obtain

$$(3.6.23') \quad E(Y^l) = \int_{-\infty}^{+\infty} \int_{-\infty}^{+\infty} y^l f(x, y) \, dx \, dy$$

$$= \int_{-\infty}^{+\infty} f_1(x) E(Y^l \mid X = x) \, dx = E[E(Y^l \mid x)].$$

Formulas (3.6.22) and (3.6.23) may be written as follows:

$$(3.6.24) \qquad E(Y^l \mid X \in S) = E[E(Y^l \mid X = x) \mid X \in S],$$

and formulas (3.6.22') and (3.6.23') may be written in the form

$$(3.6.24') \qquad\qquad E(Y^l) = E[E(Y^l \mid x)].$$

These formulas give us a new approach to the problem of conditional probabilities. Let E be the set of elementary events e and let Z be a Borel field of random events which are subsets of E. Let $A \in Z$ be a fixed random event. Let us define the random variable Y as follows:

$$Y(e) = \begin{cases} 1 & \text{if } e \in A, \\ 0 & \text{if } e \notin A. \end{cases}$$

Hence $P(Y = 1) = P(A)$. Let X be another random variable defined on the same set E of elementary events such that the right side of (3.6.24) exists. Relations (3.6.24) and (3.6.24′) are, respectively, of the form

(3.6.24″) $P(A \mid X \in S) = E[P(A \mid X = x) \mid X \in S],$

(3.6.24‴) $P(A) = E[P(A \mid x)].$

Hence $P(A \mid x)$ is a random variable.

We observe that formulas (3.6.24) to (3.6.24‴) have been derived under the assumption that the random variable (X, Y) is either of the discrete or of the continuous type and the conditions formulated in Section 2.7 are satisfied. However, in the general case, equations (3.6.24) and (3.6.24″) (see Supplement) serve as definitions of the conditional expected value and the conditional probability, respectively. These definitions are due to Kolmogorov [7] who has shown that if $E(Y^l)$ exists, equation (3.6.24) determines the random variable $E(Y^l \mid x)$ uniquely up to equivalence, that is, if another random variable U besides $E(Y^l \mid x)$ satisfies equation (3.6.24), $P[U - E(Y^l \mid x) = 0] = 1$. In particular, equation (3.6.24″) determines the conditional probability $P(A \mid x)$ uniquely up to equivalence.

The further development of these ideas can be found in the monograph by Doob [5].

D   In this section we have considered only moments of two-dimensional random variables since there is no difficulty in extending the notions introduced here to more than two dimensions. We shall mention only the conditional expected value for m-dimensional distributions $(m > 2)$.

Consider a random variable $(X_1, X_2, \ldots, X_m)$ of the continuous type and suppose that the density function $f(x_1, x_2, \ldots, x_m)$ is everywhere continuous and that the density of the marginal distribution

$$\int_{-\infty}^{\infty} f(x_1, x_2, \ldots, x_m)\, dx_1 = g(x_2, \ldots, x_n)$$

of the random variable $(X_2, \ldots, X_m)$ is everywhere continuous and positive. We can consider here the conditional moments of order l $(l = 1, 2, \ldots)$

$$(3.6.25)\ E(X_1^l \mid X_2 = x_2, \ldots, X_m = x_m) = \int_{-\infty}^{\infty} x_1^l \frac{f(x_1, x_2, \ldots, x_m)}{g(x_2, \ldots, x_m)}\, dx_1.$$

Similarly, we could consider the conditional moments of the remaining random variables or the multidimensional conditional moments of some subclasses of these random variables.

For discrete random variables the reasoning is similar.

E   If the standard deviations σ_1 and σ_2 of the random variables X and Y exist and are different from zero, we can define a very important parameter characterizing the two-dimensional distribution of (X, Y), namely, the *coefficient of correlation*. The formula for the coefficient of correlation is the following:

$$(3.6.26) \qquad \rho = \frac{E[(X - m_{10})(Y - m_{01})]}{\sqrt{E[(X - m_{10})^2]} \sqrt{E[(Y - m_{01})^2]}} = \frac{\mu_{11}}{\sigma_1 \sigma_2}.$$

The basic properties of the coefficient of correlation are given by the following two theorems.

Theorem 3.6.4. *The coefficient of correlation satisfies the double inequality*

$$(3.6.27) \qquad\qquad -1 \leqslant \rho \leqslant 1.$$

Proof. For arbitrary real numbers t and u consider the non-negative expression

$$(3.6.28) \qquad E\{[t(X - m_{10}) + u(Y - m_{01})]^2\}$$
$$= t^2 E[(X - m_{10})^2] + 2tu E[(X - m_{10})(Y - m_{01})]$$
$$+ u^2 E[(Y - m_{01})^2]$$
$$= t^2 \sigma_1^2 + 2tu\mu_{11} + u^2 \sigma_2^2.$$

Since the left-hand side of (3.6.28) is always non-negative, we must have

$$\mu_{11}^2 - \sigma_1^2 \sigma_2^2 \leqslant 0, \qquad \text{or} \qquad -\sigma_1 \sigma_2 \leqslant \mu_{11} \leqslant \sigma_1 \sigma_2,$$

and hence

$$-1 \leqslant \frac{\mu_{11}}{\sigma_1 \sigma_2} \leqslant 1.$$

Inequality (3.6.27) now follows directly from the definition of the coefficient of correlation.

If the random variables X and Y are independent, we have $\mu_{11} = 0$, and hence $\rho = 0$. As has been mentioned the converse is not true; thus if $\rho = 0$, we say that X and Y are *uncorrelated* (to distinguish this from the term independent).

We notice that theorem 3.6.3 is also true when the random variables are pairwise uncorrelated. The following theorem tells us how to interpret the relation $\rho^2 = 1$.

Theorem 3.6.5.   *The equality $\rho^2 = 1$ is a necessary and sufficient condition for the relation*

$$(3.6.29) \qquad P(Y = aX + b) = 1$$

to hold.

Before we give the proof of this theorem we shall present its interpretation. Thus, according to this theorem, if $\rho^2 = 1$, the probability that the random variable (X, Y) takes on values lying on a straight line in the plane (x, y) is equal to one. If we take this line as the coordinate axis, we have a one-dimensional random variable. In such cases we usually say, using pictorial language, that "the whole mass of probability" is concentrated on a line.

Proof.   Suppose that relation (3.6.29) is satisfied, that is, $P(Y \neq aX + b) = 0$. From formula (3.6.24) we obtain

$$
\begin{aligned}
m_{01} &= E(Y) \\
&= P(Y = aX + b)E(Y \mid Y = aX + b) \\
&\quad + P(Y \neq aX + b)E(Y \mid Y \neq aX + b) \\
&= E(aX + b) = am_{10} + b,
\end{aligned}
$$

$$
\begin{aligned}
\sigma_2{}^2 = E[(Y - m_{01})^2] &= P(Y = aX + b)E[(Y - m_{01})^2 \mid Y = aX + b] \\
&\quad + P(Y \neq aX + b)E[(Y - m_{01})^2 \mid Y \neq aX + b] \\
= E[(aX + b - m_{01})^2] &= E[a^2(X - m_{10})^2] = a^2 E[(X - m_{10})^2] = a^2 \sigma_1{}^2.
\end{aligned}
$$

Similarly, we obtain $\mu_{11} = a\sigma_1{}^2$. From these equalities it follows that $\rho^2 = 1$.

Suppose now that $\rho^2 = 1$. From the definition of the coefficient of correlation we then have

$$\mu_{20}\mu_{02} - \mu_{11}{}^2 = 0.$$

Then the quadratic form (3.6.28) takes on the value zero for some pair of values $t = t_0$ and $u = u_0$, where at least one of the values t_0 and u_0 is not zero. For these values t_0 and u_0 we have

$$E\{[t_0(X - m_{10}) + u_0(Y - m_{01})]^2\} = 0.$$

This equation is satisfied only when we have the equation

$$P[(t_0(X - m_{10}) + u_0(Y - m_{01}) = 0] = 1.$$

If, for instance, we suppose that $u_0 \neq 0$, we obtain

$$P\left(Y = -\frac{t_0}{u_0} X + \frac{m_{01}u_0 + m_{10}t_0}{u_0}\right) = 1.$$

Theorem 3.6.5 is now proved.

It follows that the case $\rho^2 = 1$ is the opposite of independence.

Example 3.6.1. The random variables X and Y have the joint density given by the formula

$$f(x, y) = \frac{1}{2\pi} \exp\left(-\frac{x^2 - 2xy + 2y^2}{2}\right).$$

This expression is a density function, since it is non-negative and

$$\int_{-\infty}^{+\infty} \int_{-\infty}^{+\infty} f(x, y)\, dx\, dy = \frac{1}{2\pi} \int_{-\infty}^{+\infty} \int_{-\infty}^{+\infty} \exp\left(-\frac{x^2 - 2xy + 2y^2}{2}\right) dx\, dy$$

$$= \frac{1}{\sqrt{2\pi}} \int_{-\infty}^{+\infty} \exp\left(-\frac{y^2}{2}\right) \frac{1}{\sqrt{2\pi}}$$

$$\int_{-\infty}^{+\infty} \exp\left[-\frac{(x - y)^2}{2}\right] dx\, dy$$

$$= \frac{1}{\sqrt{2\pi}} \int_{-\infty}^{+\infty} \exp\left(-\frac{y^2}{2}\right) dy = 1.$$

This distribution is a special case of the *two-dimensional normal distribution* (see Section 5.11).

Let us compute the moments of the first and second order.

$$m_{10} = \int_{-\infty}^{+\infty} \int_{-\infty}^{+\infty} x f(x, y)\, dx\, dy$$

$$= \frac{1}{\sqrt{2\pi}} \int_{-\infty}^{+\infty} \exp\left(-\frac{y^2}{2}\right) \frac{1}{\sqrt{2\pi}} \int_{-\infty}^{+\infty} x \exp\left[-\frac{(x - y)^2}{2}\right] dx\, dy$$

$$= \frac{1}{\sqrt{2\pi}} \int_{-\infty}^{+\infty} y \exp\left(-\frac{y^2}{2}\right) dy = 0.$$

$$m_{01} = \frac{1}{2\sqrt{\pi}} \int_{-\infty}^{+\infty} \exp\left(-\frac{x^2}{4}\right) \frac{1}{\sqrt{\pi}} \int_{-\infty}^{+\infty} y \exp\left[-\left(y - \frac{x}{2}\right)^2\right] dy\, dx$$

$$= \frac{1}{2\sqrt{\pi}} \int_{-\infty}^{+\infty} \frac{x}{2} \exp\left(-\frac{x^2}{4}\right) dx = 0.$$

Thus we have

$$\mu_{11} = m_{11} = \frac{1}{\sqrt{2\pi}} \int_{-\infty}^{+\infty} y \exp\left(-\frac{y^2}{2}\right) \frac{1}{\sqrt{2\pi}} \int_{-\infty}^{+\infty} x \exp\left[-\frac{(x - y)^2}{2}\right] dx\, dy$$

$$= \frac{1}{\sqrt{2\pi}} \int_{-\infty}^{+\infty} y^2 \exp\left(-\frac{y^2}{2}\right) dy = 1,$$

$$\mu_{20} = m_{20} = \frac{1}{2\sqrt{\pi}} \int_{-\infty}^{+\infty} x^2 \exp\left(-\frac{x^2}{4}\right) \frac{1}{\sqrt{\pi}} \int_{-\infty}^{\infty} \exp\left[-\left(y - \frac{x}{2}\right)^2\right] dy\, dx$$

$$= \frac{1}{2\sqrt{\pi}} \int_{-\infty}^{+\infty} x^2 \exp\left(-\frac{x^2}{4}\right) dx = 2,$$

$$\mu_{02} = m_{02} = \frac{1}{\sqrt{2\pi}} \int_{-\infty}^{+\infty} y^2 \exp\left(-\frac{y^2}{2}\right) \frac{1}{\sqrt{2\pi}} \int_{-\infty}^{+\infty} \exp\left[-\frac{(x-y)^2}{2}\right] dx\, dy$$

$$= \frac{1}{\sqrt{2\pi}} \int_{-\infty}^{+\infty} y^2 \exp\left(-\frac{y^2}{2}\right) dy = 1.$$

Hence the coefficient of correlation equals

$$\rho = \mu_{11}/\sigma_1\sigma_2 = \frac{1}{\sqrt{2}}.$$

Besides the coefficient of correlation there are many other measures of dependence between two random variables. Detailed information concerning this subject is found in the books by Ezekiel [1] and M. G. Kendall [1]. More recent papers are those of Gebelein [1], Hirschfeld [1], and Rényi [8] and [9]. In the last two papers the properties of some important measures of dependence of random variables are compared.
F   We now consider the n-dimensional random variable

$$(X_1, X_2, \ldots, X_n).$$

Suppose that the variances σ_i^2 $(i = 1, 2, \ldots, n)$ of the random variables X_i exist and are positive. Then the covariances of all pairs of these random variables exist also. Denote by λ_{ik} and ρ_{ik} the covariance and the coefficient of correlation, respectively, of X_i and X_k.
   The symmetric matrix

$$(3.6.30) \qquad \mathbf{M} = \begin{bmatrix} \lambda_{11} & \lambda_{12} & \cdots & \lambda_{1n} \\ \lambda_{21} & \lambda_{22} & \cdots & \lambda_{2n} \\ & & \cdots & \\ \lambda_{n1} & \lambda_{n2} & \cdots & \lambda_{nn} \end{bmatrix}$$

is called the *matrix of second-order moments*. The determinant of the matrix $\mathbf{M}$ is denoted by $|\mathbf{M}|$.
   The generalization of theorem 3.6.5 to the random variables under consideration is a theorem which provides a criterion for determining whether there are linear relations among $X_1, X_2, \ldots, X_n$.
   Theorem 3.6.6.   *The probability that the random variables $X_1, X_2, \ldots, X_n$, whose variances exist, satisfy at least one linear relation, equals 1 if and only if* $|\mathbf{M}| = 0$.
   We shall not give a detailed proof of this theorem. It is similar to the proof of theorem 3.6.5, with expression (3.6.28) replaced by the non-negative quadratic form

$$(3.6.31) \qquad E\left\{\left[\sum_{i=1}^{n} t_i(X_i - E(X_i))\right]^2\right\} = \sum_{i,k=1}^{n} \lambda_{ik} t_i t_k \geqslant 0.$$

The interpretation of theorem 3.6.6 is similar to that of theorem 3.6.5, namely, if $|\mathbf{M}| = 0$, then the "whole mass of probability" is concentrated on a hyperplane of dimension less than n.
G   In connection with theorem 3.6.6 we introduce the following definition.

Definition 3.6.4.   If the components $X_1, \ldots, X_n$ of the random vector $(X_1, \ldots, X_n)$ satisfy at least one linear relation with probability 1, the distribution of $(X_1, \ldots, X_n)$ is called *degenerate*.

If the determinant $|\mathbf{M}| \neq 0$, the distribution of $(X_1, \ldots, X_n)$ is *nondegenerate*.

Definition 3.6.5.   The determinant $|\mathbf{M}|$ is called the *generalized variance*.

To justify this name, we shall show that $|\mathbf{M}|$ may serve as a measure of dispersion. Notice that if $|\mathbf{M}| \neq 0$, the quadratic form (3.6.31) is positive definite; hence using the equality $\lambda_{ik} = \lambda_{ki}$, we see that the matrix $\mathbf{M}$ is symmetric and positive definite. It follows[1] that $|\mathbf{M}| \leqslant \sigma_1^2\sigma_2^2 \ldots \sigma_n^2$. Thus $|\mathbf{M}|$ takes on its maximum value if $\lambda_{ik} = 0$ $(i \neq k)$, that is, if the random variables $X_1, \ldots, X_n$ are uncorrelated. On the other hand, the smaller the σ_i^2 are, and hence the more concentrated the marginal distributions of the random variables X_i are, the smaller is the value of the product $\sigma_1^2 \sigma_2^2 \ldots \sigma_n^2$, which is the upper bound of the generalized variance. In the boundary case, when at least one of the σ_i^2 tends to zero, by theorem 3.6.6, the distribution of $(X_1, \ldots, X_n)$ becomes degenerate, that is, becomes a distribution of a random variable of less than n dimensions. It is natural to consider such a distribution as "highly concentrated" in n-dimensional space.

The notion of generalized variance was introduced by Wilks [1].

Consider now the matrix $\mathbf{R}$ of the correlation coefficients ρ_{ik}, taking $\rho_{ii} = 1$;

$$\mathbf{R} = \begin{bmatrix} 1 & \rho_{12} & \cdots & \rho_{1n} \\ \rho_{21} & 1 & \cdots & \rho_{2n} \\ & \cdot & \cdot & \cdot \\ \cdots & & \cdot\cdot & \cdot \\ & & & \cdot \\ \rho_{n1} & \rho_{n2} & \cdots & 1 \end{bmatrix}.$$

From the fact that $\rho_{ik} = \lambda_{ik}/\sigma_i\sigma_k$ we obtain the following relation between the determinants $|\mathbf{M}|$ and $|\mathbf{R}|$:

(3.6.32) $|\mathbf{M}| = \sigma_1^2\sigma_2^2 \ldots \sigma_n^2\,|\mathbf{R}|.$

The matrix $\mathbf{R}$ is also symmetric and its determinant satisfies the relation $|\mathbf{R}| \leqslant 1$. The determinant $|\mathbf{R}|$ has its maximum value 1 when all the

[1] See, for example, A. C. Aitken [1].

coefficients of correlation $\rho_{ik} = 0$, and its minimum value 0 if at least one of the coefficients of correlation equals 1; hence if the distribution is degenerate. It follows that the value of the determinant $|\mathbf{R}|$ may serve as a measure of degeneracy of the distribution of the random vector $(X_1, \ldots, X_n)$. This measure was introduced by Frisch [1], who gave the following definition.

Definition 3.6.6. The expression $\sqrt{|R|}$ is called the *scatter coefficient*.

3.7 REGRESSION OF THE FIRST TYPE

A   Let (X, Y) be a two-dimensional random variable of the discrete type with jump points (x_i, y_k) and jumps p_{ik}, and let $p_{i.}$ and $p_{.k}$ denote, as before, the probabilities in the marginal distributions of X and Y, respectively. Consider the conditional expectations of X and Y and denote them respectively by $m_1(y_k)$ and $m_2(x_i)$. We obtain them from (3.6.18) and (3.6.19), setting $l = 1$; thus

$$m_1(y_k) = E(X \mid Y = y_k) = \sum_i x_i \frac{p_{ik}}{p_{.k}},$$

(3.7.1)

$$m_2(x_i) = E(Y \mid X = x_i) = \sum_k y_k \frac{p_{ik}}{p_{i.}}.$$

In this way we obtain two collections of points in the plane (x, y). The first collection consists of points with coordinates

(3.7.2) $x = m_1(y_k), \qquad y = y_k,$

and the second of points with coordinates

(3.7.3) $x = x_i, \qquad y = m_2(x_i).$

Suppose now that (X, Y) is a two-dimensional random variable of the continuous type with density $f(x, y)$ and marginal densities $f_1(x)$ and $f_2(y)$. Suppose that the density functions satisfy the conditions considered in Section 2.7 so that the conditional distributions of X and Y are defined. The conditional expectations $m_1(y)$ and $m_2(x)$ are obtained from (3.6.20) and (3.6.21); thus

$$m_1(y) = E(X \mid Y = y) = \int_{-\infty}^{\infty} x \frac{f(x, y)}{f_2(y)} \, dx,$$

(3.7.1′)

$$m_2(x) = E(Y \mid X = x) = \int_{-\infty}^{\infty} y \frac{f(x, y)}{f_1(x)} \, dy.$$

Again we obtain here two collections of points in the plane (x, y) with the respective coordinates

$$(3.7.4) \qquad\qquad x = m_1(y), \qquad y,$$

$$(3.7.5) \qquad\qquad x, \qquad y = m_2(x).$$

Definition 3.7.1. The set of points of the plane (x, y) with coordinates given in (3.7.2) or (3.7.4) is called *the regression curve of the random variable X on the random variable Y*; the set of points with the coordinates given in (3.7.3) or (3.7.5) is called the *regression curve of the random variable Y on the random variable X.*

From this definition it follows that the conditional expectations of X, when Y runs over all possible values, are situated on the regression curve of X on Y. The situation is similar in regard to the regression of Y on X. These two regression curves do not usually coincide.

If all the points of the regression curve lie on a straight line, we say that there is *linear regression.*

If there is a linear dependence $Y = aX + b$ between the random variables X and Y, the random variable (X, Y) takes on only those values that correspond to the points of this line. It follows that the expected values of the random variables lie on this line, and thus both regression curves lie on this line.

If X and Y are independent,

$$m_2(x) = E(Y \mid X = x) = E(Y),$$

$$m_1(y) = E(X \mid Y = y) = E(X).$$

Here the conditional expectation $m_2(x)$ is independent of x, and the regression curve of Y on X lies on a line parallel to the x-axis; similarly, the regression curve of X on Y lies on a line parallel to the y-axis. These two lines intersect at the point with coordinates (m_1, m_2).

To distinguish the regression discussed in this section from another type of regression discussed in the next section, we call the regression just defined *regression of the first type.*

Example 3.7.1. The random variable (X, Y) can take on the pairs of values (x_k, y_l), $(k, l = 1, 2, 3, 4, 5)$, where $x_1 = 1$, $x_2 = 2$, $x_3 = 3$, $x_4 = 4$, $x_5 = 5$, $y_1 = 1$, $y_2 = 2$, $y_3 = 3$, $y_4 = 4$, $y_5 = 5$, and the probabilities p_{kl} for the particular pairs (x_k, y_l) are given in Table 3.7.1.

Besides the values of p_{kl}, the table gives in the last right-hand column the marginal distribution of Y and in the last row the marginal distribution of X.

In Table 3.7.2. are given the conditional distributions of Y under the condition that $X = x_k$, where $k = 1, 2, 3, 4, 5$, and in Table 3.7.3 are given the conditional distributions of X under the condition that $Y = y_l$, where $l = 1, 2, 3, 4, 5$.

TABLE 3.7.1

PROBABILITIES p_{kl}

y_l	x_k					Marginal Distribution of the Random Variable Y
	1	2	3	4	5	
1	$\frac{1}{12}$	$\frac{1}{24}$	0	$\frac{1}{24}$	$\frac{1}{30}$	$\frac{1}{5}$
2	$\frac{1}{24}$	$\frac{1}{24}$	$\frac{1}{24}$	$\frac{1}{24}$	$\frac{1}{30}$	$\frac{1}{5}$
3	$\frac{1}{12}$	$\frac{1}{24}$	$\frac{1}{24}$	0	$\frac{1}{30}$	$\frac{1}{5}$
4	$\frac{1}{12}$	0	$\frac{1}{24}$	$\frac{1}{24}$	$\frac{1}{30}$	$\frac{1}{5}$
5	$\frac{1}{24}$	$\frac{1}{24}$	$\frac{1}{24}$	$\frac{1}{24}$	$\frac{1}{30}$	$\frac{1}{5}$
Marginal Distribution of the Random Variable X	$\frac{1}{3}$	$\frac{1}{6}$	$\frac{1}{6}$	$\frac{1}{6}$	$\frac{1}{6}$	1

Let us compute the conditional expected value of one random variable under the condition that the second takes on a given constant value. We have

$$E(Y \mid X = 1) = \tfrac{1}{4} \cdot 1 + \tfrac{1}{8} \cdot 2 + \tfrac{1}{4} \cdot 3 + \tfrac{1}{4} \cdot 4 + \tfrac{1}{8} \cdot 5 = 2\tfrac{7}{8},$$

$$E(Y \mid X = 2) = 2\tfrac{3}{4}, \quad E(Y \mid X = 3) = 3\tfrac{1}{2}, \quad E(Y \mid X = 4) = 3,$$

$$E(Y \mid X = 5) = 3.$$

TABLE 3.7.2

y_l	x_k				
	1	2	3	4	5
1	$\frac{1}{4}$	$\frac{1}{4}$	0	$\frac{1}{4}$	$\frac{1}{5}$
2	$\frac{1}{8}$	$\frac{1}{4}$	$\frac{1}{4}$	$\frac{1}{4}$	$\frac{1}{5}$
3	$\frac{1}{4}$	$\frac{1}{4}$	$\frac{1}{4}$	0	$\frac{1}{5}$
4	$\frac{1}{4}$	0	$\frac{1}{4}$	$\frac{1}{4}$	$\frac{1}{5}$
5	$\frac{1}{8}$	$\frac{1}{4}$	$\frac{1}{4}$	$\frac{1}{4}$	$\frac{1}{5}$
Total	1	1	1	1	1

TABLE 3.7.3

y_l	x_k					Total
	1	2	3	4	5	
1	$\frac{5}{12}$	$\frac{5}{24}$	0	$\frac{5}{24}$	$\frac{1}{6}$	1
2	$\frac{5}{24}$	$\frac{5}{24}$	$\frac{5}{24}$	$\frac{5}{24}$	$\frac{1}{6}$	1
3	$\frac{5}{12}$	$\frac{5}{24}$	$\frac{5}{24}$	0	$\frac{1}{6}$	1
4	$\frac{5}{12}$	0	$\frac{5}{24}$	$\frac{5}{24}$	$\frac{1}{6}$	1
5	$\frac{5}{24}$	$\frac{5}{24}$	$\frac{5}{24}$	$\frac{5}{24}$	$\frac{1}{6}$	1

Similarly,

$$E(X \mid Y = 2) = \tfrac{5}{24} \cdot 1 + \tfrac{5}{24} \cdot 2 + \tfrac{5}{24} \cdot 3 + \tfrac{5}{24} \cdot 4 + \tfrac{1}{6} \cdot 5 = 2\tfrac{11}{12},$$

$$E(X \mid Y = 1) = 2\tfrac{1}{2}, \quad E(X \mid Y = 3) = 2\tfrac{7}{24}, \quad E(X \mid Y = 4) = 2\tfrac{17}{24},$$

$$E(X \mid Y = 5) = 2\tfrac{11}{12}.$$

In this way we obtain two collections of points in the plane (x, y). The first collection consists of the points with coordinates

$$x = x_k, \quad y = E(Y \mid X = x_k) \quad (k = 1, 2, 3, 4, 5),$$

the second consists of the points with coordinates

$$y = y_l, \quad x = E(X \mid Y = y_l) \quad (l = 1, 2, 3, 4, 5).$$

The points of the first collection form the regression curve of Y on X. Similarly, the points of the second collection form the regression curve of X on Y.

Example 3.7.2. Let us find the regression curves for the two-dimensional normal distribution given in example 3.6.1.

First we determine the marginal densities

$$f_1(x) = \int_{-\infty}^{+\infty} f(x, y)\, dy = \int_{-\infty}^{+\infty} \frac{1}{2\pi} \exp\left(-\frac{x^2 - 2xy + 2y^2}{2} \right) dy$$

$$= \frac{1}{2\sqrt{\pi}} \exp\left(-\frac{x^2}{4} \right),$$

$$f_2(y) = \int_{-\infty}^{+\infty} f(x, y)\, dx = \int_{-\infty}^{+\infty} \frac{1}{2\pi} \exp\left(-\frac{x^2 - 2xy + 2y^2}{2} \right) dx$$

$$= \frac{1}{\sqrt{2\pi}} \exp\left(-\frac{y^2}{2} \right).$$

By (2.7.6) we obtain the conditional densities

$$f(y \mid x) = \frac{1}{\sqrt{\pi}} \exp\left[-\left(y - \frac{x}{2} \right)^2 \right], \quad f(x \mid y) = \frac{1}{\sqrt{2\pi}} \exp\left[-\frac{(x - y)^2}{2} \right].$$

Hence we obtain

$$m_2(x) = \frac{1}{\sqrt{\pi}} \int_{-\infty}^{+\infty} y \exp\left[-\left(y - \frac{x}{2} \right)^2 \right] dy = \frac{x}{2},$$

$$m_1(y) = \frac{1}{\sqrt{2\pi}} \int_{-\infty}^{+\infty} x \exp\left[-\frac{(x - y)^2}{2} \right] dx = y.$$

As we see, the regression curves thus obtained are straight lines. We shall see later that for the normal distribution the regression curves are always straight lines.

B   The regression curve of the first type has an important extremal property. Thus, the regression curve of the random variable Y on the random variable X satisfies the relation

(3.7.6) $E\{[Y - m_2(X)]^2\} = \text{minimum}.$

In other words, the mean quadratic deviation of Y from a function $u(X)$ achieves its minimum when $u(X)$ equals $m_2(X)$, with probability 1.

We shall prove relation (3.7.6) for the case in which (X, Y) is a two-dimensional random variable of the continuous type with density $f(x, y)$. Then we have

$$E\{[Y - u(X)]^2\} = \int_{-\infty}^{\infty} \int_{-\infty}^{\infty} [y - u(x)]^2 f(x, y)\, dx\, dy$$

(3.7.7)

$$= \int_{-\infty}^{\infty} f_1(x)\left\{\int_{-\infty}^{\infty} [y - u(x)]^2 f(y \mid x)\, dy\right\} dx.$$

From formula (3.2.12) we obtain that the right-hand side of (3.7.7) takes on its minimal value when $u(x) = m_2(x)$.

For discrete random variables the proof of relation (3.7.6) is similar. Similarly, the regression curve of X on Y satisfies the relation

(3.7.8) $E\{[X - m_1(Y)]^2\} = \text{minimum}.$

C   For n-dimensional $(n > 2)$ random variables we can introduce the notion of the *regression surface* of the first type. We shall restrict ourselves to random variables of the continuous type. Let $f(x_1, x_2, \ldots, x_n)$ be the density function of the random variable $(X_1, X_2, \ldots, X_n)$. Suppose that the conditional moment defined by formula (3.6.25) for $l = 1$ exists. Denote this moment by $m_1(x_2, \ldots, x_n)$. Then we have

$$m_1(x_2, \ldots, x_n) = E(X_1 \mid X_2 = x_2, \ldots, X_n = x_n)$$

$$= \frac{\displaystyle\int_{-\infty}^{\infty} x_1 f(x_1, x_2, \ldots, x_n)\, dx_1}{\displaystyle\int_{-\infty}^{\infty} f(x_1, x_2, \ldots, x_n)\, dx_1}.$$

Definition 3.7.2. The set of points of the n-dimensional space $(x_1, x_2, \ldots, x_n)$ with the coordinates

$$x_1 = m_1(x_2, \ldots, x_n), \qquad x_2, \ldots, x_n$$

is called the *regression surface of the first type of the random variable X_1 on the random variables $X_2, \ldots, X_n$.*

In a similar way the regression surfaces of the remaining random variables can be defined.

The regression surface of the first type has a property similar to property (3.7.6), namely

(3.7.9) $E\{[X_1 - m_1(X_2, \ldots, X_n)]^2\} = \text{minimum};$

thus the mean quadratic deviation of X_1 from a function $u(X_2, \ldots, X_n)$ has its minimal value when the function $u(X_2, \ldots, X_n)$ equals $m_1(X_2, \ldots, X_n)$, with probability 1.

3.8   REGRESSION OF THE SECOND TYPE

A   We have defined the regression curve of the first type for the random variable Y on the random variable X as the set of points of the plane (x, y) with coordinates $[x, m_2(x)]$, where $m_2(x)$ is the conditional expectation of Y under the condition that $X = x$. In a similar way the regression curve of X on Y has been defined. It has been shown that these regression curves satisfy relations (3.7.6) and (3.7.8), respectively. In general, these curves are not straight lines. Nevertheless in practical problems it is often necessary to find a straight line such that the mean quadratic deviation of the random variable Y from that line is smaller than from any other straight line in the plane (x, y). Thus if the equation of that line has the form $y = \alpha x + \beta$, the condition

(3.8.1) $E\{[Y - (\alpha X + \beta)]^2\} = \text{minimum}$

must be satisfied.

Definition 3.8.1.   The line with the property expressed by formula (3.8.1) is called the *regression line of the second type of the random variable Y on the random variable X.*

Let us find the coefficients α and β of this line. Denoting

$$E(X) = m_{10}, \quad E(Y) = m_{01}, \quad \sigma_1 = \sqrt{\mu_{20}} = \sqrt{E[(X - m_{10})^2]},$$

$$\sigma_2 = \sqrt{\mu_{02}} = \sqrt{E[(Y - m_{01})^2]}, \quad \mu_{11} = E[(X - m_{10})(Y - m_{01})],$$

we obtain

$$E\{[Y - (\alpha X + \beta)]^2\} = E\{[Y - m_{01} - \alpha(X - m_{10}) + m_{01} - \alpha m_{10} - \beta]^2\}$$

(3.8.2)
$$= E[(Y - m_{01})^2] + \alpha^2 E[(X - m_{10})^2]$$

$$- 2\alpha E[(X - m_{10})(Y - m_{01})] + (m_{01} - \alpha m_{10} - \beta)^2$$

$$= \mu_{02} + \alpha^2 \mu_{20} - 2\alpha \mu_{11} + (m_{01} - \alpha m_{10} - \beta)^2.$$

To find the coefficients α and β which minimize expression (3.8.2), we differentiate it with respect to α and β; the partial derivatives thus obtained should vanish. Hence, we find two equations

$$2\alpha \mu_{20} - 2\mu_{11} - 2m_{10}(m_{01} - \alpha m_{10} - \beta) = 0,$$

(3.8.3)
$$m_{01} - \alpha m_{10} - \beta = 0.$$

The solution of this system is

$$\alpha = \frac{\mu_{11}}{\mu_{20}}, \qquad \beta = m_{01} - \alpha m_{10}.$$

Introducing the coefficient of correlation, we obtain

(3.8.4) $$\alpha = \rho \frac{\sigma_2}{\sigma_1}, \qquad \beta = m_{01} - \rho \frac{\sigma_2}{\sigma_1} m_{10}.$$

In the sequel we denote the coefficients α and β by α_{21} and β_{21}. Similarly, the coefficients of the regression line of X on Y are of the form

(3.8.5) $$\alpha_{12} = \rho \frac{\sigma_1}{\sigma_2}, \qquad \beta_{12} = m_{10} - \rho \frac{\sigma_1}{\sigma_2} m_{01}.$$

The number α_{21} is called the *regression coefficient of Y on X*, and the number α_{12} the *regression coefficient of X on Y*. The equation of the regression line of Y on X may be written in the form

(3.8.6) $$y - m_{01} = \rho \frac{\sigma_2}{\sigma_1} (x - m_{10}),$$

and the equation of the regression line of X on Y in the form

(3.8.7) $$y - m_{01} = \frac{1}{\rho} \frac{\sigma_2}{\sigma_1} (x - m_{10}).$$

From equations (3.8.6) and (3.8.7) it follows that if $\rho^2 = 1$, the regression lines coincide. It is easy to verify that

(3.8.8) $$\begin{aligned} E(Y - \alpha_{21} X - \beta_{21}) &= 0, \\ E\{[Y - (\alpha_{21} X + \beta_{21})]^2\} &= \sigma_2^2 (1 - \rho^2). \end{aligned}$$

The number $\sigma_2^2 (1 - \rho^2)$ is the variance of the random variable $Y - \alpha_{21} X - \beta_{21}$ and, at the same time, the minimum quadratic deviation of Y from any straight line in the plane (x, y). A simple computation leads us to the equation

(3.8.9) $$E[(X - m_{10})(Y - \alpha_{21} X - \beta_{21})] = 0.$$

Thus, the "remainder" $Y - \alpha_{21} X - \beta_{21}$ and X are not correlated. The following formulas are true:

(3.8.10) $$\begin{aligned} E(X - \alpha_{12} Y - \beta_{12}) &= 0, \\ E\{[X - (\alpha_{12} Y + \beta_{12})]^2\} &= \sigma_1^2 (1 - \rho)^2, \\ E[(Y - m_{01})(X - \alpha_{12} Y - \beta_{12})] &= 0. \end{aligned}$$

The numbers $\sigma_1^2 (1 - \rho^2)$ and $\sigma_2^2 (1 - \rho^2)$ are called the *residual variances*.

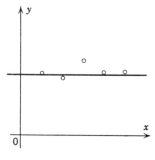 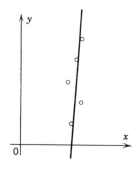

Fig. 3.8.1   Fig. 3.8.2

Example 3.8.1.   Let us find the regression lines of the second type in example 3.7.1.   To obtain them, we compute from Table 3.7.1 the following quantities:

$$m_{10} = E(X) = 2\tfrac{2}{3}, \qquad m_{01} = E(Y) = 3,$$
$$\sigma_1 = \sqrt{D^2(X)} = 1.49, \quad \sigma_2 = \sqrt{D^2(Y)} = 1.41, \quad \rho = 0.06.$$

According to formula (3.8.6), the equation of the regression line of Y on X is of the form

(3.8.11) $y - 3 = 0.056(x - 2\tfrac{2}{3}).$

According to formula (3.8.7), the equation of the regression line of X on Y is of the form

$$y - 3 = 15.7(x - 2\tfrac{2}{3}).$$

In Fig. 3.8.1 the graph of the regression line defined by formula (3.8.11) and the graph of the regression curve of the first type of Y on X are given. In Fig. 3.8.2 the graphs of the regression of X on Y are given.

Example 3.8.2.   We shall determine the regression lines of the two-dimensional normal distribution of example 3.6.1.

We have already computed $m_{01} = m_{10} = 0$, $\sigma_1 = \sqrt{2}$, $\sigma_2 = 1$, $\rho = 1/\sqrt{2}$. Substituting these values into formulas (3.8.4) and (3.8.5), we obtain

$$\alpha_{21} = \tfrac{1}{2}, \quad \beta_{21} = 0, \quad \alpha_{12} = 1, \quad \beta_{12} = 0.$$

Thus the equations of the regression lines are

$$y = \tfrac{1}{2}x \qquad \text{and} \qquad y = x.$$

The equations of the regression lines of the second type are here identical with those of the first type (see example 3.7.2). This property is true for every two-dimensional normal distribution.

B   The generalization of the notion of the regression line of the second type to an n-dimensional random variable $(n > 2)$ is the notion of the *regression hyperplane of the second type*.

Let $(X_1, X_2, \ldots, X_n)$ be an n-dimensional random variable. For simplicity suppose that $E(X_i) = 0$ $(i = 1, 2, \ldots, n)$. This assumption does not restrict the generality, since if it is not satisfied, we can consider the variables $X_i - E(X_i)$. We have shown that the regression surface of the first type of X_1 on the remaining random variables satisfies relation (3.7.9). The regression surface is not always a hyperplane.

Definition 3.8.2. The hyperplane defined by

$$(3.8.12) \qquad X_1 = \alpha_{12}X_2 + \alpha_{13}X_3 + \ldots + \alpha_{1n}X_n,$$

and satisfying the condition

$$(3.8.13) \quad E[(X_1 - \alpha_{12}X_2 - \alpha_{13}X_3 - \ldots - \alpha_{1n}X_n)^2] = \text{minimum},$$

is called the *regression hyperplane of the second type of the random variable* X_1 *on the random variables* $X_2, \ldots, X_n$.

We can suppose that the determinant of the matrix $\mathbf{M}$ defined by formula (3.6.30) is different from zero; for if it were zero we would have, with probability 1, at least one linear relation among the random variables $X_1, \ldots, X_n$, according to theorem 3.6.6; hence the hyperplane determined by this relation would satisfy condition (3.8.13).

Let us write formula (3.8.13) in the form

$$\sigma_1{}^2 + \sum_{i,k=2}^{n} \lambda_{ik}\alpha_{1i}\alpha_{1k} - 2\sum_{i=2}^{n} \lambda_{1i}\alpha_{1i} = \text{minimum}.$$

Differentiating this expression successively with respect to the α's, we obtain the system of linear equations

$$(3.8.14) \quad \begin{aligned} \lambda_{22}\alpha_{12} + \lambda_{23}\alpha_{13} + \ldots + \lambda_{2n}\alpha_{1n} &= \lambda_{12}, \\ \lambda_{32}\alpha_{12} + \lambda_{33}\alpha_{13} + \ldots + \lambda_{3n}\alpha_{1n} &= \lambda_{13}, \\ \cdot \quad \cdot \quad \cdot \quad \cdot \quad \cdot \quad \cdot \quad \cdot \quad \cdot & \\ \lambda_{n2}\alpha_{12} + \lambda_{n3}\alpha_{13} + \ldots + \lambda_{nn}\alpha_{1n} &= \lambda_{1n}. \end{aligned}$$

We have assumed that the determinant $|\mathbf{M}| \neq 0$; hence the symmetric quadratic form (3.6.31) is non-negative. It follows that every diagonal subdeterminant of the determinant $|\mathbf{M}|$ is different from zero;[1] hence the solution of the system of equations (3.8.14) is uniquely determined. We obtain

$$(3.8.15) \qquad \alpha_{1i} = -\frac{|\mathbf{M}_{1i}|}{|\mathbf{M}_{11}|} \qquad (i = 2, 3, \ldots, n),$$

where $|\mathbf{M}_{1i}|$ and $|\mathbf{M}_{11}|$ are the algebraic complements of λ_{1i} and λ_{11}, respectively, in the determinant $|\mathbf{M}|$. Denote

$$Y_1 = X_1 - \alpha_{12}X_2 - \ldots - \alpha_{1n}X_n.$$

---

[1] See, for instance A. C. Aitken [1].

Substituting the α_{1i} from formulas (3.8.15) we obtain

$$(3.8.16) \qquad Y_1 = X_1 + \frac{1}{|\mathbf{M}_{11}|} \sum_{i=2}^{n} |\mathbf{M}_{1i}| \, X_i.$$

We notice that $E(Y_1) = 0$. We have

$$E(Y_1 X_k) = E\left(X_1 X_k + \frac{1}{|\mathbf{M}_{11}|} \sum_{i=2}^{n} |\mathbf{M}_{1i}| \, X_i X_k\right)$$

$$= \lambda_{1k} + \frac{1}{|\mathbf{M}_{11}|} \sum_{i=2}^{n} \lambda_{ik} \, |\mathbf{M}_{1i}|$$

$$= \frac{1}{|\mathbf{M}_{11}|} \sum_{i=1}^{n} \lambda_{ik} \, |\mathbf{M}_{1i}|.$$

It follows that

$$(3.8.17) \qquad E(Y_1 X_k) = \begin{cases} |\mathbf{M}|/|\mathbf{M}_{11}| & \text{for} \quad k = 1, \\ 0 & \text{for} \quad k = 2, 3, \ldots, n. \end{cases}$$

Thus we find that the residual Y_1 is not correlated with any of the random variables $X_2, X_3, \ldots, X_n$.

Furthermore, we have from (3.8.17)

$$(3.8.18) \qquad D^2(Y_1) = E(Y_1^2) = E(Y_1 X_1) = \frac{|\mathbf{M}|}{|\mathbf{M}_{11}|} \, .$$

Expression (3.8.18) is called the *residual variance*.

By analogy to formula (3.8.15) we obtain for the coefficients of the regression hyperplane of X_k on the remaining random variables, the formula

$$(3.8.19) \quad \alpha_{ki} = - \frac{|\mathbf{M}_{ki}|}{|\mathbf{M}_{kk}|} \qquad (i = 1, \ldots, k-1, k+1, \ldots, n).$$

The residual Y_k is given by the formula

$$(3.8.20) \qquad Y_k = X_k + \frac{1}{|\mathbf{M}_{kk}|} \sum_{\substack{i=1 \\ i \neq k}}^{n} |\mathbf{M}_{ki}| \, X_i.$$

The random variable Y_k is not correlated with any of the random variables X_i where $i \neq k$, and the residual variance is given by the formula

$$(3.8.21) \qquad D^2(Y_k) = \frac{|\mathbf{M}|}{|\mathbf{M}_{kk}|} \, .$$

Example 3.8.3. The random variable (X_1, X_2, X_3) has a three-dimensional normal distribution with density function

$$(3.8.22) \quad f(x_1, x_2, x_3) = \frac{1}{(2\pi)^{3/2}}$$

$$\times \exp\left[-\tfrac{1}{4}(3x_1^2 + 3x_2^2 + 3x_3^2 - 2x_1 x_2 - 2x_1 x_3 - 2x_2 x_3)\right].$$

After some simple calculations we obtain for the matrix of moments of the second order

$$\mathbf{M} = \begin{bmatrix} 1 & \frac{1}{2} & \frac{1}{2} \\ \frac{1}{2} & 1 & \frac{1}{2} \\ \frac{1}{2} & \frac{1}{2} & 1 \end{bmatrix}.$$

We shall find the regression plane of the second type of X_1 on X_2 and X_3. From formula (3.8.15) we obtain

$$\alpha_{12} = \frac{\begin{vmatrix} \frac{1}{2} & \frac{1}{2} \\ \frac{1}{2} & 1 \end{vmatrix}}{\begin{vmatrix} 1 & \frac{1}{2} \\ \frac{1}{2} & 1 \end{vmatrix}} = \tfrac{1}{3}, \qquad \alpha_{13} = -\frac{\begin{vmatrix} \frac{1}{2} & 1 \\ \frac{1}{2} & \frac{1}{2} \end{vmatrix}}{\begin{vmatrix} 1 & \frac{1}{2} \\ \frac{1}{2} & 1 \end{vmatrix}} = \tfrac{1}{3}.$$

The equation of the regression plane of X_1 is then

$$X_1 = \tfrac{1}{3}X_2 + \tfrac{1}{3}X_3.$$

By formula (3.8.18), the variance of $Y_1 = X_1 - \tfrac{1}{3}X_2 - \tfrac{1}{3}X_3$ equals $\tfrac{2}{3}$. We notice that the variance of X_1 is equal to 1.

The reader can verify that the regression surface of the first type of X_1 coincides with the regression hyperplane of the second type. This property is valid for every multidimensional normal distribution.

C   The method of finding the unknown coefficients of the straight line and hyperplane which we have applied in this section is called the *method of least squares*. The basic ideas of this method were created by Legendre [1] and Gauss [1]. The further development of these ideas, so important for probability theory and mathematical statistics, is due to Markov [5]. We also recommend to the reader the paper of David and Neyman [1]. In the book by Linnik [3] a comprehensive description is given of the most important results thus far obtained in the method of least squares.

Problems and Complements

3.1. Express the central moment μ_4 as a function of the ordinary moments m_1, m_2, m_3, and m_4.

3.2. Express the ordinary moment m_4 as a function of the central moments μ_2, μ_3, and μ_4.

3.3. Find the moments μ_4 and m_4 of the random variable X of example 3.2.3.

3.4. Show that if X_1 and X_2 are independent and have the same distribution, $Y = X_1 - X_2$ has a symmetric distribution.

The results discussed in Problems 3.5 to 3.8 are due to Kolmogorov [7].

3.5. A real function $g(x)$ is non-negative and satisfies the inequality $g(x) \geqslant b > 0$ for all $x \geqslant a$. Prove that for an arbitrary variable X such that $E[g(X)]$ exists, we have

$$P(X \geqslant a) \leqslant \frac{E[g(X)]}{b}.$$

3.6.(*a*) A real function $g(x)$ is non-negative, even and, for positive values of x, nondecreasing. Prove that for an arbitrary random variable X, if $E[g(X)]$ exists, then for arbitrary positive ε

$$P(|X| \geqslant \varepsilon) \leqslant \frac{E[g(X)]}{g(\varepsilon)}.$$

(*b*) Derive the Chebyshev inequality from the last relation.

3.7. A real function $g(x)$ satisfies all the assumptions of 3.6(*a*) and the relation $|g(x)| \leqslant K$ for all x. Prove that for any $\varepsilon > 0$ we have

$$P(|X| \geqslant \varepsilon) \geqslant \frac{E[g(X)] - g(\varepsilon)}{K}.$$

3.8. A real function $g(x)$ satisfies the assumptions of 3.6(*a*). Suppose that $P(|X| \leqslant M) = 1$. Prove that for any $\varepsilon > 0$

$$P(|X| \geqslant \varepsilon) \geqslant \frac{E[g(X)] - g(\varepsilon)}{g(M)};$$

hence in particular

$$P(|X| \geqslant \varepsilon) \geqslant \frac{E(X^2) - \varepsilon^2}{M^2}.$$

3.9. Prove (Berge [1]) that for arbitrary random variables Y_1 and Y_2 with standard deviations σ_1 and σ_2 and correlation coefficient ρ, we have for arbitrary k

$$P(|Y_1 - E(Y_1)| \geqslant k\sigma_1 \quad \text{or} \quad |Y_2 - E(Y_2)| \geqslant k\sigma_2) \leqslant \frac{1 + \sqrt{1 - \rho^2}}{k^2}.$$

(The generalization to random vectors with a number of dimensions greater than two was given by Olkin and Pratt [1].)

3.10. Show that if ρ is the correlation coefficient of the random variables X_1 and X_2, ρ is also the coefficient of correlation of the random variables $Y_i = a_i X_i + b_i (i = 1, 2)$.

3.11. Show that if X and Y have the expected value zero and the same standard deviation, the random variables

$$Z = X + Y \quad \text{and} \quad U = X - Y$$

are uncorrelated. Are they necessarily independent? What if the independence of X and Y is assumed? Compare with Skitovitch's theorem in Section 5.7.

3.12. The random variables X and Y have finite second order moments and Z and U are linear functions

$$Z = aX + bY + c, \quad U = dX + eY + f.$$

Find the coefficients a, b, c, d, e, and f such that the random variables Z and U are uncorrelated.

3.13. The expression

$$\Theta_{yx} = \frac{1}{D^2(Y)} E[m_2(X) - E(Y)]^2,$$

where $m_2(X)$ is given by (3.7.1) or (3.7.1'), is called the *correlation ratio of the random variable Y on the random variable X.* (*a*) Show that:

I. $$0 \leqslant \Theta_{yx}^{\,2} \leqslant 1.$$

II. The equality $\Theta_{yx}^{\,2} = 0$ holds if and only if $m_2(X) = $ constant.
III. The equality $\Theta_{yx}^{\,2} = 1$ holds if and only if $P[Y - m_2(X) = 0] = 1$.
(*b*) What does the correlation ratio measure?

3.14. Find the correlation ratios of X on Y and Y on X in examples 2.6.1, 3.6.1, and 3.7.1.

3.15. Let

$$Y_1 = X_1 - \alpha_{13}X_3 - \ldots - \alpha_{1n}X_n,$$

$$Y_2 = X_2 - \alpha_{23}X_3 - \ldots - \alpha_{2n}X_n,$$

where $E(X_i) = 0$ ($i = 1, \ldots, n$) and the coefficients α_{1i} and α_{2i} ($i = 3, \ldots, n$) are determined from formulas analogous to (3.8.15). The expression

$$\rho_{12\cdot3\ldots n} = \frac{E(Y_1 Y_2)}{\sqrt{E(Y_1^2)E(Y_2^2)}}$$

is called the *partial correlation coefficient of X_1 and X_2 with respect to $X_3, \ldots, X_n$.* This is a measure of correlation between X_1 and X_2 after eliminating the influence of the random variables $X_3, \ldots, X_n$.

(*a*) Prove that (notation of Section 3.8)

$$\rho_{12\cdot3\ldots n} = -\frac{|M_{12}|}{\sqrt{|M_{11}||M_{22}|}}.$$

(*b*) Verify that

$$\rho_{12\cdot3} = \frac{\rho_{12} - \rho_{13}\rho_{23}}{(1 - \rho_{23}^2)(1 - \rho_{13}^2)},$$

where ρ_{12}, ρ_{13}, and ρ_{23} are the correlation coefficients of the pairs of random variables (X_1, X_2), (X_1, X_3), and (X_2, X_3), respectively.

3.16. Find all the partial correlation coefficients in example 3.8.3.

3.17. Let

$$X_1^* = \alpha_{12}X_2 + \ldots + \alpha_{1n}X_n,$$

where $E(X_i) = 0$ ($i = 1, \ldots, n$) and the α_{1i} ($i = 2, \ldots, n$) satisfy (3.8.15). The correlation coefficient of the random variables X_1^* and X_1 is called the *multiple correlation coefficient between X_1 and $(X_2, \ldots, X_n)$.* We denote it by $\rho_{1(2\ldots n)}$. Thus we have

$$\rho_{1(2\ldots n)} = \frac{E(X_1 X_1^*)}{\sqrt{E(X_1^2)E(X_1^{*2})}}.$$

Prove that among all linear combinations of the random variables $X_2, \ldots, X_n$ the random variable X_1^* is the most closely correlated with X_1.

3.18. Compute all multiple correlation coefficients in example 3.8.3.

3.19. Show that the probability that the random variables $X_1, X_2, \ldots, X_n$, whose variances exist, satisfy exactly $n - r$ linear relations, equals one if and only if r is the rank of the matrix **M** of moments of the second order. The distribution is then said to be of *rank r* (Frisch [1]).

3.20. Let $\{X_n\}$ $(n = 1, 2, 3, \ldots)$ be a sequence of random variables with mean values $E(X_n)$. Prove that if

$$\sum_{n=1}^{\infty} E(|X_n|) < \infty,$$

then

$$E\left(\sum_{n=1}^{\infty} X_n\right) = \sum_{n=1}^{\infty} E(X_n).$$

Deduce that $\sum_{n=1}^{\infty} X_n$ converges with probability one.

Hint: Use the Monotone Convergence Theorem for the sequence Y_n, where $Y_n = \sum_{k=1}^{n} |X_k|$.

See also problem 4.10.

CHAPTER 4

Characteristic Functions

4.1 PROPERTIES OF CHARACTERISTIC FUNCTIONS

In this chapter we investigate the expected value of a certain function of a random variable and obtain a method of investigation which is extremely useful in further work on probability theory and its application to statistics.

Let X be a random variable and let $F(x)$ be its distribution function.

Definition 4.1.1. The function

$$(4.1.1) \qquad \phi(t) = E(e^{itX}),$$

where t is a real number and i is the imaginary unit, is called the *characteristic function* of the random variable X or of the distribution function $F(x)$.

If X is a random variable of the discrete type with jump points x_k $(k = 1, 2, \ldots)$ and $P(X = x_k) = p_k$, the characteristic function of X has the form

$$(4.1.2) \qquad \phi(t) = E(e^{itX}) = \sum_k p_k e^{itx_k}.$$

Since $|e^{itx_k}| = 1$ and $\sum_k p_k = 1$, the series on the right-hand side of (4.1.2) is absolutely and uniformly convergent. Thus, the characteristic function $\phi(t)$, as the sum of a uniformly convergent series of continuous functions, is continuous for every real value of t.

Example 4.1.1. The random variable X can take on the values $x_1 = -1$ and $x_2 = +1$ with probabilities $P(X = -1) = P(X = +1) = 0.5$. We shall determine the characteristic function of this random variable.

By (4.1.2) we have

$$(4.1.3) \quad \phi(t) = 0.5e^{-it} + 0.5e^{it} = 0.5(\cos t - i \sin t) + 0.5 (\cos t + i \sin t)$$
$$= \cos t.$$

105

If X is a random variable of the continuous type with density function $f(x)$, its characteristic function is given by the formula

(4.1.4) $$\phi(t) = E(e^{itX}) = \int_{-\infty}^{+\infty} f(x)e^{itx}\, dx.$$

Since $$\int_{-\infty}^{+\infty} f(x)\,|e^{itx}|\, dx = \int_{-\infty}^{+\infty} f(x)\, dx = 1,$$

the integral in (4.1.4) is absolutely and uniformly convergent; hence $\phi(t)$ is a continuous function for every value of t.

Example 4.1.2. The density $f(x)$ is defined as

(4.1.5) $$f(x) = \begin{cases} 0, & \text{for} \quad x < 0, \\ 1, & \text{for} \quad 0 \leqslant x \leqslant 1, \\ 0, & \text{for} \quad x > 1. \end{cases}$$

This distribution is called *uniform* or *rectangular*. Its characteristic function is

(4.1.6) $$\phi(t) = \int_{-\infty}^{+\infty} f(x)e^{itx}\, dx = \int_0^1 f(x)e^{itx}\, dx = \left[\frac{e^{itx}}{it}\right]_0^1 = \frac{e^{it} - 1}{it}.$$

We now investigate some of the properties of characteristic functions. We have

(4.1.7) $$\phi(0) = E(e^0) = E(1) = 1.$$

Since $$|\phi(t)| = |E(e^{itX})| \leqslant E(|e^{itX}|) = 1,$$
we have

(4.1.8) $$|\phi(t)| \leqslant 1.$$

We next have

$$\phi(-t) = E(e^{-itX}) = E(\cos tX - i \sin tX) = E(\cos tX) - iE(\sin tX).$$

Since

$$\phi(t) = E(e^{itX}) = E(\cos tX + i \sin tX) = E(\cos tX) + iE(\sin tX),$$

we obtain

(4.1.9) $$\phi(-t) = \overline{\phi(t)},$$

where $\overline{\phi(t)}$ denotes the complex number conjugate to $\phi(t)$.

Every characteristic function must satisfy conditions (4.1.7), (4.1.8), and (4.1.9). These conditions are, however, not sufficient; thus not every function $\phi(t)$ satisfying these conditions is a characteristic function of some random variable.

Other necessary conditions have been given by Marcinkiewicz [1]. He has shown that a function $\phi(t)$ which is not identically constant and which, in a neighborhood of zero, can be represented in the form

$$\phi(t) = 1 + 0(t^{2+\alpha})$$

with $\alpha > 0$ cannot be a characteristic function. It follows immediately that neither the function $\phi(t) = \exp(-t^4)$ nor the function $\phi(t) = 1/(1 + t^4)$ can be a characteristic function. Further necessary conditions have been given by Lukacs [4]. We shall present without proof the theorem of Bochner [2], giving necessary and sufficient conditions for a function $\phi(t)$ to be a characteristic function.

Theorem 4.1.1. *Let the function $\phi(t)$ defined for $-\infty < t < +\infty$ satisfy condition (4.1.7). The function $\phi(t)$ is the characteristic function of some distribution function if and only if*

1. *$\phi(t)$ is continuous.*
2. *for $n = 1, 2, 3, \ldots$ and every real $t_1, \ldots, t_n$ and complex $a_1, \ldots, a_n$ we have*

$$\sum_{j,k=1}^{n} \phi(t_j - t_k) a_j \bar{a}_k \geqslant 0.$$

A relatively simple proof of this theorem is given in the monograph by Loève [4].

Let us recall that a function satisfying condition 2 of theorem 4.1.1 is called positive definite.

Another necessary and sufficient condition for the function $\phi(t)$ to be a characteristic function has been given by Cramér [5] (see Problem 4.3).

4.2 THE CHARACTERISTIC FUNCTION AND MOMENTS

A Consider a random variable X and suppose that its lth moment $m_l = E(X^l)$ exists.

Suppose that X is a random variable of the discrete type with jump points x_k. Then we can differentiate (4.1.2) l times with respect to t.

In fact, the lth derivative with respect to t of the expression under the summation sign in (4.1.2) equals $p_k i^l x_k^l e^{itx_k}$. On the other hand, from the existence of the lth moment there follows the existence of the absolute lth moment. Since

$$\sum_k |i^l p_k x_k^l e^{itx_k}| = \sum_k |p_k x_k^l| = \beta_l,$$

we can differentiate (4.1.2) l times under the summation sign. Hence we have

(4.2.1)
$$\phi^{(l)}(t) = \sum_k p_k i^l x_k^l e^{itx_k}$$
$$= E(i^l X^l e^{itX}).$$

Suppose now that $f(x)$ is the density function of a random variable X of the continuous type. Then we can differentiate (4.1.4) l times.

Indeed, the lth derivative with respect to t of the expression under the integral sign in (4.1.4) equals $i^l x^l f(x) e^{itx}$. We have

$$\int_{-\infty}^{+\infty} |i^l x^l f(x) e^{itx}| \, dx = \int_{-\infty}^{+\infty} |x^l f(x)| \, dx = \beta_l.$$

By assumption, the absolute moment β_l is finite. Thus we can differentiate the formula for $\phi(t)$ l times under the integral sign. We obtain

$$(4.2.2) \qquad \phi^{(l)}(t) = \int_{-\infty}^{+\infty} i^l x^l f(x) e^{itx} \, dx$$

$$= E(i^l X^l e^{itX}).$$

Thus we have obtained the same result as for a random variable of the discrete type

$$(4.2.3) \qquad \phi^{(l)}(t) = E(i^l X^l e^{itx}).$$

Let us compute $\phi^{(l)}(0)$ from relation (4.2.3). We have

$$\phi^{(l)}(0) = i^l E(X^l) = i^l m_l.$$

Hence

$$(4.2.4) \qquad m_l = \frac{\phi^{(l)}(0)}{i^l}.$$

Thus we have proved the following theorem.

Theorem 4.2.1. *If the lth moment m_l of a random variable exists, it is expressed by formula (4.2.4), where $\phi^{(l)}(0)$ is the lth derivative of the characteristic function $\phi(t)$ of this random variable at $t = 0$.*

Example 4.2.1. Suppose that the random variable X has a Poisson distribution, that is, it can take on the values $x_k = k$, where k is any non-negative integer, and the probability function is given by the formula

$$(4.2.5) \qquad P(X = k) = \frac{\lambda^k}{k!} e^{-\lambda},$$

where λ is a positive constant. We shall find the characteristic function of X. From (4.1.2) we obtain

(4.2.6)
$$\phi(t) = \sum_{k=0}^{\infty} e^{itk} \frac{\lambda^k}{k!} e^{-\lambda} = e^{-\lambda} \sum_{k=0}^{\infty} \frac{(\lambda e^{it})^k}{k!} = \exp(-\lambda) \exp(\lambda e^{it}) = \exp[\lambda(e^{it} - 1)].$$

Furthermore,

$$\phi'(t) = \lambda i \exp(it) \exp[\lambda(e^{it} - 1)].$$

From (4.2.4) we obtain

(4.2.7) $$m_1 = \frac{\phi'(0)}{i} = \frac{\lambda i}{i} = \lambda.$$

Similarly,

$$\phi''(t) = \lambda i^2 \exp{(it)} \exp{[\lambda(e^{it} - 1)]}[\lambda \exp{(it)} + 1].$$

Hence

(4.2.8) $$m_2 = \frac{\phi''(0)}{i^2} = \frac{i^2\lambda(\lambda + 1)}{i^2} = \lambda(\lambda + 1).$$

Thus the central moment of the second order is

(4.2.9) $$\mu_2 = \lambda(\lambda + 1) - \lambda^2 = \lambda.$$

In a similar manner we can obtain the moments of higher orders.

Example 4.2.2. We shall find the characteristic function and the moments of a normal distribution.

We have

$$f(x) = \frac{1}{\sqrt{2\pi}} e^{-x^2/2};$$

hence

(4.2.10)
$$\phi(t) = \frac{1}{\sqrt{2\pi}} \int_{-\infty}^{+\infty} \exp{(itx)} \exp{\left(-\frac{x^2}{2}\right)} dx$$
$$= \frac{1}{\sqrt{2\pi}} \int_{-\infty}^{+\infty} \exp{\left[-\frac{(x - it)^2}{2}\right]} \exp{\left(-\frac{t^2}{2}\right)} dx = \exp{\left(-\frac{t^2}{2}\right)}.$$

Since $\phi'(t) = -t \exp{(-t^2/2)}$, we have

(4.2.11) $$m_1 = \frac{\phi'(0)}{i} = 0.$$

Next, we have $\phi''(t) = (t^2 - 1) \exp{(-t^2/2)}$; hence

(4.2.12) $$m_2 = \frac{\phi''(0)}{i^2} = 1.$$

We have already obtained the same values m_1 and m_2 in examples 3.1.4 and 3.2.2. The reader can verify that all the odd order moments equal zero and that the even order moments are expressed by the formula

(4.2.13) $$m_{2l} = 1 \cdot 3 \cdot 5 \cdot \ldots \cdot (2l - 1).$$

We notice that the converse of theorem 4.2.1 is not true. An example of a random variable, whose expectation does not exist and whose characteristic function is differentiable at $t = 0$, is given in Problem 4.9. But if the characteristic function $\phi(t)$ has a finite derivative of an even order $2k$ at $t = 0$, then the moment of order $2k$ of the corresponding random variable exists (See Problem 4.8). As we know, in this case all the moments of orders smaller than $2k$ also exist.

B   Now consider the characteristic function of a linear transformation of the random variable X.

First consider the translation

$$Y = X + b.$$

Denoting by $\phi_1(t)$ the characteristic function of the random variable Y, we obtain

(4.2.14) $\phi_1(t) = E(e^{itY}) = E(e^{it(X+b)}) = E(e^{itX})e^{itb} = e^{itb}\phi(t).$

We see that when the random variable is translated by a constant b, its characteristic function is multiplied by the factor e^{itb}.

Now let

$$Y = aX.$$

We have

(4.2.15) $\phi_1(t) = E(e^{itY}) = E(e^{itaX}) = \phi(at).$

Thus, the characteristic function of the random variable aX equals the characteristic function of the random variable X at the point at. In particular, if $a = -1$, we obtain

$$\phi_1(t) = \phi(-t) = \overline{\phi(t)}.$$

Now let us consider the transformation

$$Y = aX + b.$$

Denoting the characteristic functions of the random variables X and Y by $\phi(t)$ and $\phi_1(t)$ respectively, we obtain from equations (4.2.14) and (4.2.15)

(4.2.16) $\phi_1(t) = e^{ibt}\phi(at).$

In particular, let

$$Y = \frac{X - m_1}{\sigma},$$

where m_1 and σ denote respectively the expected value and the standard deviation of X. Then

(4.2.17) $\phi_1(t) = \exp\left(-\frac{m_1 it}{\sigma}\right)\phi\left(\frac{t}{\sigma}\right).$

4.3   SEMI-INVARIANTS

Sometimes it is convenien. to deal with a set of parameters other than the set of moments. We obtain such parameters by considering the function

(4.3.1) $\psi(t) = \log \phi(t),$

where $\phi(t)$ is the characteristic function of the random variable under consideration.

Let us formally expand the function $\phi(t)$ in a power series in a neighborhood of $t = 0$,

$$(4.3.2) \qquad \phi(t) = 1 + \sum_{s=1}^{\infty} \frac{m_s}{s!} (it)^s.$$

Let us denote by z the series on the right-hand side of (4.3.2) and let us formally expand the function $\psi(t)$ into a power series

$$(4.3.3) \quad \psi(t) = \log \phi(t) = \log(1 + z) = \frac{z}{1} - \frac{z^2}{2} + \frac{z^3}{3} - \ldots = \sum_{s=1}^{\infty} \frac{\kappa_s}{s!} (it)^s.$$

From (4.3.2) and (4.3.3) we obtain the formal equation

(4.3.4)

$$\phi(t) = 1 + \sum_{s=1}^{\infty} \frac{m_s}{s!} (it)^s = \exp\left[\sum_{s=1}^{\infty} \frac{\kappa_s}{s!} (it)^s \right]$$

$$= 1 + \sum_{s=1}^{\infty} \frac{\kappa_s}{s!} (it)^s + \frac{1}{2!}\left[\sum_{s=1}^{\infty} \frac{\kappa_s}{s!} (it)^s \right]^2 + \frac{1}{3!}\left[\sum_{s=1}^{\infty} \frac{\kappa_s}{s!} (it)^s \right]^3 + \ldots.$$

Definition 4.3.1. The coefficients κ_s in (4.3.3) are called *semi-invariants*.

To express the semi-invariants in terms of the moments or the moments in terms of the semi-invariants, we compare successively the coefficients of $(it)^s$ for particular values of s in equation (4.3.4). In this way we obtain

$$(4.3.5) \qquad \begin{aligned} \kappa_1 &= m_1, \\ \kappa_2 &= m_2 - m_1^2 = \sigma^2, \\ \kappa_3 &= m_3 - 3m_1m_2 + 2m_1^3, \\ \kappa_4 &= m_4 - 3m_2^2 - 4m_1m_3 + 12m_1^2m_2 - 6m_1^4, \end{aligned}$$

$$\cdots \cdots \cdots \cdots \cdots \cdots \cdots$$

and also

$$(4.3.6) \qquad \begin{aligned} m_1 &= \kappa_1, \\ m_2 &= \kappa_2 + \kappa_1^2, \\ m_3 &= \kappa_3 + 3\kappa_1\kappa_2 + \kappa_1^3, \\ m_4 &= \kappa_4 + 3\kappa_2^2 + 4\kappa_1\kappa_3 + 6\kappa_1^2\kappa_2 + \kappa_1^4, \end{aligned}$$

$$\cdots \cdots \cdots \cdots \cdots \cdots \cdots$$

The semi-invariants can also be expressed in terms of the central moments,

$$(4.3.7) \qquad \begin{aligned} \kappa_1 &= m_1, \\ \kappa_2 &= \mu_2 = \sigma^2, \\ \kappa_3 &= \mu_3, \\ \kappa_4 &= \mu_4 - 3\mu_2^2, \end{aligned}$$

$$\cdots \cdots \cdots \cdots$$

From (4.3.5) and (4.3.6) it follows that if the moment of the lth order exists, all the semi-invariants of order not greater than l also exist.

The name semi-invariants comes from the fact that under a translation, that is, under a transformation $Y = X + b$, all semi-invariants except κ_1 remain unchanged. If we denote by $\phi(t)$ and $\phi_1(t)$ the characteristic functions of the random variables X and Y, respectively, we have, by equation (4.2.14)

$$(4.3.8) \qquad \log \phi_1(t) = bit + \log \phi(t).$$

Thus the translation changes only the coefficient of the term with it to the first power in the expansion (4.3.4); hence it changes only the semi-invariant of the first order.

Example 4.3.1. We shall compute the semi-invariants of the Poisson distribution discussed in example 4.2.1.

The characteristic function of the Poisson distribution is

$$\phi(t) = \exp [\lambda(e^{it} - 1)].$$

Hence we obtain

$$(4.3.9) \quad \psi(t) = \log \phi(t) = \lambda(e^{it} - 1) = \lambda \left(\sum_{k=0}^{\infty} \frac{(it)^k}{k!} - 1 \right) = \lambda \sum_{k=1}^{\infty} \frac{(it)^k}{k!}.$$

From formula (4.3.3), we obtain

$$(4.3.10) \qquad \kappa_k = \lambda \quad (k = 1, 2, \ldots).$$

Using the formulas for the relations between semi-invariants and moments we can obtain from formula (4.3.10) the moments of arbitrary order of the Poisson distribution.

4.4   THE CHARACTERISTIC FUNCTION OF THE SUM OF INDEPENDENT RANDOM VARIABLES

Let X and Y be two independent random variables. From the considerations of Section 2.8 it follows that the random variables e^{itX} and e^{itY} are also independent. We shall find the characteristic function of the sum

$$Z = X + Y.$$

Let $\phi(t)$, $\phi_1(t)$ and $\phi_2(t)$ denote respectively the characteristic functions of the random variables Z, X, and Y. We have

$$(4.4.1) \qquad \phi(t) = E(e^{itZ}) = E(e^{it(X+Y)}) = E(e^{itX}e^{itY}).$$

By the independence of the random variables e^{itX} and e^{itY}, we obtain from theorem 3.6.2

$$(4.4.2) \qquad \phi(t) = E(e^{itX})E(e^{itY}) = \phi_1(t)\phi_2(t).$$

This result can be generalized to an arbitrary finite number of independent random variables.

Theorem 4.4.1. *The characteristic function of the sum of an arbitrary finite number of independent random variables equals the product of their characteristic functions.*

Thus, if Z is the sum of n independent random variables,

$$Z = X_1 + X_2 + \ldots + X_n,$$

and $\phi(t)$, $\phi_1(t)$, $\phi_2(t), \ldots, \phi_n(t)$ denote the characteristic functions of $Z, X_1, X_2, \ldots, X_n$, respectively, then

(4.4.3) $$\phi(t) = \phi_1(t)\phi_2(t) \ldots \phi_n(t).$$

Example 4.4.1. Suppose two independent random variables X_1 and X_2 have Poisson distributions

$$P(X_1 = r) = \frac{\lambda_1^r}{r!} e^{-\lambda_1}, \qquad P(X_2 = r) = \frac{\lambda_2^r}{r!} e^{-\lambda_2} \ (r = 0, 1, \ldots).$$

Consider the random variable

$$Z = X_1 - X_2.$$

We shall determine the characteristic function and the semi-invariants of Z. By equation (4.2.6) the characteristic functions $\phi_1(t)$ and $\phi_2(t)$ of X_1 and X_2 have the form

$$\phi_1(t) = \exp\left[\lambda_1(e^{it} - 1)\right], \qquad \phi_2(t) = \exp\left[\lambda_2(e^{it} - 1)\right].$$

By (4.2.16), the characteristic function of $-X_2$ is

$$\phi_2(-t) = \exp\left[\lambda_2(e^{-it} - 1)\right].$$

Since X_1 and $-X_2$ are independent, we obtain by (4.4.2) for the characteristic function of the random variable Z

$$\phi(t) = \exp\left[\lambda_1(e^{it} - 1)\right] \exp\left[\lambda_2(e^{-it} - 1)\right] = \exp\left(\lambda_1 e^{it} + \lambda_2 e^{-it} - \lambda_1 - \lambda_2\right).$$

Expanding the exponents e^{it} and e^{-it} into power series, we obtain

$$\phi(t) = \exp\left[(\lambda_1 - \lambda_2)(it) + (\lambda_1 + \lambda_2)\frac{(it)^2}{2!} + (\lambda_1 - \lambda_2)\frac{(it)^3}{3!} + \ldots\right],$$

$$\psi(t) = \log \phi(t) = (\lambda_1 - \lambda_2)\frac{(it)}{1!} + (\lambda_1 + \lambda_2)\frac{(it)^2}{2!} + (\lambda_1 - \lambda_2)\frac{(it)^3}{3!} + \ldots$$

From (4.3.3) it follows that all the semi-invariants of odd order of Z equal $\lambda_1 - \lambda_2$, and all the semi-invariants of even order equal $\lambda_1 + \lambda_2$. The expected value and the variance of Z can be obtained from (4.3.5),

$$m_1 = \kappa_1 = \lambda_1 - \lambda_2, \qquad \sigma^2 = \kappa_2 = \lambda_1 + \lambda_2.$$

We notice that the converse of theorem 4.4.1 is not true; that is, the characteristic function of the sum of dependent random variables may equal the product of their characteristic functions. The following is an example of this.

Example 4.4.2. The joint distribution of the random variable (X, Y) is given by the density

$$f(x, y) = \begin{cases} \frac{1}{4}[1 + xy(x^2 - y^2)] & \text{for} \quad |x| \leqslant 1 \quad \text{and} \quad |y| \leqslant 1, \\ 0 & \text{for all other points.} \end{cases}$$

We first show that the random variables X and Y are dependent. The marginal distributions in the domains $|x| \leqslant 1$ and $|y| \leqslant 1$ are, respectively, of the form

$$f_1(x) = \int_{-1}^{+1} \tfrac{1}{4}[1 + xy(x^2 - y^2)] \, dy = \tfrac{1}{4}(y + \tfrac{1}{2}x^3 y^2 - \tfrac{1}{4}xy^4)_{-1}^{+1} = \tfrac{1}{2},$$

$$f_2(y) = \int_{-1}^{+1} \tfrac{1}{4}[1 + xy(x^2 - y^2)] \, dx = \tfrac{1}{4}(x + \tfrac{1}{4}x^4 y - \tfrac{1}{2}x^2 y^3)_{-1}^{+1} = \tfrac{1}{2}.$$

We then obtain $f_1(x)f_2(y) = \tfrac{1}{4} \neq f(x, y)$; hence the random variables X and Y are not independent.

We now find the density of the sum $Z = X + Y$. From (2.9.8) we obtain

$$f_3(z) = \int_{-\infty}^{+\infty} f(x, z - x) \, dx.$$

The endpoints of the intervals of x values, for which $f(x, z - x) > 0$, depend on z. To find them, observe that by introducing the variables x, z instead of x, y we transform the square $|x| \leqslant 1$, $|y| \leqslant 1$ into the domain defined by the inequalities

(4.4.4) $$|x| \leqslant 1, \quad x - 1 \leqslant z \leqslant x + 1.$$

This transformation is illustrated in Figs. 4.4.1 and 4.4.2. In Fig. 4.4.1 the shaded area represents the domain in the (x, y) plane defined by the inequalities $|x| \leqslant 1$, $|y| \leqslant 1$, and in Fig. 4.4.2 the corresponding domain in the (x, z) plane.

Let us write inequalities (4.4.4) in the form

$$|x| \leqslant 1, \quad z - 1 \leqslant x \leqslant z + 1.$$

Furthermore, we notice that for $z \leqslant 0$ we have

$$z - 1 \leqslant -1, \quad z + 1 \leqslant 1,$$

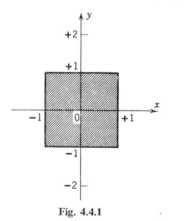

Fig. 4.4.1   **Fig. 4.4.2**

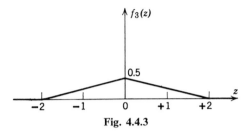

Fig. 4.4.3

and for $z > 0$ we have

$$z - 1 > -1, \qquad z + 1 > 1.$$

Thus for $z \leqslant 0$ we integrate the function $f(x, z - x)$ from -1 to $z + 1$, and for $z > 0$ from $z - 1$ to 1. After simple computations we obtain

$$
f_3(z) = \begin{cases}
\displaystyle\int_{-1}^{z+1} \tfrac{1}{4}(1 + 3z^2x^2 - 2zx^3 - z^3x)\,dx = \tfrac{1}{4}(2 + z) & \text{for } -2 \leqslant z \leqslant 0, \\[2ex]
\displaystyle\int_{z-1}^{1} \tfrac{1}{4}(1 + 3z^2x^2 - 2zx^3 - z^3x)\,dx = \tfrac{1}{4}(2 - z) & \text{for } 0 < z \leqslant 2, \\[2ex]
0 & \text{for } |z| > 2.
\end{cases}
$$

A distribution such as that of Z is called a *triangular distribution*. The graph of the function $f_3(z)$ is represented in Fig. 4.4.3.

We now determine the characteristic functions of X, Y and $Z = X + Y$. We have

$$\phi_1(t) = \frac{1}{2}\int_{-1}^{+1} e^{itx}\,dx = \frac{1}{2}\left[\frac{e^{itx}}{it}\right]_{-1}^{+1} = \frac{e^{it} - e^{-it}}{2it} = \frac{\sin t}{t}.$$

Similarly,

$$\phi_2(t) = \frac{\sin t}{t}.$$

Since the variable z takes on the values from the interval $[-2, +2]$, we find

$$\phi_3(t) = \frac{1}{4}\int_{-2}^{0}(2 + z)e^{itz}\,dz + \frac{1}{4}\int_{0}^{2}(2 - z)e^{itz}\,dz = \frac{1}{4}\left(\frac{2 - e^{2it} - e^{-2it}}{t^2}\right)$$

$$= \frac{1}{2t^2}\left(1 - \frac{e^{2it} + e^{-2it}}{2}\right) = \frac{1}{2t^2}(1 - \cos 2t) = \left(\frac{\sin t}{t}\right)^2.$$

It follows that the equality $\phi_3(t) = \phi_1(t)\phi_2(t)$ holds; nevertheless X and Y are dependent.

4.5 DETERMINATION OF THE DISTRIBUTION FUNCTION BY THE CHARACTERISTIC FUNCTION

A Formula (4.1.1) uniquely determines the characteristic function of a given distribution function. We shall prove the theorem of Lévy [3] that the converse is also true: from the characteristic function we can uniquely determine the distribution function.

Theorem 4.5.1. *Let $F(x)$ and $\phi(t)$ denote respectively the distribution function and the characteristic function of the random variable X. If $a + h$ and $a - h$ ($h > 0$) are continuity points of the distribution function $F(x)$,*

$$(4.5.1) \quad F(a + h) - F(a - h) = \lim_{T \to \infty} \frac{1}{\pi} \int_{-T}^{T} \frac{\sin ht}{t} e^{-ita} \phi(t)\, dt.$$

Before proving it we shall show how to apply theorem 4.5.1. Since the numbers a and h are arbitrary, formula (4.5.1) gives the difference $F(x_2) - F(x_1)$ for arbitrary continuity points x_1 and x_2. By the relation

$$F(x_2) - F(x_1) = P(x_1 \leqslant X < x_2),$$

if we know the characteristic function $\phi(t)$, we obtain from theorem 4.5.1 the probability that the value of X belongs to an arbitrary continuity interval (see definition 2.5.6).

Let $x = x_2$ be a given continuity point and let $x_1 \to -\infty$, where the passage to the limit is performed over the set of continuity points. Here the sequence of differences $F(x) - F(x_1)$ is determined by the characteristic function and is convergent to $F(x)$; thus the distribution function $F(x)$ is determined at every continuity point; hence it is determined everywhere. We now give the proof of theorem 4.5.1.

Proof. We give the proof only for a random variable of the continuous type with density function $f(x)$. Denote

$$(4.5.2) \qquad J = \frac{1}{\pi} \int_{-T}^{+T} \frac{\sin ht}{t} e^{-ita} \phi(t)\, dt.$$

From the definition of the characteristic function we obtain

$$J = \frac{1}{\pi} \int_{-T}^{+T} \left[\int_{-\infty}^{+\infty} \frac{\sin ht}{t} e^{-ita} e^{itx} f(x)\, dx \right] dt$$

$$= \frac{1}{\pi} \int_{-T}^{T} \left[\int_{-\infty}^{+\infty} \frac{\sin ht}{t} e^{it(x-a)} f(x)\, dx \right] dt.$$

We notice that we can interchange the order of integration since the limits of integration with respect to t are finite and the integral is absolutely convergent with respect to x. Thus

$$\int_{-\infty}^{+\infty} \left| \frac{\sin ht}{t} e^{it(x-a)} \right| f(x)\, dx = \int_{-\infty}^{+\infty} \left| \frac{\sin ht}{t} \right| f(x)\, dx \leqslant h \int_{-\infty}^{+\infty} f(x)\, dx = h.$$

We obtain

$$
\begin{aligned}
J &= \frac{1}{\pi} \int_{-\infty}^{+\infty} \left[\int_{-T}^{T} \frac{\sin ht}{t} e^{it(x-a)} f(x)\, dt \right] dx \\
&= \frac{1}{\pi} \int_{-\infty}^{+\infty} \left[\int_{-T}^{T} \frac{\sin ht}{t} \{ \cos \left[(x-a)t \right] + i \sin \left[(x-a)t \right] \} f(x)\, dt \right] dx \\
&= \frac{2}{\pi} \int_{-\infty}^{+\infty} \left\{ \int_{0}^{T} \frac{\sin ht}{t} \cos \left[(x-a)t \right] f(x)\, dt \right\} dx.
\end{aligned}
$$

By the formula

$$
\sin A \cos B = \tfrac{1}{2}[\sin (A + B) + \sin (A - B)]
$$

and the substitution $A = ht$, $B = xt - at$, we obtain

(4.5.2′)

$$
\begin{aligned}
J &= \int_{-\infty}^{+\infty} \left\{ \frac{1}{\pi} \int_{0}^{T} \frac{\sin \left[(x-a+h)t \right]}{t}\, dt \right. \\
&\qquad \left. - \frac{1}{\pi} \int_{0}^{T} \frac{\sin \left[(x-a-h)t \right]}{t}\, dt \right\} f(x)\, dx \\
&= \int_{-\infty}^{+\infty} g(x, T) f(x)\, dx,
\end{aligned}
$$

where $g(x, T)$ denotes the expression in the braces. It is known from mathematical analysis that the integral $\int_{0}^{T} (\sin x/x)\, dx$ is bounded for all $T > 0$ and converges to $\frac{1}{2}\pi$ as $T \to +\infty$. It follows that the expression $|g(x, T)|$ is bounded and

$$
\lim_{T \to \infty} \frac{1}{\pi} \int_{0}^{T} \frac{\sin \alpha t}{t}\, dt =
\begin{cases}
\frac{1}{2} & \text{for } \alpha > 0, \\
-\frac{1}{2} & \text{for } \alpha < 0.
\end{cases}
$$

Here the convergence is uniform with respect to α where $|\alpha| = |x - a \pm h| > \delta > 0$. From this fact we obtain

(4.5.3)
$$
\lim_{T \to \infty} g(x, T) =
\begin{cases}
0 & \text{for } x < a - h, \\
\frac{1}{2} & \text{for } x = a - h, \\
1 & \text{for } a - h < x < a + h, \\
\frac{1}{2} & \text{for } x = a + h, \\
0 & \text{for } x > a + h.
\end{cases}
$$

It follows that in computing $\lim_{T \to \infty} J$ we can pass to the limit under the integral sign on the right-hand side of (4.5.2′).

Hence we obtain

$$(4.5.4) \qquad \lim_{T \to \infty} J = \int_{-\infty}^{+\infty} \lim_{T \to \infty} g(x, T) f(x) \, dx$$

$$= \int_{a-h}^{a+h} f(x) \, dx = F(a + h) - F(a - h).$$

From (4.5.2) and (4.5.4) we obtain (4.5.1). Thus the theorem is proved for a random variable of the continuous type. For a random variable of the discrete type the proof is similar; it is only necessary to replace the integrals by series.

B   If the characteristic function $\phi(t)$ is absolutely integrable over the interval $(-\infty, +\infty)$, then the corresponding density function $f(x)$ can be determined[1] by $\phi(t)$. In fact, from the absolute integrability of the function $\phi(t)$ it follows that the improper integral (4.5.1) exists. Dividing both sides of (4.5.1) by $2h$, we then have

$$(4.5.5) \qquad \frac{F(x + h) - F(x - h)}{2h} = \frac{1}{2\pi} \int_{-\infty}^{+\infty} \frac{\sin ht}{ht} e^{-itx} \phi(t) \, dt,$$

where $x + h$ and $x - h$ are continuity points of $F(x)$. When $h \to 0$, the expression under the integral sign tends to $e^{-itx}\phi(t)$. Moreover, the expression under the integral sign is, in absolute value, not greater than $|\phi(t)|$, which by assumption is integrable. It follows that we can pass to the limit with $h \to 0$ under the integral sign in expression (4.5.5). Then we obtain

$$\lim_{h \to 0} \frac{F(x + h) - F(x - h)}{2h} = \frac{1}{2\pi} \int_{-\infty}^{+\infty} e^{itx} \phi(t) \, dt.$$

Since the right-hand side of this equation is a continuous function of x, we obtain

$$(4.5.6) \qquad F'(x) = f(x) = \frac{1}{2\pi} \int_{-\infty}^{+\infty} e^{-itx} \phi(t) \, dt.$$

From the absolute and uniform convergence of the last integral it follows that the density $F'(x)$ exists and is a continuous function. Thus formula (4.5.6) allows us to determine the density $f(x)$ from the characteristic function $\phi(t)$, under the assumption that $\phi(t)$ is absolutely integrable. It is apparent that formula (4.5.6) is the inverse of formula (4.1.4).

Example 4.5.1.   The characteristic function of the random variable X is given by the formula

$$\phi(t) = \exp\left(-\frac{t^2}{2}\right).$$

We shall find the density function of this random variable.

---

[1] The validity of this assertion under weaker assumptions on $\phi(t)$ has been recently stated by van der Vaart [2].

From equation (4.5.6) we obtain

$$f(x) = \frac{1}{2\pi} \int_{-\infty}^{+\infty} \exp{(-itx)} \exp{\left(-\frac{t^2}{2}\right)} dt$$

$$= \frac{1}{2\pi} \int_{-\infty}^{+\infty} \exp{\left[-\frac{(t+ix)^2}{2}\right]} \exp{\frac{(ix)^2}{2}} dt$$

$$= \frac{1}{\sqrt{2\pi}} \exp{\left(-\frac{x^2}{2}\right)} \frac{1}{\sqrt{2\pi}} \int_{-\infty}^{+\infty} \exp{\left[-\frac{(t+ix)^2}{2}\right]} dt = \frac{1}{\sqrt{2\pi}} \exp{\left(-\frac{x^2}{2}\right)}.$$

This is the density of the normal (Gauss) distribution. (Compare this result with that given by (4.2.10) in example 4.2.2, where we solved the converse problem; that is, we found the characteristic function of the normal (Gauss) distribution.)

If the random variable X is of the discrete type and can take on only integer values, then its probability function can easily be obtained from the characteristic function

For every integer k, let

$$p_k = P(X = k),$$

where, of course, not all p_k must be positive. We have

$$\phi(t) = \sum_{k=-\infty}^{\infty} p_k e^{ikt}.$$

Let k' be a fixed integer. Then we have

$$e^{-itk'}\phi(t) = \sum_{\substack{k=-\infty \\ k \neq k'}}^{+\infty} e^{-it(k'-k)}p_k + p_{k'}.$$

Integrating both sides of this equation from $-\pi$ to $+\pi$ and using the fact that for every $k \neq k'$, we have

$$\int_{-\pi}^{\pi} e^{-it(k'-k)} dt = 0,$$

we obtain, replacing k' by k,

(4.5.7) $$p_k = \frac{1}{2\pi} \int_{-\pi}^{\pi} e^{-itk}\phi(t) dt.$$

Notice the analogy between formulas (4.5.6) and (4.5.7).

C   Gnedenko [1] proved that the values of the characteristic function in a finite interval do not uniquely determine the distribution function. We show this by the following example.

Example 4.5.2. Let us find the density function of the random variable X, whose characteristic function is

(4.5.8)
$$\phi_1(t) = \begin{cases} 1 - |t| & \text{for} & |t| \leqslant 1, \\ 0 & \text{for} & |t| > 1. \end{cases}$$

It is obvious that the function $\phi_1(t)$ is absolutely integrable over the interval $(-\infty < t < +\infty)$. From formula (4.5.6) we obtain

$$f(x) = \frac{1}{2\pi} \int_{-\infty}^{+\infty} e^{-itx}\phi_1(t)\, dt = \frac{1}{2\pi} \int_{-1}^{0} (1 + t)e^{-itx}\, dt + \frac{1}{2\pi} \int_{0}^{1} (1 - t)e^{-itx}\, dt;$$

$$\int_{-1}^{0} (1 + t)e^{-itx}\, dt = \left[\frac{e^{-itx}}{-ix}(1 + t) \right]_{-1}^{0} - \frac{1}{-ix} \int_{-1}^{0} e^{-itx}\, dt$$

$$= -\frac{1}{ix} + \frac{1}{ix}\left[\frac{e^{-itx}}{-ix} \right]_{-1}^{0}$$

$$= -\frac{1}{ix} - \frac{1}{(ix)^2}(1 - e^{ix}),$$

$$\int_{0}^{1} (1 - t)e^{-itx}\, dt = \left[\frac{e^{-itx}}{-ix}(1 - t) \right]_{0}^{1} + \frac{1}{-ix} \int_{0}^{1} e^{-itx}\, dt$$

$$= \frac{1}{ix} - \frac{1}{ix}\left[\frac{e^{-itx}}{-ix} \right]_{0}^{1} = \frac{1}{ix} + \frac{1}{(ix)^2}(e^{-ix} - 1).$$

We then have

(4.5.9) $\quad f(x) = \dfrac{1}{2\pi x^2}(2 - e^{ix} - e^{-ix}) = \dfrac{1}{\pi x^2}\left(1 - \dfrac{e^{ix} + e^{-ix}}{2} \right) = \dfrac{1 - \cos x}{\pi x^2}.$

Let us now consider the random variable Y of the discrete type, with the probability function defined by the formulas

(4.5.10)
$$P(Y = 0) = \tfrac{1}{2},$$

$$P[Y = (2k - 1)\pi] = \frac{2}{(2k - 1)^2\pi^2} \quad (k = 0, \pm 1, \pm 2, \ldots).$$

The characteristic function of this random variable is

$$\phi_2(t) = \frac{1}{2} + \sum_{k=-\infty}^{\infty} \frac{2}{(2k - 1)^2\pi^2} e^{it(2k-1)\pi}$$

(4.5.11)
$$= \frac{1}{2} + \frac{2}{\pi^2} \sum_{k=-\infty}^{+\infty} \frac{\cos(2k - 1)t\pi + i\sin(2k - 1)t\pi}{(2k - 1)^2}$$

$$= \frac{1}{2} + \frac{4}{\pi^2} \sum_{k=1}^{\infty} \frac{\cos(2k - 1)t\pi}{(2k - 1)^2}.$$

We shall show that for $|t| \leqslant 1$ we have

$$\phi_1(t) = \phi_2(t).$$

Expanding the function $\psi(t) = |t|$ in the interval $|t| \leqslant 1$ in a Fourier series, we have

$$\psi(t) = \frac{a_0}{2} + \sum_{n=1}^{\infty} a_n \cos n\pi t.$$

We compute the coefficients of this expansion from the formulas

$$\frac{a_0}{2} = \int_0^1 t \, dt = \frac{1}{2},$$

$$a_n = 2 \int_0^1 t \cos n\pi t \, dt = \left[\frac{2t \sin n\pi t}{n\pi} \right]_0^1 - \frac{2}{n\pi} \int_0^1 \sin n\pi t \, dt$$

$$= -\frac{2}{n\pi} \left[\frac{-\cos n\pi t}{n\pi} \right]_0^1 = 2 \frac{\cos n\pi - 1}{\pi^2 n^2}.$$

For even n we have $a_n = 0$, and for odd n, that is, for $n = 2k - 1$, we have

$$a_{2k-1} = -\frac{4}{(2k-1)^2 \pi^2}.$$

Finally we obtain

(4.5.12) $$\psi(t) = |t| = \frac{1}{2} - \frac{4}{\pi^2} \sum_{k=1}^{\infty} \frac{\cos (2k-1)\pi t}{(2k-1)^2}.$$

From formulas (4.5.11) and (4.5.12) we have $\phi_2(t) = 1 - |t| = \phi_1(t)$, in spite of the fact that $\phi_1(t)$ and $\phi_2(t)$ are the characteristic functions of two different distributions. We observe that for $|t| > 1$ the characteristic functions $\phi_1(t)$ and $\phi_2(t)$ are not equal. In fact, from the definition we then have $\phi_1(t) \equiv 0$, whereas the function $\phi_2(t)$ is not identically zero since the values taken by this function in the interval $|t| \leqslant 1$ repeat periodically.

4.6 THE CHARACTERISTIC FUNCTION OF MULTI-DIMENSIONAL RANDOM VECTORS

A   The notion of the characteristic function of a one-dimensional random variable can be generalized to a random variable with an arbitrary finite number of dimensions. We restrict ourselves to two-dimensional random variables.

Let (X, Y) be a two-dimensional random variable and let $F(x, y)$ be its distribution function. Let t and u be two arbitrary real numbers. The *characteristic function* of the random variable (X, Y) or of the distribution function $F(x, y)$ is defined by the formula

(4.6.1) $$\phi(t, u) = E[e^{i(tX + uY)}].$$

Example 4.6.1.   The two-dimensional random variable can take on four pairs of values: $(+1, +1)$, $(+1, -1)$, $(-1, +1)$, and $(-1, -1)$ with the probabilities

$$P(X = 1, Y = 1) = \tfrac{1}{3}, \qquad P(X = 1, Y = -1) = \tfrac{1}{3},$$
$$P(X = -1, Y = 1) = \tfrac{1}{6}, \qquad P(X = -1, Y = -1) = \tfrac{1}{6}.$$

The reader can verify that X and Y are independent. For the characteristic function of the random variable (X, Y), we obtain from (4.6.1)

$$
\begin{aligned}
\phi(t, u) &= E(e^{i(tX+uY)}) = \tfrac{1}{3}e^{i(t+u)} + \tfrac{1}{3}e^{i(t-u)} + \tfrac{1}{6}e^{i(-t+u)} + \tfrac{1}{6}e^{i(-t-u)} \\
(4.6.2) \quad &= \tfrac{1}{3}e^{it}(e^{iu} + e^{-iu}) + \tfrac{1}{6}e^{-it}(e^{iu} + e^{-iu}) = \tfrac{1}{6}(e^{iu} + e^{-iu})(2e^{it} + e^{-it}) \\
&= \tfrac{1}{3}\cos u \,(3\cos t + i\sin t).
\end{aligned}
$$

We shall investigate some of the properties of characteristic functions of multidimensional random variables. We have

$$(4.6.3) \qquad \phi(0, 0) = E(e^{i(0X+0Y)}) = 1,$$

$$|\phi(t, u)| = |E(e^{i(tX+uY)})| \leqslant E(|e^{i(tX+uY)}|) = 1,$$

hence

$$(4.6.4) \qquad\qquad |\phi(t, u)| \leqslant 1,$$

$$(4.6.5) \qquad \phi(-t, -u) = E(e^{-i(tX+uY)}) = \overline{\phi(t, u)}.$$

It can be shown that, as in the one-dimensional case, if all the moments of order k of a multidimensional random variable exist, then the derivatives

$$(4.6.6) \qquad \frac{\partial^k \phi(t, u)}{\partial t^{k-l}\, \partial u^l} \qquad \text{for } l = 0, 1, 2, \ldots, k$$

exist and can be obtained from the formula

$$(4.6.7) \qquad \frac{\partial^k \phi(t, u)}{\partial t^{k-l}\, \partial u^l} = i^k E(X^{k-l} Y^l e^{i(tX + uY)}).$$

From (4.6.7) we see that the moment $m_{k-l, l}$ can be obtained from the formula

$$(4.6.8) \qquad m_{k-l, l} = E(X^{k-l}Y^l) = \frac{1}{i^k}\left[\frac{\partial^k \phi(t, u)}{\partial t^{k-l}\, \partial u^l}\right]_{\substack{t=0 \\ u=0}}.$$

For the moments of the first and second order we obtain the expressions

$$(4.6.8') \qquad m_{10} = \frac{1}{i}\left[\frac{\partial \phi(t, u)}{\partial t}\right]_{\substack{t=0 \\ u=0}}, \qquad m_{01} = \frac{1}{i}\left[\frac{\partial \phi(t, u)}{\partial u}\right]_{\substack{t=0 \\ u=0}},$$

$$m_{20} = \frac{1}{i^2}\left[\frac{\partial^2 \phi(t, u)}{\partial t^2}\right]_{\substack{t=0 \\ u=0}}, \quad m_{11} = \frac{1}{i^2}\left[\frac{\partial^2 \phi(t, u)}{\partial t\, \partial u}\right]_{\substack{t=0 \\ u=0}}, \quad m_{02} = \frac{1}{i^2}\left[\frac{\partial^2 \phi(t, u)}{\partial u^2}\right]_{\substack{t=0 \\ u=0}}.$$

We obtain the characteristic functions of the marginal distributions of the random variables X and Y from formula (4.6.1) by putting $u = 0$ or $t = 0$, respectively. Thus

$$(4.6.9) \qquad \phi(t, 0) = E(e^{itX}) = \phi_1(t).$$

This is simply the characteristic function of X. Similarly,

(4.6.10) $$\phi(0, u) = E(e^{iuY}) = \phi_2(u)$$

is the characteristic function of Y.

B We shall now give without proof the generalization of theorem 4.5.1 to two-dimensional random vectors. The proof is similar to that for a one-dimensional random variable.

Theorem 4.6.1. *Let $\phi(t, u)$ be the characteristic function of the random variable (X, Y). If the rectangle $(a - h \leqslant X < a + h, b - g \leqslant Y < b + g)$ is a continuity rectangle (see definition 2.5.6), then*

(4.6.11) $P(a - h \leqslant X < a + h, b - g \leqslant Y < b + g)$

$$= \lim_{T \to \infty} \frac{1}{\pi^2} \int_{-T}^{+T} \int_{-T}^{+T} \frac{\sin ht}{t} \frac{\sin gu}{u} \exp\left[- i(at + bu) \right] \phi(t, u) \, dt \, du.$$

Thus, if we know $\phi(t, u)$, formula (4.6.11) allows us to determine the probability

(4.6.12) $$P(x_1 \leqslant X < x_2, y_1 \leqslant Y < y_2)$$

for an arbitrary continuity rectangle. However, the probabilities (4.6.12) for continuity rectangles completely determine the probability distribution in the plane (x, y).

Theorem 4.6.2. *Let $F(x, y)$, $F_1(x)$, $F_2(y)$, $\phi(t, u)$, $\phi_1(t)$, and $\phi_2(u)$ denote the distribution functions and the characteristic functions of the random variables (X, Y), X and Y, respectively. The random variables X and Y are then independent if and only if the equation*

(4.6.13) $$\phi(t, u) = \phi_1(t)\phi_2(u)$$

holds for all real t and u.

Proof. Suppose that X and Y are independent. From theorem 3.6.2 we have, for any real t and u

$$\phi(t, u) = E(e^{i(tX + uY)}) = E(e^{itX}e^{iuY}) = E(e^{itX})E(e^{iuY}) = \phi_1(t)\phi_2(u).$$

Suppose now that equality (4.6.13) is satisfied. Then from (4.6.13), (4.6.11) and (4.5.1) we obtain the equation

(4.6.14)

$$P(X_1 \leqslant X < x_2, y_1 \leqslant Y < y_2) = P(x_1 \leqslant X < x_2)P(y_1 \leqslant Y < y_2),$$

which is valid for arbitrary continuity rectangles. From equation (4.6.14) we obtain, for arbitrary x and y

$$F(x, y) = F_1(x)F_2(y).$$

Thus theorem 4.6.2 is proved.

The following Cramér-Wold theorem [1] is useful in the theory of random vectors.

Theorem 4.6.3. *The distribution function $F(x, y)$ of a two-dimensional random variables (X, Y) is uniquely determined by the class of all one-dimensional distribution functions of $tX + uY$, where t and u run over all possible real values.*

Proof. Suppose we are given for all real t and u the characteristic functions $\phi_z(v)$ of $Z = tX + uY$,

(4.6.15) $\phi_z(v) = E\{\exp[iv(tX + uY)]\} = E\{\exp[i(vtX + vuY)]\}.$

Putting $v = 1$, we obtain for the right-hand side of (4.6.15) the expression

$$E\{\exp[i(tX + uY)]\},$$

which is the characteristic function $\phi(t, u)$ of the distribution function $F(x, y)$. According to theorem 4.6.1, the function $\phi(t, u)$ uniquely determines $F(x, y)$. Thus the theorem is proved.

Let us write

$$P(tX + uY < z) = P(X \cos \alpha + Y \sin \alpha < w),$$

where

$$\cos \alpha = \frac{t}{\sqrt{t^2 + u^2}}, \quad \sin \alpha = \frac{u}{\sqrt{t^2 + u^2}}, \quad w = \frac{z}{\sqrt{t^2 + u^2}} \ (0 \leqslant \alpha \leqslant 2\pi).$$

The Cramér-Wold theorem can now be formulated in the following way: *The distribution function $F(x, y)$ is uniquely determined by the distribution functions of the projections of (X, Y) on all straight lines passing through the origin.*

Rényi [5] showed that if, with probability 1, (X, Y) satisfies the inequality

$$X^2 + Y^2 \leqslant R^2 < \infty,$$

then the distribution function $F(x, y)$ is uniquely determined by the class of distribution functions of $X \cos \alpha + Y \sin \alpha$, where α runs over an arbitrary countable set of different values from the interval $[0, 2\pi]$. Also the papers of Gilbert [1] and Heppes [1] are devoted to the question of determining the two-dimensional distribution from the distributions of the projections on countable sets of straight lines.

Theorems 4.6.1 to 4.6.3 can obviously be generalized to random vectors with more than two dimensions.

For a comprehensive study of the theory of characteristic functions, see Lukacs [5].

4.7 PROBABILITY-GENERATING FUNCTIONS

When investigating random variables which take on only the integers $k = 0, 1, 2, \ldots$ it is simpler to deal with probability-generating functions than with characteristic functions. Let X be a random variable and let

$$p_k = P(X = k) \quad (k = 0, 1, 2, \ldots),$$

where $\sum\limits_k p_k = 1$.

Definition 4.7.1. The function defined by the formula

$$(4.7.1) \qquad \psi(s) = \sum_k p_k s^k,$$

where $-1 \leqslant s \leqslant 1$, is called the *probability generating function* of X.

We notice that $\psi(1) = \sum\limits_k p_k = 1$. Hence the series on the right-hand side of (4.7.1) is absolutely and uniformly convergent in the interval $|s| \leqslant 1$. Thus the generating function is continuous. It determines the probability function uniquely, since $\psi(s)$ can be represented in a unique way as a power series of the form (4.7.1).

Example 4.7.1. The random variable X has a binomial distribution, that is,

$$p_k = \binom{n}{k} p^k (1 - p)^{n-k} \quad (k = 0, 1, \ldots, n).$$

Therefore

$$(4.7.2) \qquad \psi(s) = \sum_{k=0}^{n} \binom{n}{k} (ps)^k (1 - p)^{n-k} = (ps + q)^n.$$

Example 4.7.2. The random variable X has a Poisson distribution, that is

$$p_k = e^{-\lambda} \frac{\lambda^k}{k!} \quad (k = 0, 1, 2, \ldots).$$

Therefore

$$(4.7.3) \qquad \psi(s) = \sum_{k=0}^{\infty} e^{-\lambda} \frac{(\lambda s)^k}{k!} = e^{-\lambda} e^{\lambda s} = e^{-\lambda(1-s)}.$$

The moments of the random variable X can be determined by the derivatives at the point 1 of the generating function. Let us, for example, determine the moments of the first and second order. We have

$$\psi'(s) = \sum_k k p_k s^{k-1},$$

$$\psi''(s) = \sum_k k(k - 1) p_k s^{k-2}.$$

Hence

$$\psi'(1) = \sum_k kp_k = E(X),$$

$$\psi''(1) = \sum_k k(k-1)p_k = E(X^2) - E(X).$$

We then obtain

$$E(X^2) = \psi''(1) + \psi'(1).$$

There is a theorem for probability generating functions similar to theorem 4.4.1. The proof of this theorem is elementary.

Problems and Complements

4.1. Find the characteristic functions of the random variables whose densities are

(a)
$$f(x) = \begin{cases} 0 & \text{for} & |x| \geqslant a > 0, \\ \dfrac{a-1}{a^2} & \text{for} & |x| < a, \end{cases}$$

(b)
$$f(x) = \frac{2 \sin^2 (ax/2)}{\pi a x^2}.$$

Hint: Compare with the derivation of formula (5.10.10).

4.2. Prove that the function

$$\phi(t) = \exp(-|t|^r)$$

with $r > 2$ is not the characteristic function of any random variable.

4.3. A bounded and continuous function $\phi(t)$ is a characteristic function of some distribution function if and only if $\phi(0) = 1$ and the function

$$\psi(x, A) = \int_0^A \int_0^A \phi(t-u) e^{ix(t-u)}\, dt\, du$$

is real and non-negative for all real x and all $A > 0$ (Cramér [5]).

4.4. Let $\phi(t)$ be the characteristic function of the random variable X. Prove that

(a) if X is of the continuous type, then $\lim_{|t| \to \infty} \phi(t) = 0$,

(b) if X is of the discrete type, then $\limsup_{|t| \to \infty} |\phi(t)| = 1$.

4.5. Prove that (a) if $\phi(t)$ is the characteristic function of a random variable, then so is $|\phi(t)|^2$; (b) if $\phi(t)$ is the characteristic function of a random variable with a continuous distribution function, then so is $|\phi(t)|^2$.

4.6. Prove that the characteristic function of a random variable X is real if and only if X has a symmetric distribution about 0.

4.7. (a) Show that for an arbitrary characteristic function $\phi(t)$ we have

$$1 - Re\, \phi(2t) \leqslant 4[1 - Re\, \phi(t)].$$

(b) Deduce that

$$1 - |\phi(2t)|^2 \leqslant 4[1 - |\phi(t)|^2].$$

4.8. Prove that if the characteristic function $\phi(t)$ of the random variable X has a derivative of an even order $\phi^{(2l)}(0)$, then the moments of the orders $1, 2, \ldots, 2l$ of the random variable X exist (Cramér [2]).

Hint: Prove, using symmetric derivatives, that

$$\phi^{(2l)}(0) = \lim_{h \to 0} \int_{-\infty}^{\infty} \left(\frac{e^{ihx} - e^{-ihx}}{2h} \right)^{2l} dF(x) = (-1)^l \lim_{h \to 0} \int_{-\infty}^{\infty} \left(\frac{\sin hx}{h} \right)^{2l} dF(x).$$

4.9. (a). Prove (Zygmund [1]) that if the characteristic function $\phi(t)$ of the distribution function $F(x)$ is differentiable at the point 0, then

$$\lim_{a \to \infty} \int_{-a}^{a} x \, dF(x) = \frac{\phi'(0)}{i}.$$

The left side of the preceding equation is called the *generalized mean value of* X.

(b) Prove (Pitman [2]) that if the generalized mean value of the random variable X exists and

$$\lim_{a \to \infty} aP(|X| > a) = 0,$$

then its characteristic function $\phi(t)$ is differentiable at $t = 0$.

(c) Show that

$$\phi(t) = C \sum_{\substack{n = -\infty \\ n \neq 0, \pm 1 \pm 2}}^{\infty} \frac{\cos nt}{n^2 \log |n|},$$

where C is a suitably chosen constant, is the characteristic function of a random variable without expected value, and that $\phi(t)$ is everywhere differentiable.

4.10. Prove Theorem 3.2.1.

Hint: Using the inequality (3.4.5) show that

$$\lim_{k \to \infty} \frac{\beta_k}{k!} r^k = 0.$$

Next, show that for $|t| < r$ the series

$$\phi(t) = \sum_{k=0}^{\infty} \frac{m_k}{k!} (it)^k$$

is convergent. Finally, apply the method of analytic continuation to show the convergence of this series for all t.

4.11. Show that the characteristic functions $\phi_1(t)$, $\phi_2(t)$, and $\phi_3(t)$ may satisfy the relation

$$\phi_1(t)\phi_2(t) = \phi_1(t)\phi_3(t)$$

in spite of the fact that $\phi_2(t)$ and $\phi_3(t)$ are not identically equal.

4.12. (a) Prove (Bochner [1]) that if x_0 is a point of discontinuity of the distribution function $F(x)$ and $F(x_0 + 0) - F(x_0) = p$, then

$$p = \lim_{T \to \infty} \frac{1}{2T} \int_{-T}^{+T} e^{-itx_0}\phi(t) \, dt,$$

where $\phi(t)$ is the characteristic function of $F(x)$.

(b) Prove (Wiener [2]) that $F(x)$ is continuous if and only if

$$\lim_{T\to\infty} \frac{1}{2T} \int_{-T}^{T} |\phi(t)|\, dt = 0.$$

(c) Deduce from 4.12(b) (Lorch and Newman [1]) that

$$\lim_{T\to\infty} \frac{1}{2T} \int_{-T}^{T} |\phi(t)|\, dt = \lim_{T\to\infty} \frac{1}{2T} \int_{-T}^{T} e^{itx_j} |\sum_j [F(x_j + 0) - F(x_j)]|\, dt,$$

where the x_j are the jump points of $F(x)$, if any.

4.13. Prove (Lévy [1]) that for an arbitrary characteristic function $\phi(t)$

$$\lim_{T\to\infty} \frac{1}{2T} \int_{-T}^{+T} |\phi(t)|^2\, dt = \sum_j [F(x_j + 0) - F(x_j)],$$

where the summation is extended over all the jump points x_j, if any, of the corresponding distribution function $F(x)$.

4.14. Show that if the characteristic function $\phi(t)$ satisfies the relation $|\phi(t_k)| = 1$ for a sequence of points $t_k \neq 0$ $(k = 1, 2, \ldots)$ which converges to zero, then $|\phi(t)| \equiv 1$.

4.15. The expression

$$m_{[l]} = \sum_k k(k-1)\ldots(k-1+l)p_k,$$

where $p_k = P(X = k)$ and $\sum_k p_k = 1$, is called the *factorial moment of order l* of X.

(a) Express $m_{[l]}$ by means of the probability generating function.

(b) Find the relations between the factorial moments and the ordinary moments.

(c) Which expression would you call the central factorial moment of order l?

4.16. Prove that if the series on the right-hand side of (4.7.1) converges for some $s_0 > 1$, then the corresponding random variable has moments of all orders.

4.17. The *moment generating function* of a random variable X or of its distribution function $F(x)$ is defined as

$$g(t) = E(e^{tX}),$$

if this expression is finite.

(a) Write the formula for $g(t)$ for random variables of the continuous and of the discrete type.

(b) Find the function $g(t)$ for the random variables of examples 4.2.1 and 4.2.2.

(c) Express the ordinary moments by means of the moment generating function.

See also Problems 5.27 and 5.28.

CHAPTER 5

Some Probability Distributions

5.1 ONE-POINT AND TWO-POINT DISTRIBUTIONS

In the previous chapters we have dealt with various probability distributions. We shall now investigate more closely some probability distributions of special importance in either theory or practice.

We begin with the one-point distribution.

Definition 5.1.1. The random variable X has a *one-point distribution* if there exists a point x_0 such that

$$(5.1.1) \qquad P(X = x_0) = 1.$$

We also say that the probability mass is concentrated at one point. It is clear that a random variable with a one-point distribution has a degenerate distribution given by definition 3.6.4.

Formula (5.1.1) gives us the probability function. The distribution function of this probability distribution is given by the formula

$$(5.1.2) \qquad F(x) = \begin{cases} 0 & \text{for } x \leqslant x_0, \\ 1 & \text{for } x > x_0. \end{cases}$$

We obtain the characteristic function of this distribution from the formula

$$(5.1.3) \qquad \phi(t) = e^{itx_0}.$$

From (4.2.4) we see that $m_1 = x_0$ and, more generally. we have $m_k = x_0{}^k$ for every k. Hence we obtain

$$D^2(X) = m_2 - m_1{}^2 = x_0{}^2 - x_0{}^2 = 0.$$

Conversely, if the variance of a random variable X equals zero, then X has a one-point distribution. To prove this suppose that

$$(5.1.4) \qquad D^2(X) = E[X - E(X)]^2 = 0.$$

Since the expression $[X - E(X)]^2$ is non-negative, equation (5.1.4) is satisfied only if

$$P[X - E(X) = 0] = 1, \quad \text{or} \quad P[X = E(X)] = 1.$$

From (5.1.1) we find that the random variable X has a one-point distribution.

Definition 5.1.2. The random variable X has a *two-point distribution* if there exist two values x_1 and x_2, such that

(5.1.5) $P(X = x_1) = p, \quad P(X = x_2) = 1 - p \quad (0 < p < 1).$

We often put $x_1 = 1$ and $x_2 = 0$. In place of (5.1.5) we then have

(5.1.6) $P(X = 1) = p, \quad P(X = 0) = 1 - p \quad (0 < p < 1).$

This distribution is called the *zero-one distribution*.

The characteristic function of distribution (5.1.6) is given by the formula

(5.1.7) $\phi(t) = pe^{it \cdot 1} + (1 - p)e^{it \cdot 0} = pe^{it} + 1 - p = 1 + p(e^{it} - 1).$

From (4.2.4) we obtain for every k

(5.1.8) $m_k = p.$

Hence

(5.1.9) $D^2(X) = m_2 - m_1^2 = p - p^2 = p(1 - p).$

From (3.2.9) and (3.2.17) we obtain

$$\mu_3 = m_3 - 3m_1 m_2 + 2m_1^3 = p - 3p^2 + 2p^3 = p(1 - p)(1 - 2p),$$

(5.1.10)
$$\gamma = \frac{\mu_3}{\mu_2^{3/2}} = \frac{p(1 - p)(1 - 2p)}{p^{3/2}(1 - p)^{3/2}} = \frac{1 - 2p}{\sqrt{p(1 - p)}}.$$

We see that if $p = 0.5$ then $\gamma = 0$ since here X has a symmetric distribution.

5.2 THE BERNOULLI SCHEME.
THE BINOMIAL DISTRIBUTION

From the following scheme of trials, called the *Bernoulli scheme*, we obtain a random variable X with binomial distribution, defined in example 3.1.3.

We perform n random experiments. Through an experiment we can obtain the event A, which we designate a success, with probability p, or the complementary A, which we designate a failure, with probability $q = 1 - p$. The results of the n experiments are independent. As a result of n random experiments, event A may occur k times $(k = 0, 1, 2, \ldots, n)$. The number of occurrences of A is a random variable X that can take on the values $k = 0, 1, \ldots, n$, where the equality $X = k$ means that in n experiments the event A has occurred k times. It is shown in example 3.1.3 that X has the binomial probability function given by the formula

$$(5.2.1) \qquad P(X = k) = \binom{n}{k} p^k (1 - p)^{n-k}.$$

The distribution function of the binomial distribution is given by the formula

$$F(x) = P(X < x) = \sum_{k < x} \binom{n}{k} p^k (1 - p)^{n-k},$$

where the summation extends over all non-negative integers less than x.

We notice that for $n = 1$ the binomial distribution is reduced to the zero-one distribution. For $n \geqslant 2$ the binomial distribution can also be obtained from the zero-one distribution as follows.

Let X_r $(r = 1, 2, \ldots, n)$ be independent random variables with the same zero-one distribution. The probability function of every X_r has the form

$$P(X_r = 1) = p, \quad P(X_r = 0) = 1 - p.$$

Consider the random variable equal to the sum of the X_r,

$$(5.2.2) \qquad X = X_1 + X_2 + \ldots + X_n.$$

The random variable X can take on the values $k = 0, 1, 2, \ldots, n$. The event $X = k$ occurs if and only if k of the n random variables X_r take on the value one and $n - k$ of them take on the value zero. For a given k this may happen in $\binom{n}{k}$ different ways. By the independence of the random variables X_r we obtain formulas (5.2.1).

From (5.1.7) and (5.2.2), for the characteristic function $\phi(t)$ of X, we obtain

$$(5.2.3) \qquad \phi(t) = [1 + p(e^{it} - 1)]^n.$$

From (4.2.4) we obtain the moments of this distribution. In particular,

$$(5.2.4) \qquad m_1 = np, \qquad m_2 = np + n(n - 1)p^2,$$

$$\mu_2 = np(1 - p), \quad \mu_3 = np(1 - p)(1 - 2p).$$

We then have

$$(5.2.5) \qquad \gamma = \frac{1 - 2p}{\sqrt{np(1 - p)}}.$$

We have already obtained the formula for μ_2 by another method in example 3.2.3.

Example 5.2.1. Table 5.2.1 gives the binomial distributions for the values $p_1 = 0.1$, $p_2 = 0.3$, and $p_3 = 0.5$ for $n = 20$. The first column gives the values $k = 0, 1, \ldots, 20$ and the remaining columns the probabilities that the random variable takes on the value k. These probabilities are given with a maximum error of 0.00005.

TABLE 5.2.1

k	$P(X = k)$			k	$P(X = k)$		
	$p_1 = 0.1$	$p_2 = 0.3$	$p_3 = 0.5$		$p_1 = 0.1$	$p_2 = 0.3$	$p_3 = 0.5$
0	0.1216	0.0008	—	11	—	0.0120	0.1602
1	0.2702	0.0068	—	12	—	0.0039	0.1201
2	0.2852	0.0278	0.0002	13	—	0.0010	0.0739
3	0.1901	0.0716	0.0011	14	—	0.0002	0.0370
4	0.0898	0.1304	0.0046	15	—	—	0.0148
5	0.0319	0.1789	0.0148	16	—	—	0.0046
6	0.0089	0.1916	0.0370	17	—	—	0.0011
7	0.0020	0.1643	0.0739	18	—	—	0.0002
8	0.0004	0.1144	0.1201	19	—	—	—
9	0.0001	0.0654	0.1602	20	—	—	—
10	—	0.0308	0.1762				

The graphs of these distributions are represented in Fig. 5.2.1.

As we can see, the closer p is to 0.5, the more symmetric is the distribution and the greater its dispersion. We might have expected these results from comparing the values of the parameters μ_2 and γ for the considered values of p with $n = 20$. These values, computed from (5.2.4) and (5.2.5) are given in Table 5.2.2.

TABLE 5.2.2

	$p_1 = 0.1$	$p_2 = 0.3$	$p_3 = 0.5$
$\sigma = \sqrt{\mu_2}$	1.34	2.05	2.24
γ	0.597	0.195	0.000

Let X and Y be two independent random variables with binomial distributions and let the characteristic functions of X and Y be, respectively

$$\phi_1(t) = [1 + p(e^{it} - 1)]^{n_1},$$

$$\phi_2(t) = [1 + p(e^{it} - 1)]^{n_2}.$$

Consider the random variable

$$Z = X + Y.$$

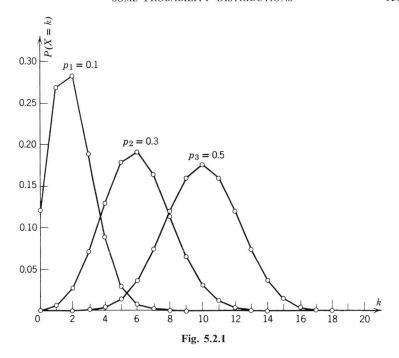

Fig. 5.2.1

Because of the independence of X and Y, the characteristic function of Z is

$$(5.2.6) \qquad \phi(t) = [1 + p(e^{it} - 1)]^{n_1 + n_2}.$$

As we see from (5.2.6), Z has the binomial distribution with $n = n_1 + n_2$. This is the *addition theorem* for the binomial distribution.

In applications we often deal with the distribution of

$$Y = \frac{X}{n},$$

where the random variable X has the binomial distribution. The random variable Y can take on the values

$$\frac{k}{n} = 0, \frac{1}{n}, \ldots, \frac{n-1}{n}, 1.$$

Since the probability that $Y = k/n$ is equal to the probability that $X = k$, the probability function of Y is given by (5.2.1)

$$P\left(Y = \frac{k}{n}\right) = P(X = k) = \binom{n}{k} p^k (1 - p)^{n-k}.$$

From (4.2.15) and (5.2.3) we obtain for the characteristic function of Y

(5.2.7) $\phi(t) = [1 + p(e^{it/n} - 1)]^n.$

From (4.2.4) we obtain the moments. In particular,

(5.2.8) $m_1 = p, \qquad m_2 = \dfrac{p}{n} + \dfrac{n-1}{n} p^2, \qquad \mu_2 = \dfrac{p(1-p)}{n}.$

5.3   THE POISSON SCHEME. THE GENERALIZED BINOMIAL DISTRIBUTION

Poisson considered the following scheme of experiments. We perform n random trials. As a result of the kth trial ($k = 1, 2, \ldots, n$), the event A (or a success) may occur with probability p_k; thus the probability of the complementary event, $q_k = 1 - p_k$. The results of the n experiments are independent. Unlike Bernoulli's scheme, here the probabilities of the occurrence of event A in individual trials are not necessarily equal.

The number of occurrences of A in n trials is a random variable. We say that this random variable has a *generalized binomial distribution*.

The random variable Z with the generalized binomial distribution can also be represented as the sum

(5.3.1) $Z = Z_1 + \ldots + Z_n,$

where the random variables Z_k ($k = 1, 2, \ldots, n$) are independent and have the zero-one distribution with the probability functions

$$P(Z_k = 1) = p_k, \qquad P(Z_k = 0) = 1 - p_k.$$

The formula for the probability function of the random variable Z is not as simple as that for the probability function of the binomial distribution. The probability that $Z = r$ can be found by the summation of the probabilities of each possible combination of r 1's and $(n - r)$ 0's.

Example 5.3.1.   We have three lots of oranges. The fraction of rotten oranges in the first lot is $p_1 = 0.02$, in the second, $p_2 = 0.05$, and in the third, $p_3 = 0.01$. We choose one orange at random from each lot. We assign the number one to the appearance of a good orange, and the number zero to the appearance of a rotten one. Here Z_1, Z_2, and Z_3 are random variables which take on the value 0 or 1, according to whether we have obtained a rotten or a good orange from the first, second, or third lot. These random variables are independent, and we have

$$P(Z_1 = 1) = 0.98, P(Z_2 = 1) = 0.95, P(Z_3 = 1) = 0.99.$$

Consider the random variable $Z = Z_1 + Z_2 + Z_3$. This random variable can take on the values $r = 0, 1, 2,$ or 3.

As a result of the independence of Z_1, Z_2, Z_3 we obtain the probabilities

$$P(Z = 0) = P(Z_1 = 0)P(Z_2 = 0)P(Z_3 = 0) = 0.00001,$$

$$P(Z = 1) = P(Z_1 = 0)P(Z_2 = 0)P(Z_3 = 1)$$
$$+ P(Z_1 = 0)P(Z_2 = 1)P(Z_3 = 0)$$
$$+ P(Z_1 = 1)P(Z_2 = 0)P(Z_3 = 0) = 0.00167,$$

$$P(Z = 2) = P(Z_1 = 0)P(Z_2 = 1)P(Z_3 = 1)$$
$$+ P(Z_1 = 1)P(Z_2 = 0)P(Z_3 = 1)$$
$$+ P(Z_1 = 1)P(Z_2 = 1)P(Z_3 = 0) = 0.07663,$$

$$P(Z = 3) = P(Z_1 = 1)P(Z_2 = 1)P(Z_3 = 1) = 0.92169.$$

The reader can verify that

$$P(Z = 0) + P(Z = 1) + P(Z = 2) + P(Z = 3) = 1.$$

The characteristic function of Z, defined by (5.3.1) to have a generalized binomial distribution, can be obtained from (5.1.7) using the independence of the Z_k, namely

$$(5.3.2) \qquad \phi(t) = \prod_{k=1}^{n} [1 + p_k(e^{it} - 1)].$$

Using (4.2.4), we compute the first two moments of Z. We have

$$(5.3.3) \quad m_1 = \sum_{k=1}^{n} p_k, \quad m_2 = \sum_{k=1}^{n} p_k + \sum_{l=1}^{n} \sum_{\substack{k=1 \\ l \neq k}}^{n} p_l p_k, \quad \mu_2 = \sum_{k=1}^{n} p_k(1 - p_k).$$

As we see, formulas (5.2.4) for m_1, m_2, and μ_2 are particular cases of the corresponding formulas (5.3.3).

Example 5.3.2. We compute the expected value and the standard deviation of the random variable Z of example 5.3.1.

We have

$$E(Z) = m_1 = (1 - p_1) + (1 - p_2) + (1 - p_3) = 2.92,$$

$$\sigma = \sqrt{\mu_2} = \sqrt{0.0196 + 0.0475 + 0.0099} = \sqrt{0.0770} = 0.28.$$

5.4 THE PÓLYA AND HYPERGEOMETRIC DISTRIBUTIONS

A   In practice we often deal with distributions which can be reduced to a scheme called the *Pólya scheme*.

Imagine that we have b white and c black balls in an urn. Let $b + c = N$. We draw one ball at random, and before drawing the next ball we replace the one we have drawn and add s balls of the same color. This procedure is repeated n times. Denote by X the random variable which takes on the value k ($k = 0, 1, \ldots, n$) if as a result of n drawings we draw a white ball k times. We shall find the probability function of X.

We notice that the probability of the successive drawing of k white balls is

$$\frac{b(b + s) \ldots [b + (k - 1)s]}{N(N + s) \ldots [N + (k - 1)s]}.$$

Similarly, the probability of drawing k white balls in turn and then $n - k$ black balls is

$$\frac{b(b + s) \ldots [b + (k - 1)s]c(c + s) \ldots [c + (n - k - 1)s]}{N(N + s) \ldots [N + (n - 1)s]}.$$

We notice that the last expression also gives the probability of drawing k white and $n - k$ black balls in any given order. The order of drawing affects only the order of the terms in the numerator of this expression. Since k white and $n - k$ black balls can be drawn in $\binom{n}{k}$ different ways, we have

(5.4.1) $P(X = k)$

$$= \binom{n}{k} \frac{b(b + s) \ldots [b + (k - 1)s]c(c + s) \ldots [c + (n - k - 1)s]}{N(N + s) \ldots [N + (n - 1)s]}.$$

Definition 5.4.1.   The random variable X with the probability distribution given by (5.4.1) has a *Pólya distribution*.

Denote

$$Np = b, \; Nq = c, \; N\alpha = s.$$

As we see, p and q are the probabilities of drawing a white and a black ball, respectively, on the first drawing. Formula (5.4.1) takes the form

(5.4.2) $P(X = k)$

$$= \binom{n}{k} \frac{p(p + \alpha) \ldots [p + (k - 1)\alpha]q(q + \alpha) \ldots [q + (n - k - 1)\alpha]}{1(1 + \alpha) \ldots [1 + (n - 1)\alpha]}.$$

It is obvious that

(5.4.3)

$$\sum_{k=0}^{n} \binom{n}{k} \frac{p(p + \alpha) \ldots [p + (k - 1)\alpha]q(q + \alpha) \ldots [q + (n - k - 1)\alpha]}{1(1 + \alpha) \ldots [1 + (n - 1)\alpha]} = 1.$$

We shall compute the first and second moments of X. The first moment is given by the formula

$$E(X) = \sum_{k=0}^{n} kP(X = k) = pn \sum_{k=1}^{n} \binom{n-1}{k-1}$$
$$\times \frac{(p+\alpha)\ldots[p+(k-1)\alpha]q(q+\alpha)\ldots[q+(n-k-1)\alpha]}{(1+\alpha)\ldots[1+(n-1)\alpha]}.$$

Putting $l = k - 1$, we obtain

(5.4.4)
$$E(X) = pn \sum_{l=0}^{n-1} \binom{n-1}{l} \frac{(p+\alpha)\ldots(p+l\alpha)q(q+\alpha)\ldots[q+(n-l-2)\alpha]}{(1+\alpha)\ldots[1+(n-1)\alpha]}.$$

It is easy to verify that the term under the summation sign in the last formula represents the probability of obtaining l white and $n - l - 1$ black balls in $n - 1$ drawings according to a Pólya scheme where at the beginning the urn contains $N + s$ balls, including $b + s$ white and c black ones. From (5.4.3) it follows that the sum on the right-hand side of (5.4.4) equals 1; hence

(5.4.5) $E(X) = np.$

For the second moment we obtain

$$E(X^2) = \sum_{k=0}^{n} k^2 P(X = k) = np \sum_{k=1}^{n} k \binom{n-1}{k-1}$$
$$\times \frac{(p+\alpha)\ldots[p+(k-1)\alpha]q(q+\alpha)\ldots[q+(n-k-1)\alpha]}{(1+\alpha)\ldots[1+(n-1)\alpha]}.$$

Putting $l = k - 1$ we obtain

$$E(X^2)$$
$$= np \sum_{l=0}^{n-1} (l+1) \binom{n-1}{l} \frac{(p+\alpha)\ldots(p+l\alpha)q(q+\alpha)\ldots[q+(n-l-2)\alpha]}{(1+\alpha)\ldots[1+(n-1)\alpha]}$$
$$= np \left\{ \sum_{l=0}^{n-1} l \binom{n-1}{l} \frac{(p+\alpha)\ldots(p+l\alpha)q(q+\alpha)\ldots[q+(n-l-2)\alpha]}{(1+\alpha)\ldots[1+(n-1)\alpha]} \right.$$
$$\left. + \sum_{l=0}^{n-1} \binom{n-1}{l} \frac{(p+\alpha)\ldots(p+l\alpha)q(q+\alpha)\ldots[q+(n-l-2)\alpha]}{(1+\alpha)\ldots[1+(n-1)\alpha]} \right\}$$
$$= np(A + B).$$

After some simple transformations we have

$$(5.4.6) \quad A = \frac{(p + \alpha)(n - 1)}{1 + \alpha} \sum_{r=0}^{n-2} \binom{n - 2}{r}$$

$$\times \frac{(p + 2\alpha) \ldots [p + (r + 1)\alpha]q \ldots [q + (n - r - 3)\alpha]}{(1 + 2\alpha) \ldots [1 + (n - 1)\alpha]},$$

$$(5.4.7) \quad B = \sum_{l=0}^{n-1} \binom{n - 1}{l} \frac{(p + \alpha) \ldots (p + l\alpha)q(q + \alpha) \ldots [q + (n - l - 2)\alpha]}{(1 + \alpha) \ldots [1 + (n - 1)\alpha]}.$$

Expression B is identical with the sum in (5.4.4); hence $B = 1$. We notice further that the term under the summation sign in (5.4.6) is the probability of drawing r white and $n - r - 2$ black balls in $n - 2$ drawings according to a Pólya scheme where the urn contains $N + 2s$ balls at the beginning, among which $b + 2s$ are white and c are black. Thus we obtain from (5.4.3)

$$A = \frac{(p + \alpha)(n - 1)}{1 + \alpha}.$$

Finally,

$$(5.4.8) \quad E(X^2) = np\left[\frac{(p + \alpha)(n - 1)}{1 + \alpha} + 1\right] = np\frac{np + q + n\alpha}{1 + \alpha}.$$

Using (5.4.5), we obtain

$$(5.4.9) \qquad\qquad D^2(X) = npq\frac{1 + n\alpha}{1 + \alpha}.$$

The Pólya scheme can be applied to such phenomena as infectious diseases where the realization of an event (appearance of the disease) causes an increase in the probability of being infected with the disease.

In the Pólya scheme s may also be negative. Since the inequalities

$$b + (k - 1)s \geqslant 1 \text{ and } c + (n - k - 1)s \geqslant 1$$

must hold, k must then satisfy the double inequality

$$\max\left(0, n - 1 + \frac{c - 1}{s}\right) \leqslant k \leqslant \min\left(n, \frac{1 - b}{s} + 1\right).$$

B   Let N, b, and c tend to infinity so that

$$(5.4.10) \qquad\qquad p = \frac{b}{N} = \text{constant}.$$

Here, of course, $q = 1 - p$ is also constant. Suppose that $\lim_{N\to\infty} \alpha = 0$.

This condition will be satisfied, in particular, if s is constant and N tends to infinity. It follows from (5.4.1) and (5.4.2) that

$$(5.4.11) \qquad \lim_{N\to\infty} P(X = k) = \binom{n}{k} p^k q^{n-k}.$$

We have proved the following theorem.

Theorem 5.4.1. *If for $N = 1, 2, \ldots$ equality (5.4.10) is satisfied and $\lim_{N\to\infty} \alpha = 0$, then the probability function of the random variable X with the Pólya distribution tends to the probability function of the binomial distribution as $N \to \infty$.*

C   A particular case of the Pólya distribution is the *hypergeometric distribution*. In this distribution $s = -1$, which simply means that we do not replace the ball which has been drawn before drawing the next ball.

The probability function of the hypergeometric distribution can be obtained from (5.4.2) by putting $\alpha = -1/N$. We obtain for k satisfying the double inequality max $(0, n - Nq) \leqslant k \leqslant$ min (n, Np)

$$(5.4.12) \quad P(X = k)$$

$$= \binom{n}{k} \frac{Np(Np - 1) \ldots (Np - k + 1)Nq \ldots (Nq - n + k + 1)}{N(N - 1) \ldots (N - n + 1)}$$

$$= \frac{\binom{Np}{k}\binom{Nq}{n-k}}{\binom{N}{n}}.$$

The expected value $E(X)$ equals np, and formula (5.4.9) for the variance takes the form

$$(5.4.13) \qquad D^2(X) = \frac{N - n}{N - 1}\, npq.$$

The hypergeometric distribution is often applied in statistical quality control of mass production. For example, let the lot under control consist of b good items, and $N - b = c$ defective items. Here a good item plays the role of a white ball; a defective item, the role of a black ball; and the lot under control, the role of the urn. From the lot we draw n items at random to determine their quality; usually the chosen items are not returned to the lot. If the numbers b and c are known, by using the formulas obtained previously we can compute the probability that among n chosen items there are k good ones. However, in practice the numbers

b and c are unknown, and the investigation of the quality of some number of items may serve to estimate these numbers. This type of question is considered in Part II of the book.

5.5 THE POISSON DISTRIBUTION

In example 4.2.1 we considered a random variable X with a Poisson distribution. Let us summarize the most important properties of such a random variable.

Such a random variable can take on the values $r = 0, 1, 2, \ldots$ Its probability function is given by the formula

$$(5.5.1) \qquad P(X = r) = \frac{\lambda^r}{r!} e^{-\lambda},$$

where λ is a positive constant. According to (4.2.6) its characteristic function has the form

$$\phi(t) = e^{\lambda(e^{it} - 1)}.$$

From (4.2.7) to (4.2.9), we obtain

$$m_1 = \lambda, \quad m_2 = \lambda(\lambda + 1), \quad \mu_2 = \lambda.$$

The probability function (5.5.1) can be obtained as the limit of a sequence of probability functions of the binomial distribution. We shall prove Poisson's theorem (Poisson [1]).

Theorem 5.5.1. *Let the random variable X_n have a binomial distribution defined by the formula*

$$(5.5.2) \qquad P(X_n = r) = \frac{n!}{r!\,(n-r)!}\, p^r(1-p)^{n-r},$$

where r takes on the values $0, 1, 2, \ldots, n$. If for $n = 1, 2, \ldots$ the relation

$$(5.5.3) \qquad p = \frac{\lambda}{n}$$

holds,[1] where $\lambda > 0$ is a constant, then

$$(5.5.4) \qquad \lim_{n \to \infty} P(X_n = r) = \frac{\lambda^r}{r!} e^{-\lambda}.$$

Since the expected value of X_n is np, condition (5.5.3) means that as n increases the expected value of X_n remains constant.

---

[1] The assertion of this theorem will still hold if relation (5.5.3) is replaced by

$$\lim_{n \to \infty} np = \lambda.$$

Proof. Let us transform formula (5.5.2) in the following way:

$$P(X_n = r) = \frac{n!}{r!\,(n-r)!}\left(\frac{\lambda}{n}\right)^r\left(1-\frac{\lambda}{n}\right)^{n-r}$$

$$= \frac{\lambda^r}{r!}\left(1-\frac{\lambda}{n}\right)^n \frac{n(n-1)\ldots(n-r+1)}{n^r} \cdot \frac{1}{\left(1-\frac{\lambda}{n}\right)^r}$$

$$= \frac{\lambda^r}{r!}\left(1-\frac{\lambda}{n}\right)^n \frac{1\cdot\left(1-\frac{1}{n}\right)\cdot\ldots\cdot\left(1-\frac{r-1}{n}\right)}{\left(1-\frac{\lambda}{n}\right)^r}.$$

Using the fact that

$$\lim_{n\to\infty}\left(1-\frac{\lambda}{n}\right)^n = e^{-\lambda} \quad \text{and} \quad \lim_{n\to\infty}\frac{1\cdot\left(1-\frac{1}{n}\right)\cdot\ldots\cdot\left(1-\frac{r-1}{n}\right)}{\left(1-\frac{\lambda}{n}\right)^r} = 1,$$

we obtain formula (5.5.4).

We return to theorem 5.5.1 in Section 6.9,C.

In Fig. 5.5.1 there are two graphs, one of the binomial distribution with $n = 5$ and $p = 0.3$, hence $\lambda = np = 1.5$, and one of the Poisson distribution with the same expected value $\lambda = 1.5$. Figure 5.5.2 represents two such graphs for $n = 10$ and $p = 0.15$; hence again $\lambda = 1.5$. For larger values of n, for instance, $n = 100$, the graphs of the binomial and Poisson distributions will almost coincide.

Often the Poisson distribution is interpreted as a distribution of a random variable which can take on many different values (the number n is large) but with small probabilities (the probability $p = \lambda/n$ is small). That is why the Poisson distribution is sometimes called the *law of small numbers*. However, as is shown by the next two examples, this name is not justified.

Bortkiewicz [1], who has investigated the Poisson distribution, has given some empirical examples of random events to which this distribution can be applied. We present one of them.

Example 5.5.1. From Prussian army data, Bortkiewicz computed the number of soldiers in ten cavalry corps who died within a period of twenty years from a kick by a horse. We consider as a random variable the number r ($r = 0, 1, 2, \ldots$) of men in one corps killed in one year by a kick from a horse. The number of observations was $10 \times 20 = 200$, that is, the observations concerned ten army corps over a period of twenty years.

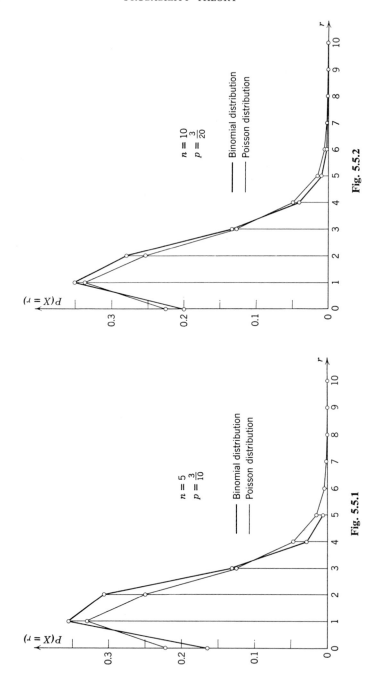

Fig. 5.5.2

Fig. 5.5.1

Bortkiewicz observed the following frequencies of appearance of values of r (Table 5.5.1):

TABLE 5.5.1
THE FREQUENCIES OF DEATH FROM A KICK BY A HORSE

r	0	1	2	3	4
Frequency	0.545	0.325	0.110	0.015	0.005
Probability	0.544	0.331	0.101	0.021	0.003

From the central row of this table we compute the mean

$$E(X) = 0 \cdot 0.545 + 1 \cdot 0.325 + 2 \cdot 0.110 + 3 \cdot 0.015 + 4 \cdot 0.005 = 0.61 \cdot$$

Let us compute the corresponding probabilities $P(X = r)$ for the Poisson distribution with $\lambda = 0.61$. Usually we find these probabilities from Poisson distribution tables, but here we compute them directly. We have

$$P(X = 0) = e^{-0.61} = 0.544,$$
$$P(X = 1) = 0.61 e^{-0.61} = 0.331,$$
$$P(X = 2) = \frac{0.61^2 e^{-0.61}}{2!} = 0.101,$$
$$P(X = 3) = \frac{0.61^3 e^{-0.61}}{3!} = 0.021,$$
$$P(X = 4) = \frac{0.61^4 e^{-0.61}}{4!} = 0.003.$$

These values are presented in the lower row of Table 5.5.1. As we see, these probabilities differ but little from the corresponding frequencies. We shall return to this example in Chapter 12 (example 12.4.3).

In many physical and technical problems we deal with distributions close to the Poisson distribution. Here we give an example from physics. In Chapter 8 we investigate more closely the models of random phenomena in which the Poisson distribution appears.

Example 5.5.2. We present here the results of the famous experiments of Rutherford and Geiger. They observed the numbers of α particles emitted by a radioactive substance in $n = 2608$ periods of 7.5 sec each. These data are presented in Table 5.5.2. In this table n_i denotes the number of periods in which the number of emitted particles was equal to i. The average number λ of particles emitted during a period of 7.5 sec is

$$\lambda = \frac{\sum n_i i}{n} = 3.87,$$

and

$$p_i = \frac{3.87^i}{i!} e^{-3.87}.$$

The reader will notice the striking closeness of the second and third columns in Table 5.5.2.

TABLE 5.5.2

i	n_i	np_i
0	57	54.399
1	203	210.523
2	383	407.361
3	525	525.496
4	532	508.418
5	408	393.515
6	273	253.817
7	139	140.325
8	45	67.882
9	27	29.189
10	16	17.075
	2608	2,608.000

Just as for the binomial distribution we can prove the *addition theorem* for independent random variables with Poisson distributions.

Let the independent random variables X_1 and X_2 have the respective Poisson distributions

$$P(X_1 = r) = \frac{\lambda_1^r}{r!} e^{-\lambda_1}, \qquad P(X_2 = r) = \frac{\lambda_2^r}{r!} e^{-\lambda_2} \qquad (r = 0, 1, 2, \ldots).$$

Consider the random variable

$$X = X_1 + X_2.$$

According to (4.2.6) the characteristic functions of X_1 and X_2 are

$$\phi_1(t) = \exp[\lambda_1(e^{it} - 1)], \qquad \phi_2(t) = \exp[\lambda_2(e^{it} - 1)].$$

By the independence of X_1 and X_2 the characteristic function of X has the form

(5.5.5) $$\phi(t) = \exp[(\lambda_1 + \lambda_2)(e^{it} - 1)].$$

Formula (5.5.5) represents the characteristic function of a random variable with the Poisson distribution having the expected value $\lambda_1 + \lambda_2$. This proves the *addition theorem* for independent random variables with Poisson distributions.

Raikov [1] has proved that the converse theorem is also true: *if X_1 and X_2 are independent and $X = X_1 + X_2$ has a Poisson distribution, then each of the random variables X_1 and X_2 has a Poisson distribution.*

Raikov's theorem is true for an arbitrary finite number of independent random variables $X_1, \ldots, X_n$.

5.6 THE UNIFORM DISTRIBUTION

A The simplest example of a random variable of the continuous type is a random variable with the uniform distribution. In example 4.1.2 we considered a particular case of the uniform distribution. The general definition is as follows.

Definition 5.6.1. The random variable X has a *uniform*, or *rectangular distribution* if its density function $f(x)$ is given by the formula

(5.6.1)

$$f(x) = \begin{cases} \dfrac{1}{2h} & \text{for } a - h \leqslant x \leqslant a + h, \text{ where } a \text{ and } h > 0 \text{ are constants,} \\ 0 & \text{otherwise.} \end{cases}$$

The distribution function $F(x)$ of this random variable is given by the formula

(5.6.2) $F(x) = \begin{cases} 0 & \text{for } x < a - h, \\ \dfrac{1}{2h} \displaystyle\int_{a-h}^{x} dx = \dfrac{x - (a - h)}{2h} & \text{for } a - h \leqslant x \leqslant a + h, \\ 1 & \text{for } x > a + h. \end{cases}$

The characteristic function of X is

(5.6.3) $$\phi(t) = \frac{1}{2h} \int_{a-h}^{a+h} e^{itx}\, dx = \frac{1}{2h} \left(\frac{e^{itx}}{it} \right)_{a-h}^{a+h}$$
$$= \frac{1}{2h} \cdot \frac{e^{it(a+h)} - e^{it(a-h)}}{it} = e^{ita} \frac{\sin th}{th}.$$

We obtain the moments directly from the formula

(5.6.4) $$m_k = \frac{1}{2h} \int_{a-h}^{a+h} x^k\, dx = \frac{1}{2h} \cdot \frac{(a + h)^{k+1} - (a - h)^{k+1}}{k + 1}.$$

In particular,

$$m_1 = a, \qquad m_2 = \tfrac{1}{3}(3a^2 + h^2).$$

Hence

(5.6.5) $$\mu_2 = m_2 - m_1^2 = \tfrac{1}{3}h^2.$$

By a linear transformation of X we can obtain a random variable with a uniform distribution in the interval $[0, 1]$. To do this we write

$$Y = \frac{X - (a - h)}{2h}.$$

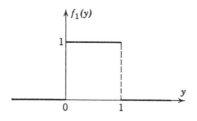

Fig. 5.6.1

The density of Y, which we shall denote by $f_1(y)$, is given by the following formula:

$$(5.6.6) \qquad f_1(y) = \begin{cases} 1 & \text{in the interval } [0, 1], \\ 0 & \text{otherwise.} \end{cases}$$

This is the rectangular distribution which we considered in example 4.1.2. The density of this distribution is shown in Fig. 5.6.1.

B In statistical problems we often deal with rectangular distributions. It is worthwhile to mention that if the distribution function $F(x)$ of the random variable X is continuous, then the random variable $Y = F(X)$ has the uniform distribution given by (5.6.6).

In fact, to every infinite interval $-\infty < X \leqslant x$ of values of the random variable X there corresponds the set of values of the random variable Y contained in the interval $0 \leqslant Y \leqslant y = F(x)$.

On the other hand, by the assumption that the distribution function $F(x)$ is continuous, to every y $(0 \leqslant y \leqslant 1)$ there corresponds at least one x satisfying the relation

$$(5.6.7) \qquad y = F(x) = P(X < x).$$

However, transformation (5.6.7) may not be one-to-one since the inverse image $F^{-1}(y)$ of some values of y may be an interval in which the function $F(x)$ is constant. Here, for a given y we can take for $x = F^{-1}(y)$ any of the values of x from the interval in which the distribution function $F(x)$ is constant, and for every such value of x we shall have $F[F^{-1}(y)] = y$; in particular we can take as $x = F^{-1}(y)$ the least x for which this equality holds.

If we denote by $F_1(y)$ the distribution function of the random variable Y, we obtain

$$F_1(y) = P(Y < y) = P[F(X) < y]$$

$$= \begin{cases} 0 & \text{for } y \leqslant 0, \\ P[X < F^{-1}(y)] = F[F^{-1}(y)] = y & \text{for } 0 < y < 1, \\ 1 & \text{for } y \geqslant 1. \end{cases}$$

From these formulas we obtain

$$F_1'(y) = f_1(y) = \begin{cases} 1 & \text{for } 0 \leqslant y \leqslant 1, \\ 0 & \text{for the remaining } y. \end{cases}$$

5.7 THE NORMAL DISTRIBUTION

In the examples we have often considered random variables with normal distributions. We now investigate the general form of the normal distribution.

Definition 5.7.1. The random variable X has a *normal distribution* if its density function is given by the formula

$$(5.7.1) \qquad f(x) = \frac{1}{\sigma\sqrt{2\pi}} \exp\left(-\frac{(x-m)^2}{2\sigma^2} \right),$$

where $\sigma > 0$.

We first verify that (5.7.1) is a density. To see this let us denote

$$(5.7.2) \qquad Y = \frac{X-m}{\sigma}.$$

We obtain

$$(5.7.3) \qquad f(y) = \frac{1}{\sqrt{2\pi}} e^{-y^2/2}.$$

Since the function $f(y)$ given by (5.7.3) (see Section 2.9) is a density, we have the equation

$$\frac{1}{\sigma\sqrt{2\pi}} \int_{-\infty}^{+\infty} \exp\left(-\frac{(x-m)^2}{2\sigma^2} \right) dx = \frac{1}{\sqrt{2\pi}} \int_{-\infty}^{+\infty} e^{-y^2/2} \, dy = 1.$$

The characteristic function $\phi(t)$ of the random variable Y has already been obtained in example 4.2.2; we have $\phi(t) = e^{-t^2/2}$.

Using equations (4.2.14), (4.2.15), and (5.7.2), we obtain the expression

$$(5.7.4) \qquad \phi_1(t) = \exp\left(itm - \tfrac{1}{2}\sigma^2 t^2 \right)$$

for the characteristic function of X.

From (5.7.4) and (4.2.4) we obtain the moments

$$(5.7.5) \qquad m_1 = m, \qquad m_2 = \sigma^2 + m^2, \qquad \mu_2 = \sigma^2.$$

As we can see from equalities (5.7.5), the constants m and σ which appear in (5.7.1) may be easily interpreted; m is the expected value of X and σ is its standard deviation. The shape of the curve of the density of the normal distribution depends on the parameter σ; this curve is called,

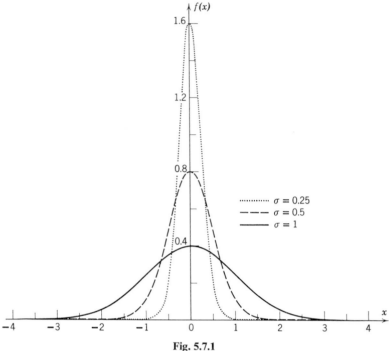

Fig. 5.7.1

in brief, the *normal curve*. It is illustrated in Fig. 5.7.1, representing three normal distributions with the same expected value $m = 0$ and different standard deviations: $\sigma = 1$, $\sigma = 0.5$ and $\sigma = 0.25$.

The normal distribution with expected value m and standard deviation σ is often denoted by $N(m; \sigma)$.

By the symmetry of the normal curve with respect to the expected value m all the central moments of odd order vanish,

(5.7.6) $\mu_{2k+1} = 0$ for every k.

It can be easily shown that

(5.7.7) $\mu_{2k} = 1 \cdot 3 \cdot \ldots \cdot (2k - 1)\sigma^{2k}.$

Formula (4.2.13) is a particular case of formula (5.7.7) for $\sigma = 1$.

There are very exact tables of the normal distribution which are used in computation. Usually we are interested in the probability that the random variable X with a normal distribution differs in absolute value from the expected value $m = E(X)$ by more than $\lambda\sigma$ ($\lambda > 0$), that is, more than a given multiple of the standard deviation. We find this probability, expressed as a function of λ, in the tables of the normal distribution giving the value of the integral

$$P(|X - m| > \lambda\sigma) = \frac{2}{\sqrt{2\pi}} \int_\lambda^\infty e^{-y^2/2} \, dy.$$

In fact,

$$P(|X - m| > \lambda\sigma) = P\left(\frac{|X - m|}{\sigma} > \lambda\right) = P(|Y| > \lambda),$$

where $Y = (X - m)/\sigma$.

We may also ask for the probability that X exceeds the expected value by more than a given multiple of the standard deviation $\lambda\sigma$, that is, the probability $P(X > m + \lambda\sigma)$. We have

$$P(X > m + \lambda\sigma) = P(Y > \lambda) = \frac{1}{\sqrt{2\pi}} \int_\lambda^{+\infty} e^{-y^2/2} \, dy.$$

Example 5.7.1. The random variable X has the distribution $N(1; 2)$. Find the probability that X is greater than 3 in absolute value.

Let us introduce the standardized random variable $Y = (X - 1)/2$. We have

$$P(|X| > 3) = P(|2Y + 1| > 3) = P(|Y + \tfrac{1}{2}| > \tfrac{3}{2})$$
$$= P(Y + \tfrac{1}{2} < -\tfrac{3}{2}) + P(Y + \tfrac{1}{2} > \tfrac{3}{2}) = P(Y < -2) + P(Y > 1).$$

By definition of Y, we have

$$P(Y < -2) = \frac{1}{\sqrt{2\pi}} \int_{-\infty}^{-2} e^{-t^2/2} \, dt = \frac{1}{\sqrt{2\pi}} \int_2^{+\infty} e^{-t^2/2} \, dt \cong 0.023,$$

$$P(Y > 1) = \frac{1}{\sqrt{2\pi}} \int_1^{+\infty} e^{-t^2/2} \, dt \cong 0.159.$$

The values of these integrals are obtained from tables of the normal distribution. Finally, we have

$$P(|X| > 3) = 0.182.$$

From tables of the normal distribution we see that, for a random variable X with the normal distribution $N(m; \sigma)$, the following equalities are satisfied:

$$P(|X - m| > \sigma) \cong 0.3173,$$
$$P(|X - m| > 2\sigma) \cong 0.0455,$$
$$P(|X - m| > 3\sigma) \cong 0.0027.$$

We see thus that the normal distribution is highly concentrated around its expected value. The probability that the value of X differs from the expected value by more than 3σ is smaller than 0.01. This property of the normal distribution has led many statisticians to apply the "three-sigma" rule, according to which for an arbitrary distribution there is small probability that the random variable differs from the expected value by more than 3σ. This rule should be applied very carefully. In fact, from the Chebyshev inequality follows only the fact that for an arbitrary random variable X whose variance exists

$$P(|X - m| \geqslant 3\sigma) \leqslant \tfrac{1}{9}.$$

The three-sigma rule can be applied only to distributions which do not differ much from the normal distribution. Thus they must be almost symmetric distributions, having only one maximum point in the neighborhood of the center of symmetry.

The *addition theorem* also holds for the normal distribution. Let X and Y be two independent random variables, and let X have the distribution $N(m_1; \sigma_1)$ and Y the distribution $N(m_2; \sigma_2)$. The respective characteristic functions of these distributions are

$$\phi_1(t) = \exp\left(m_1 it - \tfrac{1}{2}t^2\sigma_1^2\right),$$

$$\phi_2(t) = \exp\left(m_2 it - \tfrac{1}{2}t^2\sigma_2^2\right).$$

Because of the independence of X and Y, the random variable $Z = X + Y$ has the characteristic function

$$(5.7.8) \qquad \phi(t) = \exp\left[(m_1 + m_2)it - \tfrac{1}{2}(\sigma_1^2 + \sigma_2^2)t^2\right].$$

Expression (5.7.8) is the characteristic function of the normal distribution $N(m_1 + m_2; \sqrt{\sigma_1^2 + \sigma_2^2})$, which was to be proved.

Cramér [4] proved (see Problem 5.12) that the converse theorem is also true: *if X_1 and X_2 are independent and the random variable $X = X_1 + X_2$ has a normal distribution, then each of the random variables X_1 and X_2 has a normal distribution.*

Cramér's theorem is true for an arbitrary finite number of independent random variables.

Besides Cramér's theorem, many others are known which characterize the normal distribution. One of them is given in Section 9.5,C. We present here the theorem of Skitowitch [1]. *Let $X_1, \ldots, X_n$ be independent and have the same nondegenerate distribution. Then the independence of the random variables L_1 and L_2, defined by*

$$L_1 = a_1 X_1 + \ldots + a_n X_n,$$

$$L_2 = b_1 X_1 + \ldots + b_n X_n$$

with $\sum_{j=1}^{n} a_j b_j = 0$ and $\sum_{j=1}^{n} (a_j b_j)^2 \neq 0$, is a necessary and sufficient condition for the distributions of the random variables $X_1, \ldots, X_n$ to be normal. For $n = 2$ this theorem has been proved by Bernstein [4], Darmois [2], and Gnedenko [10]. We also mention the papers of Pólya [2] (see Problem

6.10), Kac [1], Marcinkiewicz [1] (see Problem 5.11), Linnik [1], Laha [3], and Basu and Laha [1].

The normal distribution is of great importance in probability theory and statistics. In nature and technology we very often deal with distributions that are close to normal. This phenomenon is an object of investigation of the theory of stochastic processes (see Section 8.9). Moreover, under rather general assumptions the normal distribution is the limiting distribution for sums of independent random variables when the number of terms increases to infinity. This question is discussed in the next chapter.

5.8 THE GAMMA DISTRIBUTION

In applications we often use a distribution associated with the gamma function, defined for $p > 0$ by the formula

$$(5.8.1) \qquad \Gamma(p) = \int_0^\infty x^{p-1} e^{-x}\, dx.$$

It is known that integral (5.8.1) is uniformly convergent with respect to p and thus $\Gamma(p)$ is a continuous function. Integrating (5.8.1) by parts, we obtain

$$\Gamma(p + 1) = \int_0^\infty x^p e^{-x}\, dx = [-e^{-x} x^p]_0^\infty + p \int_0^\infty x^{p-1} e^{-x}\, dx.$$

Hence

$$(5.8.2) \qquad \Gamma(p + 1) = p\Gamma(p).$$

In particular, if $p = n$, where n is an integer, we obtain from (5.8.2)

$$\Gamma(n + 1) = n\Gamma(n),$$

$$(5.8.3) \qquad \Gamma(n) = (n - 1)\Gamma(n - 1),$$

$$\dotsc\dotsc\dotsc\dotsc\dotsc\dotsc\dotsc$$

$$\Gamma(2) = 1\Gamma(1).$$

Since

$$\Gamma(1) = \int_0^\infty e^{-x}\, dx = -[e^{-x}]_0^\infty = 1,$$

we obtain from equalities (5.8.3)

$$(5.8.4) \qquad \Gamma(n + 1) = n!.$$

Substituting $y = x/a$ $(a > 0)$ in (5.8.1), we have

$$(5.8.5) \qquad \frac{\Gamma(p)}{a^p} = \int_0^\infty y^{p-1} e^{-ay}\, dy.$$

Equation (5.8.5) is also valid when a is a complex number $a = b + ic$, where $b > 0$. We shall not give the proof of (5.8.5) for this case.

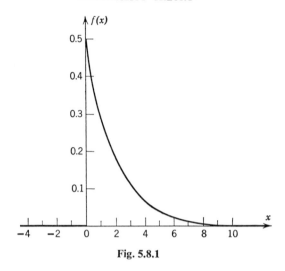

Fig. 5.8.1

Let X be a random variable with the density defined by the formula

$$(5.8.6) \qquad f(x) = \begin{cases} 0 & \text{for } x \leqslant 0, \\ \dfrac{b^p}{\Gamma(p)}\, x^{p-1} e^{-bx} & \text{for } x > 0, \end{cases}$$

where $b > 0$ and $p > 0$. The fact that (5.8.6) defines a density follows directly from (5.8.5), since

$$\int_{-\infty}^{+\infty} f(x)\, dx = \int_{0}^{+\infty} \frac{b^p}{\Gamma(p)}\, x^{p-1} e^{-bx}\, dx = 1$$

and $f(x)$ is a non-negative function.

Definition 5.8.1. If a random variable X has the density given by (5.8.6) we shall say that X has a *gamma distribution*.

Figure 5.8.1 represents such a density for $p = 1$ and $b = 0.5$ (see also Fig. 9.4.1).

We now find the characteristic function of this distribution. We have

$$(5.8.7) \qquad \phi(t) = \int_{-\infty}^{+\infty} e^{itx} f(x)\, dx = \frac{b^p}{\Gamma(p)} \int_{0}^{+\infty} x^{p-1} e^{-(b-it)x}\, dx.$$

Since, as has already been stated, equation (5.8.5) is valid when $a = b + ic$ and $b > 0$, we obtain from formula (5.8.7)

$$(5.8.8) \qquad \phi(t) = \frac{b^p}{\Gamma(p)} \cdot \frac{\Gamma(p)}{(b - it)^p} = \frac{1}{(1 - it/b)^p}.$$

The function $\phi(t)$ can be differentiated an arbitrary number of times. Its kth derivative is expressed by the formula

$$\phi^{(k)}(t) = \frac{p(p+1)\ldots(p+k-1)}{b^k} i^k \frac{1}{(1-it/b)^{p+k}} \quad \text{for } k = 1, 2, \ldots .$$

From (4.2.4) we obtain

$$(5.8.9) \qquad m_k = \frac{\phi^{(k)}(0)}{i^k} = \frac{p(p+1)\ldots(p+k-1)}{b^k}.$$

In particular, we have,

$$(5.8.10) \qquad m_1 = \frac{p}{b}, \qquad m_2 = \frac{p(p+1)}{b^2}, \qquad \mu_2 = \frac{p}{b^2}.$$

Example 5.8.1. The random variable X has the gamma distribution with the density given by the formula

$$f(x) = \begin{cases} 0 & \text{for} \quad x \leqslant 0, \\ 2e^{-2x} & \text{for} \quad x > 0. \end{cases}$$

The reader may verify that if we substitute $p = 1$ and $b = 2$ in (5.8.6) we obtain the distribution considered in this example. What is the probability that X is not smaller than two?
We have

$$P(X \geqslant 2) = 2 \int_2^\infty e^{-2x} \, dx = -[e^{-2x}]_2^\infty = e^{-4} \simeq 0.0183.$$

In more complicated cases we can make use of the tables by K. Pearson [1] to compute probabilities of the gamma distribution.

The probability distribution considered in example 5.8.1 is a particular case of the exponential distribution.

Definition 5.8.2. The random variable with density $f(x)$, defined by the formula

$$(5.8.11) \qquad f(x) = \begin{cases} 0 & \text{for } x \leqslant 0, \\ \lambda e^{-\lambda x} & \text{for } x > 0, \end{cases}$$

where $\lambda > 0$, has an *exponential distribution*.
We now show that the *addition theorem* is valid for random variables with gamma distributions.
Let X_1 and X_2 be two independent random variables with gamma distributions and with the respective characteristic functions

$$\phi_k(t) = \frac{1}{(1-it/b)^{p_k}} \qquad (k = 1, 2).$$

Consider the sum of these random variables, $X = X_1 + X_2$. From the independence of X_1 and X_2 it follows that the characteristic function $\phi(t)$ of X equals

$$\phi(t) = \frac{1}{(1 - it/b)^{p_1}} \cdot \frac{1}{(1 - it/b)^{p_2}} = \frac{1}{(1 - it/b)^{p_1 + p_2}}.$$

We see that X also has the gamma distribution, which proves the theorem.

Laha [1] and Lukacs [3] have given theorems characterizing the gamma distribution. We mention here the following quite simple theorem of Lukacs.

Let the independent random variables X and Y with nondegenerate distributions take on only positive values. Then X and Y have the gamma distribution with the same parameter b if and only if the random variables U and V, where

$$U = X + Y, \qquad V = \frac{X}{Y},$$

are independent.

5.9   THE BETA DISTRIBUTION

In the applications we also deal with a distribution associated with the function defined by the formula

$$(5.9.1) \quad B(p, q) = \int_0^1 x^{p-1}(1 - x)^{q-1}\, dx, \qquad \text{where } p > 0, q > 0.$$

In the monograph of Saks and Zygmund [1] the reader will find a proof of the following equation connecting the function $B(p, q)$ with the function Γ defined by (5.8.1):

$$(5.9.2) \qquad\qquad B(p, q) = \frac{\Gamma(p)\Gamma(q)}{\Gamma(p + q)}.$$

Definition 5.9.1.   We say that the random variable X has a *beta distribution* if its density is given by the formula

$$(5.9.3) \quad f(x) = \begin{cases} \dfrac{1}{B(p, q)} x^{p-1}(1 - x)^{q-1} & \text{for } 0 < x < 1, \\ 0 & \text{for } x \leqslant 0 \quad \text{and} \quad x \geqslant 1, \end{cases}$$

where $p > 0, q > 0$.

That the function $f(x)$ given by (5.9.3) is a density follows from formula (5.9.1) and the fact that it is non-negative.

It is convenient to obtain the moments of the beta distribution directly from the formula

$$(5.9.4) \quad m_k = \frac{\Gamma(p+q)}{\Gamma(p)\Gamma(q)} \int_0^1 x^{p+k-1}(1-x)^{q-1}\,dx = \frac{\Gamma(p+q)}{\Gamma(p)\Gamma(q)} \cdot B(p+k,q)$$

$$= \frac{\Gamma(p+q)\Gamma(p+k)}{\Gamma(p)\Gamma(p+q+k)} = \frac{p(p+1)\ldots(p+k-1)}{(p+q)(p+q+1)\ldots(p+q+k-1)}.$$

In particular,

$$(5.9.5) \qquad m_1 = \frac{p}{p+q}, \qquad m_2 = \frac{p(p+1)}{(p+q)(p+q+1)},$$

$$(5.9.6) \qquad\qquad \mu_2 = \frac{pq}{(p+q)^2(p+q+1)}.$$

Figure 5.9.1 represents the density of the beta distribution with $p = q = 2$.

Example 5.9.1. The random variable X has the beta distribution with $p = q = 2$; hence its density $f(y)$ has the form

$$f(y) = \begin{cases} 0 & \text{for} \quad y \leqslant 0 \quad \text{and} \quad y \geqslant 1, \\ \dfrac{\Gamma(4)}{\Gamma(2)\Gamma(2)}\, y(1-y) = 6y(1-y) & \text{for} \quad 0 < y < 1. \end{cases}$$

What is the probability that X is not greater than 0.2?
We have

$$P(Y \leqslant 0.2) = 6 \int_0^{0.2} y(1-y)\,dy = 6\left[\frac{y^2}{2} - \frac{y^3}{3}\right]_0^{0.2} = 0.104.$$

For computing the probabilities of the beta distribution we can use Pearson's tables [4].

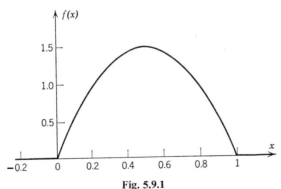

Fig. 5.9.1

5.10   THE CAUCHY AND LAPLACE DISTRIBUTIONS

Definition 5.10.1.   The random variable X has a *Cauchy distribution* if its density is given by the formula

$$(5.10.1) \qquad f(x) = \frac{1}{\pi} \cdot \frac{\lambda}{\lambda^2 + (x - \mu)^2}, \qquad \text{where } \lambda > 0.$$

The function $f(x)$ is non-negative. By substituting

$$(5.10.2) \qquad\qquad y = \frac{x - \mu}{\lambda},$$

we obtain

$$\int_{-\infty}^{+\infty} f(x)\, dx = \frac{1}{\pi} \int_{-\infty}^{+\infty} \frac{dy}{1 + y^2} = \frac{1}{\pi} \left[\arctan y \right]_{-\infty}^{+\infty} = 1.$$

To find the characteristic function of the random variable X let us first find the characteristic function of the random variable Y which is the linear transformation of X given by (5.10.2). Thus Y has the density

$$(5.10.3) \qquad \bullet \qquad f(y) = \frac{1}{\pi} \cdot \frac{1}{1 + y^2}$$

and the characteristic function

$$(5.10.4) \qquad\qquad \phi(t) = \frac{1}{\pi} \int_{-\infty}^{+\infty} e^{ity} \frac{1}{1 + y^2}\, dy.$$

To find $\phi(t)$ consider first the density

$$(5.10.5) \qquad\qquad f_1(y) = \tfrac{1}{2} e^{-|y|}.$$

The reader may verify that expression (5.10.5) is a density. The characteristic function of the random variable with the density (5.10.5) is

$$\phi_1(t) = \tfrac{1}{2} \int_{-\infty}^{+\infty} e^{ity} e^{-|y|}\, dy = \tfrac{1}{2} \int_{-\infty}^{+\infty} (\cos ty + i \sin ty) e^{-|y|}\, dy$$

$$= \int_0^{\infty} \cos tye^{-y}\, dy.$$

Integrating by parts, we obtain

$$\int_0^{\infty} \cos tye^{-y}\, dy = \left[-e^{-y} \cos ty \right]_0^{\infty} - t \int_0^{\infty} \sin tye^{-y}\, dy$$

$$= 1 - t \int_0^{\infty} \sin tye^{-y}\, dy.$$

Similarly,

$$\int_0^\infty \sin tye^{-y}\,dy = \left[-e^{-y}\sin ty\right]_0^\infty + t\int_0^\infty e^{-y}\cos ty\,dy$$

$$= t\int_0^\infty e^{-y}\cos ty\,dy.$$

Hence we obtain

$$\int_0^\infty e^{-y}\cos ty\,dy = 1 - t^2\int_0^\infty e^{-y}\cos ty\,dy.$$

Finally we have

(5.10.6) $$\phi_1(t) = \int_0^\infty e^{-y}\cos ty\,dy = \frac{1}{1+t^2}.$$

The characteristic function (5.10.6) is absolutely integrable over the interval $(-\infty, +\infty)$; by (4.5.6) its corresponding density is

(5.10.7) $$f_1(y) = \frac{1}{2\pi}\int_{-\infty}^{+\infty}\frac{e^{-ity}}{1+t^2}\,dt.$$

From (5.10.5) we obtain

$$e^{-|y|} = \frac{1}{\pi}\int_{-\infty}^{+\infty}\frac{e^{-ity}}{1+t^2}\,dt.$$

Changing e^{-ity} into e^{ity} under the integral sign (this does not affect the value of the integral) and changing the roles of t and y, we obtain

(5.10.8) $$e^{-|t|} = \frac{1}{\pi}\int_{-\infty}^{+\infty}\frac{e^{ity}}{1+y^2}\,dy.$$

The right-hand side of (5.10.8) is identical with that of (5.10.4); thus we finally obtain

(5.10.9) $$\phi(t) = e^{-|t|}.$$

Since X is a linear transformation of Y, for the characteristic function $\phi_2(t)$ of X we obtain the formula

(5.10.10) $$\phi_2(t) = \exp(i\mu t - \lambda|t|).$$

Since, as can easily be seen, the function $\phi_2(t)$ is not differentiable at $t = 0$, none of the moments of the Cauchy distribution exist.

The *addition theorem* is valid for the Cauchy distribution. In fact, let X_1 and X_2 be two independent random variables with densities

$$g_1(x) = \frac{1}{\pi}\cdot\frac{\lambda_1}{\lambda_1^2 + (x - \mu_1)^2}, \quad g_2(x) = \frac{1}{\pi}\cdot\frac{\lambda_2}{\lambda_2^2 + (x - \mu_2)^2} \quad (\lambda_1, \lambda_2 > 0).$$

The characteristic functions of these random variables are, respectively,

$$\psi_1(t) = \exp\left(i\mu_1 t - \lambda_1 |t|\right), \qquad \psi_2(t) = \exp\left(i\mu_2 t - \lambda_2 |t|\right).$$

Consider the random variable $X = X_1 + X_2$. Its characteristic function is

(5.10.11) $\psi(t) = \exp\left[i(\mu_1 + \mu_2)t - (\lambda_1 + \lambda_2)|t|\right].$

Formula (5.10.11) is also the characteristic function of a Cauchy distribution.

In the proof of (5.10.9) we considered the density defined by (5.10.5). The corresponding characteristic function is given by (5.10.6).

The random variable Y with the density (5.10.5) is a particular case of a random variable with a Laplace distribution. Namely, X has a *Laplace distribution* if $X = \lambda Y + \mu$, where Y has the density function (5.10.5). According to formula (2.4.10) the density of X is of the form

(5.10.12) $f(x) = \dfrac{1}{2\lambda} \exp\left(-\dfrac{|x - \mu|}{\lambda}\right) \qquad (\lambda > 0).$

The characteristic function of this random variable has the form

(5.10.13) $\phi(t) = \dfrac{e^{\mu t i}}{1 + \lambda^2 t^2}.$

It follows from (5.10.13) that a random variable with a Laplace distribution has moments of any order.

5.11 THE MULTIDIMENSIONAL NORMAL DISTRIBUTION

A   In example 3.6.1 we considered a particular case of the two-dimensional normal distribution. We now characterize the multidimensional normal distribution in its general form.

Definition 5.11.1. The random variable (X, Y) has a *two-dimensional normal distribution* if its density function $f(x, y)$ is given by the formula

(5.11.1) $f(x, y) = \dfrac{1}{2\pi\sigma_1\sigma_2\sqrt{1 - \rho^2}} \exp\left\{-\dfrac{1}{2(1 - \rho^2)}\left[\dfrac{(x - m_1)^2}{\sigma_1^2}\right.\right.$

$$\left.\left. - 2\dfrac{\rho(x - m_1)(y - m_2)}{\sigma_1\sigma_2} + \dfrac{(y - m_2)^2}{\sigma_2^2}\right]\right\},$$

where m_1, m_2, σ_1, σ_2, and ρ are constants.

It can be shown that

$$\int_{-\infty}^{+\infty}\int_{-\infty}^{+\infty} f(x, y)\, dx\, dy = 1.$$

The moments of the first two orders are the following:

(5.11.2) $m_{10} = m_1$, $m_{01} = m_2$, $\mu_{20} = \sigma_1^2$, $\mu_{02} = \sigma_2^2$, $\mu_{11} = \rho\sigma_1\sigma_2$.

The characteristic function of the distribution under consideration has the form

(5.11.3) $\phi(t, u) = \exp[i(m_1 t + m_2 u) - \frac{1}{2}(\sigma_1^2 t^2 + 2\rho\sigma_1\sigma_2 tu + \sigma_2^2 u^2)]$.

As we see, the function $f(x, y)$ depends on five constants. According to (3.6.5), (3.6.9), and (3.6.26) two of them, namely, m_1 and m_2, are the expected values of X and Y; the constants σ_1 and σ_2 are the standard deviations of these random variables; finally, ρ is the correlation coefficient of X and Y.

If $\rho^2 = 1$, the expression (5.11.1) is meaningless. However, according to theorem 3.6.5, from the equality $\rho^2 = 1$ it follows that there exists a linear relation $Y = aX + b$ between X and Y, with probability 1. Here we have a one-dimensional distribution which is called the *degenerate two-dimensional normal distribution*. In the sequel we shall not consider such distributions.

If the correlation coefficient $\rho = 0$, formula (5.11.1) takes the form

$$(5.11.4) \quad f(x, y) = \frac{1}{2\pi\sigma_1\sigma_2} \exp\left\{-\frac{1}{2}\left[\frac{(x - m_1)^2}{\sigma_1^2} + \frac{(y - m_2)^2}{\sigma_2^2}\right]\right\}$$

$$= \frac{1}{\sigma_1\sqrt{2\pi}} \exp\left(-\frac{(x - m_1)^2}{2\sigma_1^2}\right)\frac{1}{\sigma_2\sqrt{2\pi}} \exp\left(-\frac{(y - m_2)^2}{2\sigma_2^2}\right)$$

$$= f_1(x)f_2(y),$$

where $f_1(x)$ and $f_2(y)$ are the densities of X and Y, respectively.

From relation (5.11.4) it follows that *if two random variables with joint normal distribution are uncorrelated* ($\rho = 0$), *they are independent*. As we know, not all two-dimensional distributions have this property.

According to (2.7.6), the conditional density of Y provided that $X = x$ has the form

$$(5.11.5) \quad f(y \mid x) = \frac{1}{\sigma_2\sqrt{1 - \rho^2}\sqrt{2\pi}} \exp\left\{-\frac{1}{2\sigma_2^2(1 - \rho^2)}\right.$$

$$\left. \times \left[(y - m_2) - \frac{\rho\sigma_2}{\sigma_1}(x - m_1)\right]^2\right\}.$$

This is the density of a normal distribution, the expected value being the following function of x:

$$(5.11.6) \quad m_2(x) = m_2 + \frac{\rho\sigma_2}{\sigma_1}(x - m_1).$$

Formula (5.11.6) is the equation of the regression curve of the first type of Y on X. As we see, the regression curve in this case is a straight line.

Similarly, for the regression curve of X on Y we obtain the formula

$$(5.11.7) \qquad m_1(y) = m_1 + \frac{\rho \sigma_1}{\sigma_2}(y - m_2).$$

A particular case of the regression curves (5.11.6) and (5.11.7) has been considered in example 3.7.2.

B We now investigate the n-dimensional normal distribution for the case $n \geqslant 2$.

Definition 5.11.2. The random variable $(X_1, X_2, \ldots, X_n)$ has an n-dimensional *normal distribution* if its density function $f(x_1, x_2, \ldots, x_n)$ is of the form

$$(5.11.8) \quad f(x_1, x_2, \ldots, x_n) = \frac{1}{(2\pi)^{n/2}\sqrt{|\mathbf{M}|}} \exp\left[-\frac{1}{2|\mathbf{M}|} \right.$$

$$\left. \times \sum_{j,k=1}^{n} |\mathbf{M}_{jk}| (x_j - m_j)(x_k - m_k) \right],$$

where $|\mathbf{M}| \neq 0$ is the determinant of the matrix $\mathbf{M}$ of the second-order moments λ_{jk} defined by formula (3.6.30), and $|\mathbf{M}_{jk}|$ is the algebraic complement of the term λ_{jk} in the determinant of the matrix $\mathbf{M}$.

When the correlation coefficients of all possible pairs of random variables $X_1, X_2, \ldots, X_n$ are equal to 0, then $|\mathbf{M}_{jk}| = 0$ for $j \neq k$; hence $X_1, X_2, \ldots, X_n$ are independent.

We notice that if $|\mathbf{M}| = 0$, then, according to theorem 3.6.6, the probability that among the random variables $X_1, X_2, \ldots, X_n$ there is at least one linear relation equals 1.

The characteristic function $\phi(t_1, t_2, \ldots, t_n)$ of a random variable with the density given by (5.11.8) has the form

$$(5.11.9) \quad \phi(t_1, t_2, \ldots, t_n) = \exp\left(i\sum_{j=1}^{n} m_j t_j - \tfrac{1}{2}\sum_{j,k=1}^{n} \lambda_{jk} t_j t_k \right).$$

A theorem characterizing the multidimensional normal distribution was given by Laha [2].

Example 5.11.1. Denote by V the velocity of a particle of an ideal gas. Consider the rectangular system of coordinates x_1, x_2, x_3. Let $\vec{V}$ be the velocity vector of the particle, and V_1, V_2, V_3 its projections on the corresponding axes. The position and length of the vector $\vec{V}$ have a random character; hence

(V_1, V_2, V_3) is a three-dimensional random variable. To determine its density function, Maxwell accepted the following two assumptions:

I. $\qquad\qquad h(v_1, v_2, v_3) = f(v_1)f(v_2)f(v_3),$

II. $\qquad h(v_1, v_2, v_3) = h_1(v_1^2 + v_2^2 + v_3^2) = h_1(v^2).$

In equality (I) the functions $f(v_1)$, $f(v_2)$, $f(v_3)$ are the densities of the random variables V_1, V_2, V_3, respectively. Assumption (I) therefore means that the random variables V_1, V_2, V_3 are independent and have the same distribution. Assumption (II) means that the density $h(v_1, v_2, v_3)$ is a function of the kinetic energy of the particle, which equals $\frac{1}{2}m(v_1^2 + v_2^2 + v_3^2)$, where m is the mass of the particle.

From these equalities we obtain

$$\log h_1(v^2) = \sum_{i=1}^{3} \log f(v_i),$$

$$2\frac{h_1'(v^2)}{h_1(v^2)} \sum_{i=1}^{3} v_i \, dv_i = \sum_{i=1}^{3} \frac{f'(v_i)}{f(v_i)} \, dv_i.$$

Comparing corresponding coefficients of dv_i, we obtain

$$\frac{f'(v_i)}{v_i f(v_i)} = k = \text{constant} \qquad (i = 1, 2, 3).$$

The solution of these differential equations gives

$$f(v_i) = C \exp\left(\tfrac{1}{2}kv_i^2\right) \qquad (i = 1, 2, 3).$$

Since $f(v_i)$ is a density, we have $k < 0$; thus we can write $k = -1/\sigma^2$. By formula (I), we have

$$h(v_1, v_2, v_3) = C^3 \exp\left(-\frac{v_1^2 + v_2^2 + v_3^2}{2\sigma^2}\right).$$

Thus the random variable (V_1, V_2, V_3) has a three-dimensional normal distribution. Since $h(v_1, v_2, v_3)$ is a density, we obtain

$$C^3 = (2\pi\sigma^2)^{-3/2}.$$

The constant σ is different for different gases.

We return to the question considered in this example in example 9.4.3.
C   Now let $(X_1, X_2, \ldots, X_n)$ be an n-dimensional random variable with a normal distribution. Suppose, for the sake of simplicity, that $m_j = 0$ $(j = 1, 2, \ldots, n)$. Consider the r-dimensional random variable $(Y_1, Y_2, \ldots, Y_r)$, where

(5.11.10) $\qquad Y_l = \sum_{j=1}^{n} \alpha_{lj} X_j \qquad (l = 1, 2, \ldots, r),$

and $r \leqslant n$. Denote by μ_{lh} the covariance of the random variables Y_l and Y_h, where $l, h = 1, 2, \ldots, r$. We have

$$(5.11.11) \qquad \mu_{lh} = E(Y_l Y_h) = E\left(\sum_{j=1}^{n} \alpha_{lj} X_j \sum_{k=1}^{n} \alpha_{hk} X_k \right)$$

$$= \sum_{j,k=1}^{n} \alpha_{lj} \alpha_{hk} E(X_j X_k) = \sum_{j,k=1}^{n} \alpha_{lj} \alpha_{hk} \lambda_{jk}.$$

Now denote by $\psi(u_1, u_2, \ldots, u_r)$ the characteristic function of $(Y_1, Y_2, \ldots, Y_r)$. We have

$$\psi(u_1, u_2, \ldots, u_r) = E\left[\exp\left(i \sum_{l=1}^{r} u_l Y_l \right) \right]$$

$$= E\left[\exp\left(i u_1 \sum_{j=1}^{n} \alpha_{1j} X_j + \ldots + i u_r \sum_{j=1}^{n} \alpha_{rj} X_j \right) \right]$$

$$= E\left[\exp\left(i \sum_{j=1}^{n} v_j X_j \right) \right],$$

where $v_j = u_1 \alpha_{1j} + \ldots + u_r \alpha_{rj}$ $(j = 1, \ldots, n)$. By (5.11.9), we have

$$\psi(u_1, u_2, \ldots, u_r) = \exp\left(-\frac{1}{2} \sum_{j,k=1}^{n} \lambda_{jk} v_j v_k \right)$$

$$= \exp\left(-\frac{1}{2} \sum_{j,k=1}^{n} \lambda_{jk} \sum_{l,h=1}^{r} u_l u_h \alpha_{lj} \alpha_{hk} \right)$$

$$= \exp\left(-\frac{1}{2} \sum_{l,h=1}^{r} u_l u_h \sum_{j,k=1}^{n} \alpha_{lj} \alpha_{hk} \lambda_{jk} \right).$$

Using formula (5.11.11), we obtain

$$(5.11.12) \qquad \psi(u_1, u_2, \ldots, u_r) = \exp\left(-\frac{1}{2} \sum_{l,h=1}^{r} \mu_{lh} u_l u_h \right).$$

Comparing (5.11.12) and (5.11.9), we see that the random variable $(Y_1, Y_2, \ldots, Y_r)$ has an r-dimensional normal distribution. Thus we have obtained the following important theorem.

Theorem 5.11.1. *Let $(X_1, X_2, \ldots, X_n)$ be an n-dimensional random variable with a normal distribution and let $Y_1, Y_2, \ldots, Y_r$, where $r \leqslant n$, be linear functions of the random variables X_j $(j = 1, 2, \ldots, n)$. Then the random variable $(Y_1, Y_2, \ldots, Y_r)$ also has a normal distribution.*

In particular, it follows from this theorem that every marginal distribution of a random vector with the normal distribution is normal.

5.12   THE MULTINOMIAL DISTRIBUTION

A   Let us consider the following generalized Bernoulli scheme.
We perform n random experiments. As a result of each experiment one
of the pairwise exclusive events A_j ($j = 1, 2, \ldots, r, r + 1$) occurs. Let
$p_j = P(A_j)$, where $p_1 + \ldots + p_r + p_{r+1} = 1$. The results of the n experiments are independent. Consider the random variable $(X_1, \ldots, X_r, X_{r+1})$,
where $X_j = k_j$ means that event A_j has occurred k_j times, $k_j = 0, 1, \ldots, n$.
By reasoning similar to that used in example 3.1.3 where we deduced the
formula for the binomial distribution, we obtain the formula

$$(5.12.1) \quad P(X_1 = k_1, \ldots, X_r = k_r, X_{r+1} = k_{r+1})$$
$$= \frac{n!}{k_1! \ldots k_r! k_{r+1}!} p_1^{k_1} \ldots p_r^{k_r} p_{r+1}^{k_{r+1}},$$

where $k_1 + \ldots + k_r + k_{r+1} = n$. This formula gives us the probability
that A_1 occurs k_1 times, A_2 occurs k_2 times, $\ldots$, A_{r+1} occurs k_{r+1} times.
We notice that the random variables $X_1, \ldots, X_r, X_{r+1}$ satisfy the
linear relation $X_1 + \ldots + X_r + X_{r+1} = n$.
Let us express one of the random variables, say X_{r+1}, in terms of the
remaining ones, that is, $X_{r+1} = n - X_1 \ldots - X_r$. Then formula (5.12.1)
can be written in the form

$$(5.12.1') \quad P(X_1 = k_1, \ldots, X_r = k_r) = \frac{n!}{k_1! \ldots k_r! (n - K)!} p_1^{k_1} \ldots p_r^{k_r} q^{n-K},$$

where $K = k_1 + \ldots + k_r$ and $q = 1 - p_1 - \ldots - p_r$.
Definition 5.12.1.   The random variable $(X_1, \ldots, X_r)$ with the probability function given by formula (5.12.1') is said to have a *multinomial
distribution*.
B   Let $(Y_1^{(1)}, Y_2^{(1)}, \ldots, Y_r^{(1)})$ and $(Y_1^{(2)}, Y_2^{(2)}, \ldots, Y_r^{(2)})$ be two random
variables. Addition of two multidimensional random variables will be
understood in the vector sense, that is, we say that the random variable
$(X_1, X_2, \ldots, X_r)$ is the sum of the random variables $(Y_1^{(1)}, Y_2^{(1)}, \ldots, Y_r^{(1)})$
and $(Y_1^{(2)}, Y_2^{(2)}, \ldots, Y_r^{(2)})$, and we write

$$(X_1, X_2, \ldots, X_r) = (Y_1^{(1)}, Y_2^{(1)}, \ldots, Y_r^{(1)}) + (Y_1^{(2)}, Y_2^{(2)}, \ldots, Y_r^{(2)}),$$

if $X_j = Y_j^{(1)} + Y_j^{(2)}$ ($j = 1, 2, \ldots, r$).
Now let $(Y_1^{(m)}, Y_2^{(m)}, \ldots, Y_r^{(m)})$, where $m = 1, 2, \ldots, n$, be independent
random vectors with the same distribution, having at most one coordinate
different from zero, where for $m = 1, \ldots, n$ and $j = 1, 2, \ldots, r$

$$P(Y_j^{(m)} = 1) = p_j, \quad P(Y_j^{(m)} = 0) = 1 - p_j,$$
$$(5.12.2)$$
$$P(Y_1^{(m)} = 0, \ldots, Y_r^{(m)} = 0) = q = 1 - p_1 - \ldots - p_r.$$

It is easy to verify that the random variable $(X_1, X_2, \ldots, X_r)$ with a multinomial distribution satisfies the relation

$$(X_1, X_2, \ldots, X_r) = \sum_{m=1}^{n} (Y_1^{(m)}, Y_2^{(m)}, \ldots, Y_r^{(m)}).$$

By (4.6.1) we find that the characteristic function $\phi_m(t_1, t_2, \ldots, t_r)$ of $(Y_1^{(m)}, Y_2^{(m)}, \ldots, Y_r^{(m)})$, for $m = 1, 2, \ldots, n$, is of the form

$$\phi_m(t_1, t_2, \ldots, t_r) = \sum_{j=1}^{r} p_j e^{it_j} + q.$$

Hence for the characteristic function $\phi(t_1, t_2, \ldots, t_r)$ of $(X_1, X_2, \ldots, X_r)$ we obtain the formula

$$(5.12.3) \quad \phi(t_1, t_2, \ldots, t_r) = \prod_{m=1}^{n} \phi_m(t_1, t_2, \ldots, t_r) = \left(\sum_{j=1}^{r} p_j e^{it_j} + q \right)^n.$$

From the last formula and from formulas analogous to (4.6.8′) we obtain for $j = 1, 2, \ldots, r$

$$(5.12.4) \qquad E(X_j) = np_j, \qquad \lambda_{jj} = D^2(X_j) = np_j(1 - p_j),$$

and for $j, k = 1, 2, \ldots, r$ and $j \neq k$

$$(5.12.5) \qquad \lambda_{jk} = E[(X_j - np_j)(X_k - np_k)] = -np_j p_k.$$

5.13   COMPOUND DISTRIBUTIONS

A   In applications we often deal with a random variable X whose distribution depends on a parameter α which is a random variable with a specified distribution. Then, we say that the random variable X has a *compound distribution*. We shall investigate more closely two compound distributions, namely, a compound binomial distribution and a compound Poisson distribution.

Let the random variables X_k $(k = 1, 2, \ldots)$ be independent and have the zero-one distribution defined by the probabilities $P(X_k = 1) = p$ and $P(X_k = 0) = 1 - p$. Consider the random variable $X = X_1 + X_2 + \ldots + X_N$. For a fixed N, X has the binomial distribution

$$(5.13.1) \quad P(X = s) = \binom{N}{s} p^s (1 - p)^{N-s} \qquad (s = 0, 1, \ldots, N).$$

Let N be a random variable independent of X_k $(k = 1, 2, \ldots)$ with the Poisson distribution

$$(5.13.2) \qquad P(N = n) = \frac{\lambda^n}{n!} e^{-\lambda} \qquad (n = 0, 1, 2, \ldots).$$

As we see, here N plays the role of the parameter α mentioned previously. Consider the two-dimensional random variable (X, N). We have (see Section 2.7)

$$P(X = s, N = n) = P(X = s \mid N = n)P(N = n).$$

We are interested in the probability of the event $X = s$ for every s; in other words, we want to find the marginal distribution of X. This distribution is given by the formula

$$P(X = s) = \sum_{n=0}^{\infty} P(X = s \mid N = n)P(N = n) \qquad (s = 0, 1, 2, \ldots).$$

Considering (5.13.1), (5.13.2), and the fact that $\binom{n}{s} = 0$ for $n < s$, we obtain

$$(5.13.3) \quad P(X = s) = \sum_{n=0}^{\infty} e^{-\lambda} \frac{\lambda^n}{n!} \binom{n}{s} p^s (1 - p)^{n-s}$$

$$= \frac{e^{-\lambda} p^s}{s!} \sum_{n=s}^{\infty} \frac{\lambda^n (1 - p)^{n-s}}{(n - s)!} = \frac{e^{-\lambda} p^s \lambda^s}{s!} \sum_{n=s}^{\infty} \frac{[\lambda(1 - p)]^{n-s}}{(n - s)!}$$

$$= \frac{e^{-\lambda} p^s \lambda^s}{s!} e^{\lambda(1-p)} = \frac{e^{-\lambda p}(\lambda p)^s}{s!}.$$

We have obtained the Poisson distribution with expected value equal to $p\lambda$. This distribution is called a *compound binomial distribution*.

Example 5.13.1. The probability that a newborn baby will be a boy is $p = 0.517$ (see example 1.1.3). The number X of boys in a family with N children (N constant) is a random variable with the binomial distribution

$$P(X = s) = \binom{N}{s} 0.517^s \cdot 0.483^{N-s} \qquad (s = 0, 1, \ldots, N).$$

We might want the probability that $X = s$ for all possible values of N, that is, the probability that there will be s boys in a family with an arbitrary number of children. Here N is a random variable with a certain distribution which can be determined empirically for a given population by establishing the fraction of families with no children, with one child, and so on. If we know the distribution of N, we can calculate the probability $P(X = s)$ for every s in a manner similar to the derivation of (5.13.3).

B We now investigate a *compound Poisson distribution*.

Let the random variable X have the Poisson distribution given by the formula

$$(5.13.4) \qquad P(X = k) = \frac{\lambda^k}{k!} e^{-\lambda} \qquad (k = 0, 1, 2, \ldots),$$

and let λ be a random variable of the continuous type with the density

$$(5.13.5) \qquad f(\lambda) = \begin{cases} \dfrac{a^v}{\Gamma(v)} \lambda^{v-1} e^{-a\lambda} & \text{for } \lambda > 0, \\ 0 & \text{for } \lambda \leqslant 0, \end{cases}$$

where $v > 0$, $a > 0$.

Now consider the two-dimensional random variable (X, λ). Here one of the random variables is discrete and the other is continuous. For every $h > 0$ and $\lambda_1 > 0$ we have

$$P(X = k, \lambda_1 \leqslant \lambda \leqslant \lambda_1 + h) =$$
$$P(X = k \mid \lambda_1 \leqslant \lambda \leqslant \lambda_1 + h)P(\lambda_1 \leqslant \lambda \leqslant \lambda_1 + h).$$

Leι us divide both sides of this equality by h and pass to the limit as $h \to 0$. From (2.3.8) and (5.13.4) we obtain

$$(5.13.6) \quad \lim_{h \to 0} \frac{1}{h} P(X = k, \lambda_1 \leqslant \lambda \leqslant \lambda_1 + h) = \frac{\lambda_1^k}{k!} e^{-\lambda_1} \frac{a^v}{\Gamma(v)} \lambda_1^{v-1} e^{-a\lambda_1}.$$

Expression (5.13.6) determines the two-dimensional distribution of (X, λ). Writing on the right-hand side of (5.13.6) λ in place of λ_1, we obtain the marginal distribution of X from the formula

$$P(X = k) = \int_0^\infty \frac{\lambda^k}{k!} e^{-\lambda} \frac{a^v}{\Gamma(v)} \lambda^{v-1} e^{-a\lambda} \, d\lambda.$$

From (5.8.5) we obtain

$$(5.13.7) \quad P(X = k) = \frac{a^v}{\Gamma(v)} \int_0^\infty \frac{\lambda^{k+v-1} e^{-(a+1)\lambda}}{k!} \, d\lambda = \frac{a^v}{\Gamma(v)} \cdot \frac{1}{k!} \cdot \frac{\Gamma(k+v)}{(a+1)^{k+v}}$$

$$= \left(\frac{a}{1+a}\right)^v \frac{v(v+1)\ldots(v+k-1)}{(1+a)^k k!}.$$

For simplicity in notation we generalize the symbol $\binom{n}{r}$, which has been used only for positive integer values of n. For every x and every positive integer r we denote

$$\binom{x}{r} = \frac{x(x-1)\ldots(x-r+1)}{r!} = \frac{x^{(r)}}{r!}.$$

Further, let $p = 1/(1+a)$ and $q = 1 - p = a/(1+a)$. By assumption, we have $a > 0$, and hence the inequalities $0 < p < 1$ and $0 < q < 1$. Using this notation, we can write (5.13.7) in the form

$$(5.13.8) \quad P(X = k) = (-1)^k \binom{-v}{k} p^k q^v \qquad (k = 0, 1, 2, \ldots).$$

Definition 5.13.1. The compound Poisson distribution whose probability function is defined by formula (5.13.8), is called the *negative binomial distribution*.

Let us compute the characteristic function of this distribution,

$$\phi(t) = \sum_{k=0}^{\infty} P(X = k)e^{itk} = q^v \sum_{k=0}^{\infty} (-1)^k \binom{-v}{k} p^k e^{itk}.$$

Using Maclaurin's expansion,

$$(1 - p)^{-v} = \sum_{k=0}^{\infty} (-1)^k \binom{-v}{k} p^k, \qquad |p| < 1,$$

we obtain

(5.13.9) $$\phi(t) = q^v(1 - pe^{it})^{-v}.$$

It follows that

(5.13.10) $$m_1 = \frac{\phi'(0)}{i} = v\frac{p}{q}, \qquad m_2 = \frac{\phi''(0)}{i^2} = \left(\frac{p}{q}\right)^2 v(v + 1) + \frac{p}{q} v,$$

$$\mu_2 = m_2 - m_1^2 = v\frac{p}{q}\left(1 + \frac{p}{q}\right).$$

For the ordinary moment of order r we obtain the formula

(5.13.11) $$m_r = \sum_{l=0}^{r-1} (-1)^{r-l} \binom{r-1}{l} \left(\frac{p}{q}\right)^{r-l} (-v)_{r-l} \qquad (r = 1, 2, \ldots).$$

Greenwood and Yule [2] gave some examples of applications of the negative binomial distribution, of which one follows.

Example 5.13.2. The number of accidents among 414 machine operators was investigated for three successive months. The data are presented in Table 5.13.1. The symbol k in the first column denotes the number of accidents which happened

TABLE 5.13.1

k	Observed Frequency	Probability
0	0.715	0.722
1	0.179	0.167
2	0.063	0.063
3	0.019	0.027
4	0.010	0.012
5	0.010	0.005
6	0.002⎫	
7	0.000⎬	0.004
8	0.002⎭	

to the same operator during the period under investigation. In the second column are given the observed frequencies for the operators who had k accidents in the period under investigation, and in the third column, the corresponding probabilities calculated from formula (5.13.8). The unknown parameters v and p appearing in this formula were found in the following way: the expected value and the variance of the observed distribution were computed, and then it was assumed that they coincide with the values of $E(X)$ and μ_2 defined by (5.13.10). In this way two equations were obtained which make it possible to determine the unknown parameters.

As we see, the observed frequencies differ little from the computed probabilities. This can be explained as follows.

The probability that a machine operator will have k accidents during the period under investigation is determined by the Poisson distribution with parameter λ. The value of this parameter is influenced by many factors depending on time, such as the extent of the protective measures taken and the atmospheric conditions. We can regard λ as a random variable. Assuming that λ has a gamma distribution, it has been established that the observed and predicted frequencies are close to each other.

C   *Definition 5.13.2.*   The distribution of a random variable X given by (5.13.8), where v is an integer, is called the *Pascal distribution*.

This distribution can also be obtained from other considerations, not as a compound but as a simple distribution.

Consider a sequence of experiments. Suppose that as a result of an experiment either the event A or the event $\bar{A}$ may occur, and suppose that the results of the experiments are independent. We say there is a success if the event A occurs, and a failure if not. Suppose that $P(A) = p$; thus $P(\bar{A}) = 1 - p = q$.

Denote by X_r the number of successes following the $(r - 1)$th failure and preceding the rth failure. Thus, for instance, X_1 is the number of successive successes preceding the first failure, X_2 the number of successive successes after the first failure and before the second failure. Let us consider the random variable $X = X_1 + X_2 + \ldots + X_v$.

The event $X = k$ is the product of two events; the event that the $(k + v)$th experiment will lead to a failure and the event that among the remaining $k + v - 1$ experiments k will lead to successes. The probability of the first event is q and the probability of the second is

$$\binom{k + v - 1}{k} p^k q^{v-1}.$$

Hence

$$(5.13.12) \quad P(X = k) = \binom{k + v - 1}{k} p^k q^v = \frac{v(v + 1) \ldots (v + k - 1)}{k!} p^k q^v$$

$$= (-1)^k \binom{-v}{k} p^k q^v \quad (k = 0, 1, 2, \ldots).$$

This formula is identical with (5.13.8).

D   We now prove a theorem for the negative binomial distribution, which is analogous to theorem 5.5.1 for the binomial distribution.

Theorem 5.13.1. *If the equation*

$$E(X) = v\,\frac{p}{q} = c,$$

where c is a positive constant, is satisfied for every v, then the probability function of the negative binomial distribution tends to the corresponding function of the Poisson distribution as $v \to \infty$.

Proof. From (5.13.8) we have

$$(5.13.13)\quad \lim_{v \to \infty} P(X = k) = \lim_{v \to \infty} \frac{v(v + 1) \ldots (v + k - 1)}{k!}\, p^k(1 - p)^v$$

$$= \frac{c^k}{k!} \lim_{v \to \infty} \frac{v(v + 1) \ldots (v + k - 1)}{(v + c)^k} \left(1 - \frac{c}{v + c}\right)^v = \frac{e^{-c}c^k}{k!}\,.$$

Formula (5.13.13) allows us to apply tables of the Poisson distribution to a negative binomial distribution.

E   Consider now the random variable Y defined by the formula

$$(5.13.14)\qquad\qquad Y = \sum_{k=1}^{N} X_k,$$

where X_k ($k = 1, 2, \ldots$) and N are random variables and N takes on only positive integer values.

Theorem 5.13.2. *Let the random variable N be independent of the random variables $X_1, X_2, \ldots$. If the inequality*

$$(5.13.15)\qquad\qquad \sum_{k=1}^{\infty} P(N \geqslant k)E(|X_k|) < \infty$$

is satisfied, the expected value of the random variable Y exists and

$$(5.13.16)\qquad\qquad E(Y) = \sum_{k=1}^{\infty} P(N \geqslant k)E(X_k).$$

Proof. From (5.13.14) and (3.6.22) we obtain

$$E(Y) = \sum_{n=1}^{\infty} P(N = n)E(Y \mid N = n)$$

$$= \sum_{n=1}^{\infty} P(N = n) \sum_{k=1}^{n} E(X_k) = \sum_{k=1}^{\infty} E(X_k) \sum_{n=k}^{\infty} P(N = n)$$

$$= \sum_{k=1}^{\infty} E(X_k)P(N \geqslant k).$$

From (5.13.15) the theorem follows (see also Problem 5.23).

Suppose in addition that the random variables X_k have the same distribution. Then, assumption (5.13.15) is satisfied if $E(N)$ and the expected value $E(X)$ of X_k exist.

Here, formula (5.13.16) has the form

$$(5.13.17) \quad E(Y) = E(X) \sum_{k=1}^{\infty} P(N \geqslant k) = E(X) \sum_{k=1}^{\infty} kP(N = k) = E(X)E(N).$$

The reader will notice that the expected value of the compound binomial distribution satisfies relation (5.13.17).

Problems and Complements

5.1.(a) Show that if the random variables X_1 and X_2 have zero-one distributions and are uncorrelated, they are independent.

(b) Check whether this property holds for all two-point random variables.

5.2. The random variable X has the binomial distribution given by (5.2.1). Let

$$A_{nk} = \frac{P(X = k + 1)}{P(X = k)} \qquad (k = 0, 1, \ldots, n - 1).$$

Prove that (a) the expressions $A_{nk} - 1$ and $(n + 1)p - 1 - k$ are either both equal to zero, both positive, or both negative.

(b) $P(X = k)$ takes on its maximum value either at one point k_0, which satisfies the inequality

$$(n + 1)p - 1 < k_0 < (n + 1)p,$$

if $(n + 1)p$ is not an integer, or at two points $(n + 1)p - 1$ and $(n + 1)p$ if $(n + 1)p$ is an integer.

(c) for $k > (n + 1)p$

$$A_{nk} < \exp\left[-\frac{k + (n + 1)p}{n} \right].$$

5.3. Prove the equality

$$\sum_{l=0}^{k} \binom{n_1}{l} \binom{n_2}{k - l} = \binom{n_1 + n_2}{k}.$$

Hint: Use the addition theorem for random variables with binomial distributions.

5.4. Prove that

$$\sum_{m=k}^{n} \binom{n}{m} p^m (1 - p)^{n-m} = \frac{n!}{(k - 1)!\,(n - k)!} \int_0^p t^{k-1}(1 - t)^{n-k}\, dt.$$

Hint: Integrate by parts. Compare formulas (10.3.7) and (10.3.8).

5.5.(a) Prove that for arbitrary $\lambda_1 > 0$, $\lambda_2 > 0$, and non-negative integer k

$$\sum_{l=0}^{k} \frac{\lambda_1^l \lambda_2^{k-l}}{l!\,(k - l)!} = \frac{(\lambda_1 + \lambda_2)^k}{k!}.$$

(b) X_1 and X_2 are independent random variables with Poisson distributions with parameters λ_1 and λ_2, respectively, and $Z = X_2 - X_1$. Find $P(Z = k)$ ($k = 0, \pm1, \pm2, \ldots$), and prove that for the particular case $\lambda_1 = \lambda_2 = \lambda$

$$P(Z = k) = e^{-2\lambda}I_k(2\lambda),$$

where $I_k(2\lambda)$ is the Bessel function with a purely imaginary argument.

5.6. Let the random variable Z have a generalized binomial distribution. Show that (notation of Section 5.3) if the p_k are functions of n such that $\sum_{k=1}^{n} p_k = \lambda$ is fixed, and $\alpha_n = \max(p_1, \ldots, p_n)$ tends to zero as $n \to \infty$, then

$$\lim_{n \to \infty} P(Z = r) = e^{-\lambda}\frac{\lambda^r}{r!} \qquad (r = 0, 1, 2, \ldots).$$

(This result is due to Mises [5, 6]. For a stronger result, see Hodges and LeCam [1].)

5.7.(a) $F(x)$ is the distribution function of a random variable X with the zero-one distribution. Find the distribution function of the random variable $Y = F(X)$.

(b) Do the same for a random variable X with the binomial distribution.

(c) Do the same when X has the Poisson distribution.

5.8. The random variable X is said to have a *log normal distribution* if its density is of the form

$$f(x) = \begin{cases} 0 & (x \leqslant c), \\ \dfrac{1}{(x - c)\sigma\sqrt{2\pi}} \exp\left\{ -\dfrac{[\log(x - c) - m]^2}{2\sigma^2} \right\} & (x < c), \end{cases}$$

where c is a constant. Find $E(X)$ and $D^2(X)$.

5.9.(a) Prove that for every $x > 0$

$$\frac{1}{\sqrt{2\pi}} e^{-x^2/2}\left(\frac{1}{x} - \frac{1}{x^3}\right) < 1 - \Phi(x) < \frac{1}{x\sqrt{2\pi}} e^{-x^2/2},$$

where $\Phi(x)$ is the distribution function of the normal distribution $N(0; 1)$.

(b) Find the analogous inequality for $x < 0$.

5.10. The random variables X_i ($i = 1, 2, 3$) are independent and have the same distribution $N(0; 1)$. Find the distribution function of the random variable $Y = \max_{1 \leqslant i \leqslant 3} |X_i|$.

5.11. Let $X_1, \ldots, X_n$ be independent and have the same distribution function $F(x)$ and moments of arbitrary order. Let $a_1, \ldots, a_n$ and $b_1, \ldots, b_n$ satisfy the relations

$$\sum_{j=1}^{n} a_j = \sum_{j=1}^{n} b_j; \quad \sum_{j=1}^{n} a_j^2 = \sum_{j=1}^{n} b_j^2,$$

where the sequence $a_1, \ldots, a_n$ is not a permutation of the sequence $b_1, \ldots, b_n$. The distribution function $F(x)$ is normal if and only if the random variables L_1 and L_2 defined as

$$L_1 = \sum_{j=1}^{n} a_j X_j, \quad L_2 = \sum_{j=1}^{n} b_j X_j$$

have the same distribution. (Marcinkiewicz [1]).

5.12. Prove the theorem of Cramér [4] (see Section 5.7). *Hint:* Let $F_1(x_1)$, $F_2(x_2)$, $\phi_1(t)$, $\phi_2(t)$ denote the distribution functions and characteristic functions of the random variables X_1 and X_2, respectively, and $\Phi[(x - m)/\sigma)]$ the normal distribution function of the sum $X_1 + X_2$. From the inequality

$$F_1(x_1)F_2(x_2) \leqslant \Phi\left(\frac{x_1 + x_2 - m}{\sigma}\right),$$

we obtain for $x_1 < 0$

$$F_1(x_1) < A \exp\left[-\frac{x_1^2}{2\sigma^2} + B|x_1|\right]$$

and for $x_1 > 0$

$$1 - F_1(x_1) < A' \exp\left[-\frac{x_1^2}{2\sigma^2} + B'x_1\right].$$

From these inequalities follows the existence of the integral (see Section 6.5)

$$I = \int_{-\infty}^{+\infty} e^{x^2/4\sigma^2} \, dF_1(x).$$

Hence we find that $\phi_1(t)$, as a function of the imaginary argument t, is an entire function of order at most two. The same property holds for $\phi_2(t)$. We then use the theorem of Hadamard and the assumed normality of the random variable $X_1 + X_2$.

5.13. (A generalization of the theorem of Cramér obtained by Linnik [1] and Linnik and Zinger [1].) Let $\phi_j(t)$ $(j = 1, \ldots, n)$ be characteristic functions of nondegenerate distributions. If the equality

$$\prod_{j=1}^{n} [\phi_j(t)]^{\alpha_j} = \exp\left(imt - \tfrac{1}{2}\sigma^2 t^2\right)$$

holds for some collection of non-negative numbers $\alpha_1, \ldots, \alpha_n$, then each $\phi_j(t)$ is the characteristic function of a random variable with a normal distribution.

5.14. Let $X_1, \ldots, X_n$ be independent and have the same normal distribution. Prove that the sum $X_1 + \ldots + X_n$ is independent of the function $g(X_1, \ldots, X_n)$ if and only if $g(X_1, \ldots, X_n)$ and $g(X_1 + a, \ldots, X_n + a)$ have the same distribution for every a (Laha [3]).

5.15. Let $X_1, \ldots, X_n$ be independent and have the same nondegenerate distribution function $F(x)$. Let

$$U = X_1 + \ldots + X_n, \qquad V = \sum_{r=1}^{n} \sum_{s=1}^{n} a_{rs} X_r X_s,$$

$$B_1 = \sum_{r=1}^{n} a_{rr}, \quad B_2 = \sum_{r=1}^{n} \sum_{s=1}^{n} a_{rs}.$$

(*a*) If $B_1 \neq 0$ and $B_2 = 0$, then $F(x)$ is the normal distribution function if and only if V has constant regression (of the first type) on U, i.e., $E(V \mid u) = E(V)$, with probability 1.

(*b*) If $E(V) = 0$, $B_1 \neq 0$, and $B_2 \neq 0$, then $F(x)$ is the gamma distribution function if and only if V has constant regression on U. (See Lukacs [3].)

5.16. (Pitman [1]). If $X_1, \ldots, X_n$ are independent and have the same gamma distribution, then the sum $X_1 + \ldots + X_n$ is independent of any function

$g(X_1, \ldots, X_n)$ which is invariant under change of scale (that is, $g(X_1, \ldots X_n) = g(aX_1, \ldots, aX_n)$ for every $a > 0$).

5.17. Let $X_1, \ldots, X_n$ be independent and have the gamma distribution with $b = 1$. Prove that the sum $X_1 + \ldots + X_n$ is independent of the function $g(X_1, \ldots, X_n)$ if and only if $g(X_1, \ldots, X_n)$ is invariant under change of scale (Laha [3]).

5.18. The random variables X_1 and X_2 are independent and have the distribution $N(0; 1)$.

(a) Show that $Y = X_1/X_2$ has the Cauchy distribution.

(b) Show that this property does not characterize the normal distribution.

(c) Find the class of random variables for which the property given in (a) holds (Laha [6], Mauldon [1], Kotlarski [1]).

5.19. Give an example of a two-dimensional random variable (X, Y) which does not have the normal distribution and whose marginal distributions are normal.

(For a general discussion of multivariate distribution functions with given marginal distribution functions, see Fréchet [3] and Gumbel [3]).

5.20. Give an example of two random variables X and Y which have the normal distribution and are such that the sum $X + Y$ does not have a normal distribution. In particular, give an example such that $X + Y$ has a discrete distribution.

5.21. Prove that if the random variables $X_1, \ldots, X_n$ are independent and have the same density $f(x_1, \ldots, x_n)$ which is a function of $x_1^2 + \ldots + x_n^2$ only, then $X_1, \ldots, X_n$ have the normal distribution.

Hint: Compare with example 5.11.1.

5.22.(a) The random variable X is said to have a *Pareto distribution* if its density $f(x)$ is of the form

$$f(x) = \begin{cases} 0 & \text{for} \quad x \leqslant x_0, \\ \left(\dfrac{a}{x_0}\right)\left(\dfrac{x_0}{x}\right)^{a+1} & \text{for} \quad x > x_0, \end{cases}$$

where x_0 and $a > 0$ are constants. Find $E(X)$. Does it always exist? Find the median of X.

(b) A random variable X is said to have a *logistic distribution* if its distribution function is given by the formula

$$F(x) = [1 + e^{-(ax+b)}]^{-1}$$

with $a > 0$. Find the characteristic function and the moments of $F(x)$ and show that the density $f(x)$ of $F(x)$ satisfies the equation

$$f(x) = a F(x) [1 - F(x)].$$

(For the generalization of the logistic distribution to two dimensions, see Gumbel [4].)

5.23. Prove that if in theorem 5.13.2 we replace the assumption of independence of N and $(X_1, X_2, \ldots)$ by the assumption that the event $N = n$ is independent of the random variables X_k for $k > n$, then its assertion remains valid (Kolmogorov and Prohorov [1]).

5.24. The probability distributions whose densities $y = f(x)$ satisfy the differential equation

$$y' = \frac{x + a}{b_0 + b_1 x + b_2 x^2} y,$$

where a, b_0, b_1, and b_2 are constants, are called *Pearson probability distributions* (K. Pearson [6] to [8]).

(a) Prove that the normal, gamma, beta, and Pareto distributions are Pearson probability distributions.

(b) Show that the constants a, b_0, b_1, and b_2 can be expressed in terms of the moments (if they exist) of orders 1 to 4.

5.25. The random variable X is said to be *infinitely divisible* if, for every integer n, X can be represented as a sum $X = X_1 + \ldots + X_n$, where X_k ($k = 1$, $\ldots, n$) are independent and have the same distribution function $F_n(x)$ (which depends upon n). The distribution function of an infinitely divisible random variable is called an *infinitely divisible distribution function*.

(a) Define infinite divisibility in terms of characteristic functions.

(b) Show that random variables with the normal, gamma, Poisson, and Cauchy distribution are infinitely divisible.

(c) Is a random variable with the binomial distribution infinitely divisible?

5.26. Prove that the sum of an arbitrary finite number of independent infinitely divisible random variables is infinitely divisible.

5.27. Show that the characteristic function of an infinitely divisible random variable does not vanish.

Hint: Infinite divisibility implies that for every n $\phi(t) = [\phi_n(t)]^n$, where $\phi_n(t)$ is also a characteristic function. The functions $\phi(t)$ and $\phi_n(t)$ differ from zero in some neighborhood of the point $t = 0$, that is for $|t| \leqslant a$, where a is some positive number. Next, show that for any $\varepsilon > 0$ there exists an n such that for all $t(|t| \leqslant a)$ we have $1 - |\phi_n(t)| < \varepsilon$. Next, use the results given in Problems 4.5(a) and 4.7(b) to show that for any $a > 0$ one can find an n such that for all t with $|t| \leqslant a$ the expression $1 - |\phi_n(t)|$ is arbitrarily small. Next, extend this result to intervals $|t| \leqslant ra$ with arbitrary integer r.

5.28. Prove that if $\phi(t)$ is the characteristic function of an infinitely divisible distribution function, then $[\phi(t)]^c$ is a characteristic function for every $c > 0$.

(A comprehensive presentation of the theory of infinitely divisible distributions is given in Lévy [3] and in Gnedenko and Kolmogorov [1].)

CHAPTER 6

Limit Theorems

6.1 PRELIMINARY REMARKS

In the previous chapters we have several times dealt with limit theorems for sequences of probability functions. In theorem 5.4.1 we dealt with the convergence of a sequence of probability functions of the Pólya distribution to the corresponding function of the binomial distribution, and in theorem 5.5.1 with the convergence of a sequence of probability functions of the binomial distribution to the corresponding function of the Poisson distribution. These theorems are special cases of limit theorems which have great theoretical, as well as practical, importance.

We distinguish two kinds of limit theorems: local theorems and integral theorems. In local theorems we investigate either the limit of a sequence of probability functions of a discrete distribution, as in theorems 5.4.1 and 5.5.1, or the limit of a sequence of density functions of random variables of the continuous type. In integral limit theorems we investigate the limit of a sequence of distribution functions. In this chapter we deal almost exclusively with limit distributions for sums of independent random variables. We present some important theorems on the convergence of sequences of distribution functions of standardized sums of independent random variables to the normal distribution function. These are the theorems of de Moivre-Laplace (integral), Lindeberg-Lévy, Lapunov, and Lindeberg-Feller. We also present Gnedenko's local limit theorem for the convergence of probability functions of sums of independent random variables, which take on only values of the form $a + kh$ ($h > 0$, k an integer), to the density function of the normal distribution. This theorem contains the de Moivre-Laplace local theorem as a particular case. Besides theorems on convergence to the normal distribution, we also present theorems, called the laws of large numbers, in which the limit distribution is the one-point distribution.

175

The modern theory of limit distributions for sums of independent random variables has developed greatly during the last thirty years, due mainly to Khintchin [2], [3], Gnedenko [2] to [4], Kolmogorov [1], [2], and [5], and Lévy [1] to [3]. A uniform general theory has been developed, in which the limit theorems presented in this book are only particular cases of general theorems which give conditions for the convergence of a sequence of distribution functions of sums (much more general than the sums considered here) to a limit distribution function and establish the set of all possible limit distributions.

The reader will find a detailed discussion of this theory in the books by Lévy [3] Gnedenko and Kolmogorov [1], and Loève [4].

The question of the convergence of a sequence of distribution functions for dependent random variables is also extremely interesting. Investigations in this domain were originated by Markov [1] and [3]. Bernstein [3] has obtained some important results. We return to these questions in Section 7.5.

6.2 STOCHASTIC CONVERGENCE

A   Consider the following example.

Example 6.2.1. The random variable Y_n can take on the values

$$0, \frac{1}{n}, \frac{2}{n}, \ldots, \frac{n-1}{n}, 1$$

and its probability function is given by the formula

(6.2.1) $$P\left(Y_n = \frac{r}{n}\right) = \binom{n}{r}\frac{1}{2^n} \qquad (r = 0, 1, \ldots, n).$$

Consider the random variable X_n defined by the formula

(6.2.2) $$X_n = Y_n - \tfrac{1}{2}.$$

Thus X_n can take on the values

$$-\frac{1}{2}, \frac{2-n}{2n}, \frac{4-n}{2n}, \ldots, \frac{n-4}{2n}, \frac{n-2}{2n}, \frac{1}{2}.$$

The probability function of X_n is given by the formula

$$P\left(X_n = \frac{2r-n}{2n}\right) = \binom{n}{r}\frac{1}{2^n}.$$

Let $n = 2$. The random variable X_2 can take on the values

$$-0.5, \quad 0, \quad 0.5$$

with the respective probabilities

$$\tfrac{1}{4}, \ \tfrac{1}{2}, \ \tfrac{1}{4}.$$

Let ε be a positive number, say $\varepsilon = 0.3$. We see that

$$P(|X_2| > 0.3) = P(X_2 = -\tfrac{1}{2}) + P(X_2 = \tfrac{1}{2}) = 0.5.$$

Now let $n = 5$. The random variable X_5 can take on the values

$$-0.5, \; -0.3, \; -0.1, \, 0.1, \, 0.3, \, 0.5$$

with the respective probabilities

$$\tfrac{1}{32}, \; \tfrac{5}{32}, \; \tfrac{10}{32}, \; \tfrac{10}{32}, \; \tfrac{5}{32}, \; \tfrac{1}{32}.$$

Hence

$$P(|X_5| > 0.3) = 0.0625.$$

Now let $n = 10$. The random variable X_{10} can take on the values

$$-0.5, \; -0.4, \; -0.3, \; -0.2, \; -0.1, \, 0.0, \, 0.1, \, 0.2, \, 0.3, \, 0.4, \, 0.5$$

with the respective probabilities

$$\tfrac{1}{1024}, \; \tfrac{10}{1024}, \; \tfrac{45}{1024}, \; \tfrac{120}{1024}, \; \tfrac{210}{1024}, \; \tfrac{252}{1024}, \; \tfrac{210}{1024}, \; \tfrac{120}{1024}, \; \tfrac{45}{1024}, \; \tfrac{10}{1024}, \; \tfrac{1}{1024}.$$

Hence

$$P(|X_{10}| > 0.3) \simeq 0.02.$$

We see that for $n = 10$ the probability that X_n will exceed $\varepsilon = 0.3$ in absolute value is very small.

From the theorem which is proved in Section 6.3, it follows that in our example

(6.2.3)
$$\lim_{n \to \infty} P(|X_n| > 0.3) = 0$$

and, moreover, that for the sequence of random variables X_n defined by formula (6.2.2), relation (6.2.3) is satisfied for every $\varepsilon > 0$.

B   Before we present the theorem just mentioned, we define the notion of stochastic convergence.

Definition 6.2.1.   The sequence $\{X_n\}$ of random variables is called *stochastically convergent*[1] *to zero* if for every $\varepsilon > 0$ the relation

(6.2.4)
$$\lim_{n \to \infty} P(|X_n| > \varepsilon) = 0$$

is satisfied.

We notice that in this definition we say nothing about the convergence of the random variables X_n to zero in the sense which is understood in analysis. Thus, if the sequence $\{X_n\}$ is stochastically convergent to zero, it does not follow that for every $\varepsilon > 0$ we can find a finite n_0 such that for all $n > n_0$ the relation $|X_n| < \varepsilon$ will be satisfied. From the definition of stochastic convergence it follows only that the probability of the event $|X_n| \geq \varepsilon$ tends to zero as $n \to \infty$.

---

[1] The notion of stochastic convergence corresponds to the notion of convergence in measure in the theory of real functions.

Theorem 6.2.1. *Let $F_n(x)$ ($n = 1, 2, \ldots$) be the distribution function of the random variable X_n. The sequence $\{X_n\}$ is stochastically convergent to zero if and only if the sequence $\{F_n(x)\}$ satisfies the relation*

$$(6.2.5) \qquad \lim_{n \to \infty} F_n(x) = \begin{cases} 0 & \text{for } x \leqslant 0, \\ 1 & \text{for } x > 0. \end{cases}$$

Proof. Suppose that the sequence $\{X_n\}$ is stochastically convergent to zero.

From relation (6.2.4) it follows that for an arbitrary $\varepsilon > 0$ as $n \to \infty$ we have

$$(6.2.6) \qquad P(X_n < -\varepsilon) = F_n(-\varepsilon) \to 0,$$

$$(6.2.6') \qquad P(X_n > \varepsilon) = 1 - F_n(\varepsilon) - P(X_n = \varepsilon) \to 0.$$

Since for every $\varepsilon > 0$ we can find an ε_1 such that $0 < \varepsilon_1 < \varepsilon$, it follows from relation (6.2.4) that for an arbitrary $\varepsilon > 0$ we have $P(X_n = \varepsilon) \to 0$. Hence from (6.2.6') it follows that

$$(6.2.7) \qquad 1 - F_n(\varepsilon) \to 0.$$

Replacing ε by $-x$ in formula (6.2.6) and by x in formula (6.2.7), where $x > 0$, we obtain (6.2.5).

Suppose now that (6.2.5) is satisfied. Then for arbitrary $\varepsilon > 0$ we have

$$\lim_{n \to \infty} P(X_n < -\varepsilon) = \lim_{n \to \infty} F_n(-\varepsilon) = 0,$$

$$\lim_{n \to \infty} P(X_n > \varepsilon) \leqslant \lim_{n \to \infty} P(X_n \geqslant \varepsilon) = \lim_{n \to \infty} [1 - F_n(\varepsilon)] = 0.$$

Relation (6.2.4) follows immediately from the last two relations, which proves the theorem.

We remind the reader that the random variable X with a one-point distribution such that $P(X = 0) = 1$, has the distribution function

$$(6.2.8) \qquad F(x) = \begin{cases} 0 & \text{for } x \leqslant 0, \\ 1 & \text{for } x > 0, \end{cases}$$

and this distribution function is continuous at every point $x \neq 0$. From relations (6.2.6) and (6.2.7) it follows that for every point $x \neq 0$ the sequence of distribution functions $F_n(x)$ converges (in the usual sense) to the distribution function $F(x)$ defined by formula (6.2.8). We conclude that the sequence of distribution functions $F_n(x)$ of random variables, convergent stochastically to zero, converges to the distribution function of the one-point distribution at every point $x \neq 0$. Since the points $x \neq 0$ are continuity points of this distribution function, we can formulate the preceding result in the following way.

The sequence $\{X_n\}$ of random variables is stochastically convergent to zero if and only if the sequence $\{F_n(x)\}$ of distribution functions of these random variables is convergent (in the usual sense) to the distribution function $F(x)$ given by (6.2.8) at every continuity point of the latter.

We stress the fact that at the point of discontinuity of $F(x)$, that is, at the point $x = 0$, the sequence $\{F_n(0)\}$ may not converge to $F(0)$.

We can also consider the *stochastic convergence* of a sequence of random variables $\{X_n\}$ *to a constant* $c \neq 0$. This will mean that the sequence of random variables $\{Y_n\} = \{X_n - c\}$ is stochastically convergent to zero.

Similarly, we can define the *stochastic convergence* of a sequence of random variables $\{X_n\}$ *to a random variable* X. This will mean that the sequence of random variables $\{Z_n\} = \{X_n - X\}$ is stochastically convergent to zero.

6.3 BERNOULLI'S LAW OF LARGE NUMBERS

We now prove the theorem stated in Section 6.2, of which formula (6.2.3) is a particular case. Denote by $\{Y_n\}$ the sequence of random variables with probability functions given by the formula

$$(6.3.1) \qquad P\left(Y_n = \frac{r}{n}\right) = \binom{n}{r} p^r (1 - p)^{n-r},$$

where $0 < p < 1$ and r can take on the values $0, 1, 2, \ldots, n$. Further denote

$$(6.3.2) \qquad X_n = Y_n - p.$$

Theorem 6.3.1. *The sequence of random variables* $\{X_n\}$ *given by* (6.3.1) *and* (6.3.2) *is stochastically convergent to* 0, *that is, for any* $\varepsilon > 0$ *we have*

$$(6.3.3) \qquad \lim_{n \to \infty} P(|X_n| > \varepsilon) = 0.$$

Proof. We shall use the Chebyshev inequality in the proof. By equalities (5.2.8) we have

$$(6.3.4) \qquad E(X_n) = 0,$$

$$(6.3.5) \qquad \sigma_n = \sqrt{D^2(X_n)} = \sqrt{p(1 - p)/n}.$$

Substituting (6.3.4) and (6.3.5) into the Chebyshev inequality, we obtain

$$(6.3.6) \qquad P(|X_n| > k\sqrt{p(1 - p)/n}) \leqslant \frac{1}{k^2},$$

where k is an arbitrary positive number. Set

$$k = \varepsilon \sqrt{n/p(1 - p)}$$

in inequality (6.3.6). We then obtain the inequality

$$(6.3.7) \qquad P(|X_n| > \varepsilon) \leqslant \frac{p(1-p)}{n\varepsilon^2} < \frac{1}{n\varepsilon^2}.$$

From inequality (6.3.7) it follows that for every $\varepsilon > 0$ we have (6.3.3), which was to be proved.

The theorem just proved is called the *Bernoulli law of large numbers*. This law can be interpreted in practice as follows.

We perform n experiments according to the Bernoulli scheme, where the probability of the event A is equal to p. The law of large numbers states that, for large values of n, the probability that the observed frequency of A will differ little from p is close to one.

In the following sections we investigate other laws of large numbers.

6.4 THE CONVERGENCE OF A SEQUENCE OF DISTRIBUTION FUNCTIONS

A   In Section 6.2 we considered a sequence of distribution functions which converges to the distribution function (6.2.8) of the one-point distribution at every continuity point of this distribution.

We now investigate the question of convergence of sequences of distribution functions generally.

Definition 6.4.1. The sequence $\{F_n(x)\}$ of distribution functions of the random variables $\{X_n\}$ is called *convergent*, if there exists a distribution function $F(x)$ such that, at every continuity point of $F(x)$, the relation

$$(6.4.1) \qquad \lim_{n \to \infty} F_n(x) = F(x)$$

is satisfied. The distribution function $F(x)$ is called the *limit distribution function*.

As we see, it is not required that the sequence $\{F_n(x)\}$ converge to $F(x)$ at the discontinuity points of $F(x)$.

Let us return to example 6.2.1. By theorem 6.3.1, the sequence $\{X_n\}$ of random variables defined by formula (6.2.2) is stochastically convergent to zero; thus the sequence $\{F_n(x)\}$ of their distribution functions converges to the distribution function $F(x)$ defined by formula (6.2.8). This distribution function is discontinuous at $x = 0$. It is easy to verify that the sequence of numbers $\{F_n(0)\}$ is not convergent to $F(0)$.

Consider the subsequence of the sequence $\{F_n(0)\}$ containing only terms with the odd indices $n = 2k + 1$. The random variable X_{2k+1} can take on the values

$$-\frac{1}{2}, \frac{2-(2k+1)}{2(2k+1)}, \frac{4-(2k+1)}{2(2k+1)}, \ldots, \frac{2k+1-4}{2(2k+1)}, \frac{2k+1-2}{2(2k+1)}, \frac{1}{2}.$$

For every k, half of these terms are each less than zero, the other half greater than zero. The probability that X_{2k+1} will take on a value less than zero equals 0.5. Thus, for every k we have $P(X_{2k+1} < 0) = F_{2k+1}(0) = 0.5$. Since $F(0) = 0$, we have

$$(6.4.2) \qquad \lim_{k \to \infty} F_{2k+1}(0) = 0.5 \neq F(0).$$

From (6.4.2) it follows that $\lim_{n \to \infty} F_n(0) \neq F(0)$. Nevertheless, by the definition of the convergence of a sequence of distribution functions, the sequence of distribution functions of example 6.2.1 is convergent to the distribution function given by formula (6.2.8).

It is important to note that we speak about the convergence of a sequence of distribution functions only when it is convergent to a distribution function. This remark is important since it may happen that a sequence of distribution functions converges to a function that is not a distribution function.

Example 6.4.1. Let us consider the sequence $\{X_n\}$ of random variables with the one-point distributions given by the formula

$$P(X_n = n) = 1 \quad (n = 1, 2, \ldots).$$

The distribution function $F_n(x)$ of X_n is of the form

$$F_n(x) = \begin{cases} 0 & \text{for } x \leqslant n, \\ 1 & \text{for } x > n. \end{cases}$$

We have the relation

$$\lim_{n \to \infty} F_n(x) = 0 \quad (-\infty < x < \infty).$$

Thus the sequence $\{F_n(x)\}$ is not convergent to a distribution function.

Let the sequence $\{F_n(x)\}$ be convergent to the distribution function $F(x)$. Let a and b, where $a < b$, be two arbitrary continuity points of the limit distribution function $F(x)$. Then we have

$$(6.4.3) \qquad \lim_{n \to \infty} P(a \leqslant X_n < b) = F(b) - F(a).$$

In fact,

$$(6.4.4) \qquad P(a \leqslant X_n < b) = F_n(b) - F_n(a).$$

From the assumption that a and b are continuity points of the distribution function $F(x)$ it follows that

$$(6.4.5) \qquad F_n(b) \to F(b), \qquad F_n(a) \to F(a).$$

From (6.4.4) and (6.4.5) follows (6.4.3).

Let the sequence $\{F_n(x)\}$ be convergent to the distribution function $F(x)$. Let $P_n(S)$ and $P(S)$ denote the probability functions corresponding respectively to the distribution functions $F_n(x)$ and $F(x)$. It can be shown that for an arbitrary Borel set S on the real line R such that $P(\bar{S} \cap \overline{R - S}) = 0$ (here $\bar{A}$ denotes the closure of the set A) we have the relation (see Problem 6.7)

$$(6.4.6) \qquad \lim_{n \to \infty} P_n(S) = P(S).$$

We observe that even when the limit distribution function is everywhere continuous, Borel sets S may exist, for which (6.4.6) is not satisfied. This will happen if $P(\bar{S} \cap \overline{R - S}) > 0$. The following example is due to Robbins [2].

Example 6.4.2. The random variable $X_n (n = 1, 2, \ldots)$ has the density $f_n(x)$ given by

$$f_n(x) = \begin{cases} \dfrac{2^n}{\varepsilon} & \text{if } \dfrac{i}{n} - \dfrac{\varepsilon}{(n2^n)} < x < \dfrac{i}{n} & (i = 1, \ldots, n), \\ 0 & \text{otherwise,} \end{cases}$$

where $0 < \varepsilon < 1$. The distribution function $F_n(x)$ of X_n is then, for $i = 1, \ldots, n$, of the form

$$(6.4.7) \qquad F_n(x) = \begin{cases} 0 & \text{if } x \leqslant 0, \\[2mm] \dfrac{i - 1}{n} & \text{if } \dfrac{i - 1}{n} \leqslant x \leqslant \dfrac{i}{n} - \dfrac{\varepsilon}{(n2^n)}, \\[4mm] \dfrac{i - 1}{n} + \dfrac{2^n\left(x - \dfrac{i}{n} + \dfrac{\varepsilon}{n2^n}\right)}{\varepsilon} \\[4mm] \qquad \text{if } \dfrac{i}{n} - \dfrac{\varepsilon}{n2^n} < x < \dfrac{i}{n}, \\[4mm] 1 & \text{if } x \geqslant 1. \end{cases}$$

Thus for every x in the interval $I = [0, 1]$ we have

$$(6.4.8) \qquad 0 \leqslant x - F_n(x) \leqslant \frac{1}{n}.$$

By considering the values taken by $F_n(x)$ outside the interval I, we obtain for every real x

$$(6.4.9) \qquad \lim_{n \to \infty} F_n(x) = F(x) = \begin{cases} 0 & \text{if } x \leqslant 0, \\ x & \text{if } 0 < x < 1, \\ 1 & \text{if } x \geqslant 1. \end{cases}$$

Let us denote by S_n the set on which $f_n(x) > 0$, and by S_∞ the Borel set defined as

$$S_\infty = \sum_{n=1}^{\infty} S_n.$$

Let $P_n(S)$ and $P(S)$ denote the probability functions which correspond to the distribution functions $F_n(x)$ and $F(x)$, respectively. We have for $n = 1, 2, \ldots$

$$P_n(S_n) = \int_{S_n} f_n(x)\, dx = 1.$$

Since $S_\infty > S_n$ we obtain $P_n(S_\infty) = 1$; hence,

$$\lim_{n \to \infty} P_n(S_\infty) = 1.$$

On the other hand,

(6.4.10) $$P(S_\infty) = \int_{S_\infty} dx \leqslant \sum_{n=1}^{\infty} \left(n\, \frac{\varepsilon}{n2^n} \right) = \varepsilon < 1;$$

thus

$$\lim_{n \to \infty} P_n(S_\infty) \neq P(S_\infty),$$

despite the fact that the distribution function $F(x)$ given by (6.4.9) is everywhere continuous. This is because $P(\bar{S}_\infty \cap \overline{R - S_\infty}) > 0$. Indeed[1], we have $P(\bar{S}_\infty \cap \overline{R - S_\infty}) = P(\bar{S}_\infty \cap \overline{I - S_\infty})$, and the set $I - S_\infty$ is perfect and nowhere dense in I; hence $\overline{I - S_\infty} = \overline{I - S_\infty}$ and $\overline{I - (I - S_\infty)} = \bar{S}_\infty = I$. Thus we obtain, using (6.4.10),

$$P(\bar{S}_\infty \cap \overline{I - S_\infty}) = P(I \cap I - S_\infty) = P(I - S_\infty) \geqslant 1 - \varepsilon.$$

B   We now give the generalization of definition 6.4.1 to random vectors.

Definition 6.4.2.   The sequence of distribution functions $\{F_n(x_1, \ldots, x_k)\}$ of random vectors $(X_{n1}, X_{n2}, \ldots, X_{nk})$ is *convergent* if there exists a distribution function $F(x_1, \ldots, x_k)$ such that at every one of its continuity points

(6.4.11) $$\lim_{n \to \infty} F_n(x_1, x_2, \ldots, x_k) = F(x_1, x_2, \ldots, x_k).$$

It is not difficult to show that if (6.4.11) holds, and $P_n(S)$ and $P(S)$ denote the respective probability functions, then for every Borel set in k-dimensional Euclidean space R^k such that $P(\bar{S} \cap \overline{R^k - S}) = 0$ relation (6.4.6) holds. This relation holds, in particular, for continuity intervals (see Section 2.5, C).

C   The following theorem has important applications (Sverdrup [1], Prohorov [3]). We present it without proof.

Theorem 6.4.1.   Let $\{F_n(x_1, \ldots, x_k)\}(n = 1, 2, \ldots)$ *be a sequence of distribution functions of random vectors* $(X_{n1}, \ldots, X_{nk})$ *and let* $F(x_1, \ldots, x_k)$ *and* $P(S)$ *be the distribution function and probability function of a random vector* $(X_1, \ldots, X_k)$, *respectively. Relation* (6.4.11) *holds if and only if for*

[1] Information concerning the notions introduced here can be found in the book by Hausdorff [2].

every function $g(x_1, \ldots, x_k)$ *continuous on a set* S *satisfying the relation* $P(S) = 1$, *the equality*

$$\lim_{n \to \infty} H_n(\alpha) = H(\alpha)$$

holds at every continuity point α *of* $H(\alpha)$, *where* $H_n(\alpha)$ *and* $H(\alpha)$ *are the distribution functions of* $g(X_{n1}, \ldots, X_{nk})$ *and* $g(X_1, \ldots, X_k)$, *respectively.*

6.5 THE RIEMANN-STIELTJES INTEGRAL

In further considerations we use the theorem proved by Lévy [3] and Cramér [1] which makes it possible to investigate the convergence of a sequence of distribution functions $\{F_n(x)\}$ of random variables $\{X_n\}$ to a distribution function $F(x)$ by investigating the convergence of the sequence of characteristic functions $\phi_n(t)$ of the random variables X_n. This theorem plays an important role in probability theory. The proof of this theorem requires the notion of the Riemann-Stieltjes integral, which for simplicity is called the Stieltjes integral. It will be seen that distributions of random variables of the continuous and discrete types, considered separately, can be treated together by means of the Stieltjes integral.

We first introduce the notion of a function of *bounded variation*. Let $F(x)$ be a function defined in the interval $[a, b]$, which can be either finite or infinite. Let us take a partition of the interval $[a, b]$ with the points

$$a = x_0 < x_1 < x_2 < \ldots < x_n = b$$

and form the sum

$$T = \sum_{k=0}^{n-1} |F(x_{k+1}) - F(x_k)|.$$

The value of T may depend, of course, on the number n and on the partition into subintervals.

Definition 6.5.1. The least upper bound of the values of T is called the *total absolute variation* of the function $F(x)$ in the interval $[a, b]$.

Definition 6.5.2. If the total absolute variation of the function $F(x)$ in the interval $[a, b]$ is finite, we shall say that F is a *function of bounded variation on the interval* $[a, b]$.

It is easy to verify that every nondecreasing bounded function is of bounded variation. Indeed, here the expression $F(x_{k+1}) - F(x_k)$ is nonnegative for arbitrary k and arbitrary partition of the interval $[a, b]$, hence

$$T = \sum_{k=0}^{n-1} [F(x_{k+1}) - F(x_k)] = F(b) - F(a),$$

and our assertion then follows from the assumption that $F(b)$ and $F(a)$ are finite. It also follows that every distribution function $F(x)$ is a function of bounded variation, and the total absolute variation of $F(x)$ is $F(+\infty) - F(-\infty) = 1$.

We now introduce the notion of the Stieltjes integral. Suppose we are given two functions $g(x)$ and $F(x)$ in a finite interval $[a, b]$.

Let us form a partition of the interval $[a, b]$ into n parts with the points

$$a = x_0 < x_1 < \ldots < x_n = b.$$

Consider the sum

(6.5.1)
$$S = \sum_{k=0}^{n-1} g(x_k')[F(x_{k+1}) - F(x_k)],$$

where x_k' is an arbitrary point in the kth interval (x_k, x_{k+1}).

Definition 6.5.3. If as $n \to \infty$ and $\max (x_{k+1} - x_k) \to 0$ the sum S tends to a finite limit I independent of the choice of the points x_k' and the partition of the interval $[a, b]$, this limit is called the *Stieltjes integral of the function $g(x)$ with respect to the function $F(x)$.*

We denote this integral as follows:

(6.5.2)
$$I = \lim_{n \to \infty} S = \int_a^b g(x)\, dF(x).$$

As we see, the Stieltjes integral is a generalization of the Riemann integral, since for $F(x) = x$ formula (6.5.2) represents the Riemann integral.

When the interval of integration is infinite, we define the improper Stieltjes integral as the limit of a sequence of proper Stieltjes integrals. Thus, if

$$\lim_{\substack{a \to -\infty \\ b \to +\infty}} \int_a^b g(x)\, dF(x)$$

exists as a and b tend arbitrarily to $-\infty$ and $+\infty$, respectively, this limit is called the *improper Stieltjes integral of the function $g(x)$ with respect to the function $F(x)$.*

We give without proof some of the properties of the Stieltjes integral.

1. If c and l are constants, then

$$\int_a^b cg(x)\, d[lF(x)] = cl \int_a^b g(x)\, dF(x).$$

2. If the integrals on the right-hand side of the next two equations exist, then the integrals on the left-hand side exist and

$$\int_a^b [g_1(x) + g_2(x)]\, dF(x) = \int_a^b g_1(x)\, dF(x) + \int_a^b g_2(x)\, dF(x),$$

$$\int_a^b g(x)\, d[F_1(x) + F_2(x)] = \int_a^b g(x)\, dF_1(x) + \int_a^b g(x)\, dF_2(x).$$

If $a < b < c$ and all three integrals

$$\int_a^b g(x)\, dF(x), \qquad \int_a^c g(x)\, dF(x), \qquad \int_c^b g(x)\, dF(x)$$

exist, the equation

$$\int_a^b g(x)\, dF(x) = \int_a^c g(x)\, dF(x) + \int_c^b g(x)\, dF(x)$$

holds.

Properties 2 are satisfied for sums of an arbitrary finite number of functions $g_i(x)$ and $F_i(x)$, and property 3 is satisfied for an arbitrary finite number of points $a < c_1 < c_2 < \ldots < c_n < b$.

In the theory of real functions it is proved that if $g(x)$ is continuous and bounded over the real axis and $F(x)$ is a function of bounded variation, both proper and improper Stieltjes integrals exist. However, the Stieltjes integral may exist when $g(x)$ is not bounded.

Suppose that the function $F(x)$ has the derivative $F'(x)$ integrable in the Riemann sense over the interval $[a, b]$. By the Lagrange formula we obtain

$$\int_a^b g(x)\, dF(x) = \lim_{n \to \infty} \sum_{k=0}^{n-1} g(x_k')F'(x_k')(x_{k+1} - x_k) = \int_a^b g(x)F'(x)\, dx.$$

Under these conditions the Stieltjes integral is reduced to the Riemann integral of the function $g(x)F'(x)$. In particular, if $F(x)$ is the distribution function of a random variable of the continuous type with the density function $f(x)$,

$$(6.5.3) \qquad \int_a^b g(x)\, dF(x) = \int_a^b g(x)f(x)\, dx.$$

Suppose that $F(x)$ is the distribution function of a random variable of the discrete type with jump points x_k' and jumps p_k ($k = 1, 2, \ldots$). Then $F(x)$ has the form

$$F(x) = \sum_{x_k' < x} [F(x_k' + 0) - F(x_k')].$$

By formulas (6.5.1) and (6.5.2) we obtain for $x_k' \in [a, b]$

$$(6.5.4) \qquad \int_a^b g(x)\, dF(x) = \sum_k g(x_k')p_k.$$

From formulas (6.5.3) and (6.5.4) it follows that for a random variable X of the continuous type as well as of the discrete type, if the expected value of the random variable $Y = g(X)$ exists, it can be expressed by the formula

(6.5.5)
$$E[g(X)] = \int_{-\infty}^{+\infty} g(x)\, dF(x),$$

where $F(x)$ denotes the distribution function of X.

The dual formulas used until now, in terms of integrals for random variables of the continuous type and of series for random variables of the discrete type, can now be written in a common formula by the use of Stieltjes integrals. Thus, for instance, setting $g(x) = x^r$ in formula (6.5.5), we obtain the general expression for the moment of the rth order (See also Supplement)

$$m_r = E(x^r) = \int_{-\infty}^{+\infty} x^r\, dF(x).$$

Similarly, setting $g(x) = e^{itx}$, we obtain the characteristic function

$$\phi(t) = \int_{-\infty}^{\infty} e^{itx}\, dF(x).$$

If the Stieltjes integral (6.5.2) exists and $g(x)$ is continuous at a and b, then it can be computed by integrating by parts, according to the formula

$$\int_a^b g(x)\, dF(x) = \Big[g(x)F(x) \Big]_a^b - \int_a^b F(x)\, dg(x).$$

We can also derive the following formulas for the distribution function of the sum, difference, product, and ratio of independent random variables X and Y without the assumption that the vector (X, Y) is of the continuous type. (For the proofs of these formulas, see Supplement.)

Let $F_1(x)$, $F_2(y)$, and $F(z)$ be the distribution functions of the random variables X, Y and Z, respectively. Then:

I If $Z = X + Y$,

(6.5.6)
$$F(z) = \int_{-\infty}^{+\infty} F_2(z - x)\, dF_1(x) = \int_{-\infty}^{+\infty} F_1(z - y)\, dF_2(y).$$

II If $Z = X - Y$,

(6.5.7)
$$F(z) = \int_{-\infty}^{+\infty} F_1(z + y)\, dF_2(y).$$

III If $Z = XY$ and $P(X = 0) = P(Y = 0) = 0$,

(6.5.8)
$$F(z) = \int_{-\infty}^{0} \left[1 - F_2\left(\frac{z}{x}\right) \right] dF_1(x) + \int_0^{\infty} F_2\left(\frac{z}{x}\right) dF_1(x)$$
$$= \int_{-\infty}^{0} \left[1 - F_1\left(\frac{z}{y}\right) \right] dF_2(y) + \int_0^{\infty} \left[F_1\left(\frac{z}{y}\right) \right] dF_2(y).$$

IV. If $Z = X/Y$ and $P(Y = 0) = 0$,

$$(6.5.9) \qquad F(z) = \int_{-\infty}^{0} [1 - F_1(zy)]\, dF_2(y) + \int_{0}^{\infty} F_1(zy)\, dF_2(y).$$

6.6. THE LÉVY-CRAMÉR THEOREM

We first present the Lévy-Cramér theorem in the form of two theorems.

Theorem 6.6.1a. *If the sequence* $\{F_n(x)\}(n = 1, 2, \ldots)$ *of distribution functions is convergent to the distribution function* $F(x)$, *then the corresponding sequence of characteristic functions* $\{\phi_n(t)\}$ *converges at every point* t ($-\infty < t < +\infty$) *to the function* $\phi(t)$ *which is the characteristic function of the limit distribution function* $F(x)$, *and the convergence to* $\phi(t)$ *is uniform with respect to* t *in every finite interval on the* t-*axis.*

Proof. From the definition of a characteristic function we have

$$\phi_n(t) = \int_{-\infty}^{\infty} e^{itx}\, dF_n(x), \qquad \phi(t) = \int_{-\infty}^{\infty} e^{itx}\, dF(x).$$

Let $a < 0$ and $b > 0$ be continuity points of the distribution function $F(x)$. We have

$$(6.6.1) \qquad \phi_n(t) = \int_{-\infty}^{a} e^{itx}\, dF_n(x) + \int_{a}^{b} e^{itx}\, dF_n(x) + \int_{b}^{\infty} e^{itx}\, dF_n(x)$$

$$= I_{n1} + I_{n2} + I_{n3},$$

$$\phi(t) = \int_{-\infty}^{a} e^{itx}\, dF(x) + \int_{a}^{b} e^{itx}\, dF(x) + \int_{b}^{\infty} e^{itx}\, dF(x)$$

$$= I_1 + I_2 + I_3.$$

Consider the difference

$$I_{n2} - I_2 = \int_{a}^{b} e^{itx}\, dF_n(x) - \int_{a}^{b} e^{itx}\, dF(x).$$

Integrating by parts, we obtain

$$I_{n2} - I_2 = e^{itx}\left\{\left[F_n(x)\right]_a^b - \left[F(x)\right]_a^b\right\} - it\int_{a}^{b} [F_n(x) - F(x)]e^{itx}\, dx;$$

hence

$$|I_{n2} - I_2| \leqslant |F_n(b) - F(b)| + |F_n(a) - F(a)| + |t|\int_{a}^{b} |F_n(x) - F(x)|\, dx.$$

Let $\varepsilon > 0$ be arbitrary. By the assumption of the theorem and by the fact that a and b are continuity points of $F(x)$, we obtain, for sufficiently large n,

$$|F_n(b) - F(b)| < \frac{\varepsilon}{9}, \qquad |F_n(a) - F(a)| < \frac{\varepsilon}{9}.$$

Furthermore, by the Lebesgue theorem on passage to the limit under the integral sign (see Sikorski [1]), by the assumption of the theorem, and by the fact that $|F_n(x) - F(x)|$ is uniformly bounded in every interval, we obtain

$$\lim_{n \to \infty} \int_a^b |F_n(x) - F(x)| \, dx = \int_a^b \lim_{n \to \infty} |F_n(x) - F(x)| \, dx.$$

Since the function under the integral sign on the right-hand side of the last formula is equal to zero except at most at a countable number of points, the integral under consideration is equal to zero. Suppose now that t satisfies the inequality $T_1 < t < T_2$, where T_1 and T_2 are arbitrary fixed numbers, and let K be the greater of the numbers $|T_1|$ and $|T_2|$, that is, $K = \max(|T_1|, |T_2|)$. Then, for sufficiently large n and all t under consideration, we have

$$|t| \int_a^b |F_n(x) - F(x)| \, dx \leqslant K \int_a^b |F_n(x) - F(x)| \, dx < \frac{\varepsilon}{9}.$$

Thus we obtain

(6.6.2) $$|I_{n2} - I_2| < \frac{\varepsilon}{3}.$$

Now consider the difference

$$I_{n1} - I_1 = \int_{-\infty}^a e^{itx} \, dF_n(x) - \int_{-\infty}^a e^{itx} \, dF(x).$$

We have

$$|I_{n1} - I_1| \leqslant \int_{-\infty}^a dF_n(x) + \int_{-\infty}^a dF(x) = F_n(a) + F(a).$$

Thus, if a is sufficiently large in absolute value, then, by the assumption of the theorem and the continuity of $F(x)$ at a, we have, for sufficiently large n,

$$F_n(a) < \frac{\varepsilon}{6}, \qquad F(a) < \frac{\varepsilon}{6}.$$

Hence for all t and sufficiently large n,

(6.6.3) $$|I_{n1} - I_1| < \frac{\varepsilon}{3}.$$

Similarly, we obtain that

(6.6.4) $$|I_{n3} - I_3| < \frac{\varepsilon}{3}.$$

The theorem follows from formulas (6.6.1) to (6.6.4).

Theorem 6.6.1b. *If the sequence of characteristic functions $\{\phi_n(t)\}$ converges at every point t $(-\infty < t < +\infty)$ to a function $\phi(t)$ continuous in* *some interval $|t| < \tau$, then the sequence $\{F_n(x)\}$ of corresponding distribution functions converges to the distribution function $F(x)$ which corresponds to the characteristic function $\phi(t)$.*

Proof. In the proof we use the Helly theorem (see Sikorski [1]), which states that every sequence of distribution functions $\{F_n(x)\}$ contains a subsequence $\{F_{n_k}(x)\}$ convergent to some nondecreasing function $F(x)$. The function $F(x)$ can be changed at its discontinuity points so that it becomes continuous from the left. It does not, however, follow from the Helly theorem that $F(x)$ is a distribution function. Of course, since $F(x)$ is the limit of distribution functions, we have $0 \leqslant F(x) \leqslant 1$, but we do not know whether $F(-\infty) = 0$ and $F(+\infty) = 1$. We show that the last relations are satisfied.

Suppose that

$$(6.6.5) \qquad \alpha = F(+\infty) - F(-\infty) < 1.$$

Since $\phi_n(t) \to \phi(t)$ and $\phi_n(0) = 1$, we have $\phi(0) = 1$. By the assumption that the function $\phi(t)$ is continuous, it follows that in some neighborhood of the origin $t = 0$ it will differ little from 1; thus for sufficiently small τ we have the inequality

$$(6.6.6) \qquad \frac{1}{2\tau}\left|\int_{-\tau}^{\tau}\phi(t)\,dt\right| > 1 - \frac{\varepsilon}{2} > \alpha + \frac{\varepsilon}{2},$$

where the number ε is chosen in such a way that $\alpha + \varepsilon < 1$. Since the subsequence $\{F_{n_k}(x)\}$ converges to $F(x)$, it follows from relation (6.6.5) that we can choose $a > 4/\varepsilon\tau$ such that a and $-a$ are continuity points of the limit distribution function, and a number K such that for $k > K$

$$\alpha_k = F_{n_k}(a) - F_{n_k}(-a) < \alpha + \frac{\varepsilon}{4}.$$

On the other hand, since $\phi_n(t) \to \phi(t)$, it follows from relation (6.6.6) that for sufficiently large k the inequality

$$(6.6.7) \qquad \frac{1}{2\tau}\left|\int_{-\tau}^{\tau}\phi_{n_k}(t)\,dt\right| > \alpha + \frac{\varepsilon}{2}$$

is satisfied.

We show that this inequality is not satisfied. Indeed, we have

$$\int_{-\tau}^{\tau}\phi_{n_k}(t)\,dt = \int_{-\tau}^{\tau}\left[\int_{-\infty}^{+\infty}e^{itx}\,dF_{n_k}(x)\right]dt = \int_{-\infty}^{+\infty}\left[\int_{-\tau}^{\tau}e^{itx}\,dt\right]dF_{n_k}(x).$$

Since $|e^{itx}| = 1$, we obtain

$$(6.6.8) \qquad \left|\int_{-\tau}^{\tau}e^{itx}\,dt\right| \leqslant 2\tau.$$

Moreover,

$$(6.6.9) \quad \left| \int_{-\tau}^{\tau} e^{itx} \, dt \right| = \left| \left[\frac{e^{itx}}{ix} \right]_{-\tau}^{\tau} \right| = \frac{2}{|x|} |\sin \tau x| \leqslant \frac{2}{|x|} < \frac{2}{a} \quad \text{for } |x| > a.$$

Divide the whole axis into two parts, namely, into the interval $|x| \leqslant a$ and the complement of this interval. We have

$$\left| \int_{-\infty}^{+\infty} \left(\int_{-\tau}^{\tau} e^{itx} \, dt \right) dF_{n_k}(x) \right|$$

$$\leqslant \left| \int_{|x| \leqslant a} \left(\int_{-\tau}^{\tau} e^{itx} \, dt \right) dF_{n_k}(x) \right| + \left| \int_{|x| > a} \left(\int_{-\tau}^{\tau} e^{itx} \, dt \right) dF_{n_k}(x) \right|.$$

Using inequality (6.6.8) for $|x| \leqslant a$ and inequality (6.6.9) for $|x| > a$, we obtain

$$(6.6.10) \quad \frac{1}{2\tau} \left| \int_{-\tau}^{\tau} \phi_{n_k}(t) \, dt \right| \leqslant \left| \int_{|x| \leqslant a} dF_{n_k}(x) \right| + \frac{1}{a\tau} \left| \int_{|x| > a} dF_{n_k}(x) \right|$$

$$\leqslant \alpha_k + \frac{1}{a\tau} \leqslant \alpha_k + \frac{\varepsilon}{4} < \alpha + \frac{\varepsilon}{4} + \frac{\varepsilon}{4} = \alpha + \frac{\varepsilon}{2}.$$

The last inequality contradicts inequality (6.6.7). Hence the function $F(x)$ is a distribution function. From theorem 6.6.1a it follows that $\phi(t)$ is its characteristic function.

We now prove that not only the subsequence $\{F_{n_k}(x)\}$, but the whole sequence $\{F_n(x)\}$ converges to $F(x)$. If this were not so there would be another subsequence $\{F_n(x)\}$ convergent to a limit function $\tilde{F}(x)$ different from $F(x)$. The previous reasoning implies that $\tilde{F}(x)$ is a distribution function and from theorem 6.6.1a it follows that $\tilde{F}(x)$ has the same characteristic function as $F(x)$. Hence, by theorem 4.5.1, $\tilde{F}(x) \equiv F(x)$. Thus every subsequence of the sequence $\{F_n(x)\}$ contains a subsequence convergent to the same distribution function $F(x)$; hence the sequence $\{F_n(x)\}$ converges to $F(x)$.

From theorems 6.6.1a and 6.6.1b we obtain immediately:

Theorem 6.6.1 (Lévy-Cramér). *Let $\{X_n\}$ $(n = 1, 2, \ldots)$ be a sequence of random variables and let $F_n(x)$ and $\phi_n(t)$ be respectively the distribution function and the characteristic function of X_n. Then the sequence $\{F_n(x)\}$ is convergent to a distribution function $F(x)$ if and only if the sequence $\{\phi_n(t)\}$ is convergent at every point t $(-\infty < t < +\infty)$ to a function $\phi(t)$ continuous in some neighborhood $|t| < \tau$ of the origin. The limit function $\phi(t)$ is then the characteristic function of the limit distribution function $F(x)$ and the convergence $\phi_n(t) \to \phi(t)$ is uniform in every finite interval on the t-axis.*

We observe that theorem 6.6.1 remains true if we assume the continuity of the limit function $\phi(t)$ only at the point $t = 0$ (see Cramér [2]). We also observe that in the general case of theorem 6.6.1 we cannot replace the convergence at every point t in the interval $(-\infty, +\infty)$ by convergence in some interval on the t-axis containing the origin. If, however, all the random variables X_n are uniformly bounded from above (or below), then for the sequence $\{F_n(x)\}$ of distribution functions to converge to a distribution function $F(x)$, it is sufficient that in some interval $|t| < \tau$ the sequence $\{\phi_n(t)\}$ is convergent to a function $\phi(t)$ continuous at the origin. This theorem was proved by Zygmund [2].

6.7 THE DE MOIVRE-LAPLACE THEOREM

A   We use theorem 6.6.1b in proving the de Moivre-Laplace theorem.

Denote by $\{X_n\}$ a sequence of random variables with the binomial distribution. For every n the random variable X_n can take on the values $0, 1, \ldots, n$ and its probability function is given by the formula

$$(6.7.1) \qquad P(X_n = r) = \binom{n}{r} p^r q^{n-r},$$

where $0 < p < 1$ and $q = 1 - p$.

As we know from formulas (5.2.4), we have

$$E(X_n) = np, \qquad D^2(X_n) = npq.$$

Consider the sequence $\{Y_n\}$ of standardized random variables

$$(6.7.2) \qquad Y_n = \frac{X_n - np}{\sqrt{npq}}.$$

We shall prove a limit theorem called the *de Moivre-Laplace theorem*.

Theorem 6.7.1.   *Let $\{F_n(y)\}$ be the sequence of distribution functions of the random variables Y_n defined by (6.7.2), where the X_n have the binomial distribution given by formula (6.7.1). If $0 < p < 1$, then for every y we have the relation*

$$(6.7.3) \qquad \lim_{n \to \infty} F_n(y) = \frac{1}{\sqrt{2\pi}} \int_{-\infty}^{y} e^{-u^2/2} \, du.$$

Proof.   According to formula (5.2.3), the characteristic function $\phi_x(t)$ of X_n has the form

$$(6.7.4) \qquad \phi_x(t) = (q + pe^{it})^n.$$

Thus by equality (4.2.17) the characteristic function $\phi_y(t)$ of the random variable Y_n is given by the formula

(6.7.5)
$$\phi_y(t) = \exp\left(-\frac{npit}{\sqrt{npq}}\right)\left[q + p\exp\left(\frac{it}{\sqrt{npq}}\right)\right]^n$$
$$= \left[q\exp\left(-\frac{pit}{\sqrt{npq}}\right) + p\exp\left(\frac{qit}{\sqrt{npq}}\right)\right]^n.$$

Let us expand the function e^{iz} in the neighborhood of $z = 0$ according to the Taylor formula for k terms with the remainder in the Peano form,

$$e^{iz} = \sum_{j=0}^{k}\frac{(iz)^j}{j!} + o(z^k).$$

We obtain

$$p\exp\left(\frac{qit}{\sqrt{npq}}\right) = p + it\sqrt{pq/n} - \frac{qt^2}{2n} + o\left(\frac{t^2}{n}\right),$$

$$q\exp\left(-\frac{pit}{\sqrt{npq}}\right) = q - it\sqrt{pq/n} - \frac{pt^2}{2n} + o\left(\frac{t^2}{n}\right),$$

where for every t we have

(6.7.6)
$$\lim_{n\to\infty} no\left(\frac{t^2}{n}\right) = 0.$$

Substituting these expressions in formula (6.7.5) and considering the fact that $p + q = 1$, we obtain

$$\phi_y(t) = \left[1 - \frac{t^2}{2n} + o\left(\frac{t^2}{n}\right)\right]^n.$$

Thus

$$\log\phi_y(t) = n\log\left[1 - \frac{t^2}{2n} + o\left(\frac{t^2}{n}\right)\right] = n\log(1 + z).$$

We observe that for every fixed t for sufficiently large n, we have $|z| < 1$. Thus we can write

$$\log\phi_y(t) = -\frac{t^2}{2} + no\left(\frac{t^2}{n}\right).$$

By (6.7.6) we obtain

$$\lim_{n\to\infty}\log\phi_y(t) = -\frac{t^2}{2}.$$

Hence

$$\lim_{n\to\infty}\phi_y(t) = e^{-t^2/2}.$$

We have thus established that the sequence of characteristic functions $\phi_y(t)$ of the standardized random variables Y_n given by formula (6.7.2) converges as $n \to \infty$ to the characteristic function (see Section 5.7) of a random variable with a normal distribution whose distribution function is given by the right-hand side of formula (6.7.3). By theorem 6.6.1b we immediately obtain formula (6.7.3). We observe that the convergence in formula (6.7.3) holds for every y, since the distribution function of the normal distribution has no discontinuity points.

The de Moivre-Laplace theorem is proved.

Let y_1 and y_2 be two arbitrary points with $y_1 < y_2$. From relation (6.7.3) it follows that

$$(6.7.7) \quad \lim_{n \to \infty} P(y_1 < Y_n < y_2) = \lim_{n \to \infty} [F_n(y_2) - F_n(y_1)] = \frac{1}{\sqrt{2\pi}} \int_{y_1}^{y_2} e^{-y^2/2} \, dy.$$

We shall rewrite the de Moivre-Laplace theorem in another form. By formula (6.7.2) we have

$$P(y_1 < Y_n < y_2) = P\left(y_1 < \frac{X_n - np}{\sqrt{npq}} < y_2\right)$$

$$= P(y_1 \sqrt{npq} + np < X_n < y_2 \sqrt{npq} + np).$$

Thus we obtain

$$\lim_{n \to \infty} P(y_1 \sqrt{npq} + np < X_n < y_2 \sqrt{npq} + np) = \frac{1}{\sqrt{2\pi}} \int_{y_1}^{y_2} e^{-y^2/2} \, dy.$$

Let

$$(6.7.8) \qquad x_1 = y_1\sqrt{npq} + np, \qquad x_2 = y_2\sqrt{npq} + np.$$

We can write formula (6.7.7) in the asymptotic form

$$P(x_1 < X_n < x_2) \cong \frac{1}{\sqrt{2\pi}} \int_{y_1}^{y_2} e^{-y^2/2} \, dy,$$

where y_1 and y_2 are determined by (6.7.8).

We say that the random variable X_n has an *asymptotically normal distribution* $N(np; \sqrt{npq})$.

Replacing y_1 and y_2 with

$$y_1 + \frac{1}{2\sqrt{npq}} \quad \text{and} \quad y_2 - \frac{1}{2\sqrt{npq}},$$

respectively, we get a somewhat better approximation.

Example 6.7.1. We throw a coin $n = 100$ times. We assign the number 1 to the appearance of heads and the number 0 to the appearance of tails. The probability of each of these events is equal to $p = q = 0.5$. What is the probability that heads will appear more than 50 times and less than 60 times?

The random variable X_n can here take on values from 0 to 100. We have

$$E(X_n) = 50, \quad D^2(X_n) = 25,$$

$$P(50 < X_n < 60) = P\left(\frac{50 - 50}{5} < \frac{X_n - 50}{5} < \frac{60 - 50}{5}\right)$$

$$= P\left(0 < \frac{X_n - 50}{5} < 2\right) \cong \frac{1}{\sqrt{2\pi}} \int_{0.1}^{1.9} e^{-t^2/2}\, dt.$$

From tables of the normal distribution we obtain that the value of this integral is 0.4315.

B   From the de Moivre-Laplace limit theorem we obtain an analogous theorem for the sequence of random variables

$$U_n = \frac{X_n}{n},$$

where X_n has the binomial distribution given by formula (6.7.1).

Indeed, since $E(U_n) = p$ and $D^2(U_n) = pq/n$, we obtain the relation

$$Z_n = \frac{U_n - p}{\sqrt{pq/n}} = \frac{X_n - np}{\sqrt{npq}} = Y_n,$$

where the random variables Y_n are defined by formula (6.7.2).

Since the sequence $\{F_n(y)\}$ of distribution functions of Y_n satisfies formula (6.7.3), we obtain for the sequence $\{F_n(z)\}$ of the distribution functions of Z_n

$$\lim_{n \to \infty} F_n(z) = \frac{1}{\sqrt{2\pi}} \int_{-\infty}^{z} e^{-z^2/2}\, dz.$$

Similarly, for every pair of constants z_1 and z_2, where $z_1 < z_2$, we obtain the relation

(6.7.9) $\lim_{n \to \infty} P(z_1 < \sqrt{n/pq}(U_n - p) < z_2) = \dfrac{1}{\sqrt{2\pi}} \displaystyle\int_{z_1}^{z_2} e^{-z^2/2}\, dz.$

Letting

(6.7.10) $u_1 = z_1\sqrt{pq/n} + p, \quad u_2 = z_2\sqrt{pq/n} + p,$

we can rewrite formula (6.7.9) in the asymptotic form

(6.7.11) $P(u_1 < U_n < u_2) \cong \dfrac{1}{\sqrt{2\pi}} \displaystyle\int_{z_1}^{z_2} e^{-z^2/2}\, dz,$

where z_1 and z_2 are determined by formula (6.7.10). We say that the random variable U_n satisfying relation (6.7.11) has an *asymptotically normal distribution* $N(p; \sqrt{pq/n})$.

C   The Bernoulli law of large numbers, proved in Section 6.3, allows us to state only that for every $\varepsilon > 0$ the probability of the inequality

$$\left|\frac{X_n}{n} - p\right| > \varepsilon$$

tends to zero as $n \to \infty$. The limit theorem, which we have just proved, allows us (for large n) to compute approximately the probability that the random variable $X_n/n - p$ is contained in the interval

$$\left(z_1 \sqrt{\frac{p(1 - p)}{n}}, z_2 \sqrt{\frac{p(1 - p)}{n}}\right)$$

for arbitrary z_1 and z_2 $(z_1 < z_2)$.

Example 6.7.2. A box contains a collection of IBM cards corresponding to the workers from some branch of industry. Of the workers 20% are minors and 80% adults. We select one IBM card in a random way and mark the age given on this card. Before choosing the next card, we return the first one to the box, so that the probability of selecting the card corresponding to a minor remains 0.2. We observe n cards in this manner. What value should n have in order that the probability will be 0.95 that the frequency of cards corresponding to minors lies between 0.18 and 0.22?

Denote the frequency of the appearance of the card corresponding to a minor by U_n. We then have

$$E(U_n) = 0.2, \quad D^2(U_n) = \frac{0.16}{n}, \quad \sqrt{D^2(U_n)} = \frac{0.4}{\sqrt{n}}.$$

Consider the probability

$$P(0.18 < U_n < 0.22) = P\left(\frac{-0.02}{0.4/\sqrt{n}} < \frac{U_n - 0.2}{0.4/\sqrt{n}} < \frac{0.02}{0.4/\sqrt{n}}\right)$$

$$= P\left(-0.05\sqrt{n} < \frac{U_n - 0.2}{0.4}\sqrt{n} < 0.05\sqrt{n}\right) \simeq 0.95.$$

By formula (6.7.11) we obtain

$$0.95 \simeq \frac{1}{\sqrt{2\pi}} \int_{-0.05\sqrt{n}}^{0.05\sqrt{n}} e^{-z^2/2}\, dz.$$

From tables of the normal distribution we obtain $0.05\sqrt{n} \simeq 1.96$; consequently $n \simeq 1537$.

6.8 THE LINDEBERG-LÉVY THEOREM

A   The de Moivre-Laplace theorem is, as we shall see later, a particular case of a more general limit theorem, namely, the *Lindeberg-Lévy theorem*.

Consider a sequence $\{X_k\}$ $(k = 1, 2, \ldots)$ of equally distributed, independent random variables whose moment of the second order exists.

For every k denote

(6.8.1) $$E(X_k) = m, \qquad D^2(X_k) = \sigma^2.$$

Consider the random variable Y_n defined by the formula

(6.8.2) $$Y_n = X_1 + X_2 + \ldots + X_n.$$

We have $E(Y_n) = nm$ and, by the independence of the X_n,

$$D^2(Y_n) = n\sigma^2.$$

Let

(6.8.3) $$Z_n = \frac{Y_n - mn}{\sigma\sqrt{n}}.$$

We shall prove the following theorem (Lindeberg [1], Lévy [1]).

Theorem 6.8.1. *If $X_1, X_2, \ldots$ are independent random variables with the same distribution, whose standard deviation $\sigma \neq 0$ exists, then the sequence $\{F_n(z)\}$ of distribution functions of the random variables Z_n, given by formulas (6.8.3) and (6.8.2), satisfies, for every z, the equality*

(6.8.4) $$\lim_{n \to \infty} F_n(z) = \frac{1}{\sqrt{2\pi}} \int_{-\infty}^{z} e^{-z^2/2} \, dz.$$

Proof. Let us write equality (6.8.3) in the form

$$Z_n = \frac{1}{\sigma\sqrt{n}} \sum_{k=1}^{n} (X_k - m).$$

All the random variables $X_k - m$ have the same distribution, hence the same characteristic function $\phi_x(t)$. According to formulas (4.2.15) and (4.4.3) the characteristic function $\phi_z(t)$ of Z_n has the form

(6.8.5) $$\phi_z(t) = \left[\phi_x\left(\frac{t}{\sigma\sqrt{n}}\right) \right]^n.$$

We have assumed the existence of the first and second moments, and we have

$$E(X_k - m) = 0 \quad \text{and} \quad D^2(X_k - m) = \sigma^2.$$

Hence we can expand the function $\phi_x(t)$ in a neighborhood of the point $t = 0$ according to the MacLaurin formula as follows:

(6.8.6) $$\phi_x(t) = 1 - \tfrac{1}{2}\sigma^2 t^2 + o(t^2).$$

Substituting expression (6.8.6) in formula (6.8.5), we obtain

$$\phi_z(t) = \left[1 - \frac{t^2}{2n} + o\left(\frac{t^2}{n}\right) \right]^n,$$

where for every t we have

(6.8.7) $$\lim_{n \to \infty} n o\left(\frac{t^2}{n}\right) = 0.$$

Let

$$u = -\frac{t^2}{2n} + o\left(\frac{t^2}{n}\right).$$

We obtain

$$\log \phi_z(t) = n \log (1 + u) = n\left[-\frac{t^2}{2n} + o\left(\frac{t^2}{n}\right)\right] = -\frac{t^2}{2} + n o\left(\frac{t^2}{n}\right).$$

By relation (6.8.7) we obtain $\lim_{n \to \infty} \log \phi_z(t) = -t^2/2$. Hence

$$\lim_{n \to \infty} \phi_z(t) = e^{-t^2/2}.$$

The expression $e^{-t^2/2}$ is the characteristic function of a random variable with the normal distribution. By theorem 6.6.1b we obtain relation (6.8.4), which proves the theorem of Lindeberg-Lévy.

B   Let z_1 and z_2 be two arbitrary numbers with $z_1 < z_2$. By relation (6.8.4) we obtain

(6.8.8) $$\lim_{n \to \infty} P(z_1 < Z_n < z_2) = \lim_{n \to \infty} \left[F_n(z_2) - F_n(z_1)\right] = \frac{1}{\sqrt{2\pi}} \int_{z_1}^{z_2} e^{-z^2/2} \, dz.$$

From formula (6.8.3) we obtain

$$P(z_1 < Z_n < z_2) = P\left(z_1 < \frac{Y_n - nm}{\sigma\sqrt{n}} < z_2\right)$$

$$= P(z_1\sigma\sqrt{n} + nm < Y_n < z_2\sigma\sqrt{n} + nm).$$

Thus, we obtain from formula (6.8.8)

(6.8.9) $$\lim_{n \to \infty} P(z_1\sigma\sqrt{n} + nm < Y_n < z_2\sigma\sqrt{n} + nm) = \frac{1}{\sqrt{2\pi}} \int_{z_1}^{z_2} e^{-z^2/2} \, dz$$

Let

(6.8.10) $$y_1 = z_1\sigma\sqrt{n} + nm, \qquad y_2 = z_2\sigma\sqrt{n} + nm.$$

Now we can write formula (6.8.9) in the asymptotic form

$$P(y_1 < Y_n < y_2) \cong \frac{1}{\sqrt{2\pi}} \int_{z_1}^{z_2} e^{-z^2/2} \, dz,$$

where z_1 and z_2 are determined by relations (6.8.10). Thus the random variable Y_n defined by formula (6.8.2) has an *asymptotically normal distribution* $N(mn; \sigma\sqrt{n})$.

When a sum of random variables has an asymptotically normal distribution, we say that it *satisfies the central limit theorem*. Thus, for the sum Y_n under consideration, the central limit theorem holds.

Example 6.8.1. Suppose that the random variables $\{X_k\}$ ($k = 1, 2, \ldots$) are independent and each of them has the same two-point distribution, that is, for every k we have

$$P(X_k = 1) = p, \quad P(X_k = 0) = 1 - p, \quad \text{where} \quad 0 < p < 1.$$

Consider the random variable $Y_n = X_1 + X_2 + \ldots + X_n$. From the fact that $E(X_k) = p$ and $D^2(X_k) = pq$, we obtain by theorem 6.8.1 that Y_n has an asymptotically normal distribution $N(np; \sqrt{npq})$.

Since the random variable Y_n has the binomial distribution, this example is, strictly speaking, a new proof of the de Moivre-Laplace limit theorem, which, as we see, is a particular case of the Lindeberg-Lévy theorem.

Example 6.8.2. The random variables X_n ($n = 1, 2, \ldots$) are independent and each of them has the Poisson distribution given by the formula

$$P(X_n = r) = \frac{2^r}{r!} e^{-2} \quad (r = 0, 1, 2, \ldots).$$

Let us find the probability that the sum $Y_{100} = X_1 + X_2 + \ldots + X_{100}$ is greater than 190 and less than 210.

The random variable Y_{100} has approximately the normal distribution $N(100; 10\sqrt{2})$, since each of the random variables X_n has standard deviation $\sigma = \sqrt{2}$ and expected value $m = 2$. Thus we have

$$P(190 < Y_{100} < 210) = P\left(-0.707 < \frac{Y_{100} - 200}{10\sqrt{2}} < 0.707\right).$$

From the normal distribution tables we find that the required probability is 0.52.

C   From the Lindeberg-Lévy theorem we obtain the following:

Theorem 6.8.2. Suppose that the random variables $X_1, X_2, \ldots$ are independent and have the same distribution with standard deviation $\sigma \neq 0$. Let the random variable U_n be defined by the formula

$$U_n = \frac{X_1 + X_2 + \ldots + X_n}{n}.$$

Furthermore, let $F_n(v)$ be the distribution function of the random variable V_n defined as

$$V_n = \frac{U_n - E(U_n)}{\sqrt{D^2(U_n)}}.$$

Then the sequence $\{F_n(v)\}$ satisfies the relation

$$(6.8.11) \qquad \lim_{n \to \infty} F_n(v) = \frac{1}{\sqrt{2\pi}} \int_{-\infty}^{v} e^{-v^2/2} \, dv.$$

Proof. We have $E(U_n) = m$ and $D^2(U_n) = \sigma^2/n$. Hence

$$V_n = \frac{\dfrac{1}{n} \sum_{k=1}^{n} X_k - m}{\sigma/\sqrt{n}} = \frac{\sum_{k=1}^{n} X_k - nm}{\sigma\sqrt{n}} = Z_n,$$

where the random variables Z_n are defined by formula (6.8.3). Since the sequence $\{F_n(z)\}$ satisfies relation (6.8.4), the sequence $\{F_n(v)\}$ satisfies (6.8.11).

Now let v_1 and v_2 be two arbitrary numbers with $v_1 < v_2$. We have the relation

$$(6.8.12) \qquad \lim_{n \to \infty} P(v_1 < V_n < v_2) = \frac{1}{\sqrt{2\pi}} \int_{v_1}^{v_2} e^{-v^2/2} \, dv.$$

Let

$$(6.8.13) \qquad u_1 = \frac{v_1 \sigma}{\sqrt{n}} + m, \qquad u_2 = \frac{v_2 \sigma}{\sqrt{n}} + m.$$

Formula (6.8.12) can be written in the asymptotic form

$$P(u_1 < U_n < u_2) \simeq \frac{1}{\sqrt{2\pi}} \int_{v_1}^{v_2} e^{-v^2/2} \, dv,$$

where v_1 and v_2 are determined from relations (6.8.13). Thus the random variable U_n has an *asymptotically normal distribution* $N(m; \sigma/\sqrt{n})$. In other words, the arithmetic mean of n independent random variables with the same, although arbitrary, distribution, where it is only assumed that the moment of the second order exists, has, for large n, an asymptotically normal distribution.

Example 6.8.3. The random variables $X_1, X_2, \ldots$ are independent and have the uniform distribution defined by the density

$$f(x) = \begin{cases} 1 & \text{for } x \text{ in the interval } [0,1], \\ 0 & \text{for } x < 0 \text{ and } x > 1. \end{cases}$$

By formulas (5.6.4) and (5.6.5) we have

$$m = \frac{1}{2}, \qquad \sigma = \frac{1}{\sqrt{12}}.$$

Consider the random variable

$$Y_n = \frac{X_1 + X_2 + \ldots + X_n}{n}.$$

By theorem 6.8.2, the random variable Y_n has the asymptotically normal distribution $N(\frac{1}{2}; 1/\sqrt{12n})$. For $n = 48$ compute the probability that Y_n will be smaller than 0.4.

We have[1]

$$P(Y_n < 0.4) = P\left(\frac{Y_n - \frac{1}{2}}{1/\sqrt{576}} < \frac{0.4 - \frac{1}{2}}{1/\sqrt{576}}\right)$$

$$= P\left(\frac{Y_n - \frac{1}{2}}{\frac{1}{24}} < -2.4\right) \simeq \Phi(-2.4) \simeq 0.0082.$$

As we see, although the random variables X_k ($k = 1, 2, \ldots$) have a uniform distribution in the interval [0, 1], their arithmetic mean has, for large n, approximately a distribution in which values that are less than $m = 0.5$ by more than 0.1 appear extremely rarely.

Example 6.8.4. The random variables $X_r (r = 1, 2, \ldots)$ are independent and have the same distribution. Each of them can take on the values $k = 0, 1, 2, \ldots, 9$ with the probabilities $P(X_r = k) = 0.1$ for every k.

We have

$$m = E(X_r) = \frac{1}{10} \sum_{k=0}^{9} k = 4.5,$$

$$\sigma^2 = D^2(X_r) = \frac{1}{10} \sum_{k=0}^{9} (k - m)^2 = \frac{1}{10} \sum_{k=0}^{9} k^2 - m^2 = 28.50 - 20.25 = 8.25.$$

Thus $\sigma = 2.87$.

Consider the random variable

$$Y_{100} = \frac{X_1 + X_2 + \ldots + X_{100}}{100}.$$

What is the probability that Y_{100} will exceed 5?

By theorem 6.8.2 we know that Y_{100} has approximately the normal distribution $N(4.5; 2.87/\sqrt{100}.)$ We obtain

$$P(Y_{100} > 5) = P\left(\frac{Y_{100} - 4.5}{0.287} > \frac{5 - 4.5}{0.287}\right) = P\left(\frac{Y_{100} - 4.5}{0.287} > 1.74\right)$$

$$\simeq 1 - \Phi(1.74) \simeq 0.041.$$

Questions similar to those considered in this example appear in statistics when we use tables of random numbers discussed in Chapter 14.

D   We now show by an example that the arithmetic mean of n random variables with the same distribution may not have an asymptotically normal distribution, if their moment of the second order does not exist.

Example 6.8.5. The random variables X_k ($k = 1, 2, \ldots$) are independent and have the Cauchy distribution given by formula (5.10.3). Since the characteristic function of X_k has, for every k, the form

$$\phi_k(t) = e^{-|t|},$$

---

[1] The distribution function of the normal distribution $N(0; 1)$ is denoted by $\Phi(x)$.

the characteristic function $\phi(t)$ of the random variable

$$Y_n = \frac{X_1 + X_2 + \ldots + X_n}{n}$$

takes the form

$$\phi(t) = e^{-n|t|/n} = e^{-|t|}.$$

Hence for an arbitrary n the random variable Y_n has the Cauchy distribution. Thus Y_n does not have an asymptotically normal distribution.

We notice, however, that a random variable with a Cauchy distribution does not have a standard deviation (see Section 5.10).

E   Let the random variables X_k ($k = 1, 2, \ldots$) satisfy the assumptions of theorem 6.8.1 and let $E(X_k) = 0$. Consider for every n the partial sums

$$S_j = \sum_{k=1}^{j} X_k \qquad (j = 1, 2, \ldots, n).$$

Erdös and Kac [1, 2] have found the limit distributions for the sequences of random variables

$$\left\{ \max_{1 \leqslant j \leqslant n} \frac{S_j}{\sqrt{n}} \right\}, \quad \left\{ \max_{1 \leqslant j \leqslant n} \frac{|S_j|}{\sqrt{n}} \right\}, \quad \left\{ \frac{1}{n^2} \sum_{j=1}^{n} S_j^2 \right\}, \quad \left\{ \frac{1}{n^{3/2}} \sum_{j=1}^{n} |S_j| \right\}.$$

These papers began a series of fruitful investigations concerning the limit distributions of a large class of functionals defined on the vectors $(S_1, \ldots, S_n)$, even with much more general assumptions concerning the random variables X_k than those considered here. We shall not discuss these results. The reader will find them in the papers of Erdös and Kac mentioned here and in the papers of Donsker [1], Prohorov [2], [3], Skorohod [1], Spitzer [1], Baxter and Donsker [1], Varadarajan [1], Lamperti [1], Bartoszyński [1, 2], and Billingsley [1].

6.9   THE LAPUNOV THEOREM

In the preceding section we discussed the limit distribution of the sum of independent random variables with the same distribution, and we established that if the variance of these random variables exists, their sum has an asymptotically normal distribution. However, the distribution of a sum of independent random variables may not converge to the normal distribution if the terms do not have the same distribution, even if all the random variables have standard deviations.

We now prove the Lapunov theorem [2], which gives a sufficient condition for a sum of independent random variables to have a limiting normal distribution.

A   Consider a sequence $\{X_k\}$ of independent random variables whose moments of the third order exist.

Theorem 6.9.1 (Lapunov). *Let $\{X_k\}$ ($k = 1, 2, \ldots$) be a sequence of independent random variables whose moments of the third order exist, and let m_k, $\sigma_k \neq 0$, a_k, and b_k denote the expected value, standard deviation, central moment of the third order, and the absolute central moment of the third order of X_k, respectively. Furthermore, let*

$$B_n = \sqrt[3]{\sum_{k=1}^{n} b_k}, \qquad C_n = \sqrt{\sum_{k=1}^{n} \sigma_k^2}.$$

If the relation

(6.9.1)
$$\lim_{n \to \infty} \frac{B_n}{C_n} = 0$$

is satisfied, the sequence $\{F_n(z)\}$ of the distribution functions of the random variables Z_n, defined as

(6.9.2)
$$Z_n = \frac{\sum_{k=1}^{n} (X_k - m_k)}{C_n},$$

satisfies, for every z, the relation

(6.9.3)
$$\lim_{n \to \infty} F_n(z) = \frac{1}{\sqrt{2\pi}} \int_{-\infty}^{z} e^{-z^2/2} \, dz.$$

Proof. Let

$$Y_k = \frac{X_k - m_k}{C_n}.$$

Let $\phi_{x_k}(t)$ denote the characteristic function of the random variable $X_k - m_k$. From the fact that $E(X_k - m_k) = 0$ and that the moments σ_k^2 and a_k exist, we obtain by formula (4.3.2) the expansion of $\phi_{x_k}(t)$ into the sum

$$\phi_{x_k}(t) = 1 - \tfrac{1}{2}\sigma_k^2 t^2 + \tfrac{1}{6}a_k(it)^3 + o(a_k t^3).$$

By formula (4.2.15), the characteristic function $\phi_{y_k}(t)$ of Y_k equals

$$\phi_{y_k}(t) = \phi_{x_k}\left(\frac{t}{C_n}\right) = 1 - \frac{\sigma_k^2 t^2}{2C_n^2} + \frac{a_k}{6C_n^3}(it)^3 + o\left(\frac{a_k t^3}{C_n^3}\right) = 1 + u_k.$$

For every t we have

(6.9.4)
$$\lim_{n \to \infty} \left[o\left(\frac{a_k t^3}{C_n^3}\right) : \frac{a_k t^3}{C_n^3} \right] = 0.$$

Since by the Lapunov inequality (3.4.2) we have $\sigma_k < \sqrt[3]{b_k}$, condition (6.9.1) implies, for every t,

(6.9.5)
$$\lim_{n \to \infty} \left| \frac{-\sigma_k^2 t^2}{2C_n^2} \right| \leqslant \lim_{n \to \infty} \frac{\sqrt[3]{b_k^2}}{2C_n^2} t^2 \leqslant \lim_{n \to \infty} \frac{1}{2} \cdot \frac{B_n^2}{C_n^2} t^2 = 0.$$

Furthermore, we have

$$(6.9.6) \qquad \lim_{n \to \infty} \left| \frac{a_k}{6C_n^3} (it)^3 \right| \leqslant \lim_{n \to \infty} \frac{b_k}{6C_n^3} |t|^3 \leqslant \lim_{n \to \infty} \frac{B_n^3 |t|^3}{6C_n^3} = 0.$$

It follows from relations (6.9.5) and (6.9.6) that

$$\lim_{n \to \infty} u_k = 0,$$

and the convergence is uniform with respect to k. Hence for every t there exists a number $N = N(t)$ such that for $n > N$ and all $k \leqslant n$ we have the inequality $|u_k| < \frac{1}{2}$. Thus

$$\log \phi_{y_k}(t) = \log (1 + u_k) = u_k - \tfrac{1}{2}u_k^2 + \tfrac{1}{3}u_k^3 - \ldots$$
$$= u_k - \tfrac{1}{2}u_k^2(1 - \tfrac{2}{3}u_k + \tfrac{2}{4}u_k^2 - \ldots) = u_k - \tfrac{1}{2}u_k^2 v_k.$$

We notice that

$$|v_k| \leqslant 1 + \tfrac{2}{3}|u_k| + \tfrac{2}{4}|u_k|^2 + \ldots < 1 + |u_k| + |u_k|^2 + \ldots$$
$$< 1 + \tfrac{1}{2} + \tfrac{1}{4} + \ldots = 2.$$

Thus we can write

$$(6.9.7) \qquad \log \phi_{y_k}(t) = u_k + \vartheta_k u_k^2,$$

where $\vartheta_k = -\tfrac{1}{2}v_k$, and $|\vartheta_k| < 1$.

Denote by $\phi_z(t)$ the characteristic function of the random variable Z_n. By formula (4.4.3), we have $\phi_z(t) = \prod_{k=1}^{n} \phi_{y_k}(t)$. Hence

$$\log \phi_z(t) = \sum_{k=1}^{n} \log \phi_{y_k}(t).$$

By equality (6.9.7) we obtain

$$(6.9.8) \qquad \log \phi_z(t) = \sum_{k=1}^{n} (u_k + \vartheta_k u_k^2).$$

Next, we have

$$(6.9.9) \qquad \sum_{k=1}^{n} u_k = -\frac{t^2}{2} + \sum_{k=1}^{n} \frac{a_k(it)^3}{6C_n^3} + \sum_{k=1}^{n} o\left(\frac{a_k t^3}{C_n^3}\right).$$

We notice that for every t

$$(6.9.10) \quad \lim_{n \to \infty} \left| \sum_{k=1}^{n} \frac{a_k(it)^3}{6C_n^3} \right| \leqslant \lim_{n \to \infty} \sum_{k=1}^{n} \frac{b_k |t|^3}{6C_n^3} = \lim_{n \to \infty} \frac{B_n^3 |t|^3}{6C_n^3} = 0.$$

Hence, by formula (6.9.4) we obtain

$$(6.9.11) \quad \lim_{n \to \infty} \sum_{k=1}^{n} o\left(\frac{a_k t^3}{C_n^3}\right) = \lim_{n \to \infty} \sum_{k=1}^{n} \left\{ \frac{a_k t^3}{6C_n^3} \cdot \left[o\left(\frac{a_k t^3}{C_n^3}\right) : \frac{a_k t^3}{6C_n^3} \right] \right\} = 0.$$

From formula (6.9.9), because of formulas (6.9.10) and (6.9.11), it follows that for every t we have

$$(6.9.12) \qquad \lim_{n \to \infty} \sum_{k=1}^{n} u_k = -\frac{t^2}{2}.$$

We now find the limit of the sum

$$\sum_{k=1}^{n} u_k^2 = \sum_{k=1}^{n} \left[\frac{-\sigma_k^2 t^2}{2C_n^2} + \frac{a_k(it)^3}{6C_n^3} + o\left(\frac{a_k t}{C_n^3}\right) \right]^2.$$

By the Lapunov inequality (3.4.2) and condition (6.9.1) we obtain

$$\lim_{n \to \infty} \sum_{k=1}^{n} \frac{\sigma_k^4 t^4}{4C_n^4} \leqslant \lim_{n \to \infty} \sum_{k=1}^{n} \frac{\sqrt[3]{b_k^4}}{4C_n^4} t^4 = \lim_{n \to \infty} \sum_{k=1}^{n} \frac{b_k \sqrt[3]{b_k}}{4C_n^3 C_n} t^4$$

$$(6.9.13) \qquad \leqslant \lim_{n \to \infty} \sum_{k=1}^{n} \frac{b_k}{4C_n^3} t^4 = \lim_{n \to \infty} \frac{B_n^3}{4C_n^3} t^4 = 0,$$

$$\lim_{n \to \infty} \left| \sum_{k=1}^{n} \left[\frac{a_k(it)^3}{6C_n^3} \right]^2 \right| \leqslant \lim_{n \to \infty} \sum_{k=1}^{n} \left| \frac{a_k t^3}{6C_n^3} \right|^2$$

$$(6.9.14) \qquad \leqslant \lim_{n \to \infty} \sum_{k=1}^{n} \frac{b_k^2 t^6}{36C_n^6} \leqslant \lim_{n \to \infty} \frac{B_n^6 t^6}{36C_n^6} = 0.$$

Taking formula (6.9.4) into consideration, we obtain

$$(6.9.15) \qquad \lim_{n \to \infty} \sum_{k=1}^{n} \left[o\left(\frac{a_k t^3}{6C_n^3}\right) \right]^3 = 0.$$

Similarly, we obtain

$$(6.9.16) \qquad \lim_{n \to \infty} \left| \sum_{k=1}^{n-1} \sum_{j=k+1}^{n} \frac{\sigma_k^2 a_j i^3 t^5}{6C_n^5} \right| = 0,$$

$$(6.9.17) \qquad \lim_{n \to \infty} \left| \sum_{k=1}^{n-1} \sum_{j=k+1}^{n} \frac{t^2 \sigma_k^2}{C_n^2} o\left(\frac{a_j t^3}{C_n^3}\right) \right| = 0.$$

$$(6.9.18) \qquad \lim_{n \to \infty} \left| \sum_{k=1}^{n-1} \sum_{j=k+1}^{n} \frac{a_k i^3 t^3}{3C_n^3} o\left(\frac{a_j t^3}{C_n^3}\right) \right| = 0.$$

From formulas (6.9.13) to (6.9.18) and the fact that for every $k \leqslant n$ we have $|\vartheta_k| < 1$, we obtain

$$(6.9.19) \qquad \lim_{n \to \infty} \sum_{k=1}^{n} \vartheta_k u_k^2 = 0.$$

Using (6.9.12) and (6.9.19) we obtain from formula (6.9.8) that for every t the relation

$$\lim_{n \to \infty} \log \phi_z(t) = -\frac{t^2}{2}$$

holds. Hence

$$\lim_{n \to \infty} \phi_z(t) = e^{-t^2/2}.$$

By the last relation and by theorem 6.6.1*b*, we obtain formula (6.9.3), which proves the Lapunov theorem.

For another proof of Lapunov's theorem, see Parzen [2].

B   The Lapunov theorem only gives a sufficient condition for relation (6.9.3). We shall now present without proof the theorem of Lindeberg-Feller, giving a necessary and sufficient condition. The reader will find the proof of this theorem in the original papers of Lindeberg [1], Feller [2], or in the book by Gnedenko [6] or Loève [4].

Theorem 6.9.2 (Lindeberg-Feller). *Let* $\{X_k\}$ $(k = 1, 2, \ldots)$ *be a sequence of independent random variables whose variances exist, and let* $G_k(x)$, m_k, *and* $\sigma_k \neq 0$ *denote, respectively, the distribution function, the expected value and the standard deviation of the random variable* X_k, *and let* $F_n(z)$ *denote the distribution function of the standardized random variable* Z_n *given by formula* (6.9.2).

Then the relations

$$\lim_{n \to \infty} \max_{1 \le k \le n} \frac{\sigma_k}{C_n} = 0, \qquad \lim_{n \to \infty} F_n(z) = \frac{1}{\sqrt{2\pi}} \int_{-\infty}^{z} e^{-z^2/2} \, dz$$

hold if and only if, for every $\varepsilon > 0$,

$$(6.9.20) \qquad \lim_{n \to \infty} \frac{1}{C_n^2} \sum_{k=1}^{n} \int_{|x-m_k| > \varepsilon C_n} (x - m_k)^2 \, dG_k(x) = 0.$$

If all the X_k are of the continuous type and $g_k(x)$ is the density of X_k, then condition (6.9.20) takes the form

$$(6.9.21) \qquad \lim_{n \to \infty} \frac{1}{C_n^2} \sum_{k=1}^{n} \int_{|x-m_k| > \varepsilon C_n} (x - m_k)^2 g_k(x) \, dx = 0.$$

If, however, all the X_k are of the discrete type with jump points x_{kl} and jumps p_{kl} $(l = 1, 2, \ldots)$, formula (6.9.20) takes the form

$$(6.9.22) \qquad \lim_{n \to \infty} \frac{1}{C_n^2} \sum_{k=1}^{n} \sum_{|x_{kl} - m_k| > \varepsilon C_n} (x_{kl} - m_k)^2 p_{kl} = 0.$$

From the theorem of Lindeberg-Feller follows this theorem.

Theorem 6.9.3. *Let* $\{X_k\}$ $(k = 1, 2, \ldots)$ *be a sequence of independent, uniformly bounded random variables, that is, there exists a constant* $a > 0$ *such that for every* k

$$(6.9.23) \qquad P(|X_k| \leqslant a) = 1,$$

and suppose that $D^2(X_k) \neq 0$ for every k. Then a necessary and sufficient condition for relation (6.9.3) to hold is

(6.9.24) $$\lim_{n \to \infty} C_n^{\ 2} = \infty.$$

Proof. Suppose that (6.9.24) is satisfied. From formula (6.9.23) it follows that the random variables $X_k - m_k$ are uniformly bounded. Hence for every $\varepsilon > 0$ we can find an N such that for $n > N$ we have

$$P(|X_k - m_k| < \varepsilon C_n; \ k = 1, 2, \ldots, n) = 1.$$

Formula (6.9.20) follows immediately.

Suppose now that (6.9.3) holds, and (6.9.24) does not. Then there exists a $C < \infty$ such that $\lim_{n \to \infty} C_n^{\ 2} = C^2$. From the last relation, and from formulas (6.9.3) and (6.9.2), it follows that $\sum_{k=1}^{\infty}(X_k - m_k)$ has (see Problem 6.43) the normal distribution $N(0; C)$. Let

$$U = (X_2 - m_2) + (X_3 - m_3) + \ldots .$$

The random variables $X_1 - m_1$ and U are independent, and their sum has a normal distribution. By the Cramér theorem (see Section 5.7) both $X_1 - m_1$ and U have normal distributions. However, by hypothesis (6.9.23), the random variable $X_1 - m_1$ does not have a normal distribution. Hence (6.9.3) is not satisfied, and the theorem is proved.

In particular, it follows from this theorem that if the random variable $Y_n = \sum_{k=1}^{n} X_k$ has the generalized binomial distribution, that is, if the probability function of X_k is given by the formulas $P(X_k = 1) = p_k$, $P(X_k = 0) = q_k = 1 - p_k$ $(k = 1, 2, \ldots)$, then the divergence of the series $\Sigma p_k q_k$ is a necessary and sufficient condition for Y_k to have the asymptotically normal distribution

$$N\left(\sum_{k=1}^{n} p_k; \ \sqrt{\sum_{k=1}^{n} p_k q_k} \right) .$$

Example 6.9.1. At a construction site there are lots of bricks from five different factories. Judging by previous experience, the quality of bricks from different factories differs and the fraction of defective items is not the same for all lots. The production of the ith factory is characterized by the number p_i, giving the fraction of good bricks. The values of p_i are the following:

$$p_1 = 0.95, \quad p_2 = 0.90, \quad p_3 = 0.98, \quad p_4 = 0.92, \quad p_5 = 0.96.$$

Since the lots are very large, we assume it is certain that the defectiveness of a lot produced by the ith factory is exactly $1 - p_i$ $(i = 1, \ldots, 5)$. The probability of choosing a good brick from a given lot is thus p_i. We select 20 bricks at random from each lot. Since each lot contains many bricks, and the drawing of 20 bricks does not change practically the probability of selecting a good brick, we may assume that this probability is constant while drawing bricks and hence equals p_i. After checking the quality of all 100 selected bricks, it turned out that 11 of them were defective. This result created some doubts as to whether the assumptions about the numbers p_i were not too optimistic.

The mathematical model of this example is the following. We have 100 independent random variables X_k and each of them can take on two values; 1 when a good brick is selected and 0 when a defective one is selected. These random variables are divided into five groups. The ith group consists of those random variables which take on the value 1 with probability p_i. Let us form the random variable

$$Y_{100} = X_1 + \ldots + X_{20} + X_{21} + \ldots + X_{40} + X_{41} + \ldots + X_{60}$$
$$+ X_{61} + \ldots + X_{80} + X_{81} + \ldots + X_{100}.$$

This is a random variable with a generalized binomial distribution. We have

$$E(Y_{100}) = 20 \cdot 0.95 + 20 \cdot 0.90 + 20 \cdot 0.98 + 20 \cdot 0.92 + 20 \cdot 0.96 = 94.20,$$

$$D^2(Y_{100}) = 20 \cdot 0.05 \cdot 0.95 + 20 \cdot 0.10 \cdot 0.90 + 20 \cdot 0.02 \cdot 0.98$$
$$+ 20 \cdot 0.08 \cdot 0.92 + 20 \cdot 0.04 \cdot 0.96 = 5.382,$$

$$\sigma = 2.32.$$

Before we apply the central limit theorem, we must examine the result obtained above which gives the divergence of the series $\Sigma p_k q_k$ as a necessary and sufficient condition for the convergence of the generalized binomial distribution to the normal distribution. If, however, this series is convergent, then $p_k q_k \to 0$ as $k \to \infty$. Hence $\min(p_k; 1 - p_k) \to 0$. Thus the sequence $\{p_k\}$ must contain a subsequence convergent either to zero or to one. In the language of this example, this would mean that the series $\Sigma p_k q_k$ would converge if the bricks produced contained very often (theoretically an infinite number of times) either only good or only defective items. However, many years of practice in the production of bricks show that this is not true and thus the series $\Sigma p_k q_k$ is not convergent. Thus we can apply the central limit theorem.

According to this theorem, the random variable Y_{100} has approximately the normal distribution N (94.2; 2.32). Thus we have

$$P(Y_{100} \leqslant 89) = P\left(\frac{Y_{100} - 94.2}{2.32} \leqslant -2.25\right) \simeq \Phi(-2.25).$$

From tables of the normal distribution we find that $\Phi(-2.25)$ is rather small, about 0.01. In such cases we are inclined to accept the conclusion that our assumptions about the p_i were too optimistic.

In this example we have touched on questions which will be systematically and exhaustively considered in the second part of this book. This example was given to show that the central limit theorem is not only a beautiful mathematical achievement but can also be applied to the solution of many practical problems.

C   From the considerations of Sections 6.6 to 6.9, we see how important a role the normal distribution plays in probability theory and its applications. However, the theorem which we now present shows that under rather general assumptions a sequence of distribution function of sums of independent random variables may converge to a limit distribution function different from the normal. We remind the reader that in Section 6.1 we gave references to papers presenting solutions of many fundamental questions connected with limit distributions for sums of independent random variables.

Consider a sequence $\{Y_n\}$ $(n = 1, 2, \ldots)$ of random variables, where for every n, Y_n is the sum of n independent random variables X_{nk} $(k = 1, 2, \ldots, n)$,

$$(6.9.25) \qquad Y_n = \sum_{k=1}^{n} X_{nk}.$$

These sums are more general than the sums considered in this chapter, where we have

$$X_{nk} = X_k \quad (n = 1, 2, \ldots; \ k = 1, 2, \ldots, n).$$

We restrict ourselves to the case when, for every n, the random variables X_{nk} $(k = 1, 2, \ldots, n)$ have the same distribution[1] given by the probability function

$$(6.9.26) \qquad P(X_{nk} = x_l) = p_{nl} \qquad (l = 1, 2, \ldots, r),$$

where

$$0 < p_{nl} < 1, \ \sum_{l=1}^{r} p_{nl} = 1,$$

and r $(r \geqslant 2)$ is some natural number.

Theorem 6.9.4.   Let Y_n be defined by formula (6.9.25) and let X_{nk} $(k = 1, 2, \ldots, n)$ be independent and have the distribution defined by formula (6.9.26). Let $F_n(z)$ be the distribution function of the random variable Z_n defined as

$$Z_n = \frac{Y_n - E(Y_n)}{\sqrt{D^2(Y_n)}}.$$

Then:

I. If

$$(6.9.27) \qquad \lim_{n \to \infty} n(p_{n1}p_{n2} + p_{n1}p_{n3} + \cdots + p_{n,r-1}p_{nr}) = \infty,$$

[1] The case when the X_{nk} do not have the same distribution for all k was considered by Kubik [1] (see Problem 6.17).

the sequence $\{F_n(z)\}$ satisfies the relation

$$\lim_{n \to \infty} F_n(z) = \frac{1}{\sqrt{2\pi}} \int_{-\infty}^{z} e^{-z^2/2} \, dz.$$

II. *If the limits (finite or infinite)*

$$\lim_{n \to \infty} p_{nl} \quad and \quad \lim_{n \to \infty} n p_{nl} \quad (l = 1, 2, \ldots, r)$$

exist, and the relation

(6.9.28) $$\lim_{n \to \infty} n(p_{n1} p_{n2} + p_{n1} p_{n3} + \cdots + p_{n,r-1} p_{nr}) = \lambda,$$

where $\lambda > 0$, holds, then the sequence $\{F_n(z)\}$ converges to the distribution function of a random variable which is a linear combination of s $(1 \leqslant s \leqslant r - 1)$ independent random variables, each having a Poisson distribution.

Theorem 6.9.4 was proved by Fisz [1]. We do not present the proof here.

We notice that in theorem 5.5.1 we have dealt with sums of the form (6.9.25). Indeed, let Y_n be the number of successes in n trials in the Bernoulli scheme and let the probability of success p_n be a function of n satisfying the relation

(6.9.29) $$\lim_{n \to \infty} n p_n = \lambda,$$

where $\lambda > 0$. Then we can write

$$Y_n = \sum_{k=1}^{n} X_{nk},$$

where X_{nk} is the number of successes (equal to 0 or 1) in the kth trial $(k = 1, 2, \ldots, n)$; thus the X_{nk} are independent and have the same distribution given by the formulas

$$P(X_{nk} = 1) = p_{n1} = p_n, \qquad P(X_{nk} = 0) = p_{n2} = 1 - p_n,$$

and we have $r = 2$. From formula (6.9.29) follow the relations

$$\lim_{n \to \infty} p_{n1} = 0, \qquad \lim_{n \to \infty} p_{n2} = 1,$$

$$\lim_{n \to \infty} n p_{n1} = \lambda, \qquad \lim_{n \to \infty} n p_{n2} = \infty, \qquad \lim_{n \to \infty} n p_{n1} p_{n2} = \lambda,$$

where $\lambda > 0$. All the assumptions of assertion II of theorem 6.9.4 are satisfied; thus the sequence $\{F_n(z)\}$, where $F_n(z)$ is the distribution function of the random variable

$$Z_n = \frac{Y_n - n p_n}{\sqrt{n p_n (1 - p_n)}},$$

converges as $n \to \infty$ to the distribution function of a Poisson random variable with the parameter λ. This is the *integral Poisson theorem* (see Section 6.1), whereas theorem 5.5.1 is the *local Poisson theorem*.

D   Besides establishing the convergence of a sequence of distribution functions $F_n(z)$ of random variables Z_n defined by formula (6.9.2) to the normal distribution function $\Phi(z)$, it is important to evaluate the rate of this convergence. In other words, we are interested in evaluating the difference $|F_n(z) - \Phi(z)|$ for all n and z. The reader will find the solution of these questions well advanced, for Z_n binomially distributed, in Uspensky [1] and Feller [11] and, for the general case, in the papers of Cramér [1, 3], Esseen [1], and Berry [1] (see Problem 6.19).

6.10   THE GNEDENKO THEOREM

A   In this chapter we have considered the convergence of sequences of distribution functions of standardized random variables to a limit distribution function. However, from the convergence of distribution functions the local limit theorem does not, in general, follow. Thus the sequence of densities (in the continuous case) and the sequence of probability functions (in the discrete case) may not converge to the corresponding limit density or probability function (see Problems 6.25 and 6.26).

We shall not investigate all the aspects of this question. We restrict ourselves to the proof of the local limit theorem for a sequence of independent random variables $\{X_i\}$ with the same distribution of a special discrete type, namely, when X_i can take on only values of the form $a + kh$ with $h > 0$, where k runs through integral values (not necessarily all). Such a distribution is called a *lattice distribution*, and the number h is called a *span* of the distribution. Evidently, h is a common divisor of the differences of the pairs of values that can be taken on by X_i. The number h is called the *maximum span* if it is the greatest common divisor of such differences. At first we shall investigate only random variables which can take on only integer values.

Theorem 6.10.1 (Gnedenko [5]).   *Suppose that the independent and equally distributed random variables X_i ($i = 1, 2, \ldots$) of the discrete type can take on with positive probability only integer values, and let $E(X_i) = m$ and $D^2(X_i) = \sigma^2 > 0$. Then the relation*

$$(6.10.1) \qquad \lim_{n \to \infty} \left[\sigma\sqrt{n} P_n(k) - \frac{1}{\sqrt{2\pi}} \exp\left(-\frac{z_{nk}^2}{2} \right) \right] = 0,$$

where

$$(6.10.2) \qquad P_n(k) = P\left(\sum_{i=1}^{n} X_i = k \right),$$

$$(6.10.3) \qquad z_{nk} = \frac{k - mn}{\sigma\sqrt{n}},$$

is satisfied uniformly with respect to k in the interval $(-\infty < k < +\infty)$ *if and only if the maximum span of the distribution of* X_i *is equal to one.*

An integer valued random variable satisfying the last condition is said to satisfy condition (w).

A particular case of this theorem is the local limit theorem of de Moivre-Laplace when X_i can take the values 0 and 1, with probabilities $1 - p$ and p, respectively.

B In the proof of the theorem just stated we use the following lemma.

Lemma. *If a random variable X of the discrete type takes on with positive probability only integer values and satisfies condition* (w), *then its characteristic function satisfies the relations*

$$(6.10.4) \qquad \phi(2\pi) = 1, \qquad |\phi(t)| < 1 \qquad \text{if} \quad 0 < |t| < 2\pi.$$

Proof of the lemma. We have

$$\phi(t) = \sum_{k=-\infty}^{+\infty} p_k e^{itk},$$

where $p_k = P(X = k)$. Hence we obtain

$$\phi(2\pi) = \sum_{k=-\infty}^{+\infty} p_k e^{2\pi k i} = \sum_{k=-\infty}^{+\infty} p_k = 1.$$

Let $0 < |t_0| < 2\pi$ and let $|\phi(t_0)| = 1$. Then there exists an a such that $\phi(t_0) = e^{ia}$, or

$$\int_{-\infty}^{+\infty} e^{it_0 x} \, dF(x) = e^{ia}.$$

Multiplying both sides by e^{-ia} we obtain

$$\int_{-\infty}^{+\infty} e^{i(t_0 x - a)} \, dF(x) = 1.$$

Hence

$$\int_{-\infty}^{+\infty} \cos (t_0 x - a) \, dF(x) = \sum_{k=-\infty}^{+\infty} p_k \cos (t_0 x_k - a) = 1.$$

This condition can be satisfied only if, for all values of k, we have the equality $\cos (t_0 x_k - a) = 1$; hence only if

$$x_k = \frac{a}{t_0} + m_k \frac{2\pi}{t_0} \qquad (m_k \text{ integer}).$$

By hypothesis, the numbers x_k are integers; hence differences of any two x_k have the form sh, where s is an integer and $h = \dfrac{2\pi}{|t_0|}$ is the greatest common divisor of the differences of x_k. From the relation $|t_0| < 2\pi$ it follows that $h > 1$, which contradicts the assumption that condition (w) is satisfied. This completes the proof of the lemma.

C We now prove theorem 6.10.1. First we show that condition (w) is sufficient.

Let $\phi(t)$ and $\phi_n(t)$ denote, respectively, the characteristic functions of the random variables X_i and the random variable $Y_n = \sum_{i=1}^{n} X_i$. We have

$$\phi(t) = \sum_{k=-\infty}^{+\infty} p_k e^{itk}, \qquad \phi_n(t) = \sum_{k=-\infty}^{+\infty} P_n(k) e^{itk}.$$

By formula (4.5.7)

$$P_n(k) = \frac{1}{2\pi} \int_{-\pi}^{\pi} \phi_n(t) e^{-itk} \, dt,$$

and by formula (6.10.3) we obtain $k = \sigma\sqrt{n} z_{nk} + mn$. Thus

$$P_n(k) = \frac{1}{2\pi} \int_{-\pi}^{\pi} \phi_n(t) \exp\left[-it(\sigma\sqrt{n} z_{nk} + mn)\right] dt.$$

Letting $u = \sigma\sqrt{n} t$, we obtain

$$(6.10.5) \quad P_n(k) = \frac{1}{2\sigma\sqrt{n\pi}} \int_{-\pi\sigma\sqrt{n}}^{\pi\sigma\sqrt{n}} \phi_n\left(\frac{u}{\sigma\sqrt{n}}\right) \exp\left[-iu\left(z_{nk} + \frac{mn}{\sigma\sqrt{n}}\right)\right] du.$$

In example 4.5.1 we derived the formula

$$(6.10.6) \quad \frac{1}{\sqrt{2\pi}} \exp\left(-\frac{z^2}{2}\right) = \frac{1}{2\pi} \int_{-\infty}^{+\infty} \exp\left(-izu - \frac{u^2}{2}\right) du.$$

We now show that the difference

$$R_n = 2\pi\left[\sigma\sqrt{n}\, P_n(k) - \frac{1}{\sqrt{2\pi}} \exp\left(-\frac{z_{nk}^2}{2}\right)\right]$$

converges to zero uniformly with respect to k in the interval $-\infty < k < +\infty$. To show this, we use formulas (6.10.5) and (6.10.6) and represent R_n in the form of a sum of four integrals,

$$(6.10.7) \quad R_n = \int_{-\pi\sigma\sqrt{n}}^{\pi\sigma\sqrt{n}} \phi_n\left(\frac{u}{\sigma\sqrt{n}}\right) \exp\left[-iu\left(z_{nk} + \frac{mn}{\sigma\sqrt{n}}\right)\right] du$$

$$- \int_{-\infty}^{+\infty} \exp\left(-iuz_{nk} - \frac{u^2}{2}\right) du = I_1 + I_2 + I_3 + I_4,$$

where

$$I_1 = \int_{|u|<A} \exp(-iuz_{nk}) \left[\phi_n\left(\frac{u}{\sigma\sqrt{n}}\right) \exp\left(-\frac{imnu}{\sigma\sqrt{n}}\right) - \exp\left(-\frac{u^2}{2}\right) \right] du,$$

$$I_2 = \int_{A \leqslant |u| < \varepsilon\sigma\sqrt{n}} \phi_n\left(\frac{u}{\sigma\sqrt{n}}\right) \exp\left[-iu\left(z_{nk} + \frac{mn}{\sigma\sqrt{n}}\right) \right] du,$$

$$I_3 = \int_{\varepsilon\sigma\sqrt{n} \leqslant |u| \leqslant \pi\sigma\sqrt{n}} \phi_n\left(\frac{u}{\sigma\sqrt{n}}\right) \exp\left[-iu\left(z_{nk} + \frac{mn}{\sigma\sqrt{n}}\right) \right] du,$$

$$I_4 = -\int_{|u| \geqslant A} \exp\left(-iuz_{nk} - \frac{u^2}{2}\right) du.$$

In these integrals A and ε are positive constants which will be fixed later. We now evaluate these integrals.

1. Since, by assumption, the variance of the random variables X_i exists, it follows, from the Lindeberg-Lévy theorem (see Section 6.8) and from theorem 6.6.1a, that in every finite interval of values of u the relation

$$\phi_n\left(\frac{u}{\sigma\sqrt{n}}\right) \exp\left(-\frac{imnu}{\sigma\sqrt{n}}\right) - \exp\left(-\frac{u^2}{2}\right) \to 0$$

is satisfied uniformly with respect to u as $n \to \infty$. Thus for an arbitrary constant A we have $I_1 \to 0$ uniformly with respect to z_{nk} as $n \to \infty$.

2. From the assumption of the existence of the second moment of X_i it follows that its characteristic function is twice differentiable. By formula (4.3.2), in a sufficiently small neighborhood of $t = 0$, we have

$$\log\left[e^{-imt}\phi(t) \right] = \log\left[1 - \frac{\sigma^2 t^2}{2} + o(t^2) \right].$$

Thus if the neighborhood of $t = 0$ is sufficiently small, we have

$$\log\left[e^{-imt}\phi(t) \right] = -\frac{\sigma^2 t^2}{2} + o(t^2),$$

$$|e^{-imt}\phi(t)| \leqslant e^{-\sigma^2 t^2/4}.$$

Hence, for a sufficiently small $\varepsilon > 0$ we have, in the domain $|u| < \varepsilon\sigma\sqrt{n}$,

$$\left| \exp\left(-\frac{imnu}{\sigma\sqrt{n}}\right) \phi_n\left(\frac{u}{\sigma\sqrt{n}}\right) \right| \leqslant \exp\left(-\frac{u^2}{4}\right),$$

and we immediately obtain an estimate of the integral I_2,

$$|I_2| \leqslant 2\int_A^{\varepsilon\sigma\sqrt{n}} e^{-u^2/4}\, du < 2\int_A^\infty e^{-u^2/4}\, du = 4\sqrt{\pi}\left[1 - \Phi\left(\frac{A}{\sqrt{2}}\right) \right].$$

By choosing A sufficiently large, we can obtain for $|I_2|$ a value smaller than any given positive number.

3. We now evaluate the integral I_3. From the lemma previously proved it follows that, for $0 < |u| < 2\pi\sigma\sqrt{n}$,

$$\left| \phi\left(\frac{u}{\sigma\sqrt{n}}\right) \right| < 1;$$

hence for an arbitrary $\varepsilon > 0$ we can find a number $c > 0$ such that if $\varepsilon\sigma\sqrt{n} \leqslant |u| \leqslant \pi\sigma\sqrt{n}$, then $|\phi(u \mid \sigma\sqrt{n})| \leqslant e^{-c}$, and thus

$$\left| \phi_n\left(\frac{u}{\sigma\sqrt{n}}\right) \right| \leqslant e^{-nc}.$$

Hence we obtain

$$|I_3| \leqslant e^{-nc} \int_{\varepsilon\sigma\sqrt{n} \leqslant |u| \leqslant \pi\sigma\sqrt{n}} du < 2\pi\sigma\sqrt{n}e^{-nc},$$

and the integral I_3 converges to zero uniformly with respect to z_{nk} as $n \to \infty$.

4. The evaluation of the integral I_4 is immediately obtained,

$$|I_4| \leqslant 2 \int_A^\infty e^{-u^2/2}\,du = 2\sqrt{2\pi}[1 - \Phi(A)].$$

By choosing A sufficiently large, we can obtain for $|I_4|$ a value smaller than any given positive number.

As we see, the integrals I_1 and I_3 are convergent to zero uniformly with respect to z_{nk} as $n \to \infty$, independently of the values of A and ε, and the integrals I_2 and I_4 can be made arbitrarily small by a suitable selection of A and ε, and this estimate is independent of n. Hence R_n, as the sum of these integrals, is sufficiently small for sufficiently large values of n, which proves the sufficiency of condition (w).

Suppose now that condition (w) is not satisfied. Then the greatest common divisor h of the differences of the pairs of values which can be taken on with positive probability by the X_i is greater than 1. Since the X_i take on with positive probability only numbers of the form

$$x_k = a + kh \qquad (k = 0, \pm 1, \pm 2, \ldots),$$

the set of values taken on with positive probability by the random variable Y_n consists of numbers y_k of the form

(6.10.8) $$y_k = na + kh \qquad (k = 0, \pm 1, \pm 2, \ldots).$$

Let

$$k_n = na + \left[n\frac{m - a}{h} \right] h + 1 \qquad (n = 1, 2, \ldots).$$

([A] here denotes the greatest integer not greater than A). Since $h > 1$, the numbers k_n are not of the form (6.10.8) and we have the equality

$$(6.10.9) \qquad P_n(k_n) = 0 \qquad (n = 1, 2, \ldots).$$

It follows from the definition of the numbers z_{nk} and k_n that

$$|z_{nk_n}| \leqslant \frac{|h| + 1}{\sigma \sqrt{n}} \qquad (n = 1, 2, \ldots),$$

and hence

$$\lim_{n \to \infty} \frac{1}{\sqrt{2\pi}} \exp\left(-\frac{z_{nk_n}^2}{2}\right) = \frac{1}{\sqrt{2\pi}}.$$

From the last relation and from relation (6.10.9) it follows that relation (6.10.1) does not hold uniformly with respect to k, which proves the necessity of condition (w).

D   Suppose that the random variables X_i can take on only the values $x_k = a + kh$, where k is an integer and $h > 0$ is the maximum span of the distribution of X_i. Then the random variables $Z_i = (X_i - a)/h$ take on only integer values and satisfy condition (w). The local limit theorem proved can thus be applied to these random variables.

A generalization of the theorem of Gnedenko to random variables of not necessarily the same distribution was given by Richter [1]. Mejzler, Parasiuk, and Rvatcheva [1] and Richter [2] have generalized this theorem to random vectors.

6.11 POISSON'S CHEBYSHEV'S, AND KHINTCHIN'S LAWS OF LARGE NUMBERS

A   In Section 6.3 we obtained the Bernoulli law of large numbers. This law of large numbers, historically the oldest, is only a particular case of more general theorems which are known under the common name of *laws of large numbers*.

Consider first a sequence of random variables $\{X_k\}$ ($k = 1, 2, \ldots$); the only assumption we make is that for every k the first two moments exist, that is,

$$E(X_k) = m_k, \qquad E[(X_k - m_k)^2] = \sigma_k^2.$$

The random variables X_k may or may not be dependent.

By inequality (3.3.1) we have for every k and $\varepsilon > 0$

$$(6.11.1) \qquad P(|X_k - m_k| \geqslant \varepsilon) \leqslant \frac{\sigma_k^2}{\varepsilon^2}.$$

If the Markov condition

(6.11.2) $$\lim_{k \to \infty} \sigma_k^2 = 0$$

is satisfied, from formula (6.11.1) we obtain

$$\lim_{k \to \infty} P(|X_k - m_k| \geqslant \varepsilon) = 0.$$

We state the last result in the form of Chebyshev's theorem.

Theorem 6.11.1. Let $\{X_k\}$ $(k = 1, 2, \ldots)$ be an arbitrary sequence of random variables with variances σ_k^2. If the Markov condition (6.11.2) is satisfied, the sequence $\{X_k - m_k\}$ is stochastically convergent to zero.

Suppose that the X_k considered in this theorem are pairwise uncorrelated. Consider the random variable

(6.11.3) $$Y_n = \frac{X_1 + X_2 + \ldots + X_n}{n}.$$

We have

$$E(Y_n) = \frac{1}{n} \sum_{k=1}^{n} m_k.$$

Since the X_k are pairwise uncorrelated, we have

$$D^2(Y_n) = \frac{1}{n^2} \sum_{k=1}^{n} \sigma_k^2.$$

If

(6.11.4) $$\lim_{n \to \infty} \frac{1}{n^2} \sum_{k=1}^{n} \sigma_k^2 = 0,$$

then by the Chebyshev theorem 6.11.1 it follows that

$$\lim_{n \to \infty} P[|Y_n - E(Y_n)| \geqslant \varepsilon] = 0.$$

Thus we have obtained the following corollary of the Chebyshev theorem.

Let $\{X_k\}$ $(k = 1, 2, \ldots)$ be a sequence of random variables pairwise uncorrelated and let $E(X_k) = m_k$ and $D^2(X_k) = \sigma_k^2$. If condition (6.11.4) is satisfied, then the sequence

$$\left\{ Y_n - \frac{m_1 + m_2 + \ldots + m_n}{n} \right\} \qquad (n = 1, 2, \ldots)$$

is stochastically convergent to 0.

B   In 5.3 we considered the Poisson scheme and the generalized binomial distribution associated with it. In this scheme we consider the sum of n independent random variables X_k $(k = 1, 2, \ldots, n)$ with the zero-one distribution, where $P(X_k = 0) = 1 - p_k, P(X_k = 1) = p_k$. Since $D^2(X_k) = p_k(1 - p_k) \leqslant \frac{1}{4}$, condition (6.11.4) is satisfied. Thus the corollary of the Chebyshev theorem takes a form which could be called the *Poisson law of large numbers.*

Theorem 6.11.2. *If the random variable Y_n is the arithmetic mean of the random variables X_k in the Poisson scheme,*

$$Y_n = \frac{X_1 + X_2 + \ldots + X_n}{n},$$

then the sequence

$$\left\{ Y_n - \frac{p_1 + p_2 + \ldots + p_n}{n} \right\} \qquad (n = 1, 2, \ldots)$$

is stochastically convergent to 0.

C Let us now consider the case where the pairwise uncorrelated random variables X_k $(k = 1, 2, \ldots)$ have the same expected value and the same standard deviation. Thus for every k we can write

$$E(X_k) = m, \qquad D^2(X_k) = \sigma^2.$$

If we introduce the random variables Y_n defined by (6.11.3), we have

$$E(Y_k) = m, \qquad D^2(Y_n) = \sigma^2/n.$$

Thus

$$\lim_{n \to \infty} D^2(Y_n) = 0.$$

According to the corollary of the Chebyshev theorem, the sequence $\{Y_n - m\}$ is stochastically convergent to zero. Thus we have obtained the *Chebyshev law of large numbers* (Chebyshev [1]), which we formulate in the following way.

Theorem 6.11.3. *Let $\{X_k\}$ $(k = 1, 2, \ldots)$ be a sequence of pairwise uncorrelated random variables with the same expected value and the same standard deviation, and let Y_n be given by formula* (6.11.3). *Then the sequence $\{Y_n\}$ is stochastically convergent to the common expected value m of the random variables X_k.*

The reader can verify that the Bernoulli law of large numbers is a special case of the Chebyshev law of large numbers.

D In all laws of large numbers considered until now we have assumed that the variances of the random variables involved exist. In the following theorem proved by Khintchin [1] and called *Khintchin's law of large numbers*, no assumption is made about the existence of the variances.

Theorem 6.11.4. Let $\{X_k\}$ $(k = 1, 2, \ldots)$ be a sequence of independent random variables with the same distribution and with expected value $E(X_k) = m$. Then the sequence $\{Y_n\}$, where

$$Y_n = \frac{X_1 + X_2 + \ldots + X_n}{n},$$

is stochastically convergent to m.

Proof. Let $\phi(t)$ be the common characteristic function of the random variables X_k. By the independence of the X_k, the characteristic function of Y_n is

(6.11.5)
$$\left[\phi\left(\frac{t}{n}\right) \right]^n.$$

Since the expected value m exists, we can expand $\phi(t)$ in the neighborhood of $t = 0$ according to the MacLaurin formula,

(6.11.6)
$$\phi(t) = 1 + mit + o(t).$$

Substituting the expression (6.11.6) into (6.11.5), we obtain

$$\left[\phi\left(\frac{t}{n}\right) \right]^n = \left[1 + \frac{mit}{n} + o\left(\frac{t}{n}\right) \right]^n.$$

Proceeding as in the proof of the de Moivre-Laplace theorem, we obtain

(6.11.7)
$$\lim_{n \to \infty} \left[\phi\left(\frac{t}{n}\right) \right]^n = e^{mit}.$$

The right-hand side of formula (6.11.7) is the characteristic function of the random variable Y with the one-point distribution such that

$$P(Y = m) = 1.$$

By theorem 6.6.1b the sequence $\{F_n(y)\}$ of distribution functions of Y_n converges to the distribution function of the random variable Y; thus, by theorem 6.2.1, the sequence $\{Y_n\}$ is stochastically convergent to m.

Example 6.11.1. Let us return to example 6.8.5. There we showed that the characteristic function $\phi(t)$ of the random variable

$$Y_n = \frac{X_1 + X_2 + \ldots + X_n}{n},$$

where the random variables X_k $(k = 1, 2, \ldots)$ are independent and have the same Cauchy distribution, is of the form $\phi(t) = e^{-|t|}$. Thus the law of large numbers does not apply to the sequence $\{Y_n\}$. This result is not surprising, (see Problem 6.38).

Necessary and sufficient conditions for the validity of the law of large numbers for a sequence of independent random variables have been given by Kolmogorov [1] (see Problem 6.39) and Feller [8]

6.12   THE STRONG LAW OF LARGE NUMBERS

A   The laws of large numbers considered until now state that under certain conditions the sequence $\{Z_n\}$ of random variables defined by the formula

$$(*) \qquad\qquad Z_n = \frac{1}{n} \sum_{k=1}^{n} X_k - c_n,$$

where $c_n = (1/n) \sum_{k=1}^{n} E(X_k)$ and the random variables X_k are independent, is stochastically convergent to zero. Thus for arbitrary $\varepsilon > 0$ and $\eta > 0$ we can find an N such that, for $n > N$, we have $P(|Z_n| > \varepsilon) < \eta$. It does not follow, however, that for arbitrary $\varepsilon > 0$ and $\eta > 0$ we can find an N such that

$$(6.12.1) \qquad\qquad P(\sup_{n \geqslant N} |Z_n| > \varepsilon) < \eta.$$

We observe that relation (6.12.1) implies that the probability of occurrence of the inequality $|Z_n| > \varepsilon$ for at least one value $n \geqslant N$ is smaller than η; thus, instead of the probability of one event ($|Z_n| > \varepsilon$) we have here the probability of an alternative of events

$$(|Z_N| > \varepsilon) \cup (|Z_{N+1}| > \varepsilon) \cup (|Z_{N+2}| > \varepsilon) \cup \dots.$$

We show in Section 6.12,G that relation (6.12.1) is equivalent to the relation

$$(6.12.2) \qquad\qquad P(\lim_{n \to \infty} Z_n = 0) = 1.$$

If (6.12.2) holds, we say that the sequence $\{Z_n\}$ is *convergent to zero almost everywhere*.

The laws of large numbers considered in the previous sections are called *weak laws of large numbers*.

Definition 6.12.1.   We say that the sequence $\{X_k\}$ ($k = 1, 2, \dots$) of random variables *obeys the strong law of large numbers* if there exists a sequence of constants $\{c_n\}$ ($n = 1, 2, \dots$) such that, for the random variables Z_n defined by formula (*), relation (6.12.1) holds for all $\varepsilon > 0$ and $\eta > 0$.

B   Thus far we do not have a satisfactory solution of the problem of necessary and sufficient conditions for the validity of the strong law of large numbers for a sequence of independent random variables. Detailed information on the present state of investigations in this field can be found in the monograph by Loève [1] and in the paper of Chung [1]. The most advanced results have been obtained by Prohorov [1, 4]. We shall present the theorem of Kolmogorov [2] giving sufficient conditions for the validity of the strong law of large numbers. The proof of this theorem is based on a generalization of the Chebyshev inequality which was proved by Kolmogorov [1] and which is given in Section 6.12,C. In Section 6.12,D we prove the *Borel-Cantelli lemma* (Borel [1], Cantelli [1]), which is used in the proof of the next theorem of Kolmogorov [7], stating that for a sequence of independent, identically distributed random variables the existence of the expected value is a necessary and sufficient condition for the validity of the strong law of large numbers. Section 6.12,F is devoted to the question of existence of necessary and sufficient conditions for the validity of the strong law of large numbers, expressed in terms of variances. These results are formulated in theorems 6.12.4 and 6.12.5.

C   The Kolmogorov inequality. *Let* $X_1, X_2, \ldots, X_n$ *be independent random variables whose variances exist. Let* $U_i = X_i - E(X_i)$ *and*

$$Y_k = \sum_{i=1}^{k} U_i,$$

where $k = 1, 2, \ldots, n,$ *Then for arbitrary* $\varepsilon > 0$ *we have the inequality*

$$(6.12.3) \qquad P(\max_{1 \leqslant k \leqslant n} |Y_k| \geqslant \varepsilon) \leqslant \frac{D^2(Y_n)}{\varepsilon^2}.$$

Proof. Denote by A_0 the event $|Y_k| < \varepsilon$ for $k = 1, 2, \ldots, n$, by A_1 the event $|Y_1| \geqslant \varepsilon$, and by A_j $(j = 2, 3, \ldots, n)$ the event $|Y_k| < \varepsilon$ for $k = 1, 2, \ldots, j - 1$ and $|Y_j| \geqslant \varepsilon$. Notice that the event $\max_{1 \leqslant k \leqslant n} |Y_k| \geqslant \varepsilon$ is equal to the event $\sum_{j=1}^{n} A_j$; hence

$$P(\max_{1 \leqslant k \leqslant n} |Y_k| \geqslant \varepsilon) = P\left(\sum_{j=1}^{n} A_j\right).$$

Since the events A_j are pairwise exclusive, inequality (6.12.3) is equivalent to

$$(6.12.4) \qquad \sum_{j=1}^{n} P(A_j) \leqslant \frac{D^2(Y_n)}{\varepsilon^2}.$$

Denote by $F_n(y)$ the distribution function of the random variable Y_n. We observe that for every y we have

$$F_n(y) = P(Y_n < y) = P(Y_n < y, A_0) + \ldots + P(Y_n < y, A_n)$$

$$= P(A_0)P(Y_n < y \mid A_0) + \ldots + P(A_n)P(Y_n < y \mid A_n)$$

$$= P(A_0)F_n(y \mid A_0) + \ldots + P(A_n)F_n(y \mid A_n),$$

where $F_n(y \mid A_j)$ is the conditional distribution function of Y_n under the condition that the event A_j has occurred. Since $E(Y_n) = 0$, we obtain

$$(6.12.5) \quad D^2(Y_n) = \int_{-\infty}^{+\infty} y^2 \, dF_n(y) = \sum_{j=0}^{n} P(A_j) \int_{-\infty}^{+\infty} y^2 \, dF_n(y \mid A_j)$$

$$= \sum_{j=0}^{n} P(A_j) E(Y_n^2 \mid A_j) \geq \sum_{j=1}^{n} P(A_j) E(Y_n^2 \mid A_j).$$

Furthermore, we have

$$E(Y_n^2 \mid A_j) = E\left[\left(Y_j^2 + 2\sum_{k>j} Y_j U_k + \sum_{k>j} U_k^2 + 2\sum_{k>h>j} U_k U_h \right) \middle| A_j \right]$$

$$\geq E\left[\left(Y_j^2 + 2\sum_{k>j} Y_j U_k + 2\sum_{k>h>j} U_k U_h \right) \middle| A_j \right].$$

We observe that the event A_j concerns only the first j random variables U_i and that the random variables X_k are independent. Thus Y_j and U_k are independent if $k > j$. Similarly, if $k > h > j$, the random variables U_k and U_h are independent. Thus for $j \geq 1$ and for the k and h under consideration, we have

$$E(Y_j U_k \mid A_j) = 0, \qquad E(U_k U_h \mid A_j) = 0.$$

Thus

$$E(Y_n^2 \mid A_j) \geq E(Y_j^2 \mid A_j) \qquad (j = 1, 2, \ldots, n).$$

By the definition of the A_j and the last inequality, we obtain

$$E(Y_n^2 \mid A_j) \geq \varepsilon^2 \qquad (j = 1, 2, \ldots, n).$$

Using formula (6.12.5), we finally obtain

$$D^2(Y_n) \geq \varepsilon^2 \sum_{j=1}^{n} P(A_j).$$

Formula (6.12.4) is equivalent to the Kolmogorov inequality, which was to be proved.

The Kolmogorov inequality has been generalized (see Problem 6.41) by Hájek and Rényi [1] and quite recently by Birnbaum and Marshall [1].

Theorem 6.12.1 (Kolmogorov). *Let $\{X_k\}$ $(k = 1, 2, \ldots)$ be a sequence of independent random variables and let the variance $D^2(X_k)$ of X_k exist. If*

$$(6.12.6) \qquad \sum_{k=1}^{\infty} \frac{D^2(X_k)}{k^2} < \infty,$$

the sequence $\{X_k\}$ obeys the strong law of large numbers with

$$c_n = \frac{1}{n} \sum_{k=1}^{n} E(X_k).$$

Proof. Let

$$Z_n = \sum_{k=1}^{n} \frac{X_k - E(X_k)}{n}.$$

Let N and m_0 be two integers satisfying the inequalities

(6.12.7) $$2^{m_0} < N \leqslant 2^{m_0+1}.$$

Let

$$P_N = P(\sup_{n \geqslant N} |Z_n| > \varepsilon).$$

We shall prove that for arbitrary $\varepsilon > 0$ and $\eta > 0$ we can find an N such that $P_N < \eta$.

Let

$$P_m = P(\max_{2^m < n \leqslant 2^{m+1}} |Z_n| > \varepsilon),$$

where $m \geqslant m_0$ is an integer. The inequality

(6.12.8) $$P_N \leqslant \sum_{m=m_0}^{\infty} P_m$$

is satisfied. Replacing $D^2(X_k)$ by $D^2(X_k/2^{m+1})$ in the Kolmogorov inequality, we obtain

$$P_m \leqslant P(\max_{1 \leqslant n \leqslant 2^{m+1}} |Z_n| > \varepsilon) \leqslant \frac{1}{\varepsilon^2 2^{2m+2}} \sum_{k=1}^{2^{m+1}} D^2(X_k).$$

Hence by formula (6.12.8)

$$P_N \leqslant \sum_{m=m_0}^{\infty} \frac{1}{\varepsilon^2 2^{2m+2}} \sum_{k=1}^{2^{m+1}} D^2(X_k).$$

Changing the order of summation in the last formula, we obtain

(6.12.9) $$P_N \leqslant \frac{1}{\varepsilon^2} \sum_{k=1}^{\infty} D^2(X_k) \sum_{m=m(k)}^{\infty} \frac{1}{2^{2m+2}},$$

where $m(k) = m_0$ if $1 \leqslant k \leqslant 2^{m_0+1}$ and $m(k) = m_0 + j$ if $2^{m_0+j} + 1 \leqslant k \leqslant 2^{m_0+j+1}$ ($j = 1, 2, \ldots$). We have

$$\sum_{m=m(k)}^{\infty} \frac{1}{2^{2m+2}} = \frac{1}{2^{2m(k)} \cdot 3} \leqslant \frac{4}{3k^2}.$$

Thus

(6.12.10) $$P_N \leqslant \frac{1}{3\varepsilon^2} \left[\frac{1}{2^{2m_0}} \sum_{k=1}^{m_0} D^2(X_k) + 4 \sum_{k=m_0+1}^{\infty} \frac{D^2(X_k)}{k^2} \right].$$

Now let $\eta > 0$ be an arbitrary positive number. From conditions (6.12.6) and (6.12.7) it follows that for sufficiently large N, hence for sufficiently large m_0, we have

(6.12.11)
$$\frac{4}{3\varepsilon^2} \sum_{k=m_0=1}^{\infty} \frac{D^2(X_k)}{k^2} < \frac{\eta}{2} .$$

Moreover,

$$\frac{1}{3\varepsilon^2} \cdot \frac{1}{2^{2m_0}} \sum_{k=1}^{m_0} D^2(X_k) \leqslant \frac{1}{3\varepsilon^2 m_0^2} \sum_{k=1}^{m_0} D^2(X_k).$$

By the well-known Kronecker lemma we obtain from formula (6.12.6)

$$\lim_{m_0 \to \infty} \frac{1}{m_0^2} \sum_{k=1}^{m_0} D^2(X_k) = 0 .$$

Thus for sufficiently large N we have

(6.12.12)
$$\frac{1}{3\varepsilon^2 2^{2m_0}} \sum_{k=1}^{m_0} D^2(X_k) < \frac{\eta}{2} .$$

From formulas (6.12.10) to (6.12.12) it follows that

$$P_N < \eta,$$

which proves the Kolmogorov theorem.

From theorem 6.12.1 we obtain the following corollary. *If the variances of a sequence of independent random variables are uniformly bounded, this sequence obeys the strong law of large numbers.* In particular, if X_k is the number of successes (equal to 0 or 1) in the Bernoulli scheme, the sequence $\{X_k\}$ obeys the strong law of large numbers, which here is called the *Borel law of large numbers.* Borel discovered it in 1909 (Borel [1]). It is easy to see that the number of successes in the Poisson scheme also obeys the strong law of large numbers.

Dvoretzky [1] found (see Problem 6.30) a sufficient condition under which the number of successes, in a scheme in which the events are dependent in an arbitrary way, obeys the strong law of large numbers.

Sufficient conditions for the validity of the strong law of large numbers, expressed in terms of moments of order higher than the second, have been obtained by Brunk [1] and Prohorov [1, 4]. Bobrov [1] gave necessary (but not sufficient) conditions and sufficient (but not necessary) conditions, and these differ but little.

D   We now prove the following lemma.

Borel-Cantelli lemma. *Let* $\{A_n\}$ *(*$n = 1, 2, \ldots$*) be a sequence of events and let* $P(A_n)$ *denote the probability of the event* A_n*, where* $0 < P(A_n) < 1$. *Then:* I. *If*

(6.12.13)
$$\sum_{n=1}^{\infty} P(A_n) < \infty,$$

with probability one only a finite number of events A_n *occur.*

II. *If the events* $\{A_n\}$ $(n = 1, 2, \ldots)$ *are independent and*

(6.12.14) $$\sum_{n=1}^{\infty} P(A_n) = \infty,$$

with probability one an infinite number of events A_n *occur.*

Proof. Suppose that condition (6.12.13) is satisfied. Denote by A the event that an infinite number of events A_n occur; hence (see Problem 1.7)

$$A = \bigcap_{r=1}^{\infty} \bigcup_{n=r}^{\infty} A_n.$$

The relation

$$A \subset \bigcup_{n=r}^{\infty} A_n,$$

which is true for every integer r, implies (see Problem 1.10)

$$P(A) \leqslant P\left(\bigcup_{n=r}^{\infty} A_n\right) \leqslant \sum_{n=r}^{\infty} P(A_n).$$

It follows from formula (6.12.13) that for sufficiently large r the right-hand side of the last inequality is smaller than any given $\varepsilon > 0$. Hence $P(A) = 0$, which proves assertion I.

Suppose now that the events A_n are independent and that relation (6.12.14) is satisfied. The event $\bar{A}$, the complement of the event A, occurs if and only if at most a finite number of events A_n occur. Hence,

$$\bar{A} = \bigcup_{r=1}^{\infty} \bigcap_{n=r}^{\infty} \bar{A}_n.$$

In view of the independence of the A_n, we obtain

(6.12.15) $\quad 1 - P(A) = P(\bar{A}) = P\left(\bigcup_{r=1}^{\infty} \bigcap_{n=r}^{\infty} \bar{A}_n\right)$

$$\leqslant \sum_{r=1}^{\infty} P\left(\bigcap_{n=r}^{\infty} \bar{A}_n\right) = \sum_{r=1}^{\infty} \prod_{n=r}^{\infty} [1 - P(A_n)].$$

Formula (6.12.14) implies that the infinite product on the right-hand side of formula (6.12.15) is divergent to zero for any r. Hence $P(A) = 1$, which proves assertion II.

For an arbitrary sequence of random events A_n, Loève[3] found necessary and sufficient conditions (see Problem 6.34(a)) for the probability of occurrence of an infinite number of A_n to be zero, and Nash [1] found necessary and sufficient conditions (see Problem 6.34(b)) for this probability to be one.

Definition 6.12.2. We say that the sequences $\{X_k\}$ and $\{X_k^*\}$ $(k = 1, 2, \ldots)$ are *equivalent in the sense of Khintchin* if

$$\sum_{k=1}^{\infty} P(X_k \neq X_k^*) < \infty.$$

Theorem 6.12.2. *Let* $\{X_k\}$ *and* $\{X_k^*\}$ $(k = 1, 2, \ldots)$ *be two sequences of random variables equivalent in the sense of Khintchin. Then either both these sequences obey the strong law of large numbers, or neither does.*

Proof. Denote by A_k the event $(X_k \neq X_k^*)$. From definition 6.12.2 and assertion I of the Borel-Cantelli lemma, it follows that only a finite number of events A_k may occur. Hence for every sequence (x_k, x_k^*) of pairs of values of the random variable (X_k, X_k^*) there exists a k_0 (the number k_0 may be different for different sequences (x_k, x_k^*)) such that for $k > k_0$ we have $x_k = x_k^*$. Since the expressions

$$\frac{1}{n} \sum_{k=1}^{k_0} x_k, \qquad \frac{1}{n} \sum_{k=1}^{k_0} x_k^*$$

tend to zero as $n \to \infty$, it follows immediately from definition 6.12.1 that if either one of the sequences $\{X_k\}$, $\{X_k^*\}$ obeys the strong law of large numbers then the other one also does.

E. Kolmogorov [7] proved the following theorem concerning the validity of the strong law of large numbers for identically distributed random variables; it is called the *Kolmogorov law of large numbers*.

Theorem 6.12.3. *Let* $\{Y_i\}$ $(i = 1, 2, \ldots)$ *be a sequence of independent random variables with the same distribution function* $F(y)$. *Then the relation*

$$(6.12.16) \qquad P\left[\lim_{n \to \infty} \left(\frac{1}{n} \sum_{i=1}^{n} Y_i - c \right) = 0 \right] = 1$$

holds for some c *if and only if the expected value* $E(Y)$ *of a random variable* Y *with the distribution function* $F(y)$ *exists; here* $c = E(Y)$.

Proof. Suppose that $E(Y)$ exists. Then $E(|Y|)$ also exists. Let $X = Y - E(Y)$. Then $E(X) = 0$ and we have

$$(6.12.17) \quad \sum_{i=1}^{\infty} P(|X| > i) = \sum_{i=1}^{\infty} \sum_{k=i}^{\infty} P(k < |X| \leqslant k + 1)$$

$$= \sum_{k=1}^{\infty} k P(k < |X| \leqslant k + 1) \leqslant E(|X|) < \infty.$$

Consider the sequence of random variables $\{X_i^*\}$ $(i = 1, 2, \ldots)$ defined as follows:

$$X_i^* = \begin{cases} X_i & \text{if } |X_i| \leqslant i, \\ 0 & \text{otherwise,} \end{cases}$$

where $X_i = Y_i - E(Y)$. It follows from formula (6.12.17) that the sequences $\{X_i\}$ and $\{X_i{}^*\}$ are equivalent in the sense of Khintchin. Thus, in order to prove that the sequence $\{X_i\}$ obeys the strong law of large numbers it is sufficient to prove this (by theorem 6.12.2) for the sequence $\{X_i{}^*\}$. To do this, we shall show that inequality (6.12.6) is satisfied for the sequence $\{X_i{}^*\}$. We have

$$D^2(X_i{}^*) \leqslant E(X_i{}^{*2}) = \int_{-i}^{i} x^2 \, dF(x) \leqslant \sum_{k=0}^{i} (k+1)^2 P(k < |X| \leqslant k+1).$$

Hence

$$\sum_{i=1}^{\infty} \frac{D^2(X_i{}^*)}{i^2} \leqslant \sum_{i=1}^{\infty} \sum_{k=0}^{i} \frac{(k+1)^2}{i^2} P(k < |X| \leqslant k+1)$$

$$= P(0 < |X| \leqslant 1) \sum_{i=1}^{\infty} \frac{1}{i^2} + \sum_{k=1}^{\infty} (k+1)^2 P(k < |X| \leqslant k+1) \sum_{i=k}^{\infty} \frac{1}{i^2}.$$

Since

$$\sum_{i=k}^{\infty} \frac{1}{i^2} < \frac{1}{k^2} + \frac{1}{k} < \frac{2}{k},$$

we have

$$\sum_{i=1}^{\infty} \frac{D^2(X_i{}^*)}{i^2} < 2P(0 < |X| \leqslant 1) + 2 \sum_{k=1}^{\infty} \frac{(k+1)^2}{k} P(k < |X| \leqslant k+1)$$

$$= 2P(0 < |X| \leqslant 1 + 2 \sum_{k=1}^{\infty} kP(k < |X| \leqslant k+1)$$

$$+ 4 \sum_{k=1}^{\infty} P(k < |X| \leqslant k+1) + 2 \sum_{k=1}^{\infty} \frac{P(k < |X| \leqslant k+1)}{k}.$$

It is obvious that the first, third, and fourth terms on the right-hand side of the last formula are finite. From formula (6.12.17) it follows that the second term is finite; thus inequality (6.12.6) holds. Hence, by theorem 6.12.1 it follows that the sequence $\{X_i{}^*\}$ obeys the strong law of large numbers, and thus the same is true for the sequences $\{X_i\}$ and $\{Y_i\}$.

This completes the proof that the existence of the expected value $E(Y)$ is a sufficient condition for the validity of the strong law of large numbers for the sequence $\{Y_i\}$. Hence there exists a sequence of constants c_n such that

(6.12.18)
$$P\left[\lim_{n \to \infty} \left(\frac{1}{n} \sum_{i=1}^{n} Y_i - c_n\right) = 0\right] = 1.$$

We have still to show that we can take $c_n = E(Y)$.

We observe that from formula (6.12.18) it follows that the sequence $\left\{\frac{1}{n} \sum_{i=1}^{n} Y_i - c_n\right\}$ is stochastically convergent to zero. On the other hand, the random variables Y_i satisfy all the assumptions of theorem 6.11.4, and thus the sequence $\left\{\dfrac{1}{n}\sum_{i=1}^{n}Y_i - E(Y)\right\}$ is stochastically convergent to zero. It follows that

(6.12.19) $$\lim_{n\to\infty} c_n = E(Y).$$

From relations (6.12.18) and (6.12.19) we obtain relation (6.12.16) with $c = E(Y)$.

Suppose now that for some c relation (6.12.11) is satisfied. Let $U_i = Y_i - c$. Formula (6.12.16) then takes the form

$$P\left(\lim_{n\to\infty}\frac{1}{n}\sum_{i=1}^{n}U_i = 0\right) = 1.$$

Hence we have

$$P\left(\lim_{i\to\infty}\frac{U_i}{i} = 0\right) = 1.$$

It follows from the last formula that at most a finite number of the events $|U_i| > i$ can occur. Hence by assertion II of the Borel-Cantelli lemma, we obtain the inequality

(6.12.20) $$\sum_{i=1}^{\infty} P(|U_i| > i) < \infty.$$

Denote by U a random variable with the same distribution as U_i. By formula (6.12.20) we obtain

$$E(|U|) \leqslant \sum_{k=0}^{\infty}(k+1)P(k < |U| \leqslant k+1)$$

$$= \sum_{k=0}^{\infty}kP(k < |U| \leqslant k+1) + 1 = \sum_{i=1}^{\infty}P(|U| > i) + 1 < \infty.$$

Thus the expected value $E(U)$ exists; hence the expected value $E(Y)$ also exists. Here, as has been shown, we can substitute $c = E(Y)$.

Theorem 6.12.3 is proved.

Looking at the proof of relation (6.12.20), we see that the assumption that the distributions of the random variables U_i are the same is not relevant. Thus we have proved: If $\{U_i\}$ $(i = 1, 2, \ldots)$ is a sequence of independent random variables obeying the strong law of large numbers with $c_n = 0$, then relation (6.12.20) is satisfied.

F   We now present without proof the theorem of Prohorov [1] giving necessary and sufficient conditions, expressed in terms of variances, for the validity of the strong law of large numbers for a sequence of bounded random variables X_i satisfying the relation

(6.12.21) $$|X_i| = o\left(\frac{i}{\log\log i}\right).$$

Define, for $r = 1, 2, \ldots,$

(6.12.22) $$H_r = \frac{1}{2^{2r}} \sum_{i=2^r+1}^{2^{r+1}} D^2(X_i).$$

Theorem 6.12.4 (Prohorov). *Let $\{X_i\}$ ($i = 1, 2, \ldots$) be a sequence of independent random variables satisfying relation (6.12.21). Then the sequence $\{X_i\}$ obeys the strong law of large numbers if and only if, for an arbitrary $\varepsilon > 0$,*

(6.12.23) $$\sum_{r=1}^{\infty} \exp\left(-\frac{\varepsilon}{H_r}\right) < \infty,$$

where H_r is defined by formula (6.12.22).

Fisz [6] showed that for a sequence of independent random variables $\{X_i\}$ satisfying a relation weaker than (6.12.21), namely,

(6.12.24) $$|X_i| < i,$$

there do not exist necessary and sufficient conditions for the validity of the strong law of large numbers, expressed in terms of variances. (see also Prohorov [5])

Theorem 6.12.5. *Let $\{X_i\}$ ($i = 1, 2, \ldots$) be independent random variables satisfying relation (6.12.24). There do not exist for the sequence $\{X_i\}$ necessary and sufficient conditions for the validity of the strong law of large numbers, expressed only in terms of variances.*

This theorem may be formulated as follows: *If $W(\sigma_1^2, \sigma_2^2, \ldots)$, where $\sigma_i^2 = D^2(X_i)$, is any sufficient condition for the validity of the strong law of large numbers for $\{X_i\}$, then there exists a sequence of independent random variables satisfying condition (6.12.24) which obeys the strong law of large numbers, and does not satisfy the condition $W(\sigma_1^2, \sigma_2^2, \ldots)$.*

Proof. To prove the theorem it is sufficient to show that there exist two sequences $\{X_i\}$ and $\{Y_i\}$ of independent random variables satisfying relation (6.12.24) and the equalities

$$D^2(X_i) = D^2(Y_i) = \sigma_i^2$$

such that one of them obeys the strong law of large numbers and the other does not.

Let us define the random variables X_i and Y_i as follows:

(6.12.25) $$P\left(X_i = \frac{i}{\log i}\right) = P\left(X_i = -\frac{i}{\log i}\right) = \frac{\log i}{2i},$$

$$P\left(X_i = 0\right) = 1 - \frac{\log i}{i}.$$

(6.12.26) $$P(Y_i = \beta i) = P(Y_i = -\beta i) = \frac{1}{2\beta^2 i \log i},$$

$$P(Y_i = 0) = 1 - \frac{1}{\beta^2 i \log i},$$

where $0 < \beta < 1$, and $i = 2, 3, \ldots$. We have

$$D^2(X_i) = D^2(Y_i) = \frac{i}{\log i}.$$

It is obvious that the random variables X_i satisfy relation (6.12.21). We shall show that they also satisfy relation (6.12.23). In fact,

$$\exp\left(-\frac{\varepsilon}{H_r}\right) < \exp\left(-\frac{\varepsilon r}{2} \log 2\right) = \frac{1}{2^{\varepsilon r/2}}.$$

Hence, for arbitrary $\varepsilon > 0$,

$$\sum_r \exp\left(-\frac{\varepsilon}{H_r}\right) < \sum_r \frac{1}{2^{\varepsilon r/2}} < \infty.$$

From theorem 6.12.4 it follows that the sequence $\{X_i\}$ obeys the strong law of large numbers. However, the sequence $\{Y_i\}$ does not; for suppose that $\{Y_i\}$ obeys the strong law of large numbers for some sequence $\{c_n\}$. Then it is also satisfied[1] for $c_n = 0$ $(n = 1, 2, \ldots)$. Thus the relation

(6.12.27) $$P\left(\lim_{i \to \infty} \frac{Y_i}{i} = 0\right) = 1$$

is satisfied. However, from formula (6.12.26) it follows that

$$\sum_{i=2}^{\infty} P\left(\frac{|Y_i|}{i} = \beta\right) = \infty.$$

Thus, according to assertion II of the Borel-Cantelli lemma, an infinite number of the events $(|Y_i| = \beta i)$ occur, which contradicts relation (6.12.27).

---

[1] We arrive at this conclusion using the theorem that if the random variables Y_i are of the order $0(i)$ and the sequence $\{Y_i\}$ obeys the strong law of large numbers, then we can take $c_n = E\left(\frac{1}{n} \sum_{i=1}^{n} Y_i\right)$.

Thus the sequence $\{Y_i\}$ does not obey the strong law of large numbers, which proves theorem 6.12.5.

We observe that the fact that necessary and sufficient conditions for the validity of the strong law of large numbers, expressed in terms of variances, do not exist for sequences of random variables not satisfying any relation like (6.12.24) could have been deduced from the example given by Kolmogorov [2]. It can also be shown (Fisz [6]) that in this general case necessary and sufficient conditions expressed in terms of moments of any finite order do not exist.

G   We now prove that relations (6.12.1) and (6.12.2) are equivalent.

Denote by A_N the event that $\sup_{n \geqslant N} |Z_n| > \varepsilon$, where $\varepsilon > 0$, and by A the product of the events A_N, that is,

$$(6.12.28) \qquad A = \prod_N A_N.$$

We observe that for every N

$$A_{N+1} \subset A_N,$$

and hence by theorem 1.3.4 we have

$$(6.12.29) \qquad P(A) = \lim_{N \to \infty} P(A_N).$$

The event $\overline{A_N}$, the complement of the event A_N, occurs if and only if, for every $n \geqslant N$, we have the relation $|Z_n| \leqslant \varepsilon$. Thus we have for every N

$$\overline{A_{N+1}} \supset \overline{A_N};$$

hence by theorem 1.3.5

$$(6.12.29') \qquad P(\overline{A}) = P(\sum_N \overline{A_N}) = \lim_{N \to \infty} P(\overline{A_N}).$$

Suppose now that relation (6.12.1) is not satisfied. Then there exist $\varepsilon > 0$ and $\eta > 0$ such that for every N

$$P(A_N) \geqslant \eta.$$

From the last relation and from relation (6.12.29) we obtain the inequality

$$(6.12.30) \qquad P(A) \geqslant \eta > 0,$$

from which it follows that relation (6.12.2) is not satisfied; for if it were satisfied, then for every $\varepsilon > 0$ the probability would be zero that for every N there exists an $n \geqslant N$ such that $|Z_n| > \varepsilon$. Hence $P(A) = 0$, in contradiction to (6.12.30).

Suppose, now, that relation (6.12.2) is not satisfied. Then there exist $\varepsilon > 0$ and $\eta > 0$ such that the probability of occurrence of the event $\overline{A_N}$ is smaller than $1 - \eta$ for every N, or

(6.12.31) $P(\bar{A}) < 1 - \eta.$

It follows from the last inequality that relation (6.12.1) is not satisfied; for if it were satisfied, then for any $\varepsilon > 0$ and $\eta > 0$ there would exist an N such that $P(\overline{A_N}) \geqslant 1 - \eta$, so that from the fact that the sequence $\{\bar{A}_n\}$ is nondecreasing and from formula (6.12.29') we would obtain $P(\bar{A}) \geqslant 1 - \eta$, in contradiction to (6.12.31).

The equivalence of relations (6.12.1) and (6.12.2) is proved.

At the end of this section we give an example of a sequence of random variables which converges to zero stochastically but does not converge to zero almost everywhere (see also Problems 6.38).

Example 6.12.1. Let us consider the sequence $\{Z_n\}$ ($n = 1, 2, \ldots$) of independent random variables, where

(6.12.32) $P(Z_n = 1) = \dfrac{1}{n},$

$$P(Z_n = 0) = 1 - \frac{1}{n}.$$

The sequence $\{Z_n\}$ converges to zero stochastically, since from the equality $P(|Z_n| > \varepsilon) = P(Z_n = 1)$, which holds for every $0 < \varepsilon < 1$, we obtain, for any $\varepsilon > 0$,

$$\lim_{n \to \infty} P(|Z_n| > \varepsilon) = \lim_{n \to \infty} \frac{1}{n} = 0.$$

However, the considered sequence $\{Z_n\}$ does not satisfy relation (6.12.2); for, denoting by A_n the event $(Z_n = 1)$, it follows from (6.12.32) that

$$\sum_{n=1}^{\infty} P(A_n) = \infty.$$

From the independence of the A_n and from the Borel-Cantelli lemma it follows that the probability that an infinite number of the A_n will occur equals one; hence with probability one there will exist a subsequence of the sequence $\{Z_n\}$ which is not convergent to zero. This obviously contradicts relation (6.12.2).

6.13 MULTIDIMENSIONAL LIMIT DISTRIBUTIONS

Theorems 6.6.1a and 6.6.1b, generalized to multidimensional distributions, play a basic role in the theory of multidimensional limit distributions. We do not formulate these generalizations, as it is sufficient to replace the one-dimensional characteristic functions and distribution functions in these theorems by the corresponding multidimensional ones.

Let now $(X_1, \ldots, X_r)$ be a random variable with the multidimensional distribution defined by formula (5.12.1'). Thus

(6.13.1) $P(X_1 = k_1, X_2, = k_2, \ldots, X_r = k_r)$

$$= \frac{n!}{k_1! \, k_2! \ldots ! \, k_r! \, (n - K)!} \, p_1^{k_1} p_2^{k_2} \ldots p_r^{k_r} q^{n-K},$$

where $K = k_1 + \ldots + k_r$ and $q = 1 - p_1 - \ldots - p_r$. In Section 5.12 we showed that the characteristic function $\phi_x(t_1, \ldots, t_r)$ of the random variable $(X_1, \ldots, X_r)$ has the form

(6.13.2) $$\phi_x(t_1, t_2, \ldots, t_r) = \left(\sum_{j=1}^r p_j e^{it_j} + q \right)^n.$$

We have also given the equalities

(6.13.3) $E(X_j) = np_j, \quad \lambda_{jj} = D^2(X_j) = np_j(1 - p_j) \quad (j = 1, 2, \ldots, r),$

$$\lambda_{jk} = E[(X_j - np_j)(X_k - np_k)] = -np_j p_k$$

$$(j, k = 1, 2, \ldots, r; \; j \neq k).$$

Consider the random variable $(Z_1, \ldots, Z_r)$, where

(6.13.4) $$Z_j = \frac{X_j - np_j}{\sqrt{np_j(1 - p_j)}} \quad (j = 1, 2, \ldots, r).$$

The second-order moments η_{jk} of the random variables Z_j are

(6.13.4') $$\eta_{jj} = 1 \quad (j = 1, \ldots, r),$$

$$\eta_{jk} = -\sqrt{p_j p_k / (1 - p_j)(1 - p_k)} \quad (j, k = 1, \ldots, r; \; j \neq k).$$

The following theorem is a generalization of the integral theorem of de Moivre-Laplace to a multinomial distribution.

Theorem 6.13.1. Let $(X_1, \ldots, X_r)$ have the multinomial distribution defined by formula (6.13.1) and let $F_n(z_1, \ldots, z_r)$ be the distribution function of $(Z_1, \ldots, Z_r)$, where the Z_j $(j = 1, 2, \ldots, r)$ are given by formula (6.13.4). If $0 < q < 1$ and $0 < p_i < 1$ $(i = 1, 2, \ldots, r)$, then the sequence $\{F_n(z_1, \ldots, z_r)\}$ satisfies the relation

(6.13.5) $\lim_{n \to \infty} F_n(z_1, z_2, \ldots, z_r)$

$$= \frac{1}{(2\pi)^{r/2} \sqrt{|\mathbf{N}|}} \int_{-\infty}^{z_1} \int_{-\infty}^{z_2} \ldots \int_{-\infty}^{z_r} \exp \left(-\frac{1}{2|\mathbf{N}|} \sum_{j,k=1}^r |\mathbf{N}_{jk}| z_j z_k \right) dz_1 \, dz_2 \ldots dz_r,$$

where $|\mathbf{N}|$ *is the matrix of the moments* η_{jk} *and*

$$(6.13.6) \quad |\mathbf{N}| = \frac{q}{(1 - p_1) \ldots (1 - p_r)} \neq 0,$$

$$|\mathbf{N}_{jj}| = \frac{(1 - p_j)(q + p_j)}{(1 - p_1) \ldots (1 - p_r)} \quad (j = 1, 2, \ldots, r),$$

$$|\mathbf{N}_{jk}| = \frac{\sqrt{p_j p_k (1 - p_j)(1 - p_k)}}{(1 - p_1) \ldots (1 - p_r)} \quad (j, k = 1, 2, \ldots, r; k \neq j).$$

Proof. Denote by $\phi_z(t_1, \ldots, t_r)$ the characteristic function of the random variable $(Z_1, \ldots, Z_r)$. By formulas (6.13.3) and (6.13.4) we obtain

$$\phi_z(t_1, t_2, \ldots, t_r) = \exp\left(-\sum_{j=1}^{r} it_j \sqrt{\frac{np_j}{1 - p_j}}\right)\left\{\sum_{j=1}^{r} p_j \exp\left[\frac{it_j}{\sqrt{np_j(1 - p_j)}}\right] + q\right\}^n,$$

$$\log \phi_z(t_1, t_2, \ldots, t_r)$$

$$= -i\sum_{j=1}^{r} t_j \sqrt{\frac{np_j}{1 - p_j}} + n \log\left\{\sum_{j=1}^{r} p_j \exp\left[\frac{it_j}{\sqrt{np_j(1 - p_j)}}\right] + q\right\}$$

$$= -i\sqrt{n}\sum_{j=1}^{r} t_j \sqrt{\frac{p_j}{1 - p_j}}$$

$$\quad + n \log\left[1 + \frac{i}{\sqrt{n}}\sum_{j=1}^{r} t_j \sqrt{\frac{p_j}{1 - p_j}} - \frac{1}{2n}\sum_{j=1}^{r} \frac{t_j^2}{1 - p_j} + o\left(\frac{1}{n}\right)\right]$$

$$= -i\sqrt{n}\sum_{j=1}^{r} t_j \sqrt{\frac{p_j}{1 - p_j}} + n \log(1 + s),$$

where

$$s = \frac{i}{\sqrt{n}}\sum_{j=1}^{r} t_j \sqrt{\frac{p_j}{1 - p_j}} - \frac{1}{2n}\sum_{j=1}^{r} \frac{t_j^2}{1 - p_j} + o\left(\frac{1}{n}\right).$$

Since for any fixed $t_1, \ldots, t_r$ and for sufficiently large n we have $|s| < 1$, we can write

$$\lim_{n \to \infty}\left[n \log(1 + s) - i\sqrt{n}\sum_{j=1}^{r} t_j \sqrt{\frac{p_j}{1 - p_j}}\right] = -\frac{1}{2}\left[\sum_{j=1}^{r} \frac{t_j^2}{1 - p_j}\right.$$

$$\left. - \left(\sum_{j=1}^{r} t_j \sqrt{\frac{p_j}{1 - p_j}}\right)^2\right] = -\frac{1}{2}\left[\sum_{j=1}^{r} t_j^2 - \sum_{\substack{j,k=1 \\ j \neq k}}^{r} t_j t_k \sqrt{\frac{p_j p_k}{(1 - p_j)(1 - p_k)}}\right].$$

Finally, we have

$$(6.13.7) \quad \lim_{n \to \infty} \phi_z(t_1, t_2, \ldots, t_r)$$

$$= \exp\left\{-\frac{1}{2}\left[\sum_{j=1}^{r} t_j^2 - \sum_{\substack{j,k=1 \\ j \neq k}}^{r} t_j t_k \sqrt{\frac{p_j p_k}{(1 - p_j)(1 - p_k)}}\right]\right\}.$$

Comparing formulas (6.13.7) and (5.11.9), we see that on the right-hand side of (6.13.7) we have the characteristic function of an r-dimensional random variable with a normal distribution whose second-order moment matrix is given by formulas (6.13.4'). Finally, from the generalization of theorem 6.6.1b, mentioned at the beginning of this section, we obtain formula (6.13.5).[1] Formulas (6.13.6) can easily be derived.

In Section 5.12 we showed that the random variable $(X_1, \ldots, X_r)$ with the multinomial distribution given by formula (6.13.1) can be represented as the sum of n independent r-dimensional random variables with the same distribution. From this point of view we can treat theorem 6.13.1 as a theorem on the limit of a sequence of distribution functions of sums of independent r-dimensional random variables with the same distribution. Moreover, the Lindeberg-Lévy theorem can also be generalized to multidimensional random variables. We now formulate this theorem.

Theorem 6.13.2. *Let* $(Y_{m1}, \ldots, Y_{mr})$ *$(m = 1, 2, \ldots)$ be a sequence of independent r-dimensional random vectors with the same distribution and the second-order moment matrix* $\mathbf{M} = [\lambda_{jk}]$, *where the determinant* $|\mathbf{M}| \neq 0$. *Let* $F_n(z_1, \ldots, z_r)$ *be the distribution function of the random variable* $(Z_{n1}, \ldots, Z_{nr})$, *where*

$$
Z_{nj} = \sum_{m=1}^{n} \frac{Y_{mj} - E(Y_{mj})}{\sqrt{n\lambda_{jj}}} \qquad (j = 1, 2, \ldots, r).
$$

Then the sequence $\{F_n(z_1, \ldots, z_r)\}$ *satisfies the relation*

$$
\lim_{n \to \infty} F_n(z_1, z_2, \ldots, z_r)
$$
$$
= \frac{1}{(2\pi)^{r/2}\sqrt{|\mathbf{N}|}} \int_{-\infty}^{z_1} \int_{-\infty}^{z_2} \cdots \int_{-\infty}^{z_r} \exp\left(-\frac{1}{2|\mathbf{N}|} \sum_{j,k=1}^{r} |\mathbf{N}_{jk}| z_j z_k\right) dz_1 dz_2 \ldots dz_r,
$$

where $\mathbf{N} = [\eta_{jk}]$ *is the second-order moment matrix with*

$$
\eta_{jj} = 1 \qquad (j = 1, 2, \ldots, r),
$$

$$
\eta_{jk} = \frac{\lambda_{jk}}{\sqrt{\lambda_{jj}\lambda_{kk}}} \qquad (j, k = 1, 2, \ldots, r; \ j \neq k).
$$

We leave to the reader the proof of this theorem, which is analogous to that of theorem 6.13.1.

The generalization of the Lapunov theorem to multidimensional distributions was given by Bernstein [3]. The generalization of the Lindeberg-Feller theorem to multidimensional distributions can be found in

---

[1] The Poisson theorem can also be generalized to the multinomial distribution. The class of all possible limit distributions of the multinomial distribution was given (see Problem 6.45) by Fisz [2].

the paper of Gnedenko [9]. E. Mourier [1] gave a further generalization of the Lindeberg-Lévy theorem to random variables that take on values in a Hilbert space.

6.14 LIMIT THEOREMS FOR RATIONAL FUNCTIONS OF SOME RANDOM VARIABLES

A   We now present the following theorem, which plays an important role in applications.

Theorem 6.14.1.   Let $\{X_n\}$ $(n = 1, 2, \ldots)$ be an arbitrary sequence of random variables (dependent or not) and let the corresponding sequence of distribution functions $\{F_n(x)\}$ converge as $n \to \infty$ to the distribution function $F(x)$. Further, let $\{Y_n\}$ $(n = 1, 2, \ldots)$ be another sequence of random variables stochastically convergent to a constant a. Then

(α) the sequence of distribution functions of the random variables $X_n + Y_n$ converges to the distribution function $F(x - a)$.

(β) the sequence of distribution functions of the random variables $X_n - Y_n$ converges to the distribution function $F(x + a)$.

(γ) the sequence of distribution functions of the random variables $X_n Y_n$ converges to the distribution function $F(x/a)$ if $a > 0$ and to the distribution function $1 - F(x/a)$ if $a < 0$.

(δ) the sequence of distribution functions of the random variables X_n / Y_n converges to the distribution function $F(ax)$ if $a > 0$ and to the distribution function $1 - F(ax)$ if $a < 0$.

We restrict ourselves to the proof of (γ) for $a > 0$, since the proofs of the remaining assertions are analogous.

Denote by $\Phi_n(x)$ the distribution function of the product $X_n Y_n$. Let x/a be a continuity point of the limit distribution function $F(x)$. We want to show that the relation

(6.14.1) $$\Phi_n(x) = P(X_n Y_n < x) \to F\left(\frac{x}{a}\right)$$

holds.

Denote by S the set of all points in the plane (x_n, y_n) for which the relation $x_n y_n < x$ is satisfied. The set S can be represented as the sum of two disjoint sets, S_1 and S_2, consisting of points satisfying the inequalities

(6.14.2) $x_n y_n < x, \quad |y_n - a| \leqslant \varepsilon,$

(6.14.3) $x_n y_n < x, \quad |y_n - a| > \varepsilon,$

respectively, where $0 < \varepsilon < a$. Next denote by A and B the sets in the plane $x_n y_n$ satisfying, respectively, the inequalities

(6.14.4) $x_n(a - \varepsilon) < x, \quad |y_n - a| \leqslant \varepsilon,$

(6.14.5) $x_n(a + \varepsilon) < x, \quad |y_n - a| \leqslant \varepsilon.$

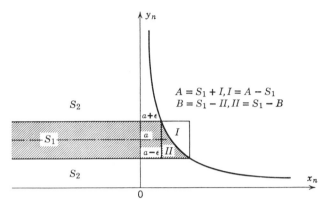

Fig. 6.14.1

In Fig. 6.14.1, where only the half-plane $y_n > 0$ is represented, the set S is situated outside the hyperbola $x_n y_n = x$. The set S_1 is shaded, the set $A - S_1$, is denoted by the symbol I, and the set $S_1 - B$ is denoted by the symbol II.

Denote by $P_n(S)$, $P_n(S_1)$, $P_n(S_2)$, $P_n(A)$ and $P_n(B)$, respectively, the probabilities that the two-dimensional random variable (X_n, Y_n) takes on a value from these sets. We have

$$(6.14.6) \qquad \Phi_n(x) = P(X_n Y_n < x) = P_n(S) = P_n(S_1) + P_n(S_2).$$

Since $B \subset S_1 \subset A$,

$$(6.14.7) \qquad\qquad P_n(B) \leqslant P_n(S_1) \leqslant P_n(A).$$

We notice that

$$F_n\left(\frac{x}{a + \varepsilon}\right) = P\left(X_n < \frac{x}{a + \varepsilon}\right)$$

$$= P\left(X_n < \frac{x}{a + \varepsilon}, |Y_n - a| \leqslant \varepsilon\right) + P\left(X_n < \frac{x}{a + \varepsilon}, |Y_n - a| > \varepsilon\right)$$

$$= P_n(B) + P\left(X_n < \frac{x}{a + \varepsilon}, |Y_n - a| > \varepsilon\right),$$

$$F_n\left(\frac{x}{a - \varepsilon}\right) = P_n(A) + P\left(X_n < \frac{x}{a - \varepsilon}, |Y_n - a| > \varepsilon\right).$$

By formulas (6.14.6) and (6.14.7) we obtain the inequality

$$(6.14.8) \quad F_n\left(\frac{x}{a + \varepsilon}\right) - P\left(X_n < \frac{x}{a + \varepsilon}, |Y_n - a| > \varepsilon\right)$$

$$\leqslant \Phi_n(x) - P_n(S_2) \leqslant F_n\left(\frac{x}{a - \varepsilon}\right) - P\left(X_n < \frac{x}{a - \varepsilon}, |Y_n - a| > \varepsilon\right).$$

We have

$$P\left(X_n < \frac{x}{a + \varepsilon}, \, |Y_n - a| > \varepsilon\right) \leqslant P(|Y_n - a| > \varepsilon),$$

$$P_n(S_2) = P(X_n Y_n < x, \, |Y_n - a| > \varepsilon) \leqslant P(|Y_n - a| > \varepsilon),$$

$$P\left(X_n < \frac{x}{a - \varepsilon}, \, |Y_n - a| > \varepsilon\right) \leqslant P(|Y_n - a| > \varepsilon).$$

From the assumption of the stochastic convergence of $\{Y_n\}$ to the constant a, it follows that the probabilities on the right-hand sides of these inequalities decrease to zero as $n \to \infty$. Thus from inequality (6.14.8) we obtain

$$F\left(\frac{x}{a + \varepsilon}\right) = \lim_{n \to \infty} F_n\left(\frac{x}{a + \varepsilon}\right) \leqslant \lim_{n \to \infty} \Phi_n(x) \leqslant \lim_{n \to \infty} F_n\left(\frac{x}{a - \varepsilon}\right) = F\left(\frac{x}{a - \varepsilon}\right),$$

where $x/(a + \varepsilon)$ and $x/(a - \varepsilon)$ are continuity points of $F(x)$. Since ε can be chosen arbitrarily small, we immediately obtain (6.14.1).

Theorem 6.14.1 presented here is due to Cramér [2]. A proof of assertion (α) was also given by Neyman [8] (see also Wolfowitz [1]). A simple proof of all the assertions of theorem 6.14.1 could also have been obtained from theorem 6.4.1 (see Sverdrup [1]).

A generalization of assertion (α) of theorem 6.14.1 to the case when the sequence of distribution functions of the random variables Y_n converges to the distribution function of a nondegenerate random variable was given by Teicher [1].

B   As a corollary of theorem 6.14.1, we obtain the following theorem of Slutsky [1].

Theorem 6.14.2.   *If the sequences* $\{X_{1k}\}, \{X_{2k}\}, \ldots, \{X_{rk}\}$ ($k = 1, 2, \ldots$) *of random variables* (r *is fixed*) *are stochastically convergent to the constant values* $a_1, a_2, \ldots, a_r$, *then an arbitrary rational function* $R(X_{1k}, X_{2k}, \ldots X_{rk})$ *is stochastically convergent to the constant* $R(a_1, a_2, \ldots, a_r)$, *provided this constant is finite.*

To prove this theorem it is sufficient to substitute for $F(x)$ in theorem 6.14.1 the distribution function of the one-point distribution and to observe that this theorem is true for a sequence of random variables obtained by performing a finite number of rational operations on random variables.

C   As an easy corollary of theorem 6.14.1 we obtain the following result.

Theorem 6.14.3.   *Let* $\{Z_n\}$ ($n = 1, 2, \ldots$) *be a sequence of random variables stochastically convergent to a random variable* Z_0. *Then the relation*

$$(6.14.9) \qquad \lim_{n \to \infty} F_n(z) = F_0(z),$$

where $F_n(z)$ *and* $F_0(z)$ *are the distribution functions of* Z_n *and* Z_0, *respectively, holds at every continuity point of* $F_0(z)$.

Proof. Suppose that the assumption is satisfied. Write $Z_n = Z_0 + (Z_n - Z_0)$. Since the sequence $\{Z_n - Z_0\}$ is stochastically convergent to zero, formula (6.14.9) follows immediately from assertion (α) of theorem 6.14.1. We observe that the converse of this theorem is not true. We know, however, (see theorem 6.2.1) that the converse is true in the particular case when Z_0 takes on a constant value with probability one.

6.15 FINAL REMARKS

Recently some new papers have appeared dealing with the limit distributions of sums of a random number of independent random variables. In Section 5.14 we considered the random variable Y defined by the formula

$$Y = X_1 + X_2 + \ldots + X_N,$$

where both X_k and N are random variables. Suppose that the distribution of N depends on some parameter λ. Then the problem arises of finding the limit distribution of the random variable Y (suitably standardized) as $\lambda \to \infty$. We restrict ourselves to mentioning some of the most important papers in this domain, namely, those of Robbins [1], Dobrushin [4], and Anscombe [1]. These results find many practical applications (see, for instance, Rényi [7] and Takács [5]). We deal with similar problems in Chapter 17. (See also Problems 6.46 and 6.47.)

Problems and Complements

6.1. The random variables X and Y are independent and X has the Poisson distribution with the parameter $\lambda = 1$.

(*a*) Applying formula (6.5.6), find the distribution function of the random variable $X + Y$, where Y has the binomial distribution with $n = 5$ and $p = \frac{1}{3}$.

(*b*) Applying formula (6.5.9), find the distribution function of the ratio $Z = X/Y$, where Y is a random variable with the normal distribution $N(0; 1)$.

6.2. Prove the theorem of Helly: If the sequence $\{F_n(x)\}$ ($n = 1, 2, \ldots$) of distribution functions converges to the distribution function $F(x)$ and $g(x)$ is bounded and continuous for all x, then

$$\lim_{n \to \infty} \int_{-\infty}^{+\infty} g(x)\, dF_n(x) = \int_{-\infty}^{+\infty} g(x)\, dF(x).$$

Hint: The proof is similar to that of theorem 6.6.1*a*

6.3. Give an example of a sequence of distribution functions $\{F_n(x)\}$ of random variables X_n with variances $D^2(X_n)$ such that

$$\lim_{n \to \infty} F_n(x) = \frac{1}{\sqrt{2\pi}} \int_{-\infty}^{x} e^{-x^2/2}\, dx,$$

whereas the relation

$$\lim_{n \to \infty} D^2(X_n) = 1$$

does not hold.

Hint: Consider first the sequence of random variables Y_n for which

$$P(Y_n = 0) = 1 - \frac{1}{n} \qquad (n = 1, 2, 3, \ldots),$$

$$P(Y_n = n) = \frac{1}{n},$$

and then use assertion (α) of theorem 6.14.1.

6.4. Prove that the sequence $\{F_n(x)\}$ converges to the distribution function $F(x)$ if and only if the relation

$$\lim_{n \to \infty} F_n(x) = F(x)$$

holds for all points x in a set S which is everywhere dense in the interval $(-\infty, +\infty)$.

6.5. Show that if the sequence of characteristic functions $\{\phi_n(t)\}$ converges to the characteristic function $\phi(t)$ and $t_n \to t_0$, then $\phi_n(t_n) \to \phi(t_0)$.

6.6. The distribution functions $F(x)$ and $G(x)$ are said to be of the *same type* if there exist constants $a > 0$ and b such that for every x

$$G(x) = F(ax + b).$$

Prove that if the sequence of distribution functions $\{F_n(x)\}$ converges as $n \to \infty$ to a nondegenerate distribution function $F(x)$, and if $F_n(a_n x + b_n)$ converges to a nondegenerate distribution function $G(x)$, then $G(x)$ is of the same type as $F(x)$. (Khintchin [8]).

Hint: Restrict consideration to the subsequences a_{n_k} and b_{n_k} which converge to the numbers a and b (finite or infinite), respectively. Show that $0 < a < \infty$ and $-\infty < b < +\infty$ and that $F_n(a_n x + b_n) \to F(ax + b)$.

The generalization of the notion of type and the preceding theorem to n-dimensional random vectors are given in Fisz [2].

6.7. Prove formula (6.4.6).

Hint: Frontier of $S = F_r(S) = \bar{S} \cap \overline{R - S}$; interior of $S = \text{Int}(S) = S - F_r(S)$. Use the equation $P_n[\text{Int}(S)] + P_n[\text{Int}(R - S)] + P_n[F_r(S)] = 1$ $(n = 1, 2, 3, \ldots)$. Show first that (6.4.1) and $P[F_r(S)] = 0$ imply $\lim_{n \to \infty} P_n[\text{Int}(S)] = P[\text{Int}(S)]$, $\lim_{n \to \infty} P_n[\text{Int}(R - S)] = P[\text{Int}(R - S)]$ and $\lim_{n \to \infty} P_n[F_r(S)] = P[F_r(S)]$.

6.8. (The converse of the Lindeberg-Lévy theorem). Let $\{X_k\}$ $(k = 1, 2, 3, \ldots)$ be a sequence of independent and identically distributed random variables. If, for some constants a and A_n $(n = 1, 2, 3, \ldots)$, the relation

$$\lim_{n \to \infty} P\left(\frac{1}{a\sqrt{n}} \sum_{k=1}^{n} X_k - A_n < z\right) = \frac{1}{\sqrt{2\pi}} \int_{-\infty}^{z} \exp\left(-\frac{z^2}{2}\right) dz$$

holds for any z, then the variance σ^2 of X_k exists. If this is so, then $a = \sigma$ and A_n may be chosen to equal $\frac{\sqrt{n}}{\sigma} E(X_k)$ (see Gnedenko and Kolmogorov [1], Section 35).

6.9.(a) The random variables X_i $(i = 1, \ldots, r)$ are independent and X_i has the Poisson distribution with the parameter $\lambda_i > 0$. Show that if $\lambda_1 + \ldots + \lambda_r \to \infty$, then $\sum_{i=1}^{r} X_i$ has the asymptotically normal distribution

$$N\left(\sum_{i=1}^{r} \lambda_i; \sqrt{\sum_{i=1}^{r} \lambda_i}\right).$$

(b) Let X_1 and X_2 be independent and have the Poisson distribution with the parameters $\lambda_1 > 0$ and $\lambda_2 > 0$, respectively. Show that if $\lim_{\lambda_1 \to \infty, \lambda_2 \to \infty} (\lambda_1/\lambda_2) = 1$, then for any $p > 0$, as $\lambda_1 \to \infty$ and $\lambda_2 \to \infty$, the random variable

$$Y = \frac{X_1 - X_2}{(X_1 + X_2)^p}$$

has the asymptotically normal distribution $N\left(\dfrac{\lambda_1 - \lambda_2}{(\lambda_1 + \lambda_2)^p} ; \dfrac{1}{(\lambda_1 + \lambda_2)^{p-0.5}} \right)$.

6.10. Let X_1, X_2 and X have the same nondegenerate distribution with variance σ^2, and let X_1 and X_2 be independent. Prove that if the equality

$$X_1 + X_2 = aX$$

holds for some $a > 0$, then the random variables X_1, X_2 and X have the normal distribution $N(0, \sigma)$, and $a = \sqrt{2}$ (Pólya [2]).

Hint: Find that $a = \sqrt{2}$ and that, for every integer k, the characteristic function $\phi(t)$ of the random variable X satisfies the condition

$$\phi^{2^k}\left(\frac{t}{\sqrt{2^k}} \right) = \phi(t).$$

Then use the Lindeberg-Lévy theorem.

(A generalization to multidimensional random variables and random vectors in Hilbert space was given by Prohorov and Fisz [1].)

6.11. Prove that the assumptions of theorem 6.9.1 imply condition (6.9.20).

6.12. Let $\{X_k\}$ $(k = 1, 2, \ldots)$ be a sequence of independent random variables, where

$$P(X_k = -1) = P(X_k = 1) = \frac{1 - 2^{-k}}{2},$$

$$P(X_k = -2^k) = P(X_k = 2^k) = \frac{1}{2^{k+1}}.$$

Check whether condition (6.9.20) is satisfied for this sequence.

6.13.(a) Show that condition (6.9.20) is satisfied for the sequences of random variables $\{X_i\}$ and $\{Y_i\}$ whose probability functions are given by (6.12.25) and (6.12.26).

(b) Do the same for the sequence $\{X_k\}$ with $P(X_k = -k^a) = P(X_k = k^a) = \frac{1}{2}$, where $a > 0$.

6.14. The random variables X_k $(k = 1, 2, \ldots)$ are independent and have the normal distribution $N(0; 2^{-k/2})$. Check whether for this sequence (a) the central limit theorem holds; (b) relation (6.9.20) is satisfied; (c) the relation $\lim_{n \to \infty} \max_{1 \le k \le n} (\sigma_k/C_n) = 0$ holds.

6.15. Let us consider the sequence of independent random variables $\{X_k\}$ $(k = 0, 1, 2, \ldots)$, where X_0 has the gamma distribution given by (5.8.6) and the X_k $(k = 1, 2, \ldots)$ have the same distributions as in the preceding problem. Does the central limit theorem hold for the sequence $\{X_k\}$ $(k = 0, 1, 2, \ldots)$?

6.16. The random variables X_k $(k = 1, 2, \ldots)$ are independent and X_k has the Poisson distribution with the parameter $\lambda_k = 2^{-k}$. Check whether the central limit theorem holds for this sequence.

6.17. Let $Y_n = \sum\limits_{k=1}^{k_n} X_{nk}$ (see Section 6.9,C), where the random variables X_{nk} $(k = 1, \ldots, k_n)$ are independent for each n and have the probability functions given by the formula $P(X_{nk} = x_{nkl}) = p_{nkl}$, where $\sum\limits_{l=1}^{n} p_{nkl} = 1$ $(n = 1, 2, \ldots,$ $k = 1, 2, \ldots, k_n,$ $l = 1, 2, \ldots, r,$ $r \geqslant 2)$. Assume that (1) the X_{nk} are *asymptotically constant*, that is, for every $\varepsilon > 0$ we have

$$\lim_{n \to \infty} \max_{1 \leq k \leq k_n} P(|X_{nk} - m_{nk}| > \varepsilon) = 0, \text{ where } m_{nk} \text{ is the median of } X_{nk},$$

$$(2)\lim_{n \to \infty} \max_{1 \leq k \leq k_n} z_{nkl} = \lim_{n \to \infty} \min_{1 \leq k \leq k_n} z_{nkl} = x_l \, (l = 1, \ldots, r),$$

where $z_{nkl} = x_{nk,l+1} - x_{nkl}$. Find the class of all possible limit distribution functions of sequences $\{F_n(y)\}$ of distribution functions of $Y_n - A_n$ for arbitrary sequences of constants $\{A_n\}$ (Kubik [1]).

6.18. Let us denote by $m_k^{(n)}$ the moment of order k of the random variable X_n with the distribution function $F_n(x)$. Prove that if, for $k = 1, 2, \ldots$, the finite limits

$$m_k = \lim_{n \to \infty} m_k^{(n)}$$

exist and, moreover, these limits uniquely determine a distribution function $F(x)$ (see Problem 4.10), the sequence $\{F_n(x)\}$ converges to $F(x)$. (This is called the *second central limit theorem.* See Fréchet and Shohat [1]).

6.19. Let the random variables $X_1, X_2, X_3, \ldots$ satisfy all the assumptions of the Lindeberg-Lévy theorem, and suppose that the moment $E|X_i|^3$ exists. Then the relation

$$|F_n(z) - \Phi(z)| \leqslant c \, \frac{E|X_i|^3}{\sigma^3} \, \frac{1}{\sqrt{n}}$$

holds, where c is a constant (Cramér [1], Esseen [1], Berry [1]).

6.20. The random variables $X_i(i = 1, 2, \ldots)$ are independent and have the same probability distribution, given by the formulas

$$P(X_i = 0) = P(X_i = 3) = P(X_i = 7) = P(X_i = 12) = \tfrac{1}{4}.$$

Check whether for this sequence the local limit theorem of Gnedenko holds.

6.21. The random variable X has the Poisson distribution with the parameter λ. Let $u_r = P(X = r) \, (r = 0, 1, \ldots), t = (r - \lambda)/\sqrt{\lambda}$ and

$$v_r = \frac{1}{\sqrt{2\pi\lambda}} \exp\left(-\frac{t^2}{2}\right),$$

$$w_r = v_r\left(1 - \frac{t}{2\sqrt{\lambda}} + \frac{t^2}{6\sqrt{\lambda}}\right).$$

Applying Stirling's formula, show that if $\lambda \to \infty$ and $r \to \infty$ in such a way that t remains bounded in absolute value, then for any $\varepsilon > 0$

$$\lim_{\lambda \to \infty} [\lambda^{1-\varepsilon}(u_r - v_r)] = 0,$$

$$\lim_{\lambda \to \infty} [\lambda^{3/2-\varepsilon}(u_r - w_r)] = 0.$$

6.22. The random variables X_i $(i = 1, 2, \ldots)$ satisfy all the assumptions of theorem 6.10.1 and condition (w). Suppose that the moment of order $v \geqslant 3$ exists. Then (notation of Section 6.10) as $n \to \infty$ the following asymptotic formula holds:

$$P_n(k) = \frac{1}{\sigma \sqrt{n}} \left\{ f(z) + \sum_{r=1}^{v-2} n^{-r/2} U_r[-f(z)] \right\} + o(n^{-(v-1)/2}),$$

where $z = z_{nk}$, $f(z) = (1/\sqrt{2\pi}) \exp(-z^2/2)$, and for every r the function U_r may be determined by comparing the coefficients with successive powers of the term $1/\sqrt{n}$ in the identity

$$\exp \left\{ \frac{\kappa_{r+2}}{(r+2)!} [-f(z)]^{r+2} n^{-r/2} \right\} = 1 + \sum_{r=1}^{\infty} U_r[-f(z)]n^{-r/2},$$

and where κ_{r+2} is the semi-invariant of order $r + 2$ of the random variable X_i/σ (Esseen [1]).

(a) Express $[-f(z)]^{(r+2)}$ in terms of Hermite polynomials.

(b) Compute the expression $P_n(k)$ for $v = 4$.

6.23. (Continued) The independent random variables X_i $(i = 1, 2)$ have Poisson distributions with parameters $\lambda_1 > 0$ and $\lambda_2 > 0$, respectively, where $\lambda_i = d_i n$ and d_1 and d_2 are positive constants.

(a) Prove that as $\lambda_i \to \infty$ $(i = 1, 2)$, for every integer $u \geqslant 1$, the following asymptotic formula holds:

$$P(X_2 - X_1 = k) = \frac{1}{\sqrt{\lambda_1 + \lambda_2}} \left\{ f(z) + \sum_{r=1}^{u} U_r[-f(z)] \right\} + o[(\lambda_1 + \lambda_2)^{-(u+1)/2}],$$

where $z = [k - (\lambda_1 + \lambda_2)]/\sqrt{\lambda_1 + \lambda_2}$ and $U_r[-f(z)]$ is determined as in Problem 6.22 with

$$\kappa_{r+2} = \begin{cases} n^{r/2}(\lambda_1 + \lambda_2)^{-r/2} & (r = 2, 4, \ldots), \\ n^{r/2}(\lambda_2 - \lambda_1)(\lambda_1 + \lambda_2)^{-(r+2)/2} & (r = 1, 3, \ldots). \end{cases}$$

(b) Use Problem 5.5(b) and find an asymptotic formula for the Bessel function $I_k(2\lambda)$ with a purely imaginary argument.

6.24. Applying Esseen's result given in Problem 6.22, find an asymptotic formula for u_r, more accurate than the formula given in Problem 6.21.

6.25. The random variables $X_1, X_2, \ldots$ are independent and have the same distribution with the density

$$f(x) = \begin{cases} 0 & \text{if} \quad |x| \geqslant \dfrac{1}{e}, \\[2ex] \dfrac{1}{2 |x| \log^2 |x|} & \text{if} \quad |x| < \dfrac{1}{e}. \end{cases}$$

Show that the Lindeberg-Lévy integral limit theorem holds for the sequence $\{X_i\}$, whereas the sequence of densities of the random variables $(X_1 + \ldots + X_n)/\sigma \sqrt{n}$ does not converge to the density of the normal distribution (Gnedenko and Kolmogorov [1]).

6.26. Let $\{X_i\}$ $(i = 1, 2, \ldots)$ be a sequence of independent random variables with the same density function $f(x)$ and let $f_n(x)$ be the density of the random variable

$$Z_n = \frac{X_1 + \ldots + X_n}{B_n} - A_n.$$

Sequences $\{A_n\}$ and $\{B_n\}$ such that the relation

$$(*) \qquad \lim_{n \to \infty} \sup_x \left[f_n(x) - \frac{1}{\sqrt{2\pi}} \exp\left(\frac{-x^2}{2}\right) \right] = 0$$

holds exist if and only if: (1) the sequence of distribution functions $\{F_n(z)\}$ of the random variables Z_n converges to $\Phi(z)$; (2) there exists an integer m such that the distribution function of the sum $X_1 + \ldots + X_m$ is absolutely continuous and its derivative is bounded (Gnedenko [2, 3]).

(a) Check whether for a sequence of independent random variables with the uniform distribution in the interval [0, 1] relation (*) holds.

(b) Do the same for X_i having a gamma distribution.

(c) Do the same for X_i having a beta distribution.

6.27. Prove that the assertion of theorem 6.11.3 remains true if: (a) the assumption that the variance is the same is replaced by the assumption that all variances are uniformly bounded; (b) the assumption that the random variables are pairwise uncorrelated is replaced by the assumption that the covariance of any pair (X_{k_1}, X_{k_2}) tends uniformly to 0 as $|k_2 - k_1| \to \infty$.

6.28. Suppose that the random variables $X_1, X_2, \ldots$ and X are uniformly bounded. Show that the sequence $\{X_k\}$ converges stochastically to X if and only if

$$\lim_{k \to \infty} E(|X_k - X|^2) = 0.$$

Hint: Use the Chebyshev inequality and the Kolmogorov inequality from Problem 3.8.

6.29. Let $\{A_k\}$ $(k = 1, 2, \ldots)$ be a sequence of random events, dependent or not. Let us consider the sequence of random variables Y_k $(k = 1, 2, \ldots)$, where

$$P(Y_k = 1) = P(A_k), \quad P(Y_k = 0) = 1 - P(A_k). \quad \text{Let} \quad X_n = \frac{1}{n}\sum_{k=1}^{n}[Y_k - P(A_k)],$$

$$p_1(n) = \frac{1}{n}\sum_{k=1}^{n}P(A_k), \quad p_2(n) = \frac{2}{n(n-1)}\sum_{1<j<k<n}P(A_j A_k), \quad d_n = p_2(n) - p_1^2(n).$$

Prove that the sequence $\{X_n\}$ converges stochastically to zero if and only if $\lim_{n \to \infty} d_n = 0$. Formulate this result in terms of the frequency of successes in n trials.

6.30. (Notation of Problem 6.29). Prove that if

$$\sum_{n=1}^{\infty} \frac{d_n}{n} < \infty,$$

then (Dvoretzky [1], also Loève) the sequence $\{X_n\}$ converges to zero with probability one. *Hint:* First prove the following lemma:

Let $a_n \geqslant 0$ $(n = 1, 2, \ldots)$ and let

$$\sum_{n=1}^{\infty} \frac{a_n}{n} < \infty.$$

Then there exists a sequence of integers n_i such that as $n_i \to \infty$, $n_{i+1} - n_i = o(n_i)$ and $\sum_{i=1}^{\infty} a_{n_i} < \infty$.

6.31. Deduce from the preceding result that the frequency of successes in the sequence of pairwise independent events A_k such that $P(A_k) = p = $ constant, converges to p with probability one. (Compare also with Problem 6.40).

6.32. The events A_k $(k = 1, 2, \ldots)$ have the same probability

$$P(A_k) = p \qquad (0 < p < 1).$$

(a) Show that $P(\limsup_{k \to \infty} A_k) \geqslant p$,

(b) Give an example of a sequence $\{A_k\}$ for which $P(\limsup_{k \to \infty} A_k) = p$.

6.33. Let us consider the sequence of events $\{A_k\}$ $(k = 1, 2, \ldots)$ with probabilities $P(A_k)$, and the random variables Y_k, where $P(Y_k = 1) = P(A_k)$, $P(Y_k = 0) = 1 - P(A_k)$. Prove that (Borel [1, 2]), if $0 < P(A_1) < 1$ and, for every k and all values $y_1, \ldots, y_k$ of $Y_1, \ldots, Y_k$

$$0 < p_k' \leqslant P(A_{k+1} \mid y_1, \ldots, y_k) \leqslant p_k'' < 1,$$

$$p_k' \leqslant P(A_k) \leqslant p_k'',$$

then

$$\sum_k p_k'' < \infty \text{ implies } P(\limsup_{k \to \infty} A_k) = 0,$$

$$\sum_k p_k' = \infty \text{ implies } P(\limsup_{k \to \infty} A_k) = 1.$$

6.34. (Continued) (a). Let us define for k larger than n

$$p_{nk} = P(A_{k+1} \mid y_n = y_{n+1} = \ldots = y_k = 0),$$

and let

$$p_{nn} = P(A_n).$$

Show that $P(\limsup_{k \to \infty} A_k) = 0$ if and only if $\lim_{n \to \infty} \sum_{k=n}^{\infty} p_{nk} = 0$ (Loève [3]).

(b) Show that $P(\limsup_{k \to \infty} A_k) = 1$ if and only if

$$\sum_{k=1}^{\infty} P(A_{k+1} \mid y_1, \ldots y_k) = \infty$$

for every sequence $y_1, y_2, \ldots$ of outcomes of trials for which only finitely many $y_k = 1$ and no $P(y_1 y_2 \ldots y_k) = 0$ (Nash [1]).

6.35. (Generalization of the Borel-Cantelli lemma, due to Erdös and Rényi [2])· Let $\{A_k\}$ $(k = 1, 2, 3, \ldots)$ be a sequence of random events such that

(*) $$\sum_{k=1}^{\infty} P(A_k) = +\infty.$$

(a) Prove that if in addition to (*) the relation

(**) $$\lim_{n \to \infty} \frac{\sum_{k=1}^{n} \sum_{l=1}^{n} P(A_k A_l)}{\left[\sum_{k=1}^{n} P(A_k) \right]^2} = 1$$

holds, then $P(\limsup_{k \to \infty} A_k) = 1$.

(b) Deduce from (a) that its assertion remains true if (*) holds, whereas (**) is replaced either by the assumption that the A_k are pairwise independent or by the assumption that for any k and l ($k \neq l$) the relation

$$P(A_k A_l) \leq P(A_k)P(A_l)$$

is true.

6.36. Let $\quad p_k = P(A_k), \quad v_k = P(A_{k+1} | \bar{A}_1, \ldots, \bar{A}_k), \quad$ where $\quad 0 < p_k < 1,$ $0 < v_k < 1 \ (k = 1, 2, 3, \ldots)$.

(a) Show that (Fisz)

$$P\left(\bigcap_{k=1}^{\infty} \bar{A}_k \right) = 0$$

if and only if

(*) $$\sum_k v_k = \infty.$$

(b) Prove that if (*) holds and moreover

(**) $$\sum_k p_k = \infty,$$

then $P(\limsup_{k \to \infty} A_k) = 1$.

(c) Show that relation (**) is not implied by relation (*).

6.37. Is it possible that the central limit theorem holds for a sequence of independent random variables in spite of the fact that this sequence does not obey the law of large numbers? (Compare Problem 6.13 (b) for $a \geq \frac{1}{2}$). What is the answer if we assume that the random variables of this sequence have the same distribution?

6.38(a) Show that the sequence $\{Y_n\}$, where $Y_n = \dfrac{1}{n} \sum_{k=1}^{n} X_k$, with independent X_k having the same distribution function $F(x)$ and characteristic function $\phi(t)$, is stochastically convergent to some constant a if and only if $\phi(t)$ is differentiable at 0 and $\phi'(0) = ai$.

Hint: Method 1. Use Kolmogorov's [7] theorem that if Y_n converges stochastically to a, then $\lim_{x \to \infty} xP(|X_k| > x) = 0$ and $a = \lim_{x \to \infty} \int_{-x}^{x} x \, dF(x)$ and then use Pitman's result [see Problem 4.9(b)].

Method 2. (Ehrenfeucht and Fisz [1]). Make use of the lemma: If $\psi(t)$ is continuous in $(-\infty, 0)$ and $(0, +\infty)$ and $\lim_{n \to \infty} \psi\left(\dfrac{t}{n}\right) = 0$ for every $t \neq 0$, then $\lim_{t \to 0} \psi(t) = 0$.

See also Dugué [2].

(b) Construct a series of random variables that obeys the weak law of large numbers [in the sense of Problem 6.38(a)] and does not obey the strong law of large numbers.

6.39. Let $\{X_k\}$ ($k = 1, 2, 3, \ldots$) be a sequence of independent random variables such that $E(X_k) = 0$ ($k = 1, 2, 3, \ldots$) Define

$$X_{nk} = \begin{cases} X_k & \text{if} \quad |X_k| < n, \\ 0 & \text{if} \quad |X_k| \geq n, \end{cases}$$

$$Y_n = \frac{1}{n} \sum_{k=1}^{n} X_k.$$

Then (Kolmogorov [1]) $\{Y_n\}$ is stochastically convergent to zero if and only if the following three relations hold:

(*)
$$\sum_{k=1}^{n} P(|X_k| \geq n) \to 0,$$

(**)
$$\frac{1}{n} \sum_{k=1}^{n} E(X_{nk}) \to 0,$$

(***)
$$\frac{1}{n^2} \sum_{k=1}^{n} D^2(X_{nk}) \to 0.$$

Remark. As shown by Breiman [1], relation (***) cannot be replaced by the relation

$$\frac{1}{n^2} \sum_{k=1}^{n} E(X_{nk}^2) \to 0.$$

6.40. Let $\{X_k\}$ be a sequence of pairwise independent random variables. If the variances $D^2(X_k)$ $(k = 1,2,3, \ldots)$ exist and are uniformly bounded, the sequence $\{X_k\}$ obeys the strong law of large numbers (Rajchman [1]).

6.41. Let X_i $(i = 1, 2, \ldots)$ be independent random variables with $E(X_i) = 0$ and $0 < D^2(X_i)$. Let c_i be a nonincreasing sequence of positive constants. Let

$$Y_k = \sum_{i=1}^{k} X_i \qquad (k = 1, \ldots, n).$$

(a) Prove that for arbitrary $\varepsilon > 0$ and j and l $(1 \leqslant j < l \leqslant n)$ the following inequality holds (Hájek and Rényi [1]):

$$P(\max_{j \leq k \leq l} c_k |Y_k| \geqslant \varepsilon) \leqslant \frac{1}{\varepsilon^2}\left[c_j^2 D^2(Y_j) + \sum_{i=j+1}^{l} c_i^2 D^2(X_i) \right].$$

(b) Using this inequality, derive theorem 6.12.1 (the proof will become considerably simpler).

In the next problem, the *Law of the Iterated Logarithm* for the binomial distribution, proved by Khintchin [8], is presented. For generalizations, see Kolmogorov [16], Erdös [1], and Feller [10].

6.42. Let $\{X_n\}$ $(n = 1, 2, \ldots)$ have the binomial distribution and let Y_n be given by (6.7.2). Then

$$P\left(\limsup_{n \to \infty} \frac{Y_n}{\sqrt{2 \log \log n}} = 1 \right) = 1,$$

$$P\left(\liminf_{n \to \infty} \frac{Y_n}{\sqrt{2 \log \log n}} = -1 \right) = 1.$$

Hint: Use the inequality of Problem 6.41.

6.43. Let $\{X_k\}$ $(k = 1,2,3, \ldots)$ be a sequence of independent random variables. Then

(a) the probability of convergence of the series

(*)
$$\sum_{k=1}^{\infty} X_k = Y$$

is equal either to one or to zero.

(*b*) the series (*) converges stochastically if and only if it converges with probability one (Lévy [3], Marcinkiewicz and Zygmund [1]).

(*c*) if both series

$$\sum_{k=1}^{\infty} E(X_k), \qquad \sum_{k=1}^{\infty} D^2(X_k)$$

converge, then the series (*) converges with probability one and, moreover,

$$E(Y) = \sum_{k=1}^{\infty} E(X_k),$$

$$D^2(Y) = \sum_{k=1}^{\infty} D^2(X_k).$$

6.44. Let $\{X_k\}$ ($k = 1, 2, 3, \ldots$) be a sequence of independent random variables. Define

$$Z_k = \begin{cases} X_k & \text{if } |X_k| \leqslant 1, \\ 0 & \text{if } |X_k| > 1. \end{cases}$$

Then the series (*) in Problem 6.43 converges with probability one if and only if the following three series converge simultaneously:

$$\sum_{k=1}^{\infty} P(|X_k| > 1), \qquad \sum_{k=1}^{\infty} E(Z_k), \qquad \sum_{k=1}^{\infty} D^2(Z_k).$$

This theorem is called the *Three Series theorem* (Khintchin and Kolmogorov [1]).

6.45. The random vector $(X_1, \ldots, X_r)$ has the distribution given by (6.13.1), where the probabilities p_j may be functions of n; we shall denote them by p_{nj}. Find the class of all possible limit distribution functions of the random vectors $(A_{n1} X_1 + B_{n1}, \ldots, A_{nr} X_r + B_{nr})$, for arbitrary sequences of constants $\{A_{nj}\}$ and $\{B_{nj}\}$. (Fisz [2]).

Hint: Use the multidimensional analogues of theorem 6.6.1*C* and of Khintchin's theorem (see Problem 6.6).

6.46. Let $\{X_k\}$ ($k = 1, 2, \ldots$) be a sequence of independent random variables with the same distribution and suppose that $m = E(X_k)$ and $\sigma^2 = D^2(X_k)$ exist. Let N be an integer-valued random variable, which is independent of $X_1, X_2, \ldots$ and such that the expected value $E(N) = c$ and $d^2 = D^2(N)$ exist, and suppose that the distribution of N depends upon the parameter λ. Let $X = \sum_{k=1}^{N} X_k$.

(*a*) Prove that $E(X) = mc$, $D^2(X) = c\sigma^2 + m^2 d^2$.

(*b*) Prove that if, as $\lambda \to \infty$, N has the asymptotically normal distribution $N(c; d)$, then X has the asymptotically normal distribution $N(mc; \sqrt{c\sigma^2 + m^2 d^2})$ (see Robbins [1]. A generalization was given by Dobrushin [4]).

(*c*) Find the limit distribution of a random variable X with the compound binomial distribution (see Section 5.13,A) as $\lambda \to \infty$.

6.47. Prove the theorem: Let the sequence $\{X_k\}$ ($k = 1, 2, \ldots$) satisfy all the assumptions of theorem 6.8.1. Let $m = 0$, $\sigma = 1$, and let N be an integer-valued random variable whose distribution depends upon $\lambda > 0$. Suppose that, as $\lambda \to \infty$, N/λ stochastically converges to a positive constant. Then the random variable $X/\sqrt{N}$, where $X = \sum_{k=1}^{N} X_k$, has the limit distribution $N(0; 1)$ (Anscombe [1]; see also Rényi [7]).

6.48. Prove that if a sequence of infinitely divisible distribution functions (see Problem 5.25) converges to a nondegenerate distribution function $F(x)$, then $F(x)$ is infinitely divisible.

6.49. Prove (Pólya [3]) that if a sequence of distribution functions $F_n(x)$ converges as $n \to \infty$ to a continuous distribution function $F(x)$, then the convergence is uniform with regard to x.

For a generalization of this result, see R. Ranga-Rao [1].

6.50. The circumference of a horizontal circular disk, pivoted at its center to a table, is marked off into $2\,m$ arcs, alternately red and black, with arcs of angle δ_1 and δ_2, respectively. Initially, an arrow fixed to the table points to a point zero on the circumference of the disk. The disk is being spun. Denote by A the random event that the disk comes to rest with the arrow pointing to a red point on the circumference, and by X the angle through which the disk turns. If $F(x)$ is the distribution function of X, then

$$P(A) = \int_{\infty}^{\infty} h(x)\, dF(x),$$

where $h(x) = 1$ for $k(\delta_1 + \delta_2) \leqslant x \leqslant k(\delta_1 + \delta_2) + \delta_1\ (k = 0,1,2,\ldots)$, and $h(x) = 0$ otherwise.

(a) If X has an arbitrary density function $f(x)$, then $\lim P(A) = \delta_1/(\delta_1 + \delta_2)$, as $m \to \infty$, $\delta_1 \to 0$, $\delta_2 \to 0$, whereas δ_1/δ_2 remains constant and $f(x)$ is held fixed (Poincaré [1], Fréchet [1], Hostinský [1]).

(b) Let now δ_1 and δ_2 be constant. Consider successive spinnings applied without touching the disk between spins. Denote by X_i the angle of the ith ($i = 1, 2, 3 \ldots$) spin, and by Y_n the proportion of occurrences of A among the first n spins. If the $X_i\ (i = 1, 2, 3, \ldots)$ have a common, arbitrary distribution function, except that X_i cannot have a lattice distribution with maximum span $h > 0$, a rational multiple of π, then (Robbins [3])

$$P[\lim_{n \to \infty} Y_n = \delta_1/(\delta_1 + \delta_2)] = 1.$$

Notice that if the disk were returned after each spin to its initial position, then by Borel's strong law of large numbers we would have

$$P[\lim_{n \to \infty} Y_n = P(A)] = 1.$$

See also Problem 14.8

CHAPTER 7

Markov Chains

7.1 PRELIMINARY REMARKS

Thus far we have mainly considered independent random events and independent random variables. In fact, in the applications of probability theory we can often assume that the random events or random variables under consideration are independent. However, there are many problems in physics, engineering, and other areas of applications of probability theory where the assumption of independence is not satisfied, not even approximately. Therefore, the investigation of dependent random events and dependent random variables is an important problem in probability theory. But to abandon the assumption of independence creates serious complications in the reasoning and in the proofs. It is a great achievement of Markov [2] that in the investigation of dependent events he distinguished a scheme of experiments, now called the scheme of events forming a Markov chain, which can be considered as the simplest generalization of the scheme of independent trials. Markov's investigations have become the starting point for the development of a new and important branch of probability theory, the theory of Markov stochastic processes. Some elements of this theory will be considered in this and the following chapters. The Markov chains considered here are also, as we shall see in Section 8.2, Markov processes, but for didactic and traditional reasons we present the theory of Markov chains in a separate chapter.

7.2 HOMOGENEOUS MARKOV CHAINS

We assume that all the conditional probabilities appearing in this and the following chapters are defined (see Section 2.7).

Imagine that we are given a sequence of experiments and as a result of each experiment there can be one and only one event from a finite or countable set of pairwise exclusive events $E_1, E_2, E_3, \ldots$. We call these events *states*. When the event E_j occurs we say that the system *passes into the state* E_j. We use the symbol $E_j^{(n)}$ to denote that at the nth trial the system passes into the state E_j; the symbol $E_j^{(0)}$ denotes that the initial state was E_j. Next we denote by $p_{ij}^{(n)}$ the conditional probability that at the nth trial the system passes into the state E_j, provided that after the $(n-1)$-st trial it was in the state E_j, that is,

$$p_{ij}^{(n)} = P(E_j^{(n)} \mid E_i^{(n-1)}).$$

Definition 7.2.1. We say that a sequence of trials forms a *Markov chain* if for any $i, j, n = 1, 2, 3, \ldots$ the equalities

$$(7.2.1) \qquad p_{ij}^{(n)} = P(E_j^{(n)} \mid E_i^{(n-1)})$$
$$= P(E_j^{(n)} \mid E_i^{(n-1)} E_{i_{n-2}}^{(n-2)} \ldots E_{i_1}^{(1)} E_{i_0}^{(0)})$$

are satisfied for arbitrary $E_{i_{n-2}}^{(n-2)}, \ldots, E_{i_1}^{(1)}, E_{i_0}^{(0)}$.

Definition 7.2.2. We say that a sequence of trials forms a *homogeneous Markov chain*, if for $i, j = 1, 2, 3, \ldots$ the probability $p_{ij}^{(n)}$ is independent of n, that is,

$$(7.2.2) \qquad\qquad p_{ij}^{(n)} = p_{ij} \qquad (n = 1, 2, \ldots).$$

The probability p_{ij} is called the *transition probability* from the state E_i to the state E_j in one trial.

We also use the time terminology, that is, we consider the trials as performed at every unit of time and, instead of saying that at the nth trial the system passes from the state E_i to the state E_j, we say that this transition is performed at the moment $t = n$. Besides this, we shall assume that at the initial moment, that is, at $t = 0$, the system may be in the state E_i with probability $P(E_i)$. In this terminology p_{ij} is the transition probability from the state E_i to the state E_j in a unit of time.

By formulas (7.2.1), (7.2.2), and (1.5.7) we obtain the following formula for the probability of the product of states $(E_{i_0} E_{i_1} \ldots E_{i_n})$ in n successive trials of a homogeneous Markov chain:

$$(7.2.3) \quad P(E_{i_0} E_{i_1} \ldots E_{i_n}) = P(E_{i_0})P(E_{i_1} \mid E_{i_0}) \ldots P(E_{i_n} \mid E_{i_{n-1}})$$
$$= P(E_{i_0}) p_{i_0 i_1} \ldots p_{i_{n-1} i_n}.$$

The reader will notice an essential difference between the last formula and formula (1.5.7).

It follows from formula (7.2.3) that the probability of every product of states is given if we know all the transition probabilities p_{ij} and all the probabilities $P(E_{i_0})$ of the initial states.

7.3 THE TRANSITION MATRIX

A   The matrix with the transition probabilities p_{ij} as elements is called the *transition matrix*. This matrix is denoted by $\mathbf{M}_1$,

$$\mathbf{M}_1 = \begin{bmatrix} p_{11} & p_{12} & p_{13} \cdots \\ p_{21} & p_{22} & p_{23} \cdots \\ p_{31} & p_{32} & p_{33} \cdots \\ \cdots\cdots\cdots\cdots \end{bmatrix}$$

We observe that all the elements p_{ij}, being probabilities, are non-negative. Suppose that the system is in the state E_i. The event that as a result of the experiment the system either remains in the state E_i or passes to any of the states E_j, where $i \neq j$, is the sure event. Since the events E_j are pairwise exclusive, for $i = 1, 2, 3, \ldots$, we obtain the formula

(7.3.1) $P[(\sum_j E_j) \mid E_i] = \sum_j p_{ij} = 1.$

Thus the sum of the terms in each row of the matrix $\mathbf{M}_1$ equals one. However, the sum of the terms in a column need not be one.

Example 7.3.1.   Consider a sequence of trials in the Bernoulli scheme. Here we have two states E_1 and E_2, and in each experiment

$$p_{11} = p_{21} = p, \qquad p_{12} = p_{22} = q.$$

Thus the transition matrix is of the form

$$\mathbf{M}_1 = \begin{bmatrix} p & q \\ p & q \end{bmatrix}.$$

It is easy to verify that in an independent sequence of trials the rows of the transition matrix are always identical.

Example 7.3.2.   Here we consider the *random walk with absorbing barriers*. It is a model of certain phenomena which often appear in physics. A particle may be at one of the points $1, 2, 3, \ldots, s$ on the x-axis. It will remain forever, with probability one, at the point $x = 1$ if it arrives there at some moment t. The same is true for the point $x = s$. The points 1 and s are called *absorbing barriers*. If at the moment t the particle comes to the point $x = i$, where $2 \leqslant i \leqslant s - 1$, then during the next unit of time the particle will pass to the point $i + 1$ with probability p and to the point $i - 1$ with probability $q = 1 - p$. Here we have a homogeneous Markov chain with s states, where the state E_i occurs if the particle has the coordinate $x = i$. In fact, the probability of passing from the state E_i to the state E_j at the moment t does not depend on the previous path of the particle and does not depend on t but only on the state at the moment t. The transition probabilities are

$$p_{11} = p_{ss} = 1,$$

and for $2 \leqslant i \leqslant s - 1$

$$p_{ij} = \begin{cases} p & \text{for } j = i + 1, \\ q = 1 - p & \text{for } j = i - 1, \\ 0 & \text{otherwise.} \end{cases}$$

Thus the transition matrix has the form

$$
\mathbf{M}_1 = \begin{bmatrix}
1 & 0 & 0 & 0 & \dots & 0 & 0 & 0 \\
q & 0 & p & 0 & \dots & 0 & 0 & 0 \\
0 & q & 0 & p & \dots & 0 & 0 & 0 \\
\multicolumn{8}{c}{\dotfill} \\
0 & 0 & 0 & 0 & \dots & q & 0 & p \\
0 & 0 & 0 & 0 & \dots & 0 & 0 & 1
\end{bmatrix}.
$$

We now give an example due to Malécot [1] of the application of Markov chains to genetics.

Example 7.3.3. In the genetics based on Mendel's laws we assume that inherited characteristics depend on the genes. Genes always appear in pairs. In the simplest case, which we consider here, every gene may be of one of two forms, A or a. If both genes of the organism being considered are of type A, we say that the organism is of genotype AA; if both genes are of type a we say that it is of genotype aa; finally, if one gene is of type A and the other of type a we say that the organism is of genotype Aa. Furthermore, we assume that the reproductive cells, or gametes, have only one gene; thus the gametes of an organism of genotype AA or aa have the gene A or a, respectively, whereas the gametes of an organism of genotype Aa may have the gene A or a with equal probability. An offspring receives one gene from each parent under the conditions of the Bernoulli scheme. This should be understood as follows: consider the set of all genes of all organisms belonging to the generation of parents of a given offspring as the population from which two genes are drawn at random under the conditions of the Bernoulli scheme. Similarly, the genotype structure of N offspring is a result of $2N$ such drawings from the set of genes under consideration.

Suppose, now, that the population under consideration consists of N elements in each generation. This may be achieved by an appropriate selection of organisms in each generation. Thus we have $2N$ genes in each generation. If in some generation $i(0 \leqslant i \leqslant 2N)$ of the genes are of the form A, we say that the generation is in the state E_i. From the assumed reproduction scheme it follows that we have here a homogeneous Markov chain with $2N + 1$ possible states: $E_0, E_1, \dots E_{2N}$. The probability of passing from the state E_i in some generation to the state E_j in the next generation is given by the formula

$$
p_{ij} = \binom{2N}{j} \left(\frac{i}{2N} \right)^j \left(1 - \frac{i}{2N} \right)^{2N - j}.
$$

We observe that the states E_0 and E_{2N} are the absorbing barriers. Indeed, if in some generation the population is in one of these states it will remain there forever; if, for instance, all the organisms are of the genotype AA, no offspring can have the gene a.

Example 7.3.4. Here we consider a model of a *random walk without absorbing barriers*, having a countable number of states. The set of states is the set of all non-negative integers and the transition probabilities are given by the formulas

$$
p_{11} = q = 1 - p,
$$

$$
p_{ij} = \begin{cases}
p & \text{for } i = 1,2,3,\dots ; \quad j = i + 1, \\
q & \text{for } i = 2,3,\dots ; \quad j = i - 1, \\
0 & \text{for the remaining pairs } (i, j).
\end{cases}
$$

The number 0 is a *reflecting barrier*.

The transition matrix M_1 is of the form

$$
M_1 = \begin{bmatrix}
q & p & 0 & 0 & 0 & 0 & \cdots \\
q & 0 & p & 0 & 0 & 0 & \cdots \\
0 & q & 0 & p & 0 & 0 & \cdots \\
\cdots\cdots\cdots\cdots\cdots\cdots\cdots
\end{bmatrix}.
$$

Example 7.3.5. Let us now return to the Pólya scheme. We use the notation of Section 5.4. We have two states, state E_1 consists of drawing a white ball and state E_2 consists of drawing a black ball, and the initial probabilities are $p_1 = b/N$ and $p_2 = 1 - p_1 = c/N$, respectively. The probability of passing from the state E_1 in the first drawing to the state E_1 in the second drawing is $(b + s)/(N + s)$. However, the probability of choosing a white ball in the third drawing if in the second drawing a white ball was drawn, equals $(b + 2s)/(N + 2s)$ provided that in the first drawing we obtained the state E_1, and it equals $(b + s)/(N + 2s)$ provided in the first drawing we obtained the state E_2. Thus the sequence of trials in the Pólya scheme is not a Markov chain.

We can, however, obtain a Markov chain in the Pólya scheme if we define the states in another way, namely, if we agree to say that after n drawings the system is in the state E_i ($i = 0, 1, 2, \ldots, n$), if i is the number of white balls obtained in n drawings. Then at the $(n + 1)$st trial the system may remain in the state E_i or pass to the state E_{i+1}, according to whether in the $(n + 1)$st trial a black or a white ball was drawn. These transition probabilities depend only on the state of the system after the nth trial and are independent of the results of the first $n - 1$ trials. However, these probabilities depend on the number of trials and we have here a nonhomogeneous Markov chain with the transition probabilities $p_{ij}^{(n+1)}$ given by the formula

$$
p_{ij}^{(n+1)} = \begin{cases}
\dfrac{c + (n - i)s}{N + ns} & \text{for } j = i, \\[2mm]
\dfrac{b + is}{N + ns} & \text{for } j = i + 1, \\[2mm]
0 & \text{for } j \neq i, i + 1.
\end{cases}
$$

B   We denote by $p_{ij}(n)$ the probability of passing in n trials from the state E_i to the state E_j in a homogeneous Markov chain. Sometimes we call it the *probability of transition in n steps*. We show how to compute the probabilities $p_{ij}(n)$ from the probabilities p_{ij}. Let us start by computing $p_{ij}(2)$. We observe that the event A of passing from the state E_i to the state E_j in two trials is the union of the pairwise exclusive events A_k, where A_k occurs if and only if the system passes from the state E_i to E_k in the first step and from E_k to E_j in the second step. Thus for every pair (i, j) we have

$$
(7.3.2) \qquad\qquad p_{ij}(2) = \sum_k p_{ik} p_{kj},
$$

where the summation is extended over all possible states.

In an analogous way we find the formulas

$$(7.3.3) \qquad p_{ij}(n) = \sum_k p_{ik}(m)p_{kj}(n-m),$$

where $n = 2, 3, 4, \ldots$ and m is an integer satisfying the condition $1 \leqslant m < n$. Equation (7.3.3) plays a basic role in the theory of homogeneous Markov chains and is called the *Markov equation*.

The matrix whose elements are the transition probabilities $p_{ij}(n)$ is called the *matrix of transition in n steps* and is denoted by the symbol M_n. It is easy to find the relation between the matrices M_n and M_1. Let us first find the relation between the matrices M_1 and M_2. From formula (7.3.2) it follows that the element of matrix M_2 at the intersection of the ith row and jth column is the sum of products of the elements of the ith row by the jth column of M_1. Thus, according to the rule of multiplication of matrices, we obtain

$$M_2 = M_1{}^2.$$

By induction and formula (7.3.3), we have

$$(7.3.4) \qquad M_n = M_1{}^n.$$

7.4 THE ERGODIC THEOREM

A We start this section with a classification (see also Problems 7.5 to 7.8) of states of Markov chains; this will allow us to interpret the assumptions of the ergodic theorem. This classification was introduced by Kolmogorov [9].

Definition 7.4.1. The state E_i is called *unintrinsic* if there exists a state E_j and an integer k such that $p_{ij}(k) > 0$ and $p_{ji}(m) = 0$ for $m = 1, 2, 3, \ldots$.

Definition 7.4.2. The state E_i is called *intrinsic* if, for every state E_j, the existence of an integer k_j such that $p_{ij}(k_j) > 0$ implies the existence of an integer m_i such that $p_{ji}(m_i) > 0$.

Definition 7.4.3. The intrinsic state E_i is called *periodic* if there exists an integer $d > 1$ such that $p_{ii}(n) = 0$ for n not a multiple of d.

In example 7.3.2 the states E_1 and E_s are intrinsic and not periodic, whereas all other states are unintrinsic. Indeed, we can pass from the state E_i ($i = 2, 3, \ldots, s - 1$) to the state E_1 in $i - 1$ steps, with positive probability, but the probability of leaving the state E_1 is 0.

We observe, however, that we cannot pass from the state E_1 to E_s, nor can we pass from E_s to E_1, despite the fact that both states are intrinsic. This remark gives rise to the following definition.

Definition 7.4.4. The set W of intrinsic states forms *one class* if for every pair of intrinsic states E_i and E_j of W there exists an integer m_{ij} such that $p_{ij}(m_{ij}) > 0$.

Examples 7.4.1 to 7.4.4, which are presented in this section, illustrate the notions just introduced.

B   We now discuss the ergodic theorem. This theorem tells how the probabilities $p_{ij}(n)$ behave as $n \to \infty$. In other words, it explains what influence the initial state E_i has on the probability $p_{ij}(n)$ after a large number of steps n. Theorem 7.4.1 gives a condition for the convergence $p_{ij}(n) \to p_j$ for a homogeneous Markov chain, where the limits p_j are independent of i, that is, are independent of the initial state E_i. The theorem given here does not give a complete solution to this problem; in particular, it does not consider Markov chains with a countable number of states. The reader can find more detailed information on this subject in the books by Feller [7], Chung [5], and Kemeny and Snell [1], and in the papers of Chung [2] and [3] (see also Problem 7.9).

Theorem 7.4.1.   *Let* $\mathbf{M}_1 = [p_{ij}]$ *be the matrix of one step transition probabilities in a homogeneous Markov chain with a finite number of states* $E_1, \ldots, E_s$. *If there exists an integer* r *such that the terms* $p_{ij}(r)$ *of the matrix* $\mathbf{M}_r$ *satisfy the relation*

$$(7.4.1) \qquad\qquad \min_{1 \leqslant i \leqslant s} p_{ij}(r) = \delta > 0$$

in s_1 ($s_1 \geqslant 1$) *columns, then the equalities*

$$(7.4.2) \qquad\qquad \lim_{n \to \infty} p_{ij}(n) = p_j \qquad (j = 1, 2, \ldots, s)$$

are satisfied, and $p_j \geqslant \delta$ *for those* j *for which relation* (7.4.1) *holds. Moreover,* $\sum_j p_j = 1$ *and*

$$(7.4.3) \qquad\qquad |p_{ij}(n) - p_j| \leqslant (1 - s_1 \delta)^{n/r - 1}.$$

As we see, one of the assumptions of this theorem requires that the elements $p_{ij}(r)$ of at least one column of the matrix $\mathbf{M}_r$ be positive. Theorem 7.4.1 is a modification of the theorem of Markov [2], which requires that for some integer r all the elements of the matrix $\mathbf{M}_r$ be positive. Then in the assertion of the theorem we have $p_j > 0$ ($j = 1, 2, \ldots, s$).

Theorem 7.4.1 is called the *ergodic theorem* and the limit probabilities p_j are called the *ergodic probabilities.*

The explanation of this name is given at the end of this section.

C   Before presenting the proof of theorem 7.4.1, we give some examples that will explain the meaning of the assumptions of the theorem.

Example 7.4.1. Let us return to example 7.3.2 and suppose, for simplicity, that $s = 3$. Then we have

$$\mathbf{M}_1 = \begin{bmatrix} 1 & 0 & 0 \\ q & 0 & p \\ 0 & 0 & 1 \end{bmatrix}.$$

Let us compute

$$\mathbf{M}_2 = \mathbf{M}_1{}^2 = \begin{bmatrix} 1 & 0 & 0 \\ q & 0 & p \\ 0 & 0 & 1 \end{bmatrix} \cdot \begin{bmatrix} 1 & 0 & 0 \\ q & 0 & p \\ 0 & 0 & 1 \end{bmatrix} = \begin{bmatrix} 1 & 0 & 0 \\ q & 0 & p \\ 0 & 0 & 1 \end{bmatrix} = \mathbf{M}_1.$$

In general, we have

$$\mathbf{M}_n = \mathbf{M}_1.$$

Thus the assumptions of the ergodic theorem are not satisfied. There does not exist an r such that the matrix $\mathbf{M}_r$ has at least one column of positive elements $p_{ij}(r)$. It is obvious that the assertion of theorem 7.4.1 is not satisfied either, since $p_{11}(n) = 1$ so that $\lim_{n \to \infty} p_{11}(n) = 1$, while $\lim_{n \to \infty} p_{21}(n) = q$ and $\lim_{n \to \infty} p_{31}(n) = 0$. The irregularity of this Markov chain is caused by the existence of two intrinsic states E_1 and E_3 such that passage from one to the other is impossible; thus the set of states does not form one class.

Example 7.4.2. Consider a homogeneous Markov chain with four states E_1, E_2, E_3, E_4 and the transition matrix

$$\mathbf{M}_1 = \begin{bmatrix} 0 & 0 & \frac{1}{2} & \frac{1}{2} \\ 0 & 0 & \frac{1}{2} & \frac{1}{2} \\ \frac{1}{2} & \frac{1}{2} & 0 & 0 \\ \frac{1}{2} & \frac{1}{2} & 0 & 0 \end{bmatrix}.$$

We obtain here

$$\mathbf{M}_2 = \begin{bmatrix} \frac{1}{2} & \frac{1}{2} & 0 & 0 \\ \frac{1}{2} & \frac{1}{2} & 0 & 0 \\ 0 & 0 & \frac{1}{2} & \frac{1}{2} \\ 0 & 0 & \frac{1}{2} & \frac{1}{2} \end{bmatrix}.$$

Generally, for $k = 1, 2, 3, \ldots$, we have

$$\mathbf{M}_{2k+1} = \mathbf{M}_1, \qquad \mathbf{M}_{2k} = \mathbf{M}_2.$$

Thus neither the assumption nor the assertion of theorem 7.4.1 is satisfied. The reader will notice the periodicity of this Markov chain. All the states are intrinsic; but they are periodic, so that, for instance, the system may return from the state E_1 to the state E_1 only in an even number of steps. This periodicity causes the observed irregularity, as a result of which the ergodic theorem is not satisfied.

Example 7.4.3. Let us return to example 7.3.4 and suppose that the number of states is 3 and the matrix $\mathbf{M}_1$ has the form

$$\mathbf{M}_1 = \begin{bmatrix} q & p & 0 \\ q & 0 & p \\ 0 & q & p \end{bmatrix}.$$

Then

$$\mathbf{M}_2 = \begin{bmatrix} q^2 + pq & qp & p^2 \\ q^2 & 2qp & p^2 \\ q^2 & pq & qp + p^2 \end{bmatrix}.$$

Thus the assumptions of theorem 7.4.1 are satisfied. We observe that all three states are intrinsic, nonperiodic, and form one class.

We show later how to compute the ergodic probabilities.

Example 7.4.4. Let us modify example 7.4.1 in such a way that the transition matrix takes the form

$$\mathbf{M}_1 = \begin{bmatrix} 1 & 0 & 0 \\ q & 0 & p \\ 0 & q & p \end{bmatrix}.$$

In this example, the state E_1 is an absorbing barrier, and the state E_3 is a reflecting barrier. We have

$$\mathbf{M}_2 = \begin{bmatrix} 1 & 0 & 0 \\ q & pq & p^2 \\ q^2 & pq & qp + p^2 \end{bmatrix}.$$

Thus the assumptions of theorem 7.4.1 are satisfied. It is easy to verify that the state E_1 is intrinsic and not periodic and the remaining two states are unintrinsic.

Later we show (see example 7.4.6) that the limit probabilities p_2 and p_3 are zero.

These examples suggest, and it can be shown that this is true, that if in a homogeneous Markov chain with a finite number of states all the intrinsic states are nonperiodic and form one class, then the assumptions of theorem 7.4.1 are satisfied. However, the possibility that some states are unintrinsic is not excluded. But if all the states are intrinsic, non-periodic, and form one class, then there exists an r such that all the elements $p_{ij}(r)$ of the matrix $\mathbf{M}_r$ are positive, hence greater than some $\delta > 0$, since there are only a finite number of them.

Let us mention here that Kaucky [1] and Konêcny [1] have given necessary and sufficient conditions for the ergodicity of homogeneous Markov chains; their conditions are expressed in terms of eigenvalues of the matrix $\mathbf{M}_1$.

D   We now give the proof of theorem 7.4.1.

Proof. For $v = 1, 2, 3, \ldots$, denote

$$(7.4.4) \qquad b_j(v) = \min_{1 \leqslant i \leqslant s} p_{ij}(v), \qquad B_j(v) = \max_{1 \leqslant i \leqslant s} p_{ij}(v).$$

Considering formula (7.3.3) for $v = 1, 2, 3, \ldots$, we obtain

$$b_j(v+1) = \min_{1 \leqslant i \leqslant s} p_{ij}(v+1) = \min_{1 \leqslant i \leqslant s} \sum_{k=1}^{s} p_{ik} p_{kj}(v)$$

$$\geqslant \min_{1 \leqslant i \leqslant s} \sum_{k=1}^{s} p_{ik} b_j(v) = b_j(v).$$

Hence

(7.4.5) $$b_j(v + 1) \geqslant b_j(v).$$

Similarly,

(7.4.6) $$B_j(v + 1) \leqslant B_j(v).$$

From formulas (7.4.5) and (7.4.6) we obtain

(7.4.7) $$b_j(1) \leqslant b_j(2) \leqslant \ldots \leqslant B_j(2) \leqslant B_j(1).$$

Suppose that r is defined as in the assumptions of theorem 7.4.1, and let $\sum_k^+$ and $\sum_k^-$ denote, respectively, the sums extended over those k for which $p_{ik}(r) \geqslant p_{mk}(r)$ and $p_{ik}(r) < p_{mk}(r)$. Then

(7.4.8) $$\sum_k^+ [p_{ik}(r) - p_{mk}(r)] + \sum_k^- [p_{ik}(r) - p_{mk}(r)] = 0.$$

Suppose that $n > r$. Consider the difference

$$B_j(n) - b_j(n) = \max_{1 \leqslant i \leqslant s} p_{ij}(n) - \min_{1 \leqslant m \leqslant s} p_{mj}(n)$$

$$= \max_{1 \leqslant i \leqslant s} \sum_{k=1}^{s} p_{ik}(r) p_{kj}(n - r) - \min_{1 \leqslant m \leqslant s} \sum_{k=1}^{s} p_{mk}(r) p_{kj}(n - r)$$

$$= \max_{1 \leqslant i, m \leqslant s} \sum_{k=1}^{s} [p_{ik}(r) - p_{mk}(r)] p_{kj}(n - r)$$

$$\leqslant \max_{1 \leqslant i, m \leqslant s} \{ \sum_k^+ [p_{ik}(r) - p_{mk}(r)] B_j(n - r) + \sum_k^- [p_{ik}(r) - p_{mk}(r)] b_j(n - r) \}.$$

Hence by formula (7.4.8)

(7.4.9)

$$B_j(n) - b_j(n) \leqslant \max_{1 \leqslant i, m \leqslant s} \sum_k^+ [p_{ik}(r) - p_{mk}(r)][B_j(n - r) - b_j(n - r)]$$

$$= [B_j(n - r) - b_j(n - r)] \max_{1 \leqslant i, m \leqslant s} \sum_k^+ [p_{ik}(r) - p_{mk}(r)].$$

Suppose that relation (7.4.1) holds for w terms of the sum $\sum_k^+$. Obviously, $w \leqslant s_1$, where s_1 is the number of columns for which (7.4.1) is satisfied. Thus

$$- \sum_k^+ p_{mk}(r) \leqslant -w\delta.$$

Next, since for $s_1 - w$ terms of the sum $\sum_k^-$ relation (7.4.1) is also satisfied, we have

$$\sum_k^+ p_{ik}(r) + (s_1 - w)\delta \leqslant 1.$$

Finally,

(7.4.10) $\sum\limits_{k}^{+}[p_{ik}(r) - p_{mk}(r)] \leqslant 1 - (s_1 - w)\delta - w\delta = 1 - s_1\delta.$

From formulas (7.4.9) and (7.4.10) follows the inequality

$$B_j(n) - b_j(n) \leqslant (1 - s_1\delta)[B_j(n - r) - b_j(n - r)].$$

Similarly, for $n > 2r$

$$B_j(n) - b_j(n) \leqslant (1 - s_1\delta)^2[B_j(n - 2r) - b_j(n - 2r)].$$

Repeating this procedure $[n/r]^1$ times, we obtain

(7.4.11)

$$B_j(n) - b_j(n) \leqslant (1 - s_1\delta)^{[n/r]}\left\{B_j\left(n - \left[\frac{n}{r}\right]r\right) - b_j\left(n - \left[\frac{n}{r}\right]r\right)\right\}.$$

We observe that from (7.4.10) and from the fact that $\delta > 0$, $s_1 \geqslant 1$ follows the inequality

$$0 \leqslant 1 - s_1\delta < 1.$$

Thus from formula (7.4.7) follows the existence of the limits of $\{b_j(n)\}$ and $\{B_j(n)\}$, and from formula (7.4.11) it follows that these limits are equal. Therefore,

(7.4.12) $\lim\limits_{n \to \infty} \max\limits_{1 \leqslant i \leqslant s} p_{ij}(n) = \lim\limits_{n \to \infty} \min\limits_{1 \leqslant i \leqslant s} p_{ij}(n) = p_j,$

which proves formula (7.4.2). Next, it is obvious that for those j for which relation (7.4.1) is satisfied, we have $p_j \geqslant \delta$. The equality $\sum\limits_{j} p_j = 1$ is also obvious.

It remains to prove relation (7.4.3). In fact, by formulas (7.4.7) and (7.4.11) we obtain

$$|p_{ij}(n) - p_j| \leqslant B_j(n) - b_j(n) \leqslant (1 - s_1\delta)^{n/r-1},$$

which completes the proof of theorem 7.4.1.

E   We now show how to calculate the ergodic probabilities p_j if they are known to exist. By formula (7.3.3) we obtain

$$p_{ij}(n) = \sum_{k=1}^{s} p_{ik}(n - 1)p_{kj}.$$

Thus, if the ergodic probabilities p_j exist, then after passage to the limit as $n \to \infty$ on both sides of the last inequality we have

(7.4.13) $p_j = \sum\limits_{k=1}^{s} p_k p_{kj}$ $(j = 1, 2, \ldots, s).$

---

[1] The symbol $[A]$ denotes here the greatest integer not exceeding A.

From these equations and from the relation

$$\sum_{j=1}^{s} p_j = 1$$

we can determine the probabilities p_j.

Example 7.4.5. Let us return to example 7.4.3 and calculate the ergodic probabilities p_j. Formula (7.4.13) gives us three linear equations

$$p_1 = p_1 q + p_2 q,$$
$$p_2 = p_1 p + p_3 q,$$
$$p_3 = p_2 p + p_3 p.$$

Hence

$$p_2 = \frac{p}{q} p_1,$$

$$p_3 = \left(\frac{p}{q}\right)^2 p_1.$$

Since $p_1 + p_2 + p_3 = 1$, we obtain

$$p_1 \left[1 + \frac{p}{q} + \left(\frac{p}{q}\right)^2 \right] = 1.$$

Thus if $p = q = \frac{1}{2}$, then $p_1 = p_2 = p_3 = \frac{1}{3}$, and thus in the limit each state has the same probability. If $p \neq q$, then

$$p_j = \frac{1 - (p/q)}{1 - (p/q)^3} \left(\frac{p}{q}\right)^{j-1} \qquad (j = 1, 2, 3).$$

If $p > q$, then the probabilities p_j increase with the number j of the state; if $p < q$ they decrease. These results agree with our intuition. Thus if $p/q = 2$ we have

$$p_1 = \tfrac{1}{7}, \qquad p_2 = \tfrac{2}{7}, \qquad p_3 = \tfrac{4}{7},$$

and if $p/q = \frac{1}{2}$

$$p_1 = \tfrac{4}{7}, \qquad p_2 = \tfrac{2}{7}, \qquad p_3 = \tfrac{1}{7}.$$

We observe that all three ergodic probabilities are positive. This is because all states are intrinsic. In a Markov chain with a countable number of states ergodic probabilities of intrinsic states may be equal zero (see also Problem 7.9).

Example 7.4.6. Let us calculate the limit probabilities in example 7.4.4. By formula (7.4.13) we have

$$p_1 = p_1 + p_2 q,$$
$$p_2 = p_3 q,$$
$$p_3 = (p_2 + p_3) p.$$

Since $p_1 + p_2 + p_3 = 1$, we obtain $p_1 = 1, p_2 = p_3 = 0$.
As has been mentioned, this is because the states E_2 and E_3 are unintrinsic.

The notion of ergodicity and conditions for the validity of the ergodic theorem for nonhomogeneous Markov chains can be found in papers of Kolmogorov [3], Sarymsakov [1], and Hajnal [1, 2].

F   We now find the relations between the ergodic probabilities and the absolute probabilities in a homogeneous Markov chain. Let us compute the absolute probability of the event that after n steps the system passes into the state E_j. Denote this probability by $c_j(n)$. We have

$$(7.4.14) \qquad c_j(n) = \sum_k P(E_k)p_{kj}(n)$$

$$= \sum_k c_k(n-1)p_{kj},$$

where $P(E_k)$ is the initial probability of the state E_k.

Definition 7.4.5.   A homogeneous Markov chain for which the equalities

$$P(E_j) = c_j(1) \qquad (j = 1, 2, \ldots)$$

are satisfied is called a *stationary chain* and the probabilities $c_j(n)$ are called *stationary absolute probabilities*.

We observe that from the last equalities and from formula (7.4.14), for $j = 1, 2, \ldots$ and $n = 1, 2, 3, \ldots$, it follows that

$$c_j(1) = c_j(2) = \ldots = c_j(n) = c_j.$$

Thus from formula (7.4.14) we obtain the equalities

$$(7.4.15) \qquad c_j = \sum_k c_k p_{kj} \qquad (j = 1, 2, \ldots).$$

Suppose that the number of states is finite and equal to s. Suppose that the assumptions of theorem 7.4.1 are satisfied; thus the ergodic probabilities p_j exist. By comparing formulas (7.4.13) and (7.4.15), it is easy to verify that $c_j = p_j$ ($j = 1, 2, 3, \ldots, s$). Thus, if the initial probabilities $P(E_j)$ are equal, for $j = 1, \ldots, s$, to the ergodic probabilities p_j, then $c_j(n) = p_j$ will be constant for $n = 1, 2, 3, \ldots$; hence the chain will be stationary. We shall have an equilibrium in the sense of the invariance of absolute probabilities. This explains the name "ergodic theorem."

However, we observe that for an arbitrary Markov chain with a finite number of states we have the following theorem.

Theorem 7.4.2.   *The limits of the absolute probabilities*

$$(7.4.16) \qquad \lim_{n \to \infty} c_j(n) = c_j \qquad (j = 1, 2, \ldots, s)$$

for a homogenous Markov chain with a finite number of states exist independently of the initial distribution if and only if the ergodic probabilities p_j exist. We then have $c_j = p_j$ ($j = 1, 2, \ldots, s$).

Proof. Suppose that (7.4.16) is satisfied and c_j does not depend on the initial distribution. Then we may put $P(E_i) = 1$ and $P(E_j) = 0$ $(i \neq j)$. Hence by formula (7.4.14) we have

$$c_j(n) = p_{ij}(n).$$

Therefore, by (7.4.16)

$$p_j = \lim_{n \to \infty} p_{ij}(n) = c_j \qquad (i, j = 1, 2, \ldots, s).$$

Conversely, suppose that the ergodic probabilities p_j exist. Then by formula (7.4.14) for an arbitrary initial distribution we obtain

$$\lim_{n \to \infty} c_j(n) = \lim_{n \to \infty} \sum_{k=1}^{s} P(E_k) p_{kj}(n) = p_j \sum_{k=1}^{s} P(E_k) = p_j.$$

7.5 RANDOM VARIABLES FORMING A HOMOGENEOUS MARKOV CHAIN

A   The considerations of the previous sections may be applied to random variables. Let $\{X_n\}$ $(n = 0, 1, 2, \ldots)$ be random variables that can take on the values x_i $(i = 1, 2, 3, \ldots)$. The values x_i correspond to the states E_i previously discussed. We now give definitions analogous to the definitions given in Section 7.2.

Definition 7.5.1.   We say that the sequence $\{X_n\}$ $(n = 0, 1, 2, \ldots)$ of random variables with possible values x_i $(i = 1, 2, 3, \ldots)$ forms a *Markov chain* if for $i, j, n = 1, 2, 3, \ldots$ the equalities

$$(7.5.1) \quad p_{ij}^{(n)} = P(X_n = x_j \mid X_{n-1} = x_i)$$
$$= P(X_n = x_j \mid X_{n-1} = x_i, X_{n-2} = x_{i_{n-2}}, \ldots, X_1 = x_{i_1}, X_0 = x_{i_0})$$

are satisfied for arbitrary $x_{i_{n-2}}, \ldots, x_{i_1}, x_{i_0}$.

Definition 7.5.2.   We say that the sequence $\{X_n\}$ $(n = 0, 1, 2, \ldots)$ of random variables with possible values x_i $(i = 1, 2, 3, \ldots)$, forms a *homogeneous Markov chain* if for $i, j, n = 1, 2, 3, \ldots$ the conditional probabilities $p_{ij}^{(n)}$ are independent of n, that is,

$$(7.5.2) \qquad\qquad p_{ij}^{(n)} = p_{ij}.$$

In the terminology of random variables the probability $p_{ij}(n)$ of transition from the state E_i to E_j in n steps is the probability that $X_n = x_j$ provided $X_0 = x_i$, which means

$$p_{ij}(n) = P(X_n = x_j \mid X_0 = x_i).$$

Formula (7.3.3) takes the form

$$(7.5.3) \quad p_{ij}(n) = \sum_{k} P(X_m = x_k \mid X_0 = x_i) P(X_n = x_j \mid X_m = x_k),$$

where $1 \leqslant m < n$. The absolute probabilities $c_j(n)$ expressed by formula (7.4.14) take the form

$$(7.5.4) \quad c_j(n) = P(X_n = x_j) = \sum_k P(X_0 = x_k)P(X_n = x_j \mid X_0 = x_k).$$

Definition 7.5.3. A sequence $\{X_n\}$ $(n = 0, 1, 2, 3, \ldots)$ of random variables with possible values x_i $(i = 1, 2, \ldots)$ forming a homogeneous Markov chain is *stationary* if for $j = 1, 2, 3, \ldots$

$$c_j(0) = P(X_0 = x_j) = P(X_1 = x_j) = c_j(1) = c_j.$$

If follows from the last equality that for $n = 0, 1, 2, \ldots$ and $j = 1, 2, 3, \ldots$

$$P(X_n = x_j) = c_j.$$

Thus a stationary sequence of random variables is a sequence of identically distributed random variables.

It is also easy to formulate the classification of states and the theorems proved previously in the terminology of random variables. We leave this to the reader.

B   It should be stated that the theory of limit distributions for random variables forming a homogeneous Markov chain is less advanced than the same theory for independent random variables.

Conditions for the validity of the central limit theorem for Markov chains with three states were found by Markov [3], and for chains with an arbitrary finite number of states by Romanovsky [1], Fréchet [2], and Onicescu and Mihoc [1, 2]. Doeblin [2] showed that for a certain class of Markov chains with a countable number of states, the question of limit theorems can be reduced to the analogous question for independent random variables. For chains with an arbitrary number of states, some results were obtained by Doeblin [3], Doob [5], Dynkin [2], and Chung [2, 3]. A quite advanced result, which is essentially a generalization of the Lindeberg-Lévy theorem to a large class of random variables forming a homogeneous Markov chain, was recently obtained by Nagayev [1]. The local limit theorem for Markov chains with a finite number of states was given by Kolmogorov [13]. Sirazhdinov [1] obtained some results concerning the rate of convergence to the limit distribution in the local and integral limit theorems for Markov chains with a finite number of states. In the paper quoted, Nagayev obtained the local central limit theorem for Markov chains with a countable number of states and estimated the rate of convergence to the normal distribution.

Some theorems concerning the laws of large numbers for random variables forming a Markov chain can be found in the book by Doob [5] and the paper of Chung [3]. Breiman [2] recently obtained a general result in this field.

We merely state here without proof the law of large numbers and the central limit theorem for random variables forming a homogeneous Markov chain with a finite number of states.

Theorem 7.5.1. Let $\{X_k\}$ ($k = 0, 1, 2, \ldots$) be a stationary sequence of random variables forming a homogeneous Markov chain with a finite number of states. If all the intrinsic states are nonperiodic and form one class, then

$$(7.5.5) \qquad P\left[\lim_{n\to\infty} \frac{1}{n+1} \sum_{k=0}^{n} X_k = E(X_0)\right] = 1.$$

Thus if the assumptions of this theorem are satisfied, then the sequence $\{X_k\}$ obeys the strong law of large numbers. Compare this theorem with the theorem of Kolmogorov 6.12.3.

Example 7.5.1. Consider a stationary sequence of random variables X_k ($k = 0, 1, 2, \ldots$) which can take on only two values x_1 and x_2 and form a homogeneous Markov chain. The number x_1 is the state E_1 and the number x_2 is the state E_2. Suppose that the transition matrix is

$$\mathbf{M}_1 = \begin{bmatrix} p_{11} & p_{12} \\ p_{21} & p_{22} \end{bmatrix},$$

where $0 < p_{12} < 1$, $0 < p_{21} < 1$. The assumptions of theorem 7.4.1 are satisfied for $r = 1$. Both states are intrinsic and nonperiodic and they form one class. By formula (7.4.13) and the relation $p_1 + p_2 = 1$ we obtain the ergodic probabilities

$$(7.5.6) \qquad p_1 = \frac{p_{21}}{p_{12} + p_{21}}, \quad p_2 = \frac{p_{12}}{p_{12} + p_{21}},$$

and $0 < p_1 < 1$, $0 < p_2 < 1$. By the assumption that $\{X_k\}$ is stationary we have $P(X_k = x_1) = p_1$ and $P(X_k = x_2) = p_2$ ($k = 0, 1, 2, \ldots$). By theorem 7.5.1 the relation

$$(7.5.7) \qquad \lim_{n=\infty} \frac{1}{n+1} \sum_{k=0}^{n} X_k = \frac{p_{21}}{p_{12} + p_{21}} x_1 + \frac{p_{12}}{p_{12} + p_{21}} x_2$$

holds with probability one.

In particular, let $x_1 = 1$ and $x_2 = 0$. Then the event $\left(\sum_{k=0}^{n} X_k = m\right)$ occurs if m times among the possible $n + 1$ times the system is in the state E_1. Relation (7.5.7) states that, with probability one, the average number of times that the system is in the state E_1 tends to $p_{21}/(p_{12} + p_{21})$. If we treat the appearance of the value $x_1 = 1$ as a success, we see that the number of successes in a sequence of trials forming a homogeneous Markov chain obeys the strong law of large numbers. The weak law of large numbers for this example was obtained by Markov [1].

We now present the central limit theorem. Let

$$Y_n = \sum_{k=0}^{n} [X_k - E(X_k)].$$

Theorem 7.5.2.   *Let $\{X_k\}$ ($k = 0, 1, 2, \ldots$) be a sequence of random variables forming a homogeneous Markov chain with a finite number of states. If all the intrinsic states are nonperiodic and form one class, and if the variance $D^2(Y_n)$, when the sequence $\{X_k\}$ is stationary, satisfies the relation*

(7.5.8) $$\lim_{n \to \infty} \frac{D^2(Y_n)}{n + 1} = \sigma^2 > 0,$$

then for an arbitrary initial distribution of the random variable X_0, the relation

(7.5.9) $$\lim_{n \to \infty} P\left(\frac{Y_n}{\sigma\sqrt{n + 1}} < y \right) = \frac{1}{\sqrt{2\pi}} \int_{-\infty}^{y} e^{-y^2/2} \, dy$$

is satisfied.

Example 7.5.2.   Let us return to example 7.5.1, in which the sequence $\{X_k\}$ is stationary, and set $x_1 = 1$, $x_2 = 0$. Let $Z_k = X_k - E(X_k)$. Since $\{X_k\}$ is stationary, we have $E(X_k) = p_1$ ($k = 0, 1, 2, \ldots$). Let us find $D^2(Y_n)$ and verify that relation (7.5.8) is satisfied. Since the sequence $\{X_k\}$ is stationary, we obtain

(7.5.10) $$D^2(Y_n) = \sum_{k=0}^{n} D^2(Z_k) + 2 \sum_{k=0}^{n-1} \sum_{m=k+1}^{n} E(Z_k Z_m)$$

$$= (n + 1)p_1 p_2 + 2 \sum_{k=0}^{n-1} \sum_{m=k+1}^{n} E(Z_k Z_m).$$

To find $E(Z_k Z_m)$, we observe that the random variable $Z_k Z_m$ can take on the following values with the respective probabilities:

$$P[Z_k Z_m = (1 - p_1)^2] = P(X_k = 1)P(X_m = 1 \mid X_k = 1) = p_1 p_{11}(m - k),$$

$$P[Z_k Z_m = (1 - p_1)(-p_1)] = P(X_k = 1)P(X_m = 0 \mid X_k = 1)$$
$$+ P(X_k = 0)P(X_m = 1 \mid X_k = 0)$$
$$= p_1 p_{12}(m - k) + p_2 p_{21}(m - k),$$

$$P(Z_k Z_m = p_1^2) = P(X_k = 0)P(X_m = 0 \mid X_k = 0) = p_2 p_{22}(m - k).$$

Hence, after some simple computations,

(7.5.11) $$E(Z_k Z_m) = p_1 p_2^2 p_{11}(m - k) - p_1^2 p_2 p_{12}(m - k)$$
$$- p_1 p_2^2 p_{21}(m - k) + p_1^2 p_2 p_{22}(m - k)$$
$$= p_1 p_2 [p_{11}(m - k) - p_{21}(m - k)] = p_1 p_2 (p_{11} - p_{21})^{m-k}.$$

Therefore, using formula (7.5.10) and letting $\delta = p_{11} - p_{21}$, we obtain

$$D^2(Y_n) = p_1 p_2 \{n + 1 + 2[n\delta + (n - 1)\delta^2 + \ldots + \delta^n]\}$$

$$= p_1 p_2 \left[n + 1 + 2\left(\sum_{j=1}^{n} \delta^j + \sum_{j=1}^{n-1} \delta^j + \ldots + \delta \right) \right]$$

$$= p_1 p_2 \left[n + 1 + \frac{2\delta n}{1 - \delta} - \frac{2\delta^2(1 - \delta^n)}{(1 - \delta)^2} \right].$$

It follows that

$$\frac{D^2(Y_n)}{n+1} = p_1 p_2 \left(1 + 2\frac{\delta}{1-\delta} \cdot \frac{n}{n+1}\right) - \frac{p_1 p_2}{n+1} \cdot \frac{2\delta^2(1-\delta^n)}{(1-\delta)^2}.$$

Finally, noting that $0 < \delta < 1$, we get

$$\lim_{n \to \infty} \frac{D^2(Y_n)}{n+1} = p_1 p_2 \frac{1+\delta}{1-\delta} = \sigma^2 > 0.$$

Condition (7.5.8) holds; thus relation (7.5.9) is satisfied.

The exact distribution of the number of successes in the homogeneous Markov chains with two states considered in this example, and the moments of this random variable were investigated by Gabriel [1]. In this paper the reader will also find an application of such a chain to meteorology.

We also observe that from formula (7.5.11) it follows that the correlation coefficient ρ_{km} of the random variables X_k and X_m is

$$\rho_{km} = (p_{11} - p_{21})^{m-k}.$$

This example of the central limit theorem is a particular case of the Markov theorem [3]; the Markov theorem concerns random variables forming a nonhomogeneous Markov chain. A stronger result was obtained by Bernstein [3]. An important result concerning the central limit theorem for random variables forming a nonhomogeneous Markov chain was obtained by Dobrushin [3].

We refer the reader to recent papers of Hunt [1], Doob [10], and Kemeny and Snell [2,3], in which new methods of research of Markov chains, based on connnections between potential theory and Markov chains, were developed.

Problems and Complements

7.1. The transition matrix of a homogeneous Markov chain with four states has the form

$$\mathbf{M_1} = \begin{bmatrix} \frac{1}{4} & \frac{1}{2} & 0 & \frac{1}{4} \\ \frac{1}{5} & 0 & \frac{1}{3} & \frac{7}{15} \\ 0 & \frac{2}{3} & \frac{1}{3} & 0 \\ \frac{1}{4} & \frac{1}{4} & \frac{1}{4} & \frac{1}{4} \end{bmatrix}$$

(a) Classify all states.
(b) Check whether the ergodic theorem holds.
(c) If so, find the ergodic probabilities.

7.2. Do the same for

$$\mathbf{M_1} = \begin{bmatrix} \frac{1}{3} & \frac{2}{3} & 0 & 0 \\ \frac{3}{4} & \frac{1}{8} & \frac{1}{8} & 0 \\ 0 & 0 & \frac{1}{2} & \frac{1}{2} \\ 0 & 0 & \frac{1}{3} & \frac{2}{3} \end{bmatrix}$$

7.3. (*Ehrenfests' model*). In physics, in problems of diffusion and in problems of recurrence and reversibility (detailed information can be found in an interesting book by Kac [3]), the following model of a homogeneous Markov chain is applied: $2R$ balls numbered $1, 2, \ldots, 2R - 1, 2R$ are placed in two urns, A and B. At each unit of time the number of a ball is drawn at random, according to the multinomial scheme with a constant probability of drawing any of the numbers $1, \ldots, 2R$ equal to $(2R)^{-1}$. When its number has been drawn, a ball changes its place from the urn in which it was during the drawing to the other urn. Let us agree to say that the system of two urns is in the state E_j ($j = 0, 1, \ldots, 2R$) if the urn A contains j balls. Thus we have a homogeneous Markov chain with $2R + 1$ states and the transition matrix

$$
\mathbf{M}_1 = \begin{bmatrix}
0 & 1 & 0 & 0 & 0 & \ldots & 0 & 0 \\
(2R)^{-1} & 0 & 1 - (2R)^{-1} & 0 & 0 & \ldots & 0 & 0 \\
0 & 2(2R)^{-1} & 0 & 1 - 2(2R)^{-1} & 0 & \ldots & 0 & 0 \\
0 & 0 & 3(2R)^{-1} & 0 & 1 - 3(2R)^{-1} & \ldots & 0 & 0 \\
\hdotsfor{8} \\
0 & 0 & 0 & 0 & 0 & \ldots & 1 & 0
\end{bmatrix}
$$

(*a*) Show that the ergodic probabilities p_j exist; hence they are equal to the stationary probabilities c_j, where

$$
c_j = \binom{2R}{j} \frac{1}{2^{2R}} \qquad (j = 0, 1, \ldots, 2R).
$$

(*b*) Notice that no matter what the initial distribution is, in the limit we always obtain the binomial distribution.

(*c*) Using Problem 5.2, find which of the states is the most probable in the limit.

7.4(*a*). Prove that for an arbitrary homogeneous Markov chain with a finite number of states the limits

$$
\lim_{n \to \infty} \frac{1}{n} \sum_{k=1}^{n} p_{ij}(k) = q_{ij}
$$

exist.

(*b*) Compute q_{ij} in examples 7.4.1 to 7.4.4.

7.5. (A continuation of Kolmogorov's [9] classification of states.) Let $K_{ij}(n)$ denote the probability of passing from the state E_i to the state E_j for the first time on the nth step and let

$$
L_{ij} = \sum_{n=1}^{\infty} K_{ij}(n), \qquad R_{ij} = \sum_{n=1}^{\infty} n K_{ij}(n).
$$

The expression R_{jj} is called the *mean recurrence time* of the state E_j.

The state E_j is called *recurrent* if $L_{jj} = 1$, *transient* if $L_{jj} < 1$. A recurrent state with $R_{jj} = \infty$ is called a *null state*. A recurrent state which is neither a null state nor periodic is called an *ergodic* state.

Show that

(*a*) $K_{ij}(n) = p_{ij}(n) - K_{ij}(1) p_{jj}(n - 1) - \ldots - K_{ij}(n - 1) p_{jj}$.

(*b*) in a homogeneous Markov chain with a finite number of states, the state E_j is recurrent (transient) if and only if it is intrinsic (unintrinsic).

7.6. Let us consider a homogeneous Markov chain with a countable number of states with the transition matrix

$$
\mathbf{M}_1 = \begin{bmatrix}
p_1 & 1 - p_1 & 0 & 0 & 0 & \cdots \\
p_2 & 0 & 1 - p_2 & 0 & 0 & \cdots \\
p_3 & 0 & 0 & 1 - p_3 & 0 & \cdots \\
\cdots\cdots\cdots\cdots\cdots\cdots\cdots\cdots\cdots\cdots
\end{bmatrix}
$$

Show that if $\sum_{j=1}^{\infty} p_j < \infty$, then all states are transient and if $\sum_{j=1}^{\infty} p_j = \infty$, then all states are recurrent. Deduce that there may exist states which are at the same time transient and intrinsic. Compare with Problem 7.5(b).

7.7. Let us denote by Ω_{ij} the probability that the system will return an infinite number of times to the state E_j if at the initial moment it was in the state E_i. Prove that

(a) if $L_{jj} = 1$, then $\Omega_{ij} = 1$.

(b) for a set of intrinsic states which form one class either all $\Omega_{ij} < 1$ or all $\Omega_{ij} = 1$.

7.8. Prove that in a set of intrinsic states which form one class, either all $L_{ij} < 1$ or all $L_{ij} = 1$.

7.9. (A generalization of the ergodic theorem). Let $\mathbf{M}_1 = [p_{ij}]$ denote the transition matrix of a homogeneous Markov chain with a countable number of states $E_1, E_2, E_3, \ldots$.

I. If all states are recurrent, non-null and nonperiodic and form one class, then for $i, j = 1, 2, \ldots$,

$$
\lim_{n \to \infty} p_{ij}(n) = p_j = 1/R_{jj},
$$

where $p_1 + p_2 + \ldots = 1$, $p_j > 0$, and $p_j = c_j$, where c_j is the stationary probability. II. If E_j is a transient or a recurrent null state, then for all i we have $\lim_{n \to \infty} p_{ij}(n) = 0$. III. If E_j is a recurrent, non-null state and has period $d > 1$, then

$$
\lim_{n \to \infty} p_{ij}(n) = d/R_{jj}.
$$

(Kolmogorov [9], Feller [7]).

7.10. Let us consider a homogeneous Markov chain with a countable number of states and with the transition matrix

$$
\mathbf{M}_1 = \begin{bmatrix}
\frac{1}{2} & \frac{1}{2} & 0 & 0 & 0 & 0 & \cdots \\
\frac{2}{3} & 0 & \frac{1}{3} & 0 & 0 & 0 & \cdots \\
\frac{3}{4} & 0 & 0 & \frac{1}{4} & 0 & 0 & \cdots \\
\frac{4}{5} & 0 & 0 & 0 & \frac{1}{5} & 0 & \cdots \\
\cdots\cdots\cdots\cdots\cdots\cdots\cdots\cdots
\end{bmatrix}
$$

Show that $\lim_{n \to \infty} p_{ij}(n) = p_j = e^{-1}/j$ $(i, j, = 1, 2, \ldots)$ (compare with Problem 7.6).

7.11. Let us consider a homogeneous Markov chain with a countable number of states and with the transition matrix

$$\mathbf{M}_1 = \begin{bmatrix} \frac{1}{2} & \frac{1}{2} & 0 & 0 & 0 & \cdots \\ \frac{1}{2} & 0 & \frac{1}{2} & 0 & 0 & \cdots \\ \frac{1}{3} & 0 & 0 & \frac{2}{3} & 0 & \cdots \\ \frac{1}{4} & 0 & 0 & 0 & \frac{3}{4} & \cdots \\ \cdots\cdots\cdots\cdots\cdots\cdots\cdots \end{bmatrix}$$

Show that all states are recurrent null states, hence $\lim_{n \to \infty} p_{ij}(n) = 0$ $(i, j, = 1, 2, \ldots)$.

CHAPTER 8

Stochastic Processes

8.1 THE NOTION OF A STOCHASTIC PROCESS

In this chapter we enlarge the domain of probability theory so far considered. This enlargement will have two aspects. The first will consist in considering not only sequences of random variables, that is, finite or countable sets of random variables, but also nondenumerable sets of random variables. The formulation of the second aspect requires some additional remarks.

In the preceding chapters we considered random variables which assign either one number or a collection of n numbers to each elementary event. In other words, a realization of such a random variable is one number or a collection of n numbers. A sequence $X_1, X_2, \ldots$ of random variables may also be treated as one "random variable" whose realizations are sequences $(x_1, x_2, \ldots)$. In practice, however, we sometimes deal with more complicated random phenomena, whose realizations are functions of a real variable. Thus, for instance, the number of telephone calls in Warsaw during the time interval $(0, t)$, for a fixed t, is a random variable, but the number of calls considered as a function of t, that is, as a function of the variable t when t runs over some interval, is a random function. Similarly, the consumption of electric power at some fixed moment in Warsaw is a random variable, but this consumption considered over some time interval as a function of time is a random function. Similarly, the water level in the Vistula under the Poniatowski bridge in Warsaw at some fixed moment of the day is a random variable; however, the level of the water in the Vistula measured at some fixed moment of the day at the distance t from the Poniatowski bridge and considered as a function of the argument t is a random function.

Thus, as we have already said, the first aspect of extending probability theory consists in considering nondenumerable sets of random variables;

271

the second aspect consists in considering random functions. Probability theory so extended is called the theory of stochastic processes.

Definition 8.1.1. A family of random variables X depending on the parameter t, where t belongs to a set I of real numbers, is called a *stochastic process*. A stochastic process is denoted by the symbol $\{X_t, t \in I\}$.

The parameter t is interpreted as time.

According to this definition, a k-dimensional random variable $(k \geqslant 1)$ is a stochastic process, and here I is the set of integers $(1, 2, \ldots, k)$. A countable sequence $\{X_k\}$ $(k = 1, 2, \ldots)$ of random variables is a stochastic process in which I is the set of all positive integers. In these two instances we speak of *stochastic processes with discrete time*. However, we have an essential extension of probability theory if we deal with nondenumerable sets I. Here we shall only consider cases in which I is a finite or infinite interval. Such processes are called *stochastic processes with continuous time*.

As already said, a stochastic process may be defined in a different way. We could consider a stochastic process as a set of random functions assigned to elementary events. At this moment we shall only mention that for a fixed t, X_t is a random variable, and for a fixed elementary event, X_t is a function of the argument t which is called a *realization* of the stochastic process. The parameter t is in many cases indeed the time during which the phenomenon under consideration takes place. Other situations are also possible; either the parameter t represents not time but some other physical quantity, such as the point in space at which we observe our phenomenon, or the parameter t does not have any physical meaning whatever.

8.2 MARKOV PROCESSES AND PROCESSES WITH INDEPENDENT INCREMENTS

A   In the previous chapter we stated that the necessity of considering dependent random variables is due to the requirements of physics and technology. Now, as we start to investigate nondenumerable sets of random variables, we are immediately led to dependent random variables, for if we assumed the independence of a nondenumerable family of random variables, we could not expect to find any regularity in this chaos of random phenomena, whereas the main task of probability theory is to find such regularities.

We give the definitions of different stochastic processes. We start with Markov processes, which are important in theory as well as in applications.

Definition 8.2.1. A stochastic process $\{X_t, t \in I\}$ is called a *Markov* *process* if for $n = 1, 2, 3, \ldots$ and arbitrary $t_m \in I$ ($m = 0, 1, 2, \ldots, n$), where $t_0 < t_1 < \ldots < t_n$, and for any real x and y the equality

$$(8.2.1) \quad P(X_{t_n} < y \mid X_{t_{n-1}} = x, X_{t_{n-2}} = x_{n-2}, \ldots, X_{t_1} = x_1, X_{t_0} = x_0)$$
$$= P(X_{t_n} < y \mid X_{t_{n-1}} = x)$$

is satisfied for every real $x_{n-2}, \ldots, x_0$.

According to this definition, the conditional distribution of the random variable X_{t_n} under the condition that the random variable $X_{t_{n-1}}$ takes on the value x does not depend on any information about the values taken on by random variables of this process in the moments preceding t_{n-1}. For any $t_1 \in I$ and $t_2 \in I$, where $t_1 < t_2$, we define the conditional distribution function $F(t_1, x, t_2, y)$ by the formula

$$(8.2.2) \qquad F(t_1, x, t_2, y) = P(X_{t_2} < y \mid X_{t_1} = x).$$

This distribution function is independent of the values of X_t at $t < t_1$.

In Section 8.8 we investigate more closely the properties of the distribution function defined by relation (8.2.2).

Definition 8.2.2. A Markov process $\{X_t, t \in I\}$ is called *homogeneous* if for arbitrary $t_1 \in I$ and $t_2 \in I$, where $t_1 < t_2$, the distribution function (8.2.2) depends only on the difference $t = t_2 - t_1$.

Here we write

$$(8.2.3) \qquad F(t_1, x, t_2, y) = F(t, x, y).$$

The stochastic processes defined by the following two definitions are of great importance.

Definition 8.2.3. A process $\{X_t, t \in I\}$ is called a *process with independent increments* if for $n = 2, 3, \ldots$ and arbitrary $t_m \in I$ ($m = 0, 1, 2, \ldots, n$), where $t_0 < t_1 < \ldots < t_n$, the random variables $X_{t_1} - X_{t_0}, X_{t_2} - X_{t_1}, \ldots, X_{t_n} - X_{t_{n-1}}$ are independent.

If a stochastic process $\{X_t, t \in I\}$ with independent increments satisfies the relation $P(X_{t_0} = b) = 1$, where t_0 is the initial moment and b is a constant, then this process is a Markov process. We observe that, with probability one,

$$X_{t_n} = b + \sum_{m=1}^{n} (X_{t_m} - X_{t_{m-1}}).$$

Thus X_{t_n} is the sum of $n + 1$ independent random variables. On the other hand, a Markov process does not need to be a process with independent increments. In fact, let $P(X_{t_0} = 0) = 1$ and let the process $\{X_t, t_0 \leqslant t < a\}$ be a Markov process. Then, for t_1 and t_2 satisfying the inequality $t_0 < t_1 < t_2 < a$,

$$P(X_{t_1} = x_1, X_{t_2} = x_1 + x_2) = P(X_{t_1} = x_1, X_{t_2} - X_{t_1} = x_2)$$
$$= P(X_{t_1} = x_1)P(X_{t_2} - X_{t_1} = x_2 \mid X_{t_1} = x_1).$$

Suppose now that this process has independent increments. Then

$$P(X_{t_1} = x_1, X_{t_2} - X_{t_1} = x_2) = P(X_{t_1} = x_1)P(X_{t_2} - X_{t_1} = x_2),$$

and the results obtained are different.

Definition 8.2.4. A stochastic process $\{X_t, t \in I\}$ is called a *process with homogeneous increments* if for any $t_1 \in I$ and $t_2 \in I$, where $t_1 < t_2$, the distribution function of $X_{t_2} - X_{t_1}$ depends only on the difference $t_2 - t_1$.

It is easy to verify that a process with independent increments which is a homogeneous Markov process is a process with homogeneous increments.

B   In Sections 8.3 to 8.7 we investigate Markov processes with random variables of the discrete type. Suppose for simplicity—and this assumption is often satisfied in practice—that the jump points of these random variables are integers. These numbers are called the *states* of the process and the process is called a *discrete process*. Here it is more convenient to characterize the process by the probability function than by the distribution function. Definitions 8.2.1 and 8.2.2 should then be suitably modified.

Definition 8.2.1'. A discrete process $\{X_t, t \in I\}$ is a *Markov process* if for $n = 1, 2, 3, \ldots$ and for arbitrary $t_m \in I (m = 0, 1, 2, \ldots, n)$, where $t_0 < t_1 < \ldots < t_n$, and for arbitrary integers i, j the equality

$$(8.2.4) \quad P(X_{t_n} = j \mid X_{t_{n-1}} = i, X_{t_{n-2}} = i_{n-2}, \ldots, X_{t_0} = i_0)$$

$$= P(X_{t_n} = j \mid X_{t_{n-1}} = i)$$

is satisfied for any integers $i_{n-2}, \ldots, i_0$.

For any $t_1 \in I$ and $t_2 \in I$, where $t_1 < t_2$, we define the *transition probability function* (or simply the *transition function*) by the formula

$$(8.2.5) \qquad\qquad p_{ij}(t_1, t_2) = P(X_{t_2} = j \mid X_{t_1} = i).$$

The function $p_{ij}(t_1, t_2)$ does not depend on the values of X_t for $t < t_1$.

The reader may compare definition 8.2.1' with definition 7.2.1.

The function $p_{ij}(t_1, t_2)$ satisfies the obvious relation

$$(8.2.6) \qquad\qquad p_{ij}(t_1, t_2) \geqslant 0.$$

Moreover, for all i, we have

$$(8.2.7) \qquad\qquad \sum_j p_{ij}(t_1, t_2) = 1.$$

The equality

$$(8.2.8) \qquad\qquad p_{ij}(t_1, t_2) = \sum_k p_{ik}(t_1, s)p_{kj}(s, t_2),$$

where $t_1 < s < t_2$, holds for every t_1 and t_2. The proof of this formula is analogous to that of (7.3.3).

Formula (8.2.8) is a generalization of the Markov equation and is called the *Chapman-Kolmogorov equation*.

By (8.2.4), (8.2.5), and (1.5.7) we obtain for $n = 1, 2, 3, \ldots$ and arbitrary $t_m \in I$ $(m = 0, 1, 2, \ldots, n)$, where $t_0 < t_1 < \ldots < t_m$, the following formula for the probability function of the random variable $(X_{t_0}, X_{t_1}, \ldots, X_{t_n})$

$$(8.2.9) \quad P(X_{t_0} = i_0, X_{t_1} = i_1, \ldots, X_{t_n} = i_n)$$
$$= P(X_{t_0} = i_0) p_{i_0 i_1}(t_0, t_1) \ldots p_{i_{n-1} i_n}(t_{n-1}, t_n).$$

Thus the probability function of the random variable X_{t_0} at the initial moment t_0 and the transition probability function determine the probability distribution of the $(n + 1)$-dimensional random variable $(X_{t_0}, X_{t_1}, \ldots, X_{t_n})$ for $n = 1, 2, 3, \ldots$ and for arbitrary values of the parameter t.

Definition 8.2.2'. A discrete Markov process $\{X_t, t \in I\}$ is *homogeneous* if for arbitrary i, j and for arbitrary $t_1 \in I$ and $t_2 \in I$ $(t_1 < t_2)$ the transition probability function depends only on the difference $t = t_2 - t_1$.

Here we write

$$(8.2.10) \qquad p_{ij}(t_1, t_2) = p_{ij}(t).$$

Relations (8.2.6) to (8.2.8) take the form

$$(8.2.11) \qquad p_{ij}(t) \geqslant 0,$$

$$(8.2.12) \qquad \sum_j p_{ij}(t) = 1,$$

$$(8.2.13) \qquad p_{ij}(t + s) = \sum_k p_{ik}(t) p_{kj}(s).$$

C Theorems 7.4.1 and 7.4.2 can be extended to homogeneous Markov processes with a finite number of states with continuous time. The proofs are essentially the same as for Markov chains. We have only to replace the parameter n by the parameter t. Therefore we present these theorems without proof.

Theorem 8.2.1. *Let* $\{X_t, 0 \leqslant t < \infty\}$ *be a homogeneous Markov process with a finite number of states* $i = 0, 1, 2, \ldots, s$. *If there exists a* t^* $(0 \leqslant t^* < \infty)$ *such that*

$$p_{ij}(t^*) > 0 \qquad (i, j = 0, 1, \ldots, s),$$

then the limits

$$(8.2.14) \qquad \lim_{t \to \infty} p_{ij}(t) = p_j > 0 \qquad (i, j = 0, 1, \ldots, s)$$

exist.

Let

$$c_j(t) = P(X_t = j) \qquad (j = 0, 1, 2, \ldots).$$

Thus $c_j(t)$ is the absolute probability that at the moment t the process will be in the state j. For a homogeneous Markov process we have

$$(8.2.15) \qquad\qquad c_j(t) = \sum_i c_i(0) p_{ij}(t).$$

Theorem 8.2.2. *Let* $\{X_t, 0 \leqslant t < \infty\}$ *be a homogeneous Markov process with a finite number of states* $i = 0, 1, \ldots, s$. *Then the limits*

$$(8.2.16) \qquad \lim_{t \to \infty} c_j(t) = c_j \qquad (j = 0, 1, \ldots, s)$$

exist independently of the initial distribution at the moment $t = 0$ *if and only if the ergodic probabilities* p_j *defined by* (8.2.14) *exist, and then* $p_j = c_j$ $(j = 0, 1, \ldots, s)$.

Fuchs [1] has generalized theorem 8.2.2 to homogeneous Markov processes with an arbitrary number of states.

We investigate further the general properties of discrete Markov processes in Section 8.7. In the succeeding sections we investigate several important particular cases of such processes.

8.3 THE POISSON PROCESS

In Section 5.5 we investigated the Poisson distribution. We showed there that the Poisson distribution can be obtained as a limit distribution of a sequence of binomial distributions, and in examples 5.5.1 and 5.5.2 we established that some observed distributions are very close to the Poisson distribution. Now we can investigate this phenomenon somewhat more deeply. However, we approach the Poisson distribution from a different point of view, namely, we formulate certain conditions which are the mathematical idealization of real properties of investigated phenomena, and we show that these properties imply the Poisson distribution. This new approach to the problem of the appearance of the Poisson distribution may be applied to many other probability distributions.

Let us return to example 5.5.2 where we deal with the emission of α particles by a radioactive substance. We first formulate the conditions that lead to the Poisson distribution, and each of these conditions is then interpreted in the terminology of this example. The results of many observations indicate that these conditions are satisfied with sufficient accuracy in this example.

Let X_t denote the number of observed random events during the time interval $[0, t)$, where $0 \leqslant t < \infty$. We have here a stochastic process $\{X_t, 0 \leqslant t < \infty\}$, where for every t the random variable X_t can take on the integer values $i = 0, 1, 2, \ldots$. Moreover, for arbitrary t_1 and t_2 $(t_1 < t_2)$, the increment $X_{t_2} - X_{t_1}$ may equal $0, 1, 2, \ldots$.

If we agree to call the observed event a *signal*, we call such a process a *signal process*. In applications this may be a process of actual signals, as in example 8.3.1 where we deal with a telephone exchange. In example 5.5.2 a "signal" consists of the emission of an α particle and X_t denotes the number of emitted particles during the time $[0, t)$.

We assume that the following conditions are satisfied:

I. *The process* $\{X_t, 0 \leqslant t < \infty\}$ *is a process with independent increments.*

Thus condition I states that the number of signals in the disjoint time intervals $[t_0, t_1), [t_1, t_2), \ldots, [t_{n-1}, t_n)$ are independent random variables. In the example under consideration the number of α particles emitted in the disjoint time intervals are supposed to be independent.

II. *The process* $\{X_t, 0 \leqslant t < \infty\}$ *is a process with homogeneous increments.*

Thus condition II states that the probability of appearance of a given number of signals is the same for all time intervals of the same length. In our example the probability of emission of a given number of α particles is supposed to be the same for time intervals of the same length.

For $t > 0$ let

$$(8.3.1) \qquad W_i(t) = P(X_t - X_0 = i) \qquad (i = 0, 1, 2, \ldots).$$

$W_i(t)$ is the probability that i signals will appear during a time interval of length t. By condition II this probability depends only on t.

III. *The following relations are satisfied:*

$$(8.3.2) \qquad \lim_{t \to 0} \frac{W_1(t)}{t} = \lambda \qquad (\lambda > 0),$$

$$(8.3.3) \qquad \lim_{t \to 0} \frac{1 - W_0(t) - W_1(t)}{t} = 0.$$

Relation (8.3.2) states that the probability of occurrence of one signal during a small interval of time of length t is $\lambda t + o(t)$, and relation (8.3.3) states that the probability of occurrence of more than one signal during the time t is of the order $o(t)$. Thus condition III states that the signals occur suddenly, instantaneously, at small intervals of time, but only as single signals and not in pairs or other multiple combinations. In the example of emission of α particles we can assume that condition III is approximately satisfied.

We now introduce the notion of a homogeneous Poisson process.

Definition 8.3.1. A discrete stochastic process $\{X_t, 0 \leqslant t < \infty\}$ with independent and homogeneous increments, and with states $i = 0, 1, 2, \ldots,$

is a *homogeneous Poisson process* if for any t $(0 \leqslant t < \infty)$ and for some $\lambda > 0$ the relation

(8.3.4) $$P(X_t = i) = \frac{(\lambda t)^i}{i!} e^{-\lambda t} \qquad (i = 0, 1, 2, \ldots)$$

holds.

Theorem 8.3.1. *A stochastic process* $\{X_t, 0 \leqslant t < \infty\}$, *where* X_t *is the number of signals in the interval* $[0, t)$, *satisfying conditions* I *to* III *and the equality* $P(X_0 = 0) = 1$, *is a homogeneous Poisson process.*

Proof. From (8.3.1) and conditions I and II we have for $\Delta t > 0$

(8.3.5) $$W_0(t + \Delta t) = W_0(t)W_0(\Delta t),$$

(8.3.6) $$W_i(t + \Delta t) = \sum_{k=0}^{i} W_{i-k}(t)W_k(\Delta t) \qquad (i = 1, 2, \ldots).$$

Indeed, if no signal occurs during the time $[0, t + \Delta t)$, then no signal occurs either in the interval $[0, t)$ or in the interval $[t, t + \Delta t)$. By conditions I and II we obtain (8.3.5). If during the time $[0, t + \Delta t)$ the signal occurs i times, then this could happen in $i + 1$ pairwise disjoint ways, namely, the signal might occur $i - k$ times in the interval $[0, t)$, and k times in the interval $[t, t + \Delta t)$, where $k = 0, 1, 2, \ldots, i$. Thus by conditions I and II we obtain (8.3.6).

Let us write formulas (8.3.2) and (8.3.3), respectively, in the following forms:

(8.3.7) $$W_1(\Delta t) = \lambda \Delta t + o(\Delta t),$$

(8.3.8) $$1 - W_0(\Delta t) - W_1(\Delta t) = \sum_{k=2}^{\infty} W_k(\Delta t) = o(\Delta t).$$

It follows that

(8.3.9) $$W_0(\Delta t) = 1 - \lambda \Delta t + o(\Delta t).$$

By the last formula we obtain from (8.3.5)

$$\frac{W_0(t + \Delta t) - W_0(t)}{\Delta t} = -\lambda W_0(t) - W_0(t) \frac{o(\Delta t)}{\Delta t}.$$

Similarly, from (8.3.6) to (8.3.9) we obtain

$$\frac{W_i(t + \Delta t) - W_i(t)}{\Delta t} = -\lambda W_i(t) + \lambda W_{i-1}(t) + \frac{o(\Delta t)}{\Delta t} \sum_{k=0}^{i} W_k(t).$$

Passing to the limit as $\Delta t \to 0$ in the last two equations we obtain the following recursive system of linear differential equations:

(8.3.10) $$W_0'(t) = -\lambda W_0(t),$$

(8.3.11) $$W_i'(t) = -\lambda W_i(t) + \lambda W_{i-1}(t) \qquad (i = 1, 2, \ldots).$$

By the condition $W_0(0) = 1$ we immediately obtain from (8.3.10)

(8.3.12) $$W_0(t) = e^{-\lambda t}.$$

To solve equations (8.3.11), let us write

(8.3.13) $$W_i(t) = U_i(t)e^{-\lambda t} \qquad (i = 0, 1, 2, \ldots).$$

By (8.3.12) we have $U_0(t) = 1$. Next we have

$$W_i'(t) = [U_i'(t) - \lambda U_i(t)]e^{-\lambda t} \qquad (i = 1, 2, \ldots).$$

Hence by (8.3.11)

$$
\begin{aligned}
[U_i'(t) - \lambda U_i(t)]e^{-\lambda t} &= -\lambda[W_i(t) - W_{i-1}(t)] \\
&= -\lambda[U_i(t) - U_{i-1}(t)]e^{-\lambda t} \qquad (i = 1, 2, \ldots).
\end{aligned}
$$

Thus the functions $U_i(t)$ satisfy the differential equations

$$U_i'(t) = \lambda U_{i-1}(t) \qquad (i = 1, 2, \ldots).$$

Integrating, we obtain

$$U_i(t) - U_i(0) = \lambda \int_0^t U_{i-1}(t)\, dt \qquad (i = 1, 2, \ldots).$$

By the definition of $U_i(t)$ we have $U_i(0) = W_i(0)$ for all i; hence by the equality $W_0(0) = 1$ we obtain $U_i(0) = 0$ for $i \geqslant 1$. Finally, we obtain

$$U_i(t) = \lambda \int_0^t U_{i-1}(t)\, dt \qquad (i = 1, 2, \ldots).$$

Hence

$$U_i(t) = \frac{(\lambda t)^i}{i!}.$$

From (8.3.13) and from the last formula it follows that

(8.3.14) $$W_i(t) = \frac{(\lambda t)^i}{i!} e^{-\lambda t} \qquad (i = 1, 2, \ldots).$$

Equalities (8.3.12) and (8.3.14) imply that for arbitrary t

$$P(X_t = i) = \frac{(\lambda t)^i}{i!} e^{-\lambda t} \qquad (i = 0, 1, 2, \ldots),$$

which proves theorem 8.3.1.

Example 8.3.1. Consider a telephone exchange. Here the signals are to be interpreted as incoming calls. Condition I is satisfied, as the number of incoming calls during the disjoint time intervals may be considered as independent with sufficient accuracy. Condition II is satisfied if we consider the same time of day, when the traffic is approximately the same. Condition III is also approximately satisfied.

The homogeneous Poisson process considered here as a process with independent increments satisfying the condition $P(X_0 = 0) = 1$ is a homogeneous Markov process. The transition probability function $p_{ij}(t)$ has the form

$$(8.3.15) \quad p_{ij}(t) = P(X_{t_2} = j \mid X_{t_1} = i) = \frac{P(X_{t_1} = i, X_{t_2} - X_{t_1} = j - i)}{P(X_{t_1} = i)}$$

$$= P(X_{t_2} - X_{t_1} = j - i) = \frac{(\lambda t)^{j-i}}{(j-i)!} e^{-\lambda t},$$

where $t = t_2 - t_1$ and $j \geqslant i$. Obviously, $p_{ij}(t) = 0$ for $j < i$.

It follows from the last formula that a homogeneous Poisson process satisfies the ergodic theorem, namely, for $i = 0, 1, 2, \ldots$ and arbitrary j

$$(8.3.16) \qquad\qquad \lim_{t \to \infty} p_{ij}(t) = 0.$$

All the ergodic probabilities equal zero since all states are transient (see Problems 7.5 and 7.9) and from each state i we can pass to each state $j > i$, but we cannot pass back to the state i from the state $j < i$.

The following relations follow easily from (8.3.15):

$$(8.3.17) \qquad q_i = \lim_{t \to 0} \frac{1 - p_{ii}(t)}{t} = \lim_{t \to 0} \frac{1 - e^{-\lambda t}}{t} = \lambda,$$

$$(8.3.18) \qquad q_{ij} = \lim_{t \to 0} \frac{p_{ij}(t)}{t} = \lim_{t \to 0} \frac{(\lambda t)^{j-i} e^{-\lambda t}}{(j-i)! \, t}$$

$$= \begin{cases} \lambda & \text{for } j = i+1, \\ 0 & \text{for } j > i+1 \quad \text{and } j < i, \end{cases}$$

$$(8.3.19) \qquad\qquad \sum_{\substack{j \\ j \neq i}} q_{ij} = q_i.$$

The expressions q_i and q_{ij} are called the *intensities* of the stochastic process. As we see, the intensities of a homogeneous Poisson process are constant (independent of t).

The realizations of a homogeneous Poisson process are nondecreasing step functions. The occurrence of a signal at the point t corresponds to a jump of the random function; this jump equals one, and in every finite time interval $[0, t)$ the number of jumps is finite. Such a realization is represented in Fig. 8.3.1.

The properties of the homogeneous Poisson process have been investigated by Lévy [3]. The Poisson process, its various generalizations, and numerous applications have been the subject of investigation by

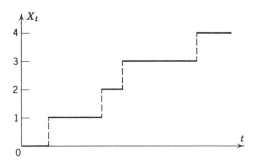

Fig. 8.3.1

many authors. We mention here the papers of Khintchin [2, 7], Florek, Marczewski, and Ryll-Nardzewski [1], Marczewski [1], Prékopa [1], Jánossy, Rényi, and Aczél [1], Rényi [1, 2, and 6], Takács [2, 3], Fortet [3], and Fisz and Urbanik [1]. One chapter of the book by Blanc-Lapierre and Fortet [1] is devoted to this subject. (See also Problems 8.1 to 8.8.)

We mention, in particular, the recent result of Pyke [1], who has given the formula for the distribution function of the lower and upper bounds of the realizations of a linear function $X_t - \alpha t$ of a Poisson process X_t in the interval $[0, t_0)$, for every positive t_0.

8.4 THE FURRY-YULE PROCESS

In spite of the many applications of the Poisson process, in a great many problems it can be treated only as a first approximation. Hence it is necessary to consider more general models. We can generalize in two directions. The first generalization consists in considering the intensity λ given by (8.3.2) as dependent on the state of the process. The second generalization consists in allowing events not only to appear but also to vanish. In this section we consider only the first generalization; the second is discussed in the following section.

A   Let X_t denote the number of events considered which occur up to and including the moment t or the number of elements of some population which exist at the moment t. Thus X_t as well as the increments of X_t in an arbitrary finite time interval can take on only the integer values $j = 0, 1, 2, \ldots$. Suppose that X_t is a homogeneous Markov process and satisfies a modified condition III of the previous section. The modification consists in considering the intensity λ given by (8.3.2) to be a function of the state j of the process at the moment t; thus instead of λ we shall write λ_j. This process is called a *birth process*.

For $j = 0, 1, 2, \ldots$ Eqs. (8.3.2) and (8.3.3) take the form

$$(8.4.1) \qquad \lim_{\Delta t \to 0} \frac{p_{j,j+1}(\Delta t)}{\Delta t} = \lambda_j,$$

$$(8.4.2) \qquad \lim_{\Delta t \to 0} \frac{1 - p_{jj}(\Delta t) - p_{j,j+1}(\Delta t)}{\Delta t} = 0.$$

In other words, for sufficiently small Δt the probability of appearance of a new element during the time interval $[t, t + \Delta t)$ equals $\lambda_j \Delta t + o(\Delta t)$, provided at the moment t the population is in the state j; the probability of two or more births during this time is $o(\Delta t)$.

Let i be a fixed integer and let $P(X_0 = i) = 1$. Let

$$(8.4.3) \qquad V_m(t) = P(X_t = i + m) \qquad (m = 0, 1, 2, \ldots).$$

By an argument similar to that in the proof of theorem 8.3.1 we obtain

$$V_0(t + \Delta t) = V_0(t)V_0(\Delta t),$$

$$V_m(t + \Delta t) = \sum_{k=0}^{m} V_{m-k}(t)V_k(\Delta t) \qquad (m = 1, 2, \ldots).$$

From (8.4.1) and (8.4.2) we obtain

$$V_1(\Delta t) = \lambda_i \Delta t + o(\Delta t),$$

$$V_0(\Delta t) = 1 - \lambda_i \Delta t + o(\Delta t).$$

The last four formulas give the following system of differential equations:

$$(8.4.4) \qquad \begin{aligned} V_0'(t) &= -\lambda_i V_0(t), \\ V_m'(t) &= -\lambda_{i+m} V_m(t) + \lambda_{i+m-1} V_{m-1}(t) \qquad (m = 1, 2, \ldots). \end{aligned}$$

The assumption $P(X_0 = i) = 1$ implies that the functions $V_m(t)$ satisfy the initial conditions

$$(8.4.5) \qquad V_m(0) = \begin{cases} 1 & \text{for} \quad m = 0, \\ 0 & \text{for} \quad m \neq 0. \end{cases}$$

Thus the first equation (8.4.4) gives

$$V_0(t) = \exp(-\lambda_i t).$$

The remaining functions $V_m(t)$ can be obtained successively from the system (8.4.4).

B   The birth process with λ_j satisfying the equalities

$$(8.4.6) \qquad \lambda_j = \lambda j \qquad (j = 1, 2, \ldots)$$

is of great importance. Such a process is called the *Furry-Yule process*. Furry [1] applied this process to problems of cosmic radiation (see example 8.4.1), and Yule [3] to biology (see example 8.4.2).

The meaning of equalities (8.4.6) is as follows. The probability that a given element will give birth to a new element during the time Δt is

$\lambda \Delta t + o(\Delta t)$. Suppose that at some moment t we have j elements, and let the random events that the existing elements give birth to new elements be independent. Then the probability of the birth of a new element during the time Δt is approximately equal to $\lambda j \Delta t + o(\Delta t)$, and the probability of two or more births is $o(\Delta t)$.

System (8.4.4) for the Furry-Yule process takes the form

(8.4.7)

$$V_0'(t) = -\lambda i V_0(t),$$

$$V_m'(t) = -\lambda(i + m)V_m(t) + \lambda(i + m - 1)V_{m-1}(t) \quad (m = 1, 2, \ldots).$$

This is a recursive system of linear differential equations. Solving these equations successively and considering the initial conditions (8.4.5), we obtain

$$(8.4.8) \quad V_m(t) = \binom{m + i - 1}{m} e^{-i\lambda t}(1 - e^{-\lambda t})^m \quad (m = 0, 1, 2, \ldots).$$

We observe that for every t

$$(8.4.9) \qquad \sum_{m=0}^{\infty} V_m(t) = 1.$$

Thus if the λ_j satisfy equalities (8.4.6), then the number of elements in an arbitrary finite time interval is finite, with probability one.

It is easy to show, as has been done in Section 8.3, that the process being considered is a homogeneous Markov process, and for arbitrary t_1 and t_2 $(0 \leqslant t_1 < t_2 < \infty)$ and arbitrary m, j, where $m \leqslant j$, we have

$$(8.4.10) \quad p_{mj}(t_1, t_2) = \binom{j - 1}{j - m} e^{-m\lambda t}(1 - e^{-\lambda t})^{j - m} = V_{j-m}(t) = p_{mj}(t),$$

where $t = t_2 - t_1$. The ergodic theorem also holds here, namely,

$$(8.4.11) \qquad \lim_{t \to \infty} p_{mj}(t) = 0 \quad (m = 1, 2, \ldots; j \geqslant m).$$

For $m = 1, 2, \ldots$ we have the formulas

$$(8.4.12) \qquad q_m = \lim_{t \to 0} \frac{1 - p_{mm}(t)}{t} = m\lambda,$$

$$(8.4.13) \quad q_{mj} = \lim_{t \to 0} \frac{p_{mj}(t)}{t} = \begin{cases} m\lambda & \text{for } j = m + 1, \\ 0 & \text{for } j \neq m, m + 1, \end{cases}$$

$$(8.4.14) \qquad \sum_{\substack{j \\ j \neq m}} q_{mj} = q_m.$$

Example 8.4.1. Cosmic radiation consists of two components, soft radiation and hard radiation. Hard radiation consists mainly of mesons and its name comes from the property that it can penetrate a sheet of lead several meters thick. Soft radiation consists mainly of negatrons and positrons (both are called electrons), and its name comes from the property that a sheet of lead 10 cm thick absorbs it completely. The theory of hard radiation and the interpretation of experimental results are not very far advanced as yet, but it is more so for soft radiation. Arley [1] has devoted a large monograph to various models of stochastic processes which could be applied to the theory of soft radiation. This example comes from this monograph. We have here two kinds of radioactive changes. The first is the following: a photon, after passing along a path of length Δt, may, with a certain probability, destroy itself by splitting into two electrons. The second radioactive change is as follows: an electron may emit a photon, losing part of its energy. Thus if at the beginning there is one photon (electron), it can initiate a chain of successive changes. If there were one electron in the first "generation," then the second "generation" would consist of a photon which in turn could create two electrons in the third "generation," and so on. Thus we have a stream of electrons or photons. In this case we speak of a *cascade* or a *cascade process*. The observations concern only electrons; thus we can observe only every second "generation."

Let X_t be the number of electrons at the depth t, thus the "time" parameter t should here be interpreted as depth. Furry [1] neglected the intermediate "generations" of photons and stated the hypothesis that an electron passing along a path of length Δt has the probability $\lambda \Delta t + o(\Delta t)$ of splitting into two electrons; since the same is true for every electron independently of any other, equality (8.4.6) holds. Thus Furry assumed that the process considered satisfies Eqs. (8.4.7). Furry's model is a step forward in comparison with that of Bhabha and Heitler [1], who assumed that this cascade process is a Poisson process. However Furry's model is not free of disadvantages, either. In particular, it does not consider the possibility of disappearance of electrons because of the loss of energy from emissions of photons. Arley [1] considered such a model and found that the observations fit the predicted results quite well. The reader will find the details in the book by Arley quoted here.

Example 8.4.2. The considered process was investigated by Yule [3] in an application to a biological population in which new elements arise by the splitting of an element into two new elements identical with the original. If we assume that the elements of the population have no influence on one another, then the splittings are independent and we may assume that the assumptions for the model are satisfied. However, a more proper model would be one where we could consider the possibility of the death of an element. Such a model was created by Feller [3].

C· We shall apply the method proposed by Arley [1] for computing the expected value $E(X_t)$ and the variance $D^2(X_t)$. Denote

$$E(X_t) = E(t), \qquad E(X_t^2) = E_2(t), \qquad D^2(X_t) = D^2(t).$$

We have

$$E(t) = \sum_{m=0}^{\infty} (i + m)V_m(t).$$

By formulas (8.4.7) we obtain[1]

$$E'(t) = \sum_{m=0}^{\infty} (i + m)V_m'(t) = \lambda \sum_{m=0}^{\infty} (i + m)V_m(t) = \lambda E(t);$$

hence considering the initial condition $E(0) = i$, we obtain

(8.4.15) $$E(t) = ie^{\lambda t}.$$

Furthermore, we have

$$E_2(t) = \sum_{m=0}^{\infty} (i + m)^2 V_m(t).$$

Hence[1]

$$[E_2(t)]' = \sum_{m=0}^{\infty} (i + m)^2 V_m'(t) = \lambda \sum_{m=0}^{\infty} (i + m)(2i + 2m + 1)V_m(t)$$
$$= 2\lambda E_2(t) + \lambda E(t).$$

Considering the initial condition $E_2(0) = i^2$, we obtain

(8.4.15') $$E_2(t) = e^{2\lambda t}[i^2 + i(1 - e^{-\lambda t})].$$

From (8.4.15) and (8.4.15') we obtain

(8.4.16) $$D^2(t) = E_2(t) - [E(t)]^2 = i(e^{2\lambda t} - e^{\lambda t}).$$

D   We have mentioned that, for an arbitrary t, equality (8.4.9) is satisfied in the Furry-Yule process. However, if equalities (8.4.6) do not hold, the functions $V_m(t)$ which are solutions of the system (8.4.4) of differential equations may not satisfy formula (8.4.9). That is, it may happen that

(8.4.17) $$\sum_{m=0}^{\infty} V_m(t) < 1.$$

This means that there is a positive probability that the number of elements of the population will increase to infinity in a finite interval of time. This can happen if the λ_j increase to infinity sufficiently fast as $j \to \infty$. This follows from the theorem which we now prove.

Theorem 8.4.1.   *The solutions $V_m(t)$ of the system (8.4.4) with the initial conditions (8.4.5) satisfy the relation (8.4.9) if and only if*

(8.4.18) $$\sum_{m=0}^{\infty} \frac{1}{\lambda_{i+m}} = \infty.$$

Proof.   Let

$$W_m(t) = \sum_{r=0}^{m} V_r(t) \qquad (m = 0, 1, 2, \ldots).$$

---

[1] It is easy to show that we can differentiate under the summation sign.

By formulas (8.4.4)

$$W_m'(t) = \sum_{r=0}^{m} V_r'(t) = -\lambda_{i+m} V_m(t).$$

Integrating both sides of the last equality and using the initial conditions, we obtain

(8.4.19) $$1 - W_m(t) = \lambda_{i+m} \int_0^t V_m(t)\, dt.$$

Since the functions $V_m(t)$ are non-negative, $1 - W_m(t)$ is, for fixed t, a nonincreasing function of the argument m; hence the limit

$$U(t) = \lim_{m \to \infty} [1 - W_m(t)]$$

exists. It follows that

$$\lambda_{i+m} \int_0^t V_m(t)\, dt \geqslant U(t).$$

Next we have

(8.4.20) $$\int_0^t W_m(t)\, dt = \sum_{r=0}^{m} \int_0^t V_r(t)\, dt \geqslant U(t) \sum_{r=0}^{m} \frac{1}{\lambda_{i+r}}.$$

Suppose that condition (8.4.18) is satisfied. We observe that the left-hand side of (8.4.20) does not exceed t, since $W_m(t) \leqslant 1$. The right-hand side of (8.4.20) goes to infinity as $m \to \infty$ if $U(t) > 0$. Thus in order that inequality (8.4.20) is satisfied, we must have $U(t) \equiv 0$ or $\lim_{m \to \infty} W_m(t) = 1$; hence (8.4.9) is satisfied.

Now, suppose that (8.4.9) is satisfied for every t. By (8.4.19), for every r

$$\int_0^t V_r(t)\, dt \leqslant \frac{1}{\lambda_{i+r}}.$$

Hence for every m and t

$$\int_0^t \sum_{r=0}^{m} V_r(t)\, dt \leqslant \sum_{r=0}^{m} \frac{1}{\lambda_{i+r}}.$$

Let us pass to the limit as $m \to \infty$ on both sides of the last inequality. By formula (8.4.9) we get the value t on the left-hand side. Since this is true for every t, the series obtained on the right-hand side must diverge. Thus relation (8.4.18) is satisfied and this completes the proof of theorem 8.4.1.

Theorem 8.4.1 is a special case of a general theorem of Feller [4] which gives conditions for equality (8.4.9) to hold for homogeneous Markov processes with a countable number of states. Lundberg [1] generalized theorem 8.4.1 to a nonhomogeneous birth process, that is, to the case when the coefficients λ_j are functions of t. We discuss another generalization of theorem 8.4.1 at the end of Section 8.5, A.

We observe that the intensities of the Furry-Yule process satisfy relation (8.4.18).

If inequality (8.4.17) is satisfied, then, as we have already said, the number of elements of the population in question can increase to infinity in a finite time interval. We then say that the process is in the state "infinity", with positive probability.

Doob [3] showed that even this exceptional case may have a probabilistic interpretation, that is, we can interpret it as a passage from one state to another in an infinite number of steps. However, it is a rather delicate question, and we refer those interested in it to Doob's paper.

8.5 BIRTH AND DEATH PROCESSES

Let us consider a further generalization of the processes considered in the previous sections, namely, processes in which the events considered may not only appear but also disappear. Thus, if in the examples given on radioactivity we consider not only the creation of some elementary particles but also their annihilation, the model obtained will be closer to reality than the model which assumes only their creation. Similarly, in the problems about the telephone exchange a model which takes into account not only the appearance of incoming calls, as in the signal process, but also the time of their duration and their termination, is a better mathematical idealization of the workings of a telephone exchange. These more general processes are called *birth and death processes*. However, we shall not consider advanced generalizations; we restrict our considerations to the simplest generalization in this direction, which allows us to explain the essential features of the problem.

A   Let us again consider a signal process $\{X_t, 0 \leqslant t < \infty\}$ satisfying conditions I to III of Section 8.3, with intensity equal to λ. Let us consider the length of "life" of a signal that is a random variable and suppose that it has the exponential distribution. If we denote this random variable by T, we have for its density $g(t)$ the formula

$$g(t) = \begin{cases} 0 & \text{for} \quad t \leqslant 0, \\ \mu e^{-\mu t} & \text{for} \quad t > 0, \end{cases}$$

where $\mu > 0$. The assumption that the "life" of a signal is exponentially distributed is a good approximation in many applications and has important theoretical meaning. In fact, under this assumption we have the equalities

$$\lim_{\Delta t \to 0} \frac{P(T < t + \Delta t \mid T \geqslant t)}{\Delta t} = \lim_{\Delta t \to 0} \frac{P(t \leqslant T < t + \Delta t)}{\Delta t P(T \geqslant t)} = \frac{g(t)}{P(T \geqslant t)} = \mu.$$

Hence, for small Δt, the probability that the signal will terminate during the time interval $[t, t + \Delta t)$, if it has lasted t units of time, is $\mu \, \Delta t + o(\Delta t)$ and is independent of t, that is, it does not depend on the length of the signal in the past. Under these conditions, the process of appearance and disappearance of signals is a Markov process. We also assume that the lengths of life of signals are independent of each other and are independent of the number of incoming signals, and we assume that the number of signals which appear and disappear during a time interval depends only on the length of this interval and not on its end points. Thus if at some moment t we have i operating signals, the probability of disappearance of one of them during the time interval $[t, t + \Delta t)$ is $\mu i \, \Delta t + o(\Delta t)$ and the probability of disappearance of two or more of them equals $o(\Delta t)$. The probability of simultaneous appearance of a new signal and the disappearance of one of the operating signals is also assumed to equal $o(\Delta t)$.

Denote by Y_t the number of signals operating at the moment t and consider the process $\{Y_t, 0 \leqslant t < \infty\}$. We say that the process is in the state j if at the moment t we have j operating signals. We find the absolute probability $c_j(t)$ that the process will be in the state j at the moment t,

$$c_j(t) = P(Y_t = j) \qquad (j = 0, 1, 2, \ldots).$$

We have

(8.5.1)
$$\begin{aligned}
c_0(t + \Delta t) &= c_0(t)[1 - \lambda \, \Delta t + o(\Delta t)] \\
&\quad + c_1(t)[\mu \, \Delta t + o(\Delta t)] + o(\Delta t),
\end{aligned}$$

(8.5.2)
$$\begin{aligned}
c_j(t + \Delta t) &= c_j(t)[1 - \lambda \, \Delta t + o(\Delta t)][1 - \mu j \, \Delta t + o(\Delta t)] \\
&\quad + c_{j-1}(t)[\lambda \, \Delta t + o(\Delta t)] \\
&\quad + c_{j+1}(t)[\mu(j + 1) \, \Delta t + o(\Delta t)] + o(\Delta t) \qquad (j = 1, 2, \ldots).
\end{aligned}$$

In fact, at the moment $t + \Delta t$ the process may be in the state zero either if it were in this state at the moment t and no signal appeared during the time $[t, t + \Delta t)$ or if it were in the state one at the moment t and the existing signal disappeared during the interval $[t, t + \Delta t)$. This proves equality (8.5.1). Next, the process may be in the state j ($j \geqslant 1$) at the moment $t + \Delta t$ either if it were in the state j at the moment t and during the time interval $[t, t + \Delta t)$ neither a new signal appeared nor any of the existing signals disappeared, or if it were in the state $j - 1$ at the moment t and during the time $[t, t + \Delta t)$ a new signal appeared, or if it were in the state $j + 1$ at the moment t and during the time $[t, t + \Delta t)$ one of the existing $j + 1$ signals disappeared. The probability of all other possibilities is $o(\Delta t)$, which proves equality (8.5.2). From these equalities we obtain the differential equations

(8.5.3)
$$\begin{aligned}
c_0'(t) &= -\lambda c_0(t) + \mu c_1(t), \\
c_j'(t) &= \lambda c_{j-1}(t) - (\lambda + j\mu)c_j(t) + \mu(j + 1)c_{j+1}(t) \qquad (j = 1, 2, \ldots).
\end{aligned}$$

Again we have obtained an infinite system of linear differential equations; it is not, however, recursive and to solve it we use the method of probability generating functions. Let

$$(8.5.4) \qquad f(t, s) = \sum_{j=0}^{\infty} c_j(t)s^j.$$

We have

$$\frac{\partial f(t, s)}{\partial t} = \sum_{j=0}^{\infty} \frac{dc_j(t)}{dt} s^j = -\lambda c_0(t)(1 - s) + \mu c_1(t)(1 - s)$$

$$- \lambda c_1(t)(1 - s)s + 2\mu c_2(t)s(1 - s) + \ldots$$

$$= -\lambda(1 - s)\sum_{j=0}^{\infty} c_j(t)s^j + \mu(1 - s)\sum_{j=1}^{\infty} jc_j(t)s^{j-1}$$

$$= -\lambda(1 - s)f(t, s) + \mu(1 - s)\frac{\partial f(t, s)}{\partial s}.$$

Thus the generating function $f(t, s)$ satisfies the linear partial differential equation of the first order

$$(8.5.5) \qquad \frac{\partial f}{\partial t} - \mu(1 - s)\frac{\partial f}{\partial s} = -\lambda(1 - s)f.$$

Suppose that at the moment $t = 0$ we have $c_i(0) = 1$. Then the function $f(t, s)$ satisfies the initial condition

$$(8.5.6) \qquad f(0, s) = s^i.$$

Let us introduce the function V by the relation

$$V(s, t, f) = 0.$$

By formula (8.5.5) and by the equations

$$\frac{\partial f}{\partial s} = -\frac{\partial V}{\partial s} : \frac{\partial V}{\partial f} ; \qquad \frac{\partial f}{\partial t} = -\frac{\partial V}{\partial t} : \frac{\partial V}{\partial f},$$

we obtain the linear homogeneous partial differential equation of the first order

$$(8.5.7) \qquad \frac{\partial V}{\partial t} - \mu(1 - s)\frac{\partial V}{\partial s} - \lambda(1 - s)f\frac{\partial V}{\partial f} = 0.$$

The following system of ordinary linear differential equations is equivalent to equation (8.5.7):

$$dt = -\frac{ds}{\mu(1 - s)},$$

$$ds = \frac{\mu}{\lambda} \cdot \frac{df}{f}.$$

The solution of this system is of the form

$$\mu t - \log (1 - s) = a_1,$$
$$\lambda s - \mu \log f = a_2.$$

Thus the solution of equation (8.5.7) is

$$\lambda s - \mu \log f = \Psi[\mu t - \log (1 - s)],$$

where Ψ is an arbitrary function. Hence

(8.5.8) $$f = \exp \left\{ \frac{1}{\mu} [\lambda s - \Psi(\mu t - \log (1 - s))] \right\}.$$

The initial condition allows us to determine the function Ψ. We have from (8.5.6) and (8.5.8)

$$\mu i \log s = \lambda s - \Psi \left(\log \frac{1}{1 - s} \right).$$

Letting $y = \log [1/(1 - s)]$, or $s = 1 - e^{-y}$, we obtain

$$\Psi(y) = \lambda(1 - e^{-y}) - i\mu \log (1 - e^{-y}).$$

Hence

$$\Psi[\mu t - \log (1 - s)] = \lambda[1 - (1 - s)e^{-\mu t}] - \mu i \log [1 - (1 - s)e^{-\mu t}].$$

Finally,

(8.5.9) $$f(t, s) = [1 - (1 - s)e^{-\mu t}]^i \exp \left[-\frac{\lambda}{\mu} (1 - s)(1 - e^{-\mu t}) \right].$$

The first term on the right-hand side of (8.5.9) is (see Section 4.7) the probability generating function of a random variable with the Poisson distribution having the expected value (hence variance) equal to $(\lambda/\mu)(1 - e^{-\mu t})$, and the second term is the probability generating function of a random variable with the binomial distribution, with the probability of a success equal to $e^{-\mu t}$, where we have a success if the signal lasts for at least t. Therefore, for every t, Y_t is the sum of two independent random variables, one of which has the Poisson distribution, and the other the binomial distribution. Hence

(8.5.10) $$c_j(t) = \exp \left[-\frac{\lambda}{\mu} (1 - e^{-\mu t}) \right]$$

$$\times \sum_{k=0}^{\min(i,j)} \binom{i}{k} \left(\frac{\lambda}{\mu} \right)^{j-k} \frac{e^{-\mu t k}(1 - e^{-\mu t})^{i+j-2k}}{(j - k)!} \quad (j = 0, 1, 2, \ldots),$$

$$E(Y_t) = \frac{\lambda}{\mu} (1 - e^{-\mu t}) + ie^{-\mu t},$$

$$D^2(Y_t) = (1 - e^{-\mu t}) \left(\frac{\lambda}{\mu} + ie^{-\mu t} \right).$$

These formulas were obtained by Palm [1]. In the particular case when $i = 0$, Y_t has the Poisson distribution with

(8.5.11)
$$c_j(t) = \exp\left[-\frac{\lambda}{\mu}(1 - e^{-\mu t})\right] \frac{\left[\frac{\lambda}{\mu} \left(1 - e^{-\mu t}\right)\right]^j}{j!} \quad (j = 0, 1, 2, \ldots).$$

We observe that $c_j(t) = \sum_l c_l(0)p_{lj}(t)$. It follows from the initial condition $c_i(0) = 1$ that

$$c_j(t) = p_{ij}(t) \quad (j = 0, 1, 2, \ldots).$$

Since formula (8.5.10) implies that

$$\lim_{t \to 0} c_j(t) = \begin{cases} 1 & \text{for } j = i, \\ 0 & \text{for } j \neq i, \end{cases}$$

using (8.5.3), we obtain the intensities

(8.5.12) $\quad q_i = \lim_{t \to 0} \dfrac{1 - p_{ii}(t)}{t} = \lim_{t \to 0} \dfrac{1 - c_i(t)}{t} = -c_i'(0) = \lambda + i\mu,$

(8.5.13) $\quad q_{ij} = \lim_{t \to 0} \dfrac{p_{ij}(t)}{t} = c_j'(0) = \begin{cases} i\mu & \text{for } j = i - 1, \\ \lambda & \text{for } j = i + 1, \\ 0 & \text{for } j \neq i - 1, i, i + 1. \end{cases}$

It is easy to verify that for every t

$$\sum_{j=0}^{\infty} c_j(t) = 1.$$

However, if the intensity λ of appearance of a signal is a function of the number of actually operating signals and is thus of the form λ_j, and the intensity of disappearance of one of the i operating signals has the form $\mu_i \neq \mu i$, then we may have

$$\sum_{j=0}^{\infty} c_j(t) < 1.$$

Necessary and sufficient conditions for the validity of the equality sign in the last series for every t have been given by Dobrushin [1] and Reuter and Lederman [1], thus generalizing theorem 8.4.1.

It can be seen from formula (8.5.10) that as $t \to \infty$ the distribution of Y_t converges independently of the initial state to the Poisson distribution with the parameter λ/μ. Thus

(8.5.14) $\quad c_j = \lim_{t \to \infty} c_j(t) = \dfrac{(\lambda/\mu)^j}{j!} \exp\left(-\dfrac{\lambda}{\mu}\right) \quad (j = 0, 1, 2, \ldots).$

Example 8.5.1. Consider a telephone exchange where the number of lines is so large that every incoming call always finds a free line. Then we can assume that the telephone exchange has an infinite number of lines. If we denote by Y_t the number of busy lines at the moment t, we see that Y_t is a process with a countable number of states. Formula (8.5.10) makes it possible to determine the absolute probability $c_j(t) = P(Y_t = j)$ for every t, and formula (8.5.14) gives the limit of this probability as $t \to \infty$.

Example 8.5.2. Let us now consider a telephone exchange with a finite number of lines. Suppose that the exchange is constructed in such a way that an incoming call that finds all the lines busy does not wait but is lost. This is called an exchange without waiting lines. The most important problem to be solved before the exchange is constructed is to determine, for any t, the probability of finding all the lines busy at t. We now investigate this question, but we shall only find the limit of this probability as $t \to \infty$.

B   Let Y_t denote, as before, the number of signals operating at the moment t. We retain all assumptions about this process with the single exception that here the number of states is finite, namely, $j = 0, 1, 2, \ldots, m$, hence $c_j(t) = 0$ for $j > m$. Equalities (8.5.1) and (8.5.2) remain unchanged for $j = 1, 2, \ldots, m - 1$, whereas for $j = m$ we have

$$c_m(t + \Delta t) = c_m(t)[1 - \mu m \, \Delta t + o(\Delta t)] + c_{m-1}(t)[\lambda \, \Delta t + o(\Delta t)] + o(\Delta t).$$

Hence we have a system of differential equations

$$c_0'(t) = -\lambda c_0(t) + \mu c_1(t),$$

$$(8.5.15) \quad c_j'(t) = \lambda c_{j-1}(t) - (\lambda + j\mu)c_j(t) + \mu(j + 1)c_{j+1}(t)$$

$$(j = 1, 2, \ldots, m - 1),$$

$$c_m'(t) = \lambda c_{m-1}(t) - \mu m c_m(t).$$

Our aim is to find the limits c_j of the functions $c_j(t)$ as $t \to \infty$; they will be independent of the initial distribution $c_j(0)$ $(j = 1, \ldots, m)$. We observe that the assumptions of theorem 8.2.1 are satisfied, that is, there exists a t^* such that the transition probabilities $p_{ij}(t^*) > 0$ $(i, j = 0, 1, \ldots, m)$; hence the ergodic theorem holds. Thus the limits $p_j = \lim_{t \to \infty} p_{ij}(t)$ exist for $j = 0, 1, 2, \ldots, m$. Then from theorem 8.2.2 it follows that the limits $c_j = \lim_{t \to \infty} c_j(t) = p_j$ exist independently of the initial distribution. Thus we have established that the limits of the right-hand sides of (8.5.15) exist as $t \to \infty$; hence the limits of the functions $c_j'(t)$ also exist. However, these limits can only equal zero since otherwise we would have $c_j(t) \to \infty$. Thus we obtain the following system of linear equations for the constants c_j:

$$-\lambda c_0 + \mu c_1 = 0,$$

$$(8.5.16) \quad \lambda c_{j-1} - (\lambda + j\mu)c_j + \mu(j + 1)c_{j+1} = 0 \quad (j = 1, 2, \ldots, m - 1),$$

$$\lambda c_{m-1} - \mu m c_m = 0.$$

These equalities imply that

$$c_j = \frac{1}{j!} \left(\frac{\lambda}{\mu}\right)^j c_0 \qquad (j = 1, 2, \ldots, m).$$

Since $c_0 + \ldots + c_m = 1$, we obtain

$$c_0 = \frac{1}{\displaystyle\sum_{k=0}^{m} \frac{1}{k!} \left(\frac{\lambda}{\mu}\right)^k}.$$

Finally,

(8.5.17) $$c_j = \frac{\dfrac{1}{j!} \left(\dfrac{\lambda}{\mu}\right)^j}{\displaystyle\sum_{k=0}^{m} \frac{1}{k!} \left(\frac{\lambda}{\mu}\right)^k} \qquad (j = 0, 1, \ldots, m).$$

These formulas are called the *Erlang formulas* [1].

Example 8.5.3. Table 8.5.1 contains the probabilities c_j computed from (8.5.17) for $m = 5$ and $m = 10$ with $\lambda = \mu$ and $\lambda = 2\mu$. It can be seen from this table that for $\lambda = 2\mu$ and for five lines the probability that all lines will be busy is relatively big, about 0.04; hence approximately four out of every hundred calls will be lost. With the same ratio $\lambda = 2\mu$, ten lines will be quite sufficient. For $\lambda = \mu$ the probability that all lines will be busy is small even for $m = 5$.

TABLE 8.5.1

VALUES OF c_j COMPUTED FROM (8.5.17)

j	$m = 5$		$m = 10$	
	$\lambda = \mu$	$\lambda = 2\mu$	$\lambda = \mu$	$\lambda = 2\mu$
0	0.36810	0.13761	0.36788	0.13534
1	0.36810	0.27523	0.36788	0.27067
2	0.18405	0.27523	0.18394	0.27067
3	0.06135	0.18349	0.06131	0.18045
4	0.01534	0.09174	0.01533	0.09022
5	0.00307	0.03670	0.00307	0.03609
6			0.00051	0.01203
7			0.00007	0.00344
8			0.00001	0.00086
9			0.00000	0.00019
10			0.00000	0.00004

C   We now consider another type of telephone exchange, one with waiting lines. This exchange is constructed so that the calls coming in when all lines are busy form a waiting line and wait until one of the lines becomes free. In such exchanges the following problems are of particular interest: (1) to find the probability function of the number of operating and waiting calls; (2) to find the distribution function of the waiting time. The first problem was treated by Kolmogorov [4]. We outline briefly the solution of this problem, where we only find the limits of the required probabilities as $t \to \infty$. The second problem was considered by Erlang [1]. The solution under some simple assumptions will be given.

We observe that problems of this type appear not only in connection with telephones but also in many other questions of technology and every-day life. Thus, for instance, the "exchange" may be a collection of ticket cashiers at a railway station and a "signal" the appearance of a passenger at the cashier's window. The process is in the state j if j passengers are waiting in a queue or are buying tickets. It is obvious that the railway company should have information about the rate of arrival of passengers as well as about the intensity of ticket selling by the cashiers at the windows in order to determine a reasonable number of windows. Processes of this kind are called *queueing processes*.

Let Y_t denote the number of signals which either operate or wait at the moment t. Suppose again that the stream of arriving signals satisfies conditions I to III of Section 8.3, with intensity λ, and the distribution of the length of life of the signals is exponential with parameter μ. We also assume that the conditions that led us to system (8.5.3) are satisfied. Again let $c_j(t) = P(Y_t = j)$, and let the number of lines be m. Then equalities (8.5.1) and (8.5.2) remain unchanged for $0 < j < m$ since for such j's at least one line remains free so that the arriving signal does not wait. For $j \geqslant m$ we have

$$c_j(t + \Delta t) = c_j(t)[1 - \mu m \, \Delta t + o(\Delta t)][1 - \lambda \, \Delta t + o(\Delta t)]$$
$$+ c_{j-1}(t)[\lambda \, \Delta t + o(\Delta t)] + c_{j+1}(t)[\mu m \, \Delta t + o(\Delta t)] + o(\Delta t).$$

The difference between the last equality and equality (8.5.2) is that on the right-hand side we have μm instead of μj or $\mu(j + 1)$. This is because the number of operating signals cannot exceed m. Finally, we obtain the following system of differential equations:

$$c_0'(t) = -\lambda c_0(t) + \mu c_1(t),$$
$$(8.5.18) \quad c_j'(t) = \lambda c_{j-1}(t) - (\lambda + j\mu)c_j(t) + \mu(j + 1)c_{j+1}(t)$$
$$(j = 1, 2, \ldots, m - 1),$$
$$c_j'(t) = \lambda c_{j-1}(t) - (\lambda + m\mu)c_j(t) + m\mu c_{j+1}(t) \quad (j \geqslant m).$$

From (8.5.18) it can be shown that

$$-\lambda c_0 + \mu c_1 = 0,$$

$$(8.5.19) \quad \lambda c_{j-1} - (\lambda + j\mu)c_j + \mu(j+1)c_{j+1} = 0 \quad (j = 1, 2, \ldots, m-1),$$

$$\lambda c_{j-1} - (\lambda + m\mu)c_j + \mu m c_{j+1} = 0 \quad (j = m, m+1, \ldots),$$

where $c_j = \lim\limits_{t \to \infty} c_j(t)$. The solution of (8.5.19) is

$$(8.5.20) \qquad c_j = \begin{cases} \dfrac{1}{j!}\left(\dfrac{\lambda}{\mu}\right)^j c_0 & (j = 1, 2, \ldots, m), \\[2ex] \dfrac{1}{m!\, m^{j-m}}\left(\dfrac{\lambda}{\mu}\right)^j c_0 & (j \geqslant m+1). \end{cases}$$

We have

$$(8.5.21) \qquad \frac{1}{c_0}\sum_{j=0}^{\infty} c_j = \sum_{j=0}^{m}\frac{1}{j!}\left(\frac{\lambda}{\mu}\right)^j + \frac{m^m}{m!}\sum_{j=m+1}^{\infty}\left(\frac{\lambda}{m\mu}\right)^j.$$

Thus if $m \leqslant \lambda/\mu$, the right-hand side of the last inequality is divergent to ∞; hence the limit distribution does not exist. In fact, the relation

$$\frac{1}{c_0}\sum_{j=0}^{\infty} c_j = \infty$$

may be satisfied either if $c_0 \neq 0$ and $\sum\limits_{j=0}^{\infty} c_j = \infty$, (however the c_j are then not probabilities), or if $c_0 = 0$. But in the latter case[1] we have, by (8.5.20), $c_j = 0$ $(j = 1, 2, \ldots)$. Then the probability that the number of waiting signals is infinite equals one.

If $\lambda/\mu < m$, then by formula (8.5.21), setting $\sum\limits_{j=0}^{\infty} c_j = 1$, we obtain

$$c_0 = \frac{1}{\displaystyle\sum_{k=0}^{m}\frac{1}{k!}\left(\frac{\lambda}{\mu}\right)^k + \frac{1}{m!}\cdot\frac{1}{m - \lambda/\mu}\left(\frac{\lambda}{\mu}\right)^{m+1}}.$$

Finally, we have

$$(8.5.22) \quad c_j = \begin{cases} \dfrac{\dfrac{1}{j!}\left(\dfrac{\lambda}{\mu}\right)^j}{\displaystyle\sum_{k=0}^{m}\dfrac{1}{k!}\left(\dfrac{\lambda}{\mu}\right)^k + \dfrac{1}{m!}\cdot\dfrac{1}{m - \lambda/\mu}\left(\dfrac{\lambda}{\mu}\right)^{m+1}} & (j = 0, 1, \ldots, m), \\[5ex] \dfrac{\dfrac{1}{m!}\cdot\dfrac{1}{m^{j-m}}\left(\dfrac{\lambda}{\mu}\right)^j}{\displaystyle\sum_{k=0}^{m}\dfrac{1}{k!}\left(\dfrac{\lambda}{\mu}\right)^k + \dfrac{1}{m!}\cdot\dfrac{1}{m - \lambda/\mu}\left(\dfrac{\lambda}{\mu}\right)^{m+1}} & (j \geqslant m+1). \end{cases}$$

[1] Here $\lim\limits_{t \to \infty} c_j(t) = 0$ $(j = 0, 1, 2, \ldots)$, but $\lim\limits_{t \to \infty}\dfrac{1}{c_0(t)}\sum\limits_{j=0}^{\infty} c_j(t) = \infty$.

Despite the fact that in the situation under consideration the number of signals is always finite, the practical consequences of $m \leqslant \lambda/\mu$ may seriously disturb the work of our "exchange" because of the many waiting signals. Thus the number of "lines" must not be too small.

Denote by Z the length of waiting time of a signal which appears at the moment t when all lines are busy, that is, when $Y_t \geqslant m$. We now find the probability $P(Z > z)$ for $z \geqslant 0$. We consider only large values of t for which we may use the limit formulas (8.5.22). Since at the moment when our signal appears Y_t may be equal to any number not smaller than m, we obtain, by the theorem of absolute probabilities,

$$(8.5.23) \qquad P(Z > z) = \sum_{j=m}^{\infty} c_j P(Z > z \mid Y_t = j).$$

If $Y_t = j$, where $j \geqslant m$, the number of waiting signals equals $j - m$. Hence if the waiting time $Z > z$, then during the time z from the moment of appearance of our signal at most $j - m$ signals terminate. Now by the assumption that the length of life of a signal has the exponential distribution with the parameter μ, and by the remaining assumptions concerning the disappearance of signals formulated in this section, and by the fact that the number of busy lines remains equal to m (since every line which becomes free is immediately taken by the next waiting signal), it follows that the stream of disappearing signals forms a Poisson process with parameter μm. Thus, according to (8.3.4), the probability that l signals terminate during the time z is

$$\frac{(\mu m z)^l}{l!} e^{-\mu m z}.$$

Hence the probability that at most $j - m$ signals terminate during the time z, or the expression $P(Z > z \mid Y_t = j)$, is given by the equality

$$(8.5.24) \qquad P(Z > z \mid Y_t = j) = e^{-\mu m z} \sum_{l=0}^{j-m} \frac{(\mu m z)^l}{l!}.$$

Let us rewrite formulas (8.5.22) for $j \geqslant m + 1$ in the form

$$(8.5.25) \qquad c_j = c_m \frac{1}{m^{j-m}} \left(\frac{\lambda}{\mu}\right)^{j-m}.$$

By (8.5.23) to (8.5.25) we obtain

$$P(Z > z) = \sum_{j=m}^{\infty} c_j \sum_{l=0}^{j-m} e^{-\mu m z} \frac{(\mu m z)^l}{l!}$$

$$= e^{-\mu m z} \sum_{j=m}^{\infty} c_m \frac{1}{m^{j-m}} \left(\frac{\lambda}{\mu}\right)^{j-m} \sum_{l=0}^{j-m} \frac{(\mu m z)^l}{l!}$$

$$= c_m e^{-\mu m z} \sum_{l=0}^{\infty} \frac{(\mu m z)^l}{l!} \sum_{j=m+l}^{\infty} \left(\frac{\lambda}{\mu m}\right)^{j-m}$$

$$= c_m e^{-\mu m z} \sum_{l=0}^{\infty} \frac{(\lambda z)^l}{l!} \sum_{j=m+l}^{\infty} \left(\frac{\lambda}{\mu m}\right)^{j-m-l}$$

$$= \frac{c_m e^{-\mu m z}}{1 - \lambda/\mu m} \sum_{l=0}^{\infty} \frac{(\lambda z)^l}{l!}$$

$$= \frac{c_m}{1 - \lambda/\mu m} e^{-(m\mu - \lambda)z}.$$

Finally,

(8.5.26) $$P(Z > z) = \frac{\dfrac{c_0}{m!} \left(\dfrac{\lambda}{\mu}\right)^m e^{-(m\mu - \lambda)z}}{1 - \lambda/\mu m}.$$

D   For further study on birth and death processes, we refer the reader to the papers of Karlin and McGregor [1] to [3], D. G. Kendall [2], and Urbanik [2]. A detailed discussion of the operation of "exchanges," in the wide sense of the term, can be found in the paper of Khintchin [7] (see also Karlin and McGregor [4]). A comprehensive review of recent mathematical developments in queueing theory is given in the books by Riordan [1], Syski [1], and Takács [6].

In the birth and death process we have considered the length of "life" of a signal. In practical applications one sometimes needs to consider some other characteristics of a signal besides the length of "life," for instance its amplitude, which can be a random variable. In other words, the original signal process originates a new stochastic process, called a *secondary process*. Under assumptions about the original signal process and the life of the signals less restrictive than ours, Takács [1] to [4] has obtained many important results.

We can also consider some more general models of birth and death processes; for instance, processes such that every existing element, let us call it a "particle," can create more than one new particle. We can also consider models with particles of different types, where the problem is to find the probability distribution of the numbers of particles of particular types at the moment t. These processes are called *branching processes*. Numerous works are devoted to the study of branching processes. We mention some of them: Kolmogorov and Dmitriev [1], Harris [1], Sevastianov [1], and Woods and Bharucha-Reid [1]. See also the book by Bharucha-Reid [2].

The concept of a branching process, with discrete or continuous time, whose random variables are of the continuous type, has been studied by Jiřina [1]. Moyal [1, 2] has studied branching processes with random variables of a quite general nature.

8.6 THE PÓLYA PROCESS

In Section 5.13 we investigated the compound Poisson distribution and we derived formula (5.13.7) for the probability function of a random variable with a negative binomial distribution. The starting point was a random variable X with a Poisson distribution given by (5.13.4), where the parameter λ had the gamma distribution given by (5.13.5). In the proof of (5.13.7) let us replace the random variable X by a Poisson process X_t, that is, let us replace (5.13.4) by

$$(8.6.1) \qquad P(X_t = k \mid \lambda) = \frac{(\lambda t)^k}{k!} e^{-\lambda t} \qquad (k = 0, 1, 2, \ldots),$$

where $\lambda > 0, t > 0$. Then instead of (5.13.7) we obtain, for $k = 1, 2, \ldots$,

$$P_k(t) = P(X_t = k) = \frac{v(v + 1) \ldots [v + (k - 1)]}{k!} \left(\frac{t}{a + t}\right)^k P_0(t),$$

where

$$P_0(t) = P(X_t = 0) = \left(\frac{a}{a + t}\right)^v.$$

Putting $a = v = 1/d$, we obtain

$(8.6.2)$

$$P_0(t) = (1 + dt)^{-1/d},$$

$$P_k(t) = \frac{1(1 + d) \ldots [1 + (k - 1) d]}{k!} \left(\frac{t}{1 + dt}\right)^k P_0(t) \quad (k = 1, 2, \ldots).$$

The functions $P_k(t)$ satisfy the initial conditions

$$P_0(0) = 1, \qquad P_k(0) = 0 \qquad (k = 1, 2, \ldots).$$

Differentiating formulas (8.6.2), we obtain the following system of differential equations for $P_k(t)$:

(8.6.3)

$$P_0'(t) = -\frac{1}{1 + td} P_0(t),$$

$$P_k'(t) = -\frac{1 + kd}{1 + td} P_k(t) + \frac{1 + (k - 1) d}{1 + td} P_{k-1}(t) \quad (k = 1, 2, \ldots).$$

This system has the same form as (8.4.4) with $i = 0$, $V_k(t) = P_k(t)$, and

(8.6.4) $$\lambda_k(t) = \frac{1 + kd}{1 + td} .$$

Thus the process is a birth process such as we considered in Section 8.4, but here the $\lambda_k(t)$ depend on t; hence if we treat t as time, the process is not homogeneous (see example 7.3.5). However, we can interpret as time the parameter τ given by the relation

$$t = \frac{e^{d\tau} - 1}{d} ,$$

and then the process being considered becomes homogeneous. In fact, after some computations we obtain from (8.6.3)

(8.6.5)

$$\frac{dP_0(\tau)}{d\tau} = -P_0(\tau),$$

$$\frac{dP_k(\tau)}{d\tau} = -(1 + kd)P_k(\tau) + [1 + (k - 1) d]P_{k-1}(\tau) \quad (k = 1, 2, \ldots).$$

Thus the coefficients $\lambda_k = 1 + kd$ are independent of τ; hence the process is homogeneous. The reasoning applied here and the change of "time" should not lead to any misunderstanding since, as we have already said, the name "time" for the parameter t is only a convention.

We can show that the solutions of (8.6.3) satisfying the initial conditions $P_0(0) = 1$, $P_k(0) = 0$ ($k \geqslant 1$) are the functions (8.6.2).

From (5.13.10) we can find the expected value $E(X_t)$ and the variance $D^2(X_t)$, namely,

(8.6.6) $$E(X_t) = \frac{1}{d} \left(\frac{dt}{1 + dt} : \frac{1}{1 + dt} \right) = t,$$

(8.6.7) $$D^2(X_t) = t(1 + dt).$$

Definition 8.6.1. A process $\{X_t, 0 \leqslant t < \infty\}$ whose random variables have the probability functions $P(X_t = k)$ given by (8.6.2) is called the *Pólya process*.

The Pólya process is a particular case of a class of stochastic processes called *compound Poisson processes*; they are obtained in the same way as the Pólya process by letting λ in formula (8.6.1) be a random variable with some distribution function $U(\lambda)$; in particular, if $U(\lambda)$ is determined by (5.13.5), we have the Pólya process. Compound Poisson processes are discussed in more detail in the monograph of Lundberg [1]. Before that, they were considered by Khintchin [2].

TABLE 8.6.1

k	$n_k/1417$	P_k
0	0.2286	0.2453
1	0.2110	0.1911
2	0.1475	0.1446
3	0.1066	0.1083
4	0.0854	0.0807
5	0.0480	0.0599
6	0.0416	0.0445
7	0.0346	0.0329
8	0.0310	0.0243
9	0.0191	0.0180
10	0.0085	0.0133
11	0.0106	0.0098
12	0.0078	0.0072
13	0.0056	0.0053
14	0.0028	0.0039
15	0.0028	0.0029
16	0.0085	0.0080
	1.0000	1.0000

The name Pólya process is connected with the Pólya distribution discussed in Section 5.4. Pólya showed that if[1] $\lim_{n \to \infty} np = t$ and $\lim_{n \to \infty} n\alpha = dt$, the right-hand side of (5.4.2) converges as $n \to \infty$ to the functions $P_k(t)$ given by (8.6.2). It is worthwhile to notice that the Pólya process can be obtained in several different ways.

Arley [1] established that the Pólya process can be applied to the theory of cosmic radiation. Lundberg [1] showed that the Pólya process has

[1] Notation of Section 5.4.

many applications in insurance statistics. The following example comes from the monograph of Lundberg just mentioned.

Example 8.6.1. The Sinkförsäkrings—A.B. Eir insurance company pays an indemnity to a client in case of complete inability to work caused by an accident or an illness. In Lundberg's data each person insured, except for a few cases, had only one policy; hence the number of insurance policies can be treated as the number of people insured. The general number of policies was 1417. Let us consider the number k of illnesses or accidents in the period of time between the third and twelfth year of insurance. These data are presented in Table 8.6.1, where n_k is the number of insured who made k reports of illness or accident during the period of time investigated and P_k is the probability computed from (8.6.2), where t and d are calculated from observations according to (8.6.6) and (8.6.7). We can see from this table that the predicted and observed frequencies match very well.

8.7 KOLMOGOROV EQUATIONS

In Sections 8.3 to 8.6 we considered homogeneous discrete processes, in which a change of state may occur only by a direct passage to a neighboring state. In this section we consider Markov discrete processes, in which the change of state may occur by a direct passage from any state to any other state.

Let $\{X_t, 0 \leqslant t < \infty\}$ be a Markov process with at most a countable number of states $i = 0, 1, 2, \ldots$. Let $p_{ij}(\tau, t)$, where $0 < \tau < t$, be the transition probability function of the process. This function satisfies relations (8.2.6) to (8.2.8). Let us write the Chapman-Kolmogorov equation in the form

$$(8.7.1) \qquad p_{ij}(\tau, t) = \sum_k p_{ik}(\tau, s)p_{kj}(s, t),$$

where $\tau < s < t$. We find a system of differential equations for $p_{ij}(\tau, t)$. Denote

$$(8.7.2) \qquad q_i(t) = \lim_{\Delta t \to 0} \frac{1 - p_{ii}(t, t + \Delta t)}{\Delta t} \qquad (i = 0, 1, 2, \ldots),$$

$$(8.7.3) \qquad q_{ij}(t) = \lim_{\Delta t \to 0} \frac{p_{ij}(t, t + \Delta t)}{\Delta t} \qquad (i, j = 0, 1, 2, \ldots; i \neq j),$$

and assume that the convergence in both relations is uniform in t. The value Δt in (8.7.2) and (8.7.3) may be negative or positive. These relations are equivalent to the following asymptotic equalities satisfied for small Δt:

$$(8.7.2') \qquad p_{ii}(t, t + \Delta t) \cong 1 - q_i(t)\,\Delta t + o(\Delta t),$$

$$(8.7.3') \qquad p_{ij}(t, t + \Delta t) \cong q_{ij}(t)\,\Delta t + o(\Delta t).$$

The functions $q_i(t)$ and $q_{ij}(t)$ are called *intensity functions*.

In Kolmogorov's paper [3], which played a basic role in the development of the theory of Markov processes, he proved theorem 8.7.1. In this theorem two systems of differential equations, called the *Kolmogorov equations*, are derived for the functions $p_{ij}(\tau, t)$. One of these systems, namely system (8.7.4), is obtained by treating j and t as constants and i and τ as variables. The second system, namely the system (8.7.6), is obtained by treating i and τ as constants and j and t as variables. Since $\tau < t$, eqs. (8.7.4) are called the *backward* (*retrospective*) *equations*, and eqs. (8.7.6) the *forward* (*prospective*) *equations*.

Theorem 8.7.1. *Let $p_{ij}(\tau, t)$ be the probability transition function of a Markov process with at most a countable number of states $i = 0, 1, 2, \ldots$. Suppose that the intensity functions $q_i(t)$ and $q_{ij}(t)$ exist and are continuous. Then:*

I. *The functions $p_{ij}(\tau, t)$ satisfy the system of differential equations*

$$(8.7.4) \qquad \frac{\partial p_{ij}(\tau, t)}{\partial \tau} = q_i(\tau) p_{ij}(\tau, t) - \sum_{k \neq i} q_{ik}(\tau) p_{kj}(\tau, t)$$

with the initial conditions

$$(8.7.5) \qquad p_{ij}(t, t) = \begin{cases} 1 & \text{for } j = i, \\ 0 & \text{for } j \neq i. \end{cases}$$

II. *If, moreover, the convergence in (8.7.3) is uniform in i for fixed j, then the functions $p_{ij}(\tau, t)$ satisfy the system of differential equations*

$$(8.7.6) \qquad \frac{\partial p_{ij}(\tau, t)}{\partial t} = -p_{ij}(\tau, t) q_j(t) + \sum_{k \neq j} p_{ik}(\tau, t) q_{kj}(t)$$

with the initial conditions

$$(8.7.7) \qquad p_{ij}(\tau, \tau) = \begin{cases} 1 & \text{for } i = j, \\ 0 & \text{for } i \neq j. \end{cases}$$

Proof. We shall prove assertion II. Let us write (8.7.1) in the form

$$p_{ij}(\tau, t + \Delta t) = \sum_k p_{ik}(\tau, t) p_{kj}(t, t + \Delta t).$$

Hence

$$\frac{p_{ij}(\tau, t + \Delta t) - p_{ij}(\tau, t)}{\Delta t}$$

$$= \frac{1}{\Delta t} \left[\sum_k p_{ik}(\tau, t) p_{kj}(t, t + \Delta t) - p_{ij}(\tau, t) \right]$$

$$= \sum_{k \neq j} p_{ik}(\tau, t) \frac{p_{kj}(t, t + \Delta t)}{\Delta t} - p_{ij}(\tau, t) \frac{1 - p_{jj}(t, t + \Delta t)}{\Delta t}.$$

As a result of (8.7.2) and (8.7.3), of the absolute convergence of the series $\sum_k p_{ik}(\tau, t)$, and of the uniform convergence in (8.7.3) with respect to the first index when the second index is fixed, we obtain

$$\frac{\partial p_{ij}(\tau, t)}{\partial t} = \lim_{\Delta t \to 0} \frac{p_{ij}(\tau, t + \Delta t) - p_{ij}(\tau, t)}{\Delta t}$$

$$= \sum_{k \neq j} p_{ik}(\tau, t) \lim_{\Delta t \to 0} \frac{p_{kj}(t, t + \Delta t)}{\Delta t} - p_{ij}(\tau, t) \lim_{\Delta t \to 0} \frac{1 - p_{jj}(t, t + \Delta t)}{\Delta t}$$

$$= -p_{ij}(\tau, t)q_j(t) + \sum_{k \neq j} p_{ik}(\tau, t)q_{kj}(t).$$

Thus we have proved assertion II. The proof of assertion I is analogous.

It can be shown that systems (8.7.4) and (8.7.6) have identical solutions $p_{ij}(\tau, t)$ satisfying (8.2.6) to (8.2.8) and the initial conditions, and that the corresponding intensity functions exist and are continuous. The proof of these facts can be found in the paper by Feller [1] for the nonhomogeneous process and in the paper by Kolmogorov already mentioned [3] for the homogeneous process.

It is obvious that if the process has only a finite number of states, then the convergence in (8.7.3) is uniform in i for fixed j.

If the process is homogeneous, then the intensity functions (8.7.2) and (8.7.3) are independent of t and we can simply write q_i and q_{ij} and call them *intensities*. Equations (8.7.4) and (8.7.6) are then of the form

(8.7.4′) $$\frac{dp_{ij}(t)}{dt} = q_i p_{ij}(t) - \sum_{k \neq i} q_{ik} p_{kj}(t),$$

(8.7.6′) $$\frac{dp_{ij}(t)}{dt} = -p_{ij}(t)q_j + \sum_{k \neq j} p_{ik}(t)q_{kj},$$

and the initial conditions are

(8.7.5′) $$p_{ij}(0) = \begin{cases} 1 & \text{for } j = i, \\ 0 & \text{for } j \neq i. \end{cases}$$

We observe that all systems of differential equations obtained in Sections 8.3 to 8.5 are particular cases of system (8.7.6′).

In theorem 8.7.1 we assumed that relations (8.7.2) and (8.7.3) are satisfied, that is, that the intensity functions $q_i(t)$ and $q_{ij}(t)$ exist and are finite. However, for homogeneous Markov processes the existence of intensities q_i and q_{ij} follows, as has been shown by Doob [2] and Kolmogorov [15], from the assumption of the continuity of $p_{ii}(t)$ at zero, that is, from the assumption that

(8.7.8) $$\lim_{t \to 0} p_{ii}(t) = 1 \qquad (i = 0, 1, 2, \ldots).$$

If in addition, we assume that the convergence in (8.7.8) is uniform in i, then the intensities q_i and q_{ij} are finite. In particular, if the number of states is finite, the existence of finite intensities follows from (8.7.8).

We observe that assumption (8.7.8) becomes quite natural if we look at the initial conditions (8.7.5).

The problem of the existence of the intensity functions $q_i(t)$ and $q_{ij}(t)$ and the properties of these functions are investigated in the papers of Austin [1, 2] and Jushkevitch [1].

8.8 PURELY DISCONTINUOUS AND PURELY CONTINUOUS PROCESSES

A   In Sections 8.3 to 8.7 we considered processes with at most a countable number of states. Now we assume that the random variables X_t of a Markov process $\{X_t, 0 \leqslant t < \infty\}$ can take on arbitrary real values. We remind the reader that for Markov processes, the conditional distribution function

$$F(t_1, x, t_2, y) = P(X_{t_2} < y \mid X_{t_1} = x)$$

is defined for every t_1 and t_2, where $0 \leqslant t_1 < t_2$, and this distribution function is independent of the values of X_t for $t < t_1$. The function $F(t_1, x, t_2, y)$, as a distribution function, is continuous at least from the left in y and satisfies the conditions

$$\lim_{y \to -\infty} F(t_1, x, t_2, y) = 0, \qquad \lim_{y \to \infty} F(t_1, x, t_2, y) = 1.$$

Let us assume that $F(t_1, x, t_2, y)$ is a continuous function with respect to t_1 and t_2 and a Baire function (see Section 2.4, A) in x. The Chapman-Kolmogorov equation takes the form

$$(8.8.1) \qquad F(t_1, x, t_2, y) = \int_{-\infty}^{\infty} F(s, z, t_2, y) \, d_z F(t_1, x, s, z),$$

where $t_1 < s < t_2$. The integral on the right-hand side of (8.8.1) is the Stieltjes integral of the function $F(s, z, t_2, y)$ with respect to the distribution function $F(t_1, x, s, z)$. If $F(t_1, x, t_2, y)$ has a density function (with respect to y) $f(t_1, x, t_2, y)$, the Chapman-Kolmogorov equation is

$$(8.8.1') \qquad F(t_1, x, t_2, y) = \int_{-\infty}^{\infty} F(s, z, t_2, y) f(t_1, x, s, z) \, dz,$$

where $t_1 < s < t_2$. Equation (8.8.1') or (8.8.1) may be interpreted in the following way. The passage from the state x at the moment t_1 to a state less than y at the moment t_2 through a state z at the intermediate moment s is a product of two passages: (1) at the moment s the process is in a state belonging to the "interval" $[z, z + dz)$ and the probability of this event is $f(t_1, x, s, z)\, dz$; (2) from the state z at the moment s the process passes to a state less than y at the moment t_2 and the probability of this event is $F(s, z, t_2, y)$. Applying the generalized theorem on absolute probability [see formula (2.7.7)], we obtain (8.8.1'). From (8.8.1') we obtain immediately

$$(8.8.2) \qquad f(t_1, x, t_2, y) = \int_{-\infty}^{\infty} f(s, z, t_2, y) f(t_1, x, s, z)\, dz.$$

It follows from the definition of the function $F(t_1, x, t_2, y)$ and from the assumption that it is continuous with respect to t_1 and t_2 that

$$(8.8.3) \qquad \lim_{t_1 \to t_2} F(t_1, x, t_2, y) = \begin{cases} 0 & \text{for } y \leqslant x, \\ 1 & \text{for } y > x, \end{cases}$$

$$(8.8.4) \qquad \lim_{t_2 \to t_1} F(t_1, x, t_2, y) = \begin{cases} 0 & \text{for } y \leqslant x, \\ 1 & \text{for } y > x, \end{cases}$$

where the convergence is from the left-hand side in (8.8.3) and from the right-hand side in (8.8.4).

B   We now introduce the notion of a purely discontinuous process. Let $\{X_t, t \in I\}$ be a Markov process, where the set of possible states may be the set of all real numbers.

We introduce two functions $q(t, x)$ and $P(t, x, y)$ given by the formulas

$$q(t, x) = \lim_{\Delta t \to 0} \frac{P(X_{t+\Delta t} - X_t \neq 0 \mid X_t = x)}{\Delta t},$$

$$P(t, x, y) = \lim_{\Delta t \to 0} P(X_{t+\Delta t} < y \mid X_t = x, X_{t+\Delta t} - X_t \neq 0).$$

The functions $q(t, x)$ and $P(t, x, y)$ are non-negative, and $P(t, x, y)$ as a function of y is a distribution function.

The process is said to be *purely discontinuous* if the changes of state can only occur by jumps; if at the moment t the process is in the state x, then the probability that at the moment $t + \Delta t$ it will be in a state different from x is equal to $q(t, x)\, \Delta t + o(\Delta t)$.

A purely discontinuous process may be defined in an analytical way as follows.

Definition 8.8.1.   A Markov process $\{X_t, t \in I\}$ is *purely discontinuous* if for every $t \in I$

(8.8.5)

$$F(t, x, t + \Delta t, y) = \begin{cases} 1 - q(t, x)\, \Delta t + q(t, x) P(t, x, y)\, \Delta t + o(\Delta t) & \text{for } x < y, \\ q(t, x) P(t, x, y)\, \Delta t + o(\Delta t) & \text{for } x \geqslant y. \end{cases}$$

In other words, if $x < y$, the process may be in a state less than y at the moment $t + \Delta t$ provided it was in the state x at the moment t, either because there was no change of state in the interval $[t, t + \Delta t)$ or the change of state did occur, but at the moment $t + \Delta t$ the process was in a state less than y. If $y \leqslant x$, the process may be in a state less than y at the moment $t + \Delta t$ only if a suitable change of state occurred.

Letting

$$E(x, y) = \begin{cases} 1 & \text{if } x < y, \\ 0 & \text{if } x \geqslant y, \end{cases}$$

we can rewrite (8.8.5) in the following form:

$$(8.8.5') \quad F(t, x, t + \Delta t, y) = [1 - q(t, x)\, \Delta t]E(x, y)$$
$$+ q(t, x)P(t, x, y)\, \Delta t + o(\Delta t).$$

Theorem 8.8.1. *Let* $\{X_t, 0 \leqslant t < \infty\}$ *be a purely discontinuous Markov process and let* $F(t_1, x, t_2, y)$ *be a Baire function in* x *and a continuous function in* t_1 *and* t_2. *Let the functions* $P(t, x, y)$ *and* $q(t, x)$ *be Baire functions in* x *and continuous in* t.

Then:

I. *The distribution function* $F(t_1, x, t_2, y)$ *satisfies the following integro-differential equations:*

$$(8.8.6) \qquad \frac{\partial F(t_1, x, t_2, y)}{\partial t_1} = q(t_1, x)\left[F(t_1, x, t_2, y) \right.$$
$$\left. - \int_{-\infty}^{+\infty} F(t_1, z, t_2, y)\, d_z P(t_1, x, z) \right],$$

$$(8.8.6') \qquad \frac{\partial F(t_1, x, t_2, y)}{\partial t_2} = -\int_{-\infty}^{y} q(t_2, z)\, d_z F(t_1, x, t_2, z)$$
$$+ \int_{-\infty}^{+\infty} q(t_2, z)P(t_2, z, y)\, d_z F(t_1, x, t_2, z).$$

II. *There exists only one solution satisfying both eqs.* (8.8.6) *and* (8.8.6'), *with initial conditions* (8.8.3) *and* (8.8.4), *and having all the properties assumed in this theorem.*

We present only the proof of (8.8.6'); the proof of (8.8.6) is analogous. The proof of assertion II is not given here. It can be found in the paper of the author of the theorem (Feller [1]).

Proof. Using (8.8.1) and (8.8.5') and the definition of $E(z, y)$, we have for $\Delta t_2 > 0$

$$F(t_1, x, t_2 + \Delta t_2, y)$$

$$= \int_{-\infty}^{+\infty} F(t_2, z, t_2 + \Delta t_2, y) \, d_z F(t_1, x, t_2, z)$$

$$= \int_{-\infty}^{+\infty} \{[1 - q(t_2, z) \Delta t_2] E(z, y) + [q(t_2, z) P(t_2, z, y) \Delta t_2] + o(\Delta t_2)\}$$

$$d_z F(t_1, x, t_2, z)$$

$$= F(t_1, x, t_2, y) - \Delta t_2 \int_{-\infty}^{y} q(t_2, z) \, d_z F(t_1, x, t_2, z)$$

$$+ \Delta t_2 \int_{-\infty}^{+\infty} q(t_2, z) P(t_2, z, y) \, d_z F(t_1, x, t_2, z) + o(\Delta t_2).$$

We observe that the existence of all these integrals follows from the assumptions about the functions q and P. From the equality of the extreme terms we obtain

$$\lim_{\Delta t_2 \to 0} \frac{F(t_1, x, t_2 + \Delta t_2, y) - F(t_1, x, t_2, y)}{\Delta t_2}$$

$$= -\int_{-\infty}^{y} q(t_2, z) \, d_z F(t_1, x, t_2, z)$$

$$+ \int_{-\infty}^{+\infty} q(t_2, z) P(t_2, z, y) \, d_z F(t_1, x, t_2, z).$$

Hence the right-hand derivative exists. Repeating this argument for $\Delta t_2 < 0$, we obtain that the left-hand derivative is equal to the right-hand derivative. It follows that the derivative $\partial F(t_1, x, t_2, y)/\partial t_2$ exists and is given by formula (8.8.6').

Equation (8.8.6) is called the *backward equation* and was derived by Feller [1]. Equation (8.8.6') is called the *forward equation* and was derived, under somewhat different assumptions, by Kolmogorov [3].

C We now discuss the purely continuous process.

Definition 8.8.2. A Markov process $\{X_t, t \in I\}$ is called *purely continuous* if for every $t \in I$ and every $\varepsilon > 0$

$$(8.8.7) \qquad \lim_{\Delta t \to 0} \frac{1}{\Delta t} \int_{|y - x| \geq \varepsilon} (y - x)^2 \, d_y F(t, x, t + \Delta t, y) = 0,$$

$$(8.8.8) \qquad \lim_{\Delta t \to 0} \frac{1}{\Delta t} \int_{|y - x| < \varepsilon} (y - x) \, d_y F(t, x, t + \Delta t, y) = m(t, x),$$

$$(8.8.9) \quad \lim_{\Delta t \to 0} \frac{1}{\Delta t} \int_{|y - x| < \varepsilon} (y - x)^2 \, d_y F(t, x, t + \Delta t, y) = \sigma^2(t, x) > 0,$$

where the convergence in the last two relations is uniform with respect to x. The value Δt in all three equations may be of arbitrary sign.

If the density $f(t_1, x, t_2, y)$ exists, then the preceding formulas take the form, respectively

$$(8.8.7') \qquad \lim_{\Delta t \to 0} \frac{1}{\Delta t} \int_{|y-x| \geqslant \varepsilon} (y - x)^2 f(t, x, t + \Delta t, y) \, dy = 0,$$

$$(8.8.8') \qquad \lim_{\Delta t \to 0} \frac{1}{\Delta t} \int_{|y-x| < \varepsilon} (y - x) f(t, x, t + \Delta t, y) \, dy = m(t, x),$$

$$(8.8.9') \quad \lim_{\Delta t \to 0} \frac{1}{\Delta t} \int_{|y-x| < \varepsilon} (y - x)^2 f(t, x, t + \Delta t, y) \, dy = \sigma^2(t, x) > 0.$$

We observe that from (8.8.7) it follows that

$$\lim_{\Delta t \to 0} \frac{1}{\Delta t} \int_{|y-x| \geqslant \varepsilon} d_y F(t, x, t + \Delta t, y) = 0,$$

and from (8.8.7) and (8.8.9) it follows that

$$\lim_{\Delta t \to 0} \frac{1}{\Delta t} \int_{-\infty}^{\infty} (y - x)^2 \, d_y F(t, x, t + \Delta t, y) = \sigma^2(t, x) > 0.$$

Let us write the last two formulas in the form, respectively

$$(8.8.10) \qquad P(|X_{t+\Delta t} - X_t| \geqslant \varepsilon \mid X_t = x) = o(\Delta t),$$

$$(8.8.11) \qquad E[(X_{t+\Delta t} - X_t)^2 \mid X_t = x] = \sigma^2(t, x) \Delta t + o(\Delta t).$$

Thus, according to (8.8.10) the probability of a big change of state in a small time interval is very small (it is of order less than Δt), and according to (8.8.11), for any Δt, no matter how small, the probability that $X_{t+\Delta t} - X_t \neq 0$ is positive. Thus, the situation is the opposite to that of the purely discontinuous process where the probability is small that in a small interval of time a change of state occurs, but if a change occurs, it may be big.

D   The processes considered in Sections 8.3 to 8.6 were purely discontinuous processes. The realizations of these processes are step functions; however, there exist purely discontinuous processes with realizations of a more complicated structure. The name "purely continuous process" might suggest that the realizations of such processes are always continuous functions; this is the situation in most cases but not in all. In the last section of this chapter we make some remarks about the properties of realizations of stochastic processes.

8.9   THE WIENER PROCESS

We now investigate one purely continuous process with an important theoretical meaning and numerous applications, particularly in physics. Let $\{X_t, 0 \leqslant t < \infty\}$ be a Markov process, where the possible states are all real numbers. Let us assume that the process is purely continuous and that for arbitrary τ and t $(0 \leqslant \tau < t)$ the conditional distribution function $F(\tau, x, t, y)$ takes the form $F(\tau, t, u)$, where $u = y - x$ and $F(\tau, t, u)$ is the probability that the increment of the state during the time interval $[\tau, t)$ is smaller than u. Suppose, in addition, that the density $f(\tau, t, u)$ exists. The function $F(\tau, t, u)$ is, by assumption, continuous with respect to τ and t. Relations (8.8.2) to (8.8.4) become, respectively

$$(8.9.1) \qquad f(\tau, t, u) = \int_{-\infty}^{\infty} f(\tau, s, u - w)f(s, t, w) \, dw,$$

where $\tau < s < t$,

$$(8.9.2) \qquad \lim_{\tau \to t} F(\tau, t, u) = \begin{cases} 0 & \text{for } u \leqslant 0, \\ 1 & \text{for } u > 0, \end{cases}$$

$$(8.9.3) \qquad \lim_{t \to \tau} F(\tau, t, u) = \begin{cases} 0 & \text{for } u \leqslant 0, \\ 1 & \text{for } u > 0. \end{cases}$$

Relations (8.8.7′) to (8.8.9′) take the form

$$(8.9.4) \qquad \lim_{\Delta t \to 0} \frac{1}{\Delta t} \int_{|u| \geqslant \varepsilon} u^2 f(t, t + \Delta t, u) \, du = 0,$$

$$(8.9.5) \qquad \lim_{\Delta t \to 0} \frac{1}{\Delta t} \int_{|u| < \varepsilon} uf(t, t + \Delta t, u) \, du = m(t),$$

$$(8.9.6) \qquad \lim_{\Delta t \to 0} \frac{1}{\Delta t} \int_{|u| < \varepsilon} u^2 f(t, t + \Delta t, u) \, du = \sigma^2(t) > 0,$$

where $\varepsilon > 0$ is arbitrary.

We assume in addition:

(A)   For all τ, t, and u, the partial derivatives

$$\frac{\partial f(\tau, t, u)}{\partial u}, \qquad \frac{\partial^2 f(\tau, t, u)}{\partial u^2}$$

exist and are continuous with respect to every argument, and for every fixed t (or τ) the derivative $\partial^2 f / \partial u^2$ is uniformly continuous and uniformly bounded in u.

Theorem 8.9.1.   *Let $\{X_t, 0 \leq t < \infty\}$ be a Markov process satisfying (8.9.1) to (8.9.6) and assumption (A). Then the density function $f(\tau, t, u)$*

satisfies the following parabolic type partial differential equation of the second order:

(8.9.7)
$$\frac{\partial f}{\partial t} = -m(t)\frac{\partial f}{\partial u} + \frac{1}{2}\sigma^2(t)\frac{\partial^2 f}{\partial u^2}.$$

Proof. Let us rewrite equation (8.9.1) in the form

(8.9.8) $f(\tau, t + \Delta t, u) = \int_{-\infty}^{\infty} f(\tau, t, u - w) f(t, t + \Delta t, w)\, dw.$

Let us expand $f(\tau, t, u - w)$ into the sum

(8.9.9) $f(\tau, t, u - w) = f(\tau, t, u) - w\frac{\partial f}{\partial u} + \frac{1}{2}w^2\frac{\partial^2 f}{\partial u^2}$

$$+ \frac{1}{2}w^2\left\{\left[\frac{\partial^2 f}{\partial u^2}\right]_{u - \vartheta w} - \frac{\partial^2 f}{\partial u^2}\right\},$$

where $0 < \vartheta < 1$. Substituting this expansion in (8.9.8), we obtain

(8.9.10) $\dfrac{f(\tau, t + \Delta t, u)}{\Delta t} = \dfrac{f(\tau, t, u)}{\Delta t}\displaystyle\int_{-\infty}^{\infty} f(t, t + \Delta t, w)\, dw$

$$- \frac{\partial f}{\partial u}\cdot\frac{1}{\Delta t}\int_{-\infty}^{\infty} wf(t, t + \Delta t, w)\, dw$$

$$+ \frac{1}{2}\cdot\frac{\partial^2 f}{\partial u^2}\cdot\frac{1}{\Delta t}\int_{-\infty}^{\infty} w^2 f(t, t + \Delta t, w)\, dw + \frac{1}{2\Delta t}J,$$

where

(8.9.11) $J = \displaystyle\int_{-\infty}^{\infty} w^2\left\{\left[\frac{\partial^2 f}{\partial u^2}\right]_{u - \vartheta w} - \frac{\partial^2 f}{\partial u^2}\right\}f(t, t + \Delta t, w)\, dw.$

We observe that

(8.9.12)
$$\int_{-\infty}^{\infty} f(t, t + \Delta t, w)\, dw = 1.$$

Next, from (8.9.4) to (8.9.6) it follows that

(8.9.13) $\displaystyle\lim_{\Delta t \to 0}\frac{1}{\Delta t}\int_{-\infty}^{\infty} wf(t, t + \Delta t, w)\, dw$

$$= \lim_{\Delta t \to 0}\frac{1}{\Delta t}\int_{|w| < \varepsilon} wf(t, t + \Delta t, w)\, dw = m(t),$$

(8.9.14) $\displaystyle\lim_{\Delta t \to 0}\frac{1}{\Delta t}\int_{-\infty}^{\infty} w^2 f(t, t + \Delta t, w)\, dw$

$$= \lim_{\Delta t \to 0}\frac{1}{\Delta t}\int_{|w| < \varepsilon} w^2 f(t, t + \Delta t, w)\, dw = \sigma^2(t).$$

Let us split the integral J into two integrals J_1 and J_2, where

$$(8.9.15) \qquad J_1 = \int_{|w|<\varepsilon} w^2 \left\{ \left[\frac{\partial^2 f}{\partial u^2} \right]_{u-\theta w} - \frac{\partial^2 f}{\partial u^2} \right\} f(t, t+\Delta t, w)\, dw,$$

$$(8.9.16) \qquad J_2 = \int_{|w|\geqslant\varepsilon} w^2 \left\{ \left[\frac{\partial^2 f}{\partial u^2} \right]_{u-\theta w} - \frac{\partial^2 f}{\partial u^2} \right\} f(t, t+\Delta t, w)\, dw.$$

By the assumption of uniform continuity with respect to u of the derivative $\partial^2 f/\partial u^2$, the expression in braces in (8.9.15) can be made arbitrarily small for every Δt if ε is sufficiently small. It follows that $J_1 = o(\Delta t)$. By the assumption of uniform boundedness in u of the derivative $\partial^2 f/\partial u^2$ and by formula (8.9.4), we obtain $J_2 = o(\Delta t)$. Hence

$$(8.9.17) \qquad J = o(\Delta t).$$

By (8.9.12) to (8.9.14) and (8.9.17), using (8.9.10), we see that the right-hand derivative $\partial f/\partial t$ exists; but the same reasoning applies to $\Delta t < 0$; hence the derivative $\partial f/\partial t$ exists. After some simple computations we obtain (8.9.7). This equation is called the *forward equation* (see Section 8.7). In an analogous way we may derive the *backward equation*. It is of the form

$$(8.9.18) \qquad \frac{\partial f}{\partial \tau} = -m(\tau) \frac{\partial f}{\partial u} - \frac{1}{2} \sigma^2(\tau) \frac{\partial^2 f}{\partial u^2}.$$

Equations (8.9.7) and (8.9.18) were found by Bachelier [1], but he did not give a precise proof. These equations form a particular case of the equations derived by Kolmogorov [3] for purely continuous processes.

To find the required density function $f(\tau, t, u)$, let us write

$$(8.9.19) \qquad v = u - \int_\tau^t m(\alpha)\, d\alpha, \qquad t^* = \int_\tau^t \sigma^2(\alpha)\, d\alpha.$$

Denoting by $g(\tau, t^*, v)$ the new density function and using (2.4.10), we obtain

$$g(\tau, t^*, v) = f[\tau, t(t^*), u(v, t^*)],$$

where $t(t^*)$ and $u(v, t^*)$ is the transformation inverse to (8.9.19). By the last equation and by (8.9.7) we obtain

$$(8.9.20) \qquad \frac{\partial g}{\partial t^*} = \frac{1}{2} \cdot \frac{\partial^2 g}{\partial v^2}.$$

Proceeding in an analogous way, we obtain from (8.9.18)

$$(8.9.21) \qquad \frac{\partial g}{\partial \tau^*} = -\frac{1}{2} \cdot \frac{\partial^2 g}{\partial v^2}.$$

Thus we have obtained the heat conductivity equations. The only solution of these equations, which is a density function and satisfies (8.9.1) to (8.9.6), is the function (see Courant and Hilbert [1])

(8.9.22)
$$g(\tau, t^*, v) = \frac{1}{\sqrt{2\pi t^*}} \exp\left\{-\frac{v^2}{2t^*}\right\}.$$

Letting

$$M(\tau, t) = \int_{\tau}^{t} m(\alpha)\, d\alpha, \qquad B^2(\tau, t) = t^*,$$

we obtain

(8.9.23)
$$f(\tau, t, u) = \frac{1}{B(\tau, t)\sqrt{2\pi}} \exp\left\{-\frac{[u - M(\tau, t)]^2}{2B^2(\tau, t)}\right\}.$$

Thus the increment $X_t - X_\tau$ has a normal distribution with expected value $M(\tau, t)$ and standard deviation $B(\tau, t)$. Since, for any $t_1, \ldots, t_n$ $(0 \leqslant t_1 < \ldots < t_n < \infty)$, we have $B^2(t_1, t_n) = \sum_{k=1}^{n-1} B^2(t_k, t_{k+1})$, the process here considered has uncorrelated increments.

We now investigate a special case of this process. We assume that the process has independent and homogeneous increments and the expected value of an increment is identically zero. Thus we have

$$M(\tau, t) \equiv 0, \qquad B^2(\tau, t) \equiv (t - \tau)\sigma^2.$$

Formula (8.9.23) has the form

(8.9.24)
$$f(\tau, t, u) = \frac{1}{\sigma\sqrt{2\pi(t - \tau)}} \exp\left[-\frac{u^2}{2(t - \tau)\sigma^2}\right].$$

Let us put $\tau = 0$ and assume that $P(X_0 = 0) = 1$. Then X_t is the increment in the interval $[0, t]$. Thus the density function $f(t, u)$ of the random variable X_t is

(8.9.25)
$$f(t, u) = \frac{1}{\sqrt{2\pi t \sigma^2}} \exp\left(-\frac{u^2}{2t\sigma^2}\right).$$

Definition 8.9.1. A process $\{X_t, 0 \leqslant t < \infty\}$ with independent and homogeneous increments, satisfying the condition $P(X_0 = 0) = 1$, and whose random variables X_t have the density function given by (8.9.25) is called the *Brownian motion process* or the *Wiener process*.

The origin of these two names is given later in this chapter.

For the Wiener process we have, for arbitrary τ and t $(0 \leqslant \tau < t < \infty)$,

(8.9.26)
$$E(X_t - X_\tau) = 0, \qquad D^2(X_t - X_\tau) = (t - \tau)\sigma^2.$$

Next, for arbitrary $n = 2, 3, \ldots$ and arbitrary t_i $(i = 1, 2, \ldots, n)$, where $t_1 < \ldots < t_n$, we have

$$X_{t_n} = X_{t_1} + (X_{t_2} - X_{t_1}) + \ldots + (X_{t_n} - X_{t_{n-1}}),$$

where the $X_{t_1}, X_{t_2} - X_{t_1}, \ldots, X_{t_n} - X_{t_{n-1}}$ are independent and have normal distributions. Then by theorem 5.11.1 the random variables $X_{t_1}, X_{t_2}, \ldots, X_{t_n}$, as linear functions of independent normally distributed random variables, have a multidimensional normal distribution with

$$E(X_{t_i}) = 0, \qquad D^2(X_{t_i}) = t_i \sigma^2 \qquad (i = 1, 2, \ldots, n),$$

$$E(X_{t_i} X_{t_j}) = t_i \sigma^2 \qquad (i, j = 1, 2, \ldots, n; \; i < j).$$

The second-order moment matrix $\mathbf{M}$ of the random variable $(X_{t_1}, X_{t_2}, \ldots, X_{t_n})$ satisfies the relation

$$\frac{\mathbf{M}}{\sigma^2} = \begin{bmatrix} t_1 & t_1 & t_1 & \cdots & t_1 \\ t_1 & t_2 & t_2 & \cdots & t_2 \\ t_1 & t_2 & t_3 & \cdots & t_3 \\ \cdots & & & & \cdots \\ t_1 & t_2 & t_3 & \cdots & t_n \end{bmatrix}$$

Example 8.9.1. Brownian motion. A microscopic particle in a liquid moves chaotically as a result of collisions with the particles of the liquid. This phenomenon, noticed by Brown in 1826, is called *Brownian motion.* Let us consider the movement of such a particle, and denote by X_t the coordinate on the x-axis of the particle at the moment t in some Cartesian coordinate system. Einstein [1] and Smoluchowski [1] showed that it is a good approximation to treat X_t as a Brownian motion process. The essential point in the explanation of this fact is that the contacts between the foreign microscopic particle and the particles of the liquid occur only at moments of collision. The collisions occur irregularly but often. Thus, if the difference $t - \tau$ is large in comparison with the time interval between two successive collisions, then the increment $X_t - X_\tau$ is the sum of a large number of small increments. Next, if the liquid is in macroscopic equilibrium we may assume that the increments depend only on the length of the time interval and hence are homogeneous, and the increments in the disjoint time intervals are independent. If we further assume that the movement is symmetric we have $E(X_t - X_\tau) = 0$. Einstein showed that $E[(X_t - X_\tau)^2] = 2D^2 t$, where D is the diffusion constant of the liquid; hence $\sigma^2 = 2D^2$ is a constant which characterizes the liquid. Thus all the assumptions that lead to the Brownian motion process may be considered as satisfied; hence we may assume that in fact we have such a process. This is confirmed by experimental data.

We observe that $X_t - X_\tau$ may be considered, as we have stated above, as the sum of a large number of independent random variables; hence also the central limit theorem suggests that $X_t - X_\tau$ has a normal distribution.

The movement of a particle of a gas under low pressure may also be treated as Brownian motion. In general, we can consider as Brownian motion the movement of any body which is subject to collision with other bodies, if the dimension of the given body is small in comparison with the dimensions of the other bodies, if the contacts occur only at moments of collision, and if these collisions are of a random character. Thus, we can consider as a body under investigation a small star, that is, a star which is small in comparison with other stars in cosmic space, and this space plays the role of the liquid in Brownian motion. By a "collision" we understand the appearance of the investigated star within the gravitational field of other stars. The orbit of such a small star is a random path and we can treat it as a model of the Brownian motion process.

Example 8.9.2. In example 5.11.1 we considered the velocity of a particle of an ideal gas and we established that each of the components of the velocity vector has a normal distribution. We can reach the same conclusion by applying the Brownian motion model. We leave this to the reader.

At the end of this section let us observe that the processes with independent and homogeneous increments considered in this chapter are particular cases of a large class of such processes. Here we only give some references to the basic works on these problems, namely, those by Cramér [1], de Finetti [1], Lévy [2], and Kolmogorov [5].

We remark also that many generalizations of the Brownian motion process (to the multidimensional case and other cases) have been considered. We mention here two papers only, those by Lévy [11] and Itô [1].

8.10   STATIONARY PROCESSES

A   In this section we discuss briefly the class of processes called *stationary processes*. The theory of these processes has been developed recently and has found more and more numerous applications. First, let us make some general statements. Stationary processes have the property that the probability distributions or at least some moments of the random variables of the process are independent of translations of the initial moment on the time axis t. Classes of Markov processes and stationary processes are not mutually exclusive. Some Markov processes may be stationary, yet many random phenomena in industry, meteorology, economics, and other areas do not satisfy the model of Markov processes and can nevertheless be treated by the theory of stationary processes.

B   In the remainder of this section we consider complex random variables, which we now define.

Definition 8.10.1. Let X and Y be two real random variables in the sense of definition 2.1.2. Then $Z = X + iY$, where i is the imaginary unit, is called a *complex random variable*.

The distribution of the random variable Z is determined by the two-dimensional distribution function $F(x, y) = P(X < x, Y < y)$.

Definition 8.10.1a. Let $\{X_t, t \in I\}$ and $\{Y_t, t \in I\}$ be two real stochastic processes in the sense of definition 8.1.1 and let $\{Z_t = X_t + iY_t, t \in I\}$. Then $\{Z_t, t \in I\}$ is a *complex stochastic process*.

Definition 8.10.2. The expression

$$D^2(Z) = E[|Z - E(Z)|^2]$$

is called the *variance of the complex random variable Z*.

The reader can verify that

$$D^2(Z) = E(|Z|^2) - |E(Z)|^2 = D^2(X) + D^2(Y).$$

Definition 8.10.3. The expression

$$E\{[Z_1 - E(Z_1)][\overline{Z_2 - E(Z_2)}]\}$$

is called the *covariance of the complex random variables Z_1 and Z_2*, the components of the vector (Z_1, Z_2).

It should be remembered that, contrary to the real case, the covariance of the components of the vector (Z_1, Z_2) is not equal to the covariance of the components of the vector (Z_2, Z_1).

C   We now introduce the notion of a stationary stochastic process.

Definition 8.10.4. A complex stochastic process $\{Z_t, -\infty < t < +\infty\}$ is called *stationary in the strict sense* if for $n = 1, 2, \ldots$ and for arbitrary real $t_1, \ldots, t_n$ and τ

$$(8.10.1) \quad F_{t_1, \ldots, t_n}(x_1, \ldots, x_n, y_1, \ldots, y_n)$$
$$= P(X_{t_1} < x_1, \ldots, X_{t_n} < x_n, Y_{t_1} < y_1, \ldots, Y_{t_n} < y_n)$$
$$= P(X_{t_1+\tau} < x_1, \ldots, X_{t_n+\tau} < x_n, Y_{t_1+\tau} < y_1, \ldots, Y_{t_n+\tau} < y_n)$$
$$= F_{t_1+\tau, \ldots, t_n+\tau}(x_1, \ldots, x_n, y_1, \ldots, y_n).$$

A real process $\{X_t, -\infty < t + \infty\}$ is *stationary in the strict sense* if for the considered n, $t_1, \ldots, t_n$ and τ

$$(8.10.1') \quad F_{t_1, \ldots, t_n}(x_1, \ldots, x_n) = F_{t_1+\tau, \ldots, t_n+\tau}(x_1, \ldots, x_n).$$

From this definition it follows in particular that for every real t we have $F_t(x, y) = F_{t-t}(x, y) = F_0(x, y)$, and, for real processes, $F_t(x) = F_0(x)$. Thus all Z_t (in the real case all X_t) have the same distribution.

All two-dimensional distribution functions depend only on the difference of values of the parameter t, namely,

$$F_{t_1, t_2}(x_1, x_2, y_1, y_2) = F_{0, t_2-t_1}(x_1, x_2, y_1, y_2),$$

and for real stationary processes

$$F_{t_1, t_2}(x_1, x_2) = F_{0, t_2-t_1}(x_1, x_2).$$

If the second-order moments of the random variables X_t and Y_t exist and $Z_t = X_t + iY_t$ is a process stationary in the strict sense, then by (8.10.1) we have, for all real t and τ,

$$E(Z_t) = E(X_t + iY_t) = m_1 + m_2 = m = \text{constant},$$
$$(8.10.2) \quad D^2(Z_t) = D^2(X_t) + D^2(Y_t) = \sigma^2 = \text{constant},$$
$$E\{[Z_{t+\tau} - E(Z_{t+\tau})][\overline{Z_t - E(Z_t)}]\} = R(\tau).$$

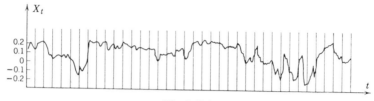

Fig. 8.10.1

The function $\overline{R(\tau)}$ is called the *covariance function*. It is easy to verify that $R(0) = D^2(Z_t)$ and $R(-\tau) = R(\tau)$. If the process is real, formulas (8.10.2) take the form

$$E(X_t) = m = \text{constant},$$

(8.10.2') $$D^2(X_t) = \sigma^2 = \text{constant},$$

$$E\{[X_{t+\tau} - E(X_{t+\tau})][X_t - E(X_t)]\} = R(\tau).$$

If the process is real and $m = 0$, $\sigma^2 = 1$, then $R(\tau)$ is called the *correlation function*.

Example 8.10.1. Let X_t denote the temperature at a fixed point of the atmosphere at the moment t. We shall consider changes of temperature in small intervals of time; hence we are concerned with microchanges of temperature of a turbulent character. These changes cause the vertical movement of heat and affect substantially various meteorological phenomena. Here, we cannot assume that X_t is a Markov process since the knowledge of the temperature at the moment t_0 does not determine the distribution of X_t for $t > t_0$. This distribution also depends on the temperatures at the earlier moments, since, for instance, the local changes of temperature are different at different times of the year. Nevertheless, for fixed weather we may assume that X_t is a stationary process. The local fluctuations of temperature have been recorded by a thermometer constructed by Kreczmar [1]. The observed realization of the process X_t is given in Fig. 8.10.1.

Among real stationary stochastic processes that of the *normal process* is of particular interest; this is a stationary process satisfying (8.10.2') with $m = 0$, $\sigma = 1$ and such that for $n = 1, 2, 3, \ldots$ and arbitrary t_m $(m = 1, 2, \ldots, n)$ the random variable $(X_{t_1}, \ldots, X_{t_n})$ has a normal distribution. By formula (5.11.9) the random variable $(X_{t_1}, \ldots, X_{t_n})$ has the characteristic function

$$\phi_{t_1, \ldots, t_n}(v_1, \ldots, v_n) = \exp\left[-\frac{1}{2}\sum_{j=1}^{n}\sum_{k=1}^{n} R(t_k - t_j)v_j v_k\right].$$

In applications it is usually difficult to deal with multidimensional distributions of the random variables X_t. Sometimes observations give us only the moments of these random variables. This fact leads us to introduce the notion of a stationary process in the wide sense.

Definition 8.10.5. A complex process $\{Z_t, -\infty < t < +\infty\}$ is called *stationary in the wide sense* (or *in the sense of Khintchin*) if the expected values, variances, and covariances of this process exist and satisfy (8.10.2). The real process $\{X_t, -\infty < t < \infty\}$ is called *stationary in the wide sense* if the first- and second-order moments of its random variables exist and satisfy (8.10.2').

It follows from this definition that a process which is stationary in the strict sense and whose first- and second-order moments exist is stationary in the wide sense. It is obvious that the real normal process considered in this section is stationary in both senses.

D   We now introduce the notion of continuity of a stationary process.

Definition 8.10.6. A process $\{Z_t, -\infty < t < +\infty\}$, stationary in the wide sense, is called *continuous* if for every real t

$$(8.10.3) \qquad \lim_{\tau \to 0} E(|Z_{t+\tau} - Z_t|^2) = 0.$$

Theorem 8.10.1. *A process stationary in the wide sense is continuous if and only if its covariance function $R(\tau)$ is continuous at zero.*

Proof. Suppose that the process $\{Z_t, -\infty < t < +\infty\}$, stationary in the wide sense, is continuous. Then after some simple transformations we obtain from (8.10.2)

$$(8.10.4) \quad |R(\tau + \Delta\tau) - R(\tau)| = |E\{(Z_{\tau+\Delta\tau} - Z_\tau)[\overline{Z_0 - E(Z_0)}]\}|$$
$$\leqslant E\{|(Z_{\tau+\Delta\tau} - Z_\tau)[\overline{Z_0 - E(Z_0)}]|\}.$$

By the Schwarz inequality (see Franklin [1]) we obtain

$$(8.10.5) \quad E\{|(Z_{\tau+\Delta\tau} - Z_\tau)[\overline{Z_0 - E(Z_0)}]|\}$$
$$\leqslant \sqrt{E(|Z_{\tau+\Delta\tau} - Z_\tau|^2)E[|Z_0 - E(Z_0)|^2]}.$$

Since $D^2(Z_0) < \infty$, it follows from (8.10.3) to (8.10.5) that

$$(8.10.6) \qquad \lim_{\Delta\tau \to 0} |R(\tau + \Delta\tau) - R(\tau)| = 0.$$

Hence $R(\tau)$ is continuous not only at zero but everywhere.

Suppose now that the function $R(\tau)$ is continuous at the origin. Since the process is stationary, we have

$$E(|Z_{t+\tau} - Z_t|^2) = E\{|Z_\tau - E(Z_\tau) - [Z_0 - E(Z_0)]|^2\}$$
$$= E[|Z_\tau - E(Z_\tau)|^2] + E[|Z_0 - E(Z_0)|^2]$$
$$- E\{[Z_0 - E(Z_0)][\overline{Z_\tau - E(Z_\tau)}]\} - E\{[Z_\tau - E(Z_\tau)][\overline{Z_0 - E(Z_0)}]\}$$
$$= 2D^2(Z_0) - R(\tau) - R(-\tau).$$

Since $R(0) = D^2(Z_0)$ and, by assumption, $R(\tau)$ is continuous at zero, the last equation gives

$$\lim_{\tau \to 0} E(|Z_{t+\tau} - Z_t|^2) = 0.$$

Hence the process is continuous.

From now on we shall always assume that in (8.10.2) and (8.10.2') we have

(8.10.7) $$m = 0, \quad \sigma = 1.$$

We now present a theorem of fundamental importance in the theory of stationary stochastic processes, which was proved by Khintchin [3].

Theorem 8.10.2. *A function $R(\tau)$ is the correlation function of a process $\{Z_t, -\infty < t < +\infty\}$ stationary in the wide sense, continuous, and satisfying (8.10.7) if and only if there exists a distribution function $F(\lambda)$ such that*

(8.10.8) $$R(\tau) = \int_{-\infty}^{+\infty} e^{i\lambda \tau} \, dF(\lambda).$$

Proof. Let $R(\tau)$ be the correlation function of a process satisfying the assumptions of the theorem. Then by (8.10.2) and (8.10.7)

(8.10.9) $$R(0) = 1.$$

Next, for $n = 1, 2, \ldots$ and arbitrary real $t_1, \ldots, t_n$ and complex $a_1, \ldots, a_n$,

$$\sum_{j,k=1}^{n} R(t_j - t_k) a_j \bar{a}_k = \sum_{j,k=1}^{n} a_j \bar{a}_k E(Z_{t_j - t_k} \bar{Z}_0)$$

$$= \sum_{j,k=1}^{n} a_j \bar{a}_k E(Z_{t_j} \bar{Z}_{t_k}) = E\left(\left| \sum_{j=1}^{n} a_j Z_j \right|^2 \right) \geqslant 0.$$

Hence $R(\tau)$ is a positive definite function.

Finally, considering (8.10.9) and the fact that, by theorem 8.10.1, $R(\tau)$ is a continuous function, we obtain from theorem 4.1.1 that there exists a distribution function $F(\lambda)$ satisfying relation (8.10.8).

Suppose now that $R(\tau)$ satisfies relation (8.10.8), where $F(\lambda)$ is a distribution function. We must show that there exists a stochastic process $\{Z_t, -\infty < t < \infty\}$ which is stationary in the wide sense, continuous, satisfying (8.10.7), and with given correlation function $R(\tau)$. To prove this, it is sufficient, of course, to give an example of such a process. For simplicity we shall do so only for the case when the function $R(\tau)$ is real. An example of such a process is the normal process, which we considered in 8.10, C. All the finite dimensional distributions of this process, that is, the distributions of the vectors $(X_{t_1}, \ldots, X_{t_n})$ for arbitrary n and $t_1, \ldots, t_n$, are determined by the second-order moment matrices, which in turn are determined by the correlation function $R(\tau)$.

The theorem is proved.

The reader has certainly noticed that under condition (8.10.7) and by theorem 8.10.2, the correlation function $R(\tau)$ has all the properties of a characteristic function.

The function $F(\lambda)$ in (8.10.8) is called the *spectral distribution function*. It follows from theorem 4.5.1 that the correlation function $R(\tau)$ uniquely determines the spectral distribution function $F(\lambda)$. If the density $f(\lambda)$ exists, the corresponding stationary process is called a *process with a continuous spectrum*. If, moreover, the function $R(\tau)$ is absolutely integrable over the whole axis, $-\infty < \tau < \infty$, the spectral density may be determined from formula (4.5.6), which here takes the form

$$f(\lambda) = \frac{1}{2\pi} \int_{-\infty}^{+\infty} e^{-i\lambda\tau} R(\tau)\, d\tau.$$

If the distribution function $F(\lambda)$ increases only at jump points λ_k, the corresponding process is called a *process with a discrete spectrum*.

We say that formula (8.10.8) gives the *spectral representation* of the correlation function.

E   Let us consider the process $\{Z_t, -\infty < t < \infty\}$ defined as

$$(8.10.10) \qquad\qquad Z_t = \sum_{k=1}^{n} W_k e^{i\lambda_k t},$$

where λ_k are real constants and the W_k $(k = 1, 2, \ldots, n)$ are complex random variables satisfying the equalities

$$(8.10.11) \qquad\qquad E(W_k) = 0, \qquad E(|W_k|^2) = \sigma_k^2,$$

$$E(W_k \overline{W}_j) = 0 \qquad (k \neq j),$$

where $\sigma_k^2 > 0\, (k = 1, 2, \ldots, n)$ and $\sum_{k=1}^{n} \sigma_k^2 = 1$. The last of the equalities (8.10.11) states that the W_k are uncorrelated.

The process $\{Z_t, -\infty < t < +\infty\}$ defined by (8.10.10) is a superposition of random vibrations with random amplitudes and random phases. Let $W_k = U_k + iV_k$, where U_k and V_k are real. Then the real part of the expression $W_k^k e^{i\lambda_k t}$ is

$$U_k \cos \lambda_k t - V_k \sin \lambda_k t = R_k \sin (\lambda_k t + \alpha_k),$$

where

$$R_k = \sqrt{U_k^2 + V_k^2}, \qquad \sin \alpha_k = \frac{U_k}{R_k}, \qquad \cos \alpha_k = -\frac{V_k}{R_k}.$$

This is a sinusoidal curve with a constant frequency λ_k, random amplitude R_k, and random phase α_k. Thus the real part of the process is the superposition of such sinusoidal curves.

The process Z_t given by (8.10.10) is a process stationary in the wide sense, continuous, and satisfying relations (8.10.7). In fact, by (8.10.10) and (8.10.11) we have

$$(8.10.12) \qquad E(Z_t) = 0, \qquad D^2(Z_t) = E(|Z_t|^2) = \sum_{k=1}^{n} \sigma_k^2 = 1,$$

$$R(\tau) = E(Z_{t+\tau}\,\bar{Z}_t) = E\left[\left(\sum_{k=1}^{n} W_k e^{i\lambda_k(t+\tau)}\right)\right.$$

$$\times \left.\left(\sum_{k=1}^{n} \bar{W}_k e^{-i\lambda_k t}\right)\right] = E\left(\sum_{k=1}^{n} W_k \bar{W}_k\, e^{i\lambda_k \tau}\right)$$

$$= \sum_{k=1}^{n} \sigma_k^2 e^{i\lambda_k \tau}.$$

Thus the process is stationary in the wide sense and satisfies relation (8.10.7). The continuity of the process follows from the continuity of $R(\tau)$.

The distribution function $F(\lambda)$, which, according to (8.10.8), corresponds to the correlation function $R(\tau)$ given by (8.10.12), is a step function, increasing only at the points λ_k with jumps equal to σ_k^2, respectively. Thus the process under consideration has a discrete spectrum.

It is easy to show that if $n = 2m$ and $2m$ terms of the sum (8.10.10) form m pairs of conjugate terms, the process is real and has the correlation function

$$R(\tau) = 2 \sum_{k=1}^{m} \sigma_k^2 \cos \lambda_k \tau.$$

F   The process given by (8.10.10) is the starting point in the harmonic analysis of stationary stochastic processes. To give even some rough information about the essential points of these problems, we must introduce a new kind of convergence, connected with the notion of continuity given in definition 8.10.6.

Definition 8.10.7.   A sequence of (real or complex) random variables $\{U_n\}$ with finite second-order moments is called *convergent to a random variable U in quadratic mean* if

$$(8.10.13) \qquad \lim_{n \to \infty} E(|U_n - U|^2) = 0.$$

Let us now consider a process $\{Z_t,\ -\infty < t < +\infty\}$ defined as

$$(8.10.14) \qquad Z_t = \lim_{n \to \infty} \sum_{k=1}^{n} W_k e^{i\lambda_k t}$$

$$= \sum_{k=1}^{\infty} W_k e^{i\lambda_k t},$$

where the W_k $(k = 1, 2, \ldots)$ are complex random variables satisfying equalities (8.10.11) and

(8.10.15) $$\sum_{k=1}^{\infty} \sigma_k^{\;2} = 1.$$

The passage to the limit in (8.10.14) should be understood in the sense of definition 8.10.7. We can show that this limit exists. (See, for instance, Doob [5], § IV, 4.)

The process Z_t is stationary in the wide sense and its correlation function has the form

(8.10.16) $$R(\tau) = \sum_{k=1}^{\infty} \sigma_k^{\;2} e^{i\lambda_k \tau}.$$

Slutsky [3] showed that every process stationary in the wide sense, continuous, and with a discrete spectrum, may be represented in the form of the series on the right-hand side of (8.10.14). This is a particular case of the *spectral representation theorem* for stationary processes. A more general theorem on the spectral representation of stationary processes which includes processes with a continuous spectrum (in which the series on the right-hand side of (8.10.14) is replaced by some specially defined integrals) was proved by Kolmogorov [10], [11]. Some theorems on the spectral representation of processes (without the assumption that they are stationary) whose random variables have finite second-order moments were given by Loève [1], [2], Karhunen [1], Getoor [1], and Cramér [6].

Kac and Siegert [1] obtained an effective formula for the representation of X_t in the form of a series when X_t is the stationary normal process considered in Section 8.10, C.

G   We now briefly discuss the interpretation of the spectral distribution function in physical problems. Let Z_t be the real stationary process discussed at the end of Section 8.10, E. Let Z_t be, for instance, the voltage fluctuation of an electric current, which can be represented as a sum of sinusoidal curves with constant frequencies and with random phases and amplitudes. Then the spectral distribution function of this process represents the distribution of electric energy carried by the components with given frequencies, provided the total energy is one. Let λ' and λ'' $(\lambda' < \lambda'')$ be two arbitrary points on the λ axis. Then

$$F(\lambda'') - F(\lambda') = \sum \frac{1}{2} \sigma_k^{\;2},$$

where the summation is extended over all λ_k in the interval $[\lambda', \lambda'')$. The increment of the spectral distribution function in the interval $[\lambda', \lambda'')$ is equal to the average energy of those components of the process which have frequencies contained in the interval $[\lambda', \lambda'')$. For processes with a continuous spectrum the interpretation is analogous, but we cannot formulate it here. The reader will find it in the paper of Yaglom [1] or in the book by Grenander and Rosenblatt [1].

H Besides stationary processes with continuous time we also consider stationary processes with discrete time, that is, stationary sequences of random variables. Here definitions 8.10.4 and 8.10.5 are the same, except that τ and t now run over the set of integers in the interval $(-\infty, +\infty)$. The notion of continuity of the process is unnecessary, since τ can take on only integer values.

We observe that if t runs over the set of integers only, we have, for integer values of j, $\exp[i(\lambda + 2\pi j)t] = \exp(i\lambda t)$; hence for stationary sequences we may consider only values of λ contained in the interval $[-\pi, +\pi]$. Theorem 8.10.2 is also satisfied for stationary sequences, with the modification that the integration in (8.10.8) is extended over the interval $[-\pi, +\pi]$. The proof is due to Herglotz [1]. It is also given in the paper of Wold [1]. The theorem on the spectral representation of processes can be extended to stationary sequences.

Example 8.10.2. Let $\{X_t\}$ ($t = 0, \pm 1, \pm 2, \ldots$) be a stationary (see definition 7.5.3) sequence of random variables with the possible states x_i ($i = 1, 2, \ldots$) forming a homogeneous Markov chain. The sequence $\{X_t\}$ is a process stationary in the strict sense, with discrete time. In fact, let $t_1, t_2, \ldots, t_{n-1}, t_n$ and τ be integers. Then by (7.2.3)

$$P(X_{t_1} = x_{i_1}, X_{t_2} = x_{i_2}, \ldots, X_{t_{n-1}} = x_{i_{n-1}}, X_{t_n} = x_{i_n})$$
$$= P(X_{t_1} = x_{i_1})p_{i_1 i_2}(t_2 - t_1) \ldots p_{i_{n-1} i_n}(t_n - t_{n-1})$$
$$= P(X_{t_1+\tau} = x_{i_1}, X_{t_2+\tau} = x_{i_2}, \ldots, X_{t_{n-1}+\tau} = x_{i_{n-1}}, X_{t_n+\tau} = x_{i_n}).$$

However, this process may not be stationary in the wide sense, since we do not assume anything about the moments of the random variables X_t.

Example 8.10.3. Let $\{X_t\}$ ($t = 0, \pm 1, \pm 2, \ldots$) be a sequence of real random variables with expected values 0, pairwise uncorrelated, and with standard deviations equal to one. Then the correlation function $R(\tau)$ is

$$(8.10.17) \qquad R(\tau) = E(X_{t+\tau} X_t) = \begin{cases} 1 & \text{for} \quad \tau = 0, \\ 0 & \text{for} \quad \tau \neq 0. \end{cases}$$

We observe that

$$R(\tau) = \int_{-\pi}^{+\pi} \frac{\cos \lambda\tau}{2\pi} \, d\lambda;$$

hence the process has the spectral density

$$(8.10.18) \qquad f(\lambda) = \tfrac{1}{2\pi}.$$

The process considered in this example has a constant spectral density. This process is called the *white noise*. This name is connected with the interpretation of the spectral distribution function in the physical problems discussed in Section 8.10, G. It comes from the fact that in the spectral distribution of white light the intensity of all frequencies is the same.

The question of the laws of large numbers for sequences of random variables forming a stochastic process stationary in the strict sense has been solved in the papers of Birkhoff [1] and Khintchin [3]. Ibragimov [1] recently obtained a result concerning the central limit theorem for such sequences.

For sequences of random variables stationary in the wide sense, we can consider a new type of laws of large numbers, where convergence is understood in the sense of definition 8.10.7. Such a law of large numbers can be found, for example, in the book by Doob [5].

I The basic problem in the theory of stationary sequences and processes is the problem of extrapolation. Let $\{X_t, t = 0, \pm 1, \pm 2, \ldots\}$ be a sequence stationary in the wide sense. The problem of extrapolation can be formulated as follows: for some integer m find a linear function

$$\sum_{j=1}^{\infty} a_j X_{t-j}$$

satisfying the relation

$$(8.10.19) \qquad E\left(\left| X_{t+m} = \sum_{j=1}^{\infty} a_j X_{t-j} \right|^2\right) = \text{minimum},$$

and find the value of this minimum. In other words, on the basis of the knowledge of the realization of X_t at the moments $t, t-1, t-2, \ldots$ we want to estimate the state of the process at the moment $t + m$ in such a way that (8.10.19) is satisfied. This is essentially the problem of finding the unknown coefficients a_j by the method of least squares. However, the direct solution by the method presented in Section 3.8 leads to a system of linear equations with an infinite number of unknowns. The solution of such a system is cumbersome even for a finite but large number of unknowns. Kolmogorov [12] and Wiener [3] have given another solution based on the relation between the spectral distribution function of the stationary stochastic process and its correlation function.

The problem of extrapolation for a process with continuous time and stationary in the wide sense can be formulated in an analogous way. Recently some papers have appeared which generalize the theory of stationary stochastic processes discussed in this section to multidimensional stationary processes. A systematic exposition of the notions and results in this domain is given in papers of Masani [1], Rozanov [1], and Wiener and Masani [1]. See also Rosenblatt [2].

8.11 MARTINGALES

We now discuss briefly the theory of another type of stochastic processes, called *martingales*. This concept has been used by Lévy [3]; however, the first systematic study of the theory of martingales and its applications is due to Doob [5, 7 to 9, 11] (see also Chow [1], Snell [1], and Billingsley [3]).

Definition 8.11.1. A sequence $\{X_n\}$ $(n = 1, 2, 3, \ldots)$ of random variables is called a *martingale* if for all $n \geqslant 1$

I. the mean value $E(X_n)$ exists (hence $E(|X_n|) < \infty$),
II. the relation (see Section 3.6, C)

(8.11.1) $$E(X_{n+1} \mid x_1, \ldots, x_n) = X_n$$

holds with probability one.

Example 8.11.1. Let $\{Y_j\}$ be a sequence of independent random variables with $E(|Y_j|) < \infty$ $(j = 1, 2, 3, \ldots)$ and $E(Y_j) = 0$ for $j > 1$. Define

$$X_n = \sum_{j=1}^{n} Y_j \qquad (n = 1, 2, 3, \ldots).$$

The sequence $\{X_n\}$ is a martingale. Indeed, in virtue of the independence of Y_{n+1} and $(X_1, \ldots, X_n)$, we get

(8.11.2) $E(X_{n+1} \mid x_1, \ldots, x_n) = E[(X_n + Y_{n+1}) \mid x_1, \ldots, x_n]$

$$= E(X_n \mid x_1, \ldots, x_n) + E(Y_{n+1}) = X_n.$$

Example 8.11.2. Let x_1 (a constant) denote the fortune of a gambler, playing some game of chance, at the beginning of the first play, and let X_{n+1} $(n = 1, 2, 3, \ldots)$ be a random variable denoting the fortune of the gambler after the nth play. It is commonly accepted to call a chance game *fair* if the mean value of the gambler's fortune after each play equals his fortune at the beginning of this play. In other words (assuming $E(|X_n|) < \infty$) a chance game is fair if

$$E(X_{n+1} \mid x_1, \ldots, x_n) = X_n \qquad (n = 1, 2, 3, \ldots)$$

with probability one. Thus a fair game can be defined as a martingale.

The following is a definition of a martingale for a stochastic process with continuous time.

Definition 8.11.2. A stochastic process $\{X_t, t \in I\}$ is called a *martingale* if $E(|X_t|) < \infty$ for all $t \in I$ and if, for $n \geqslant 1$ and arbitrary $t_1 < \ldots < t_n$ $(t_j \in I, j = 1, \ldots, n)$, the relation

$$E(X_{t_{n+1}} \mid x_{t_1}, \ldots, x_{t_n}) = X_{t_n}$$

holds with probability one.

Example 8.11.3. Let $\{X_t, 0 \leqslant t < \infty\}$ be a stochastic process with independent increments satisfying the equality $P(X_0 = 0) = 1$. If, for any $t > 0$, $E(|X_t| < \infty)$ and $E(X_t) = $ constant, the process $\{X_t, 0 \leqslant t < \infty\}$ is for $t > 0$ a martingale. Indeed, we have

$$X_{t_{n+1}} = X_{t_1} + (X_{t_2} - X_{t_1}) + \ldots + (X_{t_n} - X_{t_{n-1}}) + (X_{t_{n+1}} - X_{t_n}),$$

where the $X_{t_1}, \ldots, X_{t_{n+1}} - X_{t_n}$ are independent.
We now obtain

$$E(X_{t_{n+1}} \mid x_{t_1}, \ldots, x_{t_n}) = E\{[X_{t_n} + (X_{t_{n+1}} - X_{t_n})] \mid x_{t_1}, \ldots, x_{t_n}\}$$
$$= E(X_{t_n} \mid x_{t_1}, \ldots, x_{t_n}) + E(X_{t_{n+1}} - X_{t_n}) = X_{t_n}$$

with probability one

8.12   ADDITIONAL REMARKS

Despite its length, this chapter is only an introduction to the theory of stochastic processes. The reader who would like to have a general view of the theory of stochastic processes and its applications should read the interesting paper by Feller [6]. The reader who would like to know the fundamental results of the theory of stochastic processes and the modern mathematical apparatus for this theory is referred to the monograph by Doob [5]. In addition to the many papers on Markov processes previously mentioned, we recommend in particular two papers by Hunt [1, 2] and a paper by Blumenthal [1]. Modern treatment of the foundations and developments of the theory of Markov processes can be found in the recent monographs by Dynkin [3] and Chung [5]. Generalizations of Markov processes, namely, semi-Markov processes and renewal processes, were dealt with by Lévy [10], Smith [1], and Pyke [2, 3].

In the book by Khintchin [2] and in the papers of Erdös and Kac [1, 2], LeCam [2], Donsker [1, 2], Prohorov [2, 3], Skorohod [1], Bartoszyński [1, 2], and Billingsley [1], and also in Part 2 of this book (Section 10.11, C) the reader will find a discussion of the connections between the limit distributions for sums of random variables and the distributions which appear in the theory of stochastic processes.

We mention briefly some investigations concerning properties of the functions which are realizations of a stochastic process, namely, their continuity, the character of their discontinuities, and so on.

Wiener [1] showed that the only process $\{X_t, 0 \leqslant t < \infty\}$ with independent and homogeneous increments which satisfies (8.9.26) and $P(X_0 = 0) = 1$ whose realizations are continuous functions, with probability one, is the Brownian motion process. Further study of realizations of a Wiener process is due to Lévy. He established [8] that these realizations are irregular in nature, in particular that almost all of them are not of bounded variation. Another striking irregularity of the realizations of a Wiener process, namely, that almost all of them are nowhere increasing, has recently been established by Dvoretzky, Erdös, and Kakutani [1]. The distributions of functionals defined on the realizations of the Brownian motion process and related questions have been considered by Kac [2, 4], Doob [4], Lévy [4, 6], Fortet [1], and others. Lévy [2] has shown that for every process X_t with homogeneous and independent increments there exists a function $f(t)$ such that almost all realizations of the process $X_t - f(t)$ are bounded and have right- and left-hand limits at each point. We stress the investigations of Doeblin [1, 3], Doob [2, 3, 5], and Lévy [5, 7], which concern the properties of realizations of Markov processes (see also Chung [4]).

We mention that conditions under which almost all realizations of a Markov process are step functions were given by Doeblin [3]; conditions under which they are continuous or have right- and left-hand limits at each point were given by Dynkin [1] and Kinney [1] (see also Skorohod [2]). Urbanik [1] investigated the limit properties of realizations of Markov processes as $t \to \infty$. Conditions under which almost all realizations of a Markov process with a finite number of states have a finite expected number of discontinuity points were given by Dobrushin [2]. In the monograph by Doob [5] (Chapter XI, 9), a condition is given under which almost all realizations of a wide sense stationary process are absolutely continuous. Belayev [1] showed that almost all realizations of a Gaussian stationary process with continuous spectral function are unbounded in the interval $(-\infty, +\infty)$. For further results on the realizations of Gaussian processes, see Belayev [3], Dobrushin [6], and Ciesielski [1]. Belayev [2] has also given a condition under which almost all realizations of a wide sense stationary process are entire functions.

We mention also a paper by McKean [1] and two papers by Blumenthal and Getoor [1, 2] dealing with properties of realizations of a certain class of stochastic processes with independent increments. Properties of realizations of martingales were thoroughly studied by Doob [5, 7, 9].

For arbitrary stochastic processes, without the assumption that they are Markov processes or stationary, we note the following results: Kolmogorov (see Slutsky [2]) has given a condition under which almost all realizations of a stochastic process are continuous; Tschentsov [1] (see also Dynkin [1]) gave a condition under which almost all realizations have right- and left-hand limits; Dobrushin [5] and Fisz [7] have given conditions under which almost all realizations are either continuous or have discontinuity points with at most one one-sided limit. Fisz [7] has given conditions under which almost all realizations are step functions with a finite expected number of discontinuity points, giving at the same time the formula for the expected number of discontinuity points; finally, Loève (see Lévy [4], Complement written by Loève) and Belayev [2] have given conditions under which almost all realizations of a stochastic process whose random variables have finite second-order moments are analytic in the neighborhood of some point. (See Problems 8.21 to 8.31.)

In the preceding, we have used the equivalent formulations "almost all" and "with probability one" without a precise explanation as to what is to be understood by them; this explanation would require a discussion of the construction of probability measures in function spaces. A detailed discussion of these questions can be found in works by Kolmogorov [7] and Doob [1, 5]. Moreover, one should add that in the given information we have omitted the assumption that the processes under consideration are separable. The precise definition of this notion can be found in Doob [5] (Chapter II, 2); roughly speaking, the separability of the process X_t means that there exists a countable set of points $\{t_i\}$ $(i = 1, 2, \ldots)$ such that the supremum or infimum of almost every realization of X_t in an arbitrary open interval I is equal to the supremum or infimum of this realization over the subset of points t_i which belong to the interval I.

Problems and Complements

8.1. Show that the assertion of theorem 8.3.1 remains true if condition III is replaced by the conditions

$$\lim_{t \to 0} \frac{W_1(t)}{1 - W_0(t)} = 1 \quad \text{and} \quad \sum_{i=0}^{\infty} W_i(t) = 1 \text{ for any } t > 0.$$

8.2. Show that the assertion of theorem 8.3.1 remains true if condition III is replaced by the assumption that the number of signals in an arbitrary finite time interval $[0, t]$ is finite and that at any point at most one signal can occur.

8.3. Let $\{X_t, 0 \leqslant t < \infty\}$ be the Poisson process with probability function (8.3.4) and let τ_k denote the (random) point of appearance of the kth signal. (a) Prove that for $j = 2, 3, \ldots$ and for arbitrary $t_1, \ldots, t_j$ and T such that $0 \leqslant t_1 < t_2 < \ldots < t_j \leqslant T$

$$P(\tau_1 < t_1, \ldots, \tau_j < t_j \mid X_T = j) = \frac{j!}{T^j} \int_0^{t_1} \int_{t_1}^{t_2} \ldots \int_{t_{j-1}}^{t_j} d\tau_j \ldots d\tau_1.$$

(Compare with formula (10.7.1) when $j_k = k$ and $f(x) = 1/T$ for $0 \leqslant x \leqslant T$.)
(b) Prove that the time differences $U_k = \tau_k - \tau_{k-1}$ $(k = 1, 2 \ldots)$ are independent and have the same exponential distribution given by the formula

$$G(\tau) = P(U_k < \tau) = \begin{cases} 0 & \text{for } \tau \leqslant 0, \\ 1 - \exp(-\lambda\tau) & \text{for } \tau > 0. \end{cases}$$

Remark. Any signal process $\{X_t, 0 \leqslant t < \infty\}$ such that the U_k are independent and have a common distribution function $G(\tau)$ is called a *recurrent process*; thus the Poisson process is a particular case of a recurrent process. (See Takács [4, 6]).

8.4. Let $\{X_t, 0 \leqslant t < \infty\}$ be a signal process which satisfies conditions I and II of theorem 8.3.1 and let the realizations of this process be functions of t, continuous from the right. Show that relation (8.3.4) is satisfied if and only if almost all the realizations of this process have jumps equal only to one (Florek, Marczewski, and Ryll-Nardzewski [1].)

8.5. Let $\{X_t, 0 \leqslant t < \infty\}$ be a signal process satisfying conditions I and II of theorem 8.3.1 and let $P(X_0 = 0) = 1$. Show that (a) if, for some integer $l \geqslant 1$, we have

$$\lim_{t \to 0} \frac{W_l(t)}{t} = \lambda_l \qquad (\lambda_l > 0)$$

and

$$\lim_{t \to 0} \frac{1 - W_0(t) - W_i(t)}{t} = 0,$$

where $W_i(t)$ is given by (8.3.1), then for every $0 \leqslant t < \infty$ and $i = 0, 1, \ldots$

$$P(X_t = il) = \exp\left(-\lambda_i t\right) \frac{(\lambda_i t)^i}{i!}.$$

(b) if

$$\lim_{t \to 0} \frac{W_1(t)}{t} = \lambda_1, \qquad \lim_{t \to 0} \frac{W_2(t)}{t} = \lambda_2,$$

$$\lim_{t \to 0} \frac{1 - W_0(t) - W_1(t) - W_2(t)}{t} = 0,$$

where $\lambda_1 > 0$ and $\lambda_2 > 0$, then for every $0 \leqslant t < \infty$ and $i = 0, 1, \ldots$

$$P(X_t = i) = \exp\left[-(\lambda_1 + \lambda_2)t\right] \sum_{r_1 + 2r_2 = i} \frac{(\lambda_1 t)^{r_1}(\lambda_2 t)^{r_2}}{r_1! \, r_2!}.$$

8.6. Let $\{X_t, 0 \leqslant t < \infty\}$ be a signal process which satisfies conditions I and II of 8.3 and $P(X_0 = 0) = 1$. Suppose that for every $t > 0$

$$\sum_{i=0}^{\infty} W_i(t) = 1,$$

where $W_i(t)$ is given by (8.3.1). Show that for every $0 \leqslant t < \infty$ and $i = 0, 1, 2, \ldots$

$$W_i(t) = \exp\left(-t \sum_{k=1}^{\infty} \lambda_k\right) \sum_{\substack{r_1 + 2r_2 + \ldots + lr_l = i \\ r_1 \geqslant 0, \ldots, r_l \geqslant 0}} \frac{(\lambda_1 t)^{r_1} \ldots (\lambda_l t)^{r_l}}{r_1! \ldots r_l!}$$

and

$$\lim_{t \to 0} = \frac{W_k(t)}{t} = \lambda_k \geqslant 0 \qquad (k = 1, 2, 3, \ldots),$$

$$\lim_{t \to 0} \frac{1 - W_0(t)}{t} = \sum_{k=1}^{\infty} \lambda_k < \infty.$$

(Jánossy, Rényi, and Aczél [1].) Note that the existence of the intensities λ_k is not assumed here.

8.7. (a) Show that the process $\{X_t, 0 \leqslant t < \infty\}$ considered in the preceding problem can be represented in the form

$$X_t = \sum_{k=1}^{\infty} k \, Y_{tk},$$

where $\{Y_{tk}, 0 \leqslant t < \infty\}$ $(k = 1, 2, \ldots)$ is the Poisson process with intensity λ_k, and for every $0 \leqslant t < \infty$ the random variables $Y_{t1}, Y_{t2}, \ldots$ are independent.

(b) Show that the characteristic function $\phi_t(u)$ of the random variable X_t has the form

$$\phi_t(u) = \exp\left[t \sum_{k=1}^{\infty} \lambda_k(e^{iuk} - 1)\right] = \prod_{k=1}^{\infty} \exp\left[\lambda_k t(e^{iuk} - 1)\right].$$

8.8. Let $\{X_t, 0 \leqslant t < \infty\}$ be the signal process of Problem 8.5(b) and let the length of life of the signal have the exponential distribution. Let $\{Y_t, 0 \leqslant t < \infty\}$

denote the number of signals active at the moment t and let $c_j(t) = P(Y_t = j)$. We assume that $c_i(0) = 1$. Find: $c_j(t)$, $E(Y_t)$, and $D^2(Y_t)$.

8.9. Prove that Erlang's formulas (8.5.17) are valid if the assumption that the length of life T of the signal has the exponential distribution is replaced by the assumption that T has any distribution with finite expected value (Fortet [2], Sevastianov [2]).

8.10. Let $\{X_t, 0 \leqslant t < \infty\}$ be a Poisson process with intensity $\lambda > 0$. Let T_r denote the waiting time for the occurrence of the rth $(r = 1, 2, 3, \ldots)$ signal. Show that T_r has a gamma distribution given by (5.8.6) with x replaced by t, b by λ and p by r.

8.11. Let q_i and q_{ij} be arbitrary sequences of non-negative real numbers and let

(*)
$$q_i = \sum_{j \neq i} q_{ij} \qquad (i = 1, 2, \ldots).$$

Show that q_i and q_{ij} are the intensities of some homogeneous Markov process (Doob [3]).

8.12. Let us consider a homogeneous Markov process with a countable number of states with the transition probability function $p_{ij}(t)$ and finite intensities q_i and q_{ij} satisfying (*) of the preceding problem. Let us denote by $_np_{ij}(t)$ the probability of passing from the state i to the state j during the time t, provided n changes of state occur during that time. Show (Doob [2]) that the following formulas hold:

$$_0p_{ij}(t) = \begin{cases} \exp(-q_it) & \text{if } j = i, \\ 0 & \text{if } j \neq i, \end{cases}$$

$$_{n+1}p_{ij}(t) = \sum_{k \neq i} \int_0^t \exp[-q_j(t - s)] q_{ik} \, _np_{kj}(s) \, ds.$$

8.13. Show that if relation (8.9.4) is replaced by

$$\lim_{\Delta t \to 0} \frac{1}{\Delta t} \int_{|u| \geqslant \varepsilon} f(t, t + \Delta t, u) \, du = 0,$$

then the assertion of theorem 8.9.1 remains true.

8.14. Let $\{X_t, t \in I\}$, I finite, be a process with independent and homogeneous increments and let $P(X_0 = 0) = 1$. Then, for any $t \in I$, the random variable X_t has an infinitely divisible distribution (see Problem 5.25).

8.15. Show that if the sequence $\{U_n\}$ of random variables converges in quadratic mean to the random variable U, $\{U_n\}$ converges to U stochastically.

In Problems 8.16 and 8.17 the convergence in relations marked with (*) should be understood as convergence in quadratic mean.

8.16. Prove the following theorem: Let $\{X_t, t = 0, \pm1, \pm2, \ldots\}$ be a sequence stationary in the wide sense and let $m = 0$, $\sigma = 1$. Let $R(\tau)$ be the correlation function of the process X_t. The relation

(*)
$$\lim_{N \to \infty} \frac{1}{N + 1} \sum_{t=0}^{N} X_t = 0$$

holds if and only if (Slutsky [2])

$$\lim_{N \to \infty} \frac{1}{N + 1} \sum_{\tau=0}^{N} R(\tau) = 0.$$

8.17. Let $\{X_t, -\infty < t < +\infty\}$ be a process stationary in the wide sense and let $m = 0$, $\sigma = 1$. Let

(*) $$\int_0^T X_t \, dt = \lim_{N \to \infty} \frac{T}{N} \sum_{k=1}^N X_{kT/N}.$$

Prove that the relation

(*) $$\lim_{T \to \infty} \frac{1}{T} \int_0^T X_t \, dt = 0$$

holds if and only if

$$\lim_{T \to \infty} \frac{1}{T} \int_0^T R(\tau) \, d\tau = 0.$$

8.18. Let $\{X_t, t = 0, \pm 1, \pm 2, \ldots\}$ be a sequence stationary in the wide sense with the covariance function $R(\tau) = a^{|\tau|}$ $(\tau = 0, \pm 1, \pm 2, \ldots)$ with a real and $|a| < 1$. Find the spectral density of this process.

8.19. The spectral density of the sequence $\{X_t, t = 0, \pm 1, \pm 2, \ldots\}$, stationary in the wide sense, has the form

$$f(\lambda) = \frac{c}{2\pi} \frac{(e^{i\lambda} - b)(e^{-i\lambda} - b)}{(e^{i\lambda} - a)(e^{-i\lambda} - a)},$$

where a and b are real and $|a| < 1$, $|b| < 1$. Show that the corresponding covariance function has the form

$$R(\tau) = \begin{cases} c \, \dfrac{1 - 2ab + b^2}{1 - a^2} & \text{for } \tau = 0, \\[2ex] c \, \dfrac{(a - b)(1 - ab)}{1 - a^2} \, a^{|\tau|-1} & \text{for } \tau = \pm 1, \pm 2, \ldots \end{cases}$$

8.20. Let $\{X_t, -\infty < t < +\infty\}$ be a stationary process in the wide sense.
(a) Find the spectral density of this process if

$$R(\tau) = e^{-a|\tau|},$$

where $a > 0$.
(b) Find the spectral density of this process if

$$R(\tau) = e^{-a|\tau|} \cos b\tau \qquad (a > 0).$$

8.21. Let $\{X_t, 0 \leqslant t < \infty\}$ be a separable Markov process and let the relation

$$\lim_{\Delta t \to 0} P(X_{t+\Delta t} - X_t \neq 0 \mid X_t = x) = 0$$

hold uniformly with respect to t $(0 \leqslant t < \infty)$ and with respect to x $(-\infty < x < +\infty)$. Then almost all realizations of this process are step functions with a finite number of discontinuity points in every finite interval (Doeblin [3]).

8.22. Let $\{X_t, t \in I\}$ be a separable Markov process, where I is a closed finite interval. Let

$$C(h, \varepsilon) = \sup_{\substack{t, t+\Delta t \in I \\ -\infty < x < +\infty}} P(|X_{t+\Delta t} - X_t| > \varepsilon \mid X_t = x).$$

(a) If for every $\varepsilon > 0$

$$\lim_{h \to 0} \frac{C(h, \varepsilon)}{h} = 0,$$

then almost all realizations of the process are continuous in I (Dynkin [1], Kinney [1]. The assertion holds under a weaker assumption. See Fisz [7]).

(b) If for every $\varepsilon > 0$

$$\lim_{h \to 0} C(h, \varepsilon) = 0,$$

then almost all realizations of the process have limits from both sides at each point $t \in I$ (Kinney [1]).

8.23. (a) Let $\{X_t, 0 \leqslant t \leqslant 1\}$ be a separable stochastic process. If the relation

(*) $$E(|X_{t+\alpha} - X_t|^\beta) < C \, |\Delta t|^{1+\alpha},$$

where $\beta > 0$, $\alpha > 0$, and C are constant, holds for arbitrary points $t + \Delta t$ and t in $(0, 1]$, then for any $\gamma < \alpha/\beta$ the relation

(**) $$P\left(\lim_{\Delta t \to 0} \frac{X_{t+\Delta t} - X_t}{|\Delta t|^\gamma} = 0, \quad 0 \leqslant t \leqslant 1\right) = 1$$

holds uniformly in t. (The continuity of almost all realizations of X_t under (*) has been proved by Kolmogorov and published in Slutsky [2]. An assertion equivalent to (**) is due to Loève [4], second edition).

(b) Deduce from (**) that if $\beta < \alpha$, then

$$P(X_t = \text{constant}, \quad 0 \leqslant t \leqslant 1) = 1.$$

(c) Show that for the Wiener process relation (**) holds for any $\gamma < \tfrac{1}{2}$.

8.24. Let $\{X_t, 0 \leqslant t \leqslant 1\}$ be a separable stochastic process. Assume that for arbitrary t_1, t_2, and t_3 $(t_1 < t_2 < t_3)$ in the interval $[0, 1]$

$$E[|X_{t_1} - X_{t_2}|^p \, |X_{t_2} - X_{t_3}|^q] < C \, |t_1 - t_3|^{1+r},$$

where $p > 0, q > 0, r > 0$, and C does not depend on t. Then almost all realizations of the process have both one-sided limits at each point $t \in [0, 1]$ (Tschentsov [1]).

8.25. Let $\{X_t, t \in I_0\}$ be a separable stochastic process, where I_0 is a finite closed interval. Let X_I denote the increment of the process in the interval $I \subset I_0$. Let $a(I) = P(X_I \neq 0)$, $b(I, \varepsilon) = P(|X_I| > \varepsilon)$ and let $|I|$ denote the length of the interval I. Assume that the relation

(*) $$\lim_{|I| \to 0} b(I, \varepsilon) = 0$$

holds for every $\varepsilon > 0$, and

(**) $$\overline{\lim_{n \to \infty}} \sum_{k=1}^{n} a(I_{nk}) < \infty$$

as $\max_{1 \leqslant k \leqslant n} |I_{nk}| \to 0$, where $\{I_{nk}\}$ is a partition of the interval I_0 into the disjoint intervals I_{nk} $(k = 1, 2, \ldots n)$. Then almost all realizations of the process are step functions in I_0, and the expected value of the number of discontinuity points exists and is equal to the limit on the left-hand side of (**). Moreover, at each point $t \in I_0$ the realizations are continuous, with probability one (Fisz [3, 7]).

8.26. Let $\{X_t, t \in I_0\}$ be a separable stochastic process. Almost no realizations of the process have discontinuities of the first kind[1] if, for every $\varepsilon > 0$,

$$(*) \qquad \varlimsup_{n \to \infty} \sum_{k=1}^{n} b(I_{nk}, \varepsilon) = 0$$

as $\max_{1 \leqslant k \leqslant n} |I_{nk}| \to 0$, where $\{I_{nk}\}$ has the same meaning as in Problem 8.25 (Fisz [7]).

8.27. Let $\{X_t, t \in I_0\}$ be either a separable homogeneous Markov process or a separable martingale. If relation (*) of Problem 8.26 is satisfied, then almost all realizations of the process are continuous. (For the Markov case, see Fuchs [1]; for the martingale case, see Fisz [7].)

8.28. Let $\{X_t, t \in I_0\}$ be a separable process with independent increments and let

$$\lim_{|I| \to 0} a(I) = 0.$$

Then almost all realizations of this process satisfy all the assertions of Problem 8.25.

8.29. Almost all realizations of a real, separable, Gaussian stationary process $\{X_t, -\infty < t < +\infty\}$, whose spectral distribution function is continuous, are unbounded (Belayev [1]).

8.30. Let $\{X_t, -\infty < t < +\infty\}$ be a real, stationary, Gaussian, continuous separable process. Then (Dobrushin [6]) either almost all realizations are continuous or there exists a $\beta > 0$ such that, for every t_0, almost all realizations satisfy the relation

$$\varlimsup_{t \to t_0} X_t - \varliminf_{t \to t_0} X_t \geqslant \beta.$$

8.31. Let $\{X_t, -\infty < t < +\infty\}$ be a real, separable process, stationary in the wide sense and continuous with $E(X_t) = 0$, $D^2(X_t) = 1$. Show that if for $\tau \to 0$ the correlation function $R(\tau)$ satisfies the relation[2]

$$R(\tau) = 1 - 0(|\tau|^{1+\delta}),$$

where $\delta > 0$, then almost all realizations are continuous.

Hint: Use Problem 8.23.

[1] The function $f(t)$ has a discontinuity of the first kind at the point t if both limits $f(t - 0)$ and $f(t + 0)$ exist and are not equal.

[2] In other words, there exists a constant $c > 0$ such that for $\tau \to 0$, $|R(\tau)| < C|\tau|^{1+\delta}$.

Mathematical
Statistics

CHAPTER 9

Sample Moments and
Their Functions

9.1 THE NOTION OF A SAMPLE

In the first part of this book we discussed probability theory. We created a mathematical model in which the abstract notion of probability corresponds to the notion of frequency of random phenomena which can be observed in many domains of reality. Thus the notion of probability is related to the notion of frequency of random phenomena in the same way that the notion of geometrical figures in Euclidean geometry is related to the corresponding notion of figures in reality. Hence the significance of probability theory for statistical problems in which we investigate certain regularities of random phenomena by observing their frequency is the same as the significance of geometry for geodesics.

The aim of Part 2 is to present methods of solving many statistical problems by means of probability theory. In doing so, it will become necessary to enlarge our knowledge of probability theory, mainly in domains close to mathematical statistics.

The statistical problems we shall investigate are usually problems in which, from knowledge of some characteristics of a suitably selected part of a collection of elements, we draw conclusions about the characteristics of the unknown part.

The collection of elements under investigation is called the *population*.

We might be interested in different characteristics of the elements of the population under investigation. When we say that the population has the distribution $F(x)$, we mean that we are investigating a characteristic X of elements of this population and that this characteristic X is a random variable with the distribution function $F(x)$.

In statistics, a collection of some elements of a population is called a *sample*. We use the term *sample* to denote a sequence of values of the characteristics under investigation for some elements of the population. We now give a mathematical definition of a *random sample*.

Let the characteristic X of elements of a population be a random variable with the distribution function $F_0(x)$. Let us consider the population whose elements are all the possible j tuples ($j = 2, 3, \ldots$) that can be formed from the elements of the population. The elements of this j-dimensional population are characterized by the random vector $(X_1', \ldots, X_j')$, where X_k' ($k = 1, \ldots, j$) is the value of the characteristic X of the kth element of the j-tuple being considered. We suppose that for $k = 1, 2, \ldots$ and arbitrary $x_1, \ldots, x_k$ there exists the conditional distribution function

$$F_k(x \mid x_1, \ldots, x_k) = P(X_{k+1}' < x \mid X_1' = x_1, \ldots, X_k' = x_k).$$

Using a certain method, say M, we select from the population collections of n elements, and we observe the values $x_1, x_2, \ldots, x_n$ of the characteristic X of the chosen elements. Let $(X_1, \ldots, X_n)$ denote the *observed* random vector. Then X_k ($k = 1, 2, \ldots, n$) is a random variable, whose possible values are the observations x_k of the kth element of each of the possible samples of n elements selected from the population by the method M.

Definition 9.1.1. The method M of choosing elements of the sample is called a *random method* if

1. for every x we have

(9.1.1) $P(X_1 < x) = F_0(x).$

2. for $k = 1, 2, \ldots, n - 1$ and arbitrary $x_1, \ldots, x_k$ we have the equalities

(9.1.2) $P(X_{k+1} < x \mid X_1 = x_1, \ldots, X_k = x_k) = F_k(x \mid x_1, \ldots, x_k).$

The samples $x_1, \ldots, x_n$ obtained by a random method are called *random samples*.

Unless otherwise stated, we always understand the word sample to mean a random sample.

A geometrical interpretation is convenient; all possible values of the joint random variable $(X_1, \ldots, X_n)$ form a set of points in n-dimensional Euclidean space. A sample with n elements is a point in this n-dimensional space. The set of all possible random samples of n elements is called the *sample space*.

We return to the problems of randomness in Chapter 14.

Definition 9.1.2. A random sample is called *simple* if the random variables $X_1, \ldots, X_n$ are independent.

Example 9.1.1. Every farm in a country conducting a census has its own IBM card. One of the characteristics under investigation is the area of arable soil in acres. This characteristic is denoted by X. The set of all farms is divided into seven groups according to this characteristic. Thus X may take on one of seven values 1 to 7, according to the group to which the given farm belongs.

Here the population consists of all the IBM cards. Let N denote their number. We select $n = 10$ cards from this population in a random way, but before choosing the next card we put back the one drawn. The probability that the characteristic X takes on the value i ($i = 1, 2, \ldots, 7$) is constant while we perform the drawing. The observed values of the random variable X in ten IBM cards selected, that is, the collection of numbers $x_1, \ldots, x_{10}$, is the value of a ten-dimensional random variable. This observed value of the ten-dimensional random variable forms a sample. Here the sample space consists of all possible collections of numbers $x_1, \ldots, x_{10}$. Since x_k ($k = 1, \ldots, 10$) is one of the numbers $1, \ldots, 7$, the sample space consists of 7^{10} points.

9.2 THE NOTION OF A STATISTIC

Definition 9.2.1. A random variable which is a function of the observed random vector $(X_1, \ldots, X_n)$ is called a *statistic*.

Example 9.2.1. Let us return to example 9.1.1. Let $(X_1 \ldots, X_{10})$ denote the random vector considered in this example. The random variable $(X_1 + \ldots + X_{10})/10$ is a statistic.

The problem of finding the distributions of statistics is one of the basic problems in mathematical statistics.

We consider here two types of problems. The first is to find for every n the distribution function of some statistic $Z_n = Z(X_1, \ldots, X_n)$. Such a distribution is called an *exact distribution* of a statistic. The knowledge of exact distributions of statistics is of great importance in statistical problems where the number of observations is small. Then we speak of *small samples*.

In problems of the second type we do not investigate the distribution of a statistic for particular values of n, but we are interested only in the limit distribution of the statistic Z_n as $n \to \infty$. Limit distributions of statistics are applied to statistical investigations when the number of observations is large. Then we speak of *large samples*.

There is no general criterion which allows us to determine whether the sample under investigation is large or not. This depends on the statistics being considered. A sample which might be considered large for one statistic may appear to be insufficiently large for another.

9.3 THE DISTRIBUTION OF THE ARITHMETIC MEAN OF INDEPENDENT NORMALLY DISTRIBUTED RANDOM VARIABLES

The problem of finding the exact distributions of statistics is usually complicated. However, there are methods which provide a solution to this problem in some cases of frequent importance.

First we investigate the distributions of some statistics when the characteristic X of the population has a normal distribution with density function

(9.3.1) $$f(x) = \frac{1}{\sigma\sqrt{2\pi}} \exp\left[-\frac{(x-m)^2}{2\sigma^2}\right],$$

where m and σ denote, respectively, the expected value and the standard deviation of X.

Let us consider the *sample mean* defined as

(9.3.2) $$\bar{X} = \frac{1}{n}\sum_{k=1}^{n} X_k,$$

where the random variables X_k ($k = 1, \ldots, n$) are independent and have the same density (9.3.1). We apply the method of characteristic functions to find the distribution of this statistic. As we know from (5.7.4), the characteristic function of X is

(9.3.3) $$\phi(t) = \exp\left(itm - \frac{\sigma^2 t^2}{2}\right).$$

The random variables X_k are independent; hence by (4.4.3) and (4.2.15) the characteristic function $\phi_1(t)$ of $\bar{X}$ is

(9.3.4) $$\phi_1(t) = \exp\left(itm - \frac{\sigma^2 t^2}{2n}\right).$$

Expression (9.3.4) is the characteristic function of a random variable with normal distribution $N(m; \sigma/\sqrt{n})$; hence the density of $\bar{X}$ is

(9.3.5) $$f_1(\bar{x}) = \frac{\sqrt{n}}{\sigma\sqrt{2\pi}} \exp\left[-\frac{n}{2}\cdot\frac{(\bar{x}-m)^2}{\sigma^2}\right].$$

As we see, $\bar{X}$ has the same expected value as X, but it is more concentrated around this value.

Example 9.3.1. The random variables X_k ($k = 1, \ldots, 16$) are independent and have the same density

(9.3.6) $$f(x) = \frac{1}{2\sqrt{2\pi}} \exp\left[-\frac{1}{2}\cdot\frac{(x-1)^2}{4}\right].$$

We want to find the distribution of

$$\bar{X} = \frac{1}{16}\sum_{k=1}^{16} X_k.$$

By (9.3.5) $\bar{X}$ has the density

$$f_1(\bar{x}) = \frac{2}{\sqrt{2\pi}} \exp\left[-2(\bar{x}-1)^2\right].$$

The standard deviation of $\bar{X}$ is 0.5. Let us find the probability that $0 \leqslant \bar{X} \leqslant 2$. We have

$$P(0 \leqslant \bar{X} \leqslant 2) = P\left(-2 \leqslant \frac{\bar{X} - 1}{0.5} \leqslant 2\right).$$

From tables of the normal distribution we obtain

$$P(0 \leqslant \bar{X} \leqslant 2) \simeq 0.9544.$$

For comparison, let us compute the probability that X takes on a value from this interval. Then we have

$$P(0 \leqslant X \leqslant 2) = P\left(-\frac{1}{2} \leqslant \frac{X-1}{2} \leqslant \frac{1}{2}\right) \simeq 0.3830.$$

As we see, $\bar{X}$ is much more concentrated than X.

This example may be interpreted in practice as follows: we have a population and we are interested in the characteristic X of elements of this population. This characteristic has the normal distribution given by (9.3.6). We select from the population independent simple samples of 16 elements each, observe the values of X for the elements selected, and compute $\bar{X}$. It turns out that if we select a large series of such samples with 16 elements each, in approximately 95 cases out of 100 we obtain values of $\bar{X}$ which are non-negative and less or equal 2. If we choose a single element a large number of times, we obtain values of X contained in this interval only 38 times in 100, approximately.

From (9.3.5) we see that the distribution of the statistic $\bar{X}$ depends on n. Figure 5.7.1 shows the graphs of the densities of the normal distributions $N(0; 1)$, $N(0; 0.5)$ and $N(0; 0.25)$. At the same time they are the densities of $\bar{X}$ given by (9.3.2) for $n = 1$, $n = 4$, and $n = 16$, where the random variables X_k have densities (9.3.1) with $m = 0$, $\sigma = 1$.

9.4 THE χ^2 DISTRIBUTION

A   Let the independent random variables X_k $(k = 1, \ldots, n)$ have the same normal density

$$(9.4.1) \qquad f(x) = \frac{1}{\sigma\sqrt{2\pi}} \exp\left(-\frac{1}{2} \cdot \frac{x^2}{\sigma^2}\right).$$

The expression

$$(9.4.2) \qquad \chi^2 = \sum_{k=1}^{n} X_k^2$$

is called the *statistic χ^2*

Example 9.4.1. A characteristic X of elements of a certain population has a normal distribution with the expected value $E(X) = 0$. We choose from this population a simple sample of n elements and we observe the values of X of the elements selected.

Denote these values by $x_1, \ldots, x_n$. Compute the expression

$$\sum_{k=1}^{n} x_k^2.$$

This expression is the observed value of the statistic χ^2.

To find the distribution of χ^2, we first find the distribution of $Y = X^2$, where X has the density (9.4.1). By (2.4.14) the density of Y is

$$(9.4.3) \quad f_1(y) = \begin{cases} \dfrac{1}{2\sqrt{y}} [f(\sqrt{y}) + f(-\sqrt{y})] = \dfrac{1}{\sigma\sqrt{2\pi}} y^{-1/2} \exp\left(-\dfrac{1}{2} \cdot \dfrac{y}{\sigma^2}\right) \\ \qquad\qquad\qquad\qquad\qquad\qquad\qquad \text{for } 0 < y < \infty, \\[2mm] 0 \qquad\qquad\qquad\qquad\qquad\qquad\qquad\quad \text{for } y \leqslant 0. \end{cases}$$

Thus the random variable Y has the gamma distribution (5.8.6) with $p = \frac{1}{2}$ and $b = \frac{1}{2}\sigma^2$. Since the addition theorem holds for random variables with the gamma distribution (see Section 5.8), the random variable χ^2 given by (9.4.2) also has the gamma distribution with $p = n/2$ and $b = \frac{1}{2}\sigma^2$; hence its density is

$$(9.4.4) \quad g_n(u) = \begin{cases} \dfrac{1}{2^{n/2}\sigma^n\Gamma(\frac{1}{2}n)} \exp\left(-\dfrac{u}{2\sigma^2}\right) u^{n/2-1} & \text{for } u > 0, \\[2mm] 0 & \text{for } u \leqslant 0, \end{cases}$$

where u is the value of χ^2.

The distribution of χ^2 was obtained by Helmert [1].

From (5.8.10) we obtain the expected value and the standard deviation of χ^2,

$$(9.4.5) \qquad\qquad m_1 = n\sigma^2, \qquad \mu_2 = 2n\sigma^4.$$

The parameter n in (9.4.4) is called the *number of degrees of freedom*, which corresponds to the fact that χ^2 is the sum of n independent random variables.

For arbitrary $u_1 \geqslant 0$ and $u_2 > 0$, where $u_1 < u_2$, we have

$$P(u_1 \leqslant \chi^2 < u_2) = \int_{u_1}^{u_2} g_n(u)\, du,$$

where $g_n(u)$ is given by (9.4.4).

In the applications of the statistic χ^2, which we discuss in more detail later, we use the expression

$$(9.4.6) \quad P(\chi^2 \geqslant u_0) = \dfrac{1}{2^{n/2}\sigma^n\Gamma(\frac{1}{2}n)} \int_{u_0}^{\infty} \exp\left(-\dfrac{u}{2\sigma^2}\right) u^{n/2-1}\, du,$$

where $u_0 > 0$. The tables of the χ^2 distribution usually give the values of (9.4.6) with $\sigma = 1$ for different values of u_0 and n. To use such tables, we have to replace the expression $P(\chi^2 \geqslant u_0)$ by $P(\chi^2 \geqslant z_0\sigma^2)$, where $u_0 = z_0\sigma^2$. In fact, if we replace the variable u in (9.4.4) by $z = u/\sigma^2$, we obtain

$$(9.4.7) \qquad g_n(z) = \begin{cases} \dfrac{1}{2^{n/2}\Gamma(\frac{1}{2}n)} e^{-z/2} z^{n/2-1} & \text{for } z > 0, \\ 0 & \text{for } z \leqslant 0. \end{cases}$$

Expression (9.4.7) is the density of the statistic χ^2 for $\sigma = 1$.

In Fig. 9.4.1 the densities of χ^2 with one and six degrees of freedom are represented.

Example 9.4.2. The random variables X_k ($k = 1, \ldots, 8$) are independent and have the same normal distribution $N(0; 2)$. We consider the statistic

$$\chi^2 = \sum_{k=1}^{8} X_k^2.$$

Here the random variable χ^2 has eight degrees of freedom. By (9.4.5) the expected value and the standard deviation of this random variable are, respectively

$$m_1 = 32, \qquad \sqrt{\mu_2} = 16.$$

Let us compute the probability that χ^2 will exceed or equal 40. Since $\sigma = 2$, $40/\sigma^2 = 10$; hence the required probability is $P(\chi^2 \geqslant 10)$. From the tables of the

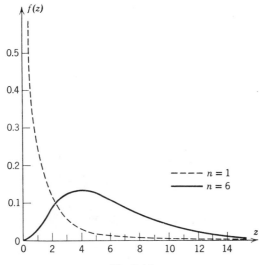

Fig. 9.4.1

χ^2 distribution for $\sigma = 1$ we find for eight degrees of freedom

$$P(\chi^2 \geqslant 9.524) = 0.30, \qquad P(\chi^2 \geqslant 11.030) = 0.20.$$

By linear interpolation we obtain

$$P(\chi^2 \geqslant 10) = 0.30 - \frac{10 - 9.524}{11.030 - 9.524} \cdot 0.10 = 0.27.$$

Example 9.4.3. In example 5.11.1 we established that the density $h(v_1, v_2, v_3)$ of the projections of the velocity vector $\vec{V}$ of a particle of an ideal gas on the axes of a rectangular coordinate system (x_1, x_2, x_3) is

$$h(v_1, v_2, v_3) = \frac{1}{(2\pi\sigma^2)^{3/2}} \exp\left(-\frac{v_1{}^2 + v_2{}^2 + v_3{}^2}{2\sigma^2}\right).$$

The random variable $V^2 = V_1{}^2 + V_2{}^2 + V_3{}^2$ has the χ^2 distribution with three degrees of freedom. By formula (9.4.4) its density $h(v^2)$ is

$$h(v^2) = \begin{cases} \dfrac{1}{\sigma^3 \sqrt{2\pi}} v \exp\left(-\dfrac{v^2}{2\sigma^2}\right) & (v > 0), \\ 0 & (v \leqslant 0). \end{cases}$$

Hence, for the density of the velocity we obtain *Maxwell's equation*

$$(9.4.8) \qquad f(v) = \begin{cases} \dfrac{\sqrt{2}}{\sigma^3 \sqrt{\pi}} v^2 \exp\left(-\dfrac{v^2}{2\sigma}\right) & (v > 0), \\ 0 & (v \leqslant 0). \end{cases}$$

We find the density of the statistic

$$Y = \frac{1}{n} \sum_{k=1}^{n} X_k{}^2,$$

where the random variables X_k are independent and have the same distribution $N(0; \sigma)$. Since $Y = \dfrac{1}{n} \chi^2$, where χ^2 is defined by (9.4.2), the density $f(y)$ of Y is

$$(9.4.9) \quad f(y) = \begin{cases} \dfrac{n^{n/2}}{2^{n/2}\sigma^n \Gamma(\frac{1}{2}n)} \exp\left(-\dfrac{ny}{2\sigma^2}\right) y^{n/2-1} & \text{for } y > 0, \\ 0 & \text{for } y \leqslant 0. \end{cases}$$

We have

$$(9.4.10) \qquad E(Y) = \sigma^2, \qquad D^2(Y) = \frac{2\sigma^4}{n}.$$

The tables of the χ^2 distribution are usually given for no more than thirty degrees of freedom. Fisher [6] showed that if the number of degrees of freedom n increases to infinity, the random variable $\sqrt{2\chi^2}$ has the asymptotically normal distribution $N(\sqrt{2n - 1}; 1)$.

For $n \geqslant 30$ we may use the tables of the normal distribution.

B   Suppose, now, that the random variables X_k ($k = 1, \ldots, n$) are independent and every X_k has the distribution $N(m_k; 1)$. Then the statistic

$$\chi^2 = \sum_{k=1}^{n} X_k^2$$

is called the *noncentral* χ^2. The density $h_n(u)$ of this statistic has the form

(9.4.11)

$$h_n(u) = \begin{cases} \dfrac{1}{\sqrt{\pi} 2^{n/2}} \exp\left[-\tfrac{1}{2}(\tau^2 + u)\right] u^{(n-2)/2} \displaystyle\sum_{j=0}^{\infty} \frac{(\tau^2 u)^j}{(2j)!} \frac{\Gamma(j + \tfrac{1}{2})}{\Gamma(j + \tfrac{1}{2}n)} & (u > 0), \\ 0 & (u \leqslant 0), \end{cases}$$

where $\tau^2 = \sum_{k=1}^{n} m_k^2$. The parameter τ^2 is called the *noncentrality parameter*, and n the *number of degrees of freedom*.

The proof of (9.4.11) can be found in the monograph by Anderson [1]. Tables of the noncentral χ^2 distribution have been prepared by Fix [1].

9.5   THE DISTRIBUTION OF THE STATISTIC $(\bar{X}, S)$

Example 9.5.1.   From a population in which the characteristic X has the normal distribution $N(0; \sigma)$, we draw a simple sample of size n. The observed values of X are $x_1, \ldots, x_n$. Let us compute the sample mean value and the *sample standard deviation s*, where

$$\bar{x} = \frac{1}{n} \sum_{k=1}^{n} x_k, \qquad s^2 = \frac{1}{n} \sum_{k=1}^{n} (x_k - \bar{x})^2 = \frac{1}{n} \sum_{k=1}^{n} x_k^2 - \bar{x}^2.$$

It is obvious that if we draw a series of independent samples of n elements from a population, in general, we obtain different pairs of values of $\bar{x}$ and s. The observed $(\bar{x}, s)$ are the observed values of the two-dimensional random variable $(\bar{X}, S)$, where

(9.5.1) $$\bar{X} = \frac{1}{n} \sum_{k=1}^{n} X_k, \qquad S^2 = \frac{1}{n} \sum_{k=1}^{n} (X_k - \bar{X})^2.$$

In applications, we often deal with problems of the following type. The values $\bar{x}$ and s are observed and found to be such that $a \leqslant \bar{x} < b$ and $s \geqslant c$. We would like to find the probability of these inequalities, or in other words, we would like to find out how often the values $\bar{x}$ of the statistic $\bar{X}$ and the values s of the statistic S satisfy these inequalities if we take a large series of observations.

To answer this question we have to find the distribution of the random variable $(\bar{X}, S)$.

A   Let X_k ($k = 1, 2, \ldots, n$) be independent random variables with the same normal density

$$f(x) = \frac{1}{\sigma\sqrt{2\pi}} \exp\left(-\frac{x^2}{2\sigma^2}\right).$$

Let $f(x_1, \ldots, x_n)$ be the density of the n-dimensional random variable $(X_1, \ldots, X_n)$. We call the expression

$$dP = f(x_1, x_2, \ldots, x_n)\, dx_1\, dx_2 \ldots dx_n$$

the *probability element* of this random variable.

Since $X_1, \ldots, X_n$ are independent, we obtain

$$(9.5.2) \qquad dP = \frac{1}{\sigma^n (2\pi)^{n/2}} \exp\left(-\frac{1}{2\sigma^2} \sum_{k=1}^{n} x_k^2\right) dx_1 \ldots dx_n$$

$$= \frac{1}{\sigma^n (2\pi)^{n/2}} \exp\left(-\frac{n\bar{x}^2 + ns^2}{2\sigma^2}\right) dx_1 \ldots dx_n.$$

Let us make the following transformation:

$$(9.5.3) \qquad x_k = \bar{x} + sz_k \qquad (k = 1, 2, \ldots, n).$$

From the relations

$$\sum_{k=1}^{n} x_k = n\bar{x}, \qquad \sum_{k=1}^{n} x_k^2 = n\bar{x}^2 + ns^2$$

follow two relations for the variables z_k,

$$(9.5.4) \qquad \sum_{k=1}^{n} z_k = 0, \qquad \sum_{k=1}^{n} z_k^2 = n.$$

Thus, two variables among the z_k, say z_n and z_{n-1}, are functions of the remaining z_k. By (9.5.4.), either

$$(9.5.5) \qquad z_{n-1} = \frac{A - B}{2}, \qquad z_n = \frac{A + B}{2},$$

or

$$(9.5.6) \qquad z_{n-1} = \frac{A + B}{2}, \qquad z_n = \frac{A - B}{2},$$

where

$$A = -\sum_{k=1}^{n-2} z_k, \qquad B = \sqrt{2n - 3\sum_{k=1}^{n-2} z_k^2 - \sum_{\substack{k,j=1 \\ k \neq j}}^{n-2} z_k z_j}.$$

Hence transformation (9.5.3) is not one-to-one, but to every system $(z_1, z_2, \ldots, z_{n-2}, \bar{x}, s)$, where $s > 0$, $\sum_{k=1}^{n-2} z_k \neq 0$, and $\sum_{k=1}^{n-2} z_k^2 < n$, there correspond two systems $(x_1, x_2, \ldots, x_n)$, namely, the system

(9.5.7)
$$x_k = \bar{x} + sz_k \quad (k = 1, 2, \ldots, n - 2),$$
$$x_{n-1} = \bar{x} + s\,\frac{A - B}{2}, \qquad x_n = \bar{x} + s\,\frac{A + B}{2},$$

and the system

(9.5.8)
$$x_k = \bar{x} + sz_k \quad (k = 1, 2, \ldots, n - 2),$$
$$x_{n-1} = \bar{x} + s\,\frac{A + B}{2}, \qquad x_n = \bar{x} + s\,\frac{A - B}{2}.$$

It is obvious that the absolute values of the Jacobians of the two transformations are equal. For transformation (9.5.7) we have ·

$$
J = \begin{vmatrix}
\dfrac{\partial x_1}{\partial \bar{x}} & \cdots & \dfrac{\partial x_n}{\partial \bar{x}} \\[2mm]
\dfrac{\partial x_1}{\partial s} & \cdots & \dfrac{\partial x_n}{\partial s} \\[2mm]
\dfrac{\partial x_1}{\partial z_1} & \cdots & \dfrac{\partial x_n}{\partial z_1} \\[2mm]
\cdots\cdots\cdots\cdots \\[1mm]
\dfrac{\partial x_1}{\partial z_{n-2}} & \cdots & \dfrac{\partial x_n}{\partial z_{n-2}}
\end{vmatrix}
=
\begin{vmatrix}
1 & 1 \ldots 1 & 1 & 1 \\[2mm]
z_1 & z_2 \ldots z_{n-2} & \dfrac{A - B}{2} & \dfrac{A + B}{2} \\[2mm]
s & 0 \ldots 0 & \dfrac{s}{2}\left(\dfrac{\partial A}{\partial z_1} - \dfrac{\partial B}{\partial z_1}\right) & \dfrac{s}{2}\left(\dfrac{\partial A}{\partial z_1} + \dfrac{\partial B}{\partial z_1}\right) \\[2mm]
\cdots\cdots\cdots\cdots \\[1mm]
0 & 0 \ldots s & \dfrac{s}{2}\left(\dfrac{\partial A}{\partial z_{n-2}} - \dfrac{\partial B}{\partial z_{n-2}}\right) & \dfrac{s}{2}\left(\dfrac{\partial A}{\partial z_{n-2}} + \dfrac{\partial B}{\partial z_{n-2}}\right)
\end{vmatrix}
$$

After a few simple computations we obtain for transformations (9.5.7) and (9.5.8)

$$|J| = ks^{n-2},$$

where $k = k(z_1, \ldots, z_{n-2})$ is a function independent of $\bar{x}$ and s. Using the fact that the density on the right-hand side of (9.5.2) has the same value for both transformations (9.5.7) and (9.5.8) and applying formula (2.9.3) to both, we obtain from (9.5.2)

(9.5.9) $\quad dP = g(\bar{x}, s, z_1, \ldots, z_{n-2})\, d\bar{x}\, ds\, dz_1 \ldots dz_{n-2}$

$$= 2\,\frac{1}{\sigma^n (2\pi)^{n/2}} \exp\left(-\frac{n\bar{x}^2 + ns^2}{2\sigma^2}\right) s^{n-2} k\, d\bar{x}\, ds\, dz_1 \ldots dz_{n-2},$$

where $g(\bar{x}, s, z_1, \ldots, z_{n-2})$ is the density of the random variable $(\bar{X}, S, Z_1, \ldots, Z_{n-2})$.

It will be easier to understand the reasoning in the proof of formula (9.5.9) by comparing it with the method of proving (2.4.14).

Let us represent formula (9.5.9) in the following form:

$$g(\bar{x}, s, z_1, \ldots, z_{n-2}) \, d\bar{x} \, ds \, dz_1 \ldots dz_{n-2}$$

$$= \frac{\sqrt{n}}{\sigma\sqrt{2\pi}} \exp\left(-\frac{n\bar{x}^2}{2\sigma^2}\right) d\bar{x} \, \frac{n^{(n-1)/2} s^{n-2} \exp\left(-\frac{ns^2}{2\sigma^2}\right) ds}{2^{(n-3)/2}\Gamma\left(\frac{n-1}{2}\right)\sigma^{n-1}}$$

$$\times \frac{\Gamma\left(\frac{n-1}{2}\right)}{n^{n/2}\pi^{(n-1)/2}} \times k(z_1, \ldots, z_{n-2}) \, dz_1 \ldots dz_{n-2}.$$

Then the probability element of $(\bar{X}, S, Z_1, \ldots, Z_{n-2})$ is a product of three factors, the first of which is the probability element of $\bar{X}$, the second is the probability element of S, and the third is the probability element of $(Z_1, \ldots, Z_{n-2})$. Hence by (2.8.5) the random variables $\bar{X}$, S and $(Z_1, \ldots, Z_{n-2})$ are independent. If we denote by $h(\bar{x}, s)$ the density of $(\bar{X}, S)$ we have

(9.5.10)

$$h(\bar{x}, s) = \begin{cases} \dfrac{\sqrt{n}}{\sigma\sqrt{2\pi}} \exp\left(-\dfrac{n\bar{x}^2}{2\sigma^2}\right) \dfrac{n^{(n-1)/2} s^{n-2} \exp\left(-ns^2/2\sigma^2\right)}{2^{(n-3)/2}\Gamma[\frac{1}{2}(n-1)]\sigma^{n-1}} & (s > 0), \\ 0 & (s \leqslant 0). \end{cases}$$

B   We now find the distribution of the statistic $U = S^2$. To do so, let us put $u = s^2$ in (9.5.10). Since $s > 0$, the transformation $u = s^2$ is one-to-one, and by (2.4.19) the density of U is

$$(9.5.11) \quad f_1(u) = \begin{cases} \dfrac{n^{(n-1)/2} u^{(n-3)/2} \exp\left(-nu/2\sigma^2\right)}{2^{(n-1)/2}\Gamma[\frac{1}{2}(n-1)]\sigma^{n-1}} & \text{for } u > 0, \\ 0 & \text{for } u \leqslant 0. \end{cases}$$

Let us now consider the random variable $Z = nS^2 = nU$. Denoting its density function by $f_2(z)$, we have

$$(9.5.12) \quad f_2(z) = \begin{cases} \dfrac{z^{(n-3)/2} \exp\left(-z/2\sigma^2\right)}{2^{(n-1)/2}\Gamma[\frac{1}{2}(n-1)]\sigma^{n-1}} & \text{for } z > 0, \\ 0 & \text{for } z \leqslant 0. \end{cases}$$

By comparing (9.5.12) with formula (9.4.4), we see that Z has the same distribution as χ^2 with $n-1$ degrees of freedom, since formula (9.5.12) can be obtained from (9.4.4) by replacing n by $n-1$. This result agrees with our intuition since, by definition, χ^2 is a sum of n independent random variables, whereas

$$nS^2 = \sum_{k=1}^{n}(X_k - \bar{X})^2$$

is a sum of n random variables satisfying the relation

$$\sum_{k=1}^{n} X_k = n\bar{X}.$$

This result gives us a better understanding of the notion of the number of degrees of freedom.

By (9.4.5)

$$E(nS^2) = (n-1)\sigma^2, \qquad D^2(nS^2) = 2(n-1)\sigma^4.$$

Hence

(9.5.13) $\qquad E(S^2) = \dfrac{n-1}{n}\sigma^2, \qquad D^2(S^2) = \dfrac{2(n-1)}{n^2}\sigma^4.$

C   Let us return to expression (9.5.10). We see that if the independent random variables X_k have the same normal distribution, then the joint density of the random variables $\bar{X}$ and S is the product of the densities of these random variables. Hence, by (2.8.5), these random variables are independent. This extremely important and interesting result was obtained by Fisher [5]. In practice it means that if we take a series of independent simple samples from a normal population and divide these samples into groups so that all the samples for which the values of s close (or equal) to one another belong to one group, then the distribution of the values of $\bar{x}$ will be approximately the same for every group. In the same way we could divide the samples into groups according to the values of $\bar{x}$, and then the distribution of the values of s would be approximately the same for every group.

The converse theorem is also true. *If the statistics $\bar{X}$ and S are independent, the random variables X_k have the normal distribution.* The proof of this theorem was given by Geary [1], Lukacs [1], Kawata and Sakamoto [1], and Zinger [1]. Later this theorem was generalized by Lukacs [2] and Basu and Laha [1].

Example 9.5.2. From a population in which the characteristic X has the normal distribution $N(1; 2)$, we draw a simple sample of size $n = 12$. We observe the following values of X:

$$x_1 = 2.0,\ x_2 = 2.5,\ x_3 = 0.5,\ x_4 = 1.0,\ x_5 = 0.0,\ x_6 = -0.9,$$
$$x_7 = 5.1,\ x_8 = -1.5,\ x_9 = 0.8,\ x_{10} = 1.1,\ x_{11} = 0.8,\ x_{12} = 0.4.$$

Hence

$$\bar{x} = \frac{1}{12} \sum_{k=1}^{12} x_k = 0.98, \qquad s^2 = \frac{1}{12} \sum_{k=1}^{12} x_k^2 - \bar{x}^2 = 2.69.$$

Thus $z = ns^2 = 12s^2 = 32.28$. What is the probability that Z will exceed or equal the value obtained, $z = 32.28$?

We know that the random variable Z has the χ^2 distribution with eleven degrees of freedom. Using the fact that the standard deviation of X is two and applying an interpolation similar to that in example 9.4.2, we find from the tables of the χ^2 distribution for eleven degrees of freedom

$$P(Z \geqslant 32.28) = P\left(\frac{nS^2}{\sigma^2} \geqslant \frac{32.28}{4}\right) = P(\chi^2 \geqslant 8.07) \cong 0.70.$$

9.6 STUDENT'S t-DISTRIBUTION

A   In Section 9.4 we considered the distribution of the statistic $\bar{X}$ given by

$$\bar{X} = \frac{1}{n} \sum_{k=1}^{n} X_k,$$

where the random variables X_k ($k = 1, \ldots, n$) are independent and have the same normal distribution $N(m; \sigma)$. We established that $\bar{X}$ has the normal distribution $N(m; \sigma/\sqrt{n})$; hence for a known m and unknown σ the distribution of $\bar{X}$ is unknown. Of course, we cannot replace σ by the value s obtained from a sample since S itself is a random variable and can take on different values in different samples. In order to be able to deduce anything about m without the knowledge of σ, we have to consider a statistic which is a function of m and with a distribution independent of σ. This problem was solved by Gosset (pseudonym: Student [1]), who introduced the statistic called *Student's t-statistic*.

Let X_k ($k = 1, \ldots, n$) be independent random variables with the same normal distribution $N(m; \sigma)$. Then *Student's t* is defined by the formula[1]

(9.6.1) $$t = \frac{\bar{X} - m}{S} \sqrt{n - 1},$$

where $\bar{X}$ and S are given by (9.5.1).

Let us first find the density $f(v)$ of $V = (\bar{X} - m)/S$. The random variable V is a ratio of two independent random variables. From (9.5.10) it

---

[1] Following tradition, we denote both Student's random variable t and the values it takes on by a small letter t, contrary to the custom used up to now.

follows that the densities of the numerator and denominator are, respectively

$$f_1(\bar{x} - m) = \frac{\sqrt{n}}{\sigma\sqrt{2\pi}} \exp\left[-\frac{n(\bar{x} - m)^2}{2\sigma^2}\right],$$

(9.6.2)

$$f_2(s) = \frac{n^{(n-1)/2}s^{n-2}\exp(-ns^2/2\sigma^2)}{2^{(n-3)/2}\Gamma[\frac{1}{2}(n-1)]\sigma^{n-1}}.$$

By (2.9.16) we obtain the density of V in the form

$$f(v) = \int_0^\infty \frac{\sqrt{n}}{\sigma\sqrt{2\pi}} \exp\left(-\frac{nv^2s^2}{2\sigma^2}\right) \frac{n^{(n-1)/2}s^{n-2}\exp(-ns^2/2\sigma^2)}{2^{(n-3)/2}\Gamma[\frac{1}{2}(n-1)]\sigma^{n-1}} s\, ds$$

$$= \frac{n^{n/2}}{\sigma^n\sqrt{\pi}2^{(n-2)/2}\Gamma[\frac{1}{2}(n-1)]} \int_0^\infty \exp\left[-\frac{ns^2(v^2+1)}{2\sigma^2}\right]s^{n-1}\, ds.$$

Introducing the variable $z = s^2$, we obtain

$$f(v) = \frac{n^{n/2}}{\sigma^n\sqrt{\pi}2^{n/2}\Gamma[\frac{1}{2}(n-1)]} \int_0^\infty \exp\left[-\frac{zn(v^2+1)}{2\sigma^2}\right]z^{(n-2)/2}\, dz.$$

Using (5.8.5) with $a = [n(v^2+1)]/2\sigma^2$, $p = n/2$ and the fact that $\Gamma(\frac{1}{2}) = \sqrt{\pi}$, we obtain, after some computation,

(9.6.3) $f(v) = \dfrac{\Gamma(\frac{1}{2}n)}{\Gamma(\frac{1}{2})\Gamma[\frac{1}{2}(n-1)]} \cdot \dfrac{1}{(v^2+1)^{n/2}}$ $(-\infty < v < \infty).$

Since

$$\frac{\Gamma(\frac{1}{2})\Gamma[\frac{1}{2}(n-1)]}{\Gamma(\frac{1}{2}n)} = B[\tfrac{1}{2}, \tfrac{1}{2}(n-1)],$$

where

$$B(p, q) = \int_0^1 x^{p-1}(1-x)^{q-1}\, dx \qquad (q > 0, p > 0),$$

we can write expression (9.6.3) in the form

(9.6.4) $f(v) = \dfrac{1}{B[\frac{1}{2}, \frac{1}{2}(n-1)]} \cdot \dfrac{1}{(v^2+1)^{n/2}}$ $(-\infty < v < \infty).$

By (9.6.4) we obtain for the density $g(t)$ of the random variable t given by (9.6.1)

(9.6.5)

$$g(t) = \frac{1}{\sqrt{n-1}B[\frac{1}{2}, \frac{1}{2}(n-1)]} \cdot \frac{1}{[1 + t^2/(n-1)]^{n/2}} \qquad (-\infty < t < \infty).$$

Thus the density of Student's t is independent of σ. As we have already mentioned, it is this fact that makes possible many applications of the

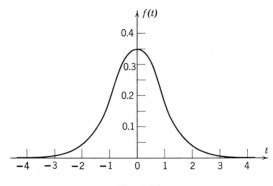

Fig. 9.6.1

t-distribution. We say that a random variable with density (9.6.5) has the *t-distribution with $n - 1$ degrees of freedom*. The density of the random variable t is symmetric with respect to $t = 0$.

Student's *t*-distribution has only moments of order $k < n - 1$. Thus, for $n = 2$ no moments exist. The reader can verify that for $n = 2$ Student's *t*-distribution is a particular case of the Cauchy distribution which, as we know (see Section 5.10), has no moments.

The probability that t belongs to the interval (t_1, t_2) can be found from the expression

$$P(t_1 < t < t_2) = \frac{1}{\sqrt{n - 1}B[\frac{1}{2}, \frac{1}{2}(n - 1)]} \int_{t_1}^{t_2} \frac{dt}{[1 + t^2/(n - 1)]^{n/2}}.$$

The graph of the density of Student's *t*-distribution for $n = 3$ is shown in Fig. 9.6.1.

If we look at the proof of (9.6.5) we can easily see that if the random variable U is defined as

$$(9.6.1') \qquad\qquad U = \frac{Z}{\sqrt{W/r}},$$

where Z and W are independent random variables, Z has the normal distribution $N(0; 1)$, and W has the χ^2 distribution with r degrees of freedom, then U has Student's *t*-distribution with r degrees of freedom.

B   If we compare the tables of Student's *t*-distribution for a number of degrees of freedom close to 30 with the tables of the normal distribution $N(0; 1)$, we see that these tables are almost identical. This is because the *t*-distribution approaches the normal distribution rapidly, as the number of degrees of freedom tends to infinity. We shall prove this.

Theorem 9.6.1. *The sequence $\{F_n(t)\}$ of distribution functions of Student's t with n degrees of freedom satisfies for every t the relation*

$$(9.6.6) \qquad \lim_{n \to \infty} F_n(t) = \frac{1}{\sqrt{2\pi}} \int_{-\infty}^{t} e^{-t^2/2} \, dt.$$

Proof. Let us write formula (9.6.1) in the form

$$t_n = \frac{[(\bar{X} - m)/\sigma]\sqrt{n}}{\sqrt{\dfrac{nS^2}{(n-1)\sigma^2}}} = \frac{Y_n}{\sqrt{Z_n}} = \frac{Y_n}{V_n}.$$

We observe that for every n the random variable Y_n has the normal distribution $N(0; 1)$. We shall prove that the sequence $\{V_n\}$ converges stochastically to one.

Indeed, the random variable nS^2 has the gamma distribution with $p = (n-1)/2$ and $b = \frac{1}{2}\sigma^2$; hence the characteristic function $\phi_n(\alpha)$ of Z_n is, by (5.8.8),

$$\phi_n(\alpha) = \left(1 - \frac{i\alpha}{\dfrac{n-1}{2}}\right)^{-(n-1)/2}$$

Hence

$$\lim_{n \to \infty} \phi_n(\alpha) = e^{i\alpha}.$$

It follows from the last equation that the sequence $\{Z_n\}$ is stochastically convergent to one; thus the sequence $\{V_n\}$ is also stochastically convergent to one. From assertion (δ) of theorem 6.14.1 equation (9.6.6) follows.

C   A statistic more general than Student's t is Student's noncentral t-statistic. It is defined in the following way.

Let X_k ($k = 1, \ldots, n$) be independent random variables with the same normal distribution $N(m; \sigma)$. The expression

$$Y = \frac{\bar{X} - x_0}{S} \sqrt{n - 1},$$

where $\bar{X}$ and S are given by (9.5.1) and x_0 is some constant, is called *Student's noncentral t*.

The difference between Student's noncentral t and Student's t is that instead of the expected value m of the random variable X in the numerator, an arbitrary constant x_0 appears. Obviously if $x_0 = m$, Y coincides with the Student t-statistic.

The density of the random variable Y can be obtained in a manner similar to that used for formula (9.6.5). This density is of the form

$$
\begin{aligned}
f(y) = {} & \frac{n^{n/2}}{\sqrt{n-1}\,\Gamma[\tfrac{1}{2}(n-1)]\sqrt{\pi}\,2^{n/2-1}} \exp\left[-\frac{n(n-1)Q^2}{2(y^2+n-1)}\right] \\
& \times \frac{1}{[1+y^2/(n-1)]^{n/2}} \int_{-Qy/\sqrt{y^2+n-1}}^{\infty} \exp\left(-\frac{n}{2}u^2\right) \\
& \times \left[u + \frac{Qy}{\sqrt{y^2+n-1}}\right]^{n-1} du,
\end{aligned}
$$

(9.6.7)

where

$$
Q = \frac{m-x_0}{\sigma}.
$$

If $x_0 = m$, that is, if $Q = 0$, formulas (9.6.7) and (9.6.5) become identical.

Large tables of the noncentral t were given by Resnikoff and Lieberman [1]; Harley [1] and Merrington and Pearson [1] gave an approximate formula for this distribution.

D   We now introduce one more statistic of great importance in applications, the distribution of which is the same as Student's t-distribution.

Let $X_1, X_2, \ldots, X_{n_1}$ and $Y_1, Y_2, \ldots, Y_{n_2}$ be independent random variables with the same normal distribution $N(m; \sigma)$. Let

$$
\bar{X} = \frac{1}{n_1}\sum_{k=1}^{n_1} X_k, \qquad \bar{Y} = \frac{1}{n_2}\sum_{l=1}^{n_2} Y_l,
$$

$$
S_1^2 = \frac{1}{n_1}\sum_{k=1}^{n_1}(X_k - \bar{Y})^2, \qquad S_2^2 = \frac{1}{n_2}\sum_{l=1}^{n_2}(Y_l - \bar{Y})^2.
$$

As we know, $\bar{X}$ and $\bar{Y}$ have, respectively, the normal distributions

$$
N\left(m; \frac{\sigma}{\sqrt{n_1}}\right) \quad \text{and} \quad N\left(m; \frac{\sigma}{\sqrt{n_2}}\right).
$$

Hence the random variable $(\bar{X} - \bar{Y})/\sigma$ has the distribution

$$
N(0; \sqrt{(n_1+n_2)/n_1 n_2}).
$$

It follows that

(9.6.8)
$$
Z = \frac{\bar{X} - \bar{Y}}{\sigma}\sqrt{n_1 n_2/(n_1+n_2)}
$$

has the distribution $N(0; 1)$. The random variable

(9.6.9) $$W = \frac{n_1 S_1^2 + n_2 S_2^2}{\sigma_2^2}$$

has the χ^2 distribution with $n_1 + n_2 - 2$ degrees of freedom. This follows from the addition theorem for χ^2, since S_1^2 and S_2^2 are independent.

Let us consider the random variable U defined as

(9.6.10)
$$U = \frac{\bar{X} - \bar{Y}}{\sqrt{n_1 S_1^2 + n_2 S_2^2}} \sqrt{\frac{n_1 n_2}{n_1 + n_2}(n_1 + n_2 - 2)} = \frac{Z}{\sqrt{W/(n_1 + n_2 - 2)}},$$

where Z and W are defined by formula (9.6.8) and (9.6.9), respectively. By comparing (9.6.10) and (9.6.1'), we see that U has Student's t-distribution with $n_1 + n_2 - 2$ degrees of freedom. Thus the distribution of U is independent of m as well as σ. This result was obtained by Fisher [5].

E   The generalization of Student's t, or more precisely t^2, to multidimensional random variables is Hotelling's T^2 (Hotelling [2]).

Let $X_k = (X_{k1}, \ldots, X_{kr})$ $(k = 1, \ldots, n)$ be independent r-dimensional random vectors with the same normal distribution, and let $m_j = E(X_{kj})$ and $\lambda_{ij} = E[(X_{kj} - m_j)(X_{ki} - m_i)]$ $(i, j = 1, \ldots, r)$, where the matrix **M** of variances and covariances has the determinant $|\mathbf{M}| \neq 0$. Let

(9.6.11) $$\bar{X}_j = \frac{1}{n} \sum_{k=1}^{n} X_{kj},$$

(9.6.12) $$W_{ji} = \frac{1}{n} \sum_{k=1}^{n} (X_{kj} - \bar{X}_j)(X_{ki} - \bar{X}_i),$$

(9.6.13) $$\mathbf{Q} = \begin{bmatrix} W_{11} & \cdots & W_{1r} \\ \cdots\cdots\cdots\cdots \\ W_{r1} & \cdots & W_{rr} \end{bmatrix}.$$

The W_{ji} are called *sample variances* (for $j = i$) and *sample covariances* (for $j \neq i$), while **Q** is called the *matrix of sample second-order moments*. The expression

(9.6.14) $$T^2 = (n - 1) \sum_{j,i=1}^{r} \left(\frac{|\mathbf{Q}_{ji}|}{|\mathbf{Q}|}\right)(\bar{X}_j - m_j)(\bar{X}_i - m_i),$$

where $|\mathbf{Q}|$ is the determinant of the matrix **Q** (different from 0, with probability one), and $|\mathbf{Q}_{ji}|$ is the algebraic complement of the term W_{ji} in the determinant of the matrix **Q**, is called *Hotelling's T^2*.

The density $g(y)$ of T^2 is given by the formula

$$(9.6.15) \quad g(y) = \begin{cases} \dfrac{1}{B\left(\dfrac{n-r}{2},\dfrac{r}{2}\right)(n-1)} \dfrac{\left(\dfrac{y}{n-1}\right)^{(r/2)-1}}{\left(1+\dfrac{y}{n-1}\right)^{n/2}} & (y > 0), \\[4mm] 0 & (y \leqslant 0). \end{cases}$$

The proof of (9.6.15) can be found in the book by Schmetterer [1] or Cramér [2].

We observe that the density (9.6.15) is independent of the second-order moments λ_{ji} of the random vectors $(X_{k1}, \ldots, X_{kr})$. We say that T^2 has $n-1$ degrees of freedom.

In mathematical statistics Hotelling's T^2 plays the same role in the multidimensional case as Student's t plays in the one-dimensional case.

9.7 FISHER'S Z-DISTRIBUTION

Let X_k $(k = 1, 2, \ldots, n_1)$ and Y_l $(l = 1, \ldots, n_2)$ be independent random variables with the same normal distribution $N(0; \sigma)$. Let

$$S_1^2 = \frac{1}{n_1} \sum_{k=1}^{n_1} (X_k - \overline{X})^2, \qquad S_2^2 = \frac{1}{n_2} \sum_{l=1}^{n_2} (Y_l - \overline{Y})^2.$$

Let us consider the statistic

$$(9.7.1) \qquad\qquad U = \frac{S_1}{S_2}.$$

By assumption, the random variables S_1 and S_2 are independent and their densities are for $i = 1, 2$

$$f_i(s_i) = \begin{cases} \dfrac{n_i^{(n_i-1)/2} s_i^{n_i-2} \exp\left(-n_i s_i^2/2\sigma^2\right)}{2^{(n_i-3)/2} \Gamma\left[\frac{1}{2}(n_i-1)\right]\sigma^{n_i-1}} & \text{for } s_i > 0, \\[4mm] 0 & \text{for } s_i \leqslant 0. \end{cases}$$

Since U is a ratio of two independent random variables, by formula (2.9.16′), the density of U is given by

$$(9.7.2)$$

$$f(u) = \begin{cases} \dfrac{n_1^{(n_1-1)/2} n_2^{(n_2-1)/2}}{2^{(n_1+n_2)/2-3}\Gamma\left[\frac{1}{2}(n_1-1)\right]\Gamma\left[\frac{1}{2}(n_2-1)\right]\sigma^{n_1+n_2-2}} \\[4mm] \quad \times \displaystyle\int_0^\infty s_2^{n_1+n_2-3} u^{n_1-2} \exp\left[-\dfrac{s_2^2}{2\sigma^2}(n_1 u^2 + n_2)\right] ds_2 & \text{for } u > 0, \\[4mm] 0 & \text{for } u \leqslant 0. \end{cases}$$

Let $y = s_2^2$. We obtain

$$f(u) = \begin{cases} \dfrac{1}{2} C_{n_1, n_2} u^{n_1 - 2} \displaystyle\int_0^\infty y^{(n_1 + n_2)/2 - 2} \exp\left(-\dfrac{n_1 u^2 + n_2}{2\sigma^2} y\right) dy & \text{for } u > 0, \\ 0 & \text{for } u \leqslant 0, \end{cases}$$

where C_{n_1, n_2} is the term preceding the integral sign in (9.7.2). Using (5.8.5) with

$$a = \frac{n_1 u^2 + n_2}{2\sigma^2}, \qquad p = \frac{n_1 + n_2}{2} - 1,$$

we obtain

$$f(u) = \begin{cases} \dfrac{n_1^{(n_1 - 1)/2} n_2^{(n_2 - 1)/2} \Gamma[\frac{1}{2}(n_1 + n_2) - 1]}{2^{(n_1 + n_2)/2 - 2} \sigma^{n_1 + n_2 - 2} \Gamma[\frac{1}{2}(n_1 - 1)] \Gamma[\frac{1}{2}(n_2 - 1)]} \\ \qquad\qquad \times \dfrac{u^{n_1 - 2} (2\sigma^2)^{(n_1 + n_2)/2 - 1}}{(n_1 u^2 + n_2)^{(n_1 + n_2)/2 - 1}} & \text{for } u > 0, \\ 0 & \text{for } u \leqslant 0. \end{cases}$$

Introducing the symbol $B(p, q)$, we obtain

(9.7.3)
$$f(u) = \begin{cases} \dfrac{2 n_1^{(n_1 - 1)/2} n_2^{(n_2 - 1)/2}}{B[\frac{1}{2}(n_1 - 1), \frac{1}{2}(n_2 - 1)]} \cdot \dfrac{u^{n_1 - 2}}{(n_1 u^2 + n_2)^{(n_1 + n_2)/2 - 1}} & \text{for } u > 0, \\ 0 & \text{for } u \leqslant 0. \end{cases}$$

Expression (9.7.3) is independent of σ. Let

(9.7.4) $$u = \left[\frac{n_2(n_1 - 1)}{n_1(n_2 - 1)}\right]^{1/2} e^z \qquad (-\infty < z < +\infty).$$

Then

$$g(z) = \frac{2 n_1^{(n_1 - 1)/2} n_2^{(n_2 - 1)/2}}{B[\frac{1}{2}(n_1 - 1), \frac{1}{2}(n_2 - 1)]} \left[\frac{n_2(n_1 - 1)}{n_1(n_2 - 1)}\right]^{(n_1 - 1)/2} e^{z(n_1 - 1)}$$

$$\times \frac{1}{\left(n_2 \dfrac{n_1 - 1}{n_2 - 1} e^{2z} + n_2\right)^{(n_1 + n_2)/2 - 1}}$$

$$= \frac{2(n_1 - 1)^{(n_1 - 1)/2} (n_2 - 1)^{(n_2 - 1)/2}}{B[\frac{1}{2}(n_1 - 1), \frac{1}{2}(n_2 - 1)]} \cdot \frac{e^{z(n_1 - 1)}}{[(n_1 - 1)e^{2z} + n_2 - 1]^{(n_1 + n_2)/2 - 1}}$$
$$(-\infty < z < \infty).$$

Letting $r_1 = n_1 - 1$ and $r_2 = n_2 - 1$, we obtain

(9.7.5) $$g(z) = \frac{2 r_1^{r_1/2} r_2^{r_2/2}}{B(\frac{1}{2}r_1, \frac{1}{2}r_2)} \cdot \frac{e^{r_1 z}}{(r_1 e^{2z} + r_2)^{(r_1 + r_2)/2}} \qquad (-\infty < z < \infty).$$

The Z-statistic, related to the statistic U by formula (9.7.4), is called *Fisher's Z* (Fisher [7]). Its density is given by (9.7.5). The pair of numbers (r_1, r_2) is called the *number of degrees of freedom* of Z.

The statistic

$$F = e^{2Z}$$

is called *Snedecor's F*. Its density can be derived from (9.7.5).

We now show by an example how we can use the statistic Z.

Example 9.7.1. We have a consignment of merchandise. We know that the characteristic X of the merchandise has a normal distribution, but the standard deviation of the distribution is unknown. We select two independent simple samples from this population. The first sample consists of $n_1 = 5$ elements and the second consists of $n_2 = 6$ elements. The standard deviation of the first sample is $s_1 = 1.3$, and of the second $s_2 = 1$. Since both samples are taken from the same population (from the same consignment), we might expect that s_1 would not differ too much from s_2. It turns out, however, that $s_1/s_2 = 1.3$. The question arises: what is the probability that in two samples from the same population, in which the characteristic under consideration has a normal distribution, the ratio S_1/S_2 will exceed or equal the observed value 1.3?

The answer to this question is given by the distribution of the statistic Z. The value $u = 1.3$ is taken as the observed value of U defined by (9.7.1). By (9.7.4)

$$z = \log \frac{u}{\sqrt{\dfrac{n_2(n_1 - 1)}{n_1(n_2 - 1)}}} = \log \frac{1.3}{\sqrt{0.96}} = 0.2826.$$

This is the observed value of Fisher's Z with (4, 5) degrees of freedom.

We ask whether the probability that $U \geqslant 1.3$ is greater than 0.05.

From Table VI at the end of the book we find that the value z_0 for which $P(Z \geqslant z_0) = 0.05$ is $z_0 = 0.8236$. Since the observed value is only 0.2826, we obtain a positive answer to our question.

If we examine the proof of (9.7.5), we see that the random variable U defined as

$$U = \tfrac{1}{2} \log \left(\frac{W_1}{r_1} \Big/ \frac{W_2}{r_2} \right),$$

where W_1 and W_2 are independent random variables and W_i $(i = 1, 2)$ has the χ^2 distribution with r_i degrees of freedom, has Fisher's Z-distribution with (r_1, r_2) degrees of freedom.

Let W_1 have the noncentral χ^2 distribution with the noncentrality parameter τ^2 and r_1 degrees of freedom (see Section 9.4, B), let W_2 have the χ^2 distribution with r_2 degrees of freedom, and let W_1 and W_2 be independent. The statistic

$$F = \frac{W_1}{r_1} \Big/ \frac{W_2}{r_2}$$

is called *Snedecor's noncentral F* and the corresponding expression $Z = \frac{1}{2} \log F$ is called *Fisher's noncentral Z*. The density $g_1(z)$ of the noncentral Z can be obtained by formulas (2.9.16), (9.4.4.), and (9.4.11). It is of the form

$$g_1(z) = \frac{2r_1^{r_1/2} r_2^{r_2/2}}{\Gamma\left(\dfrac{r_2}{2}\right)} \exp\left[-\tfrac{1}{2}\tau^2 + r_1 z\right] \sum_{j=0}^{\infty} \frac{\left(\tau^2 r_1 \dfrac{e^{2z}}{2}\right)^j \Gamma\left(\dfrac{r_1 + r_2}{2} + j\right)}{j! \, \Gamma\left(\dfrac{r_1}{2} + j\right)(r_1 e^{2z} + r_2)^{(r_1+r_2)/2 + j}}.$$

For $\tau^2 = 0$ the last formula coincides with (9.7.5).

9.8 THE DISTRIBUTION OF $\bar{X}$ FOR SOME NON-NORMAL POPULATIONS

Up to now we have considered statistics which were functions of random variables with the same normal distribution. Now we investigate some other statistics.

Let the X_k $(k = 1, \ldots, j)$ be identically distributed, independent random variables with probability function

$$P(X_k = r) = \binom{n}{r} p^r q^{n-r} \qquad (r = 0, 1, \ldots, n),$$

where $0 < p < 1$ and $q = 1 - p$. We want to determine the distribution of the arithmetic mean of these variables,

$$(9.8.1) \qquad\qquad \bar{X} = \frac{1}{j} \sum_{k=1}^{j} X_k.$$

From (5.2.3) it follows that, for every $k = 1, \ldots, j$, the characteristic function of X_k is given by the formula

$$\phi(t) = (q + pe^{it})^n.$$

By (4.2.15) and (4.4.3) we obtain for the characteristic function of $\bar{X}$

$$(9.8.2) \qquad\qquad \phi_1(t) = (q + pe^{it/j})^{nj}.$$

Expression (9.8.2) is the characteristic function of a random variable with a modified binomial distribution; $\bar{X}$ can take on the values

$$0, \frac{1}{j}, \frac{2}{j}, \ldots, \frac{nj}{j} = n,$$

and

$$P\left(\bar{X} = \frac{l}{j}\right) = \binom{nj}{l} p^l q^{nj-l} \qquad (l = 0, 1, \ldots, nj).$$

Let us now consider the independent random variables X_k $(k = 1, \ldots, j)$ with the same Poisson distribution given by

$$P(X_k = r) = e^{-\lambda} \frac{\lambda^r}{r!} \qquad (r = 0, 1, 2, \ldots),$$

where $\lambda > 0$.

To find the distribution of the statistic $\bar{X}$ defined by (9.8.1), we observe that, according to (4.2.6), the characteristic function of X_k is, for every k,

$$\phi(t) = \exp[\lambda(e^{it} - 1)].$$

By (4.2.15) and (4.4.3), the characteristic function of $\bar{X}$ is

$$\phi_1(t) = \exp[j\lambda(e^{it/j} - 1)].$$

This expression is the characteristic function of a random variable with a modified Poisson distribution; $\bar{X}$ can take on the values

$$0, \frac{1}{j}, \frac{2}{j}, \frac{3}{j}, \ldots,$$

and

$$P\left(\bar{X} = \frac{l}{j}\right) = \frac{(j\lambda)^l e^{-j\lambda}}{l!} \qquad (l = 0, 1, 2, \ldots).$$

9.9 THE DISTRIBUTION OF SAMPLE MOMENTS AND SAMPLE CORRELATION COEFFICIENTS OF A TWO-DIMENSIONAL NORMAL POPULATION

A   Let us consider a two-dimensional random variable (X, Y) with a normal distribution. Its density is

$$(9.9.1) \quad f(x, y) = \frac{1}{2\pi\sigma_1\sigma_2\sqrt{1 - \rho^2}} \exp\left\{-\frac{1}{2(1 - \rho^2)}\left[\frac{(x - m_1)^2}{\sigma_1^2} - 2\frac{\rho(x - m_1)(y - m_2)}{\sigma_1\sigma_2} + \frac{(y - m_2)^2}{\sigma_2^2}\right]\right\}.$$

In this formula, m_1 and m_2 are the expected values of X and Y, respectively. Similarly, σ_1 and σ_2 are the standard deviations of X and Y, and ρ is the correlation coefficient of X and Y.

Let us consider the statistics

$$\bar{X} = \frac{1}{n}\sum_{k=1}^{n} X_k, \qquad \bar{Y} = \frac{1}{n}\sum_{k=1}^{n} Y_k,$$

$$S_1 = \sqrt{\frac{1}{n}\sum_{k=1}^{n}(X_k - \bar{X})^2}, \qquad S_2 = \sqrt{\frac{1}{n}\sum_{k=1}^{n}(Y_k - \bar{Y})^2}$$

and

$$R = \frac{\sum_{k=1}^{n}(X_k - \bar{X})(Y_k - \bar{Y})}{nS_1 S_2},$$

where (X_k, Y_k) $(k = 1, \ldots, n)$ are independent two-dimensional random variables with the normal distribution given by (9.9.1).

Here, $\bar{X}$ and $\bar{Y}$ are the sample means of X and Y, respectively. Similarly, S_1 and S_2 are the sample standard deviations of X and Y, and R is the *sample correlation coefficient* of X and Y.

Fisher [2] and Romanovsky [1] have proved that the density function of the random vector $(\bar{X}, \bar{Y}, S_1, S_2, R)$ has the form (which we present without proof)

$$f(\bar{x}, \bar{y}, s_1, s_2, r) = u(\bar{x}, \bar{y})v(s_1, s_2, r),$$

where

(9.9.2)
$$u(\bar{x}, \bar{y}) = \frac{n}{2\pi\sigma_1\sigma_2\sqrt{1 - \rho^2}} \exp\left\{ -\frac{n}{2(1 - \rho^2)} \left[\frac{(\bar{x} - m_1)^2}{\sigma_1^2} - 2\frac{\rho(\bar{x} - m_1)(\bar{y} - m_2)}{\sigma_1\sigma_2} + \frac{(\bar{y} - m_2)^2}{\sigma_2^2} \right] \right\},$$

(9.9.3)
$$v(s_1, s_2, r) = \begin{cases} \dfrac{n^{n-1}}{\pi\Gamma(n-2)} \cdot \dfrac{s_1^{n-2}s_2^{n-2}(1 - r^2)^{(n-4)/2}}{[\sigma_1^2\sigma_2^2(1 - \rho^2)]^{(n-1)/2}} \\ \quad \times \exp\left\{ -\dfrac{n}{2\sigma_1^2\sigma_2^2(1 - \rho^2)} [\sigma_2^2 s_1^2 - 2\rho r\sigma_1\sigma_2 s_1 s_2 + \sigma_1^2 s_2^2] \right\} \\ \qquad\qquad\qquad\qquad (s_1 > 0, \quad s_2 > 0, \quad r^2 < 1), \\ 0 \qquad\qquad\qquad\qquad \text{otherwise.} \end{cases}$$

Thus $(\bar{X}, \bar{Y})$ has a two-dimensional normal distribution and, $E(\bar{X}) = m_1$, $E(\bar{Y}) = m_2$, $D^2(\bar{X}) = \sigma_1^2/n$, $D^2(\bar{Y}) = \sigma_2^2/n$, and the correlation coefficient of $\bar{X}$ and $\bar{Y}$ is ρ. From the theorem of Fisher and Romanovsky follows the important conclusion that $(\bar{X}, \bar{Y})$ is independent of (S_1, S_2, R). We observe that from (9.5.13) it follows that for $i = 1, 2$

$$E(S_i^2) = \frac{n-1}{n} \sigma_i^2, \qquad D^2(S_i^2) = \frac{2(n-1)}{n^2} \sigma_i^4.$$

Let $f(r)$ denote the density of R. Denoting, as in Section 3.6, F, $|\mathbf{M}| = \sigma_1^2\sigma_2^2(1 - \rho^2)$, we obtain from (9.9.3)

$$f(r) = \frac{n^{n-1}(1 - r^2)^{(n-4)/2}}{\pi\Gamma(n-2)|\mathbf{M}|^{(n-1)/2}} \int_0^\infty \int_0^\infty s_1^{n-2}s_2^{n-2}$$

$$\times \exp\left\{ -\frac{n}{2|\mathbf{M}|} [\sigma_2^2 s_1^2 - 2\rho r\sigma_1\sigma_2 s_1 s_2 + \sigma_1^2 s_2^2] \right\} ds_1\, ds_2.$$

Since

$$\exp\left(\frac{n}{|\mathbf{M}|}\rho r\sigma_1\sigma_2 s_1 s_2\right) = \sum_{j=0}^{\infty}\frac{1}{j!}\left(\frac{n}{|\mathbf{M}|}\rho r\sigma_1\sigma_2 s_1 s_2\right)^j,$$

we have

$$f(r) = \frac{n^{n-1}(1-r^2)^{(n-4)/2}}{\pi\Gamma(n-2)|\mathbf{M}|^{(n-1)/2}}\sum_{j=0}^{\infty}\left[\frac{1}{j!}\left(\frac{n}{|\mathbf{M}|}\rho r\sigma_1\sigma_2\right)^j \int_0^{\infty} s_1^{n-2+j}\right.$$

$$\times \left. \exp\left(-\frac{n}{2|\mathbf{M}|}\sigma_2^2 s_1^2\right) ds_1 \int_0^{\infty} s_2^{n-2+j} \exp\left(-\frac{n}{2|\mathbf{M}|}\sigma_1^2 s_2^2\right) ds_2\right].$$

Letting $z_1 = s_1^2$ and $z_2 = s_2^2$ and using (5.8.5), we obtain after integration

$$f(r) = \frac{2^{n-3}}{\pi(n-3)!}(1-\rho^2)^{(n-1)/2}(1-r^2)^{(n-4)/2}\sum_{j=0}^{\infty}\Gamma^2\left(\frac{n+j-1}{2}\right)\frac{(2\rho r)^j}{j!}.$$

Using the formula

$$\Gamma^2(p) = 2^{1-2p}\Gamma(2p)B(p,\tfrac{1}{2}),$$

we have, setting $p = (n+j-1)/2$,

$$\sum_{j=0}^{\infty}\Gamma^2\left(\frac{n+j-1}{2}\right)\frac{(2\rho r)^j}{j!}$$

$$= \frac{(n-2)!}{2^{n-2}}\int_0^1 \sum_{j=0}^{\infty}\binom{n+j-2}{j}(\rho r)^j z^{(n+j-3)/2}(1-z)^{-1/2}\,dz$$

$$= \frac{(n-2)!}{2^{n-3}}\int_0^1 \sum_{j=0}^{\infty}\binom{n+j-2}{j}(\rho r)^j x^{n-2}(1-x^2)^{-1/2}\,dx,$$

where $z = x^2$. Since

$$\sum_{j=0}^{\infty}\binom{n+j-2}{j}(\rho r x)^j = \frac{1}{(1-\rho r x)^{n-1}},$$

we have

$$\sum_{j=0}^{\infty}\Gamma^2\left(\frac{n+j-1}{2}\right)\frac{(2\rho r)^j}{j!} = \frac{(n-2)!}{2^{n-3}}\int_0^1 \frac{x^{n-2}}{(1-\rho r x)^{n-1}\sqrt{1-x^2}}\,dx.$$

Finally,

(9.9.4)

$$f(r) = \frac{n-2}{\pi}(1-\rho^2)^{(n-1)/2}(1-r^2)^{(n-4)/2}\int_0^1 \frac{x^{n-2}}{(1-\rho r x)^{n-1}\sqrt{1-x^2}}\,dx.$$

For fixed n the distribution of the sample correlation coefficient R depends only on the population correlation coefficient ρ. For different values of ρ the distribution of R differs. Tables of the distribution of R were constructed by David [1].

Let us consider in more detail the case $\rho = 0$, which in the two-dimensional normal case implies that the random variables X and Y are independent. Putting $\rho = 0$ in (9.9.4), we obtain

$$(9.9.5) \qquad f(r) = \frac{n-2}{\pi}(1-r^2)^{(n-4)/2}\int_0^1 x^{n-2}(1-x^2)^{-1/2}\,dx.$$

Introducing the variable $z = x^2$, we obtain

$$(9.9.6) \quad f(r) = \frac{n-2}{2\pi}(1-r^2)^{(n-4)/2}\int_0^1 z^{(n-1)/2-1}(1-z)^{1/2-1}\,dz$$

$$= \frac{n-2}{2\pi}B[\tfrac{1}{2}(n-1),\tfrac{1}{2}](1-r^2)^{(n-4)/2}$$

$$= \frac{1}{B[\tfrac{1}{2}(n-2),\tfrac{1}{2}]}(1-r^2)^{(n-4)/2}.$$

Let us put in (9.9.6)

$$(9.9.7) \quad t = \frac{r}{\sqrt{1-r^2}}\sqrt{n-2}; \qquad \text{thus } 1-r^2 = \frac{1}{1+\dfrac{t^2}{n-2}}.$$

We obtain

(9.9.8)

$$g(t) = \frac{1}{B[\tfrac{1}{2}(n-2),\tfrac{1}{2}]}\cdot\frac{1}{[1+t^2/(n-2)]^{(n-4)/2}}\cdot\frac{1}{\sqrt{n-2}}\cdot\frac{1}{[1+t^2/(n-2)]^{3/2}}$$

$$= \frac{1}{\sqrt{n-2}}\cdot\frac{1}{B[\tfrac{1}{2}(n-2),\tfrac{1}{2}]}\cdot\frac{1}{[1+t^2/(n-2)]^{(n-1)/2}}.$$

By comparing (9.9.8) and (9.6.5), we see that the statistic t defined as

$$t = \frac{R}{\sqrt{1-R^2}}\sqrt{n-2}$$

has, for $\rho = 0$, a density identical with that of Student's t with $n-2$ degrees of freedom.

Example 9.9.1. Two characteristics X and Y of a certain population have normal distributions, and the random variables X and Y are independent. We select from this population a simple sample of size $n = 10$. From the values of X and Y in the sample we compute the correlation coefficient $r = 0.30$. What is the probability of obtaining a value of R which exceeds the observed value 0.30?

By (9.9.7) we obtain

$$P(R > 0.3) = P\left(\frac{t}{\sqrt{8 + t^2}} > 0.3\right) = P(|t| > 0.889).$$

From tables of the t-distribution for eight degrees of freedom we find that the required probability is 0.40. In the succeeding chapters, we discuss the possible conclusions that can be drawn from such a result.

It is possible to compute the expected value and the standard deviation of the random variable R from formula (9.9.4). Here we present without proof the following approximate formulas:

$$(9.9.9) \qquad E(R) \cong \rho, \qquad D^2(R) \cong \frac{(1 - \rho^2)^2}{n}.$$

However, these approximate formulas may be applied only for large values of n, not less than 500. This is because the distribution of the random variable R is highly asymmetric, and although it converges to the normal distribution as $n \to \infty$, the convergence is very slow. It is worthwhile mentioning the fact, discovered by Fisher [3, 7], that the random variable

$$U = \tfrac{1}{2} \log \frac{1 + R}{1 - R}$$

has, even for small n, approximately the normal distribution

$$N\left(\tfrac{1}{2} \log \frac{1 + \rho}{1 - \rho} + \frac{\rho}{2(n - 1)} ; \sqrt{1/(n - 3)}\right).$$

B   Let us consider an l-dimensional ($l \geqslant 2$) random variable $(X_1, \ldots, X_l)$ normally distributed with density given by (5.11.8), with expected values $E(X_j) = m_j$ and variances and covariances $E[(X_j - m_j)(X_i - m_i)] = \lambda_{ji}$, where the matrix $\mathbf{M}$ of second-order moments has the determinant $|\mathbf{M}| \neq 0$. Let the random vectors $(X_{k1}, \ldots, X_{kl})$, where $k = 1, \ldots, n$ ($n > l$), be independent and have the same distribution as the vector $(X_1, \ldots, X_l)$. Let us now consider the sample mean values $\bar{X}_j$, the sample variances and covariances W_{ji}, and the matrix $\mathbf{Q}$, defined by (9.6.11) to (9.6.13), respectively. The joint distribution of the vector

$$(\bar{X}_1, \ldots, \bar{X}_l, W_{11}, W_{12}, \ldots, W_{l-1,l}, W_{ll})$$

was found by Wishart [1]. He established that the vectors $(\bar{X}_1, \ldots, \bar{X}_l)$ and $(W_{11}, W_{12}, \ldots, W_{l-1,l}, W_{ll})$ are independent and $(\bar{X}_1, \ldots, \bar{X}_l)$ has the l-dimensional normal distribution with $E(\bar{X}_j) = m_j$ and

$$E[(\bar{X}_j - m_j)(\bar{X}_i - m_i)] = \lambda_{ij}/n,$$

and that the density $g(w_{11}, w_{12}, \ldots, w_{l-1,l}, w_{ll})$ of the second vector is zero in the domain where the matrix $\mathbf{Q}$ is not positive definite, and is of the following form in the remaining domain:

(9.9.10)

$$g(w_{11}, w_{12}, \ldots, w_{l-1,l}, w_{ll}) = \frac{\left(\dfrac{n^l}{2^l \, |\mathbf{M}|}\right)^{(n-1)/2}}{\pi^{[l(l-1)]/4} \Gamma\left(\dfrac{n-1}{2}\right) \ldots \Gamma\left(\dfrac{n-l}{2}\right)}$$

$$\times \; |Q|^{(n-l-2)/2} \exp\left(-\frac{n}{2\,|\mathbf{M}|} \sum_{i,j=1}^{l} |M_{ji}| \, w_{ji}\right),$$

where $|\mathbf{M}_{ji}|$ is the algebraic complement of the term λ_{ji} in the determinant $|\mathbf{M}|$.

The distribution given by (9.9.10) is called the *Wishart distribution*. Besides the geometrical proof by Wishart, an algebraic proof of (9.9.10) can be found in the paper of Hsu [2].

To the notion of the generalized variance of a population (see Section 3.6,F) corresponds the *generalized sample variance* which is equal to the determinant $|\mathbf{Q}|$ of the matrix of sample variances and covariances. The distribution and the moments of the random variable $|\mathbf{Q}|$ have been investigated by Wilks [1] and Kullback [1].

9.10 THE DISTRIBUTION OF REGRESSION COEFFICIENTS

In Sections 3.7 and 3.8 we showed that for a two-dimensional normal random variable (X, Y) the regression curves of the second type are straight lines identical with the regression lines of the first type. The coefficients of these lines are given by (3.8.4) and (3.8.5), respectively. In these formulas we have constant coefficients; however, in practical statistical problems these coefficients are random variables, namely, functions of the random variables $\bar{X}$, $\bar{Y}$, S_1, S_2, and R. Thus we have the statistics

(9.10.1) $A = R\dfrac{S_2}{S_1}, \qquad B = \bar{Y} - R\dfrac{S_2}{S_1}\bar{X} = \bar{Y} - A\bar{X}.$

Similarly, for the second regression line we have

(9.10.2) $A' = R\dfrac{S_1}{S_2}, \qquad B' = \bar{X} - R\dfrac{S_1}{S_2}\bar{Y}.$

The statistics A and A' are called *sample regression coefficients*.

Let $(\bar{X}, \bar{Y}, S_1, S_2, R)$ have the distribution given by (9.9.2) and (9.9.3).

We shall find the distribution of A. To do this, let us first write down the density $g(s_1^2, s_2^2, r)$ of (S_1^2, S_2^2, R). From (9.9.3) we obtain

$$(9.10.3) \quad g(s_1^2, s_2^2, r) = \frac{n^{n-1}}{4\pi\Gamma(n-2)|\mathbf{M}|^{(n-1)/2}} s_1^{n-3} s_2^{n-3}(1 - r^2)^{(n-4)/2}$$

$$\times \exp\left[-\frac{n}{2|\mathbf{M}|}(\sigma_2^2 s_1^2 - 2\rho r\sigma_1\sigma_2 s_1 s_2 + \sigma_1^2 s_2^2)\right],$$

where $|\mathbf{M}| = \sigma_1^2\sigma_2^2(1 - \rho^2)$.

For the random variable (S_1^2, S_2^2, A), we obtain the density function

$$(9.10.4) \quad f(s_1^2, s_2^2, a) = \frac{n^{n-1}}{4\pi\Gamma(n-2)|\mathbf{M}|^{(n-1)/2}} (s_2^2 - s_1^2 a^2)^{(n-4)/2}$$

$$\times \exp\left(-\frac{n}{2|\mathbf{M}|}\sigma_1^2 s_2^2\right) s_1^{n-2} \exp\left[-\frac{n}{2|\mathbf{M}|}(\sigma_2^2 s_1^2 - 2\rho\sigma_1\sigma_2 s_1^2 a)\right].$$

Since from (9.10.1) it follows that $s_2^2 - s_1^2 a^2 \geqslant 0$, let us integrate (9.10.4) first with respect to s_2^2 in the domain $s_2^2 - s_1^2 a^2 \geqslant 0$ and then with respect to s_1^2 in the domain $(0, +\infty)$. Letting $y = s_2^2 - s_1^2 a^2$, we obtain

$$(9.10.5) \quad \int_{s_2^2 - s_1^2 a^2 \geqslant 0} (s_2^2 - s_1^2 a^2)^{(n-4)/2} \exp\left(-\frac{n}{2|\mathbf{M}|}\sigma_1^2 s_2^2\right) ds_2^2$$

$$= \exp\left(-\frac{n}{2|\mathbf{M}|}\sigma_1^2 s_1^2 a^2\right) \int_0^\infty y^{(n-4)/2} \exp\left(-\frac{n\sigma_1^2}{2|\mathbf{M}|} y\right) dy$$

$$= \frac{\Gamma[\frac{1}{2}(n-2)]2^{(n-2)/2}|\mathbf{M}|^{(n-2)/2}}{n^{(n-2)/2}\sigma_1^{2(n-2)/2}} \exp\left(-\frac{n}{2|\mathbf{M}|}\sigma_1^2 s_1^2 a^2\right).$$

Next we have

$$(9.10.6) \quad \int_0^\infty s_1^{2(n-2)/2} \exp\left[-\frac{n}{2|\mathbf{M}|}(\sigma_2^2 - 2\rho\sigma_1\sigma_2 a + \sigma_1^2 a^2)s_1^2\right] ds_1^2$$

$$= \frac{\Gamma(\frac{1}{2}n)2^{n/2}|\mathbf{M}|^{n/2}}{n^{n/2}(\sigma_2^2 - 2\rho\sigma_1\sigma_2 a + \sigma_1^2 a^2)^{n/2}}.$$

Considering the fact that

$$\Gamma(2p) = \frac{2^{2p-1}\Gamma(p)\Gamma(p + \frac{1}{2})}{\sqrt{\pi}},$$

we obtain from (9.10.4) to (9.10.6) the density function of A, namely,

$$(9.10.7) \quad h(a) = \frac{\Gamma(\frac{1}{2}n)|\mathbf{M}|^{(n-1)/2}}{\sqrt{\pi}\Gamma[\frac{1}{2}(n-1)]\sigma_1^{2(n-2)/2}(\sigma_2^2 - 2\rho\sigma_1\sigma_2 a + \sigma_1^2 a^2)^{n/2}}.$$

This formula has been given by K. Pearson [3] and Romanovsky [2]. The density function $h_1(a')$ of A' can be obtained by interchanging σ_1 and σ_2 in (9.10.7).

We leave to the reader the problem of finding the distribution of the statistic $B = \overline{Y} - A\overline{X}$. It is easy to find it if one remembers that $(\overline{X}, \overline{Y})$ has the two-dimensional normal distribution given by (9.9.2), A has the distribution given by (9.10.7), and $(\overline{X}, \overline{Y})$ and A are independent.

We now find $E(A)$ and $D^2(A)$. For simplicity let

$$K = \frac{\Gamma(\tfrac{1}{2}n)|\mathbf{M}|^{(n-1)/2}}{\sqrt{\pi}\,\Gamma[\tfrac{1}{2}(n-1)]\sigma_1^{n-2}}, \qquad V = \sigma_2^2 - 2\rho\sigma_1\sigma_2 a + \sigma_1^2 a^2$$

From (9.10.7) we obtain, for $n > 2$,

$$(9.10.8) \quad E(A) = K \int_{-\infty}^{\infty} \frac{a}{V^{n/2}}\,da$$

$$= \frac{1}{2\sigma_1^2} K \int_{-\infty}^{\infty} \frac{2\sigma_1^2 a - 2\rho\sigma_1\sigma_2}{V^{n/2}}\,da + \rho\frac{\sigma_2}{\sigma_1} K \int_{-\infty}^{\infty} \frac{da}{V^{n/2}} = \rho\frac{\sigma_2}{\sigma_1}.$$

Similarly, for $n > 3$,

$$E(A^2) = K \int_{-\infty}^{\infty} \frac{a^2}{V^{n/2}}\,da$$

$$= \frac{1}{\sigma_1^2} K \int_{-\infty}^{\infty} \frac{1}{V^{(n-2)/2}}\,da + 2\rho\frac{\sigma_2}{\sigma_1} K \int_{-\infty}^{\infty} \frac{a\,da}{V^{n/2}} - \frac{\sigma_2^2}{\sigma_1^2} K \int_{-\infty}^{\infty} \frac{1}{V^{n/2}}\,da.$$

We observe that from formula (9.10.7) with n replaced by $n - 2$ it follows that

$$\int_{-\infty}^{\infty} \frac{1}{V^{(n-2)/2}}\,da = \frac{\sqrt{\pi}\,\Gamma[\tfrac{1}{2}(n-3)]\sigma_1^{n-4}}{\Gamma[\tfrac{1}{2}(n-2)]|\mathbf{M}|^{(n-3)/2}},$$

so that after some simple computations we obtain

$$E(A^2) = \frac{\sigma_2^2}{\sigma_1^2} \cdot \frac{1 - 4\rho^2 + \rho^2 n}{n - 3}.$$

From the last equality and from (9.10.8) it follows that

$$(9.10.9) \qquad D^2(A) = \frac{1}{n - 3} \cdot \frac{\sigma_2^2}{\sigma_1^2}(1 - \rho^2).$$

The expected value and the variance of $B = \overline{Y} - A\overline{X}$ can be obtained easily, since we know the corresponding moments of $\overline{X}$, $\overline{Y}$ and A, and $(\overline{X}, \overline{Y})$ and A are independent. We obtain

$$(9.10.10) \qquad E(B) = m_2 - \rho\frac{\sigma_2}{\sigma_1} m_1,$$

$$(9.10.11) \qquad D^2(B) = \frac{\sigma_2^2}{n} \cdot \frac{n - 2 - \rho^2}{n - 3} + \frac{\sigma_2^2}{\sigma_1^2} \cdot \frac{m_1^2(1 - \rho^2)}{n - 3}.$$

Neglecting terms of order smaller than n^{-1} in the last equality, we obtain

$$(9.10.12) \qquad D^2(B) \cong \frac{\sigma_2^2}{(n-3)\sigma_1^2} [\sigma_1^2 + m_1^2(1-\rho^2)].$$

The following transformation is very convenient. Write

$$(9.10.13) \qquad t = \frac{\sigma_1\sqrt{n-1}}{\sigma_2\sqrt{1-\rho^2}} (A - \alpha),$$

where $\alpha = \rho \dfrac{\sigma_2}{\sigma_1}$. Then

$$1 + \frac{t^2}{n-1} = \frac{A^2\sigma_1^2 - 2\rho\sigma_1\sigma_2 A + \sigma_2^2}{|M|} \sigma_1^2.$$

By (9.10.7), the density function of the random variable t is

$$(9.10.14) \quad f(t) = \frac{1}{\sqrt{n-1}} \cdot \frac{1}{B[\frac{1}{2}(n-1), \frac{1}{2}]} \cdot \frac{1}{[1 + t^2/(n-1)]^{n/2}}.$$

Thus t defined by (9.10.13) has Student's t-distribution with $n - 2$ degrees of freedom.

In (9.10.13), the random variable t was defined as a function of the random variable A and of the population parameters σ_1, σ_2, and ρ. Usually, in practice these parameters are unknown; hence the following fact discovered by Bartlett [1] is of great importance. The random variable t_1 defined as

$$(9.10.15) \qquad t_1 = \frac{S_1\sqrt{n-2}}{S_2\sqrt{1-R^2}} (A - \alpha)$$

has Student's t distribution with $n - 2$ degrees of freedom. The reader will notice that if in formula (9.10.13) we replace the parameters σ_1, σ_2, and ρ by S_1, S_2, and R, respectively, and $n - 1$ by $n - 2$, we obtain t_1 given by (9.10.15).

We do not present the proof of Bartlett's theorem since it could easily be derived from (9.9.3). It is necessary first to find the joint distribution of S_1, S_2, and t_1 and then to find the marginal distribution of t_1.

9.11 LIMIT DISTRIBUTIONS OF SAMPLE MOMENTS

In the preceding sections, we discussed the exact distributions of statistics. The question of finding such a distribution is sometimes very difficult and often we obtain complicated formulas of little practical use. On the other hand, it is sometimes relatively easy to obtain the limit distributions of a large class of statistics which appear in applications. The practical application of limit distributions is based on the fact that if the size n of the sample is sufficiently big, the distribution of the statistic is close to the limit distribution obtained for $n \to \infty$. The results of this section can be applied to large sample investigations, that is, they can be applied if the size n of the sample is sufficiently big. There is no general rule valid for all statistics which would allow us to determine whether the size n of the sample is sufficiently large. This depends on the rapidity of approach of the distribution of the statistic considered to the limit distribution.

Let $X_1, X_2, \ldots, X_n$ be independent, identically distributed random variables with the same distribution as that of the characteristic X of the population. Let us assume that the moment of order k, $m_k = E(X^k)$, exists. Consider the statistic

$$A_k = \frac{1}{n} \sum_{r=1}^{n} X_r^k.$$

We observe that

$$E(A_k) = \frac{1}{n} \sum_{r=1}^{n} E(X_r^k).$$

By assumption, we have $E(X_r^k) = E(X^k)$ for $r = 1, 2, \ldots, n$, so that

$$E(A_k) = \frac{1}{n} \sum_{r=1}^{n} E(X^k) = m_k.$$

The statistic A_k is called the *sample moment* of order k. From the theorem of Khintchin it follows that for every k the sequence $\{A_k\}$ of sample moments is stochastically convergent to the population moment m_k as $n \to \infty$. By Slutsky's theorem 6.14.2, it follows that the sequence $\{B_k\}$ of *central sample moments*

$$B_k = \frac{1}{n} \sum_{r=1}^{n} (X_r - A_1)^k$$

is stochastically convergent to the population central moment μ_k as $n \to \infty$. Thus, for instance, as $n \to \infty$ the sample variance is stochastically convergent to the population variance.

If we assume the existence of the moment m_{2k} of order $2k$ of the random variable X, then it is easy to derive the formula for the variance of A_k. For $r = 1, 2, \ldots, n$ we have

$$D_2(X_r^k) = D_2(X^k) = E(X^{2k}) - [E(X^k)]^2 = m_{2k} - m_k^2,$$

and, by the independence of the random variables $X_1, \ldots, X_n$,

$$D^2(A_k) = \frac{1}{n^2} \sum_{r=1}^{n} D^2(X_r^k) = \frac{n}{n^2} D^2(X^k) = \frac{m_{2k} - m_k^2}{n}.$$

From the Lindeberg-Lévy limit theorem (see Section 6.8) it follows that as

$n \to \infty$ the sequence of distribution functions of the standardized random variables

$$Y_n = \frac{\sum_{r=1}^{n} X_r^k - nm_k}{\sqrt{n\, D^2(X^k)}} = \frac{A_k - m_k}{\sqrt{m_{2k} - m_k^2}} \sqrt{n}$$

tends to the limit distribution function $\Phi(x)$. Thus we can say that if the moment m_{2k} of the population exists, the distribution of the sample moment A_k of order k is asymptotically normal $N(m_k; \sqrt{(m_{2k} - m_k^2)/n})$.

Example 9.11.1. A collection of IBM cards corresponds to the clients of an insurance company. The insured are divided into two groups, the first group consists of those who have families, the second consists of those who are without families. To the members of the first group corresponds the number one on the IBM cards; to the members of the second group, the number zero. The fraction of insured belonging to the first category is p, hence the fraction of insured belonging to the second category is $q = 1 - p$.

From this collection of IBM cards we select a simple sample of n elements, always returning the selected card, so that the fraction p remains constant while the selection is made. Let us denote the result of the rth drawing by x_r, where x_r may take on the value zero or one. Thus x_r is the observed value of the random variable X_r with a zero-one distribution. The arithmetic mean $\bar{x}$ of the x_r is the observed value of

$$A_1 = \bar{X} = \frac{1}{n} \sum_{r=1}^{n} X_r.$$

Let n be large, say $n = 100$, and let $p = 0.3$. What is the probability that $\bar{X}$ will differ from 0.3 by not less than 0.02?

We shall use the theorem proved previously, that if the moment m_{2k} exists, then the sample moment A_k has the asymptotically normal distribution

$$N(m_k; \sqrt{(m_{2k} - m_k^2)/n}).$$

In our example

$$m_1 = m_2 = p = 0.3,$$

so that A_1 has the asymptotically normal distribution $N(0.3; 0.0458)$. We are looking for the probability

$$P(|A_1 - 0.3| \geq 0.02) = P\left(\frac{|A_1 - 0.3|}{0.0458} \geq 0.44\right).$$

From tables of the normal distribution we find that the required probability equals 0.66.

Problems and Complements

9.1.(a) Show that the density (9.6.5) of Student's t satisfies for every t the relation

$$\lim_{n \to \infty} g(t) = \frac{1}{\sqrt{2\pi}} \exp\left(-\frac{t^2}{2}\right).$$

(b) Prove (9.6.6) using Problem 9.1.(a).

9.2. Show that the mean $\bar{X}$ and the variance S^2 of simple samples drawn from a population in which the characteristic X has a symmetric distribution are uncorrelated.

9.3. Derive formula (9.6.7), proceeding in the same way as in the derivation of the density of Student's t.

9.4. Derive the formula for the density of the random variable $Y = t^2$, where t has the distribution given by (9.6.5), and compare the result with (9.6.15) for $r = 1$.

9.5. In formula (9.6.14) for Hotelling's T^2, let $r = 2$ and $n = 4$. Find values y such that (a) $P(T^2 > y) = 0.05$ (b) $P(T^2 > y) = 0.01$.

9.6. Derive the density of Snedecor's F.

9.7. Let $X_k (k = 1, \ldots, n_1)$ and $Y_l (l = 1, \ldots, n_2)$ be independent, and let X_k have the distribution $N(0; \sigma_1)$ and Y_l have the distribution $N(0; \sigma_2)$. Let

$$Z = \tfrac{1}{2} \log \frac{n_1(n_2 - 1) S_1^2}{n_2(n_1 - 1) S_2^2}.$$

Show that the density $h(z)$ of Z has the form

$$h(z) = \frac{2\sigma_1^{-r_1}\sigma_2^{-r_2}r_1^{r_1/2}r_2^{r_2/2}}{B\left(\dfrac{r_1}{2}, \dfrac{r_2}{2}\right)} \cdot \frac{e^{r_1 z}}{\left(\dfrac{r_1 e^{2z}}{\sigma_1^2} + \dfrac{r_2}{\sigma_2^2}\right)^{(r_1+r_2)/2}},$$

where $r_i = n_i - 1$ $(i = 1, 2)$.

9.8. Find the distribution of the mean $\bar{X}$ of simple samples of size n drawn from a population in which the characteristic X has the gamma distribution.

9.9. Let $\bar{X}$ be the mean of simple samples of size n drawn from a population in which the characteristic X has the uniform distribution given by (5.6.6). Prove that the density $g(x)$ of $\bar{X}$ is for $j = 0, 1, \ldots, n - 1$ of the form

$$g(x) = \frac{n^n}{(n - 1)!} \sum_{k=0}^{j} (-1)^k \binom{n}{k} \left(x - \frac{k}{n}\right)^{n-1}, \quad \left(\frac{j}{n} \leqslant x \leqslant \frac{j + 1}{n}\right).$$

9.10. Let $(X_1, \ldots, X_l)$, where $l \geqslant 3$, be a random vector and let the matrix $\mathbf{M}$ of second-order moments λ_{ji} $(j, i = 1, \ldots, l)$ have the determinant $|\mathbf{M}| \neq 0$. Let $(X_{k1}, \ldots, X_{kl})$ $(k = 1, \ldots, n)$ be independent observations of the random vector $(X_1, \ldots, X_l)$. Furthermore, let $\mathbf{Q}$ be the second-order sample moment matrix, that is, the matrix of elements W_{ji} defined by (9.6.12) and (9.6.13). Let R_{ji} $(j, i = 1, \ldots, l)$ denote the sample correlation coefficient of the random variables X_i and X_j and let $\mathbf{C}$ denote the matrix of elements R_{ij} and $|\mathbf{C}|$ its determinant (we set $R_{jj} = 1$).

Prove that if $(X_1, \ldots, X_l)$ has the normal distribution, with $\lambda_{jj} = 1$ and $\lambda_{ij} = 0$ $(j \neq i)$, then the vector $(R_{12}, R_{13}, \ldots, R_{l-1,l})$ has the density

$$\frac{\left[\Gamma\left(\dfrac{n - 1}{2}\right)\right]^{l-1}}{\pi^{\frac{l(l-1)}{4}} \Gamma\left(\dfrac{n - 2}{2}\right) \ldots \Gamma\left(\dfrac{n - l}{2}\right)} \cdot |\mathbf{C}|^{\frac{n-l-2}{2}}.$$

9.11. (Continued). The *sample partial correlation coefficient of* X_1 *and* X_2 *with respect to* $X_3, \ldots, X_l$ is defined by the formula (see Problem 3.15)

$$R_{12 \cdot 3 \cdots l} = -\frac{|\mathbf{Q}_{12}|}{\sqrt{|\mathbf{Q}_{11}| \cdot |\mathbf{Q}_{22}|}} = \frac{-|\mathbf{C}_{12}|}{\sqrt{|\mathbf{C}_{11}| \cdot |\mathbf{C}_{22}|}},$$

where $|\mathbf{Q}_{ji}|$ and $|\mathbf{C}_{ji}|$ are the algebraic complements of the term W_{ji} in the determinant $|\mathbf{Q}|$ and of the term R_{ji} in the determinant $|\mathbf{C}|$, respectively.

(a) Prove that if $(X_1, \ldots, X_l)$ has the normal distribution, then the density $f(r_{12 \cdot 3 \cdots l})$ of the random variable $R_{12 \cdot 3 \cdots l}$ is of the form (Fisher [13])

$$f(r_{12 \cdot 3 \cdots l}) = \frac{n - l}{\pi} (1 - \rho_{12 \cdot 3 \cdots l}^2)^{\frac{n-l+1}{2}} (1 - r_{12 \cdot 3 \cdots l}^2)^{(n-l-2)/2}$$

$$\times \int_0^1 \frac{x^{n-l}}{(1 - \rho_{12 \cdot 3 \cdots l} r_{12 \cdot 3 \cdots l})^{n-l+1} \sqrt{1 - x^2}} \, dx.$$

(b) Show that in the particular case when $(X_1, \ldots, X_l)$ has a normal distribution and $\rho_{12 \cdot 3 \cdots l} = 0$,

$$f(r_{12 \cdot 3 \cdots l}) = \frac{1}{B[\frac{1}{2}(n - l), \frac{1}{2}]} (1 - r_{12 \cdot 3 \cdots l}^2)^{(n-l-2)/2}.$$

(c) Compare these formulas with (9.9.4) and (9.9.6), respectively. What should we put instead of n in formulas (9.9.4) and (9.9.6) in order to obtain these formulas?

9.12. (Continued). Prove that if $\rho_{12 \cdot 3 \cdots l} = 0$, the random variable

$$\frac{R_{12 \cdot 3 \cdots l}}{\sqrt{1 - R_{12 \cdot 3 \cdots l}^2}} \sqrt{n - l}$$

has Student's t-distribution with $n - l$ degrees of freedom.

9.13. (Continued). The *sample multiple correlation coefficient between* X_1 *and* $(X_2, \ldots, X_l)$ is defined as (see Problem 3.17)

$$R_{1(2 \cdots l)} = \sqrt{1 - \frac{|\mathbf{Q}|}{W_{11} |\mathbf{Q}_{11}|}} = \sqrt{1 - \frac{|\mathbf{C}|}{|\mathbf{C}_{11}|}}.$$

Prove that if $(X_1, \ldots, X_l)$ has a normal distribution with $\rho_{ji} = 0$ $(j, i = 1, \ldots, l,$ $j \neq i)$, then $R_{1(2 \cdots l)}^2$ has the beta distribution with parameters $(l - 1)/2$ and $(n - l)/2$ (Fisher [14]. The distribution of $R_{1(2 \cdots l)}^2$ without the assumption $\rho_{ji} = 0$ was also given by Fisher [15]).

9.14. (Continued). Prove that the distribution of $nR_{1(2 \cdots l)}^2$ converges, as $n \to \infty$, to the χ^2 distribution with $l - 1$ degrees of freedom; hence it does not converge to the normal distribution.

In Problems 9.15 to 9.21, we use the notation of Section 9.11 and we assume that all the moments of the population which appear in these problems exist. For the meaning of $0(\ )$, see Problem 8.31.

9.15. Show that the following equalities hold:

$$E[(A_1 - m_1)^3] = \frac{\mu_3}{n^2},$$

$$E[(A_1 - m_1)^4] = 3\mu_2/n^2 + \frac{\mu_4 - 3\mu_2^2}{n^3}.$$

9.16. Show that for $j = 1, 2, \ldots$ as $n \to \infty$

$$E[(A_n - m_n)^{2j-1}] = 0\left(\frac{1}{n^j}\right),$$

$$E[(A_n - m_n)^{2j}] = 0\left(\frac{1}{n^j}\right).$$

9.17.(a) Show that

$$E(B_2{}^2) = \mu_2{}^2 + \frac{\mu_4 + 3\mu_2{}^2}{n} - \frac{2\mu_2 - 5\mu_2{}^2}{n^2} + \frac{\mu_4 - 3\mu_2{}^2}{n^3}.$$

(b) Using the last formula and (13.3.4), derive the formula for $D^2(B_2)$.

(c) Using formula (5.7.7) in addition to the preceding, prove that if the characteristic X of the population has the normal distribution $N(m; \sigma)$, then

$$D^2(B_2) = \frac{2(n-1)}{n^2}\sigma^4.$$

9.18. Show that

$$E\left[(A_1 - m_1)\left(B_2 \frac{n-1}{n}\mu_2\right)\right] = \frac{n-1}{n^2}\mu_3.$$

9.19. Show that

$$E(B_k) = \mu_k + 0\left(\frac{1}{n}\right),$$

$$D^2(B_k) = \frac{\mu_{2k} - 2k\mu_{k-1}\mu_{k+1} - \mu_k{}^2 + k^2\mu_2\mu_{k-1}^2}{n} + 0\left(\frac{1}{n^2}\right).$$

9.20. Show that the central moment of order k of simple samples has, as $n \to \infty$, the asymptotically normal distribution

$$N\left(\mu_k; \sqrt{\frac{\mu_{2k} - 2k\mu_{k-1}\mu_{k+1} - \mu_k{}^2 + k^2\mu_2\mu_{k-1}^2}{n}}\right).$$

9.21. Find the limit distribution as $n \to \infty$ of the random vector $[\sqrt{n}(A_1 - m_1), \sqrt{n}(A_2 - m_2)]$.

Hint: Use theorem 6.13.2.

CHAPTER 10

Order Statistics

10.1 PRELIMINARY REMARKS

In the preceding chapter, we investigated the distributions of sample moments and some of their functions. The sample moment corresponds to the population moment. In this chapter, we investigate the distributions of order statistics, which correspond to order parameters of the population. Thus we introduce the sample median and, generally, the sample quantile. The theory of distributions of these statistics, in particular the limit distributions of these statistics, is becoming more and more important in probability theory and mathematical statistics.

The close relation between the sample and the population is described by Glivenko's theorem (see Section 10.10) and the theorems of Kolmogorov-Smirnov (see Section 10.11). These theorems are among the fundamental results of the theory of order statistics and have many applications.

10.2 THE NOTION OF AN ORDER STATISTIC

In Chapter 3, we considered order parameters. Particular cases of order parameters are the median, quantile, and the smallest element. Here we consider statistics which correspond to order parameters. These statistics are called *order statistics*.

We now define this notion. Let

$$(X_1, X_2, \ldots, X_n)$$

be an n-dimensional random vector, and let the random variable $\zeta_k^{(n)}$ be a function of this random vector defined as follows.

We arrange each collection of values $(x_1, \ldots, x_n)$ of $(X_1, \ldots, X_n)$ in increasing order, that is, in such a way that we obtain a sequence $x_{r_1}, \ldots, x_{r_n}$ satisfying the inequalities

$$x_{r_1} \leqslant x_{r_2} \leqslant \ldots \leqslant x_{r_n}.$$

372

If two coordinates x_i and x_j are equal, their order is irrelevant.

Definition 10.2.1. The function of $(X_1, \ldots, X_n)$ which takes on the value x_{r_k} in each possible sequence $x_1, \ldots, x_n$ is called an *order statistic* and is denoted by $\zeta_k^{(n)}$. The number k is called the *rank of* $\zeta_k^{(n)}$.

For a given n, we can form n such order statistics, namely,

$$\zeta_1^{(n)}, \zeta_2^{(n)}, \ldots, \zeta_n^{(n)}.$$

Example 10.2.1. Each of the random variables X_1, X_2, X_3 of the discrete type can take on three values $0,1,2$. The random variable (X_1, X_2, X_3) can take on the following collections of values: $(0, 0, 0)$, $(0, 0, 1)$, $(0, 1, 0)$, $(1, 0, 0)$, $(0, 0, 2)$, $(0, 2, 0)$, $(2, 0, 0)$, $(1, 1, 0)$, $(1, 0, 1)$, $(0, 1, 1)$, $(0, 1, 2)$, $(0, 2, 1)$, $(2, 0, 1)$, $(2, 1, 0)$, $(1, 0, 2)$, $(1, 2, 0)$, $(0, 2, 2)$, $(2, 0, 2)$, $(2, 2, 0)$, $(1, 1, 1)$, $(1, 1, 2)$, $(1, 2, 1)$, $(2, 1, 1)$, $(1, 2, 2)$, $(2, 1, 2)$, $(2, 2, 1)$, $(3, 2, 2)$.

Here we can form three order statistics,

$$\zeta_1^{(3)}, \zeta_2^{(3)}, \zeta_3^{(3)}.$$

The first will take on the smallest value in each of these collections of values. The remaining two statistics are defined in a similar way.

Definition 10.2.2. The ratio k/n is called the *relative rank* of $\zeta_k^{(n)}$.

We often investigate the limit distributions of sequences $\{\zeta_k^{(n)}\}$ of order statistics as $n \to \infty$, where k may be a function of n.

We distinguish two types of sequences of order statistics.

Definition 10.2.3. If $\lim_{n \to \infty} (k/n) = 0$ or $\lim_{n \to \infty} (k/n) = 1$, the sequence of statistics $\{\zeta_k^{(n)}\}$ is called a *sequence of end rank statistics*; if $\lim_{n \to \infty} (k/n) = \lambda$, where $0 < \lambda < 1$, the sequence $\{\zeta_k^{(n)}\}$ is called a *sequence of central statistics*.

We obtain sequences of end rank statistics if for k we take a constant such as $1, 2, \ldots$ or a number of the form $n - l$, where l is a constant. When $n \to \infty$, in the first case we have $\lim_{n \to \infty} (k/n) = 0$ and in the second case $\lim_{n \to \infty} (k/n) = 1$.

Example 10.2.2. From a population select a series of samples of size n with increasing n. In each sample we observe the smallest value of the characteristic X, hence the value of the random variable $\zeta_1^{(n)}$. This is a sequence of end rank statistics.

One should not think that sequences of end rank statistics are obtained only for a constant k or for k of the form $n - l$ with constant l. We can also obtain sequences of end rank statistics if k increases with n to infinity, but in such a way that $k/n \to 0$ or l increases with n to infinity, but in such a way that $(n - l)/n \to 1$.

Example 10.2.3. Select from a population a series of samples of size n with increasing n. In each sample we observe the magnitude of the kth element, where

$k = [\sqrt{n}]$.[1] Thus for samples with

$$n = 2, 3, 4, 5, 6, 7, 8, 9, 10, 11, 12, 13, 14, 15, 16, 17, \ldots$$

elements, we have

$$k = 1, 1, 2, 2, 2, 2, 2, 3, 3, 3, 3, 3, 3, 3, 4, 4, \ldots .$$

This is a sequence of end rank statistics, since

$$\lim_{n \to \infty} (k/n) = \lim_{n \to \infty} [\sqrt{n}]/n = \lim_{n \to \infty} \sqrt{n}/n = \lim_{n \to \infty} 1/\sqrt{n} = 0.$$

Let us consider the case $k = [n/2] + 1$, that is, for odd n we take the central element (with respect to the magnitude) of the characteristic under investigation and for even $n = 2l$ we take the $(l + 1)$-st element, hence the second of two central elements. This statistic is called the *sample median*.

If we take $k = [n\lambda] + 1$, where λ is an arbitrary real number such that $0 < \lambda < 1$, then the statistic $\zeta_k^{(n)}$ is called a *sample quantile*. It is obvious that $k/n \to \lambda$ as $n \to \infty$, hence a sequence of sample quantiles is a sequence of central statistics.

If k is constant or $k = n - l + 1$ with l constant, then $\zeta_k^{(n)}$ is called a *successive element*.

Functions of order statistics are also called *order statistics*.

10.3 THE EMPIRICAL DISTRIBUTION FUNCTION

A   Let $(X_1, X_2, \ldots, X_n)$ be a random vector and let $(\zeta_1^{(n)}, \zeta_2^{(n)}, \ldots, \zeta_n^{(n)})$ and $(x_{r_1}, \ldots, x_{r_n})$ have the same meaning as in Section 10.2.

Definition 10.3.1.   The function of x $(-\infty < x < \infty)$ equal to 0 for $x \leqslant x_{r_1}$, and equal to m/n $(m = 1, \ldots, n)$ for $x > x_{r_1}$, where m is the greatest index for which

$$(10.3.1) \qquad\qquad x_{r_m} < x,$$

is called the *empirical distribution function*.

As is customary, we denote the empirical distribution function by $S_n(x)$.

Thus $nS_n(x)$ is the number of sample elements that are smaller than x. The function $S_n(x)$ may take on values between zero and one and is a non-decreasing function of x. It is easy to verify that $S_n(x)$ is continuous from the left and satisfies all the other properties of a distribution function. Hence the name empirical distribution function. However, $S_n(x)$ is a random variable for every x.

B   Now let the random variables X_r $(r = 1, \ldots, n)$ be independent and have the same continuous distribution function $F(x)$. Let $x_{r_1}, \ldots, x_{r_n}$ denote the collection of values of the observations of the random variables

---

[1] By the symbol $[A]$ we denote here the greatest integer not exceeding A.

$X_1, \ldots, X_n$ arranged in order of magnitude. From the assumption that the distribution function $F(x)$ is continuous and that the random variables $X_1, \ldots, X_n$ are independent, it follows that the probability of the appearance of two equal values x_{r_k} and $x_{r_{k+1}}$ is zero. Hence we may assume that $x_{r_1} < x_{r_2} < \ldots < x_{r_n}$. Thus we can represent the empirical distribution function in the following form:

$$(10.3.2) \qquad S_n(x) = \begin{cases} 0 & \text{for } x \leqslant x_{r_1}, \\[2mm] \dfrac{k}{n} & \text{for } x_{r_k} < x \leqslant x_{r_{k+1}} \\[2mm] & \text{and } k = 1, 2, \ldots, n - 1, \\[2mm] 1 & \text{for } x > x_{r_n}. \end{cases}$$

C   Let now $X_1, \ldots, X_n$ be independent and have the same distribution function $F(x)$. From the assumptions it follows that for every fixed x we have

$$(10.3.3) \quad P(X_r < x) = F(x) = p = \text{constant} \ (r = 1, \ldots, n).$$

Hence, for a fixed value of x, $S_n(x)$ is the frequency of successes in the Bernoulli scheme, hence the frequency of realizations of some random event in n independent trials, where the probability of the realization of this event in each particular trial is constant and equal to $p = F(x)$. Thus (see Section 5.2)

$$(10.3.4) \quad P\left[S_n(x) = \frac{m}{n}\right] = \frac{n!}{m!\,(n - m)!} [F(x)]^m [1 - F(x)]^{n - m}$$
$$(m = 0, 1, \ldots, n).$$

Let us denote the distribution function of the random variable $\zeta_k^{(n)}$ defined in the preceding section by $\Phi_{kn}(x)$,

$$(10.3.5) \qquad\qquad \Phi_{kn}(x) = P(\zeta_k^{(n)} < x).$$

We express the distribution function $\Phi_{kn}(x)$ by $S_n(x)$. The event $\zeta_k^{(n)} < x$ occurs when the kth observation, in order of magnitude, is smaller than x; thus when at least k observations are smaller than x. This is equivalent to the event that the observed frequency $s_n(x)$ is not smaller than k/n. Hence

$$(10.3.6) \quad \Phi_{kn}(x) = P(\zeta_k^{(n)} < x) = P\left[S_n(x) \geqslant \frac{k}{n}\right] = \sum_{m=k}^{n} P\left[S_n(x) = \frac{m}{n}\right].$$

By (10.3.4) and (10.3.6), we obtain

$$(10.3.7) \qquad \Phi_{kn}(x) = \sum_{m=k}^{n} \frac{n!}{m!\,(n - m)!} [F(x)]^m [1 - F(x)]^{n - m}$$

We can also obtain another form of this distribution function,

$$(10.3.8) \qquad \Phi_{kn}(x) = \frac{n!}{(k-1)!\,(n-k)!} \int_0^{F(x)} t^{k-1}(1-t)^{n-k}\,dt,$$

which can be reduced to (10.3.7) by repeated integration by parts. We do not give the details of the proof.

Let us now assume that the density $f(x) = F'(x)$ exists. Then the density $f_{kn}(x)$ of the random variable $\zeta_k^{(n)}$ exists. By differentiating (10.3.8) with respect to x, we obtain

$$(10.3.9) \quad f_{kn}(x) = \frac{n!}{(k-1)!\,(n-k)!}\,[F(x)]^{k-1}[1-F(x)]^{n-k}f(x).$$

D   Let $F(x)$ be continuous. We derive a relation between the distribution functions $\Phi_{n+k+1,n}(x)$ and $\Phi_{kn}(x)$, or between the distribution functions of the random variables $\zeta_{n-k+1}^{(n)}$ and $\zeta_k^{(n)}$. Let

$$\bar{F}(x) = P(-X < x) = P(X > -x) = 1 - F(-x).$$

Let us consider the independent random variables $-X_1, -X_2, \ldots, -X_n$ with the common distribution function $\bar{F}(x) = 1 - F(-x)$. Let $\bar{\zeta}_k^{(n)}$ denote the order statistic of the random variables $-X_1, \ldots, -X_n$ defined in the same way as the statistic $\zeta_k^{(n)}$ for the random variables $X_1, \ldots, X_n$. Let us denote its distribution function by $\bar{\Phi}_{kn}(x)$. By (10.3.8) we obtain

$$(10.3.10) \quad \bar{\Phi}_{kn}(x) = \frac{n!}{(k-1)!\,(n-k)!} \int_0^{\bar{F}(x)} t^{k-1}(1-t)^{n-k}\,dt.$$

It follows that

$$(10.3.11) \quad \bar{\Phi}_{kn}(-x) = \frac{n!}{(k-1)!\,(n-k)!} \int_0^{1-F(x)} t^{k-1}(1-t)^{n-k}\,dt.$$

By (10.3.8)

$$(10.3.12) \quad \Phi_{n-k+1,n}(x) = \frac{n!}{(n-k)!\,(k-1)!} \int_0^{F(x)} t^{n-k}(1-t)^{k-1}\,dt.$$

Substituting $y = 1 - t$ in (10.3.12), we obtain

$$(10.3.13) \quad \Phi_{n-k+1,n}(x) = -\frac{n!}{(n-k)!\,(k-1)!} \int_1^{1-F(x)} (1-y)^{n-k}y^{k-1}\,dy$$

$$= \frac{n!}{(n-k)!\,(k-1)!} \int_{1-F(x)}^1 y^{k-1}(1-y)^{n-k}\,dy.$$

Comparing (10.3.11) and (10.3.13), we have

$$(10.3.14) \qquad \Phi_{n-k+1,n}(x) = 1 - \bar{\Phi}_{kn}(-x).$$

Example 10.3.1. Suppose that

$$f(x) = \frac{1}{\sqrt{2\pi}} e^{-x^2/2}.$$

In a series of simple samples of two elements chosen from a population in which the characteristic X has the normal distribution $N(0; 1)$, let us arrange the observations obtained in order of magnitude. Thus in each sample we have two values, x_1 and x_2. Consider the distribution of the statistic $\zeta_1^{(2)}$. Suppose that $x_1 < x_2$.

In every sample the random variable $\zeta_1^{(2)}$ takes on the smaller value x_1. From (10.3.9) we find its density

$$f_{12}(x) = \frac{2}{\sqrt{2\pi}} e^{-x^2/2} \left(\frac{1}{\sqrt{2\pi}} \int_x^\infty e^{-t^2/2} \, dt \right).$$

The density of $\zeta_2^{(2)}$ is

$$f_{22}(x) = \frac{2}{\sqrt{2\pi}} e^{-x^2/2} \left(\frac{1}{\sqrt{2\pi}} \int_{-\infty}^x e^{-t^2/2} \, dt \right).$$

Example 10.3.2. Let us consider simple samples of size n from a population with the density

$$f(x) = \begin{cases} 1 & \text{for } 0 \leqslant x \leqslant 1, \\ 0 & \text{otherwise.} \end{cases}$$

Let us find the distribution of $\zeta_k^{(n)}$.
By (10.3.9) the density $f_{kn}(x)$ is

$$f_{kn}(x) = \begin{cases} \dfrac{n!}{(k-1)! \, (n-k)!} \left(\int_0^x dt \right)^{k-1} \left(\int_x^1 dt \right)^{n-k} \\ \quad = \dfrac{n!}{(k-1)! \, (n-k)!} x^{k-1}(1-x)^{n-k} \quad \text{for} \quad 0 < x < 1 \\ 0 \qquad\qquad\qquad\qquad\qquad\qquad\qquad\quad\;\; \text{otherwise.} \end{cases}$$

By comparing the last expression with (5.9.3), we see that in this example the random variable $\zeta_k^{(n)}$ has the beta distribution with $p = k$ and $q = n - k + 1$.

For instance, let $n = 5$ and $k = 1$. Then $\zeta_1^{(5)}$ has the beta distribution with $p = 1, q = 5$. Let us compute the probability that the value of $\zeta_1^{(5)}$ belongs to the interval $[0.1, 0.2]$. We have

$$P(0.1 \leqslant \zeta_1^{(5)} \leqslant 0.2) = 5 \int_{0.1}^{0.2} (1-x)^4 \, dx = [-(1-x)^5]_{0.1}^{0.2} = -0.8^5 + 0.9^5$$
$$= 0.26281.$$

In our example it was easy to compute the required probability directly. In general, to compute beta distribution probabilities we use the tables of K. Pearson [4].

10.4 STOCHASTIC CONVERGENCE OF SAMPLE QUANTILES

From now on, when considering sequences of central statistics $\{\zeta_k^{(n)}\}$ we restrict ourselves to quantiles, and when considering sequences of end rank statistics we restrict ourselves to successive elements.

Let $X_1, X_2, \ldots, X_n$ be independent random variables and let each of them have the same distribution function $F(x)$ as the random variable X. Let us consider the sequence of sample quantiles $\{\zeta_k^{(n)}\}$, $(k = [n\lambda] + 1$, $0 < \lambda < 1)$. For every λ of the interval $0 < \lambda < 1$ there exists at least one value a_λ such that

$$(10.4.1) \qquad 1 - P(X = a_\lambda) \leqslant F(a_\lambda) \leqslant \lambda.$$

As we know, this value a_λ is an order parameter, a quantile of the distribution of X.

We now prove that if relation (10.4.1) is satisfied by only one a_λ, the sequence $\{\zeta_k^{(n)}\}$ of sample quantiles is stochastically convergent to the corresponding population quantile, that is, to a_λ. In particular, the sequence of sample medians (see Section 10.2) is stochastically convergent to the population median.

Theorem 10.4.1. *Let $F(x)$ denote the distribution function of the random variable X in the population and $\zeta_k^{(n)}$ the quantile of a simple sample $(k = [n\lambda] + 1, 0 < \lambda < 1)$ taken from this population. If there is only one a_λ satisfying inequality (10.4.1), then the sequence $\{\zeta_k^{(n)}\}$ of order statistics is stochastically convergent to the order parameter a_λ.*

Proof. By the definition of stochastic convergence (see Section 6.2), we have to prove that the assumptions of the theorem imply for an arbitrary positive $\varepsilon > 0$ the relation

$$(10.4.2) \qquad \lim_{n \to \infty} P(|\zeta_k^{(n)} - a_\lambda| \geqslant \varepsilon) = 0.$$

This relation is equivalent to the following two relations:

$$\lim_{n \to \infty} P(\zeta_k^{(n)} \geqslant a_\lambda + \varepsilon) = 0,$$

$$\lim_{n \to \infty} P(\zeta_k^{(n)} \leqslant a_\lambda - \varepsilon) = 0.$$

It follows from definition 10.3.1 that these two relations are equivalent to

$$(10.4.3) \qquad \lim_{n \to \infty} P\left[S_n(a_\lambda + \varepsilon) < \frac{k}{n} \right] = 0,$$

$$(10.4.3') \qquad \lim_{n \to \infty} P\left[S_n(a_\lambda - \varepsilon) \geqslant \frac{k - 1}{n} \right] = 0.$$

We observe, however, that from the assumption that the a_λ satisfying (10.4.1) is unique, it follows that for sufficiently small $\varepsilon > 0$

$$(10.4.4) \quad \lim_{n \to \infty} \left[F(a_\lambda + \varepsilon) - \frac{k}{n} \right] = F(a_\lambda + \varepsilon) - \lambda = \delta_1 > 0,$$

$$(10.4.4') \quad \lim_{n \to \infty} \left[F(a_\lambda - \varepsilon) - \frac{k - 1}{n} \right] = F(a_\lambda - \varepsilon) - \lambda = \delta_2 < 0.$$

We prove that (10.4.4) implies (10.4.3).

Indeed, it follows from (10.4.4) that for sufficiently large n

$$F(a_\lambda + \varepsilon) - \frac{k}{n} > \tfrac{1}{2}\delta_1.$$

Hence

$$(10.4.5) \quad P\left[S_n(a_\lambda + \varepsilon) < \frac{k}{n}\right] \leqslant P[S_n(a_\lambda + \varepsilon) - F(a_\lambda + \varepsilon) < -\tfrac{1}{2}\delta_1]$$

$$\leqslant [P\,|S_n(a_\lambda + \varepsilon) - F(a_\lambda + \varepsilon)| > \tfrac{1}{2}\delta_1].$$

Since (see Section 10.3), for fixed x, the random variable $S_n(x)$ is the frequency in a Bernoulli scheme with probability of success $p = F(x)$, it follows that at the point $x = a_\lambda + \varepsilon$

$$(10.4.6) \quad E[S_n(a_\lambda + \varepsilon)] = p = F(a_\lambda + \varepsilon),$$

$$(10.4.7) \quad D^2[S_n(a_\lambda + \varepsilon)] = \frac{p(1 - p)}{n} = \frac{F(a_\lambda + \varepsilon)[1 - F(a_\lambda + \varepsilon)]}{n}.$$

By the Chebyshev inequality (3.3.4) and formula (10.4.5), we obtain

$$\lim_{n \to \infty} P\left[S_n(a_\lambda + \varepsilon) < \frac{k}{n}\right] \leqslant \lim_{n \to \infty} \frac{4F(a_\lambda + \varepsilon)[1 - F(a_\lambda + \varepsilon)]}{n\delta_1^{\,2}} \leqslant \lim_{n \to \infty} \frac{1}{n\delta_1^{\,2}} = 0,$$

which proves equality (10.4.3).

In a similar way we can prove that (10.4.4') implies (10.4.3'). Thus theorem 10.4.1 is proved.

10.5 LIMIT DISTRIBUTIONS OF SAMPLE QUANTILES

Let $\zeta_k^{(n)}$ be a sample quantile, thus, $k = [n\lambda] + 1$, where $0 < \lambda < 1$. As we know, we have $k/n \to \lambda$ as $n \to \infty$. We also have

$$(10.5.1) \quad \lim_{n \to \infty} \left(\frac{k}{n} - \lambda\right)\sqrt{n} = 0.$$

Indeed, $n\lambda < k = [n\lambda] + 1 \leqslant n\lambda + 1$; hence

$$0 < \left(\frac{k}{n} - \lambda\right)\sqrt{n} < \frac{1}{\sqrt{n}}.$$

This double inequality implies (10.5.1).

Let us now assume that $X_1, X_2, \ldots, X_n$ are independent random variables with the same distribution function $F(x)$. Usually, in practice it is not sufficient to establish that the sequence of statistics is stochastically convergent to the corresponding order parameter, since this does not allow us to determine the probability that these statistics differ from the corresponding order parameters by a given value.

Theorem 10.5.1.   *Let* $F(x)$, $f(x)$ *and* a_λ $(0 < \lambda < 1)$ *be the distribution function, density, and quantile of the random variable* X, *respectively. Let* $\zeta_k^{(n)}$ *be the quantile of a simple sample and let* $g_{kn}(y)$ *be the density of the random variable*

$$(10.5.2) \qquad Y_k^{(n)} = \sqrt{\frac{n}{\lambda(1 - \lambda)}} f(a_\lambda)(\zeta_k^{(n)} - a_\lambda).$$

If the density $f(x)$ *is continuous and positive at the point* a_λ, *then, for an arbitrary pair of real numbers* y_1, y_2 *such that* $y_1 < y_2$, *we have*

$$(10.5.3) \quad \lim_{n \to \infty} P(y_1 < Y_k^{(n)} < y_2) = \lim_{n \to \infty} \int_{y_1}^{y_2} g_{kn}(y)\, dy = \frac{1}{\sqrt{2\pi}} \int_{y_1}^{y_2} e^{-y^2/2}\, dy.$$

Proof.   By (10.3.9) the density $f_{kn}(x)$ of $\zeta_k^{(n)}$ is

$$f_{kn}(x) = \frac{n!}{(k - 1)!\,(n - k)!} [F(x)]^{k-1} [1 - F(x)]^{n-k} f(x).$$

From (2.4.10) we find for the density $g_{kn}(y)$ the formula

$$(10.5.4)$$

$$g_{kn}(y) = \frac{1}{f(a_\lambda)} \sqrt{\lambda(1 - \lambda)/n}\, f_{kn}\left[a_\lambda + \sqrt{\lambda(1 - \lambda)/n} \cdot \frac{y}{f(a_\lambda)} \right]$$

$$= \frac{1}{f(a_\lambda)} \sqrt{\lambda(1 - \lambda)/n} \cdot \frac{n!}{(k - 1)!\,(n - k)!} [F(x)]^{k-1} [1 - F(x)]^{n-k} f(x)$$

$$= \sqrt{\lambda(1 - \lambda)/n} \cdot \frac{n!}{(k - 1)!\,(n - k + 1)!} \lambda^{k-1}(1 - \lambda)^{n-k+1}$$

$$\times \frac{n - k + 1}{1 - \lambda} \frac{f(x)}{f(a_\lambda)} \left[\frac{F(x)}{\lambda} \right]^{k-1} \left[\frac{1 - F(x)}{1 - \lambda} \right]^{n-k}.$$

Let

$$A_{n1} = \sqrt{\lambda(1 - \lambda)/n} \cdot \frac{n!}{(k - 1)!\,(n - k + 1)!} \lambda^{k-1}(1 - \lambda)^{n-k+1} \frac{n - k + 1}{1 - \lambda},$$

$$A_{n2} = \frac{f\left[a_\lambda + \sqrt{\lambda(1 - \lambda)/n} \cdot \dfrac{y}{f(a_\lambda)} \right]}{f(a_\lambda)},$$

$$A_{n3} = \left[\frac{F(x)}{\lambda} \right]^{k-1} \left[\frac{1 - F(x)}{1 - \lambda} \right]^{n-k}.$$

As we see, the equality

$$(10.5.5) \qquad g_{kn}(y) = A_{n1} A_{n2} A_{n3}$$

holds. Let us consider the terms A_{n1}, A_{n2}, and A_{n3} separately.

From the Stirling formula, $m! \cong (m/e)^m \sqrt{2\pi m}$, we obtain

$$\overline{\frac{n!}{(k-1)!\,(n-k+1)!}}$$
$$\cong \frac{n^n \sqrt{2\pi n}}{\sqrt{2\pi(k-1)}\sqrt{2\pi(n-k+1)}(k-1)^{k-1}(n-k+1)^{n-k+1}}.$$

Substituting the approximate values $k-1 = n\lambda$ and $n-k+1 = n - n\lambda$, we obtain

$$\frac{n!}{(k-1)!\,(n-k+1)!} \cong \frac{1}{\sqrt{2\pi\lambda(1-\lambda)n}} \cdot \frac{n^n}{(n\lambda)^{n\lambda}(n-n\lambda)^{n-n\lambda}}$$

$$= \frac{1}{\sqrt{2\pi\lambda(1-\lambda)n}} \cdot \frac{1}{\lambda^{n\lambda}(1-\lambda)^{n-n\lambda}} \cdot \frac{n^n}{n^{n\lambda}n^{n-n\lambda}}$$

$$= \frac{1}{\sqrt{2\pi\lambda(1-\lambda)n}} \cdot \frac{1}{\lambda^{k-1}(1-\lambda)^{n-k+1}}.$$

After introducing this expression into A_{n1}, we have

$$A_{n1} \cong \sqrt{\lambda(1-\lambda)/n} \cdot \frac{1}{\sqrt{2\pi\lambda(1-\lambda)n}} \cdot \frac{n-k+1}{1-\lambda}$$

$$= \frac{1}{\sqrt{2\pi}} \cdot \frac{n-k+1}{n} \cdot \frac{1}{1-\lambda}.$$

Since $(n-k+1)/n \to 1 - \lambda$, we have

$$(10.5.6) \qquad \lim_{n\to\infty} A_{n1} = \frac{1}{\sqrt{2\pi}}.$$

By the assumption that $f(x)$ is continuous at the point $x = a_\lambda$, we have

$$(10.5.7) \qquad \lim_{n\to\infty} A_{n2} = 1.$$

We now investigate the term A_{n3}. We can expand the function $F(x)$ in the neighborhood of $x = a_\lambda$ in the form

$$(10.5.8) \quad F(x) = F(a_\lambda) + F'(a_\lambda)(x - a_\lambda) + o(x - a_\lambda)$$

$$= \lambda + \sqrt{\lambda(1-\lambda)/n}\, y + o\left[\sqrt{\lambda(1-\lambda)/n} \cdot \frac{y}{f(a_\lambda)}\right]$$

$$= \lambda + y\sqrt{\lambda(1-\lambda)/n}\,[1 + Q(n, y)],$$

where the function $Q(n, y)$ tends to zero as $n \to \infty$. Putting (10.5.8) into A_{n3}, we obtain

$$A_{n3} = \left\{ 1 + \frac{y\sqrt{\lambda(1 - \lambda)/n}[1 + Q(n, y)]}{\lambda} \right\}^{k-1}$$

$$\times \left\{ 1 - \frac{y\sqrt{\lambda(1 - \lambda)/n}[1 + Q(n, y)]}{1 - \lambda} \right\}^{n-k}.$$

Letting

$$a = \sqrt{(1 - \lambda)/\lambda}[1 + Q(n, y)], \qquad b = \sqrt{\lambda/(1 - \lambda)}[1 + Q(n, y)],$$

we obtain

(10.5.9) $$A_{n3} = \left(1 + \frac{ay}{\sqrt{n}} \right)^{k-1} \left(1 - \frac{by}{\sqrt{n}} \right)^{n-k}.$$

Hence

$$\log A_{n3} = (k - 1) \log \left(1 + \frac{ay}{\sqrt{n}} \right) + (n - k) \log \left(1 - \frac{by}{\sqrt{n}} \right).$$

Since $a/\sqrt{n}$ and $b/\sqrt{n}$ tend to zero as $n \to \infty$, for sufficiently large n we can expand the expression on the right-hand side in the following way:

(10.5.10)

$$\log A_{n3} = (k - 1) \left[\frac{a}{\sqrt{n}} y - \frac{a^2}{2n} y^2 + o\left(\frac{1}{n} \right) \right]$$

$$+ (n - k) \left[- \frac{b}{\sqrt{n}} y - \frac{b^2}{2n} y^2 + o\left(\frac{1}{n} \right) \right]$$

$$= \frac{(k - 1)a - (n - k)b}{\sqrt{n}} y - \frac{(k - 1)a^2 + (n - k)b^2}{n} \cdot \frac{y^2}{2} + no\left(\frac{1}{n} \right).$$

We have

$$\frac{(k - 1)a - (n - k)b}{\sqrt{n}}$$

$$= \frac{(k - 1)\sqrt{(1 - \lambda)/\lambda} - (n - k)\sqrt{\lambda/(1 - \lambda)}}{\sqrt{n}} [1 + Q(n, y)]$$

$$= \frac{(k - 1)(1 - \lambda) - (n - k)\lambda}{\sqrt{n}} \cdot \frac{1 + Q(n, y)}{\sqrt{\lambda(1 - \lambda)}}$$

$$= \left[\left(\frac{k}{n} - \lambda \right) \sqrt{n} + \frac{\lambda - 1}{\sqrt{n}} \right] \frac{1 + Q(n, y)}{\sqrt{\lambda(1 - \lambda)}}.$$

By (10.5.1) we obtain

(10.5.11) $$\lim_{n \to \infty} \frac{(k - 1)a - (n - k)b}{\sqrt{n}} = 0.$$

Next we have

$$(10.5.12) \quad \lim_{n \to \infty} \frac{(k-1)a^2 + (n-k)b^2}{n}$$

$$= \lim_{n \to \infty} \left[\frac{k}{n} \cdot \frac{1-\lambda}{\lambda} + \left(1 - \frac{k}{n}\right)\frac{\lambda}{1-\lambda} - \frac{a^2}{n} \right][1 + Q(n,y)]^2$$

$$= \lambda \frac{1-\lambda}{\lambda} + (1-\lambda)\frac{\lambda}{1-\lambda} = 1.$$

By (10.5.10), (10.5.11), and (10.5.12) and by the fact that $no(1/n) \to 0$ as $n \to \infty$, we obtain

$$\lim_{n \to \infty} \log A_{n3} = -\frac{y^2}{2} \; ;$$

hence

$$(10.5.13) \qquad \lim_{n \to \infty} A_{n3} = e^{-y^2/2}.$$

It follows immediately from (10.5.6), (10.5.7), and (10.5.13) that

$$(10.5.14) \qquad \lim_{n \to \infty} g_{kn}(y) = \frac{1}{\sqrt{2\pi}} e^{-y^2/2}.$$

Since the functions A_{n1}, A_{n2}, and A_{n3} are uniformly bounded in every interval (y_1, y_2), it follows from (10.5.14) that

$$\lim_{n \to \infty} P(y_1 < Y_k^{(n)} < y_2)$$
$$= \lim_{n \to \infty} \int_{y_1}^{y_2} g_{kn}(y)\, dy = \int_{y_1}^{y_2} \lim_{n \to \infty} g_{kn}(y)\, dy = \frac{1}{\sqrt{2\pi}} \int_{y_1}^{y_2} e^{-y^2/2}\, dy.$$

Thus theorem 10.5.1 is proved.

It follows from theorem 10.5.1 that if, in the population under consideration, the random variable X is of the continuous type with density $f(x)$
and if this density is positive and continuous at the point $x = a_\lambda$, where a_λ
is the quantile of the population, then the quantile of a simple sample has
the asymptotically normal distribution

$$(10.5.15) \qquad N\left(a_\lambda; \frac{1}{f(a_\lambda)}\sqrt{\lambda(1-\lambda)/n}\right).$$

Example 10.5.1. From a population, in which the characteristic X has the
normal distribution $N(0; 1)$, a simple sample of $n = 121$ elements is drawn. The
sample median is -0.5. What is the probability of obtaining a value of the
sample median smaller than -0.5?

Here we have $\lambda = 0.5$, and the median of the population is the point $a_{1/2}$ satisfying the equation

$$\frac{1}{\sqrt{2\pi}} \int_{-\infty}^{a_{1/2}} e^{-x^2/2}\, dx = 0.5.$$

Hence $a_{1/2} = 0$. Next, $f(a_{1/2}) = 1/\sqrt{2\pi}$. Thus the sample median has the asymptotically normal distribution

$$N\left(0; \frac{1}{2}\sqrt{\frac{2\pi}{121}}\right) = N(0; 0{,}114).$$

Thus

$$P(\zeta_{61}^{(121)} < -0.5) = P\left(\frac{\zeta_{61}^{(121)}}{0.114} < \frac{-0.5}{0.114}\right) = \Phi(-4.39).$$

From the tables of the normal distribution we find that the required probability is extremely small; it is approximately 0.00001.

Example 10.5.2. Let us consider a population in which the density of the characteristic X is

$$f(x) = \begin{cases} 2x & \text{for } 0 \leqslant x \leqslant 1, \\ 0 & \text{otherwise.} \end{cases}$$

The quantile $a_{1/4}$ of this population is determined by the equation

$$\int_0^{a_{1/4}} 2x\, dx = 0.25,$$

so that $a_{1/4} = 0.5$ and $f(a_{1/4}) = 1$. We draw a simple sample of 70 elements from this population and determine the sample quantile, that is, the observed value of the statistic $\zeta_k^{(n)}$, where $n = 70$ and $k = [0.25 \times 70] + 1 = 18$. What is the probability of obtaining a value of the sample quantile not smaller than 0.75?

The statistic $\zeta_{18}^{(70)}$ has approximately the normal distribution

$$N\left(0.5; \sqrt{\frac{0.25 \times 0.75}{70}}\right) = N(0.5; 0.052).$$

Hence

$$P(\zeta_{18}^{(70)} \geqslant 0.75) = P\left(\frac{\zeta_{18}^{(70)} - 0.5}{0.052} \geqslant \frac{0.25}{0.052}\right) \cong 1 - \Phi(4.81) \cong 0.0000.$$

Theorem 10.5.1 is part of the general theory of limit theorems for sequences of central order statistics, developed by Smirnov [3] for $X_1, \ldots, X_n$ identically distributed, and by Loève [5] and Cogburn [1] without the assumption that $X_1, \ldots, X_n$ have the same distribution. For a sequence of medians, theorem 10.5.1 was proved by Fogelson [1].

10.6 THE LIMIT DISTRIBUTION OF SUCCESSIVE SAMPLE ELEMENTS

At the end of Section 10.2 we defined a statistic called a successive element of a sample. It is, namely, an order statistic $\zeta_k^{(n)}$, where k is a constant, or $k = n - l + 1$ and l is a constant. In the first case we have $\lim_{n \to \infty} k/n = 0$; in the second case $\lim_{n \to \infty} k/n = 1$.

We assume that the random variables $X_1, \ldots, X_n$ are independent and are of the continuous type with the common density $f(x)$.

By (10.3.9) the density of $\zeta_k^{(n)}$ is

$$f_{kn}(x) = \frac{n!}{(k - 1)!\,(n - k)!} [F(x)]^{k-1}[1 - F(x)]^{n-k}f(x).$$

Let us introduce the random variable $Z_k^{(n)}$ defined as

(10.6.1) $Z_k^{(n)} = 2nF(\zeta_k^{(n)})$.

The random variable $Z_k^{(n)}$ can take on values only in the interval $[0, 2n]$. Let us denote its density by $h_{kn}(z)$. We have

(10.6.2)

$$h_{kn}(z) = \begin{cases} \dfrac{1}{2} \cdot \dfrac{(n - 1)!}{(k - 1)!\,(n - k)!}\left(\dfrac{z}{2n}\right)^{k-1}\left(1 - \dfrac{z}{2n}\right)^{n-k} & \text{for } 0 \leqslant z \leqslant 2n, \\ 0 & \text{for } z < 0 \quad \text{and} \quad z > 2n \end{cases}$$

Let us rewrite (10.6.2) for $0 \leq z \leq 2n$ in the form

$$h_{kn}(z) = \frac{1}{(k - 1)!} \cdot \frac{(n - 1)!}{(n - k)!\,n^{k-1}(1 - z/2n)^k 2^k}\, z^{k-1}\left(1 - \frac{z}{2n}\right)^n.$$

Since k is a constant, we have

$$\lim_{n \to \infty} \frac{(n - 1)!}{(n - k)!\,n^{k-1}} = \lim_{n \to \infty} \frac{(n - k + 1) \ldots (n - 1)}{n^{k-1}}$$

$$= \lim_{n \to \infty} \left(1 - \frac{k + 1}{n}\right) \ldots \left(1 - \frac{1}{n}\right) = 1,$$

$$\lim_{n \to \infty} \frac{1}{(1 - z/2n)^k} = 1.$$

Noting that $(k - 1)! = \Gamma(k)$ and $\lim_{n \to \infty} (1 - z/2n)^n = e^{-z/2}$, we finally obtain

(10.6.3) $$\lim_{n \to \infty} h_{kn}(z) = \begin{cases} \dfrac{z^{k-1}e^{-z/2}}{2^k\Gamma(k)} & \text{for } z \geqslant 0, \\ 0 & \text{for } z < 0. \end{cases}$$

Since the function $h_{kn}(z)$ is uniformly bounded for all n in every finite interval $a < z < b$, we have

(10.6.4) $$\lim_{n \to \infty} P(a < Z_k^{(n)} < b) = \lim_{n \to \infty} \int_a^b h_{kn}(z)\,dz = \int_a^b \lim_{n \to \infty} h_{kn}(z)\,dz$$

$$= \int_a^b \frac{z^{k-1}e^{-z/2}}{2^k\Gamma(k)}\cdot dz.$$

Thus the random variable $Z_k^{(n)}$ has asymptotically the χ^2 distribution with $m = 2k$ degrees of freedom. Similarly, we find that the random variable

$$U_k^{(n)} = 2n[1 - F(\zeta_k^{(n)})],$$

where $k = n - l + 1$, has asymptotically the χ^2 distribution with $2l$ degrees of freedom, that is

(10.6.5) $$\lim_{n \to \infty} P(c < U_k^{(n)} < d) = \int_c^d \frac{u^{l-1}e^{-u/2}}{2^l \Gamma(l)}\, du.$$

Here the statistic $\zeta_k^{(n)}$ is the lth value of the sample starting from the largest value.

The usefulness of formulas (10.6.4) and (10.6.5) in determining the asymptotic distribution of the statistic $\zeta_k^{(n)}$ is restricted by the fact that it is usually difficult to determine $\zeta_k^{(n)}$ as a function of $Z_k^{(n)}$ or $U_k^{(n)}$. We shall give only a simple application of these formulas.

Example 10.6.1.   Let the random variable X have the density

$$f(x) = \begin{cases} 1 & \text{for } 0 \leqslant x \leqslant 1, \\ 0 & \text{otherwise.} \end{cases}$$

Let us consider the statistic $\zeta_k^{(n)}$ for constant k (determined from a simple sample) and the random variable $Z_k^{(n)}$ defined by (10.6.1). In our example

$$Z_k^{(n)} = 2n\,\zeta_k^{(n)}.$$

It follows from the previous considerations that the random variable $2n\,\zeta_k^{(n)}$ has asymptotically the χ^2 distribution with $2k$ degrees of freedom. For instance, let $n = 100$ and $k = 5$. We find the probability that $\zeta_k^{(n)} > 0.0915$. We have

$$P(\zeta_k^{(n)} > 0.0915) = P(200\,\zeta_k^{(n)} > 18.3).$$

From tables of the χ^2 distribution with ten degrees of freedom we find that the required probability is approximately 0.05.

Among those who investigated the limit distributions of the largest sample element, assuming that the random variables are identically distributed, were Fréchet [1] Mises [3] and Gnedenko [8]. In his paper, Gnedenko gave the class of all limit distributions (under a suitable normalization) of the largest sample element. Gnedenko's methods were applied by Smirnov [3] to the investigation of the limit distributions of arbitrary successive sample elements. The limit distributions of the largest sample element without the assumption that the random variables $X_1, \ldots,$ X_n have the same distribution were investigated by Mejzler [1], whereas Loève [5] and Cogburn [1] investigated the limit distributions of arbitrary successive sample elements and joint limit distributions of groups of successive sample elements. (See also Gumbel's [2] monograph.)

10.7 THE JOINT DISTRIBUTION OF A GROUP OF QUANTILES

Let the random variables $X_1, X_2, \ldots, X_n$ be independent and of the continuous type with the common distribution function $F(x)$ and continuous density $f(x)$. Let us consider the order statistics $\zeta_1^{(n)}, \zeta_2^{(n)}, \ldots, \zeta_n^{(n)}$. The values taken on by these statistics are denoted respectively by $y_1, y_2, \ldots, y_n$. We want to find the distribution of the random variable

$$(\zeta_{j_1}^{(n)}, \zeta_{j_2}^{(n)}, \ldots, \zeta_{j_r}^{(n)}),$$

where $1 \leqslant j_1 < j_2 < \ldots < j_r \leqslant n$ and $1 < r \leqslant n$. Let us divide the x-axis into the segments

$$I_1 = (-\infty, y_{j_1}), \quad I_2 = (y_{j_1}, y_{j_2}), \ldots, I_{r+1} = (y_{j_r}, +\infty).$$

To find the required distribution, we find for every $(y_{j_1}, y_{j_2}, \ldots, y_{j_r})$ the probability of the event A that $j_1 - 1$ values belong to the interval I_1, one value to the "interval" $(y_{j_1}, y_{j_1} + dy_{j_1})$, next that $j_2 - j_1 - 1$ values belong to the interval I_2, one value to the "interval" $(y_{j_2}, y_{j_2} + dy_{j_2})$, and so on, and finally that $n - j_r$ values belong to the interval I_{r+1}. Let us denote by p_k the probability that X takes on a value from I_k, that is,

$$p_k = \int_{I_k} f(x)\, dx \qquad (k = 1, 2, \ldots, r + 1).$$

From (2.3.8) it follows that the probability that X takes on a value from the "interval" $(y_{j_k}, y_{j_k} + dy_{j_k})$ equals $f(y_{j_k})\, dy_{j_k}$. Thus the probability element of the random variable $(\zeta_{j_1}^{(n)}, \zeta_{j_2}^{(n)}, \ldots, \zeta_{j_r}^{(n)})$ is, according to (5.12.1'),

(10.7.1)

$$P(A) = \frac{n!}{(j_1 - 1)!\, 1!\, (j_2 - j_1 - 1)!\, 1! \ldots (j_r - j_{r-1} - 1)!\, 1!\, (n - j_r)!}$$
$$\times p_1^{j_1-1} p_2^{j_2-j_1-1} \ldots p_r^{j_r-j_{r-1}-1} p_{r+1}^{n-j_r} f(y_{j_1}) \ldots f(y_{j_r})\, dy_{j_1} \ldots dy_{j_r}$$
$$= g(y_{j_1}, \ldots, y_{j_r})\, dy_{j_1} \ldots dy_{j_r}.$$

Thus the density of the random variable $(\zeta_{j_1}^{(n)}, \zeta_{j_2}^{(n)}, \ldots, \zeta_{j_r}^{(n)})$ equals $g(y_{j_1}, y_{j_2}, \ldots, y_{j_r})$ in the domain $y_{j_1} < y_{j_2} < \ldots < y_{j_r}$ and zero in the remaining domain.

Let us consider, in particular, the distribution of the random variable $(\zeta_1^{(n)}, \zeta_n^{(n)})$. In this case we have $r = 2, j_1 = 1, j_2 = n$. From (10.7.1), for the required density function $g(y_1, y_n)$, we obtain the formula

$$(10.7.2) \quad g(y_1, y_n) = \begin{cases} n(n-1)\left[\displaystyle\int_{y_1}^{y_n} f(x)\, dx\right]^{n-2} f(y_1)f(y_n) & \text{for } y_1 < y_n, \\ 0 & \text{for } y_1 \geqslant y_n. \end{cases}$$

10.8 THE DISTRIBUTION OF THE SAMPLE RANGE

In Chapter 3 we introduced the notion of the range of a distribution, which was a parameter of the population and which can, in addition to the variance, serve as a measure of dispersion.

Definition 10.8.1. The random variable

$$(10.8.1) \qquad\qquad W = \zeta_n^{(n)} - \zeta_1^{(n)}$$

is called the *sample range*.

Since $\zeta_n^{(n)} \geqslant \zeta_1^{(n)}$, by definition, the random variable W takes on only non-negative values.

Formulas for the distribution of the sample range can be obtained from (10.7.2) but they are usually very complicated. We shall only mention that for simple samples from a normal population the probabilities connected with the distribution of the sample range are given in tables by Hartley and Pearson [1] and expressions for moments are given in the paper by Tippet [1].

Elfwing [1] investigated the asymptotic (as $n \to \infty$) distribution of the range of simple samples from a normal population. Gumbel [1] investigated the asymptotic distribution of the range of simple samples from a population in which the characteristic X is an unbounded random variable and has moments of all orders, but is otherwise unrestricted.

From (6.5.7) and (10.7.2) we can easily obtain the density $h(w)$ of the range W of simple samples of size n drawn from a population with the uniform distribution in [0, 1]. In fact, here (10.7.2) will take the form

$$g(y_1, y_n) = \begin{cases} n(n-1)(y_n - y_1)^{n-2} & (0 \leqslant y_1 < y_n \leqslant 1), \\ 0 & \text{otherwise.} \end{cases}$$

If we let $w = y_n - y_1$ and remember that $0 \leqslant y_1 + w \leqslant 1$, hence $0 \leqslant y_1 \leqslant 1 - w$, we get

$$(10.8.2)$$
$$h(w) = \begin{cases} n(n-1)\displaystyle\int_0^{1-w} w^{n-2}\,dy_1 = n(n-1)w^{n-2}(1-w) & (0 \leqslant w \leqslant 1), \\ 0 & \text{otherwise.} \end{cases}$$

10.9 TOLERANCE LIMITS

Let the random variable X have a continuous distribution function $F(x)$. Let $\zeta_1^{(n)}, \zeta_2^{(n)}, \ldots, \zeta_n^{(n)}$ be order statistics of simple samples of n elements taken from a population with the characteristic X under investigation. Let us consider the integral W defined by the formula

$$(10.9.1) \qquad W = \int_{L_1}^{L_2} dF(x),$$

where $L_i = L_i(\zeta_1^{(n)}, \zeta_2^{(n)}, \ldots, \zeta_n^{(n)})(i = 1, 2)$ are single-valued functions of the random variables $\zeta_1^{(n)}, \zeta_2^{(n)}, \ldots, \zeta_n^{(n)}$. The functions L_1 and L_2 are random variables, hence W is also a random variable. We observe that W is the probability that $L_1 \leqslant X \leqslant L_2$. Because of the random character of L_1 and L_2 this probability is a random variable. In other words, W is the fraction of elements of the population contained between the random limits L_1 and L_2.

Definition 10.9.1. The limits L_1 and L_2 in (10.9.1) are called *tolerance limits*. If the distribution of the random variable W is independent of $F(x)$, then L_1 and L_2 are called *distribution-free tolerance limits*.

Let us consider the special case when $L_1 = \zeta_1^{(n)}$ and $L_2 = \zeta_n^{(n)}$. We shall show that the limits L_1 and L_2 are then distribution-free tolerance limits. To do so, let us rewrite (10.9.1) in the form

$$W = F(\zeta_n^{(n)}) - F(\zeta_1^{(n)}).$$

From the assumption that $F(x)$ is continuous it follows (see Section 5.6) that $Y = F(X)$ has the uniform distribution in $[0, 1]$. Omitting the domains where $F(x)$ is constant (they are of no significance), it is an increasing function, hence W is the range of a simple sample drawn from a population with the uniform distribution. Thus the density $h(w)$ of W is given by (10.8.2) and thus it is independent of $F(x)$; hence $\zeta_1^{(n)}$ and $\zeta_n^{(n)}$ are distribution-free tolerance limits. Hence for an arbitrary random variable X with a continuous distribution function and for arbitrary $\alpha > 0$ we can find the least n such that

$$P(W \geqslant w) = n(n-1) \int_w^1 w^{n-2}(1-w)\, dw \geqslant a.$$

Thus, for a suitably chosen n we have probability $\geqslant \alpha$ that the fraction of elements of the population contained between the smallest and the largest elements of a sample of size n is at least w. This result was obtained by Wilks [2]. Wald [1] generalized the result to multidimensional random variables.

We observe that W has the beta distribution (see Section 5.9) with $p = n - 1$ and $q = 2$. Thus from formulas (5.9.5) and (5.9.6) we obtain

$$(10.9.2) \qquad E(W) = \frac{n-1}{n+1}, \qquad D^2(W) = \frac{n-1}{n+1} \cdot \frac{2}{(n+1)(n+2)}.$$

We see that for relatively small n the statistic W is close to one.

Example 10.9.1. Let $\alpha = 0.95$ and $w = 0.95$. Then

$$P(W \geqslant 0.95) = n(n-1) \int_{0.95}^{1} w^{n-2}(1-w)\,dw \geqslant 0.95.$$

Hence we have $n \cong 93$.

Another example of tolerance limits which is important in applications follows.

Example 10.9.2. Let $\bar{X}$ and S^2 be the mean and the standard deviation of a simple sample of size n from a population with the normal distribution $N(m; \sigma)$, where m and σ are unknown. Let

(10.9.3)
$$S' = \sqrt{n/(n-1)}\, S,$$

$$W = \frac{1}{\sigma\sqrt{2\pi}} \int_{\bar{X}-hS'}^{\bar{X}+hS'} \exp\left[-\frac{(x-m)^2}{2\sigma^2}\right] dx.$$

The limits of integration $\bar{X} - hS'$ and $\bar{X} + hS'$ are random variables; hence W is a random variable, namely, W is the fraction of elements of the population contained in the interval $\bar{X} - hS'$, $\bar{X} + hS'$. The end points of this interval are tolerance limits. They are not, however, distribution-free. Wald and Wolfowitz [3] gave a method for approximately computing the probability

$$P(W \geqslant w) = \alpha$$

as a function of w, h, n, where W is given by (10.9.3). Tables connected with computing these probabilities were constructed by Bowker [1].

It is necessary to stress the fact that, for fixed h, W is a random variable; hence it is not correct to compute W from tables of the normal distribution, as is sometimes done in practice, putting $W = P(-h \leqslant Y \leqslant +h)$, where Y has the normal distribution $N(0; 1)$.

10.10 GLIVENKO THEOREM

A   In Section 10.3, A we defined the empirical distribution function $S_n(x)$. If we choose a sample of size n, the number of observed sample elements that are smaller than x is $nS_n(x)$. Thus $S_n(x)$ can take on values in the interval $[0, 1]$. We have already established that $S_n(x)$ is a nondecreasing function, continuous from the left. It follows that $S_n(x)$, as a function of x, has all the properties of a distribution function; hence it is called the empirical distribution function to distinguish it from the distribution function $F(x)$ of the population, which is also called the *theoretical distribution function*. We have also established that for a fixed x, $S_n(x)$ is, for a simple sample, the frequency in a Bernoulli scheme, that is, the frequency of occurrences of some random event in n independent trials, where the probability of the occurrence of that event in a single trial is $p = F(x)$.

It is extremely important to find the relation between the empirical distribution function $S_n(x)$ and the theoretical distribution function $F(x)$. We might intuitively suspect that $S_n(x)$ and $F(x)$ are, roughly speaking, very closely related. In fact, such a close relation does exist.

Let us first of all state that for simple samples we have, for every fixed x,

$$P[\lim_{n \to \infty} S_n(x) = F(x)] = 1.$$

This follows from theorem 6.12.3, since for fixed x, $S_n(x)$ is the frequency in the Bernoulli scheme with $p = F(x)$.

A much stronger result is contained in the theorem of Glivenko [1].

Theorem 10.10.1 (Glivenko). *Let $S_n(x)$ be the empirical distribution function of a simple sample of n elements drawn from a population in which the characteristic X has the theoretical distribution function $F(x)$. The probability that the sequence $S_n(x)$ is convergent to $F(x)$, as $n \to \infty$, uniformly in x $(-\infty < x < +\infty)$ equals one.*

Let

$$(10.10.1) \qquad D_n = \sup_{-\infty < x < \infty} |S_n(x) - F(x)|.$$

The expression D_n is a random variable. The assertion of Glivenko's theorem states that

$$(10.10.2) \qquad P(\lim_{n \to \infty} D_n = 0) = 1.$$

Before we prove Glivenko's theorem we prove the following lemma.

Lemma. *Let $\{A_k\}$ (k = 1, 2, . . .) be a sequence of random events and let*

$$(10.10.3) \qquad P(A_k) = 1 \quad (k = 1, 2, \ldots).$$

Then

$$P\left(\prod_{k \geqslant 1} A_k\right) = 1.$$

Proof of the lemma. At first we shall prove the lemma for two events A_1 and A_2. Let $P(A_1) = P(A_2) = 1$. Then we have

$$1 = P(A_1 + A_2) = P(A_1) + P(A_2) - P(A_1 A_2)$$
$$= 1 + 1 - P(A_1 A_2).$$

Hence

$$P(A_1 A_2) = 1.$$

By induction we can extend this result to every finite number of events, that is, if relation (10.10.3) holds for $k = 1, 2, \ldots, m$, then

$$(10.10.4) \qquad P(A_1 A_2 \ldots A_m) = 1.$$

To complete the proof of the lemma, let us write

$$\prod_{k \geqslant 1} A_k = A_1(A_1A_2)(A_1A_2A_3) \ldots = \prod_{m \geqslant 1} B_m,$$

where $B_m = \prod_{k=1}^{m} A_k$. We observe that for every m we have $B_{m+1} \subset B_m$; hence B_m is a nonincreasing sequence of events. By theorem 1.3.4 we have

$$(10.10.5) \qquad P\left(\prod_{k \geqslant 1} A_k\right) = P\left(\prod_{m \geqslant 1} B_m\right) = \lim_{m \to \infty} P(B_m).$$

The lemma follows from (10.10.4) and (10.10.5).

Proof of Glivenko's theorem. Let r be a positive integer. Denote by x_{rk} $(k = 1, 2, \ldots, r)$ the smallest value of x for which

$$(10.10.6) \qquad F(x) \leqslant \frac{k}{r} \leqslant F(x + 0).$$

Let $S_n(x_{rk})$ denote the value of the empirical distribution function at the point x_{rk} and let A_0^r and A_k^r $(k = 1, 2, \ldots, r - 1)$ denote the event that the first and the second of the following relations, respectively, holds:

$$\lim_{n \to \infty} \max \left[|S_n(x_{r1}) - F(x_{r1})|, \quad |S_n(x_{rr} + 0) - F(x_{rr} + 0)| \right] = 0,$$

$$\lim_{n \to \infty} \max \left[|S_n(x_{rk} + 0) - F(x_{rk} + 0)|, \quad |S_n(x_{r,k+1}) - F(x_{r,k+1})| \right] = 0.$$

It follows from theorem 6.12.3 that

$$(10.10.7) \qquad P(A_k^r) = 1 \quad (k = 0, 1, 2, \ldots, r - 1).$$

Let us denote by A^r the product of events $A_0^r A_1^r \ldots A_{r-1}^r$. The event A^r is equivalent to the relation

$$\lim_{n \to \infty} \max_{1 \leqslant k \leqslant r} \max \left[|S_n(x_{rk}) - F(x_{rk})|, |S_n(x_{rk} + 0) - F(x_{rk} + 0)| \right] = 0.$$

From (10.10.7) and from the definition of the event A^r it follows that $P(A^r) = 1$.

Let us denote $A = A^1 A^2 A^3 \ldots$. By the lemma we have

$$(10.10.8) \quad P(A) = P\left(\prod_{r \geqslant 1} \lim_{n \to \infty} \max_{1 \leqslant k \leqslant r} \max \left[|S_n(x_{rk}) - F(x_{rk})|, \right.\right.$$

$$\left.\left. |S_n(x_{rk} + 0) - F(x_{rk} + 0)| \right] = 0\right) = 1.$$

Now let x be any number satisfying the double inequality $x_{rk} < x \leqslant x_{r,k+1}$. We have

$$S_n(x_{rk} + 0) \leqslant S_n(x) \leqslant S_n(x_{r,k+1}),$$

$$F(x_{rk} + 0) \leqslant F(x) \leqslant F(x_{r,k+1}).$$

Hence we obtain

(10.10.9)

$$S_n(x_{rk} + 0) - F(x_{r,k+1}) \leqslant S_n(x) - F(x) \leqslant S_n(x_{r,k+1}) - F(x_{rk} + 0).$$

From (10.10.6) it follows that, for $x_{rk} < x_{r,k+1}$,

$$0 \leqslant F(x_{r,k+1}) - F(x_{rk} + 0) \leqslant \frac{1}{r}$$

for $k = 1, \ldots, r - 1$, so that from (10.10.9) we obtain

$$\sup_{x_{rk} < x \leqslant x_{r,k+1}} |S_n(x) - F(x)|$$

$$\leqslant \max \left[|S_n(x_{rk} + 0) - F(x_{rk} + 0)|, |S_n(x_{r,k+1}) - F(x_{r,k+1})| \right] + \frac{1}{r}.$$

Thus

(10.10.10) $\quad \max \left[|S_n(x_{rr} + 0) - F(x_{rr} + 0)|, \displaystyle\sup_{x_{r1} \leqslant x \leqslant x_{rr}} |S_n(x) - F(x)| \right]$

$$\leqslant \max_{1 \leqslant k \leqslant r} \max \left[|S_n(x_{rk}) - F(x_{rk})|, |S_n(x_{rk} + 0) - F(x_{rk} + 0)| \right] + \frac{1}{r}.$$

From (10.10.10) we obtain

$$P\left(\lim_{n \to \infty} \sup_{-\infty < x < \infty} |S_n(x) - F(x)| = 0 \right)$$

$$\geqslant P\left(\lim_{n \to \infty} \lim_{r \to \infty} \max_{1 \leqslant k \leqslant r} \max \left[|S_n(x_{rk}) - F(x_{rk})|, \right.\right.$$

$$\left.\left. |S_n(x_{rk} + 0) - F(x_{rk} + 0)| \right] = 0 \right)$$

$$\geqslant P\left(\prod_{r \geqslant 1} \lim_{n \to \infty} \max_{1 \leqslant k \leqslant r} \max \left[|S_n(x_{rk}) - F(x_{rk})|, \right.\right.$$

$$\left.\left. |S_n(x_{rk} + 0) - F(x_{rk} + 0)| \right] = 0 \right).$$

From the last formula and from (10.10.8) we finally obtain

$$P\left(\lim_{n \to \infty} \sup_{-\infty < x < \infty} |S_n(x) - F(x)| = 0 \right) = 1.$$

Glivenko's theorem is thus proved.

B The empirical distribution function can also be written as follows. Let $\{X_r\}$ $(r = 1, 2, \ldots)$ be a sequence of random variables. Define for every x $(-\infty < x < +\infty)$ the function

(10.10.11) $$f_x(X_r) = \begin{cases} 0 & \text{for } X_r \geqslant x, \\ 1 & \text{for } X_r < x. \end{cases}$$

Then

(10.10.12) $$S_n(x) = \frac{1}{n} \sum_{r=1}^{n} f_x(X_r).$$

Glivenko's theorem has been generalized in various ways. Wolfowitz [5, 6, 8], generalized it to the case when not all the random variables X_r have the same distribution function and to the case of random vectors. (For the last case, see also Blum [1].) Fortet and Mourier [1] and Fisz [5] generalized Glivenko's Theorem to the case when the functions $f_x(X_r)$ in formula (10.10.12) are more general than the functions defined by (10.10.11). Tucker [1] generalized it to the case when the sequence X_r of random variables is stationary in the strict sense. (See also Varadarajan [2] and Ranga-Rao [1].)

10.11   THE THEOREMS OF KOLMOGOROV AND SMIRNOV

A   The following theorem gives the limit distribution of the random variable D_n defined by (10.10.1). Let us denote by $Q_n(\lambda)$ the distribution function of the random variable $D_n\sqrt{n}$; hence

(10.11.1) $$Q_n(\lambda) = \begin{cases} P(D_n\sqrt{n} < \lambda) = P\left(D_n < \dfrac{\lambda}{\sqrt{n}}\right) & \text{for } \lambda > 0, \\ 0 & \text{for } \lambda \leqslant 0. \end{cases}$$

Theorem 10.11.1.   *Let* $S_n(x)$ *be the empirical distribution function of a simple sample of size* n *drawn from a population in which the random variable* X *has a continuous distribution function* $F(x)$.
Then

(10.11.2) $$\lim_{n\to\infty} Q_n(\lambda) = Q(\lambda) = \begin{cases} \sum_{k=-\infty}^{\infty} (-1)^k \exp(-2k^2\lambda^2) & \text{for } \lambda > 0, \\ 0 & \text{for } \lambda \leqslant 0. \end{cases}$$

This theorem is due to Kolmogorov [6]. The idea of the proof is given in part C of this Section.

Table VIII at the end of the book gives the values of $Q(\lambda)$ for certain λ's. These data are taken from Smirnov's [1] tables of $Q(\lambda)$.

Tables of $Q_n(\lambda)$ for small n are given in the paper of Z. W. Birnbaum [1].

We observe that the limit distribution $Q(\lambda)$ is independent of the theoretical distribution function $F(x)$; it is only assumed that $F(x)$ is a continuous function. Thus, we say that the limit distribution of D_n is *distribution-free* (the general characterization of distribution-free statistics can be found in the papers of Z. W. Birnbaum [2], Birnbaum and Rubin [1] and Bell [1]). If, however, the theoretical distribution function $F(x)$ is not continuous, relation (10.11.2) is not valid. This situation was investigated by Schmid [1], who stated that in this case the limit of $Q_n(\lambda)$ depends on the theoretical distribution, namely on the values of $F(x)$ at the discontinuity points (see also Gihman [3]).

Example 10.11.1. We have a batch of screws which should have identical diameters of standard size. However, precise measurements show that there are some small deviations from this size. If, during the production process, there were no significant deviations from the norms characterizing this production process, we may, in general, assume that the results of the measurements x, treated as values of a random variable X, have a normal distribution. For instance, let

$$(10.11.3) \qquad f(x) = \frac{3}{\sqrt{2\pi}} \exp\left[-\frac{9}{2}(x-2)^2 \right]$$

be the density of X, that is, let X have the normal distribution $N(2; \frac{1}{3})$ with a millimeter as unit of measurement.

From the batch we draw a simple sample of $n = 80$ elements, that is, we make 80 measurements. Let the results of these measurements, arranged in increasing order, be $x_1, x_2, \ldots, x_{80}$. The least upper bound of the values $|s_{80}(x) - F(x)|$ was calculated, where $F(x)$ is the theoretical and $s_{80}(x)$ is the observed empirical distribution function. It turned out that $d_{80} = 0.21$. Since n is not small, we shall use Kolmogorov's theorem to compute the probability that D_{80} exceeds or equals the observed value.

We have $d_{80}\sqrt{80} = 0.21 \times 8.94 = 1.88 = \lambda$. In Table VIII we find

$$P(D_{80} \geqslant 0.21) = P\left(D_{80} \geqslant \frac{1.88}{8.94} \right) \cong 1 - Q(1.88) \cong 0.0017.$$

Thus this probability is of the order 0.001, and it may create some doubts as to whether the density of the random variable X has the form (10.11.3) or whether the choice of the sample elements was really random. The methods of drawing such conclusions are discussed in the following chapters.

B We now present Smirnov's theorem. Let $X_{11}, X_{12}, \ldots, X_{1n_1}$, $X_{21}, X_{22}, \ldots, X_{2n_2}$ be independent random variables with the same distribution function $F(x)$. Let $S_{1n_1}(x)$ and $S_{2n_2}(x)$ be the empirical distribution functions of the first and second group of random variables, respectively. Thus, $S_{1n_1}(x)$ and $S_{2n_2}(x)$ are the empirical distribution functions of two independent simple samples drawn from the same population, in which the characteristic X has the distribution function $F(x)$, and n_1 and n_2 are the sizes of the samples, respectively. Let us consider the random variables

$$(10.11.4) \qquad D^+_{n_1 n_2} = \max_{-\infty < x < \infty} [S_{1n_1}(x) - S_{2n_2}(x)],$$

$$(10.11.5) \qquad D_{n_1 n_2} = \max_{-\infty < x < \infty} |S_{1n_1}(x) - S_{2n_2}(x)|.$$

Let $n = n_1 n_2 / (n_1 + n_2)$ and let $Q^+_{n_1 n_2}(\lambda)$ and $Q_{n_1 n_2}(\lambda)$ be the distribution functions of the random variables $\sqrt{n} D^+_{n_1 n_2}$ and $\sqrt{n} D_{n_1 n_2}$, respectively. Thus

$$(10.11.6) \qquad Q^+_{n_1 n_2}(\lambda) = \begin{cases} P(D^+_{n_1 n_2} < \lambda/\sqrt{n}) & \text{for } \lambda > 0, \\ 0 & \text{for } \lambda \leqslant 0, \end{cases}$$

$$(10.11.7) \qquad Q_{n_1 n_2}(\lambda) = \begin{cases} P(D_{n_1 n_2} < \lambda/\sqrt{n}) & \text{for } \lambda > 0, \\ 0 & \text{for } \lambda \leqslant 0. \end{cases}$$

The following theorem holds (Smirnov [2]).

Theorem 10.11.2. *Let $S_{1n_1}(x)$ and $S_{2n_2}(x)$ be two empirical distribution functions of two independent simple samples drawn from the same population, in which the characteristic X has a continuous distribution function. Then*

$$(10.11.8) \qquad \lim_{\substack{n_1 \to \infty \\ n_2 \to \infty}} Q^+_{n_1 n_2}(\lambda) = \begin{cases} 1 - \exp(-2\lambda^2) & \text{for } \lambda > 0, \\ 0 & \text{for } \lambda \leqslant 0, \end{cases}$$

$$(10.11.9) \qquad \lim_{\substack{n_1 \to \infty \\ n_2 \to \infty}} Q_{n_1 n_2}(\lambda) = \begin{cases} \displaystyle\sum_{j=-\infty}^{\infty} (-1)^j \exp(-2j^2\lambda^2) & \text{for } \lambda > 0, \\ 0 & \text{for } \lambda \leqslant 0. \end{cases}$$

Smirnov proved this theorem under the additional assumption that $n_1/n_2 = \tau$, where $\tau > 0$, but Gihman [1] established that this assumption is not necessary.

C   The history of the various proofs of the theorems of Kolmogorov and Smirnov is very interesting. The original proofs given by the authors were very complicated. The proof by Feller [5] was simpler but still cumbersome. A short and elegant proof, based on a heuristic assertion, was given by Doob [4], and the validity of this heuristic assertion was proved by Donsker [2]. We present only the idea of the proof of theorem 10.11.1, due to Doob. Let

$$U_n(x) = \sqrt{n}[S_n(x) - F(x)].$$

Since the distribution function $F(x)$ is continuous, we may assume that the random variable X has a uniform distribution in the interval $[0, 1]$, since if it did not, we could consider the random variable $Y = F(X)$, which would then have this uniform distribution (see Section 5.6). Thus we can write

$$U_n(x) = \sqrt{n}[S_n(x) - x], \qquad D_n\sqrt{n} = \sup_{0 \leqslant x \leqslant 1} |U_n(x)| \qquad (0 \leqslant x \leqslant 1).$$

It is easy to compute

(*) $E[U_n(x)] = 0$ $(0 \leqslant x \leqslant 1)$,

$E(U_n(x_1)U_n(x_2)) = x_1(1 - x_2)$ $(0 \leqslant x_1 \leqslant x_2 \leqslant 1)$.

From the multidimensional de Moivre-Laplace theorem (theorem 6.13.1) it follows that, for $r = 1, 2, \ldots$ and arbitrary points $x_1, \ldots, x_r$, where $0 \leqslant x_1 \leqslant \ldots \leqslant x_r \leqslant 1$, the random vector $[U_n(x_1), \ldots, U_n(x_r)]$ has the asymptotically normal distribution with the expected values, variances, and covariances given by formulas (*).

Let us now consider a normal stochastic process $\{U(x),\ 0 \leqslant x \leqslant 1\}$ such that, for $r = 1, 2, \ldots$ and for arbitrary $x_1, \ldots, x_r$ $(0 \leqslant x_1 \leqslant \ldots \leqslant x_r \leqslant 1)$, the random vector $[U(x_1), \ldots, U(x_r)]$ has the normal distribution with expected values, variances, and covariances given by (*). The realizations of such a stochastic process are continuous functions, with probability one. Doob's heuristic assertion states that

$$\lim_{n \to \infty} P\left[\sup_{0 \leqslant x \leqslant 1} |U_n(x)| < \lambda \right] = P\left[\max_{0 \leqslant x \leqslant 1} |U(x)| < \lambda \right].$$

Doob [4] established that the right-hand side of the last equality is equal to the function $Q(\lambda)$ given by (10.11.2).

Smirnov's theorem can also be proved by Doob's method. However, we shall present a proof of this theorem by another method, for the particular case $n_1 = n_2 = m$. In this case $n = \frac{1}{2}m$.

First we prove a lemma which gives the exact distributions of the statistics $\sqrt{\frac{1}{2}m}D_{mm}{}^+$ and $\sqrt{\frac{1}{2}m}D_{mm}$ and then we pass to the limit as $m \to \infty$, obtaining theorem 10.11.2. This simple and ingenious proof is due to Gnedenko and Koroluk [1]. This paper originated a series of important investigations in this domain which are discussed in the paper of Gnedenko [7]. Later, Koroluk [1] showed that the method of proof given by Gnedenko and himself can be applied to the proof of Smirnov's theorem in its general form, and also to the proof of Kolmogorov's theorem.

Lemma. *Let the assumptions of theorem 10.11.2 be satisfied.*
Then

$$(10.11.10) \quad Q_{mm}{}^+(\lambda) = \begin{cases} 0 & \text{for } \lambda \leqslant 0, \\[2ex] 1 - \dfrac{\dbinom{2m}{m-c}}{\dbinom{2m}{m}} & \text{for } 0 < \lambda \leqslant \sqrt{\tfrac{1}{2}m}, \\[2ex] 1 & \text{for } \lambda > \sqrt{\tfrac{1}{2}m}, \end{cases}$$

$$(10.11.11) \quad Q_{mm}(\lambda) = \begin{cases} 0 & \text{for } \lambda \leqslant \dfrac{1}{\sqrt{2m}}, \\[2ex] \displaystyle\sum_{j=-[m/c]}^{[m/c]} (-1)^j \dfrac{\dbinom{2m}{m-jc}}{\dbinom{2m}{m}} & \text{for } \dfrac{1}{\sqrt{2m}} < \lambda \leqslant \sqrt{\tfrac{1}{2}m}, \\[2ex] 1 & \text{for } \lambda > \sqrt{\tfrac{1}{2}m}, \end{cases}$$

where $c = -[-\lambda\sqrt{2m}]$.

Proof of the lemma. Let us denote by $y_1, y_2, \ldots, y_{2m}$ the values $x_{11}, x_{12}, \ldots, x_{1m}, x_{21}, x_{22}, \ldots, x_{2m}$ arranged in increasing order, that is,

$$y_1 < y_2 < \ldots < y_{2m}.$$

Let us introduce the random variables Z_j $(j = 1, 2, \ldots, 2m)$, where $Z_j = 1$ or $Z_j = -1$, according to whether y_j is an element of the first or of the second sample. Next, let us denote

$$V_0 = 0, \qquad V_k = \sum_{j=1}^{k} Z_j \qquad (k = 1, 2, \ldots, 2m).$$

We have the equalities

$$(10.11.12) \quad mD_{mm}{}^+ = m \max_{-\infty < x < \infty} [S_{1m}(x) - S_{2m}(x)] = \max_{0 \leqslant k \leqslant 2m} V_k,$$

$$(10.11.13) \quad mD_{mm} = m \max_{-\infty < x < \infty} |S_{1m}(x) - S_{2m}(x)| = \max_{0 \leqslant k \leqslant 2m} |V_k|.$$

Equality (10.11.12) can be proved as follows. For a given x, $m[S_{1m}(x) - S_{2m}(x)]$ is the difference between the number of elements in the first sample which are less than x and the number of such elements in the second sample, and $mD_{mm}{}^+$ is the largest such difference as x runs over the interval $(-\infty < x < +\infty)$. By the definitions of Z_j and V_k, we obtain equality (10.11.12). Equality (10.11.13) can be proved in a similar way.

The following illustration will explain the further reasoning. Let us consider a random walk of a particle in the plane (t, v). At the initial moment $t = 0$ the particle was in the state $v = 0$. The particle is subject

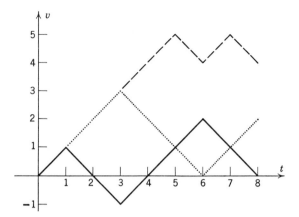

Fig. 10.11.1

to random translations at the moments $t = 1, 2, \ldots, 2m$ in such a way that at the moment j, where $1 \leqslant j \leqslant 2m$, the coordinate v may change its value by $+1$ or -1, according to whether $Z_j = 1$ or $Z_j = -1$, and among the $2m$ translations there are m positive and m negative; hence at the moment $t = 2m$ we have again $v = 0$. We next join successively the points with coordinates (t, v) by line segments. We call the broken line thus obtained a *path*. It is a random line with initial point $(0, 0)$ and end point $(2m, 0)$. Such a path is represented in Fig. 10.11.1 by the continuous line.

By (10.11.6) we have

$$Q_{mm}{}^+(\lambda) = P(\sqrt{\tfrac{1}{2}m}D_{mm}{}^+ < \lambda) = P(mD_{mm}{}^+ < \lambda\sqrt{2m}).$$

Since $mD_{mm}{}^+$ assumes only integer values, from formula (10.11.12) we have

(10.11.14) $$Q_{mm}{}^+(\lambda) = P(\max_{1 \leqslant k \leqslant 2m} V_k < c),$$

where c is a positive integer. The problem of finding the distribution function $Q_{mm}{}^+(\lambda)$ consists in finding, for every positive integer c, the probability that the observed path will always lie below the line $v = c$.

Let us now consider the set of all possible paths as the set of elementary events. Every path is determined by $2m$ possible translations, of which m are positive and m negative. In other words, every path is determined by a system of m $+1$'s and m -1's; hence the number of all possible paths is equal to the number of combinations of m elements among $2m$ elements, that is, $\binom{2m}{m}$. By the assumption that the random variables $X_{11}, \ldots, X_{1m}$,

$X_{21}, \ldots, X_{2m}$ are independent and have the same distribution, every path has the same probability; hence this probability is $1 \Big/ \binom{2m}{m}$. To find $Q_{mm}^{+}(\lambda)$, we must find the number of paths that always lie below the line $v = c$. First we find the number of paths which intersect the line $v = c$. To do this, we proceed in the following way. To each path that intersects the line $v = c$ we uniquely assign a path which is identical with the original one up to the first intersection with the line $v = c$ and from this point on is the reflection of the original line in the line $v = c$. The new path to the right of the first intersection with the line $v = 1$ is represented in Fig. 10.11.1 by the dotted line. The new path ends at the point $(2m, 2c)$; hence the number of "steps" up is $m + c$, and the number of "steps" down is $m - c$. Thus the number of such paths is $\binom{2m}{m + c} = \binom{2m}{m - c}$ and the number of paths that do not touch the line $v = c$ is $\binom{2m}{m} - \binom{2m}{m - c}$. Since the probability of each path is the same, from (10.11.14) we have, for $0 < \lambda \leqslant \sqrt{\tfrac{1}{2}m}$,

$$(10.11.15) \quad Q_{mm}^{+}(\lambda) = \frac{\binom{2m}{m} - \binom{2m}{m - c}}{\binom{2m}{m}} = 1 - \frac{\binom{2m}{m - c}}{\binom{2m}{m}}.$$

By the obvious relation $Q_{mm}^{+}(\lambda) = 0$ for $\lambda \leqslant 0$, and by the fact that if $\lambda > \sqrt{\tfrac{1}{2}m}$ then $c > m$ and that no path can intersect the line $v = m + 1$, so that $Q_{mm}^{+}(\lambda) = 1$ for $\lambda > \sqrt{\tfrac{1}{2}m}$, we obtain relation (10.11.10).

We now prove relation (10.11.11). We observe that in analogy to relation (10.11.14) we have

$$(10.11.16) \quad Q_{mm}(\lambda) = P(\sqrt{\tfrac{1}{2}m}\,D_{mm} < \lambda) = P(m D_{mm} < c)$$
$$= P(\max_{1 \leqslant k \leqslant 2m} |V_k| < c).$$

The problem of finding the distribution function $Q_{mm}(\lambda)$ consists in finding, for every positive integer c, the probability that the observed path will lie between the lines $v = -c$ and $v = c$ and will not intersect any of these lines.

Let us denote the line $v = c$ by α and the line $v = -c$ by β; next denote the set of all paths by A. The number of elements of the set A is, as we know, $\binom{2m}{m}$. Let us denote by A_0 the set of paths which lie between the lines α and β and do not intersect these lines. Let us denote by A_1 the set of paths which intersect the line α at least once; by A_2 the set of paths which intersect, at least once, first the line α and then the line β; by A_3 the set of paths which intersect the lines α and β in the order $\alpha\beta\alpha$ at least once. Generally, let A_{2i-1} denote the set of paths which intersect the lines α and β in the order $\alpha, \beta, \ldots, \alpha$ at least once (here, α appears i times and β appears $i-1$ times), while A_{2i} denotes the set of paths that intersect α and β in the order $\alpha, \beta, \ldots, \alpha, \beta$ at least once (here, both α and β appear i times). Similarly, let B_{2i-1} denote the set of paths which intersect the lines α and β in the order $\beta, \alpha, \ldots, \beta$ at least once (here, β appears i times, α appears $i-1$ times), while B_{2i} denotes the set of paths which intersect the lines α and β in the order $\beta, \alpha, \ldots, \beta, \alpha$ at least once (both α and β appear here i times).

It is easy to prove the relation

$$(10.11.17) \qquad A = A_0 + \sum_{i \geq 1} \{(A_{2i-1} - A_{2i}) + (B_{2i-1} - B_{2i})\},$$

where the sets in $\{ \ldots \}$ are disjoint for different i's . We have shown that the number of elements of A_1 (hence B_1) is equal to $\begin{pmatrix} 2m \\ m - c \end{pmatrix}$. To find the number of elements of the set A_2, let us assign uniquely to every path in the set A_2 a path which is identical with the original path up to the first intersection with the line α, and from this point up to the intersection with the line $v = 3c$ is the reflection of the original path in the line α. After reaching the line $v = 3c$, the new path is the reflection of the original line in the line $v = 3c$. The new path ends at the point $(2m, 4c)$; hence the number of "steps" up is equal to $m + 2c$, and the number of "steps" down is equal to $m - 2c$. Thus the number of elements of A_2 is $\begin{pmatrix} 2m \\ m - 2c \end{pmatrix}$. The number of elements of B_2 is the same.

In Fig. 10.11.1 the new path to the right of the point of intersection with the line $v = 3$ is represented by the broken line.

It can be shown in an analogous way that for $j \leq [m/c]$ the number of elements of the set A_j (hence B_j) is equal to $\begin{pmatrix} 2m \\ m - jc \end{pmatrix}$. Thus by (10.11.17) the number of elements of A_0 equals

$$\begin{pmatrix} 2m \\ m \end{pmatrix} - 2\sum_{i \geq 1}\left\{\begin{pmatrix} 2m \\ m - (2i-1)c \end{pmatrix} - \begin{pmatrix} 2m \\ m - 2ic \end{pmatrix}\right\} = \sum_{j=-[m/c]}^{[m/c]} (-1)^j \begin{pmatrix} 2m \\ m - jc \end{pmatrix}.$$

Since every path has the same probability of appearance, we have from (10.11.16)

$$(10.11.18) \quad Q_{mm}(\lambda) = \sum_{j=-[m/c]}^{[m/c]} (-1)^j \frac{\begin{pmatrix} 2m \\ m - jc \end{pmatrix}}{\begin{pmatrix} 2m \\ m \end{pmatrix}} \quad \left(\frac{1}{\sqrt{2m}} < \lambda \leq \sqrt{\tfrac{1}{2}m}\right).$$

We observe that if $\lambda \leqslant 1/\sqrt{2m}$, we have $c \leqslant 1$. Since $\max\limits_{1 \leqslant k \leqslant 2m} |V_k|$ is positive, we have $Q_{mm}(\lambda) = 0$ for such λ. If $\lambda > \sqrt{\tfrac{1}{2}m}$, then $c > m$, hence $Q_{mm}(\lambda) = 1$. Relation (10.11.11) is proved.

D   *Proof of Smirnov's theorem.*   We now prove theorem 10.11.2 in the case considered, $n_1 = n_2 = m$. For $j = 1, 2, \ldots$ we have

$$I_j = \binom{2m}{m-jc} \bigg/ \binom{2m}{m} = \frac{(m!)^2}{(m-jc)!\,(m+jc)!}.$$

By an argument similar to that used in Section 10.5, by Stirling's formula $r! \cong (r/e)^r \sqrt{2\pi r}$, and by the approximate equality $c = \lambda\sqrt{2m}$, we obtain

$$I_j \cong \left(1 - \frac{\lambda j\sqrt{2}}{\sqrt{m}}\right)^{j\lambda\sqrt{2m}-m} \left(1 + \frac{\lambda j\sqrt{2}}{\sqrt{m}}\right)^{-j\lambda\sqrt{2m}-m}$$

$$\times \sqrt{\frac{m}{(\sqrt{m} - j\lambda\sqrt{2})(\sqrt{m} + j\lambda\sqrt{2})}}.$$

Hence

$$\log I_j = -2j^2\lambda^2 + \sqrt{m}\,o\!\left(\frac{1}{\sqrt{m}}\right).$$

Finally,

$$(10.11.19) \qquad\qquad \lim_{m \to \infty} I_j = \exp(-2j^2\lambda^2).$$

Passing to the limit as $m \to \infty$ in (10.11.10) and using (10.11.19) for $j = 1$, we obtain (10.11.8).

To prove (10.11.9), observe that for every $\lambda > 0$ and every $\varepsilon > 0$ we can find a $j_0 > 0$ such that

$$(10.11.20) \qquad\qquad \exp(-2j_0^2\lambda^2) < \frac{\varepsilon}{16}.$$

From Leibniz's theorem we obtain

$$(10.11.21) \quad \left| \sum_{|j| > j_0} (-1)^j \exp(-2j^2\lambda^2) \right| < 2\exp[-2(j_0+1)^2\lambda^2] < \frac{\varepsilon}{8}.$$

Let

$$R = \sum_{j_0 < |j| \leqslant [m/c]} (-1)^j \frac{\binom{2m}{m-jc}}{\binom{2m}{m}}.$$

Since for $j > 0$ we have $\binom{2m}{m - jc} > \binom{2m}{m - (j+1)c}$ and for $j < 0$ we have $\binom{2m}{m - jc} > \binom{2m}{m - (j-1)c}$, we obtain

$$|R| < 4 \frac{\binom{2m}{m - j_0 c}}{\binom{2m}{m}}.$$

By (10.11.19) and (10.11.20) it follows that for sufficiently large m

$$|R| < \frac{\varepsilon}{3}.$$

The last inequality and inequality (10.11.21) imply

$$(10.11.22) \qquad \left| R - \sum_{|j| > j_0} (-1)^j \exp\left(-2\lambda^2 j^2\right) \right| < \frac{\varepsilon}{2}.$$

On the other hand, it follows from (10.11.19) that for sufficiently large m

$$(10.11.23) \qquad \left| \sum_{|j| \leqslant j_0} (-1)^j \left\{ \left[\frac{\binom{2m}{m - jc}}{\binom{2m}{m}} - \exp\left(-2j^2\lambda^2\right) \right] \right\} \right| < \frac{\varepsilon}{2}.$$

Passing to the limit as $m \to \infty$ in (10.11.11) and using (10.11.22) and (10.11.23), we obtain (10.11.9).

For small m the distribution functions $Q_{mm}^{+}(\lambda)$ and $Q_{mm}(\lambda)$ were tabulated by Massey [2]. These tables were reproduced by Sadowski [1] and by Janko [1].

It is to be emphasized that exact tables for $Q_{mm}(\lambda)$ should be used, if available, since, as has been recently established by Hodges [1], $Q_{mm}(\lambda)$ may significantly differ from the limit $Q(\lambda)$, even for fairly large m.

Example 10.11.2. Two independent simple samples of four elements each were drawn from a population in which the characteristic X has a continuous distribution function. The following values were observed:

$$x_{11} = 1, \quad x_{12} = 3, \quad x_{13} = -7, \quad x_{14} = 2,$$
$$x_{21} = 6, \quad x_{22} = -4, \quad x_{23} = 0, \quad x_{24} = 8.$$

We have here

$$y_1 = -7, \quad y_2 = -4, \quad y_3 = 0, \quad y_4 = 1, \quad y_5 = 2, \quad y_6 = 3, \quad y_7 = 6, \quad y_8 = 8.$$

Hence

$$z_1 = 1, \, z_2 = -1, \, z_3 = -1, \, z_4 = 1, \, z_5 = 1, \, z_6 = 1, \, z_7 = -1, \, z_8 = -1,$$
$$v_1 = 1, \, v_2 = 0, \, v_3 = -1, \, v_4 = 0, \, v_5 = 1, \, v_6 = 2, \, v_7 = 1, \quad v_8 = 0.$$

The observed values of the statistics $D_{44}{}^{+}$ and D_{44} are $d_{44}{}^{+} = d_{44} = \frac{1}{2}$. Let us find the probabilities of obtaining values of these random variables which exceed or equal the observed values. We have here $\lambda/\sqrt{2} = \frac{1}{2}$; hence $\lambda = 1/\sqrt{2}$ and $c = 2$. By (10.11.10) we have $Q_{44}{}^{+}(1/\sqrt{2}) = 0.6$; hence the probability of obtaining for $D_{44}{}^{+}$ a value not less than $\frac{1}{2}$ is 0.4. By (10.11.11) we have $Q_{44}(1/\sqrt{2}) = 0.229$; hence the required complement to one is 0.771.

Example 10.11.3. We select two independent simple samples from a population in which the characteristic X has a continuous distribution function. The number of elements of the first sample is $n_1 = 70$, of the second sample, $n_2 = 100$. The observed value of $D_{n_1 n_2}$ is 0.18. What is the probability that $D_{n_1 n_2}$ is smaller than this value?

We have $n = n_1 n_2/(n_1 + n_2) = 41.2$, $\sqrt{n} = 6.42$, $\lambda = 0.18 \cdot 6.42 = 1.16$; hence by the Smirnov limit theorem

$$P(D_{n_1 n_2} < 0.18) = P\left(D_{n_1 n_2} < \frac{1.16}{6.42}\right) \cong Q(1.16).$$

In Table VIII we find that $Q(1.16) \cong 0.8644$.

E   We now give some further information. Let

$$D_n{}^{+} = \sup_{-\infty < x < +\infty} [S_n(x) - F(x)],$$

$$D_{n_1 n_2}^{-} = \max_{-\infty < x < +\infty} [S_{2n_2}(x) - S_{1n_1}(x)]$$

$$= -\min_{-\infty < x < +\infty} [S_{1n_1}(x) - S_{2n_2}(x)].$$

The distribution of $D_n{}^{+}$ was given by Wald and Wolfowitz [1] and by Birnbaum and Tingey [1].

Smirnov [1, 2] showed that if the assumptions of theorem 10.11.1 are satisfied, then for $\lambda > 0$

(10.11.24) $$\lim_{n \to \infty} P(D_n{}^{+}\sqrt{n} < \lambda) = 1 - \exp(-2\lambda^2).$$

In the papers quoted, Smirnov found the limit distribution of the random vector $(D_{n_1 n_2}^{+}, D_{n_1 n_2}^{-})$, where $n = n_1 n_2/(n_1 + n_2)$.

Gnedenko and Rvatcheva [1] found the exact two-dimensional distribution of the random vector $(\sqrt{m/2}D_{mm}{}^{+}, \sqrt{m/2}D_{mm}{}^{-})$ for $n_1 = n_2 = m$, and computed the coefficient of correlation of the random variables $D_{mm}{}^{+}$ and $D_{mm}{}^{-}$.

Smirnov [2] investigated the limit distribution of the number of intersections of the line $S_{2n_2}(x) + \lambda\sqrt{(n_1 + n_2)/n_1 n_2}$ by the empirical distribution function $S_{1n_1}(x)$. Mihalevitch [1] found the exact distribution of the number of these intersections for $n_1 = n_2$.

The distribution of a statistic which is closely related to the statistic $D_n{}^{+}$ was considered by Birnbaum and Pyke [1] (see Problem 10.19).

Kac [3] investigated the distribution of the statistic D_n under the assumption that n is a random variable with a Poisson distribution.

Some results concerning the generalizations of Kolmogorov's theorem (10.11.1) and of relation (10.11.24) to random vectors were obtained by Kiefer and Wolfowitz [3].

10.12 RÉNYI'S THEOREM

A modification of Kolmogorov's theorem 10.11.1 was given by Rényi [3]. In Kolmogorov's theorem, the difference $|S_n(x) - F(x)|$ is considered without taking into account the value of $F(x)$. Thus the difference $|S_n(x) - F(x)| = 0.01$ at some point x where $F(x) = 0.5$ (a difference that is only 2% of the value of $F(x)$) is considered as significant as the same difference at the point x where $F(x) = 0.01$ and the difference is 100% of the value of $F(x)$. Rényi therefore proposed that, instead of the statistic $|S_n(x) - F(x)|$, the statistic $[S_n(x) - F(x)]/F(x)$ be considered. He showed that if the assumptions of theorem 10.11.1 are satisfied, then for every $a > 0$ the following relations are true:

$$(10.12.1) \qquad \lim_{n \to \infty} P\left[\sqrt{n} \sup_{a \leqslant F(x)} \frac{S_n(x) - F(x)}{F(x)} < \lambda \right]$$

$$= \begin{cases} 0 & \text{for } \lambda \leqslant 0, \\ \sqrt{2/\pi} \displaystyle\int_0^{\lambda \sqrt{a/(1-a)}} \exp\left(-\lambda^2/2\right) d\lambda & \text{for } \lambda > 0, \end{cases}$$

$$(10.12.2) \qquad \lim_{n \to \infty} P\left[\sqrt{n} \sup_{a \leqslant F(x)} \left| \frac{S_n(x) - F(x)}{F(x)} \right| < \lambda \right]$$

$$= \begin{cases} 0 & \text{for } \lambda \leqslant 0, \\ \dfrac{4}{\pi} \displaystyle\sum_{j=0}^{\infty} (-1)^j \dfrac{\exp\left[-\dfrac{(2j+1)^2 \pi^2}{8} \cdot \dfrac{1-a}{2\lambda^2} \right]}{2j+1} & \text{for } \lambda > 0. \end{cases}$$

The supremum on the left-hand sides of these formulas is taken with respect to values of x for which $F(x) \geqslant a$ for an arbitrary $a > 0$.

This theorem has useful practical applications. Tables for the right-hand side of (10.12.2) are given by Rényi in the paper quoted [3]. The distribution of

$$\sqrt{n} \sup_{a \leqslant F(x)} \frac{S_n(x) - F(x)}{F(x)},$$

for finite n, was given by Ishii [1] (see also Problem 10.22).

Schmid [1] generalized Rényi's theorem to the case when the distribution function $F(x)$ has discontinuity points.

Wang [1] gave formulas analogous to (10.12.1) for two empirical distribution functions $S_{1n_1}(x)$ and $S_{2n_2}(x)$ of two independent simple samples drawn from the same population with a continuous distribution function $F(x)$. He proved that if the relation

$$\frac{n_1}{n_2} \to d \leqslant 1$$

holds as $n_1 \to \infty$ and $n_2 \to \infty$, where d is a constant, then for arbitrary $a > 0$

$$\lim_{n_2 \to \infty} P\left(\sqrt{n_1 n_2/(n_1 + n_2)} \sup_{a \leqslant S_{2n_2}(x)} \frac{S_{1n_1}(x) - S_{2n_2}(x)}{S_{2n_2}(x)} < \lambda\right)$$

$$= \begin{cases} 0 & \text{for } \lambda \leqslant 0, \\ \sqrt{2/\pi} \displaystyle\int_0^{\lambda \sqrt{a/(1-a)}} \exp\left(-\tfrac{1}{2}\lambda^2\right) d\lambda & \text{for } \lambda > 0. \end{cases}$$

Chang [1] found the exact distribution of the random variables

$$\sup_{0 < F(x) \leqslant a} \frac{S_n(x)}{F(x)}, \qquad \sup_{0 < S_n(x) \leqslant a} \frac{S_n(x)}{F(x)},$$

where a is any constant in the interval $(0, 1]$.

Manija [1] investigated the distribution of the statistic

$$\sup_{a < x < b} |S_n(x) - F(x)|,$$

where a and b are constants. Kvit [1] considered the distribution of the statistic

$$\sup_{-\infty < x \leqslant a,\, b \leqslant x < +\infty} |S_n(x) - F(x)|.$$

The modifications of the statistic of Kolmogorov considered by Rényi, Chang, Manija, and Kvit are particular cases of statistics considered by Anderson and Darling [1],

$$D_n^* = \sup_{-\infty < x < +\infty} \sqrt{n}\, |S_n(x) - F(x)| \psi[F(x)],$$

$$D_n^{**} = \sup_{-\infty < x < +\infty} \sqrt{n}[S_n(x) - F(x)] \psi[F(x)],$$

where ψ is some function.

Some generalizations and modifications of the Kolmogorov and Smirnov statistics have also been considered in papers by Rvatcheva [1], Tsao [1], Kuiper [2], Vincze [1, 2] and Dwass [1].

An extensive exposition of problems connected with the Kolmogorov and Smirnov statistics can be found in the paper of Darling [1].

10.13 THE PROBLEM OF k SAMPLES

A   Theorem 10.11.1 allows us to investigate the question whether the characteristic X of the population from which the sample has been drawn has the given distribution function $F(x)$. The theorem we now present allows us to investigate the question whether the characteristic X of k ($k \geqslant 2$) populations from which k samples have been drawn has the same distribution function $F(x)$.

Let us consider k ($k \geqslant 2$) populations in which the characteristic X has the same continuous distribution function $F(x)$. From each of these populations we draw a simple sample of size n. In other words, we have nk independent random variables $X_{11}, \ldots, X_{1n}, X_{21}, \ldots, X_{2n}, \ldots, X_{k1}, \ldots, X_{kn}$ with the same distribution, divided into k groups, and for each of these random variables we have one observation. Let us denote by $S_{nj}(x)$ the empirical distribution function of the sample drawn from the jth population, $j = 1, \ldots, k$. Let

$$(10.13.1) \quad D(n, j) = \sqrt{n} \sup_{-\infty < x < \infty} |S_{nj}(x) - F(x)| \quad (j = 1, 2, \ldots, k),$$

$$M_n = \max_{1 \leqslant j \leqslant k} D(n, j).$$

The random variables $D(n, j)$ ($j = 1, \ldots, k$) are independent and have the same distribution.

Theorem 10.13.1.   *Let $S_{nj}(x)$ ($j = 1, \ldots, k$) be the empirical distribution functions of k simple samples drawn from populations in which the characteristic X has the same continuous distribution function $F(x)$.*

Then

$$(10.13.2) \quad \lim_{n \to \infty} P(M_n < \lambda) = [Q(\lambda)]^k,$$

where Q is given by (10.11.2).

Proof.   By theorem 10.11.1

$$(10.13.3) \quad \lim_{n \to \infty} P[D(n, j) < \lambda] = Q(\lambda) \quad (j = 1, 2. \ldots, k).$$

Since the event $(M_n < \lambda)$ is equivalent to the joint event $[(D(n, 1) < \lambda, \ldots, D(n, k) < \lambda]$, and the random variables $D(n, j)$ ($j = 1, \ldots, k$) are independent, by (10.13.3) we obtain

$$\lim_{n \to \infty} P(M_n < \lambda) = \lim_{n \to \infty} P[D(n, j) < \lambda; j = 1, \ldots, k]$$

$$= \prod_{j=1}^{k} \lim_{n \to \infty} P[D(n, j) < \lambda] = [Q(\lambda)]^k.$$

The theorem is proved.

It is obvious that the assumption that the number of elements is the same in every sample is not relevant. If these numbers are not equal, and the number of elements of the jth sample is n_j $(j = 1, \ldots, k)$, then the assertion of the theorem remains true if all n_j tend to infinity.

B   We now state a theorem which is a generalization of Smirnov's theorem to k samples, $k \geqslant 2$.

Let us consider k samples with the properties mentioned in Section 10.13, A and let n_j and $S_{jn_j}(x)$ $(j = 1, \ldots, k)$ denote, respectively, the number of elements and the empirical distribution function of the jth sample. Next, define, for $i = 1, \ldots, k - 1$,

$$(10.13.4) \qquad A_i^+(n_1, \ldots, n_k) = \max_{-\infty < x < +\infty} \sum_{j=1}^{k} B_{in_j}\sqrt{n_j}S_{jn_j}(x),$$

$$(10.13.5) \qquad A_i(n_1, \ldots, n_k) = \max_{-\infty < x < +\infty} \left| \sum_{j=1}^{k} B_{in_j}\sqrt{n_j}S_{jn_j}(x) \right|,$$

where B_{in_j} are constants which depend on the arguments $n_1, \ldots, n_k, i, j$.

Theorem 10.13.2.   *Let $S_{jn_j}(x)$ $(j = 1, \ldots, k)$ be the empirical distribution functions of k simple samples drawn from k populations in which the characteristic X has the same continuous distribution function. Let us assume that the following relations are satisfied:*

$$(10.13.6) \qquad \sum_{j=1}^{k} B_{in_j}\sqrt{n_j} = 0 \qquad (i = 1, \ldots, k - 1),$$

$$(10.13.7) \qquad \lim_{n_1 \to \infty, \ldots, n_k \to \infty} B_{in_j} = B_{ij} \qquad (i = 1, \ldots, k - 1, j = 1, \ldots, k),$$

and

$$(10.13.8) \qquad \sum_{j=1}^{k} B_{hj}B_{ij} = \begin{cases} 0 & \text{for } h \neq i, \\ 1 & \text{for } h = i. \end{cases}$$

Then for arbitrary positive $\lambda_1, \ldots, \lambda_{k-1}$ and λ we have

$$(10.13.9) \qquad \lim_{n_1 \to \infty, \ldots, n_k \to \infty} P[A_i^+(n_1, \ldots, n_k) < \lambda_i, i = 1, \ldots, k - 1]$$
$$= \prod_{i=1}^{k-1} [1 - \exp(-2\lambda_i^2)],$$

$(10.13.10)$

$$\lim_{n_1 \to \infty, \ldots, n_k \to \infty} P[A_i(n_1, \ldots, n_k) < \lambda_i, \; i = 1, \ldots, k - 1] = \prod_{i=1}^{k-1} Q(\lambda_i),$$

where $Q(\lambda_i)$ is given by $(10.11.2)$,

$$(10.13.11) \qquad \lim_{n_1 \to \infty, \ldots, n_k \to \infty} P\left[\max_{1 \leqslant i \leqslant k-1} A_i^+(n_1, \ldots, n_k) < \lambda \right]$$
$$= [1 - \exp(-2\lambda^2)]^{k-1},$$

$$(10.13.12) \qquad \lim_{n_1 \to \infty, \ldots, n_k \to \infty} P\left[\max_{1 \leqslant i \leqslant k-1} A_i(n_1, \ldots, n_k) < \lambda \right] = [Q(\lambda)]^{k-1}.$$

This theorem was proved by Fisz [4], Chang and Fisz [1], and Kiefer [1]. It is obvious that we can choose different coefficients B_{in_j} satisfying relations (10.13.6) to (10.13.8). Assuming in addition that $\lim_{n_1 \to \infty} (n_j/n_1) = \alpha_j > 0$ $(j = 2, \ldots, k)$, we can consider the following set of coefficients:

(10.13.13) $\quad B_{in_j} = \begin{cases} -\sqrt{n_j n_{i+1}/N_i N_{i+1}} & (j = 1, \ldots, i), \\ \sqrt{N_i/N_{i+1}} & (j = i + 1), \\ 0 & (j = i + 2, \ldots, k), \end{cases}$

where $N_r = n_1 + \ldots + n_r$. Then

$$A_i^+(n_1, \ldots, n_k) = \sqrt{n_{i+1}N_i/N_{i+1}} \, D_i^+(N_i, n_{i+1}),$$

$$A_i(n_1, \ldots, n_k) = \sqrt{n_{i+1}N_i/N_{i+1}} \, D_i(N_i, n_{i+1}),$$

where

(10.13.14)
$$D_i^+(N_i, n_{i+1}) = \max_x \left[S_{i+1,n_{i+1}}(x) - \frac{1}{N_i} \sum_{j=1}^i n_j S_{jn_j}(x) \right],$$

$$D_i(N_i, n_{i+1}) = \max_x \left| S_{i+1,n_{i+1}}(x) - \frac{1}{N_i} \sum_{j=1}^i n_j S_{jn_j}(x) \right|.$$

Thus, for instance, for $k = 4$, we have the following three random variables:

$$D_1(N_1, n_2) = \max_x |S_{2n_2}(x) - S_{1n_1}(x)|,$$

$$D_2(N_2, n_3) = \max_x \left| S_{3n_3}(x) - \frac{n_1 S_{1n_1}(x) + n_2 S_{2n_2}(x)}{n_1 + n_2} \right|,$$

$$D_3(N_3, n_4) = \max_x \left| S_{4n_4}(x) - \frac{n_1 S_{1n_1}(x) + n_2 S_{2n_2}(x) + n_3 S_{3n_3}(x)}{n_1 + n_2 + n_3} \right|.$$

As was shown by Chang and Fisz [2], (see also Kiefer [1]) the random variables $D^+(N_1, n_2), \ldots, D_{k-1}^+(N_{k-1}, n_k)$ are independent. The same is true for the random variables $D_1(N_1, n_2), \ldots, D_{k-1}(N_{k-1}, n_k)$.

Gihman [5] and Kiefer [1] have given another generalization of the Smirnov theorem to k samples. They proved the following theorem.

Theorem 10.13.5. *Let* $S_{jn_j}(x)$ *$(j = 1, \ldots, k)$ be the empirical distribution functions of k simple samples drawn from a population in which the characteristic X has the same continuous distribution function.*

Let

$$S_{n_1, \ldots, n_k}(x) = \frac{1}{N_k} \sum_{k=1}^1 n_j S_{jn_j}(x),$$

$$D_{n_1, \ldots, n_k}^2 = \max_{-\infty < x < +\infty} \sum_{j=1}^k n_j [S_{jn_j}(x) - S_{n_1, \ldots, n_k}(x)]^2.$$

Then

(10.13.15) $\lim\limits_{n_1 \to \infty, \ldots, n_k \to \infty} P(D_{n_1, \ldots, n_k} < \lambda)$

$$= \frac{4}{\Gamma\left(\dfrac{k-1}{2}\right)(2\lambda^2)^{(k-1)/2}} \sum_{s=1}^{\infty} \frac{\mu_s^{k-3}}{\left[J_{(k-3)/2}(\mu_s)\right]^2} \exp\left(-\frac{\mu_s^2}{2\lambda^2}\right),$$

where μ_s is the s-th positive root of Bessel's function $J_{(k-3)/2}(z)$.
All theorems of Section 10.13, B are given without proofs. The presentation of these proofs requires mathematical sophistication beyond the scope of this book. The idea of these proofs is close to that of the proof of Kolmogorov's theorem, presented in Section 10.11, C.

Another generalization of Smirnov's theorem to three samples, employing more elementary methods, can be found in the papers of Ozols [1] (see Problem 10.20) and H. T. David [1], and for an arbitrary finite number of samples, in Birnbaum and Hall [1]. (See also Dwass [2].)

Problems and Complements

10.1. From a population, in which the characteristic X has the normal distribution $N(0; 1)$, a simple sample of size $n = 10$ was drawn. The following values were obtained: $x_1 = 0$, $x_2 = 0.2$, $x_3 = 0.25$, $x_4 = -0.3$, $x_5 = -0.1$, $x_6 = 2$, $x_7 = 0.15$, $x_8 = 1$, $x_9 = -0.7$, $x_{10} = -1$.
(a) Find the empirical distribution function $S_n(x)$ of this sample.
(b) Compute $E[S_n(x)]$ and $D^2[S_n(x)]$ at the point $x = 0.2$.
(c) Compute $\Phi_{kn}(x)$ for $k = 6$, $n = 10$, and $x = 0.2$.

10.2. Let $\zeta_k^{(n)}$ be an order statistic of a simple sample drawn from a population in which the characteristic X has a continuous distribution function $F(x)$. Show that the random variable $F(\zeta_k^{(n)})$ has the beta distribution with parameters k and $n - k + 1$.

10.3. Prove that under the assumptions of theorem 10.4.1 its assertion can be replaced by the stronger one, that the sequence $\{\zeta_k^{(n)}\}$ converges to a_λ as $n \to \infty$, with probability one.
Hint: Show that for every $\varepsilon > 0$ the series

$$\sum_{n=1}^{\infty} P\{|\zeta_k^{(n)} - a_\lambda| > \varepsilon\} < \infty,$$

and use the Borel-Contelli lemma.

10.4. Let $\Phi_{kn}(x)$ be defined by formula (10.3.6) and suppose that k is constant. Prove that for every x we have

$$\frac{1 - a_n}{(k-1)!} \int_0^{nF(x)} e^{-x} x^{k-1} \, dx \leqslant \Phi_{kn}(x) \leqslant \frac{(1 + b_n)}{(k-1)!} \int_0^{nF(x)} e^{-x} x^{k-1} \, dx,$$

where $F(x)$ is the theoretical distribution function, $a_n > 0$, $b_n > 0$, and $\lim\limits_{n \to \infty} a_n = \lim\limits_{n \to \infty} b_n = 0$.

10.5. Let $\{\zeta_k^{(n)}\}$ be a sequence of order statistics for a fixed k, and let $F(x)$ be the theoretical distribution function.

(a) Show that the sequence $\{\zeta_k^{(n)} - A_k^{(n)}\}$, where $A_k^{(n)}$ are constants, converges stochastically to zero if and only if for every $\varepsilon > 0$

$$\lim_{n \to \infty} nF(A_k^{(n)} - \varepsilon) = 0,$$

$$\lim_{n \to \infty} nF(A_k^{(n)} + \varepsilon) = \infty.$$

Hint: Use Problem 10.4.

(b) Solve the analogous problem for $k = n - l + 1$ for a fixed l.

10.6. (Continued). (a) Show that if, for some x_0, $F(x) = 0$ for $x \leqslant x_0$ and $F(x) > 0$ for $x > x_0$, then the sequence $\{\zeta_k^{(n)} - x_0\}$, where k is constant, converges stochastically to zero as $n \to \infty$.

(b) Formulate and prove the analogous result if, for some x_0, $F(x) < 1$ for $x < x_0$ and $F(x) = 1$ for $x \geqslant x_0$.

Hint: Use Problem 10.5.

10.7. Prove that (a) if $F(x) > 0$ for every x, then the sequence $\{\zeta_k^{(n)} - A_k^{(n)}\}$, where k is constant, is, for some sequence $\{A_k^{(n)}\}$ of constants, stochastically convergent to zero if and only if for every $\varepsilon > 0$

$$\lim_{x \to -\infty} \frac{F(x - \varepsilon)}{F(x)} = 0.$$

(b) if $F(x) < 1$ for every x, then the sequence $\{\zeta_k^{(n)} - A_k^{(n)}\}$, $k = n - l + 1$, l constant, converges stochastically to zero as $n \to \infty$ if and only if for every $\varepsilon > 0$

$$\lim_{x \to +\infty} \frac{1 - F(x + \varepsilon)}{1 - F(x)} = 0.$$

(See, for $k = 1$ or $k = n$, Gnedenko [8], and in the general case Smirnov [3].)

10.8. Let $F(x)$ and a_λ $(0 < \lambda < 1)$ be the distribution function and the quantile, respectively, of the random variable X. Let $\zeta_k^{(n)}$ be the quantile of a simple sample of size n. The relation

$$\lim_{n \to \infty} P\left(\frac{\zeta_k^{(n)} - b_n}{a_n} < z \right) = \frac{1}{\sqrt{2\pi}} \int_{-\infty}^{z} e^{-z^2/2} \, dz$$

holds for some sequence of constants $a_n > 0$ and b_n if and only if (Smirnov [3]) for every x

$$\lim_{n \to \infty} \frac{F(a_n x + b_n) - \lambda}{\sqrt{\lambda(1 - \lambda)}} \sqrt{n} = x.$$

10.9. Let $F(x), f(x), a_{\lambda_1},$ and a_{λ_2} $(0 < \lambda_1 < \lambda_2 < 1)$ be the distribution function, the density, and the quantiles, respectively, of the random variable X. Let $\zeta_{k_1}^{(n)}$ and $\zeta_{k_2}^{(n)}$ $(k_1 = [n\lambda_1] + 1, k_2 = [n\lambda_2] + 1)$ be the quantiles of a simple sample. Show that if the density $f(x)$ is continuous and positive at the points $x = a_{\lambda_1}$ and $x = a_{\lambda_2}$, then the random vector $(\zeta_{k_1}^{(n)}, \zeta_{k_2}^{(n)})$ has, as $n \to \infty$, the asymptotically normal distribution with parameters

$$m_1 = a_{\lambda_1}, \qquad m_2 = a_{\lambda_2}, \qquad \sigma_1^2 = \frac{\lambda_1(1 - \lambda_1)}{n[f(a_{\lambda_1})]^2},$$

$$\sigma_2^2 = \frac{\lambda_2(1 - \lambda_2)}{n[f(a_{\lambda_2})^2}, \qquad \rho = \sqrt{\frac{\lambda_1(1 - \lambda_2)}{\lambda_2(1 - \lambda_1)}}.$$

Generalize this result to r $(r > 2)$ order statistics.

10.10. Let $\{\zeta_k^{(n)}\}$ be a sequence of order statistics for a fixed k and let $F(x)$ be the theoretical distribution function. The relation

$$\lim_{n\to\infty} P\left(\frac{\zeta_k^{(n)} - b_n}{a_n} < z\right) = \begin{cases} 0 & (z \leqslant 0), \\ \dfrac{1}{(k-1)!}\displaystyle\int_0^z e^{-z}z^{k-1}\,dz & (z > 0) \end{cases}$$

holds for some sequences of constants $a_n > 0$ and b_n if and only if

$$\lim_{n\to\infty} nF(a_n x + b_n) = \begin{cases} 0 & (x \leqslant 0), \\ x & (x > 0). \end{cases}$$

(Gnedenko [8], Smirnov [3].)

10.11. Show that the random variables $Z_k^{(n)}$ and $U_k^{(n)}$ (see Section 10.6) are, as $n \to \infty$, asymptotically independent.

10.12. Let $\zeta_1^{(n)}$, $\zeta_2^{(n)}$, . . . , $\zeta_n^{(n)}$ be order statistics of a simple sample drawn from a population in which the characteristic X has the uniform distribution in $[0, 1]$. Prove that the random variables $V_1, V_2, \ldots, V_n$, where

$$V_1 = \frac{\zeta_1^{(n)}}{\zeta_2^{(n)}}, \qquad V_2 = \frac{\zeta_2^{(n)}}{\zeta_3^{(n)}}, \ldots, V_{n-1} = \frac{\zeta_{n-1}^{(n)}}{\zeta_n^{(n)}}, \qquad V_n = \zeta_n^{(n)}, \qquad \text{are}$$

independent and V_j $j = 1, \ldots, n-1)$ has the uniform distribution in $[0.1]$.

10.13. Let $\zeta_1^{(2)}$ and $\zeta_2^{(2)}$ be order statistics of a simple sample drawn from a population in which the characteristic X has the distribution function $F(x)$ and density $f(x)$, and suppose that $F(0) = 0$. Show that the density $f(x)$ is of the form

(*) $$f(x) = \begin{cases} \dfrac{m+1}{b^{m+1}} x^m & (0 \leqslant x \leqslant b), \\ 0 & \text{otherwise} \end{cases}$$

if and only if the random variables V_1 and V_2, where

$$V_1 = \frac{\zeta_1^{(2)}}{\zeta_2^{(2)}}, \qquad V_2 = \zeta_2^{(2)},$$

are independent. If, moreover, in formula (*) for the density $f(x)$ we have $b = 1$, the distribution of V_1 also has the same density (Fisz).

10.14. Show (Thompson [1]) that if $F(x)$ is a continuous distribution function and $a_{1/2}$ is its median,

$$P(\zeta_k^{(n)} < a_{1/2} < \zeta_{n-k+1}^{(n)}) = 1 - \frac{2}{B(k, n-k+1)} \int_0^{0.5} t^{k-1}(1-t)^{n-k}\,dt.$$

Hint: Use the fact that the random variable $F(X)$ has the uniform distribution in $[0, 1]$.

10.15. Show that the density function $h(w)$ of the range W of a simple sample of size n drawn from a population, in which the characteristic X has the density functior. $f(x)$, is of the form

$$h(w) = \begin{cases} n(n-1)\displaystyle\int_{-\infty}^{+\infty} f(u)f(w+u)[F(w+u) - F(u)]^{n-2}\,du & \text{for } 0 \leqslant u < \infty, \\ 0 & \text{for } u < 0. \end{cases}$$

10.16.(a) Show that if $F(x)$ is a continuous distribution function, then the statistic $W = P(L_1 \leqslant X \leqslant L_2)$ with $L_1 = \zeta_k^{(n)}$ and $L_2 = \zeta_{n-k+1}^{(n)}$ $(k < n - k + 1)$ has the density

$$h(w) = \begin{cases} \dfrac{n!}{(n-2k)!(2k-1)!} \, w^{n-2k}(1-w)^{2k-1} & (0 \leqslant w \leqslant 1), \\ 0 & \text{otherwise.} \end{cases}$$

(b) Find the moments $E(W)$ and $D^2(W)$.

(c) Derive the formula for $h(w)$ if $L_1 = \zeta_k^{(n)}$ and $L_2 = \zeta_{n-l+1}^{(n)}$ $(k < n - l + 1)$.

10.17. Suppose that the assumptions of theorem 10.11.2 are satisfied and let $n_1 = n_2 = m$. Let C_m denote the number of points x_k for which $S_{1m}(x_k) \geqslant S_{2m}(x_k + 0)$. Show that (Chung and Feller [1], Gnedenko and Mihalevich [1])

$$P(C_m = k) = \frac{1}{m+1} \qquad (k = 0, 1, \ldots, m).$$

Hint: Use an argument similar to that in the proof of the lemma in Section 10.11.

10.18. Suppose that the assumptions of theorem 10.11.2 are satisfied and $n_1 = n_2 = m$. Let $y_1 < y_2 < \ldots < y_{2m}$ be the values of both samples arranged in increasing order. Then the probability that, for all x between y_{k_1} and y_{k_2}, where $\lim_{m \to \infty} k_1/2m = \alpha$ and $\lim_{m \to \infty} k_2/2m = \beta$, the inequality $S_{1m}(x) \leqslant S_{2m}(x)$ holds converges as $m \to \infty$ to

$$\frac{1}{\pi} \arcsin \sqrt{\frac{\alpha(1-\beta)}{\beta(1-\alpha)}}.$$

(This is the *arc sin law* (Lévy [7], Erdös and Kac [2], Gnedenko [7], Gihman [1]).

10.19. Let $x_1, \ldots, x_n$ be a simple sample drawn from a population in which $F(x) = x$ $(0 \leqslant x \leqslant 1)$ and let $y_1 < y_2 < \ldots < y_n$ be the ordered values of x. Let y^* be the value y_k at which $D_n^+ = \sup_x [S_n(x) - x]$ assumes its maximum value and let k^* denote the index k for which $y^* = y_k$. Let Y^* and K^* be the random variables whose values are y^* and k^*, respectively. Show (Birnbaum and Pyke [1]; see also Kuiper [1]) that

(a) Y^* has the uniform distribution in the interval $[0, 1]$.

(b) $P(K^* = k) = n^{-n} \sum_{i=n-k}^{n-1} \dfrac{1}{i+1} \binom{n}{i} i^i (n-i)^{n-i-1}$.

(c) How can one formulate this result if $F(x)$ is an arbitrary continuous distribution function?

10.20. Let $S_{1n}(x)$, $S_{2n}(x)$ and $S_{3n}(x)$ be the empirical distribution functions of independent simple samples drawn from a population in which the characteristic X has a continuous distribution function. Prove (Ozols [1]) that for arbitrary $\lambda > 0$

$$\lim_{n \to \infty} P(\sqrt{n/2} \max_x [S_{1n}(x) - S_{2n}(x)] < \lambda, \ \sqrt{n/2} \max_x [S_{2n}(x) - S_{3n}(x)] < \lambda)$$

$$= 1 - \exp(-2\lambda^2) + 2 \exp(-6\lambda^2) - \exp(-8\lambda^2).$$

Hint: Apply reasoning similar to that in the proof of the lemma in Section 10.11, C

10.21. Find two different systems of coefficients B_{in_j} which satisfy relations (10.13.6.) to (10.13.8.)

10.22. (Notation of Section 10.12).   Prove that

$$\lim_{n \to \infty} P\left(\sqrt{n} \sup_{0 \leqslant F(x) \leqslant 1} \frac{S_n(x) - F(x)}{F(x)} < \lambda \right) = \begin{cases} 0 & (\lambda < 0), \\ \dfrac{\lambda}{1 + \lambda} & (\lambda \geqslant 0). \end{cases}$$

(This theorem was proved first by Daniels [2] and later discovered independently by Hannan, Robbins, Chapman, and Dempster. In Dempster's [1] paper a proof is given.)

CHAPTER 11

An Outline of the Theory of Runs

11.1 PRELIMINARY REMARKS

All the statistics considered thus far are functions of the magnitude of the characteristic under investigation of the sample elements. The theory of runs, which we now investigate, allows us to draw conclusions from the order merely in which the particular elements of the sample appear.

11.2 THE NOTION OF A RUN

Let X_j ($j = 1, 2, \ldots, n$) be independent random variables with the same two-point distribution, namely, for $j = 1, 2, \ldots, n$

$$P(X_j = a_1) = p, \quad P(X_j = a_2) = 1 - p.$$

We consider sequences $x_1, x_2, \ldots, x_n$. Each sequence consists of elements a_1 and a_2 which appear in a certain order.

Definition 11.2.1. A sequence $x_j, x_{j+1}, \ldots, x_{j+l}$, where $j = 1, 2, \ldots, n$ and $l = 0, 1, \ldots, n - j$, is called a *run* if

$$(11.2.1) \qquad x_{j-1} \neq x_j = x_{j+1} = \ldots = x_{j+l} \neq x_{j+l+1}.$$

The number $l + 1$ is called the *length* of the run.

For $j = 1$ the symbol "$x_{j-1} \neq$" on the left-hand side of (11.2.1) is unnecessary, and for $j + l = n$ the symbol "$\neq x_{j+l+1}$" on the right-hand side of (11.2.1) is unnecessary.

We denote by K_{ij} the number of runs of length j consisting of elements a_i, where $i = 1$ or 2. By N_i ($i = 1, 2$) we denote the number of a_i's; hence $N_1 + N_2 = n$. We have

$$(11.2.2) \qquad \sum_j j K_{ij} = N_i \qquad (i = 1, 2).$$

415

Finally, K_i $(i = 1, 2)$ denotes the number of runs consisting of elements a_i, and K denotes the total number of runs. Thus

$$K_i = \sum_j K_{ij} \qquad (i = 1, 2),$$

(11.2.3)

$$K = K_1 + K_2.$$

Example 11.2.1.   Let us consider the sequence

$$a_1, a_1, a_2, a_2, a_2, a_1, a_2, a_1, a_1, a_1, a_1, a_2, a_2, a_2.$$

Here we have one run of a_1's of length 1, one run of a_2's of length 1, one run of a_1's of length 2, two runs of a_2's of length 3, and, finally, one run of a_1's of length 4. In this sequence the observed values k_{ij} of the random variables K_{ij} are, respectively,

$$k_{11} = k_{21} = k_{12} = k_{14} = 1, \qquad k_{23} = 2.$$

Next, we have $n_1 = n_2 = 7$. The observed values k_1, k_2, and k of K_1, K_2, and K are, respectively,

$$k_2 = k_1 = 3, \qquad k = 6.$$

11.3   THE PROBABILITY DISTRIBUTION OF THE NUMBER OF RUNS

A   Our aim is to find the distributions of the random variables introduced here. First, we observe that K_1 and K_2 may differ at most by 1; hence, for the observed values of K_1 and K_2, we have three possibilities,

$$(1)\ k_1 = k_2, \quad (2)\ k_1 = k_2 + 1, \quad (3)\ k_1 = k_2 - 1.$$

Let us introduce the function $G(k_1, k_2)$ defined as

(11.3.1) $$G(k_1, k_2) = \begin{cases} 0 & \text{for } |k_1 - k_2| > 1, \\ 1 & \text{for } |k_1 - k_2| = 1, \\ 2 & \text{for } k_1 = k_2. \end{cases}$$

The function $G(k_1, k_2)$ gives us the number of ways in which we can obtain k_1 a_1-runs and k_2 a_2-runs. Indeed, if $k_1 = k_2$, we can arrange k_1 and k_2 corresponding runs in two ways, and if $|k_1 - k_2| = 1$, this can be done in one way only.

Let us denote by $P(k_{1j}, k_{2j}, n_1, n_2)$ the probability function of the random variable $(K_{11}, \ldots, K_{1n_1}, K_{21}, \ldots, K_{2n_2}, N_1, N_2)$. We have

(11.3.2) $P(k_{1j}, k_{2j}, n_1, n_2)$

$$= P(K_{11} = k_{11}, \ldots, K_{1n_1} = k_{1n_1}, K_{21} = k_{21}, \ldots, K_{2n_2}$$

$$= k_{2n_2}, N_1 = n_1, N_2 = n_2)$$

$$= P(K_{11} = k_{11}, \ldots, K_{1n_1} = k_{1n_1}, K_{21} = k_{21}, \ldots, K_{2n_2}$$

$$= k_{2n_2} \mid N_1 = n_1, N_2 = n_2)\, P(N_1 = n_1, N_2 = n_2).$$

From the independence and the equality of the distributions of the random variables X_k it follows that, for given n_1 and n_2, every order of a_1's and a_2's has the same probability of occurring. Next we observe that k_i ($i = 1, 2$) a_i-runs, among which k_{i1} runs are of length 1, k_{i2} runs are of length 2, ..., k_{in_i} runs are of length n_i, can be obtained in

$$\frac{k_i!}{k_{i1}!\, k_{i2}!\ldots k_{in_i}!}$$

ways. This fact and the definition of $G(k_1, k_2)$ imply that the number of ways of obtaining $k_1\, a_1$-runs, among which k_{11} are of length 1, ..., k_{1n_1} are of length n_1, and $k_2\, a_2$-runs, among which k_{21} are of length 1, ..., k_{2n_2} are of length n_2, equals

$$\frac{k_1!}{k_{11}!\, k_{12}!\ldots k_{1n_1}!} \cdot \frac{k_2!}{k_{21}!\, k_{22}!\ldots k_{2n_2}!}\, G(k_1, k_2).$$

Since n terms, among which there are $n_1\, a_1$'s and $n_2 = n - n_1\, a_2$'s, can be obtained in $\binom{n}{n_1}$ ways, the conditional probability on the right-hand side of (11.3.2) equals

$$\frac{k_1!}{k_{11}!\, k_{12}!\ldots k_{1n_1}!} \cdot \frac{k_2!}{k_{21}!\, k_{22}!\ldots k_{2n_2}!}\, G(k_1, k_2)\, \frac{n_1!\, n_2!}{n!}.$$

Finally, using the fact that

$$P(N_1 = n_1, N_2 = n_2) = \frac{n!}{n_1!\, n_2!}\, p^{n_1}(1 - p)^{n_2},$$

we obtain

(11.3.3) $P(k_{1j}, k_{2j}, n_1, n_2)$

$$= \frac{k_1!}{k_{11}!\, k_{12}!\ldots k_{1n_1}!} \cdot \frac{k_2!}{k_{21}!\, k_{22}!\ldots k_{2n_2}!}\, G(k_1, k_2) p^{n_1}(1 - p)^{n_2}.$$

B   We now find the marginal probability function $P(k_{1j}, n_1, n_2)$ of the random variable $(K_{11}, \ldots, K_{1n_1}, N_1, N_2)$. We proceed int wo steps. First, we find the probability function $P(k_{1j}, k_2, n_1, n_2)$ of the random variable $(K_{11}, \ldots, K_{1n_1}, K_2, N_1, N_2)$, and then we find the required probability function.

For fixed n_2 and k_2, let us evaluate the expression

(11.3.4)
$$\sum \frac{k_2!}{k_{21}!\, k_{22}!\ldots k_{2n_2}!},$$

where the summation is extended over those $k_{21}, \ldots, k_{2n_2}$ which satisfy the equalities

$$\sum_j jk_{2j} = n_2, \qquad \sum_j k_{2j} = k_2.$$

We use the identity

$$(x + x^2 + \ldots)^{k_2} = x^{k_2}(1 + x + x^2 + \ldots)^{k_2} = \frac{x^{k_2}}{(1 - x)^{k_2}}$$

$$= x^{k_2} \sum_{m=0}^{\infty} \frac{(k_2 + m - 1)!}{(k_2 - 1)! \, m!} x^m.$$

Let us compare the coefficient of x^{n_2} on the right- and left-hand sides of this equality. On the right-hand side this coefficient is

$$(11.3.5) \qquad \binom{n_2 - 1}{k_2 - 1},$$

and on the left-hand side it is equal to expression (11.3.4). By (11.3.3) to (11.3.5), we obtain

$$(11.3.6) \quad P(k_{1j}, k_2, n_1, n_2)$$

$$= \frac{k_1!}{k_{11}! \, k_{12}! \ldots k_{1n_1}!} \, G(k_1, k_2) p^{n_1}(1 - p)^{n_2} \sum \frac{k_2!}{k_{21}! \, k_{22}! \ldots k_{2n_2}!}$$

$$= \frac{k_1!}{k_{11}! \, k_{12}! \ldots k_{1n_1}!} \binom{n_2 - 1}{k_2 - 1} G(k_1, k_2) p^{n_1}(1 - p)^{n_2}.$$

To find the required function $P(k_{1j}, n_1, n_2)$, let us add (11.3.6) for all k_2. From the relation between K_1 and K_2 and from the definition of $G(k_1, k_2)$, we obtain

$$(n_2 - 1)! \sum_{k_2=1}^{n_2} \frac{G(k_1, k_2)}{(k_2 - 1)! \, (n_2 - k_2)!}$$

$$= (n_2 - 1)! \left[\frac{2}{(k_1 - 1)! \, (n_2 - k_1)!} + \frac{1}{(k_1 - 1)! \, (n_2 - k_1 + 1)!} \right.$$

$$\left. + \frac{1}{k_1! \, (n_2 - k_1 - 1)!} \right] = \binom{n_2 + 1}{k_1}.$$

Hence, by (11.3.6)

$$(11.3.7) \quad P(k_{1j}, n_1, n_2)$$

$$= \frac{k_1!}{k_{11}! \, k_{12}! \ldots k_{1n_1}!} \, p^{n_1}(1 - p)^{n_2}(n_2 - 1)!$$

$$\times \sum_{k_2=1}^{n_2} \frac{G(k_1, k_2)}{(k_2 - 1)! \, (n_2 - k_2)!} = \frac{k_1!}{k_{11}! \, k_{12}! \ldots k_{1n_1}!} \binom{n_2 + 1}{k_1} p^{n_1}(1 - p)^{n_2}.$$

A similar expression can be found for the probability function $P(k_{2j}, n_1, n_2)$.

C   The probability function $P(k_1, k_2, n_1, n_2)$ of the random variable (K_1, K_2, N_1, N_2) is important in applications. To find this probability function, we have to add (11.3.6) for all k_{1j} in a manner analogous to that used in (11.3.3) in the derivation of (11.3.6). Thus, using the fact that expressions (11.3.4) and (11.3.5), in which k_2, n_2, and k_{2j} are replaced by k_1, n_1, and k_{1j}, respectively, are equal, from (11.3.6) we obtain

$$(11.3.8) \quad P(k_1, k_2, n_1, n_2) = \binom{n_1 - 1}{k_1 - 1}\binom{n_2 - 1}{k_2 - 1} G(k_1, k_2) p^{n_1}(1 - p)^{n_2}.$$

By adding (11.3.8) for all k_2 or k_1, we easily find the probability functions $P(k_1, n_1, n_2)$ and $P(k_2, n_1, n_2)$ of the random variables (K_1, N_1, N_2) and (K_2, N_1, N_2), respectively. Thus

$$(11.3.9) \quad \begin{aligned} P(k_1, n_1, n_2) &= \binom{n_1 - 1}{k_1 - 1}\binom{n_2 + 1}{k_1} p^{n_1}(1 - p)^{n_2}, \\ P(k_2, n_1, n_2) &= \binom{n_1 + 1}{k_2}\binom{n_2 - 1}{k_2 - 1} p^{n_1}(1 - p)^{n_2}. \end{aligned}$$

D   Finally, let us find the probability function of the total number of runs, that is, the probability function of $K = K_1 + K_2$. Here we consider the following two cases:

(a) k is an even number. Then $k_1 = k_2 = \frac{1}{2}k$. Using the fact that $G(\frac{1}{2}k, \frac{1}{2}k) = 2$, from (11.3.8) we have
(11.3.10)

$$P(K = k) = P(\tfrac{1}{2}k, \tfrac{1}{2}k, n_1, n_2) = 2\binom{n_1 - 1}{\frac{1}{2}k - 1}\binom{n_1 - 1}{\frac{1}{2}k - 1} p^{n_1}(1 - p)^{n_2}.$$

(b) k is an odd number. Then either $k_1 = \frac{1}{2}(k - 1)$ and $k_2 = \frac{1}{2}(k + 1)$, or $k_1 = \frac{1}{2}(k + 1)$ and $k_2 = \frac{1}{2}(k - 1)$. Since in the case under consideration $G(k_1, k_2) = 1$, from (11.3.8) we obtain

$$(11.3.10') \quad P(K = k) = P[\tfrac{1}{2}(k - 1), \tfrac{1}{2}(k + 1), n_1, n_2]$$
$$+ P[\tfrac{1}{2}(k + 1), \tfrac{1}{2}(k - 1), n_1, n_2]$$
$$= \left[\binom{n_1 - 1}{\frac{1}{2}(k - 3)}\binom{n_2 - 1}{\frac{1}{2}(k - 1)} + \binom{n_1 - 1}{\frac{1}{2}(k - 1)}\binom{n_2 - 1}{\frac{1}{2}(k - 3)}\right] p^{n_1}(1 - p)^{n_2}.$$

From (11.3.9) and (11.3.10) we can obtain the conditional probability functions of the random variables K_i under the condition that $N_1 = n_1$ and $N_2 = n_2$. Thus

$$(11.3.11) \quad P(K_1 = k_1 \mid N_1 = n_1, N_2 = n_2) = \binom{n_1 - 1}{k_1 - 1}\binom{n_2 + 1}{k_1} \bigg/ \binom{n}{n_1},$$

$$P(K_2 = k_2 \mid N_1 = n_1, N_2 = n_2) = \binom{n_1 + 1}{k_2}\binom{n_2 - 1}{k_2 - 1} \bigg/ \binom{n}{n_1}.$$

Similarly, for even k and odd k, respectively,

$$(11.3.12) \quad P(K = k \mid N_1 = n_1, N_2 = n_2) = 2 \binom{n_1 - 1}{\frac{1}{2}k - 1} \binom{n_2 - 1}{\frac{1}{2}k - 1} \bigg/ \binom{n}{n_1},$$

$$P(K = k \mid N_1 = n_1, N_2 = n_2) = \left[\binom{n_1 - 1}{\frac{1}{2}(k - 3)} \binom{n_2 - 1}{\frac{1}{2}(k - 1)} \right.$$

$$\left. + \binom{n_1 - 1}{\frac{1}{2}(k - 1)} \binom{n_2 - 1}{\frac{1}{2}(k - 3)} \right] \bigg/ \binom{n}{n_1}.$$

Swed and Eisenhart [1] constructed tables which give the values of the function $P(K < k \mid N_1 = n_1, N_2 = n_2)$ for n_1 and n_2 not exceeding 20.

The results of the theory of runs given here have been obtained by Bortkiewicz [2], Mood [1], Stevens [1], and others.

Example 11.3.1. We have $P(X_j = a_1) = P(X_j = a_2) = 0.5$. In a simple sample of $n = 8$ elements there are 5 a_1's and 3 a_2's in the following order:

$$a_1, a_1, a_1, a_1, a_2, a_1, a_2, a_2.$$

We have here

$$k_{11} = k_{14} = 1, \quad k_{12} = k_{13} = k_{15} = 0, \quad k_{21} = k_{22} = 1, \quad k_{23} = 0,$$
$$k_1 = k_2 = 2, \quad k = 4, \quad n_1 = 5. \quad n_2 = 3, \quad n = 8.$$

We observe that $P(N_1 = 5, N_2 = 3) = 0.2234$. Next, from (11.3.3) we compute the probability that the random variables K_{1j}, K_{2j}, N_1 and N_2 take on the values observed in this example. This probability is 0.0319. Borel's law of large numbers implies: In a large series of independent 8-element samples, in approximately 22 cases out of every 100, we shall get samples with 5 elements a_1 and 3 elements a_2; but only 3 samples out of every 100 will contain 5 elements a_1 and 3 elements a_2 with k_{ij}'s such as appeared in this example.

We return to this example in Section 14.2.

E   We quote here without proof two results concerning limit theorems in the theory of runs. Wishart and Hirschfeld [1] proved that the random variable K with the distribution given by (11.3.10) and (11.3.10') has the asymptotically normal distribution $N(2npq; 2\sqrt{npq(1 - 3pq)})$ as $n \to \infty$. Wald and Wolfowitz [2] proved that, for $n_1 = \alpha n_2$ ($\alpha > 0$) and $n_1 \to \infty$, the conditional distribution given by (11.3.12) is asymptotically normal $N(2n_1/(1 + \alpha); \sqrt{4\alpha n_1/(1 + \alpha)^3})$. This asymptotic distribution can be used for $n_1 > 20$ and $n_2 > 20$.

We observe that we can also consider runs having more than two different elements. We shall not discuss this case. Quite a few results in this domain can be found in the paper by Mood [1]. One can also consider more general runs, namely runs such that the random variables X_j are not independent but form a homogeneous Markov chain. Some results in this domain can be found in the paper of Babkin, Belayev, and Maximov [1].

11.4 THE EXPECTED VALUE AND THE VARIANCE OF THE NUMBER OF RUNS

A   We find the expected value and the variance of some of the random variables considered in the previous sections.

From (11.3.11) we obtain

$$(11.4.1) \quad E(K_1 \mid N_1 = n_1, N_2 = n_2) = \sum_{k_1=1}^{n_1} k_1 \binom{n_1 - 1}{k_1 - 1} \binom{n_2 + 1}{k_1} \Big/ \binom{n}{n_1}$$

$$= (n_2 + 1) \sum_{k_1=1}^{n_1} \binom{n_1 - 1}{k_1 - 1} \binom{n_2}{k_1 - 1} \Big/ \binom{n}{n_1}.$$

Let us consider the identity

$$(1 + x)^{n_2} \left(1 + \frac{1}{x} \right)^{n_1 - 1} \equiv \frac{(1 + x)^{n_1 + n_2 - 1}}{x^{n_1 - 1}}.$$

Let us expand both sides of this identity according to Newton's formula and compare the terms not containing x. On the right-hand side we have $\binom{n_1 + n_2 - 1}{n_1 - 1}$, and on the left-hand side

$$\sum_{k_1=1}^{n_1} \binom{n_1 - 1}{k_1 - 1} \binom{n_2}{k_1 - 1}.$$

Hence, by (11.4.1)

$$(11.4.2) \quad E(K_1 \mid N_1 = n_1, N_2 = n_2) = \frac{(n_2 + 1)(n - 1)!}{(n_1 - 1)! \, n_2!} \Big/ \binom{n}{n_1} = \frac{(n_2 + 1)n_1}{n}.$$

Replacing n_1 and n_2 in this formula, we obtain

$$(11.4.3) \qquad E(K_2 \mid N_1 = n_1, N_2 = n_2) = \frac{(n_1 + 1)n_2}{n}.$$

The last two equalities imply that

$$(11.4.4) \quad E(K \mid N_1 = n_1, N_2 = n_2)$$

$$= E(K_1 \mid N_1 = n_1, N_2 = n_2) + E(K_2 \mid N_1 = n_1, N_2 = n_2)$$

$$= \frac{2n_1 n_2 + n}{n}.$$

To find the conditional variance $D^2(K_1 \mid N_1 = n_1, N_2 = n_2)$, let us first find $E[K_1(K_1 - 1) \mid N_1 = n_1, N_2 = n_2]$. From (11.3.11) we obtain

$$(11.4.5) \quad E[K_1(K_1 - 1) \mid N_1 = n_1, N_2 = n_2]$$

$$= n_2(n_2 + 1) \sum_{k_1=1}^{n_1} \binom{n_1 - 1}{k_1 - 1}\binom{n_2 - 1}{k_1 - 2} \Big/ \binom{n}{n_1}.$$

Comparing the coefficients of x^{-1} on both sides of Newton's expansion of the identity

$$(1 + x)^{n_2 - 1}\left(1 + \frac{1}{x}\right)^{n_1 - 1} \equiv \frac{(1 + x)^{n_1 + n_2 - 2}}{x^{n_1 - 1}},$$

we have

$$\sum_{k_1=1}^{n_1} \binom{n_1 - 1}{k_1 - 1}\binom{n_2 - 1}{k_1 - 2} = \binom{n_1 + n_2 - 2}{n_1 - 2}.$$

Hence from (11.4.5) we obtain

$$E[K_1(K_1 - 1) \mid N_1 = n_1, N_2 = n_2] = n_2(n_2 + 1)\binom{n_1 + n_2 - 2}{n_1 - 2}\Big/\binom{n}{n_1}.$$

By the last formula and (11.4.2), we obtain

$$(11.4.6) \quad D^2(K_1 \mid N_1 = n_1, N_2 = n_2) = \frac{n_1(n_1 - 1)n_2(n_2 + 1)}{(n - 1)n^2}.$$

Similarly,

$$(11.4.7) \quad D^2(K_2 \mid N_1 = n_1, N_2 = n_2) = \frac{n_1(n_1 + 1)n_2(n_2 - 1)}{(n - 1)n^2}.$$

Avoiding elementary but cumbersome computations, we present without proof the formula

$$(11.4.8) \quad D^2(K \mid N_1 = n_1, N_2 = n_2) = \frac{2n_1 n_2(2n_1 n_2 - n)}{(n - 1)n^2}.$$

By (3.6.24), (11.4.2), and (11.4.6), we find

$$E(K_1) = E\left[\frac{n_1(n - n_1 + 1)}{n}\right] = np(1 - p) + p^2,$$

$$D^2(K_1) = np(1 - 4p + 6p^2 - 3p^3) + p^2(3 - 8p + 5p^2).$$

Similarly,

$$E(K_2) = nq(1 - q) + q^2,$$

$$D^2(K_2) = nq(1 - 4q + 6q^2 - 3q^3) + q^2(3 - 8q + 5q^2),$$

where $q = 1 - p$

The covariance μ_{11} of K_1 and K_2 is given by the formula (see Mood [1])

$$\mu_{11} = -npq(3pq - 1) - pq(2 - 5pq).$$

Next we have

$$E(K) = E(K_1 + K_2) = p^2 + q^2 + 2npq,$$

(11.4.9)
$$D^2(K) = D^2(K_1) + D^2(K_2) + 2\mu_{11}$$
$$= 4npq(1 - 3pq) - 2pq(3 - 10pq).$$

In particular, for $p = q = \frac{1}{2}$

(11.4.10) $\qquad E(K) = \frac{1}{2}(n + 1), \quad D^2(K) = \frac{1}{4}(n - 1).$

B At the end of this chapter, we inform the reader that the theory of runs attracted the attention of many writers in the early days of modern probability theory. Of particular importance is the book by Bortkiewicz [2] both for the formulas it presents and for the discussion with Marbe [1] about the latter's attempt to challenge the applicability of probability theory, through the interpretation of some observed runs. In that animated discussion took part, among others, Czuber [1], Mises [2], Lexis [1], and Fryde [1].

Problems and Complements

In Problems 11.1 and 11.2 apply the same reasoning as in the proof of the formulas of Section 11.4.

11.1. Prove the following equations:

$$E(K_{1j} \mid N_1 = n_1, N_2 = n_2) = \frac{(n_2 + 1)^{(2)} n_1^{(j)}}{n^{(j+1)}} \qquad (j = 1, \ldots, n_1),$$

$$E(K_{2j} \mid N_1 = n_1, N_2 = n_2) = \frac{(n_1 + 1)^{(2)} n_2^{(j)}}{n^{(j+1)}} \qquad (j = 1, \ldots, n_2),$$

where $a^{(m)} = a(a - 1) \ldots (a - m + 1)$.

11.2. For $m, j = 1, \ldots, n_1$ and $s = 1, \ldots, n_2$, prove the following equations:

$$D^2(K_{1j} \mid N_1 = n_1, N_2 = n_2) = \frac{n_2^{(2)}(n_2 + 1)^{(2)} n_1^{(2j)}}{n^{(2j+2)}} - \left[\frac{(n_2 + 1)^{(2)} n_1^{(j)}}{n^{(j+1)}} \right]^2,$$

$$\text{cov}\,(K_{1j}, K_{1m}) = \frac{n_2^{(2)}(n_2 + 1)^{(2)}\, n_1^{(j+m)}}{n^{(j+m+2)}} - \frac{[(n_2 + 1)^{(2)}]^2 n_1^{(j)} n_1^{(m)}}{n^{(j+1)} n^{(m+1)}},$$

$$\text{cov}\,(K_{1j}, K_{2s}) = \frac{n_1^{(j+2)} n_2^{(s+2)}}{n^{(j+s+2)}} + 4\,\frac{n_1^{(j+1)} n_2^{(s+1)}}{n^{(j+s+1)}} + 2\,\frac{n_1^{(j)} n_2^{(s)}}{n^{(j+s)}}$$
$$- \frac{(n_1 + 1)^{(2)}(n_2 + 1)^{(2)}\, n_1^{(j)} n_2^{(s)}}{n^{(j+1)} n^{(s+1)}}.$$

11.3. Prove that if $n_1 = \alpha n_2$ ($\alpha > 0$) and $n_1 \to \infty$, then the limit distribution of the random vector $(K_{1j_1}, \ldots, K_{1j_l}, K_{2s_1}, \ldots, K_{2s_r})$ is asymptotically normal with the expected values variances and covariances given in Problems 11.1 and 11.2 (Mood [1]).

Hint: Derive the probability function of the considered vector from formula (11.3.3). Consider the standardized random variables $[K_{ij} - E(K_{ij})]/\sqrt{D^2(K_{ij})}$, and apply Stirling's formula.

11.4. Denote by R_{ij} the number of a_i-runs of length not smaller than j ($i = 1$, $2, j = 1, \ldots, n_i$), and by $R_j = R_{1j} + R_{2j}$ the total number of runs of length not smaller than j.

(a) Find the probability function of these random variables.

(b) Show that

$$E(R_{1j} \mid N_1 = n_1, N_2 = n_2) = \frac{n_1^{(j)}(n_2 + 1)}{n^{(j)}} \qquad (j = 1, \ldots, n_1),$$

$$E(R_{2j} \mid N_1 = n_1, N_2 = n_2) = \frac{n_2^{(j)}(n_1 + 1)}{n^{(j)}} \qquad (j = 1, \ldots, n_2).$$

11.5. Show that if $n_1 = n_2 = n/2$, then

(a) $E\left(K_{ij} \mid N_1 = N_2 = \frac{n}{2}\right) \approx \frac{n}{2^{j+2}} \qquad (i = 1, 2)$,

(b) $E\left(R_{ij} \mid N_1 = N_2 = \frac{n}{2}\right) \approx \frac{n}{2^{j+1}} \qquad (i = 1, 2)$,

(c) $E\left(R_j \mid N_1 = N_2 = \frac{n}{2}\right) \approx \frac{n}{2^j}$.

CHAPTER 12

Significance Tests

12.1 THE CONCEPT OF A STATISTICAL TEST

Every conjecture concerning the unknown distribution of a random variable is called a *statistical hypothesis*.

We give some examples of statistical hypotheses.

Example 12.1.1. The characteristic X of elements of a certain population has the normal distribution $N(m; 1)$, but the expected value m is unknown. We conjecture that $m = 0$. In other words, we formulate the statistical hypothesis $m = 0$, which we denote by $H_0(m = 0)$.

In this example the hypothesis refers only to the value of the unknown parameter of the random variable, which has a given distribution.

A statistical hypothesis which refers only to the numerical values of unknown parameters of a random variable is called a *parametric hypothesis*.

Methods of verifying statistical hypotheses are called *statistical tests*. Tests of parametric hypotheses are called *parametric tests*.

If we state one hypothesis only and the aim of the statistical test is to verify whether this hypothesis is not false but not at the same time to investigate other hypotheses, then such a test is called a *significance test*.

We illustrate by some examples the procedure of testing significance tests. Let us return to example 12.1.1.

Example 12.1.1 (continued). Assume that the hypothesis $H_0(m = 0)$ is true. Imagine that we have obtained the value $\bar{x} = 1.01$ from a simple sample of ten elements drawn from the given population. This is the observed value of the statistic $\bar{X}$. The distribution of $\bar{X}$ of simple samples taken from a population with the normal distribution $N(0; 1)$ is, by (9.3.5), the normal distribution $N(0; 1/\sqrt{10})$. The probability of obtaining a value which exceeds or equals 1.01 in absolute value is

$$P(|\bar{X}| \geq 1.01) = P(|\bar{X}| \sqrt{10} \geq 1.01 \sqrt{10}) \simeq 2[1 - \Phi(3.05)]$$

$$\simeq 2(1 - 0.999) = 0.002.$$

425

Thus, if the hypothesis $H_0(m = 0)$ is true, the observed event is extremely rare in practice, since the probability of this event is about 0.002, so that in approximately two times out of every thousand we would observe values of $\bar{X}$ not smaller than 1.01 in absolute value.

Now the question arises: Can we consider it as proved that the hypothesis H_0 is not true if the probability of the event $|\bar{X}| \geqslant 1.01$ is very small? The answer is negative, for although the probability of the event $|\bar{X}| \geqslant 1.01$ is very small, provided H_0 is true, such an event may occur in practice.

At this point the statistician must make a decision; he must choose a critical probability α, namely, such that if under H_0 the probability of an outcome as extreme or more than the observed event is not greater than α, he will reject H_0. The critical probability so defined is called the *level of significance*. The critical probability α may be different in different problems and it is in general the result of pure convention. Usually, we accept $\alpha = 0.05$ or $\alpha = 0.01$. Thus, if we take $\alpha = 0.01$ (hence all the more so if we take $\alpha = 0.05$), we reject the hypothesis H_0 $(m = 0)$ of example 12.1.1.

Example 12.1.2. From a population, where X has the normal distribution $N(m; 1)$ and m is unknown, we draw a simple sample of $n = 16$ elements. We observe the value $\bar{x} = 0.1$ and we want to test the hypothesis $H_0(m = 0)$.

If H_0 is true, the random variable $\bar{X}$ has the normal distribution $N(m; \frac{1}{4})$. Let us compute the probability

$$P(|\bar{X}| \geqslant 0.1) = P(4|\bar{X}| \geqslant 0.4) = 2[1 - \Phi(0.4)] \cong 2 \cdot 0.345 = 0.690.$$

The probability of obtaining a value of $\bar{X}$ which exceeds or equals 0.1 in absolute value, is here much greater than 0.01. It does not follow, however, that H_0 can be accepted unreservedly.

In general, a significance test allows us to make decisions only in one direction, namely, if, under the assumption that H_0 is true, the probability of the observed event or worse is not greater than α, then H_0 may be rejected. If, however, the probability of the observed event or worse is greater than α, then we can only state that the experiment does not contradict H_0. It does not appear to be well based on probability theory or reasonable to accept the hypothesis tested on the basis of a single experiment which resulted in the observation of an event that can appear somewhat more often than once or even five times out of every 100, provided the hypothesis is true. It is extremely important to stress this fact; statisticians often make the fundamental mistake of equating failure to reject a hypothesis as the result of a significance test with acceptance of the hypothesis.

However, there are methods of constructing tests for both rejecting and accepting hypotheses, which we discuss in Chapter 16.

12.2 PARAMETRIC TESTS FOR SMALL SAMPLES

From the considerations of the last section it follows that the essential concept of a parametric significance test can be characterized in the following way.

The distribution function of the characteristic X of the population depends on a parameter Q which is not known, whereas the functional form of the distribution function is known. We want to test the hypothesis $H_0(Q = Q_0)$. From the population under investigation we draw a sample of size n and we determine the observed value u of some statistic U. We accept the critical probability α, where α is small, usually 0.01 or 0.05, and we find a value u_0 of the statistic U such that $P(|U| \geqslant u_0) \leqslant \alpha$, provided the hypothesis H_0 is true. If the observed value u satisfies the inequality $|u| \geqslant u_0$, we reject H_0, since we consider the observed value $|u|$ as too big to be explained by random fluctuations, if the hypothesis H_0 is true. In some problems u_0 should be selected in such ·a way that $P(U \geqslant u_0) \leqslant \alpha$ or $P(U \leqslant u_0) \leqslant \alpha$. The question of how to determine u_0 depends upon the character of the problem. A detailed discussion of this question is given in Chapter 16.

Thus we see that the most important problem here is to know the distribution of the statistic U or, at least, the value of u_0. For this reason the methods of significance tests for small samples are different from those for large samples. In the first case the tests are based upon the exact distributions and in the second upon the limit distributions of the statistics considered.

Example 12.2.1. Student [1] gave the following example: the effect of two drugs A and B has been tested on ten patients suffering from sleeplessness. The additional hours of sleep gained by a patient as a result of drug A are denoted by X and of drug B by Y. We investigate the random variable $Z = X - Y$. The values $x - y$ are the observed values of this random variable in a sample of ten elements. The observed values of X, Y, and Z are given in Table 12.2.1. We can assume that Z has the normal distribution $N(m; \sigma)$. We formulate the hypothesis $H_0(m = 0)$, without making any statements concerning the value of σ.

From Table 12.2.1 we find $\bar{z} = 1.58$. Let us perform a significance test for H_0 at the significance level $\alpha = 0.01$.

Now the problem arises of the choice of a suitable statistic. At first it might seem that we could use the distribution of Z for samples from a normal population, as in example 12.1.2. However, this is impossible, since under H_0 the random variable Z has the normal distribution $N(0; \sigma/\sqrt{10})$, the standard deviation of which is unknown as we do not know the value of σ. The sample is small and so we cannot accept the value of the sample standard deviation equal to $s = 1.167$ as the standard deviation σ. It is true that the random variable S is stochastically convergent to σ (see Section 9.11); nevertheless its observed value

TABLE 12.2.1

Patient	x	y	$z = x - y$
1	1.9	0.7	1.2
2	0.8	−1.6	2.4
3	1.1	−0.2	1.3
4	0.1	−1.2	1.3
5	−0.1	−0.1	0.0
6	4.4	3.4	1.0
7	5.5	3.7	1.8
8	1.6	0.8	0.8
9	4.6	0.0	4.6
10	3.4	2.0	1.4

for $n = 10$ may differ considerably from σ. Hence we shall use Student's t-distribution. The observed value of the statistic t in our example equals (provided the hypothesis $H_0(m = 0)$ is true)

$$t = \frac{\bar{z}}{s}\sqrt{n - 1} = \frac{1.580}{1.167}\sqrt{9} = 4.06.$$

From tables of Student's t-distribution we determine a value t_0 such that $P(|t| \geqslant t_0) = 0.01$. For nine degrees of freedom, we find $t_0 = 3.25$. The observed value t is considerably greater than t_0; hence we reject H_0, which means that drugs A and B do not have the same effect. We might suspect that A is more effective than B.

Example 12.2.2. A factory produces lots of some merchandise A and the characteristic X of the elements of these lots has the distribution $N(m; \sigma)$, where the standard deviation σ is unknown. If the production process had not been significantly disturbed, σ should have remained constant. For this reason σ plays the role of a measure of homogeneity or of a measure of accordance with the technological norms of the given production process.

We select samples from the lots produced and compute the standard deviations of the samples. We want to test the hypothesis $H_0(\sigma = 4)$ at the significance level $\alpha = 0.01$. From a simple sample of size 15 we observe $s = 4.5$. Should we reject our hypothesis, that is, should we suppose that the observed value s is significantly greater than the hypothetical value $\sigma = 4$, so that this difference cannot be explained by random fluctuations if the hypothesis H_0 is true?

We use here the distribution of the statistic $Z = nS^2$, the density of which is given by (9.5.12), the density of the χ^2 distribution with $n - 1$ degrees of freedom. Putting $\sigma = 4$, we obtain

$$\frac{z}{\sigma^2} = \frac{15 \cdot 4.5^2}{4^2} \cong 19.$$

From tables of the χ^2 distribution with fourteen degrees of freedom we find that $\alpha = 0.01$ is the probability of the inequality $z/\sigma^2 \geqslant 29.1$. Hence the observed value 19 does not lead to the rejection of H_0 at the significance level $\alpha = 0.01$.

Example 12.2.3. We have good and defective items in a lot and the proportion p of defective items is unknown. The hypothesis to be tested is $H_0(p = 0.10)$; hence that 10% of the items in the lot are defective.

From this lot we draw a simple sample of size $n = 30$. We assign the number one to the appearance of a defective item and the number zero to the appearance of a good item. Suppose that we have found four defective items in the sample. The question is: Should we reject H_0 at the significance level $\alpha = 0.05$?

Let us consider the statistic X with the binomial distribution

$$X = \sum_{k=1}^{30} X_k,$$

where the X_k are independent and each X_k has a zero-one distribution. If H_0 is true, the probability function of X is

$$P(X = r) = \binom{30}{r} 0.1^r \cdot 0.9^{30-r} \qquad (r = 0.1, \ldots, 30).$$

Hence

(12.2.1) $$P(X \geqslant 4) = \sum_{r=4}^{30} \binom{30}{r} 0.1^r \cdot 0.9^{30-r}.$$

To compute the sum (12.2.1), we find

$$P(X = 0) = 0.9^{30},$$

$$P(X = 1) = \frac{30!}{29!} 0.1 \cdot 0.9^{29} = 30 \cdot 0.1 \cdot 0.9^{29},$$

$$P(X = 2) = \frac{30!}{2!\,28!} 0.1^2 \cdot 0.9^{28} = 15 \cdot 29 \cdot 0.1^2 \cdot 0.9^{28},$$

$$P(X = 3) = \frac{30!}{3!\,27!} 0.1^3 \cdot 0.9^{27} = 5 \cdot 29 \cdot 28 \cdot 0.1^3 \cdot 0.9^{27}.$$

The sum of these four probabilities is

$$0.9^{27}[0.9^3 + 3 \cdot 0.9^2 + 4.35 \cdot 0.9 + 4.06] = 0.9^{27} \cdot 11.134 = 0.6473.$$

Hence the required probability is $1 - 0.6473 = 0.3527$. Thus this test does not lead to the rejection of H_0.

Example 12.2.4. Table 12.2.2 is reprinted from the book of Boyev [1]. Its top entries give the characteristic X of cotton thread, called the cotton number, which is the reciprocal of the diameter of thread expressed in inches. The side entries give the strength Y of thread expressed in grams.

Sixty independent sample measurements of the strength of thread of different numbers have been made. Table 12.2.2 gives the number of observed pairs (x, y).

We assume that the random variable (X, Y) has a two-dimensional normal distribution and we wish to test the hypothesis $H_0(\rho = 0)$, where ρ is the correlation coefficient of X and Y.

Let us test this hypothesis on the basis of the sample measurements. From the data presented in Table 12.2.2, we may expect that the test will lead to the rejection of H_0.

TABLE 12.2.2

x / y	4100	4300	4500	4700	4900	5100	5300	5500	Totals
6.75		1							1
6.25	1	2	2	1					6
5.75		1	3	4	2	3			13
5.25		3	5	7	1	1			17
4.75			2	5	5	3	2		17
4.25					1	2	2		5
3.75								1	1
Totals	1	7	12	17	9	9	4	1	60

Let us compute the value of the correlation coefficient of the sample. De-note by $\bar{x}$ and $\bar{y}$, respectively, the observed values of the random variables $\bar{X}$ and $\bar{Y}$. We have $\bar{x} = 4746.67$, and $\bar{y} = 5.23$.

For further computations we may use Table 12.2.3, which gives in the top entries the values of $x - \bar{x}$ and in the side entries the values of $y - \bar{y}$. The body of this table contains the values $(x - \bar{x})(y - \bar{y})$ multiplied by the corresponding number of observations given in Table 12.2.2. We obtain

$$r = \frac{\sum (x - \bar{x})(y - \bar{y})}{\sqrt{\sum (x - \bar{x})^2 \sum (y - \bar{y})^2}} = -0.61.$$

To test hypothesis H_0, we use the fact that if $\rho = 0$, the statistic

$$\frac{R}{\sqrt{1 - R^2}} \sqrt{n - 2}$$

has Student's t-distribution with $n - 2$ degrees of freedom (see Section 9.9). The observed value of this statistic is -5.86. Student's t-statistic for 58 degrees of freedom has a distribution very close to the normal $N(0; 1)$; thus the prob-ability $P(|t| \geqslant 5.86)$ is exceedingly small. Thus if the hypothesis H_0 ($\rho = 0$) were true, we should have observed an exceedingly rare event whose probability is considerably smaller than the significance level α accepted by us. Thus we reject H_0.

TABLE 12.2.3

$x - \bar{x}$ / $y - \bar{y}$	−646.67	−446.67	−246.67	−46.67	153.33	353.33	553.33	753.33
1.52		−678.94						
1.02	−659.60	−911.21	−503.21	−47.60				
0.52		−232.27	−384.81	−97.07	159.46	551.19		
0.02		−26.80	−24.67	−6.53	3.07	7.07		
−0.48			236.80	112.01	−367.99	−508.80	−531.20	
−0.98					−150.26	−692.53	1084.53	
−1.48								−1114.93

TABLE 12.2.4

Price in Zlotys
paid in September 1947 for
one kilogram

City	of potatoes	one egg
Warszawa	11.4	12.8
Łódź	12.3	15.0
Kielce	15.0	12.5
Lublin	11.5	13.5
Bialystok	5.0	11.0
Olsztyn	7.0	12.5
Gdańsk	11.0	14.0
Bydgoszcz	10.0	12.0
Szczecin	10.5	13.5
Poznań	10.5	14.3
Wroclaw	12.3	13.5
Katowice	11.8	13.5
Kraków	17.0	13.3
Rzeszów	12.0	12.0

Source: Wiadomości Statystyczne.

Example 12.2.5. Table 12.2.4 gives the price of one kilogram (2.205 lb) of potatoes and the price of one egg in September 1947 in fourteen Polish cities. Let X and Y denote, respectively, the price of one kilogram of potatoes and the price of one egg. We find from the data presented in this table the equation of the regression line of Y on X.

From Table 12.2.4 we find

$$\bar{x} = 11.24, \quad \bar{y} = 13.10, \quad s_1{}^2 = 7.79, \quad s_2{}^2 = 1.01, \quad r = 0.4.$$

From formula (9.10.1) we obtain

$$a = r\frac{s_2}{s_1} = 0.144, \quad b = \bar{y} - a\bar{x} = 11.48.$$

Thus the required regression line has the form

$$Y = 0.144X + 11.48.$$

If we assume that (X, Y) has a normal distribution, we can apply the significance test based on formula (9.10.15) to test the hypothesis H_0 that the true value of the coefficient α is 0. We have

$$t_1 = \frac{s_1}{s_2} \cdot \frac{\sqrt{n-2}}{\sqrt{1-r^2}}(a - \alpha) = \frac{2.791}{1.005} \cdot \frac{\sqrt{12}}{\sqrt{0.84}} 0.144 = 1.61.$$

Since the number of degrees of freedom here equals 12, we can find from tables of Student's t-distribution that the probability of obtaining a value of t exceeding or equal to the observed value t_1 in absolute value is greater than 0.05. Thus there is no reason to reject hypothesis H_0. This result agrees with our intuition, since eggs are not substitutes for potatoes.

Example 12.2.6. Let us return once more to example 12.2.1. The effects of two drugs were tested on every patient; hence we treated the difference $x - y$ for one patient as one observation. Now, let us imagine that the experiment was performed in another way, that drugs A and B were tested on two different groups of patients of ten persons each. Then we would have two independent samples, a sample consisting of ten observations on the random variable X and a sample of ten observations on the random variable Y. Let us suppose that X and Y have the normal distributions $N(m_1; \sigma)$ and $N(m_2; \sigma)$, respectively. The hypothesis to be tested is H_0 $(m_1 = m_2)$; hence we suspect that the effectiveness of the two drugs is the same. Let us test this hypothesis by using the statistic U given by (9.6.8).

If the hypothesis H_0 $(m_1 = m_2)$ is true, we may consider both samples as drawn from the same normal population. To find the observed value u, we first find from Table 12.2.1 the values

$$\bar{x} = \tfrac{1}{10} \sum x = 2.33, \quad \bar{y} = \tfrac{1}{10} \sum y = 0.75,$$

We obtain
$$10s_1^{\,2} = \sum (x - \bar{x})^2 = 36.1, \qquad 10s_2^{\,2} = \sum (y - \bar{y})^2 = 28.9.$$

$$u = \frac{\bar{x} - \bar{y}}{\sqrt{10(s_1^{\,2} + s_2^{\,2})}\sqrt{\dfrac{100}{20}}} \cdot 18 = 1.86.$$

As we know (see Section 9.6), the random variable U has Student's t-distribution with eighteen degrees of freedom. From tables of the t-distribution we can find that the probability of obtaining a value of t exceeding or equal to 1.86 in absolute value equals 0.08. Thus the result obtained does not contradict the hypothesis H_0 $(m_1 = m_2)$ even at the significance level $\alpha = 0.05$.

If we compare this conclusion with the conclusion formulated in example 12.2.1, we may wonder why they are different. However, this fact can be explained in a simple and intuitive manner. In the situation of example 12.2.1, we investigate the difference of the effects of two drugs on the same patient. Thus if the average effect of both drugs were the same, the differences in their effects would be small and the observed differences $x - y$ could not be explained by random fluctuations, at the accepted level of significance. In the situation of example 12.2.6, we investigate the effect of two drugs on two different groups of patients, and here we may expect bigger differences even if the hypothesis H_0 $(m_1 = m_2)$ is true.

In example 12.2.6, we investigated two normal populations with unknown expected values m_1 and m_2, and unknown standard deviations σ_1 and σ_2. We accepted as sure, that is, without the necessity of verifying, that $\sigma_1 = \sigma_2 = \sigma$, where the value of σ has not been specified, and we formulated the hypothesis H_0 ($m_1 = m_2$). Thus if hypothesis H_0 is true, then the random variables X and Y considered in this example have the same normal distribution. Hypothesis H_0 was tested on the basis of Student's t-distribution. However, in practice we often meet a situation where we need to test either the hypothesis H_1 ($m_1 = m_2$, $\sigma_1 = \sigma_2$) or the hypothesis H_2 ($m_1 = m_2$) without the assumption $\sigma_1 = \sigma_2$. Thus if we consider the hypothesis H_1, the assumption $\sigma_1 = \sigma_2$ has to be verified, and if we consider hypothesis H_2, we do not exclude the possibility $\sigma_1 \neq \sigma_2$, that is, that X and Y have different normal distributions.

The problem of testing hypothesis H_1 and H_2 has been the subject of many discussions and controversies. We mention here the papers of Fisher [7, 12], Neyman and Pearson [2], Sukhatme [2], Hsu [1], and Welch [1].

12.3 PARAMETRIC TESTS FOR LARGE SAMPLES

We show by some examples how one can obtain significance tests using limit distributions of statistics.

Example 12.3.1. Suppose that we have a simple sample of $n = 50$ elements drawn from a population in which the characteristic X has a distribution with the standard deviation $\sigma = 5$ and an unknown expected value m. From the sample we have obtained $\bar{x} = 2$. The problem is to test the hypothesis H_0 ($m = 0$).

From the existence of the standard deviation it follows that if hypothesis H_0 is true, the statistic $\bar{X}$ has the asymptotically normal distribution $N(0; 1/\sqrt{2})$. From tables of the normal distribution we find

$$P(|\bar{X}| \geq 2) = P(|\bar{X}| \sqrt{2} > 2\sqrt{2}) \cong P(|\bar{X}| \sqrt{2} \geq 2.82) < 0.01.$$

Thus we reject H_0.

In example 12.3.1 we assumed that the standard deviation of the population is known. Usually, in practical problems it is unknown. However, if the sample is large enough, the standard deviation may be determined from the sample, that is, we may put $s = \sigma$. This is possible only if the sample contains no less than 100 elements.

Example 12.3.2. Suppose we are given a simple sample of $n = 150$ elements. The mean and the standard deviation of the sample are, respectively, $\bar{x} = 0.4$ and $s = 4$. We know neither the expected value nor the standard deviation of the population. Should we reject the hypothesis H_0 ($m = 0$)?

If we put $\sigma = s$ and if H_0 is true, then $\bar{X}$ has the asymptotically normal distribution $N(0; 4/\sqrt{150})$. Hence

$$P(|\bar{X}| \geq 0.4) = P(|\bar{X}|/0.33 \geq 1.2) = 0.23.$$

Thus there is no reason to reject H_0.

As we see, for large samples we use limit distributions of statistics; moreover, sometimes we accept the values of some parameters, other than the one to which the tested hypothesis refers, as equal to the observed values of the corresponding sample statistics. However, it ought to be stressed that the problem of whether or not the sample is large enough depends upon the parameter under investigation and upon whether or not the statistic used as a basis for the test is stochastically rapidly convergent to the parameter of the population. If some other parameters are also determined from the sample (as in example 12.3.2), the same question should also be considered in relation to them.

Example 12.3.3. One of the characteristics investigated in a farm census is whether or not a given farm has a horse. We want to know the fraction of farms in counties A and B which do not have horses. From county A we have drawn (according to the Bernoulli scheme) $n_1 = 1500$ IBM cards and from county B, $n_2 = 1800$ IBM cards. We have observed 300 farms without horses in county A and 320 farms without horses in county B. Thus the corresponding fractions for counties A and B are, respectively,

$$p_1 = \frac{300}{1500} = 0.200, \qquad p_2 = \frac{320}{1800} = 0.178.$$

The hypothesis H_0 asserts that the fraction of farms without horses is the same in both counties and equals p; hence according to this hypothesis the observed difference between p_1 and p_2 in the samples is caused by random fluctuations. We shall test this hypothesis at the significance level $\alpha = 0.05$.

Let us assign the number one to the event that we draw an IBM card corresponding to a farm without a horse and the number zero to the opposite event. Thus we have the random variables X_k with the zero-one distribution. Then $p_1 = 0.200$ is the observed value of the random variable Y in the Bernoulli scheme, where $Y = \frac{1}{n_1} \Sigma X_k$ with $n_1 = 1500$, whereas $p_2 = 0.178$ is the observed value of the random variable Z in the Bernoulli scheme, where $Z = \frac{1}{n_2} \Sigma X_k$ with $n_2 = 1800$.

If hypothesis H_0 is true, $E(Y) = E(Z) = p$. The standard deviations of Y and Z are, respectively, by (5.2.8)

$$\sigma_y = \sqrt{p(1 - p)/1500}, \qquad \sigma_z = \sqrt{p(1 - p)/1800}.$$

The random variables Y and Z are independent; hence $Y - Z$ has standard deviation

$$\sigma = \sqrt{\sigma_y^2 + \sigma_z^2} = \sqrt{11p(1 - p)/9000}.$$

Thus we see that $Y - Z$ has the asymptotically normal distribution

$$N(0; \sqrt{11p(1 - p)/9000}).$$

If we knew the value of p, we could compute the exact value of σ. Since the size of the samples is large, we accept as p the weighted average of p_1 and p_2, that is,

$$p = \frac{1500p_1 + 1800p_2}{3300} = 0.188.$$

We obtain $\sigma = 0.0137$. The observed value of $Y - Z$ is $p_1 - p_2 = 0.022$. From tables of the normal distribution we find

$$P(|Y - Z| \geqslant 0.022) = P\left(\frac{|Y - Z|}{0.0137} \geqslant \frac{0.022}{0.0137}\right) = P\left(\frac{|Y - Z|}{0.0137} \geqslant 1.61\right) = 0.107.$$

As we see, the significance test at the level $\alpha = 0.05$ does not lead to the rejection of hypothesis H_0. In other words, the results of observations do not exclude the possibility that the fraction of farms without horses is the same in both counties.

Example 12.3.4. The monthly consumption of rye bread per consumption unit[1] in workers' families in Warsaw and Łódź, as compared with that in Silesia, has been explored. In Warsaw and Łódź there were $n_1 = 65$ families under investigation, and in Silesia $n_2 = 57$ families. Let us denote the bread consumption in kilograms (2.205 lb) per consumption unit by X. The collections of families under consideration can be treated as simple samples chosen from the population of workers' families in Warsaw-Łódź, and Silesia, respectively. Let us denote by m_1 the average consumption of rye bread in Warsaw and Łódź, and by m_2 that in Silesia. The observed sample means are equal, respectively, to $\bar{x}_1 = 9.95$ and $\bar{x}_2 = 10.28$. We want to test the hypothesis H_0 that the expected values m_1 and m_2 are equal.

The difference of sample means $\bar{x}_1 - \bar{x}_2$ is the observed value of the random variable $\bar{X}_1 - \bar{X}_2$ whose distribution may be considered as normal, since the samples are rather large. If hypothesis H_0 were true, then $\bar{X}_1 - \bar{X}_2$ would have the normal distribution $N(0; \sqrt{\sigma_1^2/n_1 + \sigma_2^2/n_2})$, where σ_1 and σ_2 are the standard deviations of the random variable X in Warsaw-Łódź, and Silesia, respectively. We do not know these values, but since the sample sizes are not small we take the observed values of the sample standard deviations as estimates of these values. They equal $s_1 = 4.22$ and $s_2 = 6.01$; hence the standard deviation of $\bar{X}_1 - \bar{X}_2$ is 0.95. Since in our example $\bar{x}_1 - \bar{x}_2 = -0.33$, we find from tables of the normal distribution that the probability that a random variable with the normal distribution $N(0; 0.95)$ exceeds or equals 0.33 in absolute value equals 0.7264. Thus we have no reason to reject H_0 at the significance level $\alpha = 0.05$.

Example 12.3.5. Sittig and Freudenthal [1] describe the results of anthropometric measurements of 5001 women in Holland. This experiment was carried out for the ready-made clothing industry, and its task was to set up a series of standards so that the production of shoes and clothes based on these standards would best meet the needs of Dutch women. Among other characteristics under investigation were the height X and the length of the middle finger Y. The value of the correlation coefficient of these two characteristics obtained by observation was $r = 0.5062$. We test the hypothesis H_0 $(\rho = 0)$ that the correlation coefficient of these two characteristics of the population equals zero.

Since the sample is very large, we may assume that R has a normal distribution, and use formulas (9.9.9) for $E(R)$ and $D^2(R)$, replacing the unknown value of ρ by zero. Thus, if the hypothesis is true, R has the approximately normal distribution $N(0; 0.0141)$. Thus it is easy to verify that the significance

---

[1] A man of and over 18 is considered one consumption unit. Men of other age groups and women are considered parts of a consumption unit according to a special scale.

test, even at the significance level $\alpha = 0.01$, leads to the rejection of the hypothesis; hence the correlation coefficient ρ is indeed different from zero.

12.4 THE χ^2 TEST

A   Let $F(x)$ be the unknown distribution function of the random variable X.

A hypothesis which refers to the form of an unknown distribution function is called a *nonparametric hypothesis*.

A procedure of testing a nonparametric hypothesis is called a *nonparametric test*.

The procedure of testing a nonparametric hypothesis can be described as follows. We want to test a hypothesis H_0 which specifies the unknown distribution function $F(x)$ of the characteristic X of a population. We draw a sample of size n from the population and observe the values of the characteristic X of the sample elements. We proceed then as in parametric testing, namely, we construct a statistic U and we determine a value u_0 such that if H_0 is true, $P(|U| \geqslant u_0) \leqslant \alpha$, where α is the significance level. If the value $|u|$ observed from our sample is not smaller than u_0, we reject H_0; if, however, $|u| < u_0$, we have no reason to reject H_0 and further investigations should decide whether or not it can be accepted.

B   A commonly used nonparametric test is based on the χ^2 statistic, given by K. Pearson [5]. We recall that a statistic of the same name was discussed in Section 9.4. As we shall see, the distribution of these statistics are closely related; we should not, however, forget that they are defined in completely different ways.

We now define Pearson's χ^2 statistic. Let $F(x)$ be a given distribution function of the random variable X. Let us split the x-axis into r disjoint sets S_k and let π_k $(k = 1, 2, \ldots, r)$ be the probability that X takes on a value from S_k. If we take the interval $[a_k, a_{k+1})$ as the set S_k, then

$$(12.4.1) \qquad \pi_k = F(a_{k+1}) - F(a_k).$$

The numbers π_k are called the *theoretical frequencies*. Let us imagine that we have n independent observations of the random variable X, $x_1, x_2, \ldots, x_n$. We split these observations into r groups including in the kth group those observations which belong to the set S_k. Denote the number of observations which belong to S_k by n_k $(k = 1, 2, \ldots, r)$. In different samples of size n the n_k, of course, vary. That is, for a given k, n_k is a random variable with the binomial distribution. The statistic

$$(12.4.2) \qquad \chi^2 = \sum_{k=1}^{r} \frac{(n_k - n\pi_k)^2}{n\pi_k}$$

is called *Pearson's χ^2 statistic*.

Theorem 12.4.1. *Suppose that the theoretical frequencies π_k are given. Then the sequence $\{F_n(z)\}$ of distribution functions of the statistic χ^2 defined by (12.4.2) satisfies the relation*

(12.4.3)
$$\lim_{n \to \infty} F_n(z) = \begin{cases} \dfrac{1}{2^{(r-1)/2}\Gamma(\frac{1}{2}(r-1))} \displaystyle\int_0^z z^{(r-3)/2}e^{-z/2}\,dz & \text{for } z > 0, \\ 0 & \text{for } z \leqslant 0. \end{cases}$$

Expression (12.4.3) is the distribution function of the random variable χ^2 discussed in Section 9.4 with $r - 1$ degrees of freedom.

Proof. If the assumptions of our theorem are satisfied, then the probability that among n observations, n_1 will belong to the set S_1, n_2 will belong to the set S_2, and so on, and n_r will belong to the set S_r equals

(12.4.4)
$$\frac{n!}{n_1!\,n_2!\ldots n_r!}\pi_1^{n_1}\ldots\pi_r^{n_r}.$$

This is the probability function of the multinomial distribution considered in Section 5.12. The characteristic function $\phi(t_1, \ldots, t_r)$ of this probability function is of the form

$$\phi_1(t_1, t_2, \ldots, t_r) = \left(\sum_{k=1}^r \pi_k e^{it_k}\right)^n.$$

Let us consider the random variables

$$Y_k = \frac{n_k - n\pi_k}{\sqrt{n\pi_k}} \qquad (k = 1, 2, \ldots, r).$$

We have

$$\chi^2 = \sum_{k=1}^r Y_k^2, \qquad \sum_{k=1}^r Y_k\sqrt{\pi_k} = 0.$$

From the last equality it follows that the random variables Y_k are linearly dependent. The characteristic function of $(Y_1, \ldots, Y_r)$ has the form

(12.4.5) $\quad \phi(t_1, t_2, \ldots, t_r) = \exp\left(-\sum_{k=1}^r it_k\sqrt{n\pi_k}\right)\left[\sum_{k=1}^r \pi_k \exp\left(\frac{it_k}{\sqrt{n\pi_k}}\right)\right]^n.$

Hence

$\log \phi(t_1, t_2, \ldots, t_r)$

$\displaystyle = -i\sqrt{n}\sum_{k=1}^r t_k\sqrt{\pi_k} + n\log\left[\sum_{k=1}^r \pi_k \exp\left(\frac{it_k}{\sqrt{n\pi_k}}\right)\right]$

$\displaystyle = -i\sqrt{n}\sum_{k=1}^r t_k\sqrt{\pi_k} + n\log\left[1 + \frac{i}{\sqrt{n}}\sum_{k=1}^r t_k\sqrt{\pi_k} - \frac{1}{2n}\sum_{k=1}^r t_k^2 + o\left(\frac{1}{n}\right)\right]$

$\displaystyle = -i\sqrt{n}\sum_{k=1}^r t_k\sqrt{\pi_k} + n\log(1 + z),$

where

$$z = \frac{i}{\sqrt{n}} \sum_{k=1}^{r} t_k \sqrt{\pi_k} - \frac{1}{2n} \sum_{k=1}^{r} t_k^2 + o\left(\frac{1}{n}\right).$$

Using the fact that $z \to 0$ as $n \to \infty$, we can expand $\log(1 + z)$ in a Maclaurin series. We obtain

$$n \log(1 + z) - i\sqrt{n} \sum_{k=1}^{r} t_k \sqrt{\pi_k} \to -\frac{1}{2}\left[\sum_{k=1}^{r} t_k^2 - \left(\sum_{k=1}^{r} t_k \sqrt{\pi_k}\right)^2\right].$$

Finally,

(12.4.6) $\lim_{n \to \infty} \phi(t_1, t_2, \ldots, t_r) = \exp\left\{-\frac{1}{2}\left[\sum_{k=1}^{r} t_k^2 - \left(\sum_{k=1}^{r} t_k \sqrt{\pi_k}\right)^2\right]\right\}.$

We now show that an orthogonal transformation

$$u_k = \sum_{j=1}^{r} a_{kj} t_j \qquad (k = 1, 2, \ldots, r)$$

exists, with coefficients

$$a_{rj} = \sqrt{\pi_j} \quad (j = 1, 2, \ldots, r).$$

To start with, there are r^2 unknown coefficients. By the orthogonality assumption, we obtain the following $r + \frac{1}{2}r(r - 1)$ equalities:

$$\sum_{j=1}^{r} a_{kj} a_{ij} = \begin{cases} 1 & \text{for } k = i, \\ 0 & \text{for } k \neq i. \end{cases}$$

Moreover, $r - 1$ coefficients are determined in advance (in fact, we determine r coefficients, but they must satisfy the relation $\sum_{j=1}^{r} a_{rj}^2 = 1$). Thus the number of coefficients which are to be determined equals $r^2 - (r - 1)$, and we can determine them, since

$$r^2 - (r - 1) - r - \tfrac{1}{2}r(r - 1) = \tfrac{1}{2}(r^2 - 3r + 2) \geqslant 0 \qquad \text{for } r \geqslant 2.$$

After substituting $u_1, \ldots, u_r$ in the quadratic form on the right-hand side of (12.4.6), it takes the form

$$\sum_{k=1}^{r-1} u_k^2.$$

By (12.4.6) we find that the characteristic function of the random vector $(Y_1, \ldots, Y_r)$ converges as $n \to \infty$ to the characteristic function of a random vector with $r - 1$ independent components, each of which has the normal distribution $N(0; 1)$. From a theorem which has not been proved in this book, but which is a generalization of theorem 6.6.1b to multidimensional distributions, it follows that $(Y_1, \ldots, Y_r)$ has the asymptotically normal distribution, determined by the limit characteristic function.

Finally, using theorem 6.4.1, we find that the statistic χ^2 given by (12.4.2) is asymptotically the sum of $r - 1$ squares of independent random variables with the normal distribution $N(0; 1)$; hence it has in the limit the χ^2 distribution with $r - 1$ degrees of freedom. The theorem is proved.

C   Theorem 12.4.1 serves as the foundation for a widely applied nonparametric test, which is called the χ^2 *test*. We shall illustrate this application by an example.

Example 12.4.1. From a lot a simple sample of $n = 100$ elements was selected, and $n_1 = 22$ defective items were found in the sample. We want to test on the basis of this sample the hypothesis H_0 that the probability of drawing a defective item was constant and equal to $p = 0.20$, at the significance level $\alpha = 0.1$. In other words, according to this hypothesis, $n_1 = 22$ is the observed value of a random variable with the binomial distribution with $p = 0.20$ and $n = 100$.

We apply the χ^2 test. Since hypothesis H_0 specifies the distribution completely, we can apply theorem 12.4.1.

We split the whole sample into $n_1 = 22$ defective items and $n_2 = n - n_1$ good items. The theoretical frequencies of these groups are, respectively, $p = 0.2$ and $q = 1 - p = 0.8$. From formula (12.4.2) we obtain

$$\chi^2 = \frac{(n_1 - np)^2}{np} + \frac{(n_2 - nq)^2}{nq} = \frac{2^2}{20} + \frac{2^2}{80} = \frac{1}{4}.$$

Here we have two groups, that is, $r = 2$; thus we consider the χ^2 distribution with one degree of freedom. From tables of the χ^2 distribution we find that $P(\chi^2 \geqslant 6.6) = 0.01$; hence the probability that χ^2 exceeds or equals $\frac{1}{4}$ is considerably greater than 0.01. Thus there is no reason to reject H_0.

In applying the χ^2 test, one should remember that it is based upon a limit theorem. Therefore, the numbers n_k which appear in this test cannot be too small. One can give some sort of rule, namely that the observations should be split into such groups that $n\pi_k \geqslant 10$, that is, that the expected number of observations in each group is not smaller than ten. Some authors accept the rule that the number of observations in each group should not be smaller than five. A deeper investigation on the number of groups into which the observations should be split is contained in the paper of Mann and Wald [1]. In this paper Mann and Wald advise the splitting of the x-axis into intervals with the same theoretical frequencies; hence if the number of intervals equals r, then $\pi_k = 1/r$ $(k = 1, 2, \ldots, r)$. In connection with the paper of Mann and Wald, Williams [1] gives numerical data concerning the number and lengths of these intervals. An extensive discussion of the use of the χ^2 test is given in Cochran's [2] paper. It is worthwhile to mention that Vessereau [1], comparing, for $\pi_k = 1/r$ and small n (about 15 to 20), the exact distribution of χ^2 defined by (12.4.2) with the limit distribution χ^2 with $r - 1$ degrees of freedom, established that without fear of making a big mistake one can apply the limit χ^2

distribution for the χ^2 test, at the significance level $\alpha = 0.05$ or $\alpha = 0.01$, even if $n_k = 1$ $(k = 1, 2, \ldots, r)$.

Tumanian [1] and Gihman [4] have investigated the χ^2 limit distribution when the number of intervals into which we split the axis $(-\infty, +\infty)$ tends to infinity as $n \to \infty$ (See Problem 12.7).

D   Thus far we have considered only the case when the hypothesis H_0 completely determines the distribution. If, however, the hypothesis H_0 specifies the functional form of the distribution function $F(x)$ but does not specify the values of certain parameters $\lambda_1, \ldots, \lambda_m$, then the π_k defined by (12.4.1) are not given either. Thus we cannot apply the test based on theorem 12.4.1.

However, Fisher [4] showed that here one has only to modify theorem 12.4.1 slightly to be able to apply it. To present Fisher's theorem, we have to introduce the notion of the likelihood function. Here we consider this question only generally, but return to it in the next chapter.

Suppose that the functional form of the distribution function $F(x)$ of the characteristic X of some population is known, but the values of the parameters $\lambda_1, \ldots, \lambda_m$ are unknown. From this population we have drawn a simple sample of n elements and observed the values $x_1, \ldots, x_n$.

We consider separately random variables of the discrete type and of the continuous type.

Suppose that the random variable X is of the discrete type. Let p_k denote the probability that $X = x_k$ $(k = 1, \ldots, n)$. It is evident that the p_k are functions of $\lambda_1, \ldots, \lambda_m$ and hence we shall write $p_k(\lambda_1, \ldots, \lambda_m)$.

Definition 12.4.1.   For a random variable of the discrete type, the product

$$(12.4.7) \quad L = p_1(\lambda_1, \lambda_2, \ldots, \lambda_m)p_2(\lambda_1, \lambda_2, \ldots, \lambda_m) \ldots p_n(\lambda_1, \lambda_2, \ldots, \lambda_m)$$

is called the *likelihood function*.

Suppose now that the random variable X is of the continuous type and has the density $f(x, \lambda_1, \ldots, \lambda_m)$.

Definition 12.4.2.   For a random variable of the continuous type, the product

$$(12.4.7') \quad L = f(x_1, \lambda_1, \lambda_2, \ldots, \lambda_m)f(x_2, \lambda_1, \lambda_2, \ldots, \lambda_m) \ldots$$
$$f(x_n, \lambda_1, \lambda_2, \ldots, \lambda_m)$$

is called the *likelihood function*.

The likelihood function is a function of the parameters $\lambda_1, \ldots, \lambda_m$. The values $\lambda_1^*, \ldots, \lambda_m^*$ of these parameters determined by the system of equations

$$(12.4.8) \qquad \partial \log L/\partial \lambda_i = 0 \qquad (i = 1, 2, \ldots, m)$$

are said to be obtained by the *maximum likelihood method*.

Let us now imagine that we have split the x-axis into r intervals and have grouped the n observations into r groups, and let π_k and n_k have the same meaning as in (12.4.2). It is obvious that the π_k's are functions of the unknown parameters $\lambda_1, \ldots, \lambda_m$. Suppose that $m < r$. Let us consider the expression

$$(12.4.9) \qquad L = \pi_1^{n_1}(\lambda_1, \ldots, \lambda_m) \ldots \pi_r^{n_r}(\lambda_1, \ldots, \lambda_m).$$

This expression is the likelihood function *after the grouping of observations*. The theorem of Fisher mentioned previously, which was made precise by Cramér ([2], § 30,3), states:

Theorem 12.4.2. *Let all the $\pi_k(\lambda_1, \ldots, \lambda_m) > 0$ and let the continuous partial derivatives* $\dfrac{\partial \pi_k}{\partial \lambda_i}, \dfrac{\partial^2 \pi_k}{\partial \lambda_i \partial \lambda_j}$ *$(k = 1, \ldots, r, i, j = 1, \ldots, m)$ exist, and let the matrix with the terms* $\dfrac{\partial \pi_k}{\partial \lambda_i}$ *$(k = 1, \ldots, r;\ i = 1, \ldots, m)$ have rank m. Then, if the unknown parameters $\lambda_1, \ldots, \lambda_m$ are determined by the maximum likelihood method after the grouping of observations, that is, from the system of equations (12.4.8), where L is given by (12.4.9), the distribution of the statistic χ^2 defined by (12.4.2) tends as $n \to \infty$ to the χ^2 distribution with $r - m - 1$ degrees of freedom.*

The proof of this theorem can be found in Cramér [2].

The system (12.4.8) with L given by (12.4.9) is usually very difficult to solve. It is usually much easier to solve system (12.4.8) with the likelihood function L determined by (12.4.7) or (12.4.7'), that is, when the likelihood function is computed from the observations before grouping. However, theorem (12.4.2) is not true in this case, as was noted by Fisher [9]. Chernoff and Lehmann [1] showed that if the functions $\pi_k(\lambda_1, \ldots, \lambda_m)$ satisfy the assumptions of theorem 12.4.2 and the density $f(x, \lambda_1, \ldots, \lambda_m)$, or the probability function $p(\lambda_1, \ldots, \lambda_m)$, satisfies the assumptions of theorem 13.7.3 or, rather, the assumptions of the generalization of theorem 13.7.3 to several unknown parameters (see Section 13.7), then χ^2 defined by (12.4.2) has as $n \to \infty$ the same limit distribution as $U + a_1 Y_1^2 + \ldots + a_m Y_m^2$, where U has the χ^2 distribution with $r - m - 1$ degrees of freedom, $Y_1, \ldots, Y_m$ have the normal distribution $N(0; 1)$, $U, Y_1, \ldots, Y_m$ are independent, and $a_1, \ldots, a_m$ are certain numbers between zero and one (the method of computing $a_1, \ldots, a_m$ is given by Chernoff and Lehmann [1]). If the number of groups r is large, then the assertion of theorem 12.4.2 may be applied even if L is computed from formula (12.4.7) or (12.4.7'). If, however, r is not large, the results of such an application of theorem 12.4.2 should be used very carefully.

Watson [1, 2], considering the case when the x-axis is split into intervals of constant theoretical frequencies, hence when $\pi_k = 1/r$, reached conclusions analogous to those of Chernoff and Lehmann [1]. We recommend the papers of Neyman and Pearson [1] and Neyman [8]. In the last paper some modifications and generalizations of the χ^2 test are considered; a method of determining the unknown parameters from the sample, a particular case of which is the maximum likelihood method, is also given.

Example 12.4.2 (from the book by Boev [2]). Three hundred balls of cotton thread chosen from a consignment of cotton thread have been investigated. The subject of investigation was their strength, expressed in kilograms. The observed values x_k have been grouped into 13 groups, and the number of observations in the ith group denoted by n_i $(i = 1, \ldots, 13)$. These data are given in Table 12.4.1.

We now test the hypothesis H_0 that the characteristic X has a normal distribution. This hypothesis does not specify the expected value nor the standard deviation of X. It only states that the random variable X has a normal distribution. Thus we must determine the values of these parameters from the sample. We use the method of maximum likelihood before the grouping of observations. The density of the random variable X is of the form

$$f(x, m, \sigma) = \frac{1}{\sigma\sqrt{2\pi}} \exp\left[-\frac{(x - m)^2}{2\sigma^2}\right],$$

where the parameters m and σ are unknown. According to (12.4.7′), the likelihood function is

$$L = \left[\frac{1}{\sigma\sqrt{2\pi}}\right]^n \exp\left[-\frac{1}{2\sigma^2}\sum_{k=1}^{n}(x_k - m)^2\right] \quad (n = 300).$$

Hence

$$\log L = -\frac{1}{2\sigma^2}\sum_{k=1}^{n}(x_k - m)^2 - \tfrac{1}{2}n\log\sigma^2 - \tfrac{1}{2}n\log 2\pi.$$

Applying (12.4.8), we obtain

$$\frac{\partial \log L}{\partial m} = \frac{1}{\sigma^2}\sum_{k=1}^{n}(x_k - m) = 0,$$

(12.4.10)

$$\frac{\partial \log L}{\partial \sigma^2} = \frac{1}{2\sigma^4}\sum_{k=1}^{n}(x_k - m)^2 - \frac{n}{2\sigma^2} = 0.$$

TABLE 12.4.1

i	x	n_i	i	x	n_i
1	0.5–0.64	1	8	1.48–1.62	53
2	0.64–0.78	2	9	1.62–1.76	25
3	0.78–0.92	9	10	1.76–1.90	19
4	0.92–1.06	25	11	1.90–2.04	16
5	1.06–1.20	37	12	2.04–2.18	3
6	1.20–1.34	53	13	2.18–2.38	1
7	1.34–1.48	56			

The solution of system (12.4.10) is

$$m^* = \frac{1}{n} \sum_{k=1}^{n} x_k = \bar{x},$$

(12.4.11)

$$\sigma^{*2} = \frac{1}{n} \sum_{k=1}^{n} (x_k - \bar{x})^2 = s^2.$$

Hence the sample mean and sample variance are equal to m^* and σ^* obtained by the maximum likelihood method.

Since the intervals in Table 12.4.1 are rather narrow, we may assume, when computing $\bar{x}$ and s, that the observations are placed at the midpoints of the intervals; we denote the midpoint of each interval by x_i. Thus we obtain from Table 12.4.1 the following Table 12.4.2:

TABLE 12.4.2

i	x_i	n_i	i	x_i	n_i
1	0.57	1	8	1.55	53
2	0.71	2	9	1.69	25
3	0.85	9	10	1.83	19
4	0.99	25	11	1.97	16
5	1.13	37	12	2.11	3
6	1.27	53	13	2.28	1
7	1.41	56			

From this table we obtain

$$m^* = 1.41, \qquad \sigma^* = 0.26.$$

Since the tail groups in Table 12.4.1 are too small, we pool the two highest and two lowest groups into one group. Using tables of the normal distribution, we find the theoretical frequencies π_k. The end points of the intervals are placed in the middle between neighboring points x_i and x_{i+1}. For the random variable X with the normal distribution $N(1.41; 0.26)$, we have

$$\pi_1 = P(X < 0.78) + P(X \geqslant 2.04)$$

$$= P\left(\frac{X - 1.41}{0.26} < -2.42\right) + P\left(\frac{X - 1.41}{0.26} \geqslant 2.42\right) = 0.0156,$$

$$\pi_2 = P(0.78 \leqslant X < 0.92) = P\left(-2.42 \leqslant \frac{X - 1.41}{0.26} < -1.88\right) = 0.0223,$$

$$\pi_3 = P(0.92 \leqslant X < 1.06) = P\left(-1.88 \leqslant \frac{X - 1.41}{0.26} < -1.35\right) = 0.0584,$$

$$\pi_4 = P(1.06 \leqslant X < 1.20) = P\left(-1.35 \leqslant \frac{X - 1.41}{0.26} < -0.81\right) = 0.1205,$$

$$\pi_5 = P(1.20 \leqslant X < 1.34) = P\left(-0.81 \leqslant \frac{X - 1.41}{0.26} < -0.27\right) = 0.1846,$$

$$\pi_6 = P(1.34 \leqslant X < 1.48) = P\left(-0.27 \leqslant \frac{X - 1.41}{0.26} < 0.27\right) = 0.2128,$$

$$\pi_7 = P(1.48 \leqslant X < 1.62) = P\left(0.27 \leqslant \frac{X - 1.41}{0.26} < 0.81\right) = 0.1846,$$

$$\pi_8 = P(1.62 \leqslant X < 1.76) = P\left(0.81 \leqslant \frac{X - 1.41}{0.26} < 1.35\right) = 0.1205,$$

$$\pi_9 = P(1.76 \leqslant X < 1.90) = P\left(1.35 \leqslant \frac{X - 1.41}{0.26} < 1.88\right) = 0.0584,$$

$$\pi_{10} = P(1.90 \leqslant X < 2.04) = P\left(1.88 \leqslant \frac{X - 1.41}{0.26} < 2.42\right) = 0.0223.$$

We compute

$$\chi^2 = \sum_{j=1}^{10} \frac{(n_j - n\pi_j)^2}{n\pi_j} = 22.07.$$

Since two parameters were determined from the sample, the number of degrees of freedom is seven. From tables of the χ^2 distribution we find that $P(\chi^2 \geqslant 22.07) < 0.01$. Thus we reject H_0.

Example 12.4.3. In example 5.5.1 we established that the empirical distribution of the number of deaths from a kick by a horse coincides rather well with the Poisson distribution whose parameter λ equals the mean value of the empirical distribution.

In fact, the hypothesis stated was that the random variable considered has the Poisson distribution and, without any test, it was accepted that the results of observations do not contradict the hypothesis. We now apply the χ^2 test at the significance level $\alpha = 0.05$ to test this hypothesis.

Let us rewrite Table 5.5.1 in such a form that instead of the frequencies and probabilities we give the observed and the expected numbers. In this way we obtain Table 12.4.3.

Two last groups are pooled together because of the small number of observations they contain. We have

$$\chi^2 = \sum_{i=1}^{4} \frac{(n_i - np_i)^2}{np_i} = 0.3160.$$

TABLE 12.4.3

	0	1	2	3	4
Observed number n_i	109	65	22	3	1
Expected number np_i	108.8	66.2	20.2	4.2	0.6

Since one parameter[1] has been determined from the sample, the statistic χ^2 has two degrees of freedom. From tables of the χ^2 distribution we find that $P(\chi^2 \geq 0.3160)$ is much greater than $\alpha = 0.05$; hence there is no reason to reject the hypothesis being tested.

12.5. TESTS OF THE KOLMOGOROV AND SMIRNOV TYPE

A   Kolmogorov's theorem 10.10.1 may also serve as a basis for a nonparametric test to verify a hypothesis H_0 which specifies the unknown continuous distribution function $F(x)$. Let us recall this theorem.

Let $S_n(x)$ be the empirical distribution function of a simple sample of size n. Let

$$(12.5.1) \qquad D_n = \sup_{-\infty < x < +\infty} |F(x) - S_n(x)|.$$

According to Kolmogorov's theorem, we have for $\lambda > 0$

$$(12.5.2) \qquad \lim_{n \to \infty} Q_n(\lambda) = \lim_{n \to \infty} P\left(D_n < \frac{\lambda}{\sqrt{n}}\right) = Q(\lambda),$$

where

$$(12.5.3) \qquad Q(\lambda) = \sum_{k=-\infty}^{+\infty} (-1)^k e^{-2k^2\lambda^2}.$$

Suppose that under the hypothesis H_0 the characteristic X of the population has the continuous distribution function $F(x)$. We draw a simple sample of sufficiently large size from this population, and we compute the empirical distribution function $S_n(x)$ and the statistic D_n. To test our hypothesis at the significance level α, we find from Table VIII at the end of the book a value λ_0 such that $Q(\lambda_0) = 1 - \alpha$. If the observed value of the statistic D_n exceeds or equals $\lambda_0/\sqrt{n}$, we reject hypothesis H_0, whereas in the contrary case the test does not lead to the rejection of H_0.

Example 12.5.1. According to Wiszniewski [1], Table 12.5.1 gives the distribution of the average monthly temperature in January in Warsaw for the years 1779–1947 (except 1945). Thus the number of observations is $n = 168$.

Let X denote the average monthly temperature in January in Warsaw. The x_k in Table 12.5.1 denote the observed values of the random variable X, and n_k denotes the number of observations x_k. From this table we compute

$$\bar{x} = -4.22, \qquad s = 3.57,$$

where $\bar{x}$ and s are the sample mean and sample standard deviation, respectively. We assume the hypothesis H_0 that the random variable X has the normal distribution $N(-4.22; 3.57)$, and we test this hypothesis at the significance level

---

[1] In example 13.7.1, we prove that $\bar{x}$ is the maximum likelihood estimate of the unknown parameter λ of the Poisson distribution.

TABLE 12.5.1

x_k	n_k	$s_n(x_k)$	$F(x_k)$	x_k	n_k	$s_n(x_k)$	$F(x_k)$
−17.8	1	0.000	0.000	−4.2	1	0.440	0.504
−14.1	1	0.006	0.003	−4.1	3	0.446	0.512
−13.4	1	0.012	0.005	−4.0	2	0.464	0.524
−13.2	1	0.018	0.006	−3.9	6	0.476	0.536
−13.0	1	0.024	0.007	−3.8	2	0.511	0.548
−12.2	1	0.030	0.013	−3.6	3	0.523	0.567
−11.6	1	0.036	0.019	−3.5	2	0.541	0.579
−11.3	1	0.042	0.024	−3.4	1	0.553	0.591
−10.9	1	0.048	0.031	−3.3	2	0.559	0.603
−10.1	1	0.054	0.049	−3.2	3	0.571	0.614
−9.7	1	0.060	0.063	−3.1	1	0.589	0.622
−9.4	2	0.065	0.074	−3.0	3	0.595	0.633
−9.3	3	0.077	0.078	−2.9	4	0.612	0.644
−9.2	2	0.095	0.082	−2.8	4	0.636	0.655
−9.1	1	0.107	0.085	−2.6	1	0.660	0.674
−9.0	1	0.113	0.090	−2.5	1	0.666	0.684
−8.7	1	0.119	0.106	−2.4	3	0.672	0.695
−8.6	1	0.125	0.109	−2.3	2	0.690	0.705
−8.0	1	0.131	0.145	−2.1	1	0.702	0.722
−7.8	3	0.136	0.159	−2.0	4	0.708	0.732
−7.7	1	0.154	0.166	−1.9	2	0.731	0.742
−7.4	3	0.160	0.187	−1.7	3	0.743	0.761
−7.3	2	0.178	0.195	−1.6	2	0.761	0.767
−7.2	1	0.190	0.203	−1.4	1	0.773	0.785
−7.1	1	0.196	0.209	−1.3	1	0.779	0.794
−7.0	1	0.202	0.218	−1.2	4	0.785	0.802
−6.8	2	0.208	0.236	−1.1	3	0.809	0.808
−6.7	2	0.220	0.245	−1.0	3	0.827	0.816
−6.6	1	0.232	0.251	−0.8	2	0.844	0.832
−6.4	2	0.238	0.271	−0.6	1	0.856	0.844
−6.3	1	0.250	0.281	−0.3	1	0.862	0.864
−6.1	1	0.255	0.298	−0.2	4	0.868	0.871
−6.0	2	0.261	0.308	−0.1	2	0.892	0.875
−5.9	2	0.273	0.319	0.0	1	0.904	0.881
−5.4	2	0.285	0.371	0.1	1	0.910	0.887
−5.3	4	0.297	0.382	0.3	1	0.916	0.898
−5.2	2	0.321	0.394	0.6	1	0.922	0.911
−5.1	2	0.333	0.401	0.7	3	0.928	0.916
−5.0	1	0.345	0.413	0.9	1	0.946	0.924
−4.9	2	0.351	0.425	1.2	1	0.952	0.936
−4.8	1	0.363	0.436	1.3	2	0.958	0.939
−4.7	2	0.369	0.448	1.4	1	0.969	0.942
−4.6	2	0.380	0.456	1.6	1	0.975	0.948
−4.5	1	0.392	0.468	1.9	1	0.981	0.956
−4.4	4	0.398	0.480	2.6	1	0.987	0.972
−4.3	3	0.422	0.492	2.8	1	0.993	0.976

$\alpha = 0.05$, using the Kolmogorov theorem. To do this, we find both the values of the normal distribution function $N(-4.22; 3.57)$ and the values of the empirical distribution function at the points x_k. These data are presented in Table 12.5.1. The observed value d_n of the random variable D_n equals the largest of the observed values $|s_n(x_{k+1}) - F(x_k)|$ and $|s_n(x_k) - F(x_k)|$. Thus

$$d_n = |s_n(-5.4) - F(-5.4)| = 0.086.$$

From Table VIII we find

$$P(D_n \geqslant 0.086) = P(D_n \sqrt{168} \geqslant 1.115) \cong 0.1663.$$

Thus the test does not lead to the rejection of H_0.

We draw the reader's attention to the fact that the reasoning applied in this example is not quite correct, for Gihman [2] has shown that if the unknown distribution function depends upon unknown parameters which are estimated from the sample, theorem 10.10.1 of Kolmogorov is no longer valid. Thus the results of applying Kolmogorov's theorem in the given situation should be very carefully considered.

In connection with example 12.5.1, let us also make the following remarks. In practice, the observed values are measured only up to a certain degree of accuracy. Thus, in example 12.5.1, the temperature was measured with an accuracy up to $0.05°$ C, hence the values in the interval $(x_k - 0.05°$ C, $x_k + 0.05°$ C$)$ are represented as the point x_k. It is true, as has been shown by Gihman [1, 3], that the grouping has an influence upon the limit of the sequence $\{Q_n(\lambda)\}$. However, Gihman established that if the lengths of grouping intervals tend to zero uniformly as $n \to \infty$, Kolmogorov's theorem may be applied. According to the practical interpretation of this condition, Kolmogorov's theorem may be applied if all the grouping intervals are small. This remark also applies to Smirnov's theorem.

B   Smirnov's theorem 10.11.2 may serve as a basis for testing the hypothesis H_0 that two simple samples have been drawn from populations with the same continuous distribution function. We stress the fact that the hypothesis does not specify the unknown distribution function of these populations. If $S_{1n_1}(x)$ and $S_{2n_2}(x)$ denote the empirical distribution functions of simple samples of sizes n_1 and n_2, respectively, drawn from the considered populations, then, according to Smirnov's theorem, the sequence of distribution functions $\{Q_{n_1 n_2}(\lambda)\}$ defined as

$$(12.5.4) \qquad Q_{n_1 n_2}(\lambda) = P\left(D_{n_1 n_2} < \frac{\lambda}{\sqrt{n}}\right),$$

where $n = n_1 n_2/(n_1 + n_2)$ and

$$(12.5.5) \qquad D_{n_1 n_2} = \max_{-\infty < x < +\infty} |S_{1n_1}(x) - S_{2n_2}(x)|,$$

tends to the distribution function $Q(\lambda)$ given by (12.5.3) as $n \to \infty$.

Now suppose that we have two simple samples drawn from two populations. We formulate the hypothesis that the characteristic X has the same
continuous distribution function in both populations. If the observed value
of $D_{n_1 n_2}$ exceeds or equals $\lambda_0/\sqrt{n}$, where $Q(\lambda_0) = 1 - \alpha$, we reject H_0.

Example 12.5.2. In example 12.3.4, we considered the average consumption
of bread per consumption unit in two samples. One sample consisted of $n_1 = 65$
workers' families in Warsaw and Łódź, the other of $n_2 = 57$ families in Silesia.
We established that, as a result of a parametric test at the significance level
$\alpha = 0.05$, there is no reason to reject the hypothesis H_0 that the average consumption of rye bread in workers' families in Warsaw-Łódź, and Silesia is the same.

We now investigate a much stronger hypothesis, that not only the average
consumption, but the distribution of workers' families according to their bread
consumption is identical in Warsaw-Łódź, and Silesia. In other words, we shall
test the hypothesis H_0 that both samples have been taken from populations with
the same distribution. We apply the λ-Smirnov test.

In Table 12.5.2, x_k denotes the monthly average bread consumption per consumption unit. The numbers r_k denote the number of families whose consumption equals x_k, and $s_{1n_1}(x_k)$ and $s_{2n_2}(x_k)$ denote the values of the empirical
distribution functions at the points x_k. From Table 12.5.1 we find

$$d_{n_1 n_2} = \max_x \left| s_{1n_1}(x) - s_{2n_2}(x) \right| = 0.105.$$

Since $n = n_1 n_2/(n_1 + n_2) = 30.37$, we find from Table VIII that

$$P(D_{n_1 n_2} \geqslant 0.105) = P(D_{n_1 n_2} \sqrt{30.37} \geqslant 0.58) \simeq 0.8896.$$

Thus the λ test does not lead to the rejection of H_0.

Example 12.5.3. From three populations with continuous distribution
functions three simple samples have been drawn of sizes $n_1 = 100, n_2 = 150$, and
$n_3 = 200$, respectively. The aim of the sample investigation is to test the hypothesis H_0 that the characteristic X in all the populations considered has the same
distribution. The following values of the statistics $D_i(N_i, n_{i+1})$ given by (10.13.14)
were observed:

$$d_1(N_1, n_2) = 0.20, \qquad d_2(N_2, n_3) = 0.17.$$

We test H_0 at the significance level $\alpha = 0.05$. We have

$$\sqrt{\frac{100 \cdot 150}{250}}\, 0.2 = 1.55, \quad \sqrt{\frac{250 \cdot 200}{450}}\, 0.17 = 1.79.$$

Thus the larger of the observed numbers is 1.79. According to theorem 10.13.2,
for large samples we have the approximate equality

$$P\left(\max_{1 \leqslant i \leqslant 2} \sqrt{\frac{N_i n_{i+1}}{N_{i+1}}}\, D_i(N_i, n_{i+1}) \geqslant 1.79 \right) \simeq 1 - [Q(1.79)]^2.$$

From Table VIII we find that this probability equals 0.0066; hence we reject H_0.

TABLE 12.5.2

x_k	Warsaw and Łódź	Silesia	$s_{1n_1}(x_k)$	$s_{2n_2}(x_k)$	$\|s_{1n_1}(x_k) - s_{2n_2}(x_k)\|$
1	0	4	0.000	0.000	0.000
2	0	2	0.000	0.070	0.070
3	3	1	0.000	0.105	0.105
4	2	1	0.046	0.123	0.077
5	3	2	0.077	0.140	0.063
6	5	4	0.123	0.175	0.052
7	7	5	0.200	0.246	0.046
8	5	6	0.308	0.333	0.025
9	9	7	0.384	0.439	0.055
10	8	2	0.523	0.561	0.038
11	5	3	0.645	0.596	0.049
12	5	3	0.723	0.649	0.074
13	0	1	0.800	0.702	0.098
14	1	4	0.800	0.719	0.081
15	2	0	0.815	0.789	0.026
16	1	4	0.846	0.789	0.057
17	6	2	0.861	0.860	0.001
18	1	0	0.954	0.895	0.059
20	2	1	0.969	0.895	0.074
22	0	1	1.000	0.912	0.088
23	0	3	1.000	0.930	0.070
24	0	1	1.000	0.982	0.018
>24	0	0	1.000	1.000	0.000

12.6 THE WALD-WOLFOWITZ AND WILCOXON-MANN-WHITNEY TESTS

A An ingenious method of testing the hypothesis that two samples have been drawn from populations with the same continuous distribution function was constructed by Wald and Wolfowitz [2]. This test is based upon the theory of runs, discussed in the previous chapter.

Let x_{1l} $(l = 1, \ldots, n_1)$ and x_{2j} $(j = 1, \ldots, n_2)$ be two independent simple samples drawn from two populations in which the characteristic X is a random variable with the same continuous distribution function. Since $P(X_{1l} - X_{1l'} = 0) = P(X_{2j} - X_{2j'} = 0) = P(X_{1l} - X_{2j} = 0) = 0$, we can assume that $x_{1l} \neq x_{1l'}$ $(l \neq l')$, $x_{2j} \neq x_{2j'}$ $(j \neq j')$ and $x_{1l} \neq x_{2j}$ $(l = 1, \ldots, n_1, j = 1, \ldots, n_2)$. Let us arrange the values x_{1l} and x_{2j} in increasing order, that is, let us form from the sequences x_{1l} and x_{2j} the sequence z_m $(m = 1, \ldots, n_1 + n_2)$, where

(12.6.1) $z_1 < z_2 < \ldots < z_{n_1+n_2}.$

Let us now assign the number zero to those z_m which are elements of the first sample, and to the remaining z_m the number one. Thus, we obtain a zero-one sequence $\{y_m\}$ containing n_1 0's and n_2 1's. If the hypothesis H_0 that both samples were taken from populations with the same continuous distribution function is true, all the combinations of n_1 0's and n_2 1's have the same probability of appearance. Thus, the theory of runs may serve to test such a hypothesis.

We now have only to decide how to apply the theory of runs. Thus, we are inclined to reject hypothesis H_0 if the total number of runs is significantly smaller than the expected number of runs, since here long runs appear; hence the elements of either sample have a tendency to appear in concentration, which would indicate that the characteristic X does not have the same distribution in both populations. Generally, the bigger the number of runs and the shorter the runs, the more we may expect that our hypothesis is true. Thus the test procedure consists in choosing the largest integer value k_0 such that $P(K \leqslant k_0) \leqslant \alpha$, where α is the significance level, and rejecting the hypothesis H_0 if the observed $k \leqslant k_0$.

Example 12.6.1. Two independent simple samples of size four have been drawn from two populations in order to test the hypothesis H_0 that the characteristic X has the same continuous distribution function in both populations. The following values were obtained:

$$x_{11} = 3, \qquad x_{12} = 3.5, \qquad x_{13} = 6.5, \qquad x_{14} = 2.5,$$
$$x_{21} = 7, \qquad x_{22} = 2.9, \qquad x_{23} = 4.0, \qquad x_{24} = 6.0.$$

We obtain

$$z_1 = 2.5, \qquad z_2 = 2.9, \qquad z_3 = 3, \qquad z_4 = 3.5,$$
$$z_5 = 4, \qquad z_6 = 6, \qquad z_7 = 6.5, \qquad z_8 = 7.$$

Thus we have the following zero-one sequence $\{y_m\}$:

$$0, 1, 0, 0, 1, 1, 0, 1.$$

Using the notation of Chapter 11, we have

$$k_{11} = 2, \qquad k_{12} = 1, \qquad k_{13} = k_{14} = 0,$$
$$k_{21} = 2, \qquad k_{22} = 1, \qquad k_{23} = k_{24} = 0,$$
$$k_1 = k_2 = 3, \qquad k = 6.$$

We observe that the number of elements in the particular samples are not random variables; thus, to compute the required probabilities, we use formulas (11.3.12). We have

$$P(K = 2 \mid N_1 = 4, N_2 = 4) = 0.029,$$
$$P(K = 3 \mid N_1 = 4, N_2 = 4) = 0.086.$$

If we test hypothesis H_0 at the significance level $\alpha = 0.05$, we find that the critical value $k_0 = 2$. Since $k = 6$ was observed, there is no reason to reject H_0. We also observe that, according to (11.4.4), the expected value $E(K \mid N_1 = 4$, $N_2 = 4) = 5$, which is close to the observed k.

B   Another nonparametric test, with many applications, has been constructed by Wilcoxon [1] and Mann and Whitney [1].

Let X_1 and X_2 be independent random variables and let the distribution function $F_i(x)$ of X_i $(i = 1, 2)$ be continuous. Let us first assume that for every real x we have $F_1(x) = F_2(x)$. Further, let x_{1l} $(l = 1, \ldots, n_1)$, x_{2j} $(j = 1, \ldots, n_2)$ and y_m $(m = 1, \ldots, n_1 + n_2)$ have the same meaning as in Section 12.6, A. Let us compute how many 0's of the zero-one sequence $\{y_m\}$ are preceded by each 1, and next find the sum of these numbers. Denote this sum by u. In other words, u is the number of pairs (x_{1l}, x_{2j}) for which $x_{2j} < x_{1l}'$. This is the observed value u of the *Mann-Whitney statistic U.*

It is very cumbersome to find the exact distribution of the statistic U. Mann and Whitney [1] set up tables of the distribution of U for $n_1 \leqslant n_2 \leqslant 8$, using the recursion formula which we now derive.

Let us denote by $\bar{p}_{n_1 n_2}(u)$ and $p_{n_1 n_2}(u)$, respectively, the number of sequences $\{y_m\}$ with n_1 0's and n_2 1's in which all the 1's precede u 0's and the probability of such a sequence. Since the total number of sequences of n_1 0's and n_2 1's equals $\binom{n_1 + n_2}{n_1}$ and each sequence has the same probability, we have the relation

$$(12.6.2) \qquad p_{n_1 n_2}(u) = \frac{n_1! \, n_2!}{(n_1 + n_2)!} \, \bar{p}_{n_1 n_2}(u).$$

We observe that $\bar{p}_{n_1 n_2}(u)$ satisfies the following recursive relation:

$$(12.6.3) \qquad \bar{p}_{n_1 n_2}(u) = \bar{p}_{n_1-1, n_2}(u - n_2) + \bar{p}_{n_1, n_2-1}(u),$$

where

$$\bar{p}_{n_1 n_2}(u) = 0 \qquad \text{for } u < 0 \qquad \text{and} \quad u > n_1 n_2.$$

Formula (12.6.3) can be proved as follows: Let us split the sequences being considered into two groups. The first group consists of those sequences which have 0 in the last place on the right, the second, of those which have 1 in this place. Let us denote by $A_{n_1 n_2}(u)$ and $B_{n_1 n_2}(u)$ the number of sequences of the first and second groups, respectively, in which the 1's precede u 0's. We have

$$(12.6.4) \qquad \bar{p}_{n_1 n_2}(u) = A_{n_1 n_2}(u) + B_{n_1 n_2}(u).$$

Let us delete the last 0 from the sequences of the first group. Now these sequences have $(n_1 - 1)$ 0's and n_2 1's each, hence each 1 precedes one less 0 than in the original sequence. Since each sequence has n_2 1's,

$$(12.6.5) \qquad A_{n_1 n_2}(u) = \bar{p}_{n_1-1, n_2}(u - n_2).$$

Now let us delete the last 1 from the sequences of the second group. These sequences will then have n_1 0's and $(n_2 - 1)$ 1's. Since taking away the number 1 from the last place on the right does not change the value of u, we have

$$(12.6.6) \qquad B_{n_1 n_2}(u) = \bar{p}_{n_1, n_2-1}(u).$$

Formula (12.6.3) follows from (12.6.4) to (12.6.6).

Using the method proposed by van Dantzig [1], we now derive formulas for $E(U)$ and $D^2(U)$ without the assumption that $F_1(x) \equiv F_2(x)$.

Let us set

$$(12.6.7) \qquad p = P(X_2 - X_1 < 0) = P(X_2 - X_1 \leqslant 0).$$

From (6.5.7) we have

$$(12.6.8) \qquad p = \int_{-\infty}^{+\infty} F_2(x)\, dF_1(x).$$

Let us now define the functions $g(z)$ and ξ_{lj} as follows:

$$(12.6.9) \qquad g(z) = \begin{cases} 1 & \text{for } z > 0, \\ 0 & \text{for } z \leqslant 0, \end{cases}$$

$$(12.6.9') \qquad \xi_{lj} = g(x_{1l} - x_{2j}), \ (l = 1, \ldots, n_1, j = 1, \ldots, n_2).$$

Thus ξ_{lj} equals one if $x_{1l} > x_{2j}$; it follows that the Mann-Whitney statistic can be written in the form

$$(12.6.10) \qquad U = \sum_{l=1}^{n_1} \sum_{j=1}^{n_2} \xi_{lj}.$$

From (12.6.7) to (12.6.9') we obtain

$$(12.6.11) \qquad E(\xi_{lj}) = E(\xi_{lj}^2) = p.$$

Since, for $l \neq l_1$, and $j \neq j_1$, ξ_{lj} and $\xi_{l_1 j_1}$ are independent,

$$(12.6.12) \qquad E(\xi_{lj} \xi_{l_1 j_1}) = p^2 \qquad (l \neq l_1, j \neq j_1).$$

Let us next define

$$(12.6.13) \qquad \gamma^2 = \int_{-\infty}^{+\infty} [F_2(x) - p]^2\, dF_1(x),$$

$$(12.6.14) \qquad \phi^2 = \int_{-\infty}^{+\infty} [F_1(x) - (1 - p)]^2\, dF_2(x).$$

Suppose now that $j \neq j_1$. The product $\xi_{lj} \xi_{lj_1}$ equals one if and only if $x_{2j} - x_{1l} < 0$ and $x_{2j_1} - x_{1l} < 0$; in the contrary case $\xi_{lj}\xi_{lj_1} = 0$. Let us find the probability of this event. We have

$$P(X_{2j} < X_{1l}, X_{2j_1} < X_{1l}) = \int_{-\infty}^{+\infty} \left[\int_{-\infty}^{x} dF_2(t) \right]^2 dF_1(x)$$

$$= \int_{-\infty}^{+\infty} [F_2(x)]^2 \, dF_1(x).$$

As we see from (12.6.8) and (12.6.13), the integral on the right-hand side of the last equality is $\gamma^2 + p^2$; hence

(12.6.15) $\qquad E(\xi_{lj}\xi_{lj_1}) = \gamma^2 + p^2 \quad (j \neq j_1).$

Similarly,

(12.6.16) $\qquad E(\xi_{lj}\xi_{l_1j}) = \phi^2 + p^2 \quad (l \neq l_1).$

It follows from (12.6.10) to (12.6.12), (12.6.15), and (12.6.16) that

(12.6.17) $\quad E(U) = n_1 n_2 p,$

$$E(U^2) = \sum_{l=1}^{n_1} \sum_{l_1=1}^{n_1} \sum_{j=1}^{n_2} \sum_{j_1=1}^{n_2} E(\xi_{lj}\xi_{l_1j_1})$$

$$= n_1 n_2 [p + (n_2 - 1)(\gamma^2 + p^2)$$

$$+ (n_1 - 1)(\phi^2 + p^2) + (n_1 - 1)(n_2 - 1)p^2].$$

From the last formula and (12.6.11) we obtain

(12.6.18) $\quad D^2(U) = n_1 n_2 [(n_1 - 1)\phi^2 + (n_2 - 1)\gamma^2 + p(1 - p)].$

If $F_1(x) \equiv F_2(x)$, then, applying integration by parts in (12.6.8), (12.6.13), and (12.6.14), we obtain

$$p = \tfrac{1}{2}, \qquad \gamma^2 = p^2 = \tfrac{1}{12}.$$

Then from (12.6.27) and (12.6.18) we have

(12.6.19) $\qquad\qquad E(U) = \dfrac{n_1 n_2}{2},$

$$D^2(U) = \dfrac{n_1 n_2 (n_1 + n_2 + 1)}{12}.$$

Let us consider the statistic

(12.6.20) $\qquad\qquad V = \dfrac{U - E(U)}{\sqrt{D^2(U)}}.$

Mann and Whitney [1] established that in the particular case $F_1(x) \equiv F_2(x)$ the statistic V has the asymptotically normal distribution $N(0; 1)$

as $n_1 \to \infty$, $n_2 \to \infty$. Lehmann [2] proved the same assertion for the general case $F_1(x) \not\equiv F_2(x)$, assuming that $0 < p < 1$ and $n_2/n_1 \to s \ (s > 0)$ as $n_1 \to \infty$. Tables by Mann and Whitney [1] show that if $F_1(x) \equiv F_2(x)$, the approximation is sufficiently good for $n_1 = n_2 = 8$.

We now discuss a test based upon the statistic U. To do this, we introduce the following definition.

Definition 12.6.1. Let the random variable X_i $(i = 1, 2)$ have the distribution function $F_i(x)$. We say that X_2 is *stochastically bigger* than X_1 if for every real x such that $0 < F_1(x) < 1$,

$$(12.6.21) \qquad\qquad F_1(x) \geqslant F_2(x).$$

Let us now consider two random variables X_1 and X_2 and suppose that we want to check whether X_2 is stochastically bigger than X_1. We assume the hypothesis H_0 that the equation $F_1(x) = F_2(x)$ holds for every real x. This hypothesis will be tested by Mann-Whitney's statistic U in the following way. We draw n_i observations from the ith population $(i = 1, 2)$. Suppose at first that n_1 and n_2 do not exceed 8. From the tables of Mann-Whitney [1] we find the value u_α such that if H_0 is true, then $P(U \leqslant u_\alpha) \leqslant \alpha$, where α is the significance level. Next, from the samples obtained we find the value u of the statistic U. If $u \leqslant u_\alpha$, we reject H_0, since the fact that the number of observed pairs (x_{1i}, x_{2j}) for which $x_{2j} < x_{1i}$ is small seems to indicate that X_2 is stochastically bigger than X_1.

If n_1 and n_2 exceed 8, we use the normal approximation, that is, we compute the observed value v of V given by (12.6.20), where $E(U)$ and $D^2(U)$ are given by (12.6.19), and we reject H_0 if $v \leqslant v_\alpha$, where $\Phi(v_\alpha) = \alpha$ and α is the significance level.

Example 12.6.2. Table 12.6.1 gives the areas in acres (1 acre = 0.40468 ha) with wheat in 1936 and 1937 in thirty-four villages of Lucknow county in India. These villages form a simple sample drawn from the population of villages in Lucknow county. The hypothesis to be tested is that the random variable "area with wheat" has the same distribution in 1936 and 1937. To test this hypothesis, we apply the Mann-Whitney test. From Table 12.6.1, after arranging the numbers x_{1l} and x_{2j} in increasing order and assigning the numbers one and zero, respectively, to the observations for 1936 and 1937, we obtain the zero-one sequence $\{y_m\} = 0110110100000111011011011100001001110100110001$ 0001111011011000110010.

Hence we have $u = 596$ and

$$v = \frac{596 - 17 \cdot 34}{\sqrt{(34 \cdot 34 \cdot 69)/12}} = \frac{18}{17\sqrt{23}} = 0.221.$$

Since $\Phi(0.221) \cong 0.5875$, there is no reason to reject the verified hypothesis.

We stress the fact that in Table 12.6.1 some numbers appear several times. This

TABLE 12.6.1

Village number	1936 x_{1l}	1937 x_{2j}	Village number	1936 x_{1l}	1937 x_{2j}	Village number	1936 x_{1l}	1937 x_{2j}
1	75	52	13	125	111	25	129	103
2	163	149	14	5	6	26	192	179
3	326	289	15	427	399	27	663	330
4	442	381	16	78	79	28	236	219
5	254	278	17	78	105	29	73	62
6	125	111	18	45	27	30	62	79
7	559	634	19	564	515	31	71	60
8	254	278	20	238	249	32	137	100
9	101	112	21	92	85	33	196	141
10	359	355	22	247	221	34	255	265
11	109	99	23	134	133			
12	481	498	24	131	144			

is because the measurements were made with an accuracy up to $\frac{1}{2}$ acre. The numerical data of this example are taken from the book by Sukhatme [3].

We return to the Mann-Whitney test in Chapter 16.

Wilcoxon [1], before Mann and Whitney, considered, for $n_1 = n_2$, a statistic which is linearly related to the statistic U of Mann-Whitney, namely, the statistic $S = n_2(2n_1 + n_2 + 1)/2 - U$. Thus in the literature the test discussed here is called the *Wilcoxon-Mann-Whitney test*. Let us remark that $S = m_1 + \ldots + m_{n_2}$, where the $m_r (r = 1, \ldots, n_2)$ are the ranks of those y_m which correspond to x_{2j}.

Significance probabilities for the Wilcoxon-Mann-Whitney test have been prepared by Fix and Hodges [1].

Generalizations of this test to k samples $(k > 2)$ have been given by Kruskal [1], Rijkoort [1], and Pfanzagl [1].

C   The theory of nonparametric tests has developed very much during recent years. Fraser [1] recently wrote a large monograph on this theory. We also mention here the papers of Wolfowitz [4], Schmetterer [3], and Gihman, Gnedenko, and Smirnov [1], which contain much information on important new research in this domain.

Besides pure-parametric and pure-nonparametric tests, there are considered in the literature tests of a mixed character, tests such that the functional forms of the distribution functions considered are unknown, but the subject of the hypothesis to be tested is not the functional form of the distribution functions but the comparison of the values of unknown parameters of several populations. Such tests, concerning the expected values and variances of several populations with unknown distribution functions, have been constructed by Mosteller [1], Sadowski [1], Rosenbaum [1], and Lukaszewicz and Sadowski [1]. (Paulson [1] considered similar problems for several normal populations).

12.7 INDEPENDENCE TESTS BY CONTINGENCY TABLES

Let us imagine that the elements of a population have been classified according to two characteristics X and Y, and that the domains of values of these characteristics are split into r and s groups, respectively. Let us denote by n the sample size, and by n_{ik} the number of sample elements that belong to the ith group according to the characteristic X ($i = 1, \ldots, r$) and to the kth group according to the characteristic Y ($k = 1, \ldots, s$). Let us further define

$$(12.7.1) \qquad n_{i.} = \sum_{k=1}^{s} n_{ik},$$

$$(12.7.2) \qquad n_{.k} = \sum_{i=1}^{r} n_{ik}.$$

We have the obvious equality

$$n = \sum_{i=1}^{r} \sum_{k=1}^{s} n_{ik}.$$

The elements of the sample, classified in this way, may be represented in the form of Table 12.7.1, which is called a *contingency table*.

It is our aim to test the hypothesis H_0 that the characteristics X and Y of the population from which our sample was drawn are independent. Thus, if p_{ik} denotes the probability that an element chosen at random

TABLE 12.7.1

i	k				Totals
	1	2	...	s	$n_{i.}$
1	n_{11}	n_{12}	...	n_{1s}	$n_{1.}$
2	n_{21}	n_{22}	...	n_{2s}	$n_{2.}$
.					.
.	...		...		.
.					.
r	n_{r1}	n_{r2}	...	n_{rs}	$n_{r.}$
Totals					
$n_{.k}$	$n_{.1}$	$n_{.2}$	...	$n_{.s}$	n

belongs to the ith group of values of X and to the kth group of values of Y, and if $p_{i.}$ and $p_{.k}$ denote, respectively, the marginal probabilities, then according to H_0, for every pair (i, k),

(12.7.3) $$p_{ik} = p_{i.}p_{.k},$$

where

(12.7.4) $$\sum_{i=1}^{r} p_{i.} = \sum_{k=1}^{s} p_{.k} = 1.$$

This hypothesis does not specify the values of the $r + s$ unknown parameters, that is, the numbers $p_{i.}$ and $p_{.k}$. Since two of them may be determined from (12.7.4), there remain $r + s - 2$ unknown parameters. We determine the values of the unknown parameters by the maximum likelihood method. We have

$$L = \prod_{i=1}^{r} \prod_{k=1}^{s} p_{ik}^{n_{ik}} = \prod_{i=1}^{r} \prod_{k=1}^{s} (p_{i.}p_{.k})^{n_{ik}} = \prod_{i=1}^{r} \prod_{k=1}^{s} p_{i.}^{n_{ik}} p_{.k}^{n_{ik}}.$$

Substituting (12.7.1) and (12.7.2), we obtain

$$L = \prod_{i=1}^{r} p_{i.}^{n_{i.}} \prod_{k=1}^{s} p_{.k}^{n_{.k}}.$$

From (12.7.4) we determine p_r and $p_{.s}$, namely,

$$p_{r.} = 1 - \sum_{i=1}^{r-1} p_{i.}, \qquad p_{.s} = 1 - \sum_{k=1}^{s-1} p_{.k}.$$

Hence

$$L = \left(1 - \sum_{i=1}^{r-1} p_{i.}\right)^{n_{r.}} \left(1 - \sum_{k=1}^{s-1} p_{.k}\right)^{n_{.s}} \prod_{i=1}^{r-1} p_{i.}^{n_{i.}} \prod_{k=1}^{s-1} p_{.k}^{n_{.k}}$$

and

$$\log L = n_{r.} \log \left(1 - \sum_{i=1}^{r-1} p_{i.}\right) + n_{.s} \log \left(1 - \sum_{k=1}^{s-1} p_{.k}\right)$$

$$+ \sum_{i=1}^{r-1} n_{i.} \log p_{i.} + \sum_{k=1}^{s-1} n_{.k} \log p_{.k}.$$

We have the system of equations

$$\frac{\partial \log L}{\partial p_{i.}} = - \frac{n_{r.}}{1 - \sum\limits_{i=1}^{r-1} p_{i.}} + \frac{n_{i.}}{p_{i.}} = \frac{n_{i.}}{p_{i.}} - \frac{n_{r.}}{p_{r.}} = 0 \qquad (i = 1, 2, \ldots, r - 1),$$

$$\frac{\partial \log L}{\partial p_{.k}} = - \frac{n_{.s}}{1 - \sum\limits_{k=1}^{s-1} p_{.k}} + \frac{n_{.k}}{p_{.k}} = \frac{n_{.k}}{p_{.k}} - \frac{n_{.s}}{p_{.s}} = 0 \qquad (k = 1, 2, \ldots, s - 1).$$

Let $A = n_{r.}/p_{r.}$, $B = n_{.s}/p_{.s}$. Then

$$p_{i.} = n_{i.}/A \quad (i = 1, 2, \ldots, r),$$
$$p_{.k} = n_{.k}/B \quad (k = 1, 2, \ldots, s).$$

From equalities (12.7.4) we obtain

$$\sum_{i=1}^{r} p_{i.} = \sum_{i=1}^{r} n_{i.}/A = n/A = 1,$$

$$\sum_{k=1}^{s} p_{.s} = \sum_{k=1}^{s} n_{.k}/B = n/B = 1;$$

hence $A = B = n$. Finally,

(12.7.5) $$p_{i.} = n_{i.}/n \quad (i = 1, 2, \ldots, r),$$
$$p_{.k} = n_{.k}/n \quad (k = 1, 2, \ldots, s).$$

Considering (12.4.2), (12.7.3), and (12.7.5), we obtain

(12.7.6) $$\chi^2 = n \sum_{i=1}^{r} \sum_{k=1}^{s} \frac{(n_{ik} - n_{i.}n_{.k}/n)^2}{n_{i.}n_{.k}}.$$

Since $r + s - 2$ parameters have been determined from the sample, it follows that, according to Fisher's theorem (see Section 12.4), the random variable χ^2 defined by (12.7.6) has approximately the χ^2 distribution with

$$rs - (r + s - 2) - 1 = rs - r - s + 1 = (r - 1)(s - 1)$$

degrees of freedom. Thus we may apply the χ^2 test.

If $r = s = 2$, formula (12.7.6) for χ^2 takes the form

(12.7.7) $$\chi^2 = \frac{n(n_{11}n_{22} - n_{12}n_{21})^2}{n_{1.}n_{2.}n_{.1}n_{.2}}.$$

Example 12.7.1. The hypothesis H_0 asserts that the color of eyes X and the color of hair Y of men are independent. We test this hypothesis by using the data on the color of eyes and hair for 6800 men, which are given in Table 12.7.2. These data are taken from the book by Kendall [1]. The sample is split into three groups according to the color of eyes, and four groups according to the color of hair.

TABLE 12.7.2

i (eyes) \ k (hair)	1 (fair)	2 (brown)	3 (black)	4 (red)	Totals $n_{i.}$
1 (blue)	1768	807	189	47	2811
2 (hazel or green)	946	1387	746	53	3132
3 (brown)	115	438	288	16	857
Totals $n_{.k}$	2829	2632	1223	116	6800

From this table we compute

$$\chi^2 = 6800 \sum_{i=1}^{3} \sum_{k=1}^{4} \frac{(n_{ik} - n_{i.}n_{.k}/6800)}{n_{i.}n_{.k}} = 1075.2.$$

The number of degrees of freedom equals $(r - 1)(s - 1) = 2 \times 3 = 6$. From tables of the χ^2 distribution we find that for six degrees of freedom $P(\chi^2 \geqslant 1075.2) < 0.000001$. Thus we reject H_0.

Problems and Complements

12.1. In two simple samples from two normal populations with distributions $N(0; \sigma_1)$ and $N(0; \sigma_2)$, where σ_1 and σ_2 are unknown, we observed $s_1 = 1$ and $s_2 = 3$. The sample sizes are $n_1 = 9$ and $n_2 = 7$, respectively. At the significance level $\alpha = 0.02$, test the hypotheses
(a) H_0 $(\sigma_1 = \sigma_2 = 1)$, (b) H_0 $(\sigma_1 = 1, \sigma_2 = 2)$ (c) H_0 $(\sigma_1 = \frac{1}{2}, \sigma_2 = 1)$.
Hint: Use Problem 9.7.

12.2. The following sample correlation coefficients were obtained from a simple sample of size 7 drawn from a three-dimensional normal population: $r_{12} = 0.2$, $r_{13} = 0.4$, $r_{23} = -0.35$. Test the hypothesis $H_0(\rho_{12} = \rho_{13} = \rho_{23} = 0)$ at the significance level $\alpha = 0.05$.
Hint: Use Problem 9.10.

12.3. The sample partial correlation coefficient $r_{12\cdot3} = 0.26$ was obtained from a simple sample of size 6 drawn from a three-dimensional normal population. Test, at the significance level $\alpha = 0.05$, the hypothesis that the population partial correlation coefficient $\rho_{12\cdot3}$ equals zero.
Hint: Use Problem 9.12.

12.4. From a simple sample of size 10 drawn from a four-dimensional normal population we obtained $r_{1(234)} = 0.3$. Does this result contradict, at the significance level $\alpha = 0.01$, the hypothesis that $\rho_{12} = \rho_{13} = \rho_{14} = \rho_{23} = \rho_{24} = 0$?
Hint: Use Problem 9.13.

12.5. Svedberg, observing a thin layer of gold solution, registered at equal time intervals the number of gold particles in the microscope area. In the following table, n_j denotes the number of time intervals during which Svedberg registered j gold particles. Treating the n_j as values of a random variable N, test, at the significance level $\alpha = 0.01$, the hypothesis that N has the Poisson distribution with parameter λ equal to the observed average number of particles per unit of time.

j	0	2	3	4	5	6	7	8	Total
n_j	112	168	130	68	32	5	1	1	517

12.6. Show that the statistic χ^2 defined by (12.4.2) has the following moments:

$$E(\chi^2) = r - 1,$$
$$D^2(\chi^2) = 2(r - 1) + \frac{A - (r^2 + 2r - 2)}{n},$$

where $A = \sum_{k=1}^{r} \frac{1}{\pi_k}$. In particular, if $\pi_k = 1/r$ $(k = 1, \ldots, r)$,

$$D^2(\chi^2) = \frac{2(r - 1)(n - 1)}{n}.$$

12.7. Let r in (12.4.2) increase to infinity as $n \to \infty$. Show that if

$$\lim_{n \to \infty} \min_{1 \leqslant k \leqslant r} n\pi_k = \infty,$$

then the statistic $(\chi^2 - r)/\sqrt{2r}$ has the asymptotically normal distribution $N(0; 1)$ (Tumanian [1], generalized by Gihman [4]).

12.8. Knee jerk measurements were performed on ten patients under the following two conditions: (I) the subjects squeezed a hand dynamometer just before the stimulus struck the knee, (II) the knee jerk was obtained in a relaxed posture. The magnitude of the knee jerk measured in angular units is given in the following table (from Guilford [1]):

I	19	19	26	15	18	30	18	30	26	28
II	14	19	30	7	13	20	17	29	18	21

Applying the Mann-Whitney test, at the significance level $\alpha = 0.05$, check whether the magnitude of the knee jerk has the same distribution under conditions (I) and (II).

12.9. Each of the elements of a population of size N may have both characteristics A and B, have A and not B, have B and not A, and have neither A nor B. The sizes of the groups are N_{11}, N_{12}, N_{21}, and N_{22}, respectively. The expression

$$q = \frac{N_{11}N_{22} - N_{12}N_{21}}{N_{11}N_{22} + N_{12}N_{21}}$$

is called the *association coefficient* (Yule [1, 2]).

Treating the possession of characteristic A as a random event A, and the possession of characteristic B as a random event B, show that if events A and B are independent, then $q = 0$. How should we interpret the case $q = 1$ and $q = -1$?

12.10. (Continued). Let M_{11}, M_{12}, M_{21}, and M_{22} denote the number of elements of the respective categories in simple samples of size n drawn from the population of Problem 12.9. The *sample association coefficient* has the form

$$S = \frac{M_{11}M_{22} - M_{12}M_{21}}{M_{11}M_{22} + M_{12}M_{21}}.$$

Show that

$$D^2(S) = \frac{(1 - q^2)^2}{4} \left(\frac{1}{N_{11}} + \frac{1}{N_{12}} + \frac{1}{N_{21}} + \frac{1}{N_{22}} \right).$$

12.11. Compute the sample association coefficient between inoculation against cholera and exemption from attack from the following table (Greenwood and Yule [1]):

	Not affected	Affected	Totals
Inoculated	276	3	279
Not inoculated	473	66	539
Totals	749	69	818

Compute $D^2(S)$, replacing the theoretical by the observed sizes.

CHAPTER 13

The Theory of Estimation

13.1 PRELIMINARY NOTIONS

In Sections 12.4 and 12.7 we dealt with estimates of unknown parameters of the population on the basis of samples and we mentioned a method of obtaining estimates. The general theory dealing with such·matters is called the *theory of estimation.*

The procedure of estimating an unknown parameter on the basis of a sample is the following: On the basis of a sample we determine the value u of a certain statistic U whose distribution depends upon this parameter and we take this value u as an estimate of the unknown parameter. We speak then of *point estimation.* Both the statistic U and its observed value u are called the *estimate of* the unknown parameter. It will usually be clear from the context whether the term estimate pertains to U or to u.

We may also look for some intervals that contain the unknown parameter. We speak then of *interval estimation.*

We start our detailed discussion with a classification of estimates.

13.2 CONSISTENT ESTIMATES

Let $F(x, Q)$ be the distribution function of the characteristic X of elements of a population with unknown value of the parameter Q.

Let us consider the estimate U_n of the parameter Q, where n denotes the size of the sample. It is obvious that, in general, the estimate of the unknown parameter computed from the sample will differ from the true value of the parameter. However, it is desirable that, as the size of the sample increases to infinity, the probability of the estimate being close to the true value should tend to one.

Definition 13.2.1. We say that the sequence $\{U_n\}$ $(n = 1, 2, \ldots)$ of estimates of the unknown parameter Q is *consistent* if $\{U_n\}$ converges stochastically to Q as $n \to \infty$.

If U_n is a term of a consistent sequence $\{U_n\}$ of estimates of parameter Q, then we say, in brief, that U_n is a consistent estimate of Q.

By the definition of stochastic convergence (see Section 6.2), if the sequence $\{U_n\}$ is consistent, then for every $\varepsilon > 0$ we have

(13.2.1) $$\lim_{n \to \infty} P(|U_n - Q| > \varepsilon) = 0.$$

We often denote the estimate of Q by Q^*. We now give some examples of consistent estimates.

Example 13.2.1. The characteristic X of elements of a population has the normal distribution $N(m; 1)$ with unknown expected value m. We take as an estimate of m from simple samples of size n the statistic

$$\bar{X} = \frac{1}{n} \sum_{k=1}^{n} X_k,$$

where the random variables X_k are independent and have the same distribution as the characteristic X. As we know (see Section 9.11), $\bar{X}$ is stochastically convergent to m, hence it is a consistent estimate of m.

Example 13.2.2. Another consistent estimate of the unknown expected value m of a normal population is the sample median, the observed value y of the order statistic $\zeta_k^{(n)}$, where $k = [\frac{1}{2} n] + 1$. The median is a consistent estimate, since for the normal distribution the expected value coincides with the median and, by theorem 10.4.1, the sample median converges stochastically to the median of the population.

From these two examples we see that the median, as well as the sample mean, is a consistent estimate of m.

We could ask which of them is better, that is, which of them more often gives a value of m^* close to the true value m. We could also ask which of them is better in the sense that its expected value $E(m^*)$ is closer to the true value m. We now investigate these problems.

13.3 UNBIASED ESTIMATES

A   The consistency of an estimate expresses its limit properties; hence it can serve as a characterization of the estimate only if the sample is large. We now present the notion of an unbiased estimate, which refers to estimates from samples of arbitrary size.

In a particular sample, the estimate of an unknown parameter may differ from the true value of the parameter. However, if we have estimates from a series of samples, then it is desirable that the average value of these estimates should be close to the true value of the unknown parameter.

Definition 13.3.1. Let D be a class of distribution functions $F(x, Q)$.

The estimate U_n of the parameter Q is an *unbiased estimate for the class D* if for every distribution function $F(x, Q)$ of D

$$(13.3.1) \qquad E(U_n) = Q \qquad (n = 1, 2, 3, \ldots).$$

The difference $E(U_n) - Q$ is called the *bias* of the estimate.

The estimate U_n of Q is an *asymptotically unbiased estimate for the class D* if for every distribution function $F(x, Q)$ of D

$$(13.3.1') \qquad \lim_{n \to \infty} E(U_n) = Q.$$

Example 13.3.1. Let us consider a population in which the characteristic X has an arbitrary distribution whose first moment exists. We want to estimate the unknown expected value $E(X) = m$.

Let us take a simple sample of size n and put m equal to the value $\bar{x}$ of the statistic

$$\bar{X} = \frac{1}{n} \sum_{k=1}^{n} X_k,$$

where the random variables X_k have the same distribution as X.

The statistic $\bar{X}$ is an unbiased estimate of the parameter m, since

$$(13.3.2) \qquad E(\bar{X}) = E\left(\frac{1}{n} \sum_{k=1}^{n} X_k\right) = \frac{1}{n} \sum_{k=1}^{n} E(X_k) = m.$$

Thus the statistic $\bar{X}$ is an unbiased estimate of the expected value m for the class D of all distribution functions whose first moment exists.

Sample moments of order higher than one are not necessarily unbiased estimates of the population moments. We show this by an example (see also Problems 9.17 to 9.21).

Example 13.3.2. Consider the class D of distribution functions of random variables X whose second moments exist. Let the variance σ^2 of X be unknown. As an estimate of this parameter, we take the statistic

$$(13.3.3) \qquad S^2 = \frac{1}{n} \sum_{k=1}^{n} X_k^2 - \bar{X}^2,$$

where the random variables X_k are independent and have the same distribution as X. We have

$$E(S^2) = E\left(\frac{1}{n} \sum_{k=1}^{n} X_k^2\right) - E(\bar{X}^2) = E(X^2) - E(\bar{X}^2).$$

Since

$$E(\bar{X}^2) = E\left[\left(\frac{1}{n} \sum_{k=1}^{n} X_k\right)^2\right] = \frac{1}{n^2} E\left(\sum_{k=1}^{n} X_k^2 + \sum_{\substack{l,j \\ l \neq j}} X_l X_j\right)$$

$$= \frac{1}{n} E(X^2) + \frac{n-1}{n} [E(X)]^2;$$

we have

$$(13.3.4) \qquad E(S^2) = \frac{n-1}{n} E(X^2) - \frac{n-1}{n} [E(X)]^2 = \frac{n-1}{n} \sigma^2.$$

It follows that S^2 is not an unbiased estimate of the parameter σ^2. For instance, for $n = 2$ the expected value $E(S^2)$ equals $\frac{1}{2}\sigma^2$. However, S^2 is an asymptotically unbiased estimate of the parameter σ^2 for the class D of distribution functions being considered.

It is easy to see that the statistic

$$(13.3.5) \qquad S_1^2 = \frac{n}{n-1} S^2$$

is an unbiased estimate of the parameter σ^2 for the class of distribution functions whose second-order moments exist. However, we cannot conclude that in every particular experiment the observed value s_1^2 of the statistic S_1^2 gives a more correct estimate of the parameter σ^2 than the value s^2. Unbiasedness is an average property; hence if we want to obtain from a series of experiments an estimate which is good on the average, we should apply formula (13.3.5). However, in one experiment s^2 may be closer to σ^2 than s_1^2.

Of course, if n is large, the values of S^2 and S_1^2 differ but little.

It is interesting to note that if the expected value $E(X) = m$ is known, then the estimate

$$(13.3.6) \qquad S_0^2 = \frac{1}{n} \sum_{k=1}^{n} (X_k - m)^2,$$

computed from simple samples, is an unbiased estimate of the parameter σ^2. Indeed,

$$(13.3.7) \qquad E(S_0^2) = E\left(\frac{1}{n} \sum_{k=1}^{n} X_k^2\right) - m^2 = E(X^2) - m^2 = \sigma^2.$$

If U is an unbiased estimate of the parameter Q, it does not follow that every function $f(U)$ is an unbiased estimate of the parameter $f(Q)$. We illustrate this by an example.

Example 13.3.3. The characteristic X of elements of a certain population has the normal distribution $N(m; \sigma)$, where the expected value m is known but the standard deviation σ is not known. As we know, the statistic S_0^2 is an unbiased estimate of the parameter σ^2. We show that S_0 is not an unbiased estimate of σ.

Indeed, by (9.4.9) and (5.8.5),

$$(13.3.8) \qquad E(S_0) = \frac{n^{n/2}}{(2\sigma^2)^{n/2}\Gamma(\frac{1}{2}n)} \int_0^\infty \exp\left(-\frac{ns_0^2}{2\sigma^2}\right)(s_0^2)^{(n-1)/2}\, ds_0^2$$

$$= \frac{\sqrt{2}\,\Gamma(\frac{1}{2}(n+1))}{\sqrt{n}\,\Gamma(\frac{1}{2}n)}\,\sigma \neq \sigma.$$

It is easy to see that the statistic

$$\frac{\sqrt{n}\,\Gamma(\frac{1}{2}n)}{\sqrt{2}\,\Gamma(\frac{1}{2}(n+1))}\,S_0,$$

computed from simple samples, is an unbiased estimate of the parameter σ for the class D of normal distribution functions.

B   The notion of an unbiased estimate was introduced by Gauss (see also David and Neyman [1]). Unbiased estimates of parameters do not always exist, and, on the other hand, there may exist many unbiased estimates of the same parameter. The existence and uniqueness of unbiased estimates have been considered in papers of Halmos [1], Lehmann and Scheffé [1], Kolmogorov [14], Stein [2], and Schmetterer [2].

In the remainder of this book we use the shorter term "unbiased estimate," unless there is a definite reason to specify the class D for which the estimate is unbiased.

13.4   THE SUFFICIENCY OF AN ESTIMATE

The notion of the sufficiency of an estimate, introduced by Fisher [6], requires deeper consideration.

In Section 12.4 we defined the likelihood function L for random variables of the continuous and discrete types.

We now consider more closely random variables of the continuous type. Let $f(x, Q)$ be the density of the random variable X, where Q is an unknown parameter. Let X_k $(k = 1, 2, \ldots, n)$ be independent random variables with the same distribution as X. Let $(x_1, x_2, \ldots, x_n)$ be the observed value of the random variable $(X_1, X_2, \ldots, X_n)$; thus $(x_1, x_2, \ldots, x_n)$ is a point in n-dimensional sample space.

The likelihood function is defined by the formula

(13.4.1) $$L = f(x_1, Q)f(x_2, Q) \ldots f(x_n, Q).$$

Let us now introduce the variables

$$(u, y_1, y_2, \ldots, y_{n-1})$$

defined by the formulas

(13.4.2) $$\begin{aligned} u &= \Psi(x_1, x_2, \ldots, x_n), \\ y_j &= \Psi_j(x_1, x_2, \ldots, x_n) \quad (j = 1, 2, \ldots, n - 1), \end{aligned}$$

where $u, y_1, \ldots, y_{n-1}$ gives a one-to-one continuous transformation of $(x_1, \ldots, x_n)$ such that the partial derivatives

$$\frac{\partial u}{\partial x_k} \quad \text{and} \quad \frac{\partial y_j}{\partial x_k} \quad (j = 1, 2, \ldots, n - 1; \ k = 1, 2, \ldots, n)$$

are everywhere continuous.

The density function of the random variable $(U, Y_1, \ldots, Y_{n-1})$ is of the form

(13.4.3) $$f(x_1, Q)f(x_2, Q) \ldots f(x_n, Q) \cdot |J|,$$

where the variables $x_1, \ldots, x_n$ are functions of $u, y_1, \ldots, y_{n-1}$, and J is the Jacobian of the transformation inverse to transformation (13.4.2). Let us denote by $g(u, Q)$ the density of U.

Let us consider the conditional distribution of the random vector (Y_1, $Y_2, \ldots, Y_{n-1}$) under the condition that U takes on a given value u. Let us denote the density of this distribution by

(13.4.4) $$h(y_1, y_2, \ldots, y_{n-1} \mid u, Q).$$

By (13.4.3), after taking into consideration a formula analogous to (2.7.6), we obtain

(13.4.5) $f(x_1, Q) f(x_2, Q) \ldots f(x_n, Q) \cdot |J|$
$$= g(u, Q) h(y_1, y_2, \ldots, y_{n-1} \mid u, Q).$$

We take the statistic U defined by (13.4.2) as an estimate of the parameter Q. If in a certain interval A of values of Q, the partial derivatives $\partial f/\partial Q$, $\partial g/\partial Q$, and $\partial h/\partial Q$ exist, and

$$\left| \frac{\partial f}{\partial Q} \right| < F_0(x), \quad \left| \frac{\partial g}{\partial Q} \right| < G_0(u), \quad \left| \frac{\partial h}{\partial Q} \right| < H_0(y_1, y_2, \ldots, y_{n-1}, u),$$

where the integrals

$$\int_{-\infty}^{+\infty} F_0(x)\, dx, \quad \int_{-\infty}^{+\infty} G_0(u)\, du, \quad \int_{-\infty}^{+\infty} u G_0(u)\, du$$

and

$$\int_{-\infty}^{+\infty} \cdots \int_{-\infty}^{+\infty} H_0(y_1, y_2, \ldots, y_{n-1}, u)\, dy_1 \ldots dy_{n-1}$$

are finite, we say that U is a *regular* estimate of the parameter Q. From this point on, an estimate will mean a regular estimate, unless otherwise stated (see Problem 13.6).

We now present the notion of a sufficient estimate.

Definition 13.4.1. Let D be a class of distribution functions $F(x, Q)$. The statistic U is a *sufficient estimate of the parameter Q for the class D* if for every distribution function $F(x, Q)$ of D the density (13.4.4) of the conditional distribution of ($Y_1, Y_2, \ldots, Y_{n-1}$), provided $U = u$, is independent of Q (when $g(u, Q) > 0$).

From now on we say briefly *sufficient estimate* and omit the class D for which the considered estimate is sufficient, unless it is unclear from the context which class D is being considered.

Thus, if an estimate U is sufficient, the likelihood function L has the form

(13.4.6) $$L = g(u, Q) h(y_1, y_2, \ldots, y_{n-1} \mid u).$$

Let U be a sufficient estimate. For a fixed u, the conditional distribution is concentrated on the sample points of the hypersurface $U = u$. Since the conditional distribution on this hypersurface does not depend on Q, information about the precise position of a point on the hypersurface does not give any new information about the parameter Q. In other words, if U is a sufficient statistic, the conditional distribution of any other statistic U_1, provided $U = u$, is independent of Q; hence it does not give any information about Q.

We observe that the property of a sufficient estimate here stated was used by Fisher as a definition of this notion and it also explains the name *sufficient estimate*. Definition 13.4.1 given here is equivalent to Fisher's definition, as has been shown by Neyman [3] and Darmois [1].

If U is a sufficient estimate of Q, then, as can be seen from formula (13.4.6), the maximum likelihood estimate of Q depends only on the value of U, and does not depend on which set of values of the random variables $X_1, \ldots, X_n$ gave rise to the value of U.

Example 13.4.1. The characteristic X of elements of a population has the normal distribution $N(m; \sigma)$ with unknown expected value m. Take as an estimate of m

$$U = \frac{1}{n} \sum_{k=1}^{n} X_k = \bar{X},$$

where the random variables X_k are independent and have the same distribution $N(m; \sigma)$. For the interval A, where the estimate U is supposed to be regular, we may take any interval $a < m < b$. It follows from (9.5.10) that

$$(13.4.7) \quad L = f(x_1, m)f(x_2, m) \ldots f(x_n, m)$$

$$= 2 \frac{\sqrt{n}}{\sigma \sqrt{2\pi}} \exp\left[-\frac{n}{2} \cdot \frac{(\bar{x} - m)^2}{\sigma^2} \right] \frac{n^{(n-1)/2} \exp\left(-ns^2/2\sigma^2 \right) s^{n-2}}{(2\sigma^2)^{(n-1)/2} \Gamma[\frac{1}{2}(n-1)]}.$$

As we see, the likelihood function can be represented in the form of a product, where the second term on the right-hand side of (13.4.7) is the density of the random variable S (see Section 9.5) and is independent of the parameter m. Thus the statistic U is a sufficient estimate of m for the class of normally distributed random variables.

The notions introduced here for random variables of the continuous type can easily be defined for random variables of the discrete type as well.

The notion of a sufficient estimate can also be generalized to the case when several parameters are unknown (see Cramér [2], Schmetterer [1], Barankin and Katz [1], and Problems 13.11 to 13.14). Deeper and, at the same time, more general foundations of the theory of sufficient estimates are presented in the paper of Halmos and Savage [1].

13.5 THE EFFICIENCY OF AN ESTIMATE

A An important property of an estimate and more intuitive than its sufficiency is its efficiency, which was also introduced by Fisher [2, 6].

An unbiased estimate is a random variable with the expected value equal to the unknown parameter. It is obvious that the smaller the variance of the estimate, the more concentrated are its possible values around its expected value; hence also the greater is the probability that the estimate of the unknown parameter will be close to its true value. Thus it is desirable that the variance of the estimate be as small as possible. This is the property of the *most efficient estimate*.

First we derive a certain important inequality, called the Rao-Cramér inequality (Rao [1], Cramér [2]). We shall formulate it separately for random variables of the continuous type and for random variables of the discrete type.

The Rao-Cramér Inequality. *Let U be a regular unbiased estimate of the parameter Q from simple samples of n-elements.*

If the characteristic X of the population is of the continuous type with the density $f(x, Q)$,

$$(13.5.1) \quad D^2(U) = E[(U - Q)^2] \geqslant \frac{1}{n \int_{-\infty}^{+\infty} \left[\dfrac{\partial \log f(x, Q)}{\partial Q} \right]^2 f(x, Q)\, dx}.$$

If the characteristic X of the population is of the discrete type with the jump points $x_1, x_2, \ldots$ and the probability function $P(X = x_i) = p_i(Q)$ $(i = 1, 2, \ldots)$,

$$(13.5.1') \quad D^2(U) = E[(U - Q)^2] \geqslant \frac{1}{n \sum_i \left[\dfrac{d \log p_i(Q)}{dQ} \right]^2 p_i(Q)}.$$

Proof. We restrict ourselves to the proof of (13.5.1). From our assumptions concerning the regularity of the estimate U (see Section 13.4), it follows that we can differentiate the following equation with respect to Q under the integral sign:

$$\int_{-\infty}^{+\infty} f(x, Q)\, dx = \int_{-\infty}^{+\infty} \cdots \int_{-\infty}^{+\infty} h(y_1, \ldots, y_{n-1} \,|\, u, Q)\, dy_1 \ldots dy_{n-1} = 1.$$

Thus

$$(13.5.2) \quad \int_{-\infty}^{+\infty} \frac{\partial \log f(x, Q)}{\partial Q} f(x, Q)\, dx$$

$$= \int_{-\infty}^{+\infty} \cdots \int_{-\infty}^{+\infty} \frac{\partial \log h(y_1, \ldots, y_{n-1} \,|\, u, Q)}{\partial Q}$$

$$\times h(y_1, \ldots, y_{n-1} \,|\, u, Q)\, dy_1 \ldots dy_{n-1} = 0.$$

If we take the logarithmic derivative of both sides of Eq. (13.4.5), treating the left-hand side as a function of the variables x_k $(k = 1, 2, \ldots, n)$, then we have

$$(13.5.3) \quad \sum_{k=1}^{n} \frac{\partial \log f(x_k, Q)}{\partial Q} = \frac{\partial \log g(u, Q)}{\partial Q} + \frac{\partial \log h(y_1, \ldots, y_{n-1} \mid u, Q)}{\partial Q}.$$

Let us square both sides of equation (13.5.3) and, next, multiply its left and right sides by the respective sides of equation (13.4.5). We obtain

$$(13.5.4) \quad \left\{ \sum_{k=1}^{n} \left[\frac{\partial \log f(x_k, Q)}{\partial Q} \right]^2 + \sum_{\substack{l,j \\ l \neq j}} \frac{\partial \log f(x_l, Q)}{\partial Q} \cdot \frac{\partial \log f(x_j, Q)}{\partial Q} \right\}$$

$$\times f(x_1, Q) f(x_2, Q) \ldots f(x_n, Q)$$

$$= \left[\left(\frac{\partial \log g}{\partial Q} \right)^2 + \left(\frac{\partial \log h}{\partial Q} \right)^2 + 2 \frac{\partial \log g}{\partial Q} \cdot \frac{\partial \log h}{\partial Q} \right]$$

$$\times g(u, Q) h(y_1, \ldots, y_{n-1} \mid u, Q).$$

We have

$$(13.5.5) \quad \int_{-\infty}^{+\infty} \ldots \int_{-\infty}^{+\infty} \sum_{k=1}^{n} \left[\frac{\partial \log f(x_k, Q)}{\partial Q} \right]^2 f(x_1, Q) \ldots f(x_n, Q) \, dx_1 \ldots dx_n$$

$$= n \int_{-\infty}^{+\infty} \left[\frac{\partial \log f(x, Q)}{\partial Q} \right]^2 f(x, Q) \, dx.$$

The left-hand side of (13.5.2) gives

$$(13.5.6) \quad \int_{-\infty}^{+\infty} \ldots \int_{-\infty}^{+\infty} \sum_{\substack{l,j \\ l \neq j}} \frac{\partial \log f(x_l, Q)}{\partial Q} \cdot \frac{\partial \log f(x_j, Q)}{\partial Q}$$

$$\times f(x_1, Q) f(x_2, Q) \ldots f(x_n, Q) \, dx_1 \ldots dx_n = 0.$$

The right-hand side of (13.5.2) gives

$$(13.5.7) \quad \int_{-\infty}^{+\infty} g(u, Q) \frac{\partial \log g(u, Q)}{\partial Q} \, du$$

$$\times \int_{-\infty}^{+\infty} \ldots \int_{-\infty}^{+\infty} \frac{\partial \log h}{\partial Q} h(y_1, \ldots, y_{n-1} \mid u, Q) \, dy_1 \ldots dy_{n-1} = 0.$$

Integrating both sides of equation (13.5.4) and using formulas (13.5.5) to (13.5.7), we obtain

$$(13.5.8) \quad n \int_{-\infty}^{+\infty} \left[\frac{\partial \log f(x, Q)}{\partial Q} \right]^2 f(x, Q) \, dx$$

$$= \int_{-\infty}^{+\infty} \left[\frac{\partial \log g(u, Q)}{\partial Q} \right]^2 g(u, Q) \, du + \int_{-\infty}^{+\infty} g(u, Q) \, du$$

$$\times \int_{-\infty}^{+\infty} \ldots \int_{-\infty}^{+\infty} \left(\frac{\partial \log h}{\partial Q} \right)^2 h(y_1, \ldots, y_{n-1} \mid u, Q) \, dy_1 \ldots dy_{n-1}$$

$$\geq \int_{-\infty}^{+\infty} \left[\frac{\partial \log g(u, Q)}{\partial Q} \right]^2 g(u, Q) \, du.$$

By assumption, the statistic U is an unbiased estimate of the parameter Q, hence

(13.5.9) $$E(U) = \int_{-\infty}^{+\infty} u g(u, Q)\, du = Q.$$

Besides, by the assumption of regularity, we can differentiate expression (13.5.9) with respect to Q under the integral sign. Hence we obtain

(13.5.10) $$1 = \int_{-\infty}^{+\infty} u\, \frac{\partial g(u, Q)}{\partial Q}\, du = \int_{-\infty}^{+\infty} (u - Q)\, \frac{\partial g(u, Q)}{\partial Q}\, du^{1}$$

$$= \int_{-\infty}^{+\infty} (u - Q)\sqrt{g(u, Q)}\, \frac{\partial \log g(u, Q)}{\partial Q} \sqrt{g(u, Q)}\, du.$$

Thus

$$1 = \left[\int_{-\infty}^{+\infty} (u - Q)\sqrt{g(u, Q)}\, \frac{\partial \log g(u, Q)}{\partial Q} \sqrt{g(u, Q)}\, du \right]^{2}.$$

From the Schwarz inequality, which can be found in the book of Franklin [1], for instance, we obtain the inequality

(13.5.11) $$1 \leqslant \int_{-\infty}^{+\infty} (u - Q)^{2} g(u, Q)\, du \int_{-\infty}^{+\infty} \left[\frac{\partial \log g(u, Q)}{\partial Q} \right]^{2} g(u, Q)\, du.$$

Let us write formula (13.5.11) in the form

$$D^{2}(U) = \int_{-\infty}^{+\infty} (u - Q)^{2} g(u, Q)\, du \geqslant \frac{1}{\displaystyle\int_{-\infty}^{+\infty} [\partial \log g(u, Q)/\partial Q]^{2} g(u, Q)\, du}.$$

By (13.5.8) we obtain

$$D^{2}(U) \geqslant \frac{1}{n \displaystyle\int_{-\infty}^{+\infty} [\partial \log f(x, Q)/\partial Q]^{2} f(x, Q)\, dx}.$$

Thus inequality (13.5.1) is proved.

According to this inequality, the variance of the estimate U of the parameter Q of a population is not smaller than a certain value which is constant for fixed n and is determined by the distribution of X in this population.

B   We now present the concept of the most efficient estimate.

Definition 13.5.1.   An unbiased estimate U of the parameter Q is said to be the *most efficient for the class D of distribution functions* if for every

---

[1] This transformation follows from $\int_{-\infty}^{+\infty} g(u, Q)\, du = 1$ and the fact that this integral may be differentiated under the integral sign; hence $\int_{-\infty}^{+\infty} \frac{\partial g(u, Q)}{\partial Q}\, du = 0$.

distribution function $F(x, Q)$ of D the variance $D^2(U)$ has the minimal value given by (13.5.1) or (13.5.1').

For brevity we use the short term *most efficient estimate*.

It follows from this definition that, in general, the values of the unbiased, most efficient estimate are most concentrated around the value $E(U) = Q$; hence we have the greatest possible probability that the observed value u of this estimate will be close to the true value of the unknown parameter Q.

Theorem 13.5.1. *An unbiased estimate U of the parameter Q is the most efficient if and only if*

(1) *the estimate U is sufficient,*

(2) *for $g(u, Q) > 0$, the density $g(u, Q)$ almost everywhere[1] satisfies the relation*

$$(13.5.12) \qquad \frac{\partial \log g(u, Q)}{\partial Q} = c(u - Q),$$

where the number c is independent of u (it may depend on Q).

Proof. First we show the sufficiency of these conditions. Suppose that conditions (1) and (2) are satisfied. By condition (1), the conditional density h [see formula (13.4.4)] is independent of Q; hence, using condition (2), formula (13.5.8) takes the form

$$n \int_{-\infty}^{+\infty} \left[\frac{\partial \log f(x, Q)}{\partial Q} \right]^2 f(x, Q)\, dx = c^2 \int_{-\infty}^{+\infty} (u - Q)^2 g(u, Q)\, du = c^2 D^2(U),$$

and formula (13.5.10) takes the form

$$1 = c \int_{-\infty}^{+\infty} (u - Q)^2 g(u, Q)\, du = c D^2(U).$$

Eliminating c from the last two equalities, we obtain

$$(13.5.13) \qquad D^2(U) = \frac{1}{n \int_{-\infty}^{+\infty} \left[\dfrac{\partial \log f(x, Q)}{\partial Q} \right]^2 f(x, Q)\, dx}.$$

We now show the necessity of conditions (1) and (2). Suppose that equality (13.5.13) is satisfied. It follows from (13.5.8) that the conditional density h is independent of Q, since otherwise, by (13.5.11), we would have the inequality

$$D^2(U) > \frac{1}{n \int_{-\infty}^{+\infty} [\partial \log f(x, Q)/\partial Q]^2 f(x, Q)\, dx},$$

[1] That is everywhere, except possibly a set of values of u with Lebesgue measure zero.

which would contradict relation (13.5.13). Thus we have shown that the estimate U is sufficient; hence condition (1) is satisfied.

Let us now rewrite formula (13.5.8) in the form

$$\frac{1}{D^2(U)} = \int_{-\infty}^{+\infty} \left[\frac{\partial \log g(u, Q)}{\partial Q} \right]^2 g(u, Q)\, du.$$

This relation is identical with relation (13.5.11) with the sign $\leqslant$ replaced by the equality sign. It is known that in expression (13.5.11), derived from the Schwarz inequality, we have the equality sign only if there exist two constants w and v, at least one different from zero, for which the equation

$$(13.5.14) \quad \int_{-\infty}^{+\infty} \left[w(u - Q)\sqrt{g(u, Q)} + v \frac{\partial \log g(u, Q)}{\partial Q} \sqrt{g(u, Q)} \right]^2 du = 0$$

is satisfied.

The expression under the integral sign is non-negative; hence this equation is satisfied only if almost everywhere

$$(13.5.15) \quad w(u - Q)\sqrt{g(u, Q)} + v \frac{\partial \log g(u, Q)}{\partial Q} \sqrt{g(u, Q)} = 0.$$

The expression $(u - Q)\sqrt{g(u, Q)}$ cannot vanish almost everywhere, since in this case the variance of the estimate U would have to be zero; thus for $g(u, Q) > 0$ equation (13.5.15) holds only if equality (13.5.12) holds. Thus condition (2) is also proved.

C Let U be the unbiased and most efficient estimate of the parameter Q and let U_1 be another unbiased estimate of this parameter. As a *measure of efficiency* of U_1 the expression

$$(13.5.16) \qquad\qquad e = \frac{D^2(U)}{D^2(U_1)}$$

may serve. For the most efficient unbiased estimate we have $e = 1$.

For estimates with variances proportional to $1/n$, we can interpret e as follows: The most efficient estimate U of the unknown parameter Q from a sample of n elements has the same quality, from the point of view of efficiency, as the estimate U_1 of efficiency e from a sample of n/e elements.

In practice, it is desirable to apply the most efficient estimates as they are the most economical, but unfortunately they seldom exist.

Example 13.5.1. We showed in example 13.4.1 that the statistic $U = \bar{X}$, determined from a simple sample, is a sufficient estimate of the parameter m of the normal distribution $N(m; \sigma)$ with known standard deviation σ.

We now show that this estimate is also the most efficient.

It is sufficient to prove that condition (13.5.12) is satisfied. Indeed, by (13.4.7), we have

(13.5.17)
$$g(\bar{x}, m) = \frac{\sqrt{n}}{\sigma \sqrt{2\pi}} \exp\left[-\frac{n}{2} \cdot \frac{(\bar{x} - m)^2}{\sigma^2} \right].$$

Hence

(13.5.18)
$$\frac{\partial \log g(\bar{x}, m)}{\partial m} = \frac{n}{\sigma^2} (\bar{x} - m).$$

Thus the mean of simple samples drawn from a normal population is the most efficient estimate of the unknown expected value of this population.

As we see from (13.5.17),

(13.5.19)
$$D^2(X) = \frac{\sigma^2}{n}.$$

Thus (13.5.19) is the minimal variance of a regular and unbiased estimate of the parameter m from simple samples of size n drawn from a normal population.

Example 13.5.2. The characteristic X of elements of a population has the normal distribution with the density

(13.5.20)
$$f(x, \sigma^2) = \frac{1}{\sigma \sqrt{2\pi}} \exp\left[-\frac{(x-m)^2}{2\sigma^2} \right],$$

where m is known but σ unknown.

We want to estimate σ^2 from a simple sample of n elements and we take as an estimate of σ^2 the statistic

(13.5.21)
$$S_0^2 = \frac{1}{n} \sum_{k=1}^{n} (X_k - m)^2.$$

This estimate is regular in an arbitrary interval $a < \sigma^2 < b$ with $a > 0$. As we know (see Section 13.3), S_0^2 is an unbiased estimate of the variance σ^2 for the class of all distribution functions of random variables whose second moments exist. We show that this estimate is the most efficient for the class of all distribution functions of normally distributed random variables.

From (9.4.9) we obtain

(13.5.22)
$$g(s_0^2, \sigma^2) = \frac{n^{n/2} s_0^{n-2} \exp(-n s_0^2 / 2\sigma^2)}{(2\sigma^2)^{n/2} \Gamma(\frac{1}{2}n)}.$$

Hence

(13.5.23)
$$\frac{\partial \log g(s_0^2, \sigma^2)}{\partial \sigma^2} = \frac{n}{2\sigma^4} (s_0^2 - \sigma^2).$$

Thus condition (13.5.12) is satisfied. On the other hand, we have

(13.5.24)
$$L = f(x_1, \sigma^2) f(x_2, \sigma^2) \ldots f(x_n, \sigma^2) = \frac{1}{\sigma^n (2\pi)^{n/2}} \exp\left[-\frac{\sum_{k=1}^{n} (x_k - m)^2}{2\sigma^2} \right]$$

$$= \frac{1}{\sigma^n (2\pi)^{n/2}} \exp\left(-\frac{n s_0^2}{2\sigma^2} \right) = g(s_0^2, \sigma^2) \frac{\Gamma(\frac{1}{2}n)}{\pi^{n/2} n^{n/2} s_0^{n-2}}.$$

As we see, the second term on the right-hand side of (13.5.24) is independent of

σ^2. Thus conditions (1) and (2) are satisfied and S_0^2 is the most efficient estimate of σ^2.

The minimal variance of a regular and unbiased estimate of the parameter σ^2 from simple samples can be found from formulas (13.5.13) and (13.5.20),

$$(13.5.25) \quad D^2(S_0^2) = \frac{1}{n \displaystyle\int_{-\infty}^{+\infty} [\partial \log f(x, \sigma^2)/\partial \sigma^2]^2 f(x, \sigma^2)\, dx}$$

$$= \frac{1}{n \displaystyle\int_{-\infty}^{+\infty} [(x-m)^2/2\sigma^4 - 1/2\sigma^2]^2 f(x, \sigma^2)\, dx} = \frac{2\sigma^4}{n}.$$

Example 13.5.3. The characteristic X of elements of a population has the normal distribution $N(m; \sigma)$. We want to estimate from simple samples of n elements the variance σ^2 and we take as an estimate the statistic

$$(13.5.26) \qquad U = \frac{n}{n-1} S^2 = \frac{\displaystyle\sum_{k=1}^{n} (X_k - \bar{X})^2}{n-1}.$$

The estimate U is unbiased (see Section 13.3) and regular in an arbitrary interval $a < \sigma^2 < b$, where $a > 0$. This estimate, however, is not sufficient, and hence is not the most efficient. This follows directly from formula (13.4.7), since the standard deviation σ appears in both terms on the right-hand side. It is easy to verify that

$$D^2\left(\frac{n}{n-1} S^2\right) = \frac{2\sigma^4}{n-1}.$$

By (13.5.16), the efficiency of this estimate equals

$$e = \frac{D^2(S_0^2)}{D^2\left(\dfrac{n}{n-1} S^2\right)} = \frac{n-1}{n} < 1.$$

Example 13.5.4. Under the same conditions as in example 13.5.3, we wish to estimate the standard deviation using the statistic

$$(13.5.27) \qquad U = \frac{\sqrt{n}\,\Gamma(\tfrac{1}{2}n)}{\sqrt{2}\,\Gamma(\tfrac{1}{2}(n+1))} S_0.$$

As we know (see example 13.3.3), U is an unbiased estimate of σ. By the same reasoning, as in the proof of the sufficiency of the statistic S_0^2, it is easy to show that U is a sufficient estimate of σ for the class of distribution functions of normally distributed random variables. However, U is not the most efficient estimate of this parameter.

Indeed, we have

$$(13.5.28) \qquad g(u, \sigma) = \frac{2K_n^{n-1}}{\sigma^n} u^{n-1} \exp\left(-\frac{K_n^2 u^2}{\sigma^2}\right),$$

where

$$K_n = \frac{\Gamma(\tfrac{1}{2}(n+1))}{\Gamma(\tfrac{1}{2}n)}.$$

Hence

$$\frac{\partial \log g(u, \sigma)}{\partial \sigma} = \frac{2K_n}{\sigma^3}\left(u^2 - \frac{n\sigma^2}{2K_n}\right)$$

and condition (13.5.12) is not satisfied.

The minimal value of the variance of a regular and unbiased estimate of σ from simple samples can be found from (13.5.1). It equals

$$(13.5.29) \quad \frac{1}{n\displaystyle\int_{-\infty}^{+\infty}\left[\frac{(x - m)^2}{\sigma^3} - \frac{1}{\sigma}\right]^2 \frac{1}{\sigma\sqrt{2\pi}} \exp\left[-\frac{(x - m)^2}{2\sigma^2}\right] dx} = \frac{\sigma^2}{2n}.$$

We can show that the variance of the estimate U defined by (13.5.27) is, as was to be expected, greater than $\sigma^2/2n$.

Example 13.5.5. The characteristic X of elements of a population has the normal distribution $N(m; \sigma)$, with unknown σ and m known. As an estimate of this parameter, we take the mean deviation of simple samples of size n multiplied by $\sqrt{0.5\pi}$, that is, the statistic

$$(13.5.30) \qquad U = \frac{1}{n}\sqrt{0.5\pi}\sum_{k=1}^{n}|X_k - m|,$$

where the random variables X_k are independent and have the same distribution as X. Let us find the efficiency of U. First, we verify that U is an unbiased estimate of σ.

We have

$$E[|X - m|] = \frac{1}{\sigma\sqrt{2\pi}}\int_{-\infty}^{+\infty}|x - m| \exp\left[-\frac{(x - m)^2}{2\sigma^2}\right] dx$$

$$= \frac{2}{\sigma\sqrt{2\pi}}\int_{m}^{+\infty}(x - m) \exp\left[-\frac{(x - m)^2}{2\sigma^2}\right] dx = \sigma\sqrt{2/\pi}.$$

Hence

$$(13.5.31) \quad E(U) = \frac{1}{n}\sqrt{\pi/2}\sum_{k=1}^{n}E[|X_k - m|] = \sqrt{\pi/2}\, E[|X - m|] = \sigma.$$

Similarly, we obtain

$$(13.5.32) \qquad D^2(U) = \frac{\pi - 2}{2n}\sigma^2.$$

Thus the estimate U is not the most efficient. Its efficiency is

$$(13.5.33) \qquad e = \frac{\sigma^2}{2n}\bigg/\frac{\pi - 2}{2n}\sigma^2 = \frac{1}{\pi - 2} \cong 0.876.$$

D   We now find some most efficient estimates from samples drawn from populations, in which the characteristic considered is a random variable of the discrete type.

Example 13.5.6. The characteristic X of elements of a population has the binomial distribution, that is,

$$(13.5.34) \quad p_k = P(X = k) = \binom{r}{k} p^k q^{r-k} \quad (q = 1 - p, \quad 0 < p < 1,$$

$$k = 0, 1, \ldots, r),$$

where p is an unknown parameter which is to be estimated from a simple sample of n elements drawn from this population.

Let us compute the minimal variance of a regular and unbiased estimate U of p from simple samples. Using formula (13.5.1'), we obtain

$$(13.5.35) \quad \frac{1}{D^2(U)} = n \sum_{k=0}^{r} \left(\frac{d \log p_k}{dp} \right)^2 p_k = n \sum_{k=0}^{r} \left(\frac{k}{p} - \frac{r-k}{1-p} \right)^2 p_k$$

$$= n \sum_{k=0}^{r} \left(\frac{k - pr}{pq} \right)^2 p_k = \frac{n}{p^2 q^2} D^2(X) = \frac{n}{p^2 q^2} pqr = \frac{nr}{pq}.$$

As an estimate of p, we take the statistic

$$(13.5.36) \qquad U = \frac{\bar{X}}{r} = \frac{1}{rn} \sum_{j=1}^{n} X_j,$$

where the random variables X_j are independent and have the distribution given by (13.5.34). We have

$$(13.5.37) \qquad E(U) = \frac{nrp}{nr} = p, \qquad D^2(U) = \frac{npqr}{r^2 n^2} = \frac{pq}{nr}.$$

It follows that the statistic U is the unbiased and most efficient estimate of p for the class of binomial distributions.

Let us consider the particular case $r = 1$, that is, the case when X has the zero-one distribution and $P(X = 1) = p$, $P(X = 0) = q$. According to (13.5.36), the most efficient estimate of the unknown parameter is $U = p^* = \bar{X}$.

Suppose that the population contains good and defective items, where the fraction p of defective items is unknown. Let us assign the number one to the appearance of a defective item and the number zero to the appearance of a good item. If in a simple sample of n elements we find n_1 defective items and $n - n_1$ good items, then as an estimate of p we take

$$\frac{n_1 \cdot 1 + (n - n_1) \cdot 0}{n} = \frac{n_1}{n},$$

or the fraction of defective items in the sample. Thus we conclude that the frequency of appearance of items of a given category in a simple sample is the most efficient estimate of the fraction of this category of elements in the population.

Example 13.5.7. The characteristic X of elements of a population has the Poisson distribution

$$(13.5.38) \qquad p_k = P(X = k) = \frac{\lambda^k}{k!} e^{-\lambda} \quad (k = 0, 1, 2, \ldots),$$

where the parameter λ is unknown.

The minimal variance of a regular and unbiased estimate U of λ from simple samples of n elements can be found from formula (13.5.1),

$$(13.5.39) \qquad \frac{1}{D^2(U)} = n \sum_{k=0}^{\infty} \left(\frac{d \log p_k}{d\lambda} \right)^2 p_k = n \sum_{k=0}^{\infty} \left(\frac{k}{\lambda} - 1 \right)^2 p_k$$

$$= \frac{n}{\lambda^2} \sum_{k=0}^{\infty} (k - \lambda)^2 p_k = \frac{n\lambda}{\lambda^2} = \frac{n}{\lambda}.$$

It is easy to check that $\lambda^* = U = \bar{X}$ is the most efficient unbiased estimate of λ. Indeed, we have

$$(13.5.40) \qquad E(U) = \lambda, \qquad D^2(U) = \frac{\lambda}{n}.$$

E We now give some information about recent results concerning the efficiency of unbiased estimates and a theorem of Blackwell.

Let D be a certain class of distribution functions. Let us consider all those estimates U of the parameter Q, unbiased for the class D, whose variances $D^2(U)$ exist for every distribution function $F(x, Q)$ of D. Then, as was shown by Lehmann and Scheffé [1], at most one of these estimates is the most efficient for the class D under consideration. A generalization of this result was given by Schmetterer [2]. The problem considered by Lehmann and Scheffé was considered from another point of view in the paper of Stein [3].

Blackwell [2] proved a theorem that allows, having an unbiased estimate V of the parameter Q and another estimate U of Q which is sufficient, to construct another unbiased estimate of Q which is at least as efficient as V. This follows from the following theorem due to Blackwell.

Theorem 13.5.2. *Let V be an unbiased estimate of the parameter Q, and let U be a sufficient estimate of Q. Then the random varialbe $E(V \mid u)$ (see Section 3.6, C) is an unbiased estimate of Q. If, moreover, the variance $D^2(V)$ exists, the inequality*

$$(13.5.41) \qquad D^2[E(V \mid u)] \leqslant D^2(V)$$

holds. The equality in (13.5.41) holds only if $E(V \mid u) = V$, with probability one.

Proof. From the assumed sufficiency of U, it follows that $E(V \mid u)$ does not depend on Q. Since V is unbiased, we obtain, using formula (3.6.24′),

$$E(V) = E[E(V \mid u)] = Q;$$

hence $E(V \mid u)$ is an unbiased estimate of Q.

We now give the proof of (13.5.41) for the case when all the random variables involved are of the continuous type. (See Section 2.7, B.)

Let $f(v, u)$ and $g(u)$, respectively, denote the density of (V, U) and U, and let $h(v \mid u)$ denote the conditional density of V, provided $U = u$. Using (3.6.23'), we have

$$(13.5.42) \quad D^2(V) = E(V^2) - Q^2 = \int_{-\infty}^{\infty} g(u) \int_{-\infty}^{\infty} v^2 h(v \mid u) \, dv \, du$$

$$- \int_{-\infty}^{\infty} g(u) \left[\int_{-\infty}^{\infty} vh(v \mid u) \, dv \right]^2 du + \int_{-\infty}^{\infty} g(u) \left[\int_{-\infty}^{\infty} vh(v \mid u) \, dv \right]^2 du$$

$$- \left[\int_{-\infty}^{\infty} g(u) \int_{-\infty}^{\infty} vh(v \mid u) \, dv \, du \right]^2 = A + B,$$

where A is the sum of the first and the second integral on the right-hand side of (13.5.42), and B is the sum of the remaining two integrals.

We can write

$$(13.5.43) \quad A = \int_{-\infty}^{\infty} g(u) \left\{ \int_{-\infty}^{\infty} v^2 h(v \mid u) \, dv - \left[\int_{-\infty}^{\infty} vh(v \mid u) \, dv \right]^2 \right\} du,$$

whereas B can be written in the form

$$(13.5.44) \quad B = E\{[E(V \mid u)]^2\} - \{E[E(V \mid u)]\}^2 = D^2[E(V \mid u)].$$

The expression inside the braces on the right-hand side of (13.5.43) is the conditional variance of V, provided $U = u$; hence A is non-negative. It follows thus from formulas (13.5.42) to (13.5.44) that

$$D^2(V) = A + B \geqslant B = D^2[E(V \mid u)];$$

hence (13.5.41) holds.

Now the equality in (13.5.41) holds only if $A = 0$, hence if (see Section 5.1) the conditional distribution of V, provided $U = u$, is degenerate, with probability one. In other words, the random variable V takes on, for any given value u, one (random) value, with probability one; hence $E(V \mid u) = V$, with probability one.

Barankin [1] has extended Blackwell's theorem to arbitrary central moments of estimates of an unknown parameter.

F   The notion of the most efficient estimate can be generalized (see Cramér [2]) to estimates of several unknown parameters. The concept of *jointly most efficient estimates* has then to be introduced. The Rao-Cramér inequality can also be generalized to the case considered. This generalization was obtained by the authors of this inequality. Another generalization of the Rao-Cramér inequality is mentioned in Section 17.10, A.

13.6 ASYMPTOTICALLY MOST EFFICIENT ESTIMATES

Often we deal with estimates which are not most efficient but their efficiency satisfies the condition

$$(13.6.1) \qquad \lim_{n \to \infty} e = 1,$$

and they are at least asymptotically unbiased. Such estimates are called *asymptotically most efficient estimates*. From the practical point of view, they are most efficient estimates from large samples.

Relation (13.6.1) is not always satisfied, since the efficiency e may converge to a limit $e_0 \neq 1$ as $n \to \infty$. The limit value of the number e is called the *asymptotic efficiency* of the estimate. (For another definition of asymptotic efficiency and related questions, see A. Birnbaum [1].)

Let us mention here that there are estimates for which no asymptotic efficiency exists.

Example 13.6.1. In example 13.5.3 we showed that the unbiased estimate U of the parameter σ^2 from simple samples selected from a normal population, where

$$U = \frac{n}{n-1} S^2,$$

is not the most efficient estimate of σ^2. Its efficiency is

$$e = \frac{n-1}{n} < 1.$$

It is easy to see that $\lim_{n \to \infty} e = 1$; hence the considered estimate is asymptotically most efficient.

Example 13.6.2. Suppose that the characteristic X of elements of a population has the normal distribution $N(m; \sigma)$ with m unknown. As an estimate of m, we take the median of simple samples of size n. Let us find the asymptotic efficiency of this estimate.

By (13.5.19), we know that the minimal value of the variance of an estimate U of the parameter m equals σ^2/n. From (10.5.15) we find that the expected value and variance of the median of simple samples from a normal population are approximately m and $(\pi/2n)\sigma^2$, respectively. Hence

$$(13.6.2) \qquad e \cong \frac{\sigma^2}{n} : \frac{\pi}{2n} \sigma^2 = \frac{2}{\pi} \cong 0.64.$$

As we see, the sample median is not an asymptotically most efficient estimate of m.

Example 13.6.3. The characteristic X of elements of a population has the normal distribution $N(m; \sigma)$. Denote by $F(x)$ the distribution function of the random variable X and by $f(x)$ its density. Consider the quantiles a_p and a_{1-p} defined by the equalities

(13.6.3) $$F(a_p) = p, \qquad F(a_{1-p}) = 1 - p,$$

where $0 < p < \frac{1}{2}$. In Fig. 13.6.1 are represented the quantiles a_p and a_{1-p} for $p = 0.1$ (for $m = 0$ and $\sigma = 1$).

The statistics $\zeta_k^{(n)}$ and $\zeta_{n-k+1}^{(n)}$, where $k = [np] + 1$, were defined in Section 10.2 as sample quantiles. By theorem 10.5.1, if the samples are simple, these statistics have the asymptotically normal distributions

$$N(a_p; \ \sqrt{D^2[\zeta_k^{(n)}]}) \quad \text{and} \quad N(a_{1-p}; \ \sqrt{D^2[\zeta_{n-k+1}^{(n)}]}),$$

where

(13.6.4) $$D^2(\zeta_k^{(n)}) = D^2(\zeta_{n-k+1}^{(n)}) = \frac{p(1 - p)}{n[f(a_p)]^2}.$$

It is easy to show that the covariance μ_{11} of $\zeta_k^{(n)}$ and $\zeta_{n-k-1}^{(n)}$ is

(13.6.5) $$\mu_{11} = \frac{p^2}{n[f(a_p)]^2}.$$

Let $a_{1-p} = m + b\sigma$. Then we have $a_p = m - b\sigma$. Hence

(13.6.6) $$\sigma = \frac{a_{1-p} - a_p}{2b}, \qquad m = \frac{a_{1-p} + a_p}{2}.$$

Expressions (13.6.6) suggest that we take as estimates of σ and m the respective statistics

$$U_1 = \frac{\zeta_{n-k+1}^{(n)} - \zeta_k^{(n)}}{2b}, \qquad U_2 = \frac{\zeta_{n-k+1}^{(n)} + \zeta_k^{(n)}}{2}.$$

It follows from theorem 10.5.1 that for large samples the statistics U_1 and U_2 are unbiased estimates of the parameters σ and m. Let us compute the asymptotic efficiency of these estimates.

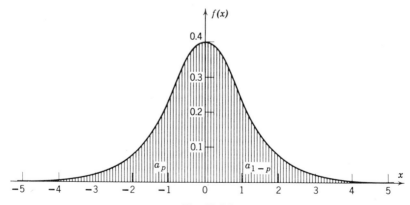

Fig. 13.6.1

From (13.6.6) and (13.6.5) we obtain

(13.6.8) $$D^2(U_1) = \frac{D^2(\zeta_{n-k+1}^{(n)}) + D^2(\zeta_k^{(n)}) - 2\mu_{11}}{4b^2} \simeq \frac{p(1-2p)}{2nb^2[f(a_p)]^2}.$$

Similarly,

(13.6.9) $$D^2(U_2) \simeq \frac{p}{2n[f(a_p)]^2}.$$

The asymptotic efficiency of U_1 is

$$e_1 \simeq \frac{\sigma^2}{2n} : \frac{p(1-2p)}{2nb^2[f(a_p)]^2} = \frac{\sigma^2 b^2[f(a_p)]^2}{p(1-2p)}.$$

Since

$$f(a_p) = \frac{1}{\sigma\sqrt{2\pi}} e^{-b^2/2},$$

we have

(13.6.10) $$e_1 \simeq \frac{b^2}{p(1-2p)} \cdot \frac{1}{2\pi} e^{-b^2}.$$

The efficiency of U_2 is

(13.6.11) $$e_2 \simeq \frac{\sigma^2}{n} : \frac{p}{2n[f(a_p)]^2} = \frac{e^{-b^2}}{p\pi}.$$

From tables of the normal distribution we find that expression (13.6.10) has its maximum $e_1 = 0.65$ for $p = 0.0692$, and expression (13.6.11) has its maximum $e_2 = 0.81$ for $p = 0.2703$. Thus, if we wish to estimate the standard deviation σ on the basis of a large sample by means of the estimate U_1, it is best to use the quantile $\zeta_k^{(n)}$ for $k = [0.0692n] + 1$, and if we wish to estimate the expected value m on the basis of a large sample using the statistic U_2, it is best to use the quantile $\zeta_k^{(n)}$ for $k = [0.2703n] + 1$.

The reader may compare these results with example 13.6.2, where we used the sample median as an estimate of m, that is, we used the sample quantile for $p = 0.5$. The asymptotic efficiency in that case is 0.64, and is much smaller than the maximum efficiency 0.81.

Let us find the asymptotic efficiency of the semi-interquartile range as an estimate of the standard deviation. Putting $p = 0.25$ into formula (13.6.10), we find $e_1 = 0.368$. Thus it is an estimate with small efficiency and is, therefore, not used in applications.

The problem discussed in example 13.6.3 was investigated by K. Pearson [6] and other authors.

Example 13.6.4. In practice, particularly in statistical quality control, it is customary to take the sample range as an estimate of the standard deviation. We do not consider this problem in detail; we merely make the following remarks.

Let us consider the statistic

(13.6.12) $$U_n = \frac{W_n}{d_n},$$

where W_n is the sample range of a simple sample of size n drawn from a population in which the characteristic X has the normal distribution $N(m; \sigma)$ and d_n is the expected value of the range of a simple sample of n-elements drawn from a population with the distribution $N(0;1)$. The values of d_n can be found in Table XXII of Volume II of the tables of K. Pearson [2].

The statistic U_n is an unbiased estimate of the parameter σ. However, the efficiency of this estimate decreases rather rapidly to zero as $n \to \infty$. This is illustrated in Table 13.6.1, taken from the paper of Davies and Pearson [1].

TABLE 13.6.1

EFFICIENCY OF SAMPLE RANGE AS AN ESTIMATE OF STANDARD DEVIATION

n	2	3	4	5	6	10	15	20	50	100
e	1.00	0.99	0.98	0.96	0.93	0.85	0.77	0.70	0.49	0.34

This result is extremely intuitive since the larger the sample, the smaller is the information contained in the smallest and the largest elements alone. Thus it is reasonable to take the sample range as an estimate of σ only in small samples.

From the viewpoint of efficiency it is convenient to split the sample of n elements in a random way into k smaller samples of $l = n/k$ elements each (called *subsamples*), and to take as an estimate of σ the statistic

$$\sigma^* = \frac{\overline{W}}{d_l},$$

where

(13.6.13)
$$\overline{W} = \frac{1}{k} \sum_{i=1}^{k} W_i,$$

and W_i $(i = 1, \ldots, k)$ is the range of the ith subsample.

Grubbs and Weaver [1] investigated the following more general and more efficient method. Split the sample of size n into k subsamples. Let the ith subsample have n_i elements $(i = 1, 2, \ldots, k)$. Consider the statistic

(13.6.13')
$$V = \sum_{i=1}^{k} a_i W_i,$$

where the coefficients a_i and the numbers k and n_i are chosen in such a way that the relations $E(V) = \sigma$ and $D^2(V) = $ minimum are satisfied.

Grubbs and Weaver worked out tables that allow us to find the numbers k, a_i, and n_i for $n = 2, \ldots, 100$. The details can be found in the paper quoted. We present only some particular results. If $n = 24$, then it is best (from the point of view of efficiency) to take $k = 3$ and $n_1 = n_2 = n_3 = 8$; if $n = 13$, then it is best to take $k = 2$, $n_1 = 6$, and $n_2 = 7$.

Example 13.6.5. The characteristic X of elements of a population has the Cauchy distribution given by the density

(13.6.14)
$$f(x) = \frac{1}{\pi[1 + (x - Q)^2]},$$

where Q is an unknown parameter.

The minimal variance of a regular and unbiased estimate U of Q from simple samples can be found from formula (13.5.1), namely,

$$(13.6.15) \qquad D^2(U) = \cfrac{1}{n\displaystyle\int_{-\infty}^{+\infty} \left[\dfrac{\partial \log f(x, Q)}{\partial Q}\right]^2 f(x, Q)\,dx}$$

$$= \cfrac{1}{n\displaystyle\int_{-\infty}^{+\infty} \dfrac{1}{\pi} \cdot \dfrac{4(x - Q)^2}{[1 + (x - Q)^2]^3}\,dx}$$

$$= \dfrac{2}{n}.$$

As we know (see example 6.8.5), the arithmetic mean of n independent random variables with the same Cauchy distribution also has the Cauchy distribution; hence neither the expected value nor the variance exists. It follows that X is not asymptotically unbiased, and, of course, is not an asymptotically most efficient estimate of Q.

Let us now take as an estimate of Q the median of a simple sample. It follows from formula (10.5.15) that the median has the asymptotically normal distribution $N(Q;\ \pi/2\sqrt{n})$. Thus the sample median is an asymptotically unbiased estimate of Q and the variance of this estimate is asymptotically equal to $\pi^2/4n$; hence

$$e \cong \frac{2}{n} : \frac{\pi^2}{4n} = \frac{8}{\pi^2} = 0.81.$$

This is the asymptotic efficiency of this estimate.

Example 13.6.6. For the birth and death process considered in Section 8.5, we assumed that the length of life, T, of a signal has the exponential distribution with density

$$(13.6.16) \qquad g(t) = \begin{cases} 0 & \text{for} \quad t \leqslant 0, \\ \mu e^{-\mu t} & \text{for} \quad t > 0. \end{cases}$$

Since the expected value of the random variable T equals $1/\mu$, as an estimate of the parameter $Q = 1/\mu$ we take the arithmetic mean $\bar{T}$ of the sample, where the elements of the sample are the observed independent lengths of life of signals, $t_1, t_2, \ldots, t_n$. Since the exponential distribution is a particular case of the gamma distribution, we find from (13.6.16) and (5.8.10) that the variance of T equals $1/\mu^2$; hence the variance of $\bar{T}$ equals $1/n\mu^2$. From (13.5.1) we easily find that the minimal variance of an estimate of the parameter $Q = 1/\mu$ is $1/n\mu^2$; hence $\bar{T}$ is the unbiased and most efficient estimate of Q.

Now let us consider as parameter the median of the population, $a_{1/2}$. Suppose that the sample is large. The sample median has the asymptotically normal distribution with the expected value $a_{1/2}$ and with the variance given by (10.5.15). To find this variance, we have to find the value of the function $g(t)$ at the point $t = a_{1/2}$. From the formula

$$\mu \int_0^{a_{1/2}} e^{-\mu t}\,dt = 1 - e^{-\mu a_{1/2}} = \frac{1}{2},$$

we find that $a_{1/2} = \log 2/\mu$ and $g(a_{1/2}) = \frac{1}{2}\mu$. Hence, using (10.5.15), we find that the variance of the median of the sample asymptotically equals $1/n\mu^2$; thus the sample median is an unbiased and asymptotically most efficient estimate of the population median.

13.7 METHODS OF FINDING ESTIMATES

A   In the preceding sections, we have dealt with the classification of estimates according to certain properties. We now briefly discuss methods of constructing estimates.

The oldest method is the *method of moments*. In this method, the sample moment serves as an estimate of the corresponding population moment, and a function of sample moments serves as an estimate of the population parameter which is the corresponding function of the population moments. Thus, the kth sample moment is an estimate of the kth population moment, the correlation coefficient for a two-dimensional population is estimated by the sample correlation coefficient, and so on. However, as Fisher [2] showed, such estimates may be of small efficiency and even small asymptotic efficiency (see Problem 13.15). From this viewpoint, the estimates obtained by the method of maximum likelihood are better. This method has been proposed and developed by Fisher [2, 6], [8, 10, and 11]. These estimates were defined in Section 12.4. We here repeat the definition briefly.

Let $F(x)$ be the distribution function of a characteristic X of elements of a population, and let $F(x)$ depend upon m unknown parameters λ_1, $\lambda_2, \ldots, \lambda_m$, which we wish to estimate on the basis of a sample of n elements, $x_1, \ldots, x_n$. If X is a random variable of the discrete type and $P(X = x_k) = p_k$ $(k = 1, 2, \ldots, n)$, the likelihood function is given by the formula

$$(13.7.1) \qquad\qquad L = p_1 p_2 \ldots p_n,$$

where the p_k $(k = 1, \ldots, n)$ are functions of the parameters $\lambda_1, \ldots, \lambda_m$.

If X is a random variable of the continuous type with the density $f(x)$, then

$$(13.7.1') \qquad\qquad L = f(x_1) f(x_2) \ldots f(x_n),$$

where the $f(x_k)$ $(k = 1, 2, \ldots, n)$ are functions of the parameters $\lambda_1, \ldots, \lambda_m$.

The *maximum likelihood estimates* of the parameters $\lambda_1, \ldots, \lambda_m$ are obtained by solving the system of m equations

$$(13.7.2) \qquad\qquad \frac{\partial L}{\partial \lambda_i} = 0 \qquad (i = 1, 2, \ldots, m).$$

The function L given by (13.7.1) is the probability of obtaining the observed sample (if X is of the continuous type, we consider the probability element $L\,dx_1, \ldots, dx_n$). Thus the principle of the method of maximum likelihood is to take estimates of the unknown parameters that maximize the probability of obtaining the observed sample. Now, since L and $\log L$ assume their maximum values for the same values of $\lambda_1, \ldots, \lambda_m$, the system of equations (13.7.2) may be replaced by the system

(13.7.3) $$\frac{\partial \log L}{\partial \lambda_i} = 0 \qquad (i = 1, 3, \ldots, m),$$

which is usually more convenient in computations.

In example 12.4.2 and in Section 12.7, we dealt with estimates of the unknown parameters of a population by the method of maximum likelihood. We now present one more example.

Example 13.7.1.   The random variable X has the Poisson distribution

$$P(X = k) = \frac{\lambda^k}{k!} e^{-\lambda} \qquad (k = 0, 1, \ldots),$$

where the parameter λ is unknown. We have n independent observations of X,

$$x_1 = k_1, \qquad x_2 = k_2, \qquad \ldots, \qquad x_n = k_n.$$

We estimate λ by the method of maximum likelihood.

We have

$$L = \frac{\lambda^{k_1}}{k_1!} e^{-\lambda} \frac{\lambda^{k_2}}{k_2!} e^{-\lambda} \cdots \frac{\lambda^{k_n}}{k_n!} e^{-\lambda} = \frac{\lambda^{k_1 + k_2 + \ldots + k_n}}{k_1! \, k_2! \ldots k_n!} e^{-\lambda},$$

$$\log L = -n\lambda + (k_1 + k_2 + \ldots + k_n) \log \lambda - \sum_{i=1}^{n} \log k_i!.$$

Hence

$$\frac{\partial \log L}{\partial \lambda} = -n + \frac{k_1 + k_2 + \ldots + k_n}{\lambda}$$

and, finally,

$$\lambda^* = \frac{k_1 + k_2 + \ldots + k_n}{n} = \bar{x}.$$

B   We now give two theorems which show the importance of estimates obtained by the method of maximum likelihood.

Theorem 13.7.1.   *Let the distribution function $F(x)$ depend upon one parameter Q, that is, $m = 1$. If there exists a sufficient estimate U of the parameter Q, then the solution of equation (13.7.3) is a function of u only.*

Proof.   If U is a sufficient estimate of Q, then by (13.4.6) we have

$$L = g(u, Q)h(y_1, y_2, \ldots, y_{n-1} \mid u),$$

where the conditional density h is independent of Q. Then

$$(13.7.4) \quad \frac{\partial \log L}{\partial Q} = \frac{\partial \log g(u, Q)}{\partial Q} + \frac{\partial \log h}{\partial Q} = \frac{\partial \log g(u, Q)}{\partial Q} = 0.$$

The solution of equation (13.7.4) is thus a function of u only.

Theorem 13.7.2. *If U is the most efficient estimate of the parameter Q, then, almost everywhere in the domain where $g(u, Q) > 0$, u is the solution of equation* (13.7.3).

Proof. If condition (13.5.12) is satisfied, then formula (13.7.4) has the form

$$(13.7.5) \qquad \frac{\partial \log L}{\partial Q} = c(u - Q) = 0,$$

where c is independent of u; the theorem follows immediately.

Among those who have studied the asymptotic properties of maximum likelihood estimates are: Hotelling [1], Doob [6], Dugué [1], Wald [5], and Wolfowitz [3]. The following theorem is presented without proof; a proof can be found in the monograph of Cramér [2]. This theorem is formulated only for random variables of the continuous type. An analogous theorem holds for random variables of the discrete type.

Theorem 13.7.3. *Let $f(x, Q)$ be the density function of the random variable X. Suppose that*

I. *the partial derivatives $\partial \log f/\partial Q$, $\partial^2 \log f/\partial Q^2$, $\partial^3 \log f/\partial Q^3$ exist for every value of the parameter Q in some interval A and for every value of x.*

II. *for every Q in A,*

$$|\partial f/\partial Q| < F_1(x), \quad |\partial^2 f/\partial Q^2| < F_2(x), \quad |\partial^3 f/\partial Q^3| < F_3(x),$$

where the functions $F_1(x)$ and $F_2(x)$ are integrable over the axis $(-\infty, +\infty)$ and the function $F_3(x)$ satisfies the relation

$$\int_{-\infty}^{\infty} F_3(x)f(x, Q)\, dx < M,$$

where M is independent of Q.

III. *for every Q in A,*

$$0 < E\left[\left(\frac{\partial \log f}{\partial Q}\right)^2\right] = \int_{-\infty}^{\infty} \left(\frac{\partial \log f}{\partial Q}\right)^2 f(x, Q)\, dx < \infty.$$

Then there exists a solution Q^* of Eq. (13.7.3) *which is stochastically convergent to Q as $n \to \infty$ and has an asymptotically normal distribution* $N(Q; 1/\sqrt{nE[(\partial \log f/\partial Q)^2]})$.

From this theorem, and from formulas (13.5.1) and (13.6.1), we see that the maximum likelihood estimate is an asymptotically unbiased and asymptotically most efficient estimate of the unknown parameter.

Theorem 13.7.3 can be generalized to maximum likelihood estimates of several parameters. The formulation and proof of such a generalization can be found in the monograph of Schmetterer [1], for instance.

C The concept of maximum likelihood estimates can be considered not only for independent observations, but also for arbitrary dependent random variables. We present only the definition of maximum likelihood estimates for Markov processes. We also give some important applications (see Lange [1]).

Let $\{X_t, t \in I\}$ be a discrete Markov process with at most a countable number of states j ($j = 0, 1 \ldots$) and with the transition probability function $p_{ij}(\tau, t) = P(X_t = j \mid X_\tau = i)$, where $\tau < t$. Let the function $p_{ij}(\tau, t)$ depend upon the parameters $\lambda_1, \lambda_2, \ldots, \lambda_m$ whose values are unknown. Thus we write $p_{ij}(\tau, t, \lambda_1, \lambda_2, \ldots, \lambda_m)$. Let $t_0 \in I$, $t_1 \in I, \ldots$, $t_n \in I$ ($t_0 < t_1 < \ldots < t_n$) and let $j_0, j_1, \ldots, j_n$ be the states of the process observed at the moments $t_0, t_1, \ldots, t_n$, respectively.

Definition 13.7.1. The product

$$(13.7.6) \quad L = c_{j_0}(t_0, \lambda_1, \lambda_2, \ldots, \lambda_m) \prod_{k=0}^{n-1} p_{j_k j_{k+1}}(t_k, t_{k+1}, \lambda_1, \lambda_2, \ldots, \lambda_m),$$

where $c_j(t_0, \lambda_1, \lambda_2, \ldots, \lambda_m) = P(X_{t_0} = j)$ is the absolute probability function of the random variable X_{t_0}, is called the *likelihood function of the discrete Markov process* $\{X_t, t \in I\}$.

Let $\{X_t, t \in I\}$ be a purely continuous Markov process (see Section 8.8) with the conditional density function $f(\tau, x, t, y, \lambda_1, \lambda_2, \ldots, \lambda_m)$ dependent upon the parameters $\lambda_1, \ldots, \lambda_m$ whose values are unknown. Let x_0, $x_1, \ldots, x_n$ be the observed states of the process at the moments $t_0 \in I$, $t_1 \in I, \ldots, t_n \in I$ ($t_0 < t_1 < \ldots < t_n$), respectively.

Definition 13.7.2. The product

$$(13.7.6') \quad L = f(t_0, x_0, \lambda_1, \lambda_2, \ldots, \lambda_m) \prod_{k=0}^{n-1} f(t_k, x_k, t_{k+1}, x_{k+1}, \lambda_1, \lambda_2, \ldots, \lambda_m),$$

where $f(t_0, x_0, \lambda_1, \ldots, \lambda_m)$ is the density of the absolute distribution of the random variable X_{t_0}, is called the *likelihood function of the purely continuous Markov process* $\{X_t, t \in I\}$.

Solving the system of equations

$$(13.7.7) \quad \frac{\partial \log L}{\partial \lambda_i} = 0 \quad (i = 1, 2, \ldots, m),$$

we obtain the *maximum likelihood estimates* $\lambda_1^*, \lambda_2^*, \ldots, \lambda_m^*$ of the unknown parameters.

The definitions of Section 12.4 are specializations of the preceding ones to cases when the t_k are replaced by integers k, and the random variables X_k $(k = 0, 1, \ldots, n)$ are independent.

Example 13.7.2. Let us consider the Poisson process (see Section 8.3), where

$$p_{ij}(\tau, t, \lambda) = \frac{[\lambda(t - \tau)]^{j-i}}{(j - i)!} \exp\left[-\lambda(t - \tau)\right],$$

$$c_j(0, \lambda) = \begin{cases} 1 & \text{for} \quad j = 0, \\ 0 & \text{for} \quad j \neq 0. \end{cases}$$

Here the parameter λ is unknown. Put $t_0 = 0$. We have n observations at the moments $t_1, t_2, \ldots, t_n$, where $0 < t_1 < t_2 < \ldots < t_n$. We have observed the states $j_1, j_2, \ldots, j_n$, respectively. Thus the likelihood function is of the form

$$L = \prod_{k=0}^{n-1} \frac{[\lambda(t_{k+1} - t_k)]^{j_{k+1}-j_k}}{(j_{k+1} - j_k)!} \exp\left[-\lambda(t_{k+1} - t_k)\right].$$

Hence

$$\log L = -\lambda t_n + j_n \log \lambda + G,$$

where G does not depend on λ. Solving (13.7.7), we obtain

$$(13.7.8) \qquad\qquad \lambda^* = \frac{j_n}{t_n}.$$

Thus λ^* is the average number of signals per unit of time. It is obvious that the estimate λ^* depends only upon the last observation. It is easy to show that $E(\lambda^*) = \lambda$ and $D^2(\lambda^*) = \lambda/t_n$.

The reader may compare (13.7.8) with the formula for λ^* in example 13.7.1.

Example 13.7.3. Let us consider the Furry-Yule birth process, where the transition probability function given by (8.4.10) has the form

$$p_{ij}(\tau, t, \lambda) = \binom{j - 1}{j - i}\{1 - \exp\left[-\lambda(t - \tau)\right]\}^{j-i} \exp\left[-i\lambda(t - \tau)\right].$$

Suppose that at the moment t_0 we have $P(X_{t_0} = j_0) = 1$. The parameter λ is unknown. At the moments $t_0, t_1, \ldots, t_n$ $(t_0 < t_1 < \ldots < t_n)$, we have observed the states $j_0, j_1, \ldots, j_n$, respectively. The likelihood function has the form

$$L = \prod_{k=0}^{n-1} \binom{j_{k+1} - 1}{j_{k+1} - j_k}\{1 - \exp\left[-\lambda(t_{k+1} - t_k)\right]\}^{j_{k+1}-j_k} \exp\left[-\lambda j_k(t_{k+1} - t_k)\right].$$

Hence

$$\log L = -\lambda \sum_{k=0}^{n-1} j_k(t_{k+1} - t_k) + \sum_{k=0}^{n-1} (j_{k+1} - j_k)$$
$$\times \log\{1 - \exp\left[-\lambda(t_{k+1} - t_k)\right]\} + G,$$

where G does not depend upon λ. Suppose that the observations have been made at equal time intervals, that is, $t_{k+1} - t_k = a\ (k = 0, 1, \ldots, n - 1)$. We then obtain a simple expression for $\log L$, namely,

$$\log L = -\lambda a \sum_{k=0}^{n-1} j_k + j_n \log (1 - e^{-\lambda a}).$$

Thus

$$\frac{\partial \log L}{\partial \lambda} = -a \sum_{k=0}^{n-1} j_k + j_n \frac{ae^{-\lambda a}}{1 - e^{-\lambda a}}.$$

From the equation $\partial \log L / \partial \lambda = 0$, we obtain

$$(13.7.9) \qquad \lambda^* = \frac{1}{a}\left(\log \sum_{k=0}^{n} j_k - \log \sum_{k=0}^{n-1} j_k\right).$$

This formula was obtained by D. G. Kendall [1].

Example 13.7.4. Let us consider the Brownian motion process (see Section 8.9), where

$$f(\tau, x, t, y, m, \sigma) = \frac{1}{\sigma \sqrt{2\pi(t - \tau)}} \exp\left\{-\frac{[y - x - m(t - \tau)]^2}{2\sigma^2(t - \tau)}\right\}$$

and $P(X_0 = 0) = 1$. We want to estimate the parameters m and σ^2. We have obtained the observations $x_0 = 0, x_1, \ldots, x_n$ at the moments $t_0, t_1, \ldots, t_n$, respectively, where $0 = t_0 < t_1 < \ldots < t_n$. From formula (13.7.6') we obtain for the likelihood function the expression

$$L = \frac{1}{\sigma^n (2\pi)^{n/2}} \prod_{k=0}^{n-1} \frac{1}{\sqrt{t_{k+1} - t_k}} \exp\left\{-\frac{[x_{k+1} - x_k - m(t_{k+1} - t_k)]^2}{2(t_{k+1} - t_k)\sigma^2}\right\}.$$

By computing $\log L$ and solving the system of equations $\partial \log L / \partial m = 0$ and $\partial \log L / \partial \sigma^2 = 0$, we obtain .

$$(13.7.10) \qquad m^* = \frac{x_n}{t_n},$$

$$\sigma^{*2} = \frac{1}{n} \sum_{k=0}^{n-1} \frac{[x_{k+1} - x_k - m^*(t_{k+1} - t_k)]^2}{t_{k+1} - t_k}.$$

The reader may compare (13.7.10) with formulas (12.4.11).

For other estimation problems for stochastic processes with independent increments, see Rubin and Tucker [1].

Many papers recently published deal with estimation and testing hypotheses for Markov chains. We mention some of them: Good[1], Hoel[1], Bharucha-Reid [1], Goodman [1, 2], Stepanov [1], and Anderson and Goodman [1]. We recommend, in particular, Billingsley's [2] paper.

Of great importance are estimation problems for stationary stochastic processes, particularly the estimation of the spectrum of the process. We mention some papers in this field; Grenander [1], Rosenblatt [1], Parzen

[1], Lommicki and Zaremba [1], Hajek [2], and Jenkins and Priestley [1]. For a more systematic treatment of these questions, see the books by Grenander and Rosenblatt [1] and Hannan [1].

D   Kiefer and Wolfowitz [2] investigated the applications of the maximum likelihood method to nonparametric estimation, that is, when it is necessary to estimate not only the unknown parameters but also the unknown functional form of the distribution function.

We mention briefly other methods of constructing estimates. Neyman [8] gave a method of constructing a large class of estimates which are asymptotically unbiased, asymptotically most efficient, and have the limit normal distribution. Neyman called them *best asymptotically normal estimates* (see also Ferguson [1]). Maximim likelihood estimates are particular cases of such estimates.

A systematic method of reduction (for large samples) of problems of estimation (and testing hypotheses) to similar problems on normal distributions has been developed by Le Cam [1].

Wolfowitz in [7] and in other papers developed a method of constructing estimates which he called the *minimum distance method*. This method can be applied to problems of parametric as well as nonparametric estimation and allows us to obtain estimates which converge not only stochastically, as consistent estimates do, but also with probability one to the unknown parameter or to the unknown distribution function. The method of minimum distance makes essential use of the empirical distribution function. The description of this method would lead us beyond the scope of this book.

In Section 15.3,C we mention estimation by the method of least squares.

13.8 CONFIDENCE INTERVALS

A   Thus far we have dealt only with methods of finding estimates and their classification. We now deal, generally speaking, with methods of finding the probability concerning the mutual position of the estimate of an unknown parameter and its true value.

Suppose that the characteristic X of elements of a population has the density $f(x, Q)$ or the probability function $P(X = x_k) = p_k(Q)$, where Q is an unknown parameter. Let $\zeta = (X_1, \ldots, X_n)$ be the observed random vector; ζ is an n-dimensional random variable whose distribution depends upon the parameter Q. Let $\underline{U}(\zeta)$ and $\bar{U}(\zeta)$ be functions of ζ satisfying the inequality
$$\underline{U}(\zeta) \leqslant \bar{U}(\zeta).$$

The functions $\underline{U}(\zeta)$ and $\bar{U}(\zeta)$ are also random variables dependent on Q; hence the interval $[\underline{U}(\zeta), \bar{U}(\zeta)]$ is a random interval dependent on Q such that its end points as well as its length can be random variables dependent on Q.

Let α be a given number such that $0 < \alpha < 1$. If, whatever may be the value of Q, the relation

(13.8.1) $$P(\underline{U}(\zeta) \leqslant Q \leqslant \bar{U}(\zeta)) \geqslant 1 - \alpha$$

holds, the interval $[\underline{U}(\zeta), \bar{U}(\zeta)]$ is called a *confidence interval*, and $1 - \alpha$ is called the *confidence coefficient*. In most cases, if $\underline{U}(\zeta)$ and $\bar{U}(\zeta)$ are random variables of the continuous type, relation (13.8.1) may be written in the form

(13.8.1') $$P[\underline{U}(\zeta) \leqslant Q \leqslant \bar{U}(\zeta)] = 1 - \alpha.$$

According to Borel's law of large numbers, equality (13.8.1') should be interpreted as follows.

The interval $[\underline{U}(\zeta), \bar{U}(\zeta)]$ is a random interval; hence in general it will be different in different samples. Some of these intervals will contain the true value of parameter Q, some of them will not. However, in a long series of independent samples the frequency of cases in which this interval contains the true value of Q will be approximately equal to $1 - \alpha$. Thus, if we take as Q a value from the confidence interval, the frequency of wrong estimates is approximately equal to α. Inequality (13.8.1) should be interpreted similarly.

We do not speak about the probability that the unknown value of the parameter is contained in some constant interval. Such a statement would have meaning only if the unknown parameter were a random variable. We speak only about the probability that, whatever may be the true value of the unknown parameter Q, the random confidence interval contains it.

It is obvious that for a given α there exist an infinite number of intervals satisfying relation (13.8.1) or (13.8.1'). For obvious reasons, we try to choose the shortest possible interval for a given α.

The notion of a confidence interval was introduced and developed by Neyman in his papers [2, 4, and 5]. In the last one the reader will find theorems concerning the existence of confidence intervals and the method of constructing them. We merely present some examples.

Example 13.8.1. In example 13.4.1, we took the mean value $\bar{X}$ of simple samples as an estimate of the unknown expected value m of the normal distribution $N(m; \sigma)$, where the parameter σ is known. Now we find a confidence interval for m with confidence coefficient $1 - \alpha = 0.95$.

We put

(13.8.2)
$$\underline{U}(\zeta) = \underline{U}(X_1, X_2, \ldots, X_n) = \bar{X} - \frac{1.96}{\sqrt{n}}\sigma,$$

$$\bar{U}(\zeta) = \bar{U}(X_1, X_2, \ldots, X_n) = \bar{X} + \frac{1.96}{\sqrt{n}}\sigma.$$

The random variable $\bar{X}$ has the normal distribution $N(m; \sigma/\sqrt{n})$. From tables of the normal distribution we find

(13.8.3)

$$P\left(-\frac{1.96\sigma}{\sqrt{n}} \leqslant \bar{X} - m \leqslant \frac{1.96\sigma}{\sqrt{n}}\right) = P\left(\bar{X} - \frac{1.96\sigma}{\sqrt{n}} \leqslant m \leqslant \bar{X} + \frac{1.96\sigma}{\sqrt{n}}\right) = 0.95.$$

Here, the probability that the estimate of the unknown parameter m has an error greater in absolute value than $1.96\sigma/\sqrt{n}$ equals 0.05. From Borel's law of large numbers, it follows that in a large series of independent simple samples we shall make errors greater than $1.96\sigma/\sqrt{n}$ in the estimation of m approximately five times out of every 100. We observe that in this example the length of the confidence interval is constant and equals $3.92\sigma/\sqrt{n}$, whereas the end points of the confidence interval are random.

Example 13.8.2.   Suppose that in example 13.8.1 the standard deviation σ is also unknown. We cannot apply the method of example 13.8.1, since we are not able to compute the functions $\underline{U}(\zeta)$ and $\bar{U}(\zeta)$. We use another function of the random variable ζ, Student's t (see Section 9.6), where

(13.8.4) $$t = \frac{\bar{X} - m}{S} \sqrt{n - 1}.$$

We write

(13.8.5) $$\underline{U}(\zeta) = -St_1/\sqrt{n - 1} + \bar{X}, \quad \bar{U}(\zeta) = St_1/\sqrt{n - 1} + \bar{X},$$

where t_1 is found from tables of Student's t distribution with $n - 1$ degrees of freedom as that value t_1 which satisfies the relation

(13.8.6) $$P(-t_1 \leqslant t \leqslant t_1) = 1 - a.$$

From (13.8.4) to (13.8.6) we obtain

(13.8.7) $$P[\underline{U}(\zeta) \leqslant m \leqslant \bar{U}(\zeta)] = 1 - a.$$

If, for instance, $n = 5$ and $1 - \alpha = 0.95$, then from (13.8.6) we obtain $t_1 = 2.78$; hence

$$P\left(\bar{X} - 2.78\frac{S}{2} \leqslant m \leqslant \bar{X} + 2.78\frac{S}{2}\right) = 0.95.$$

In example 13.8.2 not only are the end points of the confidence interval random variables but so is its length, which equals $2.78\,S$. For this problem of estimating the expected value m of the normal population $N(m; \sigma)$ with unknown m and σ, Stein [1] gave a method of constructing confidence intervals with constant length given in advance. Stein's method was generalized to other cases by Barnard [1] and Chapman [1].

Example 13.8.3.   In example 13.6.2, we considered the median $\zeta_k^{(n)}(k = [\frac{1}{2}n] + 1)$ of large simple samples as an estimate of the expected value m of a population, where the characteristic X has the normal distribution $N(m; \sigma)$. We shall find a confidence interval for m in this case.

As we know (see Section 10.5), the median of simple samples from a normal population has the asymptotically normal distribution $N(m;\ \sigma\sqrt{\pi/2n})$. If we disregard the error caused by the asymptoticity of this distribution, we obtain for $1 - \alpha = 0.95$ the following confidence interval:

$$(13.8.8) \qquad P(\zeta_k^{(n)} - 1.96\sigma\sqrt{\pi/2n} \leqslant m \leqslant \zeta_k^{(n)} + 1.96\sigma\sqrt{\pi/2n}) = 0.95.$$

Let us compare the length of the interval obtained, which equals $3.92\sigma\sqrt{\pi/2n}$, with the length of the confidence interval given by (13.8.3), which equals 3.92 $\sigma/\sqrt{n}$. The second confidence interval is shorter. Now, the random variable $\bar{X}$ is the most efficient estimate of m, whereas by (13.6.2) the asymptotic efficiency of the sample median equals only 0.64. Thus, we see the relation between the length of the confidence interval and the efficiency of the estimate; the more efficient is the estimate of the parameter, the shorter is the confidence interval.

Example 13.8.4. In example 13.5.6, we established that if the elements of a population are split into two groups, and the fraction of the first group equals p, and that of the second group $1 - p$, then the most efficient estimate of the unknown p from simple samples of size n equals

$$p^* = \frac{m}{n},$$

where m denotes the number of elements of the first group. Using this estimate, we find the confidence interval for p with $1 - \alpha = 0.95$. We restrict ourselves to large samples.

As we know (see Section 6.7), the estimate p^* has the asymptotically normal distribution

$$N(p;\ \sqrt{p(1 - p)/n}).$$

Hence

$$(13.8.9) \qquad P(p - 1.96\sqrt{p(1 - p)}/n \leqslant p^* \leqslant p + 1.96\sqrt{p(1 - p)}/n) = 0.95.$$

After some simple computations, it turns out that relation (13.8.9) is equivalent to

$$(13.8.9') \qquad P\left\{ \frac{n}{n + 1.96^2} \left[p^* + \frac{1.96^2}{2n} - 1.96\sqrt{\frac{p^*(1 - p^*)}{n} + \left(\frac{1.96}{2n}\right)^2} \right] \right.$$

$$\left. \leqslant p \leqslant \frac{n}{n + 1.96^2} \left[p^* + \frac{1.96^2}{2n} + 1.96\sqrt{\frac{p^*(1 - p^*)}{n} + \left(\frac{1.96}{2n}\right)^2} \right] \right\} = 0.95.$$

This solution is only approximate, since p^* is only asymptotically normal. Tables containing confidence intervals for small n have been constructed by Hald [1]. They are reprinted in the tables of Sadowski [1].

B Using Kolmogorov's theorem 10.11.1 we can, for large simple samples, construct confidence intervals or, rather, confidence domains for an unknown continuous distribution function $F(x)$.

Let $S_n(x)$ be the empirical distribution function of a simple sample of size n (n large) drawn from a population with the unknown continuous distribution function $F(x)$. Suppose that the confidence coefficient is $1 - \alpha$.

Applying the Kolmogorov theorem, we find the number λ_α satisfying the relation

$$P(D_n\sqrt{n} < \lambda_\alpha) = P\left(\sup_x |F(x) - S_n(x)| < \frac{\lambda_\alpha}{\sqrt{n}}\right) \cong 1 - \alpha.$$

This relation means that, with probability $1 - \alpha$, the inequality $|F(x) - S_n(x)| < \lambda_\alpha/\sqrt{n}$ is satisfied for all x in the interval $(-\infty, +\infty)$. In other words, whatever the distribution function $F(x)$, so long as it is continuous, the random domain between the lines $S_n(x) - \lambda_\alpha/\sqrt{n}$ and $S_n(x) + \lambda_\alpha/\sqrt{n}$ contains this distribution function completely, with probability $1 - \alpha$. We can write it in the form

$$P\left(S_n(x) - \frac{\lambda_\alpha}{\sqrt{n}} < F(x) < S_n(x) + \frac{\lambda_\alpha}{\sqrt{n}} \; ; \; -\infty < x < +\infty\right) \cong 1 - \alpha.$$

Thus, we have obtained a confidence domain for a problem of nonparametric estimation. Wald and Wolfowitz [1] considered more general confidence domains containing the unknown distribution function. They considered the case for which λ_α is not constant. Some details can be found in the paper of Malmquist [2].

In all these considerations, concerning the confidence domains for an unknown distribution function, an essential role is played by the empirical distribution function. Some new points, concerning the importance of the empirical distribution function as an estimate of an unknown distribution function, are discussed in the paper of Dvoretzky, Kiefer, and Wolfowitz [2].

Information about recent results on nonparametric estimation are also contained in the paper of Gihman, Gnedenko, and Smirnov [1].

13.9   BAYES THEOREM AND ESTIMATION

A   The parametric problems discussed in the previous section by the method of confidence intervals could also be solved in another way, by the application of Bayes theorem. This theorem for random events was given in Section 1.6. Now we present it for random variables.

We consider the following four cases:

1. The random variable X is of the discrete type and can take on values $x_1, x_2, \ldots$ with probabilities dependent upon the parameter Q, and this parameter is a random variable of the discrete type, which can take on the values $q_1, q_2, \ldots$.

Thus

(13.9.1) $P(X = x_k) = p_k(q_i)$ $(k = 1, 2, \ldots ; i = 1, 2 \ldots)$.

We have observed the event $X = x_k$. What can be said from this observation about the probability function of the random variable Q? In other words: what is the probability $P(Q = q_i \mid X = x_k)$ for $i = 1, 2, \ldots$? From Bayes theorem we obtain

(13.9.2) $P(Q = q_i \mid X = x_k)$

$$= \frac{P(Q = q_i)P(X = x_k \mid Q = q_i)}{\sum_j P(Q = q_j)P(X = x_k \mid Q = q_j)} \qquad (i = 1, 2, 3, \ldots).$$

This conditional distribution of Q under the condition that $X = x_k$ is called the *a posteriori distribution* of Q, whereas the marginal distribution of Q given by the expression $P(Q = q_i)$ for $i = 1, 2, \ldots$ is called the *a priori distribution of Q*.

2. For every value $q_i(i = 1, 2, \ldots)$ of the parameter Q, which is a random variable of the discrete type, the random variable X has the conditional density $f(x \mid q_i)$. Let us consider the event $(Q = q_i \mid x \leqslant X < x + h)$, where $h > 0$.

As before, we obtain

(13.9.3) $P(Q = q_i \mid x \leqslant X < x + h)$

$$= \frac{P(Q = q_i)P(x \leqslant X < x + h \mid Q = q_i)}{\sum_j P(Q = q_j)P(x \leqslant X < x + h \mid Q = q_j)}.$$

If we divide the numerator and denominator on the right-hand side of (13.9.3) by h and pass to the limit as $h \to 0$, we obtain Bayes theorem in the form

(13.9.4) $$P(Q = q_i \mid X = x) = \frac{P(Q = q_i)f(x \mid q_i)}{\sum_j P(Q = q_j)f(x \mid q_j)}.$$

3. The random variable X is of the discrete type and has a probability function dependent upon the parameter Q which is a random variable of the continuous type. By reasoning similar to that in Section 2.7, we obtain

(13.9.5) $$\frac{P(q \leqslant Q < q + h \mid X = x_k)}{h}$$

$$= \frac{P(q \leqslant Q < q + h)P(X = x_k \mid q \leqslant Q < q + h)}{hP(X = x_k)}.$$

Let us denote by $f(q)$ the density of the random variable Q and by $g(q \mid x_k)$

the conditional density of the random variable Q, provided $X = x_k$. We assume that as $h > 0$ the limit

$$\lim_{h \to 0} P(X = x_k \mid q \leqslant Q < q + h) = P(X = x_k \mid Q = q)$$

exists. Since

$$(13.9.6) \qquad g(q \mid x_k) = \lim_{h \to 0} \frac{P(q \leqslant Q < q + h \mid X = x_k)}{h},$$

$$f(q) = \lim_{h \to 0} \frac{P(q \leqslant Q < q + h)}{h},$$

we obtain from (13.9.5)

$$(13.9.7) \qquad g(q \mid x_k) = \frac{f(q)P(X = x_k \mid Q = q)}{P(X = x_k)}.$$

Since

$$P(X = x_k) = \int_{-\infty}^{+\infty} f(q)P(X = x_k \mid Q = q)\, dq,$$

Bayes theorem takes the form

$$(13.9.8) \qquad g(q \mid x_k) = \frac{f(q)P(X = x_k \mid Q = q)}{\displaystyle\int_{-\infty}^{+\infty} f(q)P(X = x_k \mid Q = q)\, dq}.$$

4. Now let (X, Q) be a two-dimensional random variable of the continuous type. Denote by $f(x, q)$, $g(q \mid x)$, and $h(x \mid q)$, respectively, the joint density and the conditional densities of X and Q. Assume that the joint density and the marginal densities $f_1(x)$ and $f_2(q)$ of X and Q satisfy all the conditions (see Section 2.7) sufficient for the existence of conditional densities.

From (2.7.6) we have

$$(13.9.9) \qquad g(q \mid x) = \frac{f(x, q)}{f_1(x)}, \qquad h(x \mid q) = \frac{f(x, q)}{f_2(q)}.$$

From (13.9.9) we obtain

$$(13.9.10) \qquad g(q \mid x) = \frac{f_2(q)h(x \mid q)}{f_1(x)}.$$

Since

$$f_1(x) = \int_{-\infty}^{+\infty} f(x, q)\, dq = \int_{-\infty}^{+\infty} f_2(q)h(x \mid q)\, dq,$$

we obtain Bayes theorem in the form

$$(13.9.11) \qquad g(q \mid x) = \frac{f_2(q)h(x \mid q)}{\displaystyle\int_{-\infty}^{+\infty} f_2(q)h(x \mid q)\, dq}.$$

The terms *a priori distribution* and *a posteriori distribution*, which were mentioned in case 1, also apply to cases 2 to 4.

In the literature, the a posteriori probability which appears in Bayes theorem is also called the *inverse probability*.

B In the problems of estimation so far considered, the parameter Q has not been a random variable but some unknown constant. However, in practice we sometimes deal with cases when the unknown parameter is a random variable.

For instance, let p denote the fraction of defective items in a lot or consignment. Let us consider a collection of lots called, in statistical quality control, a *superlot*. Even if all the lots of this superlot have been produced under the same technological conditions, the parameter p may be different in different lots; hence it may be considered to be a random variable.

If we know the a priori distribution of Q, then from Bayes theorem we can find the conditional distribution of Q provided the event $X = x_k$ has been observed. However, the a priori distribution of the unknown parameter is, in general, unknown. This is the key problem in the applicability of Bayes theorem to the theory of estimation. Usually, we' assume that the a priori distribution of the unknown parameter is uniform in a certain interval, if it is a random variable of the continuous type. If the unknown parameter can take on only a finite number of values, then we assume that each of these values has the same probability. However, the applicability of these assumptions has been the subject of much controversy.

We show by some examples how one can apply Bayes theorem to estimation theory.

Example 13.9.1. From a lot with an unknown fraction of defective items p, we select a simple sample of n elements. Suppose that p is a random variable with density

$$f(p) = \begin{cases} 1 & \text{for} \quad 0 \leqslant p \leqslant 1, \\ 0 & \text{for} \quad p < 0 \text{ and for } p > 1. \end{cases}$$

We have observed m defective items in the sample. What is the probability that p satisfies the inequality $p \leqslant 0.1$?

Let us assign to the appearance of a defective item the number one and to a good item the number zero. Let us denote by X the frequency of defective items in samples of n elements chosen from the lot. The random variable X is of the discrete type with probability function

$$(13.9.12) \quad P\left(X = \frac{m}{n}\right) = \binom{n}{m} p^m (1-p)^{n-m} \quad (m = 0, 1, 2, \ldots, n).$$

The parameter p is a random variable of the continuous type. Thus we use Bayes theorem in the form (13.9.8), where we put p in place of q. We obtain

$$(13.9.13) \quad g\left(p \left| \frac{m}{n} \right.\right) = \frac{\binom{n}{m} p^m (1-p)^{n-m}}{\int_0^1 \binom{n}{m} p^m (1-p)^{n-m} \, dp} = \frac{p^m (1-p)^{n-m}}{B(m+1, n-m+1)}.$$

It follows that the required probability equals

$$\frac{1}{B(m + 1, n - m + 1)} \int_0^{0.10} p^m (1 - p)^{n-m} \, dp.$$

For instance, let $n = 30$ and $m = 2$. From the tables by K. Pearson [2] of the incomplete beta function we find that the required probability equals 0.611.

According to Borel's law of large numbers, we can give the following frequency interpretation of the result obtained. From the superlot we independently choose a series of lots, and from each lot a simple sample of $n = 30$ elements. In approximately 61.1% of the lots such that the samples selected from them contained $m = 2$ defective items, the fraction of defective items is not greater than 0.10.

Let us now look for that value of p which maximizes expression (13.9.13). We find this value from the equation

(13.9.14)
$$\frac{d \log g \left(p \left| \frac{m}{n} \right. \right)}{dp} = \frac{m}{p} - \frac{n - m}{1 - p} = 0.$$

The solution is

(13.9.15) $p^* = m/n.$

We observe that in example 13.5.6 we showed that p^* given by (13.9.15) is the most efficient estimate of p.

Example 13.9.2. The random variable X has the normal distribution $N(m; 1)$, where the parameter m is a random variable. Let us assume that $f(m) = \frac{1}{2}a$ in the interval $(-a, a)$, $a > 0$, and $f(m) = 0$ outside this interval. In a simple sample of n elements we have observed the mean value $\bar{x}$. We find the a posteriori distribution of the random variable m.

We use here the form (13.9.11) of Bayes theorem. Considering the fact that $\bar{X}$ has the normal distribution $N(m; 1/\sqrt{n})$, we have

$$f(m \mid \bar{x}) = \frac{\dfrac{\sqrt{n}}{\sqrt{2\pi}} \exp \left[-\dfrac{n(\bar{x} - m)^2}{2} \right]}{\displaystyle\int_{-a}^{a} \dfrac{\sqrt{n}}{\sqrt{2\pi}} \exp \left[-\dfrac{n(\bar{x} - m)^2}{2} \right]}.$$

Passing to the limit as $a \to \infty$, we obtain

(13.9.16) $f(m \mid \bar{x}) = \dfrac{\sqrt{n}}{\sqrt{2\pi}} \exp \left[\dfrac{-n(\bar{x} - m)^2}{2} \right].$

This is the normal distribution $N(\bar{x}; 1/\sqrt{n})$. The result can be interpreted as in example 13.9.1.

From (13.9.16) we find that

(13.9.17) $P(\bar{x} - 1.96/\sqrt{n} \leqslant m \leqslant \bar{x} + 1.96/\sqrt{n}) = 0.95.$

If we compare the last formula with (13.8.3), it might seem that they are identical.

This is not so. In formula (13.8.3), $\bar{X}$ is a random variable; hence the interval is a random interval and m is a constant; in formula (13.9.17), $\bar{x}$ is constant; hence the interval is a constant interval and m is a random variable. The frequency interpretation of these intervals is, of course, also different.

As has already been said, the applicability of Bayes theorem has been the subject of great controversy among statisticians. The main controversial points are the question of treating the unknown parameter as a random variable and whether the uniformity assumptions can be made. When new ideas in the theory of estimation appeared, mainly due to Neyman and Fisher, there was a tendency to completely abandon applications of Bayes theorem. However, recently the discussion about the Bayesian approach is getting more and more intense and appears to be quite far from being exhausted. This controversy became deeply enframed into the more general discussion on fundamentals of statistical inference, and even, by some writers, on foundations of probability theory. We restrict ourselves to mentioning just a few references on this subject: Mises [4], Steinhaus [2], Jeffreys [1], LeCam [3], Robbins [4], L. J. Savage [4], and A. Birnbaum [2].

Let us also mention that in the general theory of Wald [6], that includes both the theory of hypotheses testing and the theory of estimation, one can also find the influence of the ideas of Bayes theorem on some basic concepts of that theory. Some general information on Wald's theory is given in the last section of Chapter 17.

Problems and Complements

13.1. The characteristic X of elements of a population has the Laplace distribution given by (5.10.12), with λ known but μ unknown. Show that for simple samples of size n the statistic $U = (\zeta_1^{(n)} + \zeta_n^{(n)})/2$ is an unbiased estimate of μ but is not consistent.

13.2. The characteristic X of elements of a population has the Cauchy distribution given by (5.10.1), where μ is unknown and λ is known. Show that (a) $E(\zeta_1^{(n)})$ and $E(\zeta_n^{(n)})$ do not exist; (b) $D^2(\zeta_2^{(n)})$ and $D^2(\zeta_{n-1}^{(n)})$ do not exist; (c) the statistic $(\zeta_k^{(n)} + \zeta_{n-k+1}^{(n)})/2$ is, for $k > 1$, an unbiased estimate of μ but is not consistent.

13.3. The characteristic X of elements of a population has the gamma distribution given by (5.8.6), with p known and b unknown. Show that for $np > 2$

(a) the statistic $U_1 = (np - 1)/n\bar{X}$ is an unbiased, consistent, and sufficient estimate of b.

(b) the statistic U_1 is not the most efficient estimate of b and its efficiency equals $e = (np - 2)/np < 1$.

(c) the statistic U_1 is an asymptotically most efficient estimate of b.

13.4. (Continued). In the preceding problem, let us consider $Q = 1/b$ as the unknown parameter.

(a) Show that the statistic $U_2 = \bar{X}/p$ is a consistent, unbiased, and sufficient estimate of Q.

(b) Find the efficiency of the estimate U_2 of Q.

(c) Find the maximum likelihood estimates of the parameter b in the preceding problem and of the parameter Q in this problem.

Remark. A table in Chapman [2] simplifies the estimation of the parameters of the gamma distribution, when using maximum likelihood estimates.

13.5. The characteristic X of elements of a population has the uniform distribution with density

$$f(x) = \begin{cases} 1/b & (0 \leqslant x \leqslant b), \\ 0 & \text{otherwise.} \end{cases}$$

Show that for simple samples of size n from this population

(a) the statistic $U_1 = (\zeta_k^{(n)} + \zeta_{n-k+1}^{(n)})/2$ is a consistent and unbiased estimate of the parameter $Q_1 = b/2$.

(b) the statistic $U_2 = \dfrac{n+1}{n-1}[\zeta_n^{(n)} - \zeta_1^{(n)}]$ is a consistent and unbiased estimate of the population range $Q_2 = b$.

13.6. (Continued). (a) Show that the Rao-Cramér inequality does not hold for the estimates U_1 and U_2 of the preceding problem. (Such estimates are called *superefficient*.)

(b) Show that U_1 and U_2 are not regular estimates of Q_1 and Q_2, respectively. (Note that the density $f(x)$ is not continuous at the point b of which the parameters Q_1 and Q_2 are functions.)

13.7. Show that if U_1 and U_2 are unbiased estimates of a parameter Q, U_1 is the most efficient estimate of Q and the efficiency e of U_2 satisfies the inequality $0 < e < 1$, then the correlation coefficient of U_1 and U_2 is equal to $\sqrt{e}$.

13.8. Show that if U_1 and U_2 are unbiased and most efficient estimates of the parameter Q, then they are equal, with probability one.

13.9. Show that if U is a sufficient estimate of a parameter Q, and $f(Q)$ is some function of Q, then $f(U)$ is a sufficient estimate of $f(Q)$.

13.10. Show that if we omit the assumption that the estimate U of the parameter Q is unbiased, the right-hand side of the Rao-Cramér inequality should be multiplied by $(1 + dK/dQ)^2$, where K is the bias of the estimate U. (Note that here $E[(U - Q)^2] \neq D^2(U)$.)

13.11. (Notation of Section 13.4). Let the density of the random variable X be $f(x, Q_1, \ldots, Q_r)$, where the parameters $Q_1, \ldots, Q_r$ are unknown. Let the functions

$$u_i = \phi_i(x_1, \ldots, x_n) \qquad (i = 1, \ldots, r),$$
$$y_j = \psi_j(x_1, \ldots, x_n) \qquad (j = 1, \ldots, n - r)$$

satisfy all the assumptions of Section 13.4 with respect to all arguments. The *likelihood function* has the form

$$L = \prod_{k=1}^{n} f(y_k, Q_1, \ldots, Q_r) |J| = g(u_1, \ldots, u_r)$$
$$\times h(y_1, \ldots, y_{n-r} \mid u_1, \ldots, u_r, Q_1, \ldots, Q_r).$$

The vector $(U_1, \ldots, U_r)$ is an estimate of the vector $(Q_1, \ldots, Q_r)$. The notion of regularity can be generalized in an obvious way. A regular estimate $(U_1, \ldots, U_r)$ of the vector $(Q_1, \ldots, Q_r)$ is said to be *jointly sufficient* if the conditional density h does not depend upon $Q_1, \ldots, Q_r$.

Show that for simple samples drawn from the normal population $N(m; \sigma)$, where both parameters are unknown, the vector (U_1, U_2), where $U_1 = \bar{X}$, $U_2 = [n/(n - 1)]S^2$, is a jointly sufficient estimate of the vector (m, σ^2). (Remember that U_2 itself is not a sufficient estimate of σ^2.)

13.12.(a)   Generalize the notion of joint sufficiency of an estimate to random vectors (X, Y) when the unknown parameters are $Q_1, \ldots, Q_r$.

(b) Show that for simple samples drawn from a two-dimensional normal population the vector $(\bar{X}, \bar{Y}, [n/(n - 1)]S_1^2, [n/(n - 1)]RS_1S_2, [n/(n - 1)] S_2^2)$ is a jointly sufficient estimate of the vector $(m_1, m_2, \sigma_1^2, \mu_{11}, \sigma_2^2)$.

13.13. Show that if the characteristic X of elements of a population has the Cauchy distribution given by (5.10.1), then a sufficient estimate of neither of the parameters λ nor μ exists if the other is known, nor does a jointly sufficient estimate of the vector (λ, μ) exist if both of them are unknown (Koopman [1]).

13.14. Show that if the characteristic X of elements of a population has the beta distribution given by (5.9.3), then, for simple samples from this population, the vector (U_1, U_2), where $U_1 = \sqrt[n]{X_1 \ldots X_n}$, $U_2 = \sqrt[n]{(1 - X_1) \ldots (1 - X_n)}$, is a jointly sufficient estimate of (p, q) (U_1 is called the *geometric mean* of $X_1, \ldots, X_n$).

13.15. The characteristic X of elements of a population has the gamma distribution given by (5.8.6) with $b = 1$; hence $E(X) = p$.

(a) Show that the efficiency of the estimate of the unknown parameter p by the method of moments, that is, the efficiency of the estimate $\bar{X}$, is smaller than one for every p and decreases to zero as $p \to 0$. Show, moreover, that it does not depend on n; hence the estimate considered is not asymptotically most efficient.

(b) Find the maximum likelihood estimate of the parameter p, and show that it is asymptotically most efficient.

Hint: Use theorem 13.7.3 or, directly, the Lindeberg-Lévy theorem.

13.16. Show that if U is a sufficient estimate of a parameter Q, the statistic U may serve for the construction of a confidence interval for Q.

13.17. (see Problems 13.3 and 13.4).   Show that for the parameter $Q = 1/b$ of the gamma distribution with known p, we obtain a confidence interval with confidence coefficient $1 - \alpha$, by putting

$$\underline{U}(\zeta) = n\bar{X}/y_1, \quad \bar{U}(\zeta) = n\bar{X}/y_2,$$

where y_1 and y_2 ($y_1 > y_2$) satisfy the relations $P(Y \geqslant y_1) = \alpha_1$, $P(Y \leqslant y_2) = \alpha_2$ ($\alpha_1 + \alpha_2 = \alpha$), and Y has the gamma distribution with density

$$f(y) = y^{np-1} e^{-y}/\Gamma(np).$$

13.18. The characteristic X has the uniform distribution in the interval $[0, Q]$, where Q is unknown. Show (Neyman [4]) that if simple samples of size two are drawn from this population, a confidence interval for Q with confidence coefficient $1 - \alpha$ is given by

$$\underline{U}(\zeta) = \frac{2\bar{X}}{1 + \sqrt{\alpha}}, \quad \bar{U}(\zeta) = \frac{2\bar{X}}{1 - \sqrt{\alpha}}.$$

13.19. (Continued). Another (Neyman [4]) confidence interval, with confidence coefficient $1 - \alpha$, for Q is given by

$$\underline{U}(\zeta) = \zeta_{\frac{1}{2}}^{(2)}, \quad \bar{U}(\zeta) = \frac{\zeta_{\frac{1}{2}}^{(2)}}{\sqrt{\alpha}}.$$

13.20. The characteristic X has a Poisson distribution with unknown parameter λ. Show that for large simple samples we obtain a confidence interval for λ with confidence coefficient $1 - \alpha = 0.95$, by putting

$$\underline{U}(\zeta) = \bar{X} + \frac{1.92}{\sqrt{n}} - \sqrt{\frac{3.84}{n}\bar{X} + \frac{3.69}{n^2}},$$

$$\bar{U}(\zeta) = \bar{X} + \frac{1.92}{\sqrt{n}} + \sqrt{\frac{3.84}{n}\bar{X} + \frac{3.69}{n^2}}.$$

13.21. Using the statistic S^2, construct a confidence interval for the unknown variance σ^2 of the normal population $N(0; \sigma)$.

13.22. The characteristic X has a continuous distribution function $F(x)$. Using the results of Problem 10.14, construct a confidence interval for its median.

13.23. The random variables X_1 and X_2 have continuous distribution functions $F_1(x)$ and $F_2(x)$, respectively, where $F_1(x)$ is known and $F_2(x)$ is not. Let us consider the parameter $p = P(X_2 < X_1)$ and the statistic

$$U = \frac{1}{n} \sum_{j=1}^{n} F_1(X_{2j}),$$

where $X_{21}, \ldots, X_{2n}$ are independent and have the distribution function $F_2(x)$. (a) Show that $E(U) = p$. (b) Show that $P(p < U + \lambda/\sqrt{n}) \geqslant P(D_n^+ < \lambda/\sqrt{n})$, where D_n^+ has been defined in Section 10.11,E. (c) Construct for $n = 100$ a one-sided (upper) confidence interval for p, with confidence coefficient $1 - \alpha = 0.95$. (Z. W. Birnbaum [4], Birnbaum and McCarty [1]).

CHAPTER 14

Methods and Schemes of Sampling

14.1 PRELIMINARY REMARKS

A random sample was defined (see Section 9.1) as a sequence of values of the characteristic X chosen by a random method from the population. The sampling method M was called a random method, if the probability distribution of the observed random vector $(X_1, \ldots, X_n)$ satisfied relations (9.1.1) and (9.1.2).

A random sample was called simple if the random variables $X_1, \ldots, X_n$ were independent.

The theory of estimation and of testing statistical hypotheses previously discussed can be applied only to random samples. In practical problems we often meet the question of whether a given sample or the method of its choice is random. The importance of randomness is explained by an example.

Example 14.1.1. Suppose that we want to determine by a sample investigation the distribution, according to sex, of people employed in Warsaw in 1951. Suppose that we take as our sample all employees in one of the big offices in Warsaw. This sample is obviously bad and its distribution according to sex cannot be considered as an approximation to the distribution of all employees in Warsaw according to sex, since the fraction of women employed in offices is considerably larger than in factories.

Suppose, now, that the aim of our sample investigation is to determine the fraction of all employees in Warsaw who have been vaccinated against typhoid fever and suppose that the vaccination has been performed in all offices and factories. Then it is quite possible that the sample consisting of all employees in one big office is a random sample.

Let us consider in this example the two dimensional distribution of all people employed in Warsaw according to two characteristics, sex (male and female) and place of employment (office and factory). These two characteristics are not independent; thus one of these characteristics cannot serve as a criterion of randomness of the sample, if we want to investigate the distribution of the population by the second characteristic.

503

A sampling method is random only if the characteristic which serves as a criterion for selecting sample elements is independent of the characteristic under investigation. It follows that a sampling method may be random with respect to one characteristic and not random with respect to another.

In the mathematical statistics literature one very frequently speaks of an investigation by the *representative method*, but this notion is not defined precisely. By the term representative method, we shall mean a random sampling method, where the sample may or may not be simple.

An important notion is that of a *sampling unit*. A sampling unit may be *individual* if it consists of only one element of the population, or *collective* if it consists of a collection of elements of the population.

Example 14.1.2. Suppose that we investigate the results of a census concerning sex, age, and family status, using the representative method. The population consists of all people covered by the census. If we take as a sampling unit a single person, the sampling unit is individual. If, however, we take as a sampling unit not a single person but a household, then the sampling unit is collective.

The results of a sampling investigation may contain some errors, which means that there may be some differences between the sample estimates and the true population values of the characteristics investigated. These errors are called *random errors* if the sampling method is random. The magnitude of error is then the result of the random selection of the sample. If, however, the sampling method is not random, the errors of such an investigation are called *systematic errors*. Thus, for instance, in example 14.1.1 the errors in estimating the distribution of the population of employees in Warsaw according to sex are systematic, since, no matter which office was used for the sample, the results would always contain errors. Moreover, these errors would be one-sided, that is, whichever office was used, the fraction of women employed would almost always be greater than the fraction of all the women employed in Warsaw.

14.2   METHODS OF RANDOM SAMPLING

A   In the first part of example 14.1.1 it was easy to predict in advance that the samping method applied there was not random with respect to the sex of employees in Warsaw. However, it is not always so easy to decide the question of randomness of a sampling method; sometimes special investigations are necessary. Usually these investigations consist of applying significance tests, which have been discussed in Chapter 12. Thus, the randomness of a sampling method is a hypothesis that has to be tested by a suitably constructed test. We illustrate it by an example.

Example 14.2.1. ZUS (The Social Insurance Agency of Poland) has investigated the distribution of the whole population of people insured according to age and type of employment; in the investigation, the first-letter sampling method was applied, that is, the sample consisted of all people whose family names started with the letter P. The analysis of the results of this experiment, performed by the author, showed that this sampling method cannot be considered as a random method with respect to the characteristics considered. We present a part of this analysis.

TABLE 14.2.1

	N_i	n_i	p_i	np_i
Workers other than those in the coal and steel industries	1,778,446	152,812	0.64504	148,638
Workers in the coal and steel industries	250,397	22,493	0.09082	20,928
White-collar and professional workers	564,147	44,040	0.20461	47,149
Civil servants	164,141	11,088	0.05953	13,718
Total	2,757,131	230,433	1.00000	230,433

The whole population of workers was split into four groups, as is shown in the second column of Table 14.2.1. The number N_i denotes the exact numbers of elements of each group in the whole population, and n_i denotes the number of elements of the ith group in the sample. N and n denote, respectively, the size of the population and the size of the sample. Let

$$p_i = \frac{N_i}{N}.$$

We test the hypothesis H_0 that the first-letter sampling method is a random method. We apply the χ^2 test. Thus, if H_0 is true, then by (5.2.4), for the random variables n_i we have the relation[1] $E(n_i) = np_i$. From Table 14.2.1 we compute

$$\chi^2 = \sum_{i=1}^{4} \frac{(n_i - np_i)^2}{np_i} = 943.$$

From tables of the χ^2 distribution we find that for three degrees of freedom the probability of obtaining a value of χ^2 not smaller than the observed value is extremely small (much smaller than 0.01). Thus we reject H_0, although it was difficult to predict at first sight the existence of any dependence between the kind of employment and the first letter of the family name.

B Tables of *random numbers*, of which there are several, serve as a device for drawing simple samples. The first such collection, the tables of random numbers by Tippet [2], was published in 1927. In addition, there

---

[1] Up to now, we have, in general, followed the rule of denoting random variables by capital letters and their possible values by small letters. Starting from this chapter, we do not obey this rule, and, without fear of being misunderstood, we mostly denote by small letters the random variables as well as their possible values.

are tables of random numbers of Fisher and Yates [1], Kadyrov [1], Kendall and Smith [1], Steinhaus [3], and Vielrose [1]. The techniques of obtaining these numbers vary, but it is not the technique of obtaining the random numbers that determines their quality, but whether they satisfy certain conditions of randomness. We mention some of them.

1. In sufficiently large sections of these tables each of the digits 0 to 9 should appear with approximately the same frequency, and the deviations from the theoretical frequency should not be too large, so that they can be explained as a result of random fluctuations.

2. This also applies to pairs, triples, and so on, consisting of the same digits.

3. Successive digits of tables of random numbers, treated as random variables, should be independent.

The authors of tables of random numbers performed various tests to show that these conditions are satisfied in their tables. Thus, for instance, Kendall and Smith performed the following tests on each thousand random digits in their tables:

I. They computed how many times each of the digits 0 to 9 appears. If the hypothesis H_0 that the tables are random is true, one might expect that each digit will appear 100 times. The deviations between observed and theoretical frequencies were tested by the χ^2 test at the significance level 0.05.

II. Among the four-digit random numbers, they computed the number of appearances of four identical digits, three identical and one different digits, two identical and two different digits, and four different digits. This test was performed to determine whether some digits have a tendency to form long runs. The deviations were tested by the χ^2 test at the significance level 0.05.

III. The numbers of digits between successive zeros were computed and compared with the expected ones. This test was performed to determine whether some digits have a tendency to appear too rarely at some places. The observed deviations were tested by the χ^2 test at the significance level 0.05.

Tests II and III are independence tests. In general, one would say that tables of random numbers are undoubtedly good, if we want to choose at random a series of samples (of arbitrary size), or one big sample. If, however, we want to choose a single small sample, tables of random numbers may give a sample which does not characterize the population properly. However, using tables of random numbers we may at least be sure that we have not had any personal influence upon the tendentious choice of the sample.

To use tables of random numbers, we have to assign numbers to all sampling units of the population. The choice is then performed in the following way. Starting from a randomly chosen digit, we choose a suitable number of digits from the tables of random numbers, and we include in the sample those elements of the population that correspond to the numbers chosen from the tables.

C   A test for randomness of a sample can also be constructed using the theory of runs. We illustrate this by example 11.3.1, to which we now return.

Example 14.2.2. From a population, in which the characteristic x may take on the values a_1 and a_2, each with probability 0.5, a simple sample has been drawn. The sample consists of the following eight elements chosen in the order shown:

$$a_1, a_1, a_1, a_1, a_2, a_1, a_2, a_2.$$

We wish to determine whether the sample is chosen randomly, that is, whether the characteristic which serves as a basis for the choice of the sample is independent of the characteristic under investigation. Let the hypothesis H_0 state that the sample is chosen at random. We test H_0 at the significance level 0.05. We have

$$n_1 = 5, \quad n_2 = 3, \quad k = 4.$$

From formulas (11.3.10) and (11.3.10′), we obtain

$$P(k = 1) = 0, \quad P(k = 2) = 0.0078,$$
$$P(k = 3) = 0.0234, \quad P(k = 4) = 0.0625;$$

hence $P(k \leqslant 4) = 0.0937$. Thus we have no reason to reject the hypothesis, in spite of the fact that the appearance of a run of terms a_1 of length four may create some doubt as to whether the sample has been chosen at random. We also observe that from formulas (11.4.10) we obtain $E(k) = 4.5$ and $D^2(k) = \frac{7}{4}$; hence the observed value $k = 4$ differs from the expected value by even less than one standard deviation.

Other applications of the theory of runs to testing randomness of samples can be found in Levene's [1] paper.

D   As has already been stated, it is not the technique of construction that determines the quality of tables of random numbers but the important question of whether they satisfy certain conditions of randomness. Steinhaus [3] constructed tables of random numbers by systematic shuffling of natural numbers. These tables, called tables of *shuffled numbers*, contain all quadruples of numbers from 0000 to 9999, and each quadruple appears exactly once. The starting point of the construction was the sequence of the four last digits of numbers of the form $4567n$ ($n = 1, 2, 3, \ldots$). The permutation of all four-digit numbers obtained in such a way was then shuffled several times, so that it is practically impossible to find any relationship between consecutive numbers. These tables, if we read them from top to bottom, have all the properties of tables of random numbers.

Further details of the construction and methods of use of these tables is given in the paper of Steinhaus [3].

Example 14.2.3. We use the theory of runs to test the randomness of Steinhaus' tables of shuffled numbers. We selected at random the first and second columns in Table XVII. Let us rewrite these columns in the form of rows

6 4 2 1 4 6 0 7 1 4 5 3 2 4 4 7 4 6 9 7 7 8 5 7 8 7 7

2 2 4 0 4 9 9 6 7 3 8 9 1 3 4 2 5 1 9 8 3 9 8 8 5 3 8

3 9 9 9 0 9 8 5 1 2 5 8 5 2 1 6 5 8 5 3 0 9 1 4 9 8.

Let us assign the number zero to an even number, and the number one to an odd number. Next, we split these numbers into ten groups of eight elements each. We obtain the following zero-one sequences

(*)   (1) 00010001,   (2) 10110001,   (3) 00111011,   (4) 01100000,

      (5) 11011011,   (6) 10011101,   (7) 10011011,   (8) 11010110,

      (9) 10101010, (10) 11011010.

From the description of the construction of the tables, it follows that the number of odd digits and of even digits are equal; hence the probability of appearance of one in these sequences is the same as the probability of appearance of zero; thus both equal 0.5.

TABLE 14.2.2

Successive number of sequence (*)	n_0	n_1	k	p
1	6	2	4	0.07
2	4	4	5	0.17
3	3	5	4	0.09
4	6	2	3	0.03
5	2	6	5	0.11
6	3	5	5	0.16
7	3	5	5	0.16
8	3	5	6	0.20
9	4	4	8	0.27
10	3	5	6	0.20

In Table 14.2.2, k denotes the number of runs, and n_0 and n_1 denote, respectively, the number of zeros and ones in the particular sequences of (*). The last column gives the probability p of appearance of a number of runs which does not exceed the number of runs observed, where this probability is obtained from formulas (11.3.10) and (11.3.10′). As we see in Table 14.2.2, p is smaller than 0.05 only for one sequence of (*) and for all sequences of (*) p is greater than 0.01. Thus, both the odd numbers and the even numbers do not show any tendency to form runs which cannot be explained by random fluctuations. Thus the test performed does not give any reason to reject the hypothesis that Steinhaus' tables of shuffled numbers are random.

E Another method of constructing tables of random numbers has been proposed by Horton [1] and Dvoretzky and Wolfowitz [1]. The starting point of this method is the observation of a sequence of random variables x_k which take on only positive integer values. Let us denote $a_n = x_1 + x_2 + \ldots + x_n$. If we wish to obtain tables of random digits containing all digits $0, \ldots, 9$, then, for large values of n, we reduce the observed values a_n, modulo 10, and write successively the numbers obtained in this way.

We remind the reader that reduction of a number a_n, modulo 10, means that if $a_n = j + 10r$, where j is one of the numbers $0, \ldots, 9$, and r is any non-negative integer, then instead of a_n we take j. As the authors mentioned have shown, tables of digits obtained in this way satisfy, under very general assumptions concerning the distributions of the random variables x_k, the randomness conditions 1 and 2 formulated in part **B** of this section.

14.3 SCHEMES OF INDEPENDENT AND DEPENDENT RANDOM SAMPLING

For a simple sample, the probability of choosing an element of the sample is constant during the process of sampling. The scheme of simple sampling is also called the *scheme of independent sampling*. We have the scheme of independent sampling if the population from which we draw the sample has an infinite number of copies of each of the different elements contained in it. However, in practical sample investigations, we always have populations containing only a finite number of elements, but the sampling scheme is still approximately a scheme of independent sampling if the population is large in comparison with the size of the sample. If we draw a sample from a finite population which is not sufficiently large in comparison with the size of the sample, then, to obtain a scheme of independent sampling, we should return each drawn element to the population before drawing the next one. Such a scheme is called *scheme of sampling with replacement*.

If the probability of drawing the particular elements changes during the process of drawing the sample, we say that a *scheme of dependent sampling* is applied.

If we draw elements from a finite population and do not return the element drawn to the population before drawing the next one, we have a *scheme of sampling without replacement*. This is a particular case of dependent sampling.

We restrict our considerations to a comparison of the efficiency of sampling with replacement and sampling without replacement in estimating the expected value m of the characteristic x of a population of size N, whose every element has the same probability $1/N$ of being selected.

Let $\bar{x}_1$ and $\bar{x}_2$ be estimates of m when the scheme of sampling with replacement and without replacement, respectively, was applied. We compute the variances $\sigma_{\bar{x}_1}{}^2$ and $\sigma_{\bar{x}_2}{}^2$ of these estimates. We have

$$(14.3.1) \qquad m = \frac{1}{N} \sum_{j=1}^{N} x_j,$$

where x_j is the value of x of the jth element of the population. In the population, the variance of x is

$$(14.3.2) \quad \sigma^2 = \frac{1}{N} \sum_{j=1}^{N} (x_j - m)^2 = \frac{1}{N} \sum_{j=1}^{N} x_j{}^2 - m^2 = E(x^2) - m^2.$$

By formula (3.2.14) and theorem 3.6.3, we obtain for the scheme of sampling with replacement the variance

$$(14.3.3) \qquad \sigma_{\bar{x}_1}{}^2 = \frac{\sigma^2}{n}.$$

The variance for the scheme of sampling without replacement is

$$(14.3.4) \qquad \sigma_{\bar{x}_2}{}^2 = E[\bar{x}_2 - E(\bar{x}_2)]^2 = E(\bar{x}_2{}^2) - [E(\bar{x}_2)]^2.$$

Let us denote by z_k the value of x of the kth element of the sample when sampling without replacement was applied. We have

$$(14.3.5) \qquad \bar{x}_2 = \sum_{k=1}^{n} \frac{z_k}{n},$$

$$E(\bar{x}_2) = \sum_{k=1}^{n} E\left(\frac{z_k}{n}\right) = E(z_k) = m.$$

We find

$$(14.3.6) \quad E(\bar{x}_2{}^2) = E\left(\sum_{k=1}^{n} \frac{z_k}{n}\right)^2 = E\left(\frac{1}{n^2} \sum_{k=1}^{n} z_k{}^2\right) + \frac{2}{n^2} E\left(\sum_{k=1}^{n-1} \sum_{l=k+1}^{n} z_k z_l\right).$$

Since

$$E(z_k{}^2) = E(x^2) = \sigma^2 + m^2,$$

we have

$$(14.3.7) \qquad E\left[\frac{1}{n^2} \sum_{k=1}^{n} z_k{}^2\right] = \frac{1}{n^2} \sum_{k=1}^{n} E(z_k{}^2) = \frac{1}{n} (\sigma^2 + m^2).$$

By (14.3.1),

$$\sum_{j=1}^{N} (x_j - m) = 0.$$

Squaring this expression, we obtain

$$\sum_{j=1}^{N} (x_j - m)^2 + 2 \sum_{s=1}^{N-1} \sum_{r=s+1}^{N} (x_s - m)(x_r - m) = 0.$$

Hence after considering (14.3.2), we have

$$2 \sum_{s=1}^{N-1} \sum_{r=s+1}^{N} (x_s - m)(x_r - m) = - \sum_{j=1}^{N} (x_j - m)^2 = -N\sigma^2.$$

Thus, for $s \neq r$,

$$(14.3.8) \qquad E[(x_s - m)(x_r - m)] = - \frac{\sigma^2}{N-1}.$$

Since $E(z_k z_l) = E[(z_k - m)(z_l - m)] + m^2$, we obtain, using (14.3.8),

$$(14.3.9) \quad \frac{2}{n^2} E\left(\sum_{k=1}^{n-1} \sum_{l=k+1}^{n} z_k z_l \right) = \frac{2}{n^2} \sum_{k=1}^{n-1} \sum_{l=k+1}^{n} E(z_k z_l)$$

$$= \frac{2}{n^2} \sum_{k=1}^{n-1} \sum_{l=k+1}^{n} \{ E[(z_k - m)(z_l - m)] + m^2 \}$$

$$= \frac{n-1}{n} m^2 - \frac{n-1}{n(N-1)} \sigma^2.$$

From (14.3.6), (14.3.7), and (14.3.9), we obtain

$$E(\bar{x}_2^{\,2}) = \frac{N-n}{N-1} \cdot \frac{\sigma^2}{n} + m^2.$$

Finally,

$$(14.3.10) \qquad \sigma_{\bar{x}_2}^{\,2} = \frac{N-n}{N-1} \cdot \frac{\sigma^2}{n}.$$

Since the ratio $(N-n)/(N-1)$ is smaller than one for $n > 1$, comparing formulas (14.3.3) and (14.3.10), we see that

$$(14.3.11) \qquad \sigma_{\bar{x}_2}^{\,2} < \sigma_{\bar{x}_1}^{\,2}.$$

Hence the estimate $\bar{x}_2$ is more efficient than $\bar{x}_1$.

We have shown that the scheme of sampling without replacement is more efficient than the scheme of sampling with replacement for the estimation of the parameter m. If the size of the population N is large in comparison with n, the difference in efficiency between these two estimates is negligible.

Suppose that the characteristic x of elements of a population may take on only two values, $x_1 = 1$ and $x_2 = 0$, and that the number of elements of the first group, for which $x = 1$, is Np and the number of elements of the second group, for which $x = 0$, is Nq, where $0 < p < 1$ and $q = 1 - p$; hence

$$(14.3.12) \qquad m = p, \qquad \sigma^2 = pq.$$

Suppose that we choose a sample of size n from this population, applying the scheme of sampling without replacement. Let us denote by $z_1, \ldots, z_n$ the values of the characteristic x of the sample elements and by $\bar{x}$ the sample mean

$$\bar{x} = \frac{1}{n} \sum_{k=1}^{n} z_k.$$

Using combinatorial considerations, we compute the probability that $\bar{x} = r/n$, where $\max(0, n - Nq) \leqslant r \leqslant \min(n, Np)$. The equality $\bar{x} = r/n$ means that we have r elements of the first kind and $n - r$ elements of the second kind in the sample. The number of possible ways of selecting n out of N elements (without considering the order of their appearance) is $\binom{N}{n}$. Similarly, the number of possible ways of choosing r elements among Np elements is $\binom{Np}{r}$, and the number of possible ways of selecting $n - r$ elements out of Nq elements is $\binom{Nq}{n-r}$. Hence

$$(14.3.13) \qquad P\left(\bar{x} = \frac{r}{n}\right) = \binom{Np}{r}\binom{Nq}{n-r} \Big/ \binom{N}{n}.$$

This is the probability function of the hypergeometric distribution. Formula (14.3.13) was obtained in a different way in Section 5.4.

From (14.3.5) and (14.3.12), we obtain

$$(14.3.14) \qquad E(\bar{x}) = p,$$

and from (14.3.10) and (14.3.12), we have

$$(14.3.15) \qquad \sigma_{\bar{x}}^2 = \frac{N - n}{N - 1} \cdot \frac{pq}{n}.$$

14.4 SCHEMES OF UNRESTRICTED AND STRATIFIED RANDOM SAMPLING

A   If we draw a random sample from the whole population without dividing it into parts before drawing, we speak of *unrestricted random sampling*. However, if before the drawing we divide the population into parts and draw random samples from each part separately, we speak of *stratified random sampling*.

The aim of stratified sampling is to increase the efficiency of the sample investigation; hence the dividing of the population must be performed in a suitable way.

Suppose that we divide the population into k strata. Denote, for $i = 1, 2, \ldots, k$,

$N_i =$ the number of elements of the ith population stratum,

$N = \sum_{i=1}^{k} N_i$, the size of the population,

$n_i =$ the size of the sample drawn from the ith stratum,

$n = \sum_{i=1}^{k} n_i$, the size of the whole sample,

$x_{ij} =$ the value of the characteristic x under investigation of the jth element of the ith stratum of the population,

$m_i =$ the expected value of x inside the ith stratum of the population,

$m =$ the expected value of x in the population,

$\sigma_i^2 =$ the variance of x inside the ith stratum of the population,

$\hat{\sigma}^2 =$ the variance of the m_i between the strata of the population,

$\sigma_0^2 =$ the variance of x in the whole population.

Thus we have the equalities

$$(14.4.1) \qquad m_i = \frac{\sum_{j=1}^{N_i} x_{ij}}{N_i},$$

$$(14.4.2) \qquad m = \sum_{i=1}^{k} \sum_{j=1}^{N_i} \frac{x_{ij}}{N},$$

$$(14.4.3) \qquad \sigma_i^2 = \frac{1}{N_i} \sum_{j=1}^{N_i} (x_{ij} - m_i)^2,$$

$$(14.4.4) \qquad \hat{\sigma}^2 = \frac{1}{N} \sum_{i=1}^{k} N_i (m_i - m)^2,$$

$$(14.4.5) \qquad \sigma_0^2 = \frac{1}{N} \sum_{i=1}^{k} \sum_{j=1}^{N_i} (x_{ij} - m)^2.$$

Expression (14.4.5) may be transformed in the following way:

(14.4.6)

$$\sigma_0^2 = \frac{1}{N} \sum_{i=1}^{k} \sum_{j=1}^{N_i} (x_{ij} - m_i + m_i - m)^2$$

$$= \frac{1}{N} \sum_{i=1}^{k} \left[\sum_{j=1}^{N_i} (x_{ij} - m_i)^2 + 2(m_i - m) \sum_{j=1}^{N_i} (x_{ij} - m_i) + N_i (m_i - m)^2 \right].$$

By (14.4.1), we have $\sum_{i=1}^{N_i} (x_{ij} - m_i) = 0$; hence

$$(14.4.7) \qquad \sigma_0^2 = \frac{1}{N} \sum_{i=1}^{k} \left[\sum_{j=1}^{N_i} (x_{ij} - m_i)^2 + N_i (m_i - m)^2 \right].$$

Thus, using (14.4.3) and (14.4.4), we obtain

(14.4.8)
$$\sigma_0^2 = \frac{1}{N} \sum_{i=1}^{k} N_i \sigma_i^2 + \hat{\sigma}^2.$$

As we see from (14.4.8), the total variance of the population may be represented as a sum of two terms, one of which represents the weighted average of variances inside the individual strata, and the second represents the variance between the strata of the populations. Since the variance σ_0^2 of the population is constant, any increasing of the variance between strata decreases the variances inside the individual strata. In other words, if we divide the whole population into strata which differ greatly in the characteristic under investigation, the variance $\hat{\sigma}^2$ will be large and the variances σ_i^2 will be suitably small.

Let us denote by $\bar{z}_i$ the sample mean of the characteristic x inside the ith stratum of the sample, and let $\bar{x}_1$, $\bar{x}_2$, $\sigma_{\bar{x}_1}^2$, and $\sigma_{\bar{x}_2}^2$ have the same meaning as in the previous section. Let us take as an estimate of the parameter m the expression

(14.4.9)
$$\bar{x} = \sum_{i=1}^{k} \frac{N_i}{N} \bar{z}_i.$$

It is easy to prove that $\bar{x}$ is an unbiased estimate of m.

By the theorem that the variance of a sum of independent random variables equals the sum of their variances, and using formulas (14.3.3) and (14.3.10), we obtain for the scheme of sampling with replacement

(14.4.10)
$$_w\sigma_{\bar{x}_1}^2 = \frac{1}{N^2} \sum_{i=1}^{k} \frac{N_i^2 \sigma_i^2}{n_i},$$

and for the scheme of sampling without replacement

(14.4.11)
$$_w\sigma_{\bar{x}_2}^2 = \frac{1}{N^2} \sum_{i=1}^{k} \frac{N_i - n_i}{N_i - 1} \cdot \frac{N_i^2 \sigma_i^2}{n_i}.$$

The letter w on the left-hand side of (14.4.10) and (14.4.11) indicates that we are dealing here with stratified sampling.

If we apply unrestricted sampling, we obtain the following formulas, which correspond to formulas (14.4.10) and (14.4.11):

(14.4.12)
$$_n\sigma_{\bar{x}_1}^2 = \frac{\sigma_0^2}{n} = \frac{1}{n} \left[\frac{1}{N} \sum_{i=1}^{k} N_i \sigma_i^2 + \hat{\sigma}^2 \right],$$

(14.4.13)
$$_n\sigma_{\bar{x}_2}^2 = \frac{N - n}{n(N - 1)} \sigma_0^2 = \frac{N - n}{n(N - 1)} \left[\frac{1}{N} \sum_{i=1}^{k} N_i \sigma_i^2 + \hat{\sigma}^2 \right].$$

The letter n on the left-hand side of (14.4.12) and (14.4.13) indicates that these formulas hold for unrestricted sampling

B   We now show that the variances (14.4.12) and (14.4.13) are, in general, bigger than the corresponding variances (14.4.10) and (14.4.11); hence, in general, the scheme of stratified random sampling is more efficient than the scheme of unrestricted random sampling.

We now make a detailed analysis of the conditions under which such an inequality will be satisfied. We restrict ourselves to sampling without replacement. Thus we compare (14.4.11) with (14.4.13).

Let us first consider the case when N and N_i are large enough to replace $N - 1$ and $N_i - 1$ by N and N_i, respectively. Expressions (14.4.11) and (14.4.13) then take the forms

$$(14.4.11') \qquad {}_w\sigma_{\bar{x}_2}{}^2 = \frac{1}{N^2} \sum_{i=1}^{k} \frac{N_i - n_i}{N_i} \cdot \frac{N_i^2 \sigma_i^2}{n_i},$$

$$(14.4.13') \qquad {}_n\sigma_{\bar{x}_2}{}^2 = \frac{N - n}{nN} \left[\frac{1}{N} \sum_{i=1}^{k} N_i \sigma_i^2 + \hat{\sigma}^2 \right].$$

Let us consider the difference

$$(14.4.14)$$

$$\begin{aligned}
{}_n\sigma_{\bar{x}_2}{}^2 - {}_w\sigma_{\bar{x}_2}{}^2 &= \frac{N - n}{nN^2} \sum_{i=1}^{k} N_i \sigma_i^2 + \frac{N - n}{nN} \hat{\sigma}^2 - \frac{1}{N^2} \sum_{i=1}^{k} \frac{N_i - n_i}{n_i N_i} N_i^2 \sigma_i^2 \\
&= \frac{1}{N^2} \sum_{i=1}^{k} N_i \sigma_i^2 \left(\frac{N - n}{n} - \frac{N_i - n_i}{n_i} \right) + \frac{N - n}{nN} \hat{\sigma}^2 \\
&= \frac{1}{N^2} \sum_{i=1}^{k} N_i \sigma_i^2 \left(\frac{N}{n} - \frac{N_i}{n_i} \right) + \frac{N - n}{nN} \hat{\sigma}^2 = A + B,
\end{aligned}$$

where

$$A = \frac{1}{N^2} \sum_{i=1}^{k} N_i \sigma_i^2 \left(\frac{N}{n} - \frac{N_i}{n_i} \right), \qquad B = \frac{N - n}{nN} \hat{\sigma}^2.$$

Expression B in (14.4.14) is always non-negative. It is equal to zero only if $\hat{\sigma}^2 = 0$; hence if the expected values of x inside individual strata are equal. Expression A may be positive as well as negative. As has already been said, for a given σ_0, the larger $\hat{\sigma}$ is (hence so is B), the smaller the σ_i-s are. Therefore, if B is large enough, A is small in absolute value, and $A + B > 0$, and then stratified sampling is more efficient than the unrestricted one. On the other hand, if B is small and $-A > B$,

$$ {}_n\sigma_{\bar{x}_2}{}^2 - {}_w\sigma_{\bar{x}_2}{}^2 < 0.$$

Then unrestricted sampling is more efficient than stratified sampling. Consequently, when planning a stratified sample investigation, we should avoid dividing the population into strata such that the expected values of the characteristic investigated inside the individual strata are close or equal.

If $N/n = N_i/n_i$ for every i, that is, if we include the same part of each stratum in the sample, then we speak of *proportional stratified random sampling*. Here, expression A in (14.4.14) equals zero. It follows that if $\hat{\sigma}^2 \neq 0$, proportional stratified sampling is more efficient than unrestricted sampling.

This result is valid only approximately, since we have replaced $N - 1$ and $N_i - 1$ by N and N_i, respectively. If, however, we do not make this replacement, expression (14.4.14) becomes

$$(14.4.15) \quad {}_n\sigma_{\bar{x}_2}^2 - {}_w\sigma_{\bar{x}_2}^2 = \sum_{i=1}^{k} N_i\sigma_i^2 \left[\frac{N - n}{nN(N - 1)} - \frac{(N_i - n_i)N_i}{n_i(N_i - 1)N^2} \right]$$

$$+ \frac{N - n}{n(N - 1)} \hat{\sigma}^2 = A' + B',$$

where

$$A' = \sum_{i=1}^{k} N_i\sigma_i^2 \left[\frac{N - n}{nN(N - 1)} - \frac{(N_i - n_i)N_i}{n_i(N_i - 1)N^2} \right], \quad B' = \frac{N - n}{n(N - 1)} \hat{\sigma}^2.$$

From the equality

$$\frac{N}{n} = \frac{N_i}{n_i},$$

it follows that

$$\frac{N - n}{n} = \frac{N_i - n_i}{n_i}.$$

Thus

$$(14.4.16) \quad A' = \frac{N - n}{nN^2} \sum_{i=1}^{k} N_i\sigma_i^2 \left[\frac{N}{N - 1} - \frac{N_i}{N_i - 1} \right]$$

$$= \frac{N - n}{nN^2} \sum_{i=1}^{k} N_i\sigma_i^2 \frac{(N_i - N)}{(N - 1)(N_i - 1)} < 0.$$

Thus it is not excluded that $A' + B' < 0$. In other words, it is possible that unrestricted sampling is more efficient than stratified proportional sampling. This was noticed by Armitage [1].

C Since in sample investigations we wish to maximize the efficiency, let us choose, for a fixed sample size n, numbers n_i such that expression (14.4.11) assumes its minimum value.

The scheme of stratified random sampling in which the numbers n_i are selected in this way is called the *optimal scheme of random sampling*. The n_i are called *optimal numbers*. The idea of the optimal scheme of random sampling is due to Tschuprov [1] and Neyman [1].

Let us rewrite formula (14.4.11) in the following way

$$(14.4.17) \quad {}_w\sigma_{\bar{x}_2}^2 = \frac{1}{N^2} \sum_{i=1}^{k} \frac{N_i^3\sigma_i^2}{n_i(N_i - 1)} - \frac{1}{N^2} \sum_{i=1}^{k} \frac{N_i^2\sigma_i^2}{N_i - 1}.$$

Since the sample size n is fixed, the n_i satisfy the equation

$$\sum_{i=1}^{k} n_i = n.$$

From this equation we express n_k in terms of the remaining n_i,

(14.4.18)
$$n_k = n - \sum_{i=1}^{k-1} n_i.$$

From (14.4.17) and (14.4.18), we obtain

(14.4.19)

$$_w\sigma_{\bar{x}_2}^2 = \frac{1}{N^2} \sum_{i=1}^{k-1} \frac{N_i^3 \sigma_i^2}{N_i - 1} \cdot \frac{1}{n_i} + \frac{1}{N^2} \cdot \frac{N_k^3 \sigma_k^2}{N_k - 1} \cdot \frac{1}{n - \sum\limits_{i=1}^{k-1} n_i} - \frac{1}{N^2} \sum_{i=1}^{k} \frac{N_i^2 \sigma_i^2}{N_i - 1}.$$

Hence

$$\frac{\partial\, _w\sigma_{\bar{x}_2}^2}{\partial n_i} = -\frac{1}{N^2} \cdot \frac{N_i^3 \sigma_i^2}{N_i - 1} \cdot \frac{1}{n_i^2} + \frac{1}{N^2} \cdot \frac{N_k^3 \sigma_k^2}{N_k - 1} \cdot \frac{1}{\left(n - \sum\limits_{i=1}^{k-1} n_i\right)^2}$$

$$= -\frac{1}{N^2}\left(\frac{N_i^3 \sigma_i^2}{N_i - 1} \cdot \frac{1}{n_i^2} - \frac{N_k^3 \sigma_k^2}{N_k - 1} \cdot \frac{1}{n_k^2}\right) \qquad (i = 1, 2, \ldots, k - 1).$$

The collection of numbers n_i which minimize expression (14.4.11) can be obtained by solving the system of k equations, namely Eq. (14.4.18) and $k - 1$ equations of the form

(14.4.20)

$$\frac{N_i^3 \sigma_i^2}{N_i - 1} \cdot \frac{1}{n_i^2} - \frac{N_k^3 \sigma_k^2}{N_k - 1} \cdot \frac{1}{n_k^2} = 0 \qquad (i = 1, 2, \ldots, k - 1).$$

This system gives

$$\frac{N_i^3 \sigma_i^2}{N_i - 1} \cdot \frac{1}{n_i^2} = a \qquad (i = 1, 2, \ldots, k),$$

where a is some constant; hence

(14.4.21)
$$n_i^2 = Q^2 \frac{N_i^3 \sigma_i^2}{N_i - 1} \qquad (i = 1, 2, \ldots, k),$$

where $Q^2 = 1/a$.

If we can replace $N_i - 1$ by N_i in this formula, then

(14.4.22)
$$n_i = Q N_i \sigma_i.$$

Formula (14.4.22) is very interesting. From it we see that the optimal numbers n_i are proportional to the size of the stratum and to the standard deviation within the stratum of the characteristic under investigation. Relation (14.4.22) agrees with our intuition, for if the dispersion of the characteristic within a certain stratum is bigger, we should take a larger proportion of the sample from that stratum.

To find the proportionality coefficient Q, let us note that from (14.4.22) we have $n = Q \sum\limits_{i=1}^{k} N_i \sigma_i$; hence

$$Q = n \Big/ \sum_{i=1}^{k} N_i \sigma_i.$$

In statistical practice, we do not usually know the standard deviations σ_i before starting the sample investigation. To find the σ_i, we usually perform a preliminary sampling of relatively small samples from each stratum in order to determine approximately the values of the σ_i. The influence of the approximate determination of the σ_i from the preliminary investigation on the results of the sample survey was discussed by Sukhatme [1].

If we base the choice of the sizes of the samples taken from the individual strata on the standard deviations σ_i of some one characteristic, then the efficiency of the sample survey may be decreased for other characteristics, and we rarely perform a sample survey in order to investigate only one characteristic. It is obvious that the variances of different characteristics inside the strata may require different collections of numbers n_i. This difficulty is usually avoided by determining the collection of optimal n_i for a characteristic which is either very important from the point of view of our investigation or is highly correlated with other characteristics. This problem is considered in the works of Dalenius [1, 2] and Ghosh [1].

We observe that the optimal numbers n_i are determined in such a way that the variance of the estimate $\bar{x}_2$ of m assumes its minimum value. If, however, the aim of the sample investigation is also to determine the expected values m_i of the characteristic x inside the individual strata, then it may happen that the sizes n_i of the samples are too small in some strata. It is then desirable to increase the sample size n. If such an increase is not possible, and the results of the investigation in individual strata are important, we have to reject the scheme of optimal sampling.

Formula (14.4.22) for the optimal numbers n_i is only approximate and can be applied only to large populations, for which $N - 1$ and $N_i - 1$ can be replaced, respectively, by N and N_i. By formula (14.4.21), for small populations, the n_i are of the form

$$(14.4.23) \qquad n_i = Q \sqrt{N_i/(N_i - 1)} N_i \sigma_i.$$

We remark that for small populations that are divided into strata in such a way that $\hat{\sigma}^2 = 0$ and the expressions $\sqrt{N_i/(N_i - 1)}\, \sigma_i$ are equal for all i, the optimal stratified sampling is even less efficient than unrestricted sampling.

Indeed, suppose that $N_i \sigma_i^2/(N_i - 1) = c^2$. Then from (14.4.23) it follows that $n_i = QcN_i$; hence the sampling is proportional and the term A' in (14.4.15) is negative. The assumption $\hat{\sigma}^2 = 0$ implies that $B' = 0$; hence

$$_n\sigma_{\bar{x}_2}^2 - {_w}\sigma_{\bar{x}_2}^2 = A' < 0.$$

The negative effect of such stratification could, by the way, be foreseen, since it requires investigating not one population but several almost identical populations with the same expected values and almost the same standard deviations; moreover, the investigation has to be performed with the use of several small samples instead of one large sample as in unrestricted sampling.

D   The following scheme of sampling is convenient and frequently applied in practice. Let us imagine that the sampling units which make up the population are numbered. Suppose that the number of sampling units in the population is N and suppose that the planned number of sampling units to be included in the sample is n. Suppose, for simplicity, that $N = nr$ where r is an integer. Let us select, using tables of random numbers, a number $j \leqslant r$ and let us include in the sample all those sampling units of the population which have numbers $j + hr$, where $h = 0, 1, \ldots, n - 1$. Such a scheme of sampling is called a *scheme of systematic sampling*. At first, it might seem that this scheme is simply a scheme of stratified sampling in which the population is divided into n strata and one element is drawn from each stratum. However, this is not so, since in the scheme under consideration, unlike stratified sampling, the drawing of one element from the first stratum uniquely determines the elements included in the sample from the remaining strata.

It is intuitively obvious that if the sampling units of the population whose numbers differ by r, that is, the elements with the numbers $j + hr$ for the selected j, are similar to each other with respect to the values of the characteristic x under investigation, then the scheme of systematic sampling gives a sample that does not characterize the population. Thus, if the characteristic x and the number j are correlated, the efficiency of systematic sampling is small, even smaller than the efficiency of unrestricted sampling. On the other hand, if x and j are not correlated, or are weakly correlated, but consecutive elements are correlated, then the scheme of systematic sampling, which guarantees that the numbers of any two sample elements differ by at least r, is more efficient than the scheme of stratified sampling in which we take one element from each stratum.

The preceding intuitive reasoning may be formalized; hence it may be expressed in the form of formulas for the variances. This was done first by W. G. Madow and L. H. Madow [1] and Cochran [1]. From among further papers dealing with systematic sampling we mention those of Hájek [1] and Zubrzycki [1]. (See also Problems 14.3 to 14.7.)

E In this chapter, we have restricted ourselves to a discussion of only the most important sampling schemes. Further details may be found in special books devoted to these problems, the books by Deming [1], Sukhatme [3], and Dalenius [2], for instance. See also the papers of Mahalanobis [1], Lieberman, and Solomon [1], and of Derman, Litauer, and Solomon [1].

14.5 RANDOM ERRORS OF MEASUREMENTS

The random errors so far considered, concerning the results of sample surveys, were due to divergence, with respect to the characteristic under investigation, between the elements included in the sample and the elements of the whole population. However, in practice, the results of sample investigations may have additional random errors which arise because the measurements of the characteristic under investigation or their records may have random errors. For instance, if we select a certain collection of women and make anthropometric measurements (see example 12.3.5), we may obtain results which have random errors because the group of women selected differs from the whole population of women who buy ready-made clothes and shoes, with respect to the characteristics under investigation. In addition, we must also consider random errors which can be made by the person who performs the measurements. As we shall see in example 14.5.1, the influence of such additional errors may be considerable.

Let z denote the observed value of the characteristic x under investigation. Hence z carries the random error which we shall denote by u. Thus we have

(14.5.1) $$z = x + u.$$

Suppose that x and u are independent, u has the normal distribution $N(0; h)$, and that the characteristic x has the expected value m and standard deviation σ. From formula (14.5.1) we obtain

(14.5.2) $$E(z) = E(x) = m, \qquad \sigma^2 = D^2(z) - h^2.$$

Hence the variance of x is smaller than the variance of z. If we denote $D^2(z)/h^2 = c$, we obtain from formula (14.5.2)

(14.5.3) $$\sigma^2 = D^2(z)\frac{c-1}{c}.$$

It can be seen from this formula that if c is large, we can equate $D^2(z)$ with σ^2.

We notice that the value of $D^2(z)$ can be estimated by the sample variance, and the value of h can be estimated from the results of repeated measurements of the same element of the sample, performed by the same people with the same tools for measurement.

Now let x and y be two random variables with the coefficient of correlation r_{xy} and let z and v be, respectively, the observed values of the random variables x and y, which carry the random errors u and w with the distributions $N(0; h_1)$ and $N(0; h_2)$, respectively. Suppose that u and w are independent, and let r_{zv} be the coefficient of correlation of z and v. By (14.5.3), we obtain

$$(14.5.4) \quad r_{xy} = \frac{E\{[x - E(x)][y - E(y)]\}}{\sqrt{D^2(x)D^2(y)}} = \frac{E\{[z - E(z)][v - E(v)]\}}{\sqrt{D^2(z)D^2(v)\dfrac{c_1 - 1}{c_1} \cdot \dfrac{c_2 - 1}{c_2}}}$$

$$= r_{zv}\sqrt{\frac{c_1 c_2}{(c_1 - 1)(c_2 - 1)}},$$

where $D^2(z)/h_1{}^2 = c_1$ and $D^2(v)/h_2{}^2 = c_2$. Thus, as intuition suggests, the coefficient of correlation of the characteristics x and y is bigger in absolute value than the coefficient of correlation between the observed random variables z and v.

Example 14.5.1. Table 14.5.1 gives the values of $\sqrt{D^2(z)}$ and the values of σ given by (14.5.3) for some characteristics of the anthropometric investigation performed in Holland (see example 12.3.5). Table 14.5.2 gives the coefficients of

TABLE 14.5.1

Characteristic	Unit	$\sqrt{D^2(z)}$	σ
Weight	kg	10.91	10.91
Height	cm	6.50	6.48
Shoulder to waist in front	cm	2.79	2.42
Shoulder to waist in back	cm	2.38	2.08
Length of sleeve	cm	3.04	2.76
Height of knee	cm	2.70	2.40
Length of middle finger	$\frac{1}{2}$ cm	0.99	0.91

correlation r_{xy} and r_{zv} for some pairs of the characteristics under investigation. We see from these tables that in some cases the influence of random errors of measurements may be considerable.

TABLE 14.5.2

Pairs of characteristics	r_{zv}	r_{xy}
Weight-hip measurement	0.9130	0.9247
Bust measurement-waist	0.9058	0.9180
Length of middle finger-height	0.5062	0.5493
Circumference of palm-size of fist	0.7007	0.8714

Problems and Complements

A *two-stage sampling scheme* proceeds as follows: The elements of a population are grouped into N collections, where the jth collection ($j = 1, \ldots, N$) consists of K_j individuals. Let us call the collections *first stage sampling units* (*fssu*), and the individuals *second stage sampling units* (*sssu*). We draw n *fssu*'s and from each of the chosen *fssu* we draw k_i ($i = 1, \ldots, n$) *sssu*'s. Let

$$x_{js}(j = 1, \ldots, N, s = 1, \ldots, K_j)$$

denote the value of the characteristic x of the sth *sssu* which belongs to the jth *fssu*. We denote the observations by the letter y with corresponding indices. We assume that each sampling unit of each stage has the same probability of being selected.

14.1. At both stages, let the sampling be independent and unrestricted and let

$$K_j = K (j = 1, \ldots, N) \text{ and } k_i = k(i = 1, \ldots, n). \text{ Let } m_{j.} = \frac{1}{K} \sum_{s=1}^{K} x_{js},$$

$$m = \frac{1}{KN} \sum_{j=1}^{N} \sum_{s=1}^{K} x_{js}, \quad \bar{y} = \frac{1}{kn} \sum_{i=1}^{n} \sum_{h=1}^{k} y_{ih}.$$

Show that

(a) $E(\bar{y}) = m$.

(b) $D^2(\bar{y}) = \left(\frac{1}{n} - \frac{1}{N}\right)\sigma_1^2 + \left(\frac{1}{k} - \frac{1}{K}\right)\frac{\sigma_2^2}{n}$,

where $\sigma_1^2 = \dfrac{1}{N-1} \sum_{j=1}^{N} (m_{j.} - m)^2$, $\sigma_2^2 = \dfrac{1}{N(K-1)} \sum_{j=1}^{N} \sum_{s=1}^{K} (x_{js} - m_{j.})^2$.

14.2. (Continued). Consider the cases (a) $N = n$, (b) $K = k$, (c) $\dfrac{N-n}{N} \approx 1$, (d) $\dfrac{K-k}{K} \approx 1$, (e) $\dfrac{N-n}{N} \approx 1$, $\dfrac{K-k}{K} \approx 1$. Compare the efficiency of the two-stage sampling scheme with the efficiency of other sampling schemes discussed in Chapter 14.

In Problems 14.3 to 14.7, we use the notation of Section 14.4, D. In addition, x_{jh} denotes the sampling unit of the population with the number $j + hr$, and

$$m_{.h} = \frac{1}{r} \sum_{j=1}^{r} x_{jh}, \qquad m_{j.} = \frac{1}{n} \sum_{h=0}^{n-1} x_{jh},$$

$$m = \frac{1}{nr} \sum_{j=1}^{r} \sum_{h=0}^{n-1} x_{jh},$$

$$\sigma_1^2 = \frac{1}{r} \sum_{j=1}^{r} (m_{j.} - m)^2, \qquad \sigma_2^2 = \frac{1}{nr} \sum_{j=1}^{r} \sum_{h=0}^{n-1} (x_{jh} - m_{.h})^2,$$

$$\sigma^2 = \frac{1}{nr} \sum_{j=1}^{r} \sum_{h=0}^{n-1} (x_{jh} - m)^2, \qquad \rho = \frac{1}{(n-1)N\sigma^2} \sum_{j=1}^{r} \sum_{\substack{h,h'=0 \\ h \neq h'}} (x_{jh} - m)(x_{jh'} - m).$$

The number ρ is called the *intraclass correlation coefficient*. The observations are denoted by y with corresponding indices.

14.3. Compare the preceding expresssions with formulas (14.4.1) to (14.4.6) and give their probabilistic interpretation.

14.4. (Continued) (a) Show that $\bar{y}_{.j} = \frac{1}{n} \sum_{h=0}^{n-1} y_{jh}$ is an unbiased estimate of the parameter m.

(b) Show that $D^2(\bar{y}_{j.}) = \sigma_1^2$.

14.5. (Continued). Show that for estimating m the systematic sampling scheme is more, less or equally efficient than the scheme of unrestricted sampling without replacement, according to whether the relation $\cdot \rho < -1/(rn - 1)$, $\rho > -1/(rn - 1)$ or $\rho = -1/(rn - 1)$ holds, respectively.

14.6. (Continued). Let the value of the characteristic x increase linearly with the number of the sampling unit, that is, $x_{jh} = a + (j + hr)$, where $j = 1, \ldots, r$, $h = 0, \ldots, n - 1$. Show that (a)

$$\sigma^2 = \frac{n^2 r^2 - 1}{12}, \qquad \sigma_1^2 = \sigma_2^2 = \frac{r^2 - 1}{12}.$$

(b) For estimating m such systematic sampling is more efficient than unrestricted sampling without replacement.

14.7. Let us treat the sampling units with the same index h as strata, and from each of the n strata considered let us draw one element. Show that in this case stratified sampling is more efficient for estimating m than systematic sampling.

14.8. (Notation of Section 14.3). Consider random samples of size n drawn without replacement from a finite population of size N. Let $n \leqslant N/2$ and let $n \to \infty$ as $N \to \infty$.
Prove (Erdös and Rényi [1]) that if the relation

$$\lim_{N \to \infty} \frac{1}{\sigma^2 N} \sum_{\substack{j=1 \\ |x_j - m| > \varepsilon n \sigma_{\bar{x}_2}}}^{N} (x_j - m)^2 = 0$$

holds for any $\varepsilon > 0$, then $\bar{x}_2$ is asymptotically normal, that is, for any real z

$$\lim_{N \to \infty} P\left(\frac{\bar{x}_2 - m}{\sigma_{\bar{x}_2}} < z \right) = \frac{1}{\sqrt{2\pi}} \int_{-\infty}^{z} \exp\left(-z^2/2\right) dz.$$

CHAPTER 15

An Outline of Analysis of Variance

15.1 ONE-WAY CLASSIFICATION

A   The statistical method known as analysis of variance, an outline of which will be presented in this chapter, was introduced by Fisher [7] for use in agricultural experiments; later it was applied to other branches as well.

Let $x_1, x_2, \ldots, x_n$ be independent, normally distributed random variables. Suppose that the standard deviation of all these random variables is the same but unknown, and that they may be classified into r groups in such a way that the random variables of the ith group ($i = 1, 2, \ldots, r$) have the same expected value m_i; hence the random variables of the ith group have the distribution $N(m_i; \sigma)$. Let us denote by n_i the number of random variables of the ith group.

This mathematical model corresponds to the following sample investigation: We have r populations in which the characteristic x has a normal distribution with the same standard deviation σ and the expected value m_i, respectively, and we draw a simple sample from each population.

Let us imagine that the aim of the sample investigation is to test the hypothesis $H_0(m_1 = m_2 = \ldots = m_r)$, that is, the hypothesis that all the expected values m_i are equal, hence that all the random variables considered have the same distribution.

If we had only two groups, the problem would be reduced to testing the hypothesis $H_0(m_1 = m_2)$, which was considered in example 12.2.6.

Let us denote the jth random variable of the ith group by x_{ij}. Furthermore, let

(15.1.1) $$\bar{x}_i = \frac{1}{n_i} \sum_{j=1}^{n_i} x_{ij},$$

(15.1.2) $$\bar{x} = \frac{1}{n} \sum_{i=1}^{r} \sum_{j=1}^{n_i} x_{ij} = \frac{1}{n} \sum_{i=1}^{r} n_i \bar{x}_i.$$

As we see, $\bar{x}_i$ is the arithmetic mean of the random variables x_{ij} of the ith group and $\bar{x}$ is the arithmetic mean of all the x_{ij}. We call these means the *group means* and the *total mean*, respectively (they should not be confused with the expected values m_i). We can also say that $\bar{x}_i$ is the sample mean of the ith sample and $\bar{x}$ is the sample mean of the whole sample

Since, by (15.1.1) and (15.1.2),

$$2\sum_{i=1}^{r}\sum_{j=1}^{n_i}(x_{ij} - \bar{x}_i)(\bar{x}_i - \bar{x}) = 0,$$

we obtain

$$(15.1.3) \quad \sum_{i=1}^{r}\sum_{j=1}^{n_i}(x_{ij} - \bar{x})^2 = \sum_{i=1}^{r}\sum_{j=1}^{n_i}(x_{ij} - \bar{x}_i + \bar{x}_i - \bar{x})^2$$

$$= \sum_{i=1}^{r}\sum_{j=1}^{n_i}(x_{ij} - \bar{x}_i)^2 + \sum_{i=1}^{r}\sum_{j=1}^{n_i}(\bar{x}_i - \bar{x})^2$$

$$= \sum_{i=1}^{r}\sum_{j=1}^{n_i}(x_{ij} - \bar{x}_i)^2 + \sum_{i=1}^{r}n_i(\bar{x}_i - \bar{x})^2.$$

As we see from (15.1.3), the sum of squares of the deviations of x_{ij} from $\bar{x}$ can be represented as a sum of two terms. The first is the sum of squares of the deviations of the random variables from their group means; we call this term the *sum of deviations within the groups*. The second term is the sum of deviations of the group means from the total mean; we call this term the *sum of deviations between groups*.

If the hypothesis H_0 ($m_1 = m_2 = \ldots = m_r$) is true, all the random variables x_{ij} have the same normal distribution and, since they are independent, by (9.5.12) the expression[1]

$$(15.1.4) \quad \sum_{i=1}^{r}\sum_{j=1}^{n_i}(x_{ij} - \bar{x})^2 = (n - 1)s^2$$

has the χ^2 distribution with $n - 1$ degrees of freedom. Similarly, we conclude that the expression

$$\sum_{j=1}^{n_i}(x_{ij} - \bar{x}_i)^2$$

has the χ^2 distribution with $n_i - 1$ degrees of freedom. Hence, using the addition theorem for the χ^2 distribution (see Section 5.8), it follows that the expression

$$(15.1.5) \quad \sum_{i=1}^{r}\sum_{j=1}^{n_i}(x_{ij} - \bar{x}_i)^2 = (n - r)s_2^2$$

has the χ^2 distribution with $\sum_{i=1}^{r}(n_i - 1) = n - r$ degrees of freedom.

The term $(n - r)s_2^2$ is called the *residual sum of squares*.

[1] The expression s^2 which appears in this formula is not the sample variance defined by (9.5.1).

Theorem 15.1.1.   *If the hypothesis H_0 is true, the expression*

$$(15.1.6) \qquad \sum_{i=1}^{r} n_i(\bar{x}_i - \bar{x})^2 = (r - 1)s_1^2$$

is independent of $(n - r)s_2^2$ and has the χ^2 distribution with $r - 1$ degrees of freedom.

The expression $(r - 1)s_1^2$ is called the *sum of squares between groups*.

In the proof of this theorem we use Fisher's [2] lemma.

Lemma (Fisher).   *Let $z_1, z_2, \ldots, z_k$ be independent random variables with the same normal distribution $N(0; \sigma)$. Let $y_1, y_2, \ldots, y_p$ $(p < k)$ be linear functions of $z_1, \ldots, z_k$,*

$$(15.1.7) \quad y_i = c_{i1}z_1 + c_{i2}z_2 + \ldots + c_{ik}z_k \quad (i = 1, 2, \ldots, p),$$

where the coefficients c_{ij} satisfy the orthogonality conditions

$$(15.1.8) \qquad \sum_{j=1}^{k} c_{ij}c_{lj} = \begin{cases} 1 & \text{for } i = l, \\ 0 & \text{for } i \neq l, \end{cases}$$

for $i = 1, 2, \ldots, p$ and $l = 1, 2, \ldots, p$. Then the expression

$$(15.1.9) \qquad \sum_{j=1}^{k} z_j^2 - \sum_{i=1}^{p} y_i^2$$

is independent of $y_1, y_2, \ldots, y_p$, hence also of y_i^2, and has the χ^2 distribution with $k - p$ degrees of ffreedom.

Proof of the lemma.   First, we prove that the random variables y_i $(i = 1, \ldots, p)$ are independent and each of them has the $N(0; \sigma)$ distribution. To do this, let us consider the characteristic function of the p-dimensional random variable $(y_1, \ldots, y_p)$. Let us write

$$(15.1.10)$$

$$\phi(t_1, t_2, \ldots, t_p)$$
$$= E\{\exp (it_1y_1 + it_2y_2 + \ldots + it_py_p)\}$$
$$= E\{\exp [it_1(c_{11}z_1 + \ldots + c_{1k}z_k) + \ldots + it_p(c_{p1}z_1 + \ldots + c_{pk}z_k)]\}$$
$$= E\{\exp [iz_1(c_{11}t_1 + \ldots + c_{p1}t_p)]\} \ldots E\{\exp [iz_k(c_{1k}t_1 + \ldots + c_{pk}t_p)]\}$$
$$= E\{\exp (iv_1z_1)\} \ldots E\{\exp (iv_kz_k)\},$$

where $v_j = c_{1j}t_1 + c_{2j}t_2 + \ldots + c_{pj}t_p$. Since the random variables z_j have the same normal distribution $N(0; \sigma)$, we have

$$(15.1.11) \qquad \phi(t_1, t_2, \ldots, t_p) = \exp \left(-\tfrac{1}{2}\sigma^2 \sum_{j=1}^{k} v_j^2 \right).$$

Using the fact that the transformation (15.1.7) is orthogonal, we finally obtain

$$(15.1.12) \quad \phi(t_1, t_2, \ldots, t_p) = \exp\left(-\tfrac{1}{2}\sigma^2 \sum_{i=1}^{p} t_i^2\right) = \prod_{i=1}^{p} \exp\left(-\tfrac{1}{2}\sigma^2 t_i^2\right).$$

This is the characteristic function of a p-dimensional random vector whose components are independent and have the same normal distribution $N(0; \sigma)$.

We now show that for the given p random variables defined by (15.1.7) we can select additional $k - p$ random variables $y_{p+1}, y_{p+2}, \ldots, y_k$ which are linear functions of the random variables $z_1, \ldots, z_k$ of the form (15.1.7) and such that the orthogonality conditions (15.1.8) are satisfied for $i = 1, 2, \ldots, p, p + 1, \ldots, k$ and $l = 1, 2, \ldots, p, p + 1, \ldots, k$. In fact, in order to determine $k - p$ such random variables, we have to determine $(k - p)k$ unknown coefficients c_{ij} which, by the assumption of orthogonality, must satisfy $\tfrac{1}{2}(k - p)(1 + k + p)$ equations. The unknown coefficients can be determined if $\tfrac{1}{2}(1 + p + k) \leq k$, thus, if

$$(15.1.13) \qquad\qquad 1 + p \leqslant k.$$

However, by assumption, $p < k$ and p and k are integers; hence inequality (15.1.13) is satisfied.

It is easy to show by the preceding method that all the random variables y_i $(i = 1, \ldots, k)$ are independent and each of them has the normal distribution $N(0; \sigma)$. It follows from the orthogonality conditions that

$$(15.1.14) \qquad\qquad \sum_{j=1}^{k} z_j^2 = \sum_{i=1}^{k} y_i^2.$$

Thus, let us rewrite formula (15.1.9) in the form

$$(15.1.15) \qquad y_{p+1}^2 + \ldots + y_k^2 = \sum_{j=1}^{k} z_j^2 - \sum_{i=1}^{p} y_i^2.$$

The left-hand side of (15.1.15) is independent of $\sum_{i=1}^{p} y_i^2$ because of the independence of the y_i $(i = 1, \ldots, k)$. Since it is a sum of squares of $k - p$ independent random variables with the distribution $N(0; \sigma)$, it has the χ^2 distribution with $k - p$ degrees of freedom. This completes the proof of Fisher's lemma.

Using this lemma, we can show that $(r - 1)s_1^2$ is independent of $(n - r)s_2^2$ and has the χ^2 distribution with $r - 1$ degrees of freedom.

Indeed, the expression $(n - 1)s^2$ defined by (15.1.4) can be represented in the following way:

$$(n - 1)s^2 = \sum_{i=1}^{r} \sum_{j=1}^{n_i} (x_{ij} - \bar{x})^2 = \sum_{i=1}^{r} \sum_{j=1}^{n_i} (x_{ij} - m + m - \bar{x})^2$$

$$= \sum_{i=1}^{r} \sum_{j=1}^{n_i} (x_{ij} - m)^2 - n(\bar{x} - m)^2$$

$$= \sum_{i=1}^{r} \sum_{j=1}^{n_i} (x_{ij} - m)^2 - \left[\sum_{i=1}^{r} \sum_{j=1}^{n_i} (x_{ij} - m)/\sqrt{n} \right]^2 = \sum_{i=1}^{r} \sum_{j=1}^{n_i} \xi_{ij}^2 - \eta^2,$$

where the $\xi_{ij} = x_{ij} - m$ are independent and have the same distribution $N(0; \sigma)$, and η is an orthogonal transformation of the ξ_{ij}.

As was established in the proof of Fisher's lemma, we can find $n - 1$ independent random variables $z_1, \ldots, z_{n-1}$ which are also an orthogonal transformation of the random variables ξ_{ij} and such that

$$(n - 1)s^2 = \sum_{i=1}^{r} \sum_{j=1}^{n_i} \xi_{ij}^2 - \eta^2 = \sum_{l=1}^{n-1} z_l^2.$$

We can also conclude that

$$(n - r)s_2^2 = \sum_{i=1}^{r} \sum_{j=1}^{n_i} (x_{ij} - \bar{x}_i)^2 = \sum_{q=1}^{n-r} y_q^2,$$

where the y_q are an orthogonal transformation of the random variables ξ_{ij} and at the same time, as can be shown by rather long, although not too complicated, computations, an orthogonal transformation of the random variables $z_1, z_2, \ldots, z_{n-1}$. It follows, then, from Fisher's lemma that the expression

$$(r - 1)s_1^2 = \sum_{l=1}^{n-1} z_l^2 - \sum_{q=1}^{n-r} y_q^2$$

has the χ^2 distribution with $(n - 1) - (n - r) = r - 1$ degrees of freedom and is independent of $(n - r)s_2^2$. Thus the theorem is proved.

B   We represent the sums of squares obtained and their number of degrees of freedom in the form of the following table.

TABLE 15.1.1

	Sum of squares	Number of degrees of freedom	Ratio
Between the means	$(r - 1)s_1^2 = \sum_{i=1}^{r} n_i(\bar{x}_i - \bar{x})^2$	$r - 1$	s_1^2
Residual	$(n - r)s_2^2 = \sum_{i=1}^{r} \sum_{j=1}^{n_i} (x_{ij} - \bar{x}_i)^2$	$n - r$	s_2^2
Total	$(n - 1)s^2 = \sum_{i=1}^{r} \sum_{j=1}^{n_i} (x_{ij} - \bar{x})^2$	$n - 1$	s^2

We observe that from (9.4.10) it follows that s_1^2, s_2^2, and s^2 are unbiased estimates of σ^2, that is,

$$(15.1.16) \qquad E(s_1^2) = E(s_2^2) = E(s^2) = \sigma^2.$$

Since $(n - 1)s^2$ has been split into two independent terms such that relation (15.1.16) holds, it is customary to call this splitting *analysis of variance*, although actually it was not the variance that was split into terms, since, as can be seen from Table 15.1.1, $s_1^2 + s_2^2 \neq s^2$.

Since $(n - r)s_2^2$ and $(r - 1)s_1^2$ are independent and have χ^2 distributions with $n - r$ and $r - 1$ degrees of freedom, respectively, the random variable z, defined as

$$(15.1.17) \qquad z = \log \frac{s_1}{s_2} = \frac{1}{2} \log \left[\frac{n - r}{r - 1} \cdot \frac{\sum\limits_{i=1}^{r} n_i(\bar{x}_i - \bar{x})^2}{\sum\limits_{i=1}^{r} \sum\limits_{j=1}^{n_i} (x_{ij} - \bar{x}_i)^2} \right],$$

has Fisher's Z-distribution (see Section 9.7) with $(r - 1, n - r)$ degrees of freedom. We observe that z equals half the difference of the logarithms of two independent unbiased estimates of σ^2.

To test hypothesis H_0, we may apply a significance test, as in Chapter 12, namely, for a given number of degrees of freedom and significance level α, we find a z_0 such that the relation $P(|z| \geqslant z_0) = \alpha$ holds if H_0 is true. Then we reject H_0 if the observed value of z exceeds or equals z_0.

Example 15.1.1. Table 15.1.2 gives twenty independent observations of the crop produced by a certain kind of plant grown in earthen pots. The observations are split into five groups of four observations each, according to the fertilizer used, and in one group there was no fertilizer at all. Thus we have

$$n = 20, \quad r = 5, \quad n_i = 4 \quad (i = 1, 2, 3, 4, 5).$$

We assume that the random variables x_{ij} have the normal distribution $N(m_i; \sigma)$. We test the hypothesis $H_0(m_1 = m_2 = m_3 = m_4 = m_5)$ according to which neither the kind of fertilizer nor the very presence of a fertilizer affects the average crop. Thus, if this hypothesis is true, the observed deviations between the values $\bar{x}_i$ would be due only to random fluctuations.

TABLE 15.1.2

i	Fertilizer	x_{ij}				$\bar{x}_i$
1	No fertilizer	67	67	55	42	57.75
2	$K_2O + N$	98	96	91	66	87.75
3	$K_2O + P_2O_5$	60	69	50	35	53.50
4	$N + P_2O_5$	79	64	81	70	73.50
5	$K_2O + P_2O_5 + N$	90	70	79	88	81.75

We compute from Table 15.1.2

$$\bar{x} = 70.85,$$

$$s_1^2 = \frac{1}{r-1}\sum_{i=1}^{r} n_i(\bar{x}_i - \bar{x})^2 = \sum_{i=1}^{5}(\bar{x}_i - x)^2 = 884.075,$$

$$s_2^2 = \frac{1}{n-r}\sum_{i=1}^{r}\sum_{j=1}^{n_i}(x_{ij} - \bar{x}_i)^2 = \frac{1}{15}\sum_{i=1}^{5}\sum_{j=1}^{4}(x_{ij} - \bar{x}_i)^2 = 144.150.$$

Hence

$$z = \frac{1}{2}\log\frac{884,075}{144,150} \simeq \frac{1}{2}\log 6.13 \simeq 0.9066.$$

Since s_1^2 has four degrees of freedom and s_2^2 has fifteen degrees of freedom, we find from tables of Fisher's Z-distribution that the value z_0 for which the relation $P(z \geqslant z_0) = 0.01$ holds equals 0.7939. As we see, $z > z_0$, and we should reject H_0 if we test it at the significance level 0.01.

One can continue the analysis of the data in Table 15.1.2. Thus, after rejecting the hypothesis that neither the fact of fertilizing nor the kind of fertilizer affects the average crop, we can test the weaker hypothesis $H_1(m_2 = m_3 = m_4 = m_5)$, that is, the hypothesis that the kind of fertilizer applied does not affect the average crop. To test this hypothesis, we use only the data in the last four rows of Table 15.1.2.

We have here $n = 16$, $r = 4$, and

$$\bar{x} = 74.125,$$

$$s_1^2 = \frac{1}{3}\sum_{i=2}^{5}4(\bar{x}_i - \bar{x})^2 = 892.55,$$

$$s_2^2 = \frac{1}{12}\sum_{i=2}^{5}\sum_{j=1}^{4}(x_{ij} - \bar{x}_i)^2 = 141.625.$$

Hence

$$z = \frac{1}{2}\log\frac{892.55}{141.625} \simeq \frac{1}{2}\log 6.30 \simeq 0.9203.$$

Since s_1^2 has three degrees of freedom and s_2^2 has twelve degrees of freedom, we find from tables of Fisher's Z-distribution that the value z_0 such that $P(z \geqslant z_0) = 0.01$ is 0.8919. However, the value of z obtained exceeds z_0; thus we reject H_1.

Finally, let us test the hypothesis H_2 that the addition of fertilizer P_2O_5 to the two remaining fertilizers does not affect the average crop. In other words, our hypothesis states that $m_2 = m_5$.

We have here $n = 8$, $r = 2$, and

$$\bar{x} = 84.75,$$

$$s_1^2 = 4 \cdot 3^2 + 4 \cdot 3^2 = 72,$$

$$s_2^2 = \frac{1}{6}\sum_{j=1}^{4}(x_{2j} - x_2)^2 + \frac{1}{6}\sum_{j=1}^{4}(x_{5j} - \bar{x}_5)^2 = 151.58.$$

In this example $s_1^2 < s_2^2$; hence in order to use Z-distribution tables, we interchange the roles of s_1^2 and s_2^2. Thus we obtain

$$z = \frac{1}{2}\log\frac{151.58}{72} = 0.3894.$$

Since s_2^2 has six degrees of freedom and s_1^2 has one, we find that the value z_0 such that $P(z \geqslant z_0) = 0.01$ is 4.3379. Thus the experiment performed does not lead to the rejection of H_2.

The last hypothesis could be treated as the hypothesis that two normal populations with the same unknown standard deviation σ, from which two independent simple samples have been drawn, have the same expected value; hence their distributions are identical. Such a hypothesis could be tested with the use of Student's t-distribution, as was shown in example 12.2.6.

The data for this example were taken from the book by Romanovsky [3].

C   The hypothesis H_0 considered in Section 15.1, A is a particular case of a linear hypothesis which we now define.

Let $x_1, \ldots, x_n$ be independent, normally distributed random variables with common, but unknown, standard deviation σ, and unknown mean values m_j $(j = 1, \ldots, n)$. It is assumed to be known that the m_j satisfy $n - r$ linear equations

$$(15.1.18) \qquad \sum_{j=1}^{n} a_{ij} m_j = 0 \qquad (i = 1, \ldots, n - r)$$

with given a_{ij}. In other words, the m_j lie in an r-dimensional linear subspace containing the origin. Any hypothesis asserting that the m_j satisfy, in addition to equations (15.1.18), $s(s \leqslant r)$ more linear equations

$$(15.1.19) \qquad \sum_{j=1}^{n} b_{kj} m_j = 0 \qquad (k = 1, \ldots, s)$$

with given b_{kj}, is called a *linear hypothesis*.

The concept of a linear hypothesis was introduced by Kolodziejczyk [1]. The theory of testing linear hypotheses is presented extensively in Anderson's [1] and Lehmann's [4] books.

15.2 MULTIPLE CLASSIFICATION

In the preceding section, we classified the observations x_{ij} of the random variable x into groups according to one criterion. In applications, we often deal with classifications of n observations of a random variable according to several criteria. For instance, if the subject under investigation is the wheat harvest from experimental fields, we can divide all the observations obtained according to the kind of wheat planted and the kind of fertilizer applied. Suppose that the observations are divided into r groups according to criterion A, and into v groups according to criterion B. Thus we have rv groups. We deal only with the case where the number of observations in each group is the same; for simplicity, we may assume that we have one observation in each group; hence $n = rv$.

Let us denote by x_{kl} the observation which belongs to the kth group according to criterion A and to the lth group according to criterion B. Furthermore, let

(15.2.1)
$$\bar{x}_{k.} = \frac{1}{v} \sum_{l=1}^{v} x_{kl},$$

(15.2.2)
$$\bar{x}_{.l} = \frac{1}{r} \sum_{k=1}^{r} x_{kl},$$

(15.2.3)
$$\bar{x} = \frac{1}{rv} \sum_{k=1}^{r} \sum_{l=1}^{v} x_{kl}.$$

This double classification is represented in Table 15.2.1.

TABLE 15.2.1

A	B k	l 1	2	$\ldots$	v	$\bar{x}_{k.}$
	1	x_{11}	x_{12}	$\ldots$	x_{1v}	$\bar{x}_{1.}$
k	2	x_{21}	x_{22}	$\ldots$	x_{2v}	$\bar{x}_{2.}$
	.	.	.	.	.	.
	.	.	.	.	.	.
	.	.	.	.	.	.
	r	x_{r1}	x_{r2}	$\ldots$	x_{rv}	$x_{r.}$
$\bar{x}_{.l}$		$\bar{x}_{.1}$	$\bar{x}_{.2}$	$\ldots$	$\bar{x}_{.v}$	$\bar{x}$

After making some simple computations, using formulas (15.2.1) to (15.2.3), we obtain the identities

(15.2.4)
$$\sum_{k=1}^{r} \sum_{l=1}^{v} (x_{kl} - \bar{x})^2$$

$$= \sum_{k=1}^{r} \sum_{l=1}^{v} (x_{kl} - \bar{x}_{k.} - \bar{x}_{.l} + \bar{x} + \bar{x}_{k.} - \bar{x} + \bar{x}_{.l} - \bar{x})^2$$

$$= \sum_{k=1}^{r} \sum_{l=1}^{v} (x_{kl} - \bar{x}_{k.} - \bar{x}_{.l} + \bar{x})^2 + v \sum_{k=1}^{r} (\bar{x}_{k.} - \bar{x})^2 + r \sum_{l=1}^{v} (\bar{x}_{.l} - \bar{x})^2.$$

Let us assume that the random variable x_{kl}, where $k = 1, \ldots, r$ and $l = 1, \ldots, v$, has the normal distribution $N(m_{kl}; \sigma)$, where σ is unknown. Let the aim of our sample investigation be to test the hypothesis that these rv observations are homogeneous, that is, the hypothesis H_0 that all the expected values m_{kl} are equal. If H_0 is true, we can show, in a manner similar to that applied in the preceding section, that the expression

$$(15.2.5) \qquad \sum_{k=1}^{r} \sum_{l=1}^{v} (x_{kl} - \bar{x})^2 = (rv - 1)s^2$$

has the χ^2 distribution with $rv - 1$ degrees of freedom. Similarly, the expression

$$(15.2.6) \qquad v \sum_{k=1}^{r} (\bar{x}_{k.} - \bar{x})^2 = (r - 1)s_1^2$$

has the χ^2 distribution with $r - 1$ degrees of freedom. The expression $(r - 1)s_1^2$ is called the *sum of deviations in rows*.

The expression

$$(15.2.7) \qquad r \sum_{l=1}^{v} (\bar{x}_{.l} - \bar{x})^2 = (v - 1)s_2^2$$

has the χ^2 distribution with $v - 1$ degrees of freedom. We call it the *sum of deviations in columns*.

Finally, the expression

$$(15.2.8) \qquad \sum_{k=1}^{r} \sum_{l=1}^{v} (x_{kl} - \bar{x}_{k.} - \bar{x}_{.l} + \bar{x})^2 = (r - 1)(v - 1)s_3^2$$

has the χ^2 distribution with $(r - 1)(v - 1)$ degrees of freedom. It is called the *residual sum of squares*.

As in the preceding section, we can show that if H_0 is true, the random variables s_1^2, s_2^2, and s_3^2 are independent, and each of them is an unbiased estimate of σ^2.

To test the hypothesis H_0, we may again apply Fisher's Z-distribution, considering either half the difference between the logarithms of s_1^2 and s_3^2, or half the difference between the logarithms of s_2^2 and s_3^2.

Table 15.2.2 corresponds to Table 15.1.1.

TABLE 15.2.2

	Sum of squares	Number of degrees of freedom	Ratio
Between the averages in rows	$(r - 1)s_1^2 = v \sum_{k=1}^{r} (\bar{x}_{k.} - \bar{x})^2$	$r - 1$	s_1^2
Between the averages in columns	$(v - 1)s_2^2 = r \sum_{l=1}^{v} (\bar{x}_{.l} - \bar{x})^2$	$v - 1$	s_2^2
Residual	$(r - 1)(v - 1)s_3^2 = \sum_{k=1}^{r} \sum_{l=1}^{v} (x_{kl} - \bar{x}_{k.} - \bar{x}_{.l} + \bar{x})^2$	$(r - 1)(v - 1)$	s_3^2
Total	$(rv - 1)s^2 = \sum_{k=1}^{r} \sum_{l=1}^{v} (x_{kl} - \bar{x})^2$	$rv - 1$	s^2

Example 15.2.1. The following data, presented in Table 15.2.3, are taken from Daniels [1]. They give the weight in grams of 100 pieces of woolen thread, each 95 yards long. These pieces are divided into 4 groups of 25 each, and each group is taken from a different ball.

In this example, we have $v = 25$, $r = 4$.

We assume that each of the 100 observations is the observed value of a random variable with normal distribution $N(m_{kl};\ \sigma)$, and we want to test the hypothesis that all the m_{kl} are identical. To do this, we apply an analysis of variance to the data presented in Table 15.2.3. This analysis is presented in Table 15.2.4.

TABLE 15.2.3

Piece	Ball 1	Ball 2	Ball 3	Ball 4	Weight	Average weight
1	7.50	7.23	7.50	7.53	29.76	7.44
2	7.52	7.81	7.77	8.05	31.15	7.79
3	7.70	7.94	7.83	8.16	31.63	7.91
4	7.93	7.94	7.96	7.76	31.59	7.90
5	7.78	7.89	8.02	7.85	31.54	7.89
6	7.73	8.23	7.99	8.14	32.09	8.02
7	8.07	8.27	8.25	8.26	32.85	8.21
8	8.01	8.54	8.24	8.54	33.33	8.33
9	8.22	8.24	8.37	8.10	32.93	8.23
10	8.24	8.35	8.43	8.15	33.17	8.29
11	8.17	8.29	8.46	8.38	33.30	8.33
12	8.09	8.54	8.33	8.47	33.43	8.36
13	8.11	8.45	8.27	8.38	33.21	8.30
14	7.96	8.43	8.24	8.60	33.23	8.31
15	8.09	8.47	8.12	8.45	33.13	8.28
16	8.04	8.33	8.14	8.43	32.94	8.24
17	7.78	8.47	8.19	8.57	33.01	8.25
18	8.11	8.63	8.36	8.38	33.48	8.37
19	8.17	8.31	8.31	8.16	32.95	8.24
20	8.12	8.31	8.47	8.41	33.31	8.33
21	8.13	8.10	8.19	8.27	32.69	8.17
22	8.01	8.01	8.37	7.96	32.35	8.09
23	8.17	7.92	8.27	8.08	32.44	8.11
24	8.05	8.27	8.07	8.16	32.55	8.14
25	7.91	7.92	8.28	8.52	32.63	8.16
Weight	199.61	204.89	204.43	205.76	816.69	
Average weight	7.98	8.20	8.18	8.23		8.15

TABLE 15.2.4

	Sum of squares	Number of degrees of freedom	Ratio
Between the segments	4.637814	24	$s_1^2 = 0.1932$
Between the balls	0.916707	3	$s_2^2 = 0.3056$
Residual	1.670418	72	$s_3^2 = 0.0232$
Total	7.224939	99	$s^2 = 0.0730$

We compute

$$z_1 = \frac{1}{2} \log \frac{0.1932}{0.0232} = \frac{1}{2} \log 8.33 = 1.0599,$$

$$z_2 = \frac{1}{2} \log \frac{0.3056}{0.0232} = \frac{1}{2} \log 13.1 = 1.2863.$$

Since s_1^2 has 24 degrees of freedom and s_3^2 has 72, we find from tables of Fisher's Z-distribution that the value z_0 which satisfies the relation $P(z \geqslant z_0) = 0.01$ lies between 0.3339 and 0.3746. As we see, the value z_1 observed in our example is much greater than z_0. Next, since s_2^2 has 3 degrees of freedom and s_3^2 has 72, we find in a similar way that the value z_0 for which the relation $P(z \geqslant z_0) = 0.01$ holds lies between 0.6867 and 0.7086. Hence z_2 is much greater than this value.

We reach the conclusion that the observed values z_1 and z_2 cannot be explained as a result of random fluctuations (at the significance level 0.01), if hypothesis H_0 is true. Thus we reject H_0.

As the name of this chapter indicates, it contains only an outline of analysis of variance and its applications, and, in fact, it amounts only to an introduction to the subject. The restricted scope of this book does not allow further presentation of this theory. A more detailed discussion can be found in the books by M. G. Kendall [1], Mann [1], Kempthorne [1], and in the comprehensive book by Scheffé [1], recently published. Many examples of applications can also be found in the book by Romanovsky [3].

15.3 A MODIFIED REGRESSION PROBLEM

A   In practice we often meet a problem which is related to the problem of regression of the second type (see Section 3.8).

Let y be a random variable which depends on a parameter x; hence here x is not a random variable. Suppose that for every value of x the random variable y has the normal distribution $N(ax + b; \sigma)$, where a, b, and $\sigma > 0$ are unknown constants. We have n pairs of observations (x_k, y_k), where $y_1, y_2, \ldots, y_n$ are independent. We shall determine the constants a, b, and σ by the method of maximum likelihood. Let us denote by $f(y, a, b, \sigma)$ the density of y. Thus

$$f(y, a, b, \sigma) = \frac{1}{\sigma\sqrt{2\pi}} \exp\left[-\frac{(y - ax - b)^2}{2\sigma^2}\right].$$

Hence for the likelihood function L we have the formula

$$L = \left(\frac{1}{\sigma\sqrt{2\pi}}\right)^n \exp\left[-\frac{1}{2\sigma^2}\sum_{k=1}^{n}(y_k - ax_k - b)^2\right].$$

We obtain

$$\frac{\partial \log L}{\partial a} = \frac{1}{\sigma^2}\sum_{k=1}^{n}(y_k - ax_k - b)x_k,$$

$$\frac{\partial \log L}{\partial b} = \frac{1}{\sigma^2}\sum_{k=1}^{n}(y_k - ax_k - b),$$

$$\frac{\partial \log L}{\partial \sigma^2} = \frac{1}{2\sigma^4}\sum_{k=1}^{n}(y_k - ax_k - b)^2 - \frac{n}{2\sigma^2}.$$

Setting these partial derivatives equal to 0, we obtain

(15.3.1) $$a^* = r\frac{s_2}{s_1}, \qquad b^* = \bar{y} - r\frac{s_2}{s_1}\bar{x},$$

$$\sigma^{*2} = \frac{1}{n}\sum_{k=1}^{n}(y_k - a^*x_k - b^*)^2 = s_2^2(1 - r^2).$$

Here, the expressions $\bar{x}$, $\bar{y}$, s_1, s_2, and r have the same form as in Section 9.9, but $\bar{x}$ and s_1 are not random variables and r is not a correlation coefficient. Formulas (15.3.1) are formally similar to formulas (3.8.4) and (3.8.8), but their meaning is different.

Let us find the distribution of the random variables a^*, b^*, and σ^{*2}. Using the fact that

$$r = \frac{1}{ns_1s_2}\sum_{k=1}^{n}(x_k - \bar{x})(y_k - \bar{y}),$$

we obtain

$$a^* = \frac{1}{ns_1^2}\sum_{k=1}^{n}(x_k - \bar{x})y_k;$$

hence a^*, as a sum of independent random variables with normal distributions, has a normal distribution. From theorems 3.6.1 and 3.6.3, we obtain

(15.3.2) $$E(a^*) = a,$$

$$D^2(a^*) = \frac{\sigma^2}{ns_1^2}.$$

Let us rewrite b^* in the form

$$b^* = \frac{1}{ns_1^2} \sum_{k=1}^{n} (s_1^2 - x_k\bar{x} + \bar{x}^2)y_k.$$

As we see, b^* also has a normal distribution, and

(15.3.3) $E(b^*) = b,$

$$D^2(b^*) = \frac{\sigma^2}{n}\left(1 + \frac{\bar{x}^2}{s_1^2}\right).$$

To find the distribution of the random variable σ^{*2}, note that we have the identity

(15.3.4) $\displaystyle\sum_{k=1}^{n} (y_k - a^*x_k - b^*)^2$

$$= \sum_{k=1}^{n}(y_k - ax_k - b)^2 - n[b^* - b + (a^* - a)\bar{x}]^2 - ns_1^2(a^* - a)^2.$$

Let

(15.3.5) $\xi_k = y_k - ax_k - b,$

$$\eta_1 = \sqrt{n}[b^* - b + (a^* - a)\bar{x}] = \frac{1}{\sqrt{n}} \sum_{k=1}^{n}\xi_k,$$

$$\eta_2 = s_1\sqrt{n}(a^* - a) = \frac{1}{s_1\sqrt{n}} \sum_{k=1}^{n}(x_k - \bar{x})\xi_k.$$

It is obvious that the random variables $\xi_k(k = 1, \ldots, n)$ are independent and each of them has the distribution $N(0; \sigma)$. From (15.3.1), (15.3.4), and (15.3.5), we obtain

(15.3.6) $n\sigma^{*2} = \displaystyle\sum_{k=1}^{n}\xi_k^2 - \eta_1^2 - \eta_2^2.$

Note that η_1 and η_2 are linear functions of $\xi_1, \ldots, \xi_n$, satisfying the orthogonality conditions (15.1.8); hence η_1 and η_2 are independent, and, by Fisher's lemma (see Section 15.1), the random variable $n\sigma^{*2}/\sigma^2$ is independent of $\eta_1^2 + \eta_2^2$ and has the χ^2 distribution with $n - 2$ degrees of freedom. From (9.4.5) it follows that $E(n\sigma^{*2}) = (n - 2)\sigma^2$. Hence, by (15.3.2) and (15.3.3), it follows that a^*, b^*, and $n\sigma^{*2}/(n - 2)$ are unbiased estimates of the parameters a, b, and σ^2, respectively, and that the random variables

(15.3.7) $t = s_1 \dfrac{a^* - a}{\sigma^*} \sqrt{n - 2},$

$$t_1 = \frac{b^* - b + (a^* - a)\bar{x}}{\sigma^*} \sqrt{n - 2}$$

have Student's t-distribution with $n - 2$ degrees of freedom.

Example 15.3.1. Let us denote by x the dose of heparine, the drug which delays coagulation of the blood, and by y the time of appearance of coagulation in the blood. The values of x are considered to be "sure", that is, we assume that the determination of the values of x is so precise that we have no doubt about their correctness. However, for a given x, y is a random variable which depends upon the individual patient who has been given the dose x. As the result of multiple observations, it has been established that y has the normal distribution $N(ax + b; \sigma)$. Table 15.3.1 gives some of the data which were observed in the Institute of Drugs in Warsaw and which were elaborated by Bażańska [1]. Three different doses were applied and for each dose four independent observations were made. Thus the total number of observations equaled twelve. From Table 15.3.1 we obtain

$$\bar{x} = 1.63, \quad \bar{y} = 5.65, \quad s_1 = 0.295, \quad s_2 = 6.36, \quad r = 0.45, \quad a^* = 9.35.$$

To test the hypothesis $H_0(a = 0)$, we compute the value of t from formulas (15.3.7) and (15.3.1). We obtain $t = 15.3$. From tables of Student's t-distribution we find that the probability of obtaining a value of t which equals or exceeds in absolute value the observed one, for ten degrees of freedom, is much smaller than 0.01. Thus we reject H_0.

TABLE 15.3.1

x		y		
1.28	2.50	2.92	2.50	2.66
1.60	4.92	4.92	4.50	4.66
2.00	9.50	9.75	9.00	10.00

B   Let

$$v_1 = a^*x + b^* - h\sigma^*\sqrt{n/(n-2)},$$

$$v_2 = a^*x + b^* + h\sigma^*\sqrt{n/(n-2)},$$

where a^*, b^*, and σ^* are given by (15.3.1). Next let us write

$$w = P(v_1 \leqslant y \leqslant v_2),$$

where y is the random variable considered in Section 15.3, A. We stress the fact, as we did in example 10.9.2, that w is a random variable, namely w is the fraction of elements of the population that lie between the random tolerance limits v_1 and v_2. An approximate method for computing the probability distribution of w was given by Wallis [2].

C   Let us modify the problem considered in Section 15.3, A by omitting the assumption of normality of the random variable y, and replacing the assumption that $y_1, \ldots, y_n$ are independent by the assumption that they are pairwise uncorrelated. Thus, for every value of x the standard deviation $\sigma > 0$ of y is the same, and the expected value is $E(y) = ax + b$. Having n pairwise uncorrelated observations y_k, where y_k corresponds to

x_k ($k = 1, \ldots, n$), we estimate the unknown parameters a and b by the *method of least squares*, that is, we find estimates a^* and b^* of the parameters a and b that satisfy the relation

$$(15.3.8) \qquad \sum_{k=1}^{n}(y_k - a^*x_k - b^*)^2 = \text{minimum.}$$

Differentiating the left-hand side of formula (15.3.8) with respect to a^* and b^* and setting these derivatives equal to zero, we again obtain for a^* and b^* expressions (15.3.1). The least squares estimate of σ^2 is σ^{*2} given by (15.3.1). Since the y_k are pairwise uncorrelated, we see that formulas (15.3.2) and (15.3.3) are also satisfied. Since

$$E[(y_k - ax_k - b)^2] = \sigma^2 \qquad (k = 1, \ldots, n),$$

$$E[\bar{x}(b^* - b)(a^* - a)] = -\frac{\sigma^2 \bar{x}^2}{ns_1{}^2},$$

we obtain from identity (15.3.4), after using formulas (15.3.2) and (15.3.3), the formula

$$(15.3.9) \qquad E(n\sigma^{*2}) = E\left[\sum_{k=1}^{n}(y_k - a^*x - b^*)^2\right] = (n - 2)\sigma^2.$$

We reach the conclusion, analogous to that in Section 15.3, A, that the statistics a^*, b^*, and $n\sigma^{*2}/(n - 2)$ are unbiased estimates of the parameters a, b, and σ^2, respectively. However, it is now possible that the random variables a^* and b^* do not have a normal distribution, and $n\sigma^{*2}/\sigma^2$ does not have the χ^2 distribution. Similarly, it is possible that t and t_1 defined by (15.3.7) do not have Student's t-distribution.

Note that a^* and b^* are linear functions of the observations y_k and, because of that, are called *linear estimates* of a and b. We observe that, according to Markov's theorem [5] (see also David and Neyman [1]), an unbiased linear estimate $\alpha a^* + \beta b^*$ of the expression $\alpha a + \beta b$, where α and β are arbitrary real numbers and a^* and b^* are obtained by the method of least squares from relation (15.3.8), is the most efficient among all unbiased linear estimates of $\alpha a + \beta b$. Thus, putting successively $\alpha = 1$, $\beta = 0$, and $\alpha = 0$, $\beta = 1$, we see that the variance of any unbiased linear estimate of a is not smaller than $D^2(a^*)$, given by (15.3.2), and that of b is not smaller than $D^2(b^*)$, given by (15.3.3).

We note that the question of whether the linearity can be omitted from the statement of Markov's theorem has been discussed by Anderson [3]. **D**   Multidimensional generalizations of problems discussed in this section, namely, when the one-dimensional random variables and the one-dimensional parameters are replaced by multidimensional ones, are frequently used in psychology and economics. We restrict ourselves to referring to Frisch [2], Guilford [1], Reiersöl [1], and Tintner [1].

Problems and Complements

15.1. We use the notation of Section 15.1. Assume that all the x_{ij} are independent and have the distributions $N(m_i; \sigma)$. Show that even if hypothesis H_0 is not true

(a) the random variable s_2^2 has the χ^2 distribution with $n - r$ degrees of freedom and $E(s_2^2) = \sigma^2$.

(b) for every pair i_1, i_2 ($i_1, i_2 = 1, \ldots, r, i_1 \neq i_2$), the random variable

$$t = \sqrt{\frac{n_{i_1} n_{i_2}}{n_{i_1} + n_{i_2}}} \frac{\bar{x}_{i_1} - \bar{x}_{i_2} - (m_{i_1} - m_{i_2})}{s_2}$$

has Student's t-distribution with $n - r$ degrees of freedom (Irwin [1]).

15.2. (Continued). Show that if $n_{i_1} + n_{i_2} - 2 < n - r$, then, to test the hypothesis concerning $m_{i_1} - m_{i_2}$ (or to obtain a confidence interval), it is better to apply the statistic t from the preceding problem than the statistic U (suitably modified) given by (9.6.10).

15.3. We use the notation and assumptions of Section 15.2, except that we do not assume that the hypothesis H_0 (m_{kl} = constant) is satisfied.

Show that s_3^2, defined by (15.2.8), is an unbiased estimate of the unknown parameter σ^2 if and only if $m_{jk} - m_{k\cdot} - m_{\cdot l} + m = 0$, where

$$m_{k\cdot} = \frac{1}{v} \sum_{l=1}^{v} m_{kl}, \qquad m_{\cdot l} = \frac{1}{r} \sum_{k=1}^{r} m_{kl}, \qquad m = \frac{1}{rv} \sum_{k=1}^{r} \sum_{l=1}^{v} m_{kl}.$$

15.4. (Continued). Show that if $x_{kl} = a_k + b_l + \xi_{kl}$, where a_k is the effect on x_{kl} of the kth group of criterion A, b_l the effect of the lth group of criterion B, and ξ_{kl} has the normal distribution $N(0; \sigma)$, then s_3^2 is an unbiased estimate of σ^2 and has the χ^2 distribution with $(r - 1)(v - 1)$ degrees of freedom.

CHAPTER 16

Theory of Hypotheses Testing

16.1 PRELIMINARY REMARKS

Considerations concerning the construction and application of tests presented thus far form only a small part of the theory of hypotheses testing, which plays a central role in mathematical statistics. We start with an example.

Example 16.1.1. From a lot of merchandise containing N elements, a simple sample of size $n = 20$ has been drawn. The lot contains Np defective items, where p is unknown. Among the elements drawn, there were r defective items.

Imagine that we want to test the hypothesis H_0 ($p = 0.1$) and we adopt the following rule: We reject H_0 if $r \geqslant 5$ and we accept it if $r < 5$. The random variable r has a binomial distribution, and we can compute the probability that $r \geqslant 5$ provided H_0 is true as well as the probability that $r < 5$ provided p equals some number different from 0.1, and both these probabilities are positive if p is neither zero nor one.

From Table 5.2.1, we find that if $p = 0.1$, the probability of the relation $r \geqslant 5$ is 0.0433. We see that the probability of rejecting H_0 is positive even when this hypothesis is true. From Borel's law of large numbers, it follows that, for a large series of simple samples, the frequency of rejecting H_0 when it is true is close to 0.0433.

Suppose now that $p = 0.3$; hence H_0 is not true. From Table 5.2.1, we find that here the probability of the relation $r < 5$ is 0.2374. Now if $r < 5$, then, according to our rule, we accept H_0, although it is not true. Thus we make a wrong decision, and the probability of an error of this kind is 0.2374.

This example shows that the procedure of testing a hypothesis cannot be restricted to the hypothetical value of the unknown parameter; it is necessary to take into account other values of the unknown parameter.

16.2 THE POWER FUNCTION AND THE OC FUNCTION

A   As has been said in Section 12.2, a parametric significance test (if only one parameter Q is unknown) can be characterized in the following

541

way: A parameter Q of the distribution of the characteristic x of elements of a population is unknown, but the functional form of this distribution function is known. We want to test the hypothesis H_0 $(Q = Q_0)$. From a sample drawn from the population under investigation, we determine the value of a certain statistic u. We find a critical value u_0 such that if H_0 is true, then $P(|u| \geqslant u_0) \leqslant \alpha$, and we reject H_0 at the significance level α if the observed value of u satisfies the relation $|u| \geqslant u_0$.

The motivation is the following: If the hypothesis tested is true, the probability of obtaining an observed value of u no less than u_0 in absolute value is at most α. Since α is usually small, the acceptance of H_0 in this case would be equivalent to the statement that we have observed an event which has very small probability, hence an event which is very rare. Thus we reject H_0.

If u is a random variable of the continuous type, we determine u_0 from the relation $P(|u| \geqslant u_0) = \alpha$. In future considerations, we deal with this case unless otherwise stated.

Some questions arise in connection with this testing procedure. The first question is that of the probability of making wrong decisions. We have already dealt with this in the preceding section; we now formulate it generally. According to the definition of α, if the hypothesis H_0 is true, the relation $|u| \geqslant u_0$ is satisfied with probability α; hence for a series of independent samples, the procedure of rejecting H_0 just discussed would make the fraction of wrong decisions equal approximately α.

The second question is the following: Let us denote by $[u_1, u_2]$ the interval such that $P(u_1 \leqslant u < u_2) = \alpha$. Is it possible to test H_0 by rejecting it if the observed value of u falls within the interval $[u_1, u_2]$?

Thus, in example 12.1.1, we rejected the hypothesis H_0 $(m = 0)$ since it turned out that if H_0 is true, the probability of obtaining a value of $\bar{x}$ which, in absolute value, exceeds or equals the value observed is smaller than α. Could we have taken any interval I such that the probability that $\bar{x}$ belongs to I is α, and reject the hypothesis H_0 if the observed value of $\bar{x}$ belongs to the interval I? In particular, is it possible to take $I = [u_0, \infty)$, where $P(\bar{x} \geqslant u_0) = \alpha$?

Note that in each of these procedures the probability of rejecting H_0 when it is true is the same.

Which of these methods is better?

Finally, the third question to be answered is how to choose the statistic that serves as basis for the test. We may choose many different statistics as basis for testing the same hypothesis H_0, and for all of these tests the probability of rejecting H_0, although it is true, will be the same. How can we decide on a choice of a statistic as basis of the test?

As we said at the end of Section 16.1, to find an answer to these questions, we must consider not only the tested hypothetical value of the parameter. In this connection, we introduce certain notions.

We are to test the hypothesis H_0 ($Q = Q_0$). Suppose that besides the tested value Q_0 of the parameter Q some other values, say, $Q_1, Q_2, \ldots$ of this parameter may be true. Then the hypotheses H_0 ($Q = Q_0$), H_1 ($Q = Q_1$), H_2 ($Q = Q_2$), ... are called the *admissible hypotheses*. To distinguish it from these other hypotheses, the hypothesis H_0 ($Q = Q_0$) is called the *null hypothesis*. Every other admissible hypothesis is called an *alternative hypothesis*.

In the following considerations, we first consider only the case of two admissible hypotheses. Let H_0 ($Q = Q_0$) be the null hypotheses. Let the alternative hypothesis be H_1 ($Q = Q_1$). We test the hypothesis H_0 by observing the value of a certain statistic u. Let v be a set of values of the statistic u such that if the hypothesis H_0 ($Q = Q_0$) is true, then $P(u \in v) = \alpha$. To the set v under consideration, there corresponds a set w in the sample space such that

(16.2.1) $$P(\zeta \in w \mid Q = Q_0) = \alpha,$$

where $\zeta = (x_1, \ldots, x_n)$ denotes the random vector being observed. The set w is called the *critical region*.

Suppose that if ζ belongs to w, we reject the hypothesis H_0 and accept the alternative hypothesis H_1. In the opposite case, we accept H_0. We have already discussed the probability of committing an error which consists in rejecting H_0 when it is true. Such an error is called an *error of the first kind*. Its probability is α; hence the probability of an error of the first kind is equal to the significance level of the test. We may, however, make an error in the opposite direction; we may accept the hypothesis H_0 although it is false, that is, $Q = Q_1$. Such an error is called an *error of the second kind*. We make this error when ζ does not belong to the critical region w although $Q = Q_1$. If we denote the probability of this event by β, we have

(16.2.2) $$P(\zeta \notin w \mid Q = Q_1) = \beta.$$

B   Thus far, we have considered only two admissible values of the unknown parameter. If besides Q_0 and Q_1 there are other admissible values of the unknown parameter Q, then the problem arises of finding the probability of rejecting the hypothesis H_0 ($Q = Q_0$) and accepting the alternative hypothesis H_1 ($Q = Q_1$) for various values of the parameter Q, which may take on a value different from both Q_0 and Q_1. Considered as a function of Q for all admissible values of Q, this probability is called the *power function* of the test, and denoted by $M(w, Q)$. The value of $M(w, Q)$ for a given $Q_1 \neq Q_0$ is called the *power* of the test against the alternative Q_1.

Since we reject H_0 when ζ belongs to the set w, we have

(16.2.3)
$$M(w, Q) = P(\zeta \in w \mid Q).$$

From the definition of α and β, that is, from relations (16.2.1) and (16.2.2), we obtain in particular

(16.2.4)
$$M(w, Q_0) = \alpha, \qquad M(w, Q_1) = 1 - \beta.$$

The following function, called the *operating characteristic function*, is closely related to the power function of the test. We denote this function by $L(w, Q)$ and define it as

(16.2.5)
$$L(w, Q) = 1 - P(\zeta \in w \mid Q).$$

For brevity, we call the operating characteristic function the *OC function*.

The OC function expresses the probability of accepting the hypothesis H_0 ($Q = Q_0$), when applying the test w, as a function of Q.

The power function and the OC function are related in the following way:

(16.2.6)
$$L(w, Q) = 1 - M(w, Q).$$

Hence in particular

(16.2.7)
$$L(w, Q_0) = 1 - \alpha, \qquad L(w, Q_1) = \beta.$$

The ideal test would, of course, be a test such that its OC function satisfies the relations

$$L(w, Q_0) = 1 \qquad \text{for } Q = Q_0,$$

$$L(w, Q) = 0 \qquad \text{for } Q \neq Q_0.$$

For such a test, the probability of accepting H_0 equals one when this hypothesis is true, and equals zero when this hypothesis is false. However, such an ideal test cannot be obtained.

The basic ideas and results of the theory of testing statistical hypotheses, which are formulated in Sections 16.2 to 16.5 of this chapter, are due to Neyman and Pearson [1, 3 to 5]. The presentation of these methods is given in the works of Neyman [6] and [7].

We now present some examples of OC functions.

Example 16.2.1. The defectiveness p of a consignment of merchandise is unknown. According to the statement made by the producer, the defectiveness p does not exceed 0.1. The customer agrees to accept a consignment with this defectiveness but he decides to reject the lot if the defectiveness exceeds or equals 0.2. The two parties may make an additional agreement concerning the procedure in case the defectiveness is greater than 0.1 but smaller than 0.2.

Disregarding the case of $p < 0.1$ and of $p > 0.2$, we have the null hypothesis H_0 ($p = 0.1$) and the alternative hypothesis H_1 ($p = 0.2$). Let us fix $\alpha = 0.08$, that is, the probability of rejecting a lot with defectiveness equal to 0.1 should not exceed[1] 0.08. To test this hypothesis, we find the number of defective items in a simple sample of size $n = 10$ chosen from the lot considered.

First, we should construct the test, that is, we should decide how many defective items found in the sample will cause the lot to be rejected. Thus, we want to find a number m such that if the number of defective items in the sample of size n exceeds m we shall reject H_0. The condition required is that if H_0 is satisfied, then the probability of obtaining more than m defective items in the sample should not exceed 0.08.

Let us assign the number one to the appearance of a defective item, and the number zero to that of a good item. If $p = 0.1$, the probability of r defective elements among ten chosen is

$$\binom{10}{r} \cdot 0.1^r \cdot 0.9^{10-r} \quad (r = 0, 1, \ldots, 10).$$

Computation yields

$$P(r = 0) = 0.3481, \quad P(r = 1) = 0.3874, \quad P(r = 2) = 0.1937;$$

hence

$$P(r > 2) = 1 - 0.9292 = 0.0708.$$

Thus, we reject the consignment if we find more than two defective items in the sample.

Let us find the OC function of this test. Thus, we have to express the probability of obtaining no more than two defective items as a function of p. This OC function, which we denote for simplicity by $L(p)$, is given by the formula

$$(16.2.8) \quad L(p) = P(r = 0) + P(r = 1) + P(r = 2) = (1 - p)^8(1 + 8p + 36p^2).$$

Hence in particular

$$\beta = L(0.2) = 0.67779.$$

The graph of this OC function is represented in Fig. 16.2.1.

The faster the graph of the OC function decreases to the right of the critical value, the better is the test. We can increase considerably the effectiveness of the investigation by a suitable selection of the numbers n and m. We shall not discuss these questions in more detail. For a detailed study of statistical problems in quality control, the reader is referred to Grant [1].

Example 16.2.2. The characteristic x has the normal distribution $N(m; \sigma)$, where m is unknown and σ is known. We have to test the hypothesis H_0 ($m = m_0$). The alternative hypothesis is H_1 ($m = m_1$), or H_1 ($|m - m_0|/\sigma = \lambda_1$), where λ_1 is a constant positive number.

We test this hypothesis at the significance level $\alpha = 0.05$. To do this, we determine the sample mean $\bar{x}$ from a simple sample of size n. Let us take as the critical region w the set of points of n-dimensional sample space for

---

[1] Since we deal here with a random variable of the discrete type, the probability of error is expressed by an inequality; thus we used the term "not exceed."

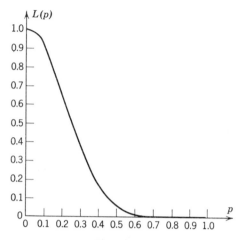

Fig. 16.2.1

which

$$\frac{|\bar{x} - m_0|}{\sigma} \sqrt{n} \geqslant 1.96.$$

Thus, if the observed value of $\bar{x}$ satisfies this relation, we reject H_0; in the contrary case we accept it.

We find the OC function of this test, as a function of the parameter λ defined by the formula

$$(16.2.9) \qquad \lambda = \frac{|m - m_0|}{\sigma},$$

where λ is an arbitrary non-negative number. From the construction of the critical region in this example, we obtain

$$L(w, m) = P\left(-1.96 < \frac{\bar{x} - m_0}{\sigma} \sqrt{n} < 1.96 \mid m\right)$$

$$= P\left(-1.96 - \frac{m - m_0}{\sigma} \sqrt{n} < \frac{\bar{x} - m}{\sigma} \sqrt{n} < 1.96 - \frac{m - m_0}{\sigma} \sqrt{n} \mid m\right).$$

By the symmetry of the normal curve, we find

$$L(w, m) = P\left(-1.96 + \frac{|m - m_0|}{\sigma} \sqrt{n} < \frac{\bar{x} - m}{\sigma} \sqrt{n}\right.$$

$$\left. < \frac{|m - m_0|}{\sigma} \sqrt{n} + 1.96 \mid m\right).$$

Finally, using formula (16.2.9), we have

$$(16.2.10) \quad L(w, m) = P\left(-1.96 + \lambda\sqrt{n} < \frac{\bar{x} - m}{\sigma} \sqrt{n} < \lambda\sqrt{n} + 1.96 \mid m\right).$$

As we see, $L(w, m_0)$ is independent of n, and equals $1 - \alpha = 0.95$, whereas for $m \neq m_0$ the OC function depends on n. We graph the OC function as a function of $\lambda \geqslant 0$. The sharper the OC function decreases in the neighborhood of $\lambda = 0$, the better is the test.

Figure 16.2.2 represents graphs of the OC function for some values of n. As we see, all these curves meet at the point ($\lambda = 0, L = 0.95$) since the significance level is constant and equals 0.05. For all $\lambda > 0$, the OC function is smaller than 0.95 and decreases as we move to the right from the point $\lambda = 0$, that is, as the difference between the true value of the parameter m and m_0 increases.

The function decreases more rapidly for greater n. Thus, for instance, if we take $\beta = 0.05$, then from Fig. 16.2.2 we find that λ is about 2.5 for $n = 2$, that is, in approximately 5 cases out of every 100, we accept H_0 ($m = m_0$) when the true value m differs from m_0 by 2.5σ. For $n = 10$, the value of λ slightly exceeds one, and, for $n \geqslant 50$, it is smaller than 0.5.

We observe, however, that although for small values of n an increase in the size of the sample causes a rapid improvement in the behavior of the OC function, for values of $n \geqslant 50$ an increase in sample size does not improve the behavior very much, hence does not increase the guarantee of not making a wrong decision. In sample investigations where the cost of sampling is quite great, it·may not be profitable to increase the sample size to over 50 when we test a hypothesis concerning the expected value of a normal population.

C We now introduce some further ideas.

Let $F(x, Q_1, \ldots, Q_k)$ be the distribution function of a random variable x whose functional form is known and which depends upon some

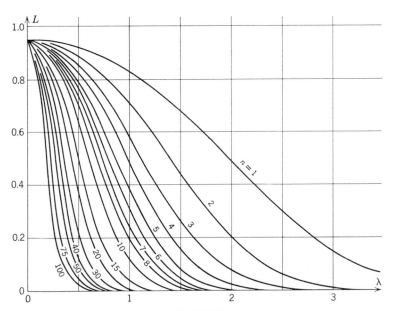

Fig. 16.2.2

unknown parameters $Q_1, \ldots, Q_k$. Let us denote by q the vector $(Q_1, \ldots, Q_k)$ and by Ω the k-dimensional space of all admissible q. The space Ω is called *parameter space*.

Definition 16.2.1. The hypothesis $H(Q_1 = Q_{10}, \ldots, Q_k = Q_{k0})$ which specifies the values of all the unknown parameters is called a *simple hypothesis*. The hypothesis $H(q \in \omega)$, where ω is a subset of Ω which contains more than one point, is called a *composite hypothesis*.

Thus, if we have k $(k > 1)$ unknown parameters and the hypothesis specifies only l $(l < k)$ among them, then the hypothesis is a composite hypothesis, and the set ω which appears in definition 16.2.1 is a $(k - l)$-dimensional hyperplane in the space Ω. If k $(k \geqslant 1)$ parameters are unknown, and the hypothesis does not specify the value of any of them, but only states that they satisfy some functional relation, then we also have a composite hypothesis, and the set ω in the space Ω is determined by the functional relation. It is obvious that the null hypothesis as well as the alternative hypothesis may be simple or composite.

Definition 16.2.1 can be formulated in an analogous way for multidimensional random variables.

The reader can verify that the hypotheses considered in examples 12.1.1, 12.2.3, 12.3.3, and both the null and alternative hypotheses in examples 16.1.1, 16.2.2 are simple hypotheses, whereas the hypotheses considered in examples 12.2.1, 12.2.4, 12.2.5, 12.2.6, 15.1.1, and 15.2.1 are composite hypotheses.

Relation (16.2.1) holds for all the tests just mentioned. In connection with the notion of a composite hypothesis, it is necessary to introduce the following definition.

Definition 16.2.2. The set w of points in the sample space is, for a given significance level α, *similar to the whole sample space with respect to the set ω of Ω* if the relation

$$(*) \qquad\qquad P(\zeta \in w \mid q \in \omega) = \alpha$$

is satisfied for all points $q \in \omega$.

The expression "similar to the sample space" is due to the fact that, for $\alpha = 1$, relation $(*)$ is satisfied for the whole sample space for all points of the parameter space Ω.

It is obvious that a set w may serve as basis for testing a composite hypothesis $H(q \in \omega)$, at a given significance level, only if this set is similar to the whole sample space with respect to the set ω. Neyman and Pearson [3] gave a condition for the existence of sets similar to the whole sample space for testing composite hypotheses. Feller [9] showed that in many cases such sets do not exist. (See also Problem 16.10.)

Note that (see the examples quoted) the set w in example 12.2.1, determined by the relation $|t| \geqslant t_0$, is similar to the sample space with respect to the set ω of Ω, where Ω and ω consist of the points (m, σ) satisfying the conditions $(-\infty < m < +\infty, 0 < \sigma < \infty)$ and $(m = 0, 0 < \sigma < \infty)$, respectively. This is because the hypothesis specifies the value of the parameter m, and the distribution of the statistic t, which serves as basis for the test, does not depend on σ. In example 12.2.4, we deal with a two-dimensional normal population for which the parameters $m_1, m_2, \sigma_1, \sigma_2$, and ρ are unknown, and the hypothesis H_0 ($\rho = 0$) does not specify the values of the remaining parameters. The parameter space Ω consists of points $(m_1, m_2, \sigma_1, \sigma_2, \rho)$ with $-\infty < m_1 < +\infty$, $-\infty < m_2 < +\infty$, $0 < \sigma_1 < \infty, 0 < \sigma_2 < \infty, -1 \leqslant \rho \leqslant 1$, and the set ω is determined by the hyperplane $\rho = 0$ in the space Ω. The critical region here is similar to the whole sample space with respect to the set ω, because the hypothesis specifies the value of ρ and the distribution of the statistic R, which serves as basis for the test, does not depend on the remaining parameters. In a similar way, the reader can analyze the similarity to the whole sample space of the critical regions considered in examples 12.2.5, 12.2.6, 15.1.1, and 15.2.1.

We now present an example of the power function of a test for a composite hypothesis.

Example 16.2.3. The characteristic x of the population under consideration has the distribution $N(m; \sigma)$, where m and σ are unknown. We want to test the hypothesis H_0 ($m = m_0$) against the alternative hypothesis H_1 ($m = m_1$). Both hypotheses H_0 and H_1 are composite since neither of them specifies the value of σ. We now rewrite H_1 in the form $H_1(|m - m_0|/\sigma = \lambda_1)$, where $\lambda_1 > 0$. The hypothesis H_0 will be tested by using Student's t at the significance level α. From simple samples, we compute the value of the statistic

$$(16.2.11) \qquad t = \frac{\bar{x} - m_0}{s} \sqrt{n - 1},$$

and if the observed t satisfies the relation $|t| \geqslant t_\alpha$, where t_α is determined from the relation

$$P(-t_\alpha < t < t_\alpha) = 1 - \alpha,$$

we reject H_0 and accept H_1. Thus the critical region w consists here of those points of the sample space for which t, computed from relation (16.2.11), satisfies the relation $|t| \geqslant t_\alpha$. As we have already said, this critical set is similar to the whole sample space with respect to the set ω in parameter space, determined by the relations $m = m_0, 0 < \sigma < \infty$.

Let us find the OC function of this test. If the hypothesis H_0 is true, the statistic t defined by (16.2.11) has Student's t-distribution. From the definition of t_α, we have

$$(16.2.12) \qquad L(w, m_0) = 1 - \alpha,$$

where w is the critical region. If, however, the hypothesis H_0 is not true, the statistic t defined by (16.2.11) has a noncentral t distribution whose density $f(y, Q)$, given by (9.6.7), depends upon the parameter

$$Q = \frac{m - m_0}{\sigma}.$$

Let $\lambda = |Q|$. Since for two values of Q which differ only in sign the distributions of the noncentral t are symmetric, we obtain

$$(16.2.13) \quad L(w, Q) = P(t \notin w \mid Q) = P(-t_\alpha \leqslant t \leqslant t_\alpha \mid \lambda) = \int_{-t_\alpha}^{t_\alpha} f(y, \lambda)\,dy = L(w, \lambda).$$

Formulas (16.2.12) and (16.2.13) determine the OC function of this test.

Figure 16.2.3 shows graphs of the OC function of the test t for some n at the significance level $\alpha = 0.05$.

These graphs give values of the function $L(w, \lambda)$ as a function of $\lambda = |Q|$. The shapes of these curves are, in general, similar to those in Fig. 16.2.2. However, for small n the OC function in example 16.2.2 is much better than the OC function of Student's t-test. Thus, we established in example 16.2.2 that if $\beta = 0.05$, then for $n = 2$ we have $\lambda = 2.5$ approximately, whereas from the graph in Fig. 16.2.3 we find that for $n = 2$ the probability of accepting the hypothesis H_0 ($m = m_0$) if $|m - m_0| = 2.5\sigma$ is about 0.77. This result agrees with our intuition, since in example 16.2.2 the value of σ is assumed to be known, hence our information is considerably larger than in this example. As n increases, the difference between the power of these tests decreases to zero.

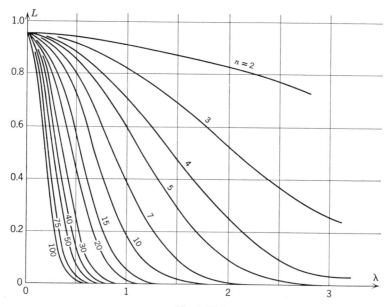

Fig. 16.2.3

Figures 16.2.2 and 16.2.3 are taken from the paper of Ferris, Grubbs, and Weaver [1]; there, the reader will find other examples of OC functions of widely applied tests.

The power function of Student's t-test has been the subject of investigations by Neyman and Tokarska [1], and Johnson and Welch [1].

In statistical practice, we often deal with populations in which the characteristic investigated does not have exactly a normal distribution, but only a distribution which is approximately normal. The question arises whether it is then possible to apply Student's t-test for testing the hypothesis concerning the unknown expected value, and, if so, what the power function of this test is. This problem has been investigated by many authors. We mention only the most recent paper in this domain, that of Srivastava [1], which also contains information on earlier papers.

We observe that the power functions of the tests considered in examples 16.2.2 and 16.2.3 depend upon the parameter σ. G. B. Dantzig [1] has proved that for a constant sample size there does not exist any test for testing the hypothesis concerning the unknown expected value of a normal population whose power function is independent of the standard deviation σ. The idea of the proof of Dantzig's theorem is the following: Let w be a set of points in the sample space such that the power function of the test, for testing the hypothesis concerning the unknown expected value m of a random variable with distribution $N(m; \sigma)$, based on the critical region w is independent of σ. In other words, the set w is such that for every fixed m the expression $P(\zeta \in w \mid m)$ is independent of σ. Then the power of this test does not depend on m either. Therefore such a test would as often lead to the rejection of the considered hypothesis when it is false as when it is true.

For this problem, Stein [1] constructed a test whose power function is independent of σ, but the number of elements of the sample is not a constant but a random variable.

Stein [1] generalized the result of Dantzig, proving that there does not exist a test for testing a linear hypothesis (see Section 15.1, C) whose power function is independent of σ.

Tang [1] has found power functions of tests used in analysis of variance which were discussed in Sections 15.1 and 15.2. Further results in this domain were obtained by Hsu [3]. Nicholson [1] gave a formula which is useful in computing the power functions of these tests. The power functions of tests used in analysis of variance, and more generally, of tests used in testing linear hypotheses, is based on Fisher's noncentral Z-distribution (see Section 9.7).

D   To complete the considerations of this section, we recall the point of view expressed in Section 12.1, that to accept a hypothesis on the basis of one test if as a result of this test the hypothesis has not been rejected is not well founded. If, however, the test is constructed so that the probabilities of an error of the first and second kind are small, we can accept or reject the hypothesis tested on the basis of one experiment. Then, acceptance of the tested hypothesis is based on the argument that since β is small (for instance, equals 0.01), if we were not to accept the hypothesis, we would be forced to state that we had observed a rare event. We recall that this is exactly the kind of reasoning we applied in Chapter 12 to justify rejecting the hypothesis tested on the basis of a single experiment.

16.3 MOST POWERFUL TESTS

A   Let Q be the unknown parameter, let $H_0 (Q = Q_0)$ be the null hypothesis, and let $H_1 (Q = Q_1)$, where $Q_1 \neq Q_0$ is a given number, be the alternative hypothesis.

Let us formulate a rule for choosing the set w which serves as basis of the test. Among all sets w for which equality (16.2.1) holds, we choose as basis of the test a set w (if one exists) for which β, satisfying equality (16.2.2), has its minimum value. In other words, from among all sets w which may serve as basis for testing the null hypothesis $H_0 (Q = Q_0)$ against the alternative hypothesis $H_1 (Q = Q_1)$, for which the probability of an error of the first kind is constant and equals α, we choose a set w, if such exists, for which the probability β of an error of the second kind has its minimum value.

Such a test is called a *most powerful test*. It does not always exist, but it can be constructed for a large class of distribution functions.

In the sequel, we often say briefly "test w" instead of "the test whose basis is the critical set w."

Let $\zeta = (x_1, \ldots, x_n)$ be a random variable of the continuous type whose density depends on Q. This density is denoted briefly by $f(\zeta, Q)$.

Theorem 16.3.1. *If the point ζ in sample space is a random variable of the continuous type with density $f(\zeta, Q)$, where the parameter Q is unknown and $H_0 (Q = Q_0)$ and $H_1 (Q = Q_1)$ $(Q_0 \neq Q_1)$ are the null and alternative hypotheses, respectively, then a most powerful test exists.*

Proof. Let $f(\zeta, Q_0)$ and $f(\zeta, Q_1)$ denote the density of the random point ζ if the true hypothesis is H_0 or H_1, respectively. Suppose that the densities $f(\zeta, Q_0)$ and $f(\zeta, Q_1)$ are everywhere defined and finite. This assumption is not an essential restriction. For every non-negative number c, denote by w_c the set of points ζ for which the inequality

(16.3.1) $f(\zeta, Q_1) \geqslant cf(\zeta, Q_0)$

is satisfied. Furthermore, let

$$\psi(c) = P(\zeta \in w_c \mid Q = Q_0).$$

That is, $\psi(c)$ is the probability that the point ζ will belong to the set w_c when $Q = Q_0$. Observe that $\psi(c)$ is a nonincreasing function, for if $c_1 < c_2$, then every point ζ for which inequality (16.3.1) is satisfied with $c = c_2$ belongs to the set w_c for $c = c_1$; hence $\psi(c_2) \leqslant \psi(c_1)$. Next, $\psi(0) = 1$, since a density cannot be negative.

Observe that

$$(16.3.2) \qquad\qquad 0 \leqslant \psi(c) \leqslant \frac{1}{c}.$$

The left-hand side of this inequality is obvious, and the right-hand side can easily be proved. Indeed, if we had $c\psi(c) > 1$, then by (16.3.1) we would obtain

$$\int_{w_c} f(\zeta, Q_1) \, d\zeta \geqslant c \int_{w_c} f(\zeta, Q_0) \, d\zeta = c\psi(c) > 1,$$

where the integrals[1] are integrated over the set w_c. However, this is impossible, since the integral on the left-hand side represents a probability and cannot exceed one. It follows from (16.3.2) that

$$\lim_{c \to \infty} \psi(c) = 0.$$

Let us assume first that there exists a c' for which

$$(16.3.3) \qquad\qquad \psi(c') = \alpha,$$

where α is the given significance level of the test. We shall show that the set $w_c' = w'$, defined by (16.3.1) with c' satisfying (16.3.3), is most powerful.

Indeed, we have

$$(16.3.4) \qquad \psi(c') = P(\zeta \in w' \mid Q = Q_0) = \int_{w'} f(\zeta, Q_0) \, d\zeta = \alpha.$$

Let w be another test with the same significance level α, that is,

$$(16.3.5) \qquad P(\zeta \in w \mid Q = Q_0) = \int_{w} f(\zeta, Q_0) \, d\zeta = \alpha.$$

Denoting by $w'w$ the set of elements common to w' and w, we obtain from (16.3.4) and (16.3.5)

(16.3.6)

$$P(\zeta \in w' \mid Q = Q_0) - P(\zeta \in w'w \mid Q = Q_0)$$
$$= \alpha - P(\zeta \in w'w \mid Q = Q_0)$$
$$= P(\zeta \in w \mid Q = Q_0) - P(\zeta \in w'w \mid Q = Q_0).$$

---

[1] $\int_{w_c} f(\zeta, Q) \, d\zeta$ is a simplified way of denoting the n-dimensional integral.

It follows from the definition of w' that if ζ does not belong to this set, we have

(16.3.7) $$c'f(\zeta, Q_0) > f(\zeta, Q_1).$$

From (16.3.6) and (16.3.7), we obtain

$$P(\zeta \in w' \mid Q = Q_1) - P(\zeta \in w'w \mid Q = Q_1)$$
$$\geqslant c'[P(\zeta \in w' \mid Q = Q_0) - P(\zeta \in w'w \mid Q = Q_0)]$$
$$= c'[P(\zeta \in w \mid Q = Q_0) - P(\zeta \in ww' \mid Q = Q_0)]$$
$$\geqslant P(\zeta \in w \mid Q = Q_1) - P(\zeta \in ww' \mid Q = Q_1).$$

Hence

(16.3.8) $P(\zeta \in w' \mid Q = Q_1) - P(\zeta \in w'w \mid Q = Q_1)$
$$\geqslant P(\zeta \in w \mid Q = Q_1) - P(\zeta \in ww' \mid Q = Q_1).$$

By adding $P(\zeta \in ww' \mid Q = Q_1)$ to both sides of the last inequality, we obtain

(16.3.9) $$P(\zeta \in w' \mid Q = Q_1) \geqslant P(\zeta \in w \mid Q = Q_1);$$

hence the power of the test based on the critical region w' is at least as great as the power of the test based on any other critical region w. Thus the test w' is most powerful.

If a constant c' satisfying (16.3.3) does not exist, then there does exist a constant c^* such that $\psi(c^* - 0) \geqslant \alpha$ and $\psi(c^* + 0) \leqslant \alpha$. Here we have

$$\psi(c^* - 0) - \psi(c^* + 0) = \int_M f(\zeta, Q_0)\, d\zeta,$$

where M is the set of points ζ for which

$$f(\zeta, Q_1) = c^*f(\zeta, Q_0).$$

Then a most powerful test is the set $w'' = w_{c^*-0} - B$, where B is a subset of M chosen so that $P(\zeta \in w'' \mid Q = Q_0) = \alpha$. The proof is similar to that of the preceding case.

Theorem 16.3.1 is proved. This theorem simultaneously provides a method for constructing the set w'.

Theorem 16.3.1 is a particular case of the *Fundamental Lemma* of Neyman and Pearson [4].

The question of whether the assumptions of theorem 16.3.1 are necessary was answered in the paper of Dantzig and Wald [1], to which the reader is referred. The problem of generalizing this theorem to discrete random variables is discussed in Section 16.3, D (see also Tocher [1]).

Example 16.3.1. The expected value m of a population, where the characteristic x has the normal distribution $N(m; 1)$, is unknown. We want to test the hypothesis H_0 ($m = 0$) against the alternative hypothesis H_1 ($m = 1$). We find a most powerful test for testing this hypothesis based on a simple sample.

If the hypothesis H_0 ($m = 0$) is true, the density $f(\zeta, Q)$ is of the form

(16.3.10) $$f(\zeta, 0) = \frac{1}{(\sqrt{2\pi})^n} \exp\left(-\frac{1}{2}\sum_{i=1}^{n} x_i^2\right).$$

If the hypothesis H_1 ($m = 1$) is true, then

(16.3.11) $$f(\zeta, 1) = \frac{1}{(\sqrt{2\pi})^n} \exp\left(-\frac{1}{2}\sum_{i=1}^{n} (x_i - 1)^2\right).$$

We find the set w' of points ζ for which inequality (16.3.1) is satisfied. We obtain

(16.3.12) $$\frac{f(\zeta, 1)}{f(\zeta, 0)} = \exp\left\{-\frac{1}{2}\sum_{i=1}^{n}[(x_i - 1)^2 - x_i^2]\right\} = \exp\left(n\bar{x} - \frac{n}{2}\right) \geqslant c'.$$

Thus the set w' consists of points ζ which satisfy the relation

(16.3.13) $$\bar{x} \geqslant \frac{n + 2\log c'}{2n} = A.$$

Since, if H_0 ($m = 0$) is true, the random variable $\bar{x}$ has the distribution $N(0; 1/\sqrt{n})$, we find for a given significance level α the number A (hence the number c') from the formula

(16.3.14) $$\frac{\sqrt{n}}{\sqrt{2\pi}} \int_A^\infty \exp\left(-\frac{n\bar{x}^2}{2}\right) d\bar{x} = \alpha.$$

For instance, we can determine A from tables of the normal distribution and then determine c'. The value of the power function for $m = 1$ can be found from the formula

(16.3.15) $$P(\bar{x} \geqslant A \mid m = 1) = \frac{\sqrt{n}}{\sqrt{2\pi}} \int_A^\infty \exp\left(-\frac{n(\bar{x} - 1)^2}{2}\right) d\bar{x};$$

hence the probability of an error of the second kind is

$$\beta = 1 - P(\bar{x} \geqslant A \mid m = 1).$$

Thus, for instance, if $\alpha = 0.05$ and $n = 30$, $\bar{x}$ has the distribution $N(0; 1/\sqrt{30})$ provided H_0 is true. From tables of the normal distribution, we find

$$A = 1.65 \cdot \frac{1}{\sqrt{30}} \cong 0.3.$$

Hence

$$\beta = 1 - P(\bar{x} \geqslant 0.3 \mid m = 1) = 0.0001.$$

B   In example 16.3.1, the sample size n and the probability α of an error of the first kind were given. The given conditions of the problem, that is, the hypotheses H_0 and H_1 and the parameters n and α, allow us to determine a most powerful test w' and the probability β of an error of the second kind. In practice, the problem is often formulated in a different way: the hypotheses H_0 and H_1 are given, the parameters α and β are also given, and it is required to find a most powerful test w and a sample size n such that, when testing the null hypothesis H_0 against the alternative H_1, the probabilities of an error of the first and second kind are α and β, respectively. We show by an example how to solve such a problem.

Example 16.3.2. The expected value m of a population, where the characteristic x has the normal distribution $N(m; 1)$, is unknown. We wish to test the hypothesis H_0 ($m = m_0$) against the alternative hypothesis H_1 ($m = m_1$), where $m_1 > m_0$, by applying a most powerful test based on a simple sample for given α and β. What value should n have?

The most powerful test for testing the hypothesis under consideration is[1] the set w' of points in the n-dimensional sample space for which

$$\bar{x} \geqslant A.$$

Two values are unknown here, A and n. For given α and β, we have the following two equalities for determining these values:

(16.3.16)
$$\frac{\sqrt{n}}{\sqrt{2\pi}} \int_A^\infty \exp\left(-\frac{n(\bar{x} - m_0)^2}{2}\right) d\bar{x} = \alpha,$$

(16.3.16′)
$$\frac{\sqrt{n}}{\sqrt{2\pi}} \int_{-\infty}^A \exp\left(-\frac{n(\bar{x} - m_1)^2}{2}\right) d\bar{x} = \beta.$$

Let us write $A = m_0 + k_\alpha/\sqrt{n}$, where k_α is chosen in such a way that for a random variable y with the normal distribution $N(0; 1)$, $P(y \geqslant k_\alpha) = \alpha$. From (16.3.16′), we have $A = m_1 + k_\beta/\sqrt{n}$, where k_β is chosen in such a way that $P(y \leqslant k_\beta) = \beta$. From the equality

$$m_0 + \frac{k_\alpha}{\sqrt{n}} = m_1 + \frac{k_\beta}{\sqrt{n}},$$

we obtain

(16.3.17)
$$n = \frac{(k_\alpha - k_\beta)^2}{(m_1 - m_0)^2},$$

(16.3.18)
$$A = \frac{k_\alpha m_1 - k_\beta m_0}{k_\alpha - k_\beta}.$$

For instance, let $m_0 = 0$, $m_1 = 1$, $\alpha = \beta = 0.05$. Then from equality (16.3.17), we obtain $\sqrt{n} = 3.3$, hence $n = 10.89$. Since n must be an integer, we take $n = 11$.

C   Theorem 16.3.1 is also valid when the hypotheses H_0 and H_1 concern two or more parameters. The proof of theorem 16.3.1 for this case is analogous to the proof presented previously.

Example 16.3.3. The parameters m and σ of a characteristic x, which has the normal distribution $N(m; \sigma)$, are unknown. On the basis of a simple sample, we

[1] See Section 16.4, formula (16.4.5).

test the hypothesis H_0 ($m = 0$, $\sigma = 1$) against the alternative H_1 ($m = m_1$, $\sigma = \sigma_1$).

Let us find the set w' of points ζ for which the inequality

$$(16.3.19) \quad \frac{f(\zeta, m_1, \sigma_1)}{f(\zeta, 0, 1)} = \frac{1}{\sigma_1^n} \exp\left\{-\left[\sum_{l=1}^n \frac{(x_l - m_1)^2}{2\sigma_1^2} - \sum_{l=1}^n \frac{x_l^2}{2}\right]\right\} \geqslant c'$$

is satisfied. Inequality (16.3.19) is the analogue of inequality (16.3.1). Since

$$n\bar{x} = \sum_{l=1}^n x_l, \qquad n(\bar{x}^2 + s^2) = \sum_{l=1}^n x_l^2,$$

from (16.3.19), we obtain

$$(16.3.20) \quad (\sigma_1^2 - 1)(\bar{x}^2 + s^2) + 2m_1\bar{x} \geqslant 2\frac{\sigma_1^2}{n} \log\left(c'\sigma_1^n\right) + m_1^2.$$

If $\sigma_1^2 > 1$, the critical region w' determined by (16.3.20) forms the domain outside a certain circle in the plane ($\bar{x}$, s). The constant c', the radius of this circle, can be found from the condition

$$P(\zeta \in w' \mid m = 0, \sigma = 1) = \alpha.$$

If $\sigma_1^2 < 1$, the set w' lies inside a certain circle in the plane ($\bar{x}$, s).

The determination of the critical region in this example requires approximate integration.

D   Now let the characteristic x of elements of a population be a random variable of the discrete type with the jump points x_k, and let its probability function $p_k(Q) = P(x = x_k \mid Q)$ depend upon the unknown parameter Q. Let the null hypothesis be H_0 ($Q = Q_0$) and the alternative be H_1 ($Q = Q_1$).

We denote in this subsection by $y_1, \ldots, y_n$ the observations of the random variable x and $\zeta = (y_1, \ldots, y_n)$. For convenience, let us denote by $f(\zeta, Q_1)$ and $f(\zeta, Q_0)$, respectively, the probability function of the observed random vector, if the hypothesis H_1 or H_0 is true. Note that, since x is of the discrete type, equality (16.2.1) may not be satisfied, for a given α, for any set w in the sample space. Thus theorem 16.3.1 can not be directly extended to random variables of the discrete type. Here a modified variant of that theorem holds. To formulate it, we need the concept of a *randomized test*.

The tests used until now (call them *nonrandomized*) divided the sample space into two parts, a rejection region w and its complement, in such a way that we had to reject H_0 whenever the observed sample point $\zeta \in w$. A randomized test assigns to any point ζ the probability $\phi(\zeta)$ of rejecting H_0 when ζ has been observed. In other words, $\phi(\zeta)$ is the conditional probability of rejecting H_0, provided ζ has been observed. The function $\phi(\zeta)$ is called *test function* or *critical function*. The *power* of the test against the alternative H_1, hence the probability of rejecting H_0 when H_1 is true, equals $E[\phi(\zeta) \mid Q = Q_1]$. Evidently, if, for some set w, we have $\phi(\zeta) = 1$ for $\zeta \in w$ and $\phi(\zeta) = 0$ for $\zeta \notin w$, the randomized test becomes equivalent to the nonrandomized one, with probability one.

We can now formulate the concept of a most powerful randomized test. We say that a test function $\phi(\zeta)$ is a *most powerful randomized test* of H_0 against H_1 at significance level α if $E[\phi(\zeta) \mid Q = Q_0] = \alpha$ and the power of $\phi(\zeta)$ against H_1 is the largest among all possible tests. Now, the *Fundamental Lemma* states that the test function $\phi(\zeta)$, defined as

(16.3.21)
$$\phi(\zeta) = \begin{cases} 1 & \text{if } f(\zeta, Q_1)/f(\zeta, Q_0) > c, \\ \gamma & \text{if } f(\zeta, Q_1)/f(\zeta, Q_0) = c, \\ 0 & \text{if } f(\zeta, Q_1)/f(\zeta, Q_0) < c \end{cases}$$

with γ and c determined from $E[\phi(\zeta) \mid Q = Q_0] = \alpha$, is a most powerful test function for testing H_0 against H_1.

Note that according to the Fundamental Lemma there exists a most powerful randomized test for any testing problem with both hypotheses H_0 and H_1 simple. The question of whether a most powerful test is unique is answered in Lehmann [4].

Example 16.3.4. The characteristic x has a Poisson distribution with unknown parameter λ. We have to test, from a simple sample of size 10, the hypothesis H_0 ($\lambda = 0.1$) against the hypothesis H_1 ($\lambda = 1$), at level $\alpha = 0.01$. According to (16.3.21), a most powerful test is given by the test function

$$\phi(\zeta) = \begin{cases} 1 & \text{if } e^{-9}10^{y_1 + \cdots + y_{10}} > c, \\ \gamma & \text{if } e^{-9}10^{y_1 + \cdots + y_{10}} = c, \\ 0 & \text{if } e^{-9}10^{y_1 + \cdots + y_{10}} < c. \end{cases}$$

In other words,

$$\phi(\zeta) = \begin{cases} 1 & \text{if } y_1 + \ldots + y_{10} > A, \\ \gamma & \text{if } y_1 + \ldots + y_{10} = A, \\ 0 & \text{if } y_1 + \ldots + y_{10} < A \end{cases}$$

with $A = (\log c + 9)/\log 10$. Denote $y = y_1 + \ldots + y_{10}$. Now, if H_0 is true, y has the Poisson distribution with $E(y) = 1$. From $P(y > A) + \gamma P(y = A) = 0.01$, we find that $A = 4$ and $\gamma = 0.413$. Therefore, we reject H_0 whenever we observe $y \geqslant 5$, and we accept H_0 whenever $y \leqslant 3$. If, however, we observe $y = 4$, we reject H_0 with probability 0.413, hence approximately in 41 cases out of 100, and accept H_0 with probability 0.587. The power of this test equals $P(y > 4) + 0.413 \, P(y = 4)$, where y has the Poisson distribution with $E(y) = 10$. We find that the power equals 0.97856.

16.4   UNIFORMLY MOST POWERFUL TEST

We have introduced the notion of a most powerful test of the null hypothesis H_0 ($Q = Q_0$) against a given single alternative H_1 ($Q = Q_1$).

It sometimes happens that the same test is most powerful for testing the hypothesis H_0 against any other admissible hypothesis H, where the admissible values of the parameter Q are those $Q \neq Q_0$ which belong to a given set ω. Such a test is called *uniformly most powerful with respect to ω*. If ω consists of all $Q \neq Q_0$, such a test is called *uniformly most powerful*.

Let us consider example 16.3.1 in its general form. The expected value m of a population with the normal distribution $N(m; \sigma)$ is unknown, but σ is known. We wish to test the hypothesis H_0 $(m = m_0)$ against the alternative H_1 $(m = m_1)$, where m_1 is an arbitrary number satisfying the inequality

(16.4.1) $$m_1 > m_0.$$

Let α be the significance level. For simple samples of size n, we find a most powerful test for testing the hypothesis under consideration. If the hypothesis H_0 $(m = m_0)$ is true, the density $f(\zeta, m_0)$ is of the form

(16.4.2) $$f(\zeta, m_0) = \frac{1}{(\sigma\sqrt{2\pi})^n} \exp\left\{ -\frac{\sum\limits_{i=1}^{n}(x_i - m_0)^2}{2\sigma^2} \right\}.$$

If the hypothesis H_1 $(m = m_1)$ is true, then

(16.4.3) $$f(\zeta, m_1) = \frac{1}{(\sigma\sqrt{2\pi})^n} \exp\left\{ -\frac{\sum\limits_{i=1}^{n}(x_i - m_1)^2}{2\sigma^2} \right\}.$$

A most powerful test satisfies relation (16.3.1); hence

(16.4.4) $$\frac{f(\zeta, m_1)}{f(\zeta, m_0)} = \exp\left\{ -\frac{\sum\limits_{i=1}^{n}[(x_i - m_1)^2 - (x_i - m_0)^2]}{2\sigma^2} \right\}$$

$$= \exp\left\{ \frac{n\bar{x}(m_1 - m_0)}{\sigma^2} \right\} \exp\left\{ \frac{-n(m_1^2 - m_0^2)}{2\sigma^2} \right\} \geqslant c'.$$

In view of formula (16.4.1), it follows that the set w' consists of those points ζ for which

(16.4.5) $$\bar{x} \geqslant \frac{\sigma^2 \log c'}{n(m_1 - m_0)} + \frac{m_1 + m_0}{2} = A.$$

The value of A (or of c') can be found from the equality

(16.4.6) $$\frac{\sqrt{n}}{\sigma\sqrt{2\pi}} \int_A^\infty \exp\left\{ -\frac{n(\bar{x} - m_0)^2}{2\sigma^2} \right\} d\bar{x} = \alpha;$$

hence simply from the fact that the significance level of the test equals α. However, as we can easily see from formula (16.4.6), the value of A does not depend on m_1. Thus we see that the most powerful test for testing the hypothesis H_0 ($m = m_0$) against the alternative H_1 ($m = m_1$) is the same for all values of m_1 greater than m_0. This test (critical region) consists of those points ζ for which inequality (16.4.5) is satisfied. Thus, if the admissible alternative hypotheses are H_1 ($m = m_1$), where $m_1 > m_0$, and otherwise arbitrary, then the test w' determined by (16.4.5) and (16.4.6) will be uniformly most powerful with respect to the set ($m_1 > m_0$).

Similarly, we can determine a uniformly most powerful test with respect to the admissible alternative hypotheses H_1 ($m = m_1$), where $m_1 < m_0$.

If, however, all alternative hypotheses H_1 ($m = m_1$), where $m_1 \neq m_0$, are admissible, then, as may be seen from the preceding reasoning, a uniformly most powerful test does not exist.

It should be stated that uniformly most powerful tests seldom exist. We present here without proof the theorem of Neyman [2], which allows us to determine sometimes whether a uniformly most powerful test exists

Theorem 16.4.1. *Let Q be the unknown parameter of the distribution of the characteristic x of elements of a population. Furthermore, let H_0 ($Q = Q_0$) be the null hypothesis. If the hypotheses H_1 ($Q = Q_1$) are admissible, where both $Q_1 < Q_0$ and $Q_1 > Q_0$, and, moreover, the density of x (or its probability function if x is of the discrete type) is differentiable with respect to Q, then there is no uniformly most powerful test for testing H_0 ($Q = Q_0$) against H_1 ($Q = Q_1$).*

It is easy to observe that, according to this theorem, in the example considered in this section, a uniformly most powerful test does not exist if we take the set of all hypotheses H_1 ($m = m_1$) with $m_1 \neq m_0$ as the set of admissible alternatives.

It should be noted that the test whose OC function we considered in example 16.2.2 is not uniformly most powerful. We return to this test in the next section.

16.5 UNBIASED TESTS

A   As was established in the previous section, uniformly most powerful tests seldom exist. This means that it is rare that we succeed in constructing a test that is most powerful among all possible tests, uniformly with respect to all possible alternatives to the null hypothesis.

In the previous section, we saw that if we restrict ourselves to a subset of all alternatives, a uniformly most powerful test with respect to this subset may exist, although a uniformly most powerful test with respect to the set of all possible alternatives does not exist.

Sometimes, we can obtain a uniformly most powerful test by another procedure, by restriction to the most powerful test among a certain subset of the set of all possible tests, but uniformly most powerful with respect to all possible alternatives.

We shall consider one such subset of tests, the unbiased tests. We explain this notion by the following example.

Example 16.5.1. The expected value m of the characteristic x of a population with the normal distribution $N(m; \sigma)$ is unknown, and we assume that σ is known. We want to test the hypothesis H_0 $(m = m_0)$ against the alternative H_1 $(m = m_1)$, where $m_1 \neq m_0$. Let us imagine that we take as the critical region the set w' of points ζ which satisfy relation (16.4.5), where A is determined from (16.4.6). As we see, the significance level of the test equals α. Thus, if the value of $\bar{x}$ observed from the sample is not smaller than A, we reject the hypothesis H_0 $(m = m_0)$ and accept the alternative H_1 $(m = m_1)$, and if $\bar{x}$ lies to the left of the point A, we accept H_0. Let us find the power function of this test.

From the definition of α, we have

(16.5.1) $\qquad M(w', m_0) = P(\zeta \in w' \mid m = m_0) = \alpha.$

If the true value of the parameter equals m_1, we have

(16.5.1') $\quad M(w', m_1) = P(\zeta \in w' \mid m = m_1) = \dfrac{\sqrt{n}}{\sqrt{2\pi}} \displaystyle\int_A^\infty \exp\left[-\dfrac{n}{2} \cdot \dfrac{(\bar{x} - m_1)^2}{\sigma^2} \right] d\bar{x}.$

If m_1 is considerably larger than m_0 (theoretically, if $m_1 = \infty$), the power of the test equals one. As m_1 decreases, the power decreases, and, for $m_1 = m_0$, it becomes α. As m_1 decreases further, the power continues to decrease; hence it becomes smaller than α, and, for $m_1 = -\infty$, it becomes equal to zero.

The graph of the power function of this test is represented in Fig. 16.5.1, where we take $n = 25$, $m_0 = 0$, $\sigma = 5$, $\alpha = 0.05$.

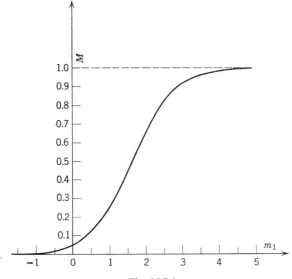

Fig. 16.5.1

Here the power function may be smaller than α; hence the probability of rejecting the hypothesis H_0 may be α when this hypothesis is true, and may be smaller than α when it is false.

If, for a test w of the hypothesis $H_0(Q = Q_0)$, there exists a $Q_1 \neq Q_0$ such that the inequality

(16.5.2) $P(\zeta \in w \mid Q = Q_1) < P(\zeta \in w \mid Q = Q_0)$

is satisfied, then w is called a *biased test*.[1] On the other hand, a test whose power function achieves its minimum value for $Q = Q_0$ is called *unbiased*.

Note that the tests considered in examples 16.2.2 and 16.2.3 are unbiased. Their *OC* functions, as may be seen from Figs. 16.2.2 and 16.2.3, assume their maximum values when H_0 is true.

Although a most powerful unbiased test does not always exist, it can be constructed for a large class of distributions.

Theorem 16.5.1. *Let the point $\zeta = (x_1, \ldots, x_n)$ in sample space be a random variable of the continuous type with density $f(\zeta, Q)$, where the parameter Q is unknown, and its admissible values belong to a certain interval K (finite or infinite). Let $H_0(Q = Q_0)$ and $H_1(Q = Q_1)$ be the null and alternative hypothesis, respectively, and let the point Q_0 belong to the interior of the interval K. If the n-dimensional density $f(\zeta, Q)$ has at each interior point Q of the interval K the partial derivative*

$$f_1(\zeta, Q) = \frac{\partial f(\zeta, Q)}{\partial Q}$$

which satisfies the inequality

(16.5.3) $|f_1(\zeta, Q)| < g(\zeta),$

where the n-dimensional integral

$$\int_{-\infty}^{+\infty} \cdots \int_{-\infty}^{+\infty} g(x_1, x_2, \ldots, x_n) \, dx_1 \ldots dx_n$$

is finite, then among the unbiased tests there exists a most powerful test.

Proof. From the assumptions, it follows that for an arbitrary set w in sample space the power $M(w, Q)$ of the test w may be differentiated as follows:

(16.5.4)

$$\frac{\partial M(w, Q)}{\partial Q} = \frac{\partial}{\partial Q} \int_w f(\zeta, Q) \, d\zeta = \int_w \frac{\partial f(\zeta, Q)}{\partial Q} \, d\zeta = \int_w f_1(\zeta, Q) \, d\zeta.$$

---

[1] Lehmann [1] has given a general definition of unbiasedness which includes as particular cases the notions of an unbiased estimate and of an unbiased test.

Since we assume that the test w is unbiased, its power function $M(w, Q)$ takes on its minimal value for $Q = Q_0$; hence the partial derivative of the power function with respect to Q takes on the value zero at the point $Q = Q_0$, or

$$(16.5.5) \qquad \left[\frac{\partial M(w, Q)}{\partial Q}\right]_{Q=Q_0} = 0.$$

Let us denote by w_{cc_1} the set of those points ζ which satisfy the inequality

$$(16.5.6) \qquad f(\zeta, Q_1) \geqslant cf(\zeta, Q_0) + c_1 f_1(\zeta, Q_0),$$

where $Q_1 \neq Q_0$ and $c \geqslant 0$. Furthermore, let

$$\psi(c, c_1) = P(\zeta \in w_{cc_1} \mid Q = Q_0) = M(w_{cc_1}, Q_0).$$

Assume[1] that there exist constants $c' \geqslant 0$ and c_1' such that for $w_{c'c_1'} = w'$ equality (16.5.5) and equality

$$(16.5.7) \qquad M(w', Q_0) = \alpha$$

are satisfied, where α is the significance level. Let w be an arbitrary unbiased test at the significance level α, that is, let

$$(16.5.8) \qquad P(\zeta \in w \mid Q = Q_0) = \alpha.$$

Denoting by ww' the set of elements common to w and w', we obtain from formulas (16.5.7) and (16.5.8)

$$(16.5.9) \qquad P(\zeta \in w' \mid Q = Q_0) - P(\zeta \in ww' \mid Q = Q_0)$$
$$= P(\zeta \in w \mid Q = Q_0) - P(\zeta \in w'w \mid Q = Q_0).$$

From (16.5.5), we obtain

$$(16.5.10) \qquad \left[\frac{\partial M(w', Q)}{\partial Q}\right]_{Q=Q_0} - \left[\frac{\partial M(w'w, Q)}{\partial Q}\right]_{Q=Q_0}$$
$$= -\left[\frac{\partial M(w'w, Q)}{\partial Q}\right]_{Q=Q_0}$$
$$= \left[\frac{\partial M(w, Q)}{\partial Q}\right]_{Q=Q_0} - \left[\frac{\partial M(w'w, Q)}{\partial Q}\right]_{Q=Q_0}.$$

From the definition of w', it follows that if ζ does not belong to w', then

$$(16.5.11) \qquad c'f(\zeta, Q_0) + c_1'f_1(\zeta, Q_0) > f(\zeta, Q_1).$$

---

[1] If such constants c' and c_1' do not exist, we should modify slightly the construction of w' in a manner analogous to that used in Section 16.3 for a most powerful test when c' does not exist.

From (16.5.9) to (16.5.11), we obtain

$$P(\zeta \in w' \mid Q = Q_1) - P(\zeta \in w'w \mid Q = Q_1)$$
$$\geqslant c'[P(\zeta \in w' \mid Q = Q_0) - P(\zeta \in ww' \mid Q = Q_0)]$$
$$+ c_1'\left\{\left[\frac{\partial M(w', Q)}{\partial Q}\right]_{Q=Q_0} - \left[\frac{\partial M(w'w, Q)}{\partial Q}\right]_{Q=Q_0}\right\}$$
$$= c'[P(\zeta \in w \mid Q = Q_0) - P(\zeta \in ww' \mid Q = Q_0)]$$
$$+ c_1'\left\{\left[\frac{\partial M(w, Q)}{\partial Q}\right]_{Q=Q_0} - \left[\frac{\partial M(w'w, Q)}{\partial Q}\right]_{Q=Q_0}\right\}$$
$$\geqslant P(\zeta \in w \mid Q = Q_1) - P(\zeta \in ww' \mid Q = Q_1).$$

Hence

(16.5.12) $$M(w', Q_1) = P(\zeta \in w' \mid Q = Q_1) \geqslant P(\zeta \in w \mid Q = Q_1)$$
$$= M(w, Q_1),$$

so that the test w' given by relation (16.5.6) and satisfying (16.5.5) and (16.5.7) is a most powerful unbiased test, at the significance level α, of the hypothesis H_0 ($Q = Q_0$) against H_1 ($Q = Q_1$), where Q_1 is a given value of Q, different from Q_0.

It may happen that we obtain the same unbiased test w' for all $Q \neq Q_0$. Then w' is a *uniformly most powerful unbiased test*.

Example 16.5.2. We find a most powerful unbiased test for testing from simple samples the hypothesis H_0 ($m = m_0$) against the alternative H_1 ($m = m_1$), where m is the unknown expected value of a normal $N(m; 1)$ population, and m_1 is an arbitrary number different from m_0. We have

$$f(\zeta, m_1) = \frac{1}{(\sqrt{2\pi})^n} \exp\left[-\frac{1}{2}\sum_{i=1}^{n}(x_i - m_1)^2\right],$$

$$f(\zeta, m_0) = \frac{1}{(\sqrt{2\pi})^n} \exp\left[-\frac{1}{2}\sum_{i=1}^{n}(x_i - m_0)^2\right],$$

$$f_1(\zeta, m_0) = \left[\frac{\partial f(\zeta, m)}{\partial m}\right]_{m=m_0} = \frac{1}{(\sqrt{2\pi})^n} \exp\left[-\frac{1}{2}\sum_{i=1}^{n}(x_i - m_0)^2\right]\sum_{i=1}^{n}(x_i - m_0)$$
$$= f(\zeta, m_0)n(\bar{x} - m_0).$$

Inequality (16.5.6) takes the form

(16.5.13)
$$\exp\left\{-\frac{1}{2}\sum_{i=1}^{n}[(x_i - m_1)^2 - (x_i - m_0)^2]\right\}$$
$$= \exp\{n\bar{x}(m_1 - m_0)\}\exp\{-\tfrac{1}{2}n(m_1^2 - m_0^2)\}$$
$$= \exp\{n(\bar{x} - m_0)(m_1 - m_0)\}\exp\{-\tfrac{1}{2}n(m_1 - m_0)^2\}$$
$$\geqslant c' + c_1'n(\bar{x} - m_0).$$

On the left-hand side of inequality (16.5.13), we have an exponential function of the difference $\bar{x} - m_0$, and on the right-hand side, a linear function of this difference. Thus the set w' consists of those points ζ for which the value of the exponential function exceeds or equals the value of the linear function. Since, for suitably chosen c' and c_1', the exponential curve intersects the straight line at two points (let us denote their abscissas by b_1 and b_2, where $b_1 < b_2$), the set w' consists of those points ζ for which either $\bar{x} - m_0 \leqslant b_1$ or $\bar{x} - m_0 \geqslant b_2$. Since, however, by relation (16.5.5)

$$\int_{w'} f_1(\zeta, m_0)\, d\zeta = \int_{w'} f(\zeta, m_0) n(\bar{x} - m_0)\, d\zeta = 0,$$

the set w' must consist of two parts symmetric with respect to m_0, hence $b_1 = -b_2$. Using (16.5.7), we obtain that w' is the set satisfying the condition

(16.5.14)
$$|\bar{x} - m_0| \geqslant b_2 = \frac{B}{\sqrt{n}},$$

where

$$\frac{1}{\sqrt{2\pi}} \int_B^{\infty} e^{-x^2/2}\, dx = \frac{\alpha}{2}.$$

Of course, we can choose the parameters $c' > 0$ and c_1' of the straight line $c' + c_1' n(\bar{x} - m_0)$ in such a way that this line will intersect the exponential curve considered here at the points

$$\bar{x} = m_0 \pm \frac{B}{\sqrt{n}}.$$

Thus the set w' defined by relation (16.5.14) is the required most powerful unbiased test. As we see from (16.5.14), the set w' is independent of m_1; thus this test is at the same time uniformly most powerful with respect to all $m \neq m_0$. This is a commonly used significance test which we have already applied several times.

Figure 16.5.2 shows three graphs. Graph I gives the power function of the uniformly most powerful unbiased test defined by (16.5.14). Graph II gives the power function of the uniformly most powerful test defined by (16.4.5), that is, when we restrict ourselves to parameter values $m_1 > m_0$. Graph III corresponds to graph II, where we restrict ourselves to the values $m_1 < m_0$. These graphs are given for $n = 16$, $m_0 = 0$, $\alpha = 0.05$. As we see, graph I for $m_1 > m_0$ is below graph II, and for $m_1 < m_0$ below graph III. This is to be expected since graph I represents a most powerful test among all unbiased tests, but not a most powerful among all tests.

The answer to the question of which of these tests should be applied in practice has to be found in the practical aspects of the problem. If we test the hypothesis H_0 $(m = m_0)$, and the only practically admissible alternatives are H_1 $(m = m_1)$ for $m_1 > m_0$, we should apply the uniformly most powerful test (graph II), in spite of the fact that it is biased. This is

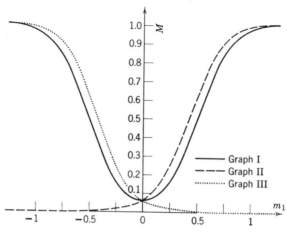

Fig. 16.5.2

because the bias concerns only the values $m_1 < m_0$, which are not of interest to us. If all the values $m_1 \neq m_0$ are practically admissible, there is a risk of accepting the hypothesis H_0 ($m = m_0$) when the true hypothesis is H_1 ($m = m_1$), where m_1 may be either larger or smaller than m_0. Then we should apply the uniformly most powerful unbiased test. The power of this test is not greatest, but, when applying this test, we minimize the risk of making an error of the second kind for all admissible alternative values of the unknown parameter.

B   The necessity of the assumptions of theorem 16.5.1 was investigated by Dantzig and Wald [1]. The problem of unbiased tests when several parameters are unknown has been the subject of investigations by Neyman and Pearson [4] and Isaacson [1].

16.6   THE POWER AND CONSISTENCY OF NONPARAMETRIC TESTS

A   Studies on the power of nonparametric tests are relatively not very far advanced at present. In this domain, as opposed to the theory of parametric tests, there are only a few general results which are completely worked out and ready to be applied. There are usually great mathematical difficulties connected with finding the power of nonparametric tests. It is also difficult, from the practical point of view, to select the alternative hypothesis from among all possible alternative hypotheses different from the hypothesis tested.

We introduce one more criterion of goodness of tests, and this new criterion can just as well be applied to parametric tests.

Let us denote by $\mathscr{G}$ the set of all admissible distribution functions of the considered random variable, and by g a subset of the set $\mathscr{G}$. Let $F(x)$ denote the unknown distribution function.

Definition 16.6.1. Let H_0 be the null hypothesis and let the alternative hypothesis be $H_1 [F(x) \in g]$. Let w_n (n denotes the sample size) be a test of H_0 against H_1. We say that *the sequence* $\{w_n\}$ *of tests is consistent with respect to the set* g if the power of w_n against H_1 converges to one as $n \to \infty$.

The term w_n of a consistent sequence of tests will be called briefly a *consistent test*.

Thus, according to this definition, a consistent test will, for large n, almost surely lead to the rejection of the null hypothesis if the alternative hypothesis is true.

The notion of consistency of a test was introduced by Wald and Wolfowitz [2].

We now investigate the power function and consistency of some non-parametric tests.

B   Let the distribution function $F(x)$ of the random variable x be continuous and strictly increasing everywhere and let the form of the distribution function $F(x)$ be unknown. We want to test the hypothesis $H_0 [F(x) \equiv F_0(x)]$ against the alternative hypothesis $H_1 [F(x) \equiv G(x)]$, where $F_0(x)$ and $G(x)$ are given functions. We use the test based upon the distribution of the statistic

$$D_n^- = \sup_{-\infty < x < +\infty} [F_0(x) - S_n(x)],$$

where $S_n(x)$ was defined in Section 10.3. Thus, we compute, according to formula (10.3.2), the empirical distribution function of a simple sample of size n drawn from the population considered. Next, we reject H_0 at the significance level α and we accept the alternative H_1 if the observed value d_n^- of D_n^- satisfies the inequality $\sqrt{n}\, d_n^+ \geqslant \lambda_{n,\alpha}$, where $\lambda_{n,\alpha}$ is chosen in such a way that, provided hypothesis H_0 is true, we have

$$P(\sqrt{n}\, D_n^- \geqslant \lambda_{n,\alpha} \mid F_0(x)) = \alpha.$$

In other words, we reject H_0 and accept H_1 if for at least one value of x the inequality $F_0(x) - S_n(x) \geqslant \lambda_{n,\alpha}/\sqrt{n}$ holds. The values of $\lambda_{n,\alpha}$ for different n and α are given in Table 16.6.1, taken from the paper of Birnbaum and Tingey [1]. For $n > 50$, we can use the limit distribution of the statistic D_n^- given by (10.11.24) in finding $\lambda_{n,\alpha}$.

We determine the power of this test. To do so, we have to find the probability

(16.6.1) $M = P[\sqrt{n}\, D_n^- \geqslant \lambda_{n,\alpha} \mid G(x)].$

Note that the event $(\sqrt{n}D_n^- < \lambda_{n,\alpha})$ occurs if and only if for $k = 1, \ldots, n$ the inequalities [see formula (10.3.2)]

(16.6.2) $F_0(y_k) - \dfrac{k-1}{n} < \dfrac{\lambda_{n,\alpha}}{\sqrt{n}}$

hold, where $y_1 < y_2 < \ldots < y_n$ are the ordered values of the observations $x_1, x_2, \ldots, x_n$. In fact, it is obvious that the relation $\sqrt{n}\, d_n^- < \lambda_{n,\alpha}$ implies inequality (16.6.2) for $k = 1, \ldots, n$. On the other hand, since $F_0(x)$ is strictly increasing and $S_n(x)$ is a nondecreasing step function, we see that if (16.6.2) is satisfied for $k = 1, \ldots, n$, then $\sqrt{n}\, d_n^- < \lambda_{n,\alpha}$. Hence, by (16.6.1), we obtain

(16.6.3) $1 - M = P\left(F_0(y_k) - \dfrac{k-1}{n} < \dfrac{\lambda_{n,\alpha}}{\sqrt{n}}\, ; \quad k = 1, \ldots, n \mid G(x)\right)$

$= P\left(y_k < F_0^{(-1)}\left(\dfrac{k-1}{n} + \dfrac{\lambda_{n,\alpha}}{\sqrt{n}}\right); \quad k = 1, \ldots, n \mid G(x)\right)$

$= P\left(G(y_k) < G\left[F_0^{(-1)}\left(\dfrac{k-1}{n} + \dfrac{\lambda_{n,\alpha}}{\sqrt{n}}\right)\right]; \quad k = 1, \ldots, n \mid G(x)\right),$

where $F_0^{(-1)}$ is the function inverse to $F_0(x)$. We recall (see Section 5.6, B) that if the distribution function of the random variable y is a continuous function, say $G(y)$, then the random variable $z = G(y)$ has the uniform distribution in the interval $[0, 1]$. Hence we can write $z_k = G(y_k) = \zeta_k^{(n)}$ $(k = 1, \ldots, n)$, where the $\zeta_k^{(n)}$ are order statistics (see Section 10.2) of n independent observations drawn from a population with the uniform distribution in the interval $[0, 1]$. Using (10.7.1) and putting $f(z_k) = 1$ in it, we obtain

(16.6.4) $1 - M =$

$n! \displaystyle\int_0^{R\left(\frac{\lambda_{n,\alpha}}{\sqrt{n}}\right)} \int_{z_1}^{R\left(\frac{1}{n} + \frac{\lambda_{n,\alpha}}{\sqrt{n}}\right)} \cdots \int_{z_{n-1}}^{R\left(\frac{n-1}{n} + \frac{\lambda_{n,\alpha}}{\sqrt{n}}\right)} dz_n \cdots dz_2\, dz_1,$

where the function $R(v)$ is given by the formula

(16.6.5) $R(v) = \begin{cases} \lim\limits_{0 < v \to 0} G[F_0^{(-1)}(v)] & \text{for } v \leqslant 0, \\[2mm] G[F_0^{(-1)}(v)] & \text{for } 0 < v < 1, \\[2mm] \lim\limits_{1 > v \to 1} G[F_0^{(-1)}(v)] & \text{for } v \geqslant 1. \end{cases}$

Expression (16.6.4) allows us to compute the power of the test considered for given distribution functions $F_0(x)$ and $G(x)$. Very often, however, we cannot compute the exact value of the integral appearing in this expression and must use numerical approximation methods.

Formula (16.6.4) was given by Z. W. Birnbaum [3]. If $G(x) \equiv F_0(x)$, then $G[F_0^{-1}(v)] = v$ and formula (16.6.4) becomes considerably simpler. We observe that, as was established by Birnbaum, formula (16.6.4) also holds without the assumption that $F_0(x)$ is strictly increasing.

TABLE 16.6.1

n	$\lambda_{n,\alpha}/\sqrt{n}$	
	$\alpha = 0.05$	$\alpha = 0.01$
5	0.5094	0.6271
8	0.4096	0.5065
10	0.3687	0.4566
20	0.2647	0.3285
40	0.1891	0.2350
50	0.1696	0.2107

Van der Waerden [1] showed that if the expected value m of the normal $N(m; 1)$ population is unknown and, for testing the hypothesis H_0 ($m = 0$) against the alternative H_1 ($m = m_1 \neq m_0$), we apply the test based on the distribution of the statistic D_n^+, then the power of this test is much smaller than the power of the classical test described in example 16.5.2. This phenomenon is quite consistent with our intuition. In fact, if the functional form of the distribution function is known, and only a parameter is unknown, by applying the nonparametric test, we do not, roughly speaking, make use of all our information, since we proceed in the same way as we would if we were testing a hypothesis concerning the functional form of the distribution function. The test considered (and nonparametric tests in general) should be applied only when we do not know the functional form of the distribution function.

C   Let $F(x)$ satisfy all the assumptions of Section 16.6, B. Let the alternative of the null hypothesis H_0 [$F(x) \equiv F_0(x)$] be H_2 [$F(x) \equiv G(x)$], where $F_0(x)$ is a given function and the only assumption about $G(x)$ is that it satisfies the relation

$$(16.6.6) \qquad \sup_{-\infty < x < +\infty} [F_0(x) - G(x)] = \delta > 0,$$

and that we know $G(x_0)$, where

$$(16.6.7) \qquad F_0(x_0) - G(x_0) = \delta.$$

Thus the hypothesis H_2 does not specify the distribution function $G(x)$, but merely states that $G(x)$ belongs to the set of continuous distribution functions satisfying (16.6.6) and (16.6.7). To test H_0 against H_2, we apply the test discussed in Section 16.6, B. It is obvious that here we cannot compute the value of the integral in (16.6.4). Z. W. Birnbaum [3] found the greatest lower bound for the power of this test, that is, he established that the power M of this test satisfies the inequality

(16.6.8) $M \geqslant 1 - I_{u_0} (j + 1, n - j)$,

where $I_{u_0} (j + 1, n - j)$ is the incomplete beta function,

(16.6.9) $u_0 = G(x_0), \qquad v_0 = F_0(x_0), \qquad j = \left[n \left(v_0 - \dfrac{\lambda_{n,\alpha}}{\sqrt{n}} \right) \right].$

Here the symbol $[A]$ denotes the greatest integer not exceeding A.

To prove this inequality, let us consider the function

(16.6.10) $G^*(x) = \begin{cases} F_0(x_0) - \delta & (x \leqslant x_0), \\ 1 & (x > x_0). \end{cases}$

By (16.6.5), (16.6.7), (16.6.9), and (16.6.10), we obtain

(16.6.11) $R^*(v) = \begin{cases} u_0 & (0 < v \leqslant v_0), \\ 1 & (v_0 < v < 1). \end{cases}$

Next, we have for an arbitrary distribution function $G(x)$ under consideration

(16.6.12) $R(v_0) = u_0.$

By formulas (16.6.9), for j and for $k - 1 \leqslant j$, we obtain

$$\frac{k - 1}{n} + \frac{\lambda_{n,\alpha}}{\sqrt{n}} \leqslant \frac{j}{n} + \frac{\lambda_{n,\alpha}}{\sqrt{n}} \leqslant v_0.$$

Since $R(v)$ is a nondecreasing function, from the last inequality and from (16.6.12), it follows that for $k - 1 \leqslant j$

$$R \left(\frac{k - 1}{n} + \frac{\lambda_{n,\alpha}}{\sqrt{n}} \right) \leqslant R(v_0) = u_0.$$

It is clear that if in integral (16.6.4) we replace the upper limits of integration $R(v)$ by $R^*(v)$ given by (16.6.11), then the value of this integral will not decrease. Hence

(16.6.13)

$$1 - M \leqslant n! \int_0^{u_0} \cdots \int_{z_j}^{u_0} \int_{z_{j+1}}^{1} \cdots \int_{z_{n-1}}^{1} dz_n \ldots dz_j \, dz_{j-1} \ldots dz_1.$$

Computing the integral on the right-hand side of (16.6.13), we obtain

$$M \geqslant \sum_{k=1}^{j+1} \binom{n}{k-1} u_0^{k-1} (1 - u_0)^{n-k+1}$$

$$= \sum_{i=1}^{j} \binom{n}{i} u_0^{i} (1 - u_0)^{n-i} = 1 - I_{u_0}(j + 1, n - j).$$

Inequality (16.6.8) is proved.

Birnbaum [3] also gave the least upper bound for the power of this test. Chapman [3] gave the minimum and maximum power for a variety of related tests. In the same paper, Chapman proposed some new criteria of goodness of nonparametric tests.

D   From formula (16.6.8), it is easy to obtain an asymptotic expression for large n for the greatest lower bound of the power of the test under consideration, and it turns out that this greatest lower bound can be expressed in terms of the normal distribution. However, we arrive at the same conclusion using another method, given by Massey [1]. Following this method, we find the greatest lower bound for the power of the test based upon the statistic $D_n = \sup_{-\infty < x < +\infty} |F_0(x) - S_n(x)|$. The same method could be applied to the test based upon the statistic D_n^{+}.

Again, let the continuous distribution function $F(x)$ be unknown; let the null hypothesis be $H_0 [F(x) \equiv F_0(x)]$ and the alternative $H_3 [F(x) \equiv G(x)]$, where $F_0(x)$ is a given function; let $G(x)$ satisfy the relation

(16.6.14) $$\sup_{-\infty < x < +\infty} |F_0(x) - G(x)| = \delta > 0,$$

and let $G(x_0)$, where

(16.6.15) $$|F_0(x_0) - G(x_0)| = \delta,$$

be given.

We apply Kolmogorov's λ test. Thus we reject H_0 and accept the alternative if the observed value $d_n \geqslant \lambda_\alpha / \sqrt{n}$, where λ_α satisfies the relation

(16.6.16) $$P(\sqrt{n} D_n \geqslant \lambda_\alpha) = \alpha.$$

The number λ_α can be found from Table VIII. The power M of this test is given by the expression

$$P(\sqrt{n} D_n \geqslant \lambda_\alpha \,|\, G(x)).$$

We have

$$M \geqslant P(\sqrt{n} |S_n(x_0) - F_0(x_0)| \geqslant \lambda_\alpha \,|\, G(x))$$

$$= 1 - P\left(F_0(x_0) - \frac{\lambda_\alpha}{\sqrt{n}} < S_n(x_0) < F_0(x_0) + \frac{\lambda_\alpha}{\sqrt{n}} \,\bigg|\, G(x) \right).$$

Let us put

$$F_0(x_0) = G(x_0) \pm \delta.$$

Then, after some simple transformations, we have

$$(16.6.17) \quad M \geqslant 1 - P\left(\frac{-\lambda_\alpha \pm \delta\sqrt{n}}{\sqrt{G(x_0)[1 - G(x_0)]}}\right.$$

$$< \frac{[S_n(x_0) - G(x_0)]\sqrt{n}}{\sqrt{G(x_0)[1 - G(x_0)]}} < \left.\frac{\lambda_\alpha \pm \delta\sqrt{n}}{\sqrt{G(x_0)[1 - G(x_0)]}}\right).$$

We observe that if H_3 is true, $S_n(x_0)$ has, according to the de Moivre-Laplace theorem, the asymptotically normal distribution $N(G(x_0);$ $\sqrt{G(x_0)(1 - G(x_0))}/\sqrt{n})$. Thus, we finally obtain from (16.6.17)

$$M \geqslant 1 - \frac{1}{\sqrt{2\pi}} \int_a^b \exp\left(-\frac{t^2}{2}\right) dt,$$

where

$$a = \frac{-\lambda_\alpha \pm \delta\sqrt{n}}{\sqrt{G(x_0)[1 - G(x_0)]}},$$

$$b = \frac{\lambda_\alpha \pm \delta\sqrt{n}}{\sqrt{G(x_0)[1 - G(x_0)]}}.$$

E   We shall show that Kolmogorov's λ test is consistent with respect to the set of continuous distribution functions satisfying (16.6.14). Indeed, let the function $G(x)$ satisfy this relation and be the true distribution function. Then we have

$$\delta = \sup_{-\infty < x < +\infty} |G(x) - F_0(x)| = \sup_{-\infty < x < +\infty} |G(x) - S_n(x) + S_n(x) - F_0(x)|$$

$$\leqslant \sup_{-\infty < x < +\infty} |G(x) - S_n(x)| + \sup_{-\infty < x < +\infty} |S_n(x) - F_0(x)|$$

or

$$\sup_{-\infty < x < +\infty} |S_n(x) - F_0(x)| \geqslant \delta - \sup_{-\infty < x < +\infty} |G(x) - S_n(x)|$$

By Glivenko's theorem (see Section 10.10), we obtain, remembering that $G(x)$ is the true distribution function,

$$(16.6.18) \quad P\left(\lim_{n \to \infty} \sup_{-\infty < x < +\infty} |S_n(x) - F_0(x)| \geqslant \delta\right) = 1.$$

Let us now denote by $\lambda_{n,\alpha}$ the number satisfying the equality

$$P(\sqrt{n}\, D_n \geqslant \lambda_{n,\alpha}) = \alpha.$$

Thus

$$\lim_{n \to \infty} M = \lim_{n \to \infty} P\left[\sup_{-\infty < x < +\infty} \left| S_n(x) - F_0(x) \right| \geqslant \frac{\lambda_{n,\alpha}}{\sqrt{n}} \left| G(x) \right| \right].$$

From Kolmogorov's theorem (see Section 10.10), it follows that

$$\lim_{n \to \infty} \lambda_{n,\alpha} = \lambda_\alpha.$$

Hence for sufficiently large n, we have $\lambda_{n,\alpha}/\sqrt{n} < \delta$, and, by relation (16.6.18), we obtain

$$\lim_{n \to \infty} M = 1,$$

so that Kolmogorov's λ test is consistent with respect to the set of alternative distribution functions considered here.

By a method similar to that used for Kolmogorov's test, we could obtain, for large n, an asymptotic expression for the greatest lower bound of the power of the test based upon Smirnov's theorem (see Section 10.11). Here, we test the hypothesis $H_0\,[F_1(x) \equiv F_2(x)]$ against the alternative $H_4\left[\sup_{-\infty < x < +\infty} |F_1(x) - F_2(x)| = \delta > 0 \right]$, where $F_1(x)$ and $F_2(x)$ are unknown continuous distribution functions. Applying the same reasoning, we can also show that Smirnov's λ test is consistent with respect to the set of continuous distribution functions satisfying the relation

$$\sup_{-\infty < x < +\infty} |F_1(x) - F_2(x)| = \delta > 0.$$

F   We now discuss the power of the Wilcoxon-Mann-Whitney test. We use the notation of Section 12.6, B. Thus the random variables X_1 and X_2 are independent, and X_i $(i = 1, 2)$ has a continuous distribution function $F_i(x)$. The null hypothesis is $H_0\,[F_1(x) \equiv F_2(x)]$.

First, we have to formulate the hypothesis alternative to H_0. The considerations of Section 12.6, B and the structure of the test suggest that we take for the alternative the hypothesis H_1 that for every real x for which $0 < F_1(x) < 1$ we have the inequality

(16.6.19) $F_2(x) < F_1(x).$

However, we consider the more general alternative hypothesis H_2, according to which

(16.6.20) $0 < p = P(X_2 - X_1 < 0) < \frac{1}{2}.$

The hypothesis H_2 is, in fact, more general than H_1 since if inequality

(16.6.19) holds for every x, relation (16.6.20) is also satisfied. Indeed, we have then (see Section 12.6, B)

$$p = \int_{-\infty}^{+\infty} F_2(x)\, dF_1(x) < \int_{-\infty}^{+\infty} F_1(x)\, dF_1(x) = \tfrac{1}{2}.$$

We have n_1 and n_2 independent observations of the random variables X_1 and X_2, namely, x_{1l} $(l = 1, \ldots, n_1)$ and x_{2j} $(j = 1, \ldots, n_2)$. From these observations, we compute the observed value u of Mann-Whitney's statistic U, and we reject H_0 and accept its alternative H_2 if $u \leqslant u_\alpha$, where α is the significance level and u_α satisfies the relation

(16.6.21) $$\psi = P(U \leqslant u_\alpha \mid H_0) \leqslant \alpha.$$

To compute the power of this test, we restrict ourselves to large samples and we use the fact, discovered by Lehmann [1] (see Section 12.6, B), that the statistic U has an asymptotically normal distribution; that is, the statistic

(16.6.22) $$V = \frac{U - E(U)}{\sqrt{D^2(U)}}$$

has for $n_2 = sn_1$ $(s > 0)$ and $n_1 \to \infty$ the normal distribution $N(0; 1)$, where $E(U)$ and $D^2(U)$ are given by (12.6.17) and (12.6.18), respectively. Thus we reject H_0 if the observed value $(u - m_0)/\sigma_0 \leqslant v_\alpha$, where $\Phi(v_\alpha) = \alpha$, and

$$m_0 = E(U \mid H_0) = \tfrac{1}{2}n_1 n_2,$$
$$\sigma_0{}^2 = D^2(U \mid H_0) = \tfrac{1}{12}n_1 n_2(n_1 + n_2 + 1).$$

The numbers u_α and v_α are related in the following way. Let $u_\alpha = m_0 - c_\alpha \sigma_0$. Then

$$\lim_{n_1 \to \infty} \psi = \alpha, \qquad \lim_{n_1 \to \infty} c_\alpha = -v_\alpha.$$

Thus the power of the test considered is obtained from the formula

(16.6.23) $$M = P\left(\frac{U - m_0}{\sigma_0} \leqslant v_\alpha \mid H_2\right) = \Phi\left(\frac{v_\alpha \sigma_0 + m_0 - E(U)}{\sqrt{D^2(U)}}\right).$$

Example 16.6.1. Let $F_1(x)$ and $F_2(x)$ be two continuous distribution functions. Let us test, by the method just described, the hypothesis H_0 $[F_1(x) \equiv F_2(x)]$ against the alternative hypothesis H_1 $[F_2(x) \equiv F_1{}^k(x)]$, where k is some integer greater than one. From (12.6.8), (12.6.17), and (12.6.18), we find

$$p = \frac{1}{k+1}, \qquad E(U) = \frac{n_1 n_2}{k+1},$$
$$D^2(U) = \frac{n_1 n_2}{(k+1)^2}\left(\frac{n_1 - 1}{k+2}k + \frac{n_2 - 1}{2k+1}k^2 + 1\right).$$

To obtain numerical results, let us put $k = 2, n_1 = n_2 = 25, \alpha = 0.05$. Then we have $m_0 = 312.5$, $\sigma_0^2 = 2656.25$. Thus we reject H_0 and accept the alternative H_1 if the observed value u satisfies the inequality $(u - 312.5)/\sqrt{2656.25} \leqslant -1.64$, hence if $u \leqslant 230$. If the alternative hypothesis is true, then $E(U) = 208\frac{1}{3}$, $D^2(U) = 2180\frac{5}{9}$. From (16.6.23), we find $M = \Phi(0.485) = 0.6866$.

If in this example we had taken $n_1 = n_2 = 60$, then the value of M would have been about 0.95 (see Lehmann [3]).

The papers of van der Vaart [1], van der Waerden [2], van Dantzig [1], and Lehmann [3] deal with the power of the Wilcoxon-Mann-Whitney test and with that of other nonparametric tests. The results of these investigations seem to indicate that the power of the Wilcoxon-Mann-Whitney test is greater than that of other nonparametric tests considered by us.

G We now show that if we test the hypothesis $H_0 [F_1(x) \equiv F_2(x)]$ against the alternative given by (16.6.20), the Wilcoxon-Mann-Whitney test is consistent. In the proof, we use the following two inequalities which can be derived immediately from (12.6.13) and (12.6.14), respectively:

$$0 \leqslant \gamma^2 = \int_{-\infty}^{+\infty} F_2^2(x)\, dF_1(x) - p^2 \leqslant \int_{-\infty}^{+\infty} F_2(x)\, dF_1(x) - p^2$$

$$= p - p^2 = p(1 - p),$$

$$0 \leqslant \phi^2 = \int_{-\infty}^{+\infty} [1 - F_1(x)]^2\, dF_2(x) - p^2 \leqslant \int_{-\infty}^{+\infty} [1 - F_1(x)]\, dF_2(x) - p^2$$

$$= 1 - (1 - p) - p^2 = p(1 - p).$$

It follows from the last inequalities and from (12.6.18) that

$$(16.6.24) \quad \frac{D^2(U)}{\sigma_0^2} = 12\frac{(n_1 - 1)\phi^2 + (n_2 - 1)\gamma^2 + p(1 - p)}{n_1 + n_2 + 1}$$

$$\leqslant \frac{12p(1 - p)(n_1 + n_2 - 1)}{n_1 + n_2 + 1} < 12p(1 - p).$$

Let us now find the power of the test. We have

$$(16.6.25) \quad 1 - M = P(U > u_\alpha \mid H_2)$$

$$= P(U > m_0 - c_\alpha \sigma_0 \mid H_2)$$

$$= P[U - n_1 n_2 p > (\tfrac{1}{2} - p)n_1 n_2 - c_\alpha \sigma_0 \mid H_2].$$

Note that, according to (16.6.20), $\tfrac{1}{2} - p$ is positive. Since c_α is bounded and $n_2 = s n_1 \ (s > 0)$, we have, for sufficiently large n_1,

$$(\tfrac{1}{2} - p)n_1 n_2 - c_\alpha \sigma_0 = (\tfrac{1}{2} - p)s n_1^2 - \frac{c_\alpha}{12} n_1 \sqrt{s(2n_1 + 1)} > 0.$$

Applying the Chebyshev inequality (see Section 3.3) to the right-hand side of (16.6.25), we obtain

$$1 - M \leqslant \frac{D^2(U)}{[(\frac{1}{2} - p)n_1 n_2 - c_\alpha \sigma_0]^2} = \frac{D^2(U)}{\sigma_0^2} \left[(\tfrac{1}{2} - p) \sqrt{\frac{12 n_1 n_2}{n_1 + n_2 + 1}} - c_\alpha \right]^{-2}.$$

By (16.6.24), the expression $D^2(U)/\sigma_0^2$ is bounded. Again in view of (16.6.20), the relation $n_2 = sn_1$, and of the fact that c_α is bounded, we see that the expression inside the square brackets on the right-hand side of the last formula diverges to infinity as $n_1 \to \infty$. Hence

$$\lim_{n_1 \to \infty} (1 - M) = 0.$$

Finally, for $p < \frac{1}{2}$

$$\lim_{n_1 \to \infty} P(U \leqslant u_\alpha \mid H_2) = 1.$$

Thus the consistency of the Wilcoxon-Mann-Whitney test is proved.

Note that if we test the hypothesis H_0 [$F_1(x) \equiv F_2(x)$] against the alternative H' [$F_1(x) \not\equiv F_2(x), p \geqslant \frac{1}{2}$], the Wilcoxon-Mann-Whitney test is not consistent. Indeed, from (16.6.25), we have

$$M = P(U \leqslant u_\alpha \mid H') = P(U - n_1 n_2 p < -[(p - \tfrac{1}{2})n_1 n_2 + c_\alpha \sigma_0] \mid H').$$

Since $p \geqslant \frac{1}{2}$ and c_α is positive for $\alpha < \frac{1}{2}$ (except for very small n_1 and n_2), we have $(p - \frac{1}{2})n_1 n_2 + c_\alpha \sigma_0 > 0$. Then we obtain, by the Chebyshev inequality,

$$M \leqslant \frac{D^2(U)}{[(p - \tfrac{1}{2})n_1 n_2 + c_\alpha \sigma_0]^2}$$

$$= \frac{D^2(U)}{\sigma_0^2} [(p - \tfrac{1}{2}) \sqrt{12 n_1 n_2/(n_1 + n_2 + 1)} + c_\alpha]^{-2}.$$

By the boundedness of $D^2(U)/\sigma_0^2$ and by the fact that the expression inside the square brackets increases to infinity as $n_1 \to \infty$, we obtain

$$\lim_{n_1 \to \infty} M = 0.$$

Thus in the case considered, $p \geqslant \frac{1}{2}$, the test is not consistent, which completely agrees with our intuition.

The results concerning the consistency of the Wilcoxon-Mann-Whitney test presented here were obtained by van Dantzig [1] (see also Lehmann [2] and Rényi [10]). The consistency of this test with respect to the set of distribution functions satisfying (16.6.19) was shown by Mann and Whitney [1].

The following example, due to Gnedenko [11], illustrates the inconsistency of the Wilcoxon-Mann-Whitney test in the case $p = \frac{1}{2}$.

Example 16.6.2. Let the distribution functions $F_1(x)$ and $F_2(x)$ be continuous and satisfy the following conditions:

$$F_1(x) = \begin{cases} 0 & (x \leqslant b), \\ 1 & (x \geqslant c), \end{cases}$$

$$F_2(x) = \begin{cases} 0 & (x < a), \\ \frac{1}{2} & (b \leqslant x \leqslant c), \\ 1 & (x \geqslant d), \end{cases}$$

where $a < b < c < d$. We have

$$p = \int_{-\infty}^{+\infty} F_2(x) \, dF_1(x) = \frac{1}{2}[F_1(c) - F_1(b)] = \frac{1}{2}.$$

If we draw samples from each of these populations, then (using the same notations as above) we have

$$E(U) = \frac{1}{2}n_1 n_2,$$
$$D^2(U) = \frac{1}{4}n_1 n_2.$$

It is obvious that for large n_1 and n_2 half of the observations x_{2j} $(j = 1, \ldots, n_2)$ belong to the interval $[a, b]$; hence for them we have $x_{2j} < x_{1l}$. Thus we shall observe a value of the statistic U close to $\frac{1}{2}n_1 n_2$. Hence if we tested the hypothesis $H_0 [F_1(x) \equiv F_2(x)]$ against the alternative $H_1 [F_1(x) \not\equiv F_2(x), p = \frac{1}{2}]$ using the Wilcoxon-Mann-Whitney test, this test would lead us to accept the hypothesis H_0, when it is false for arbitrarily large n_1 and n_2.

H The Wilcoxon-Mann-Whitney test considered thus far is one-sided, since the alternative hypothesis is given by relation (16.6.19) or (16.6.20). We can also consider a two-sided Wilcoxon-Mann-Whitney test, when the null hypothesis remains $H_0 [F_1(x) \equiv F_2(x)]$, but the alternative is $H_1 [F_1(x) \not\equiv F_2(x)]$ or $H_1' (p \neq \frac{1}{2})$. Then we proceed in the following way: For small n_1 and n_2, we determine two numbers u_1 and u_2 such that, provided the hypothesis H_0 is true, the Mann-Whitney statistic U satisfies the relation

$$P(U \leqslant u_1) \leqslant \frac{\alpha}{2}, \qquad P(U \geqslant u_2) \leqslant \frac{\alpha}{2},$$

where α is the significance level. We reject H_0 if the observed value u of the statistic U satisfies one of the relations $u \leqslant u_1$ or $u \geqslant u_2$. For large n_1 and n_2, we use the normal approximation and we reject the hypothesis H_0 if the observed value v of the statistic V given by (16.6.22) satisfies the relation $|v| \geqslant v_{\alpha/2}$, where $v_{\alpha/2}$ is determined from $\Phi(v_{\alpha/2}) = \alpha/2$.

The considerations concerning the power and consistency of the one-sided Mann-Whitney test can easily be applied to the two-sided Mann-Whitney test. In particular, the two-sided Mann-Whitney test is consistent if we test the null hypothesis $H_0 [F_1(x) \equiv F_2(x)]$ against the alternative $H_2 [F_1(x) \not\equiv F_2(x), p \neq \frac{1}{2}]$.

16.7 ADDITIONAL REMARKS

In the papers of Wolfowitz [4], Hoeffding [1], and Lehmann [3], the reader will find much information about the power and consistency of the Wald-Wolfowitz test and of other nonparametric tests. The paper of Mann and Wald [1] is devoted to the power of Pearson's χ^2 test.

The papers of Barton, David, and Mallows [1] and Barton and David [1] discuss the applicability of the Wilcoxon-Mann-Whitney test and of the Wald-Wolfowitz test to testing the hypothesis of randomness of a sequence of observations containing elements of two kinds (see example 14.2.3) and the power of these tests for various alternatives.

Problems and Complements

16.1. The parameter σ of a population, where the characteristic x has the normal distribution $N(0; \sigma)$, is unknown. The null hypothesis is $H_0 (\sigma = \sigma_0)$ and the alternative $H_1 (\sigma = \sigma_1)$. Show that for this problem a most powerful test w' for testing H_0 against H_1 from simple samples is determined by

$$\bar{x}^2 + s^2 \leqslant b \quad \text{if} \quad \sigma_0 > \sigma_1,$$
$$\bar{x}^2 + s^2 \geqslant b \quad \text{if} \quad \sigma_0 < \sigma_1,$$

where the constant b is determined by the condition $P(\zeta \in w' \mid \sigma_0) = \alpha$. Compare with example 16.3.3.

16.2. A characteristic x has the distribution with density

$$f(x) = \begin{cases} \beta \exp [-\beta(x - \gamma)] & \text{for} \quad x - \gamma \geqslant 0, \\ 0 & \text{for} \quad x - \gamma < 0. \end{cases}$$

The parameters β and γ are unknown. Let us consider the hypotheses $H_0 (\beta = \beta_0, \gamma = \gamma_0)$ and $H_1 (\beta = \beta_1, \gamma = \gamma_1)$, where the domain of admissible hypotheses is determined by the inequalities $\gamma_1 \leqslant \gamma_0, \beta_1 \geqslant \beta_0$. Show (Neyman and Pearson [5]), that a most powerful test for testing the hypotheses considered is the set $w = w_1' + w_2'$, where w_1' consists of the points $x < \gamma_0$ and w_2' consists of the sample points satisfying the inequality

$$\bar{x} \leqslant \frac{1}{\beta_1 - \beta_0} \left(\gamma_1 \beta_1 - \gamma_0 \beta_0 - \frac{1}{n} \log c + \log \frac{\beta_1}{\beta_0} \right) = A,$$

where the constant A is determined from the accepted significance level α.

Note that this test is uniformly most powerful with respect to the set of admissible hypotheses.

Consider the particular case when x has the exponential distribution with $\gamma = 0$.

16.3. The density $f(\zeta, q)$ depends on the unknown vector $q = (Q_1, \ldots, Q_r)$. We denote by q_0 and q_1 the vectors $(Q_{10}, \ldots, Q_{r0})$ and $(Q_{11}, \ldots, Q_{r1})$. We wish to test the simple null hypothesis $H_0 (q = q_0)$ against the simple alternative $H_1 (q = q_1)$. Show that if a uniformly most powerful test with respect to the class of all alternatives exists, then it may be constructed on the basis of the estimate $q^*(\zeta) = [Q_1^*(\zeta), \ldots, Q_r^*(\zeta)]$ of the vector q obtained by the maximum likelihood method.

16.4. (For notation, see Section 16.3.D.) Let $f(\zeta, Q)$, with Q any real number, denote either the density or the probability function of ζ, as the random variable ζ is either of the continuous or of the discrete type. Assume that $f(\zeta, Q)$ can be presented as

$$(*) \qquad f(\zeta, Q) = C(Q) \exp [g(Q)T(\zeta)]h(\zeta),$$

where $g(Q)$ is strictly increasing. Show that for testing the hypothesis H_0 ($Q \leqslant Q_0$) against the alternative H_1 ($Q > Q_0$) there exists a uniformly most powerful randomized test, given by the test function

$$(**) \qquad \phi(\zeta) = \begin{cases} 1 & \text{as} & T(\zeta) > c, \\ \gamma & \text{as} & T(\zeta) = c, \\ 0 & \text{as} & T(\zeta) < c, \end{cases}$$

where γ and c are determined from

$$P[T(\zeta) > c \mid Q = Q_0] + \gamma P[T(\zeta) = c \mid Q = Q_0] = \alpha.$$

If $g(Q)$ is strictly decreasing, whereas all other conditions remain unchanged, the inequalities in $(**)$ should be reversed.

16.5. Let x have the binomial probability function given by (5.2.1) and let p be unknown. Present the right side of (5.2.1) in form $(*)$ of Problem 16.4 and find a uniformly most powerful test for testing H_0 ($p \leqslant p_0$) against H_1 ($p > p_0$) at level α.

16.6. Let x be normally distributed $N(m; \sigma)$. (a) Assume that σ is known while m is unknown. Use Problem 16.4 to find a uniformly most powerful test for testing H_0 ($m \leqslant m_0$) against H_1 ($m > m_0$). Compare the test obtained with the test discussed in Section 16.4.

(b) Assume that m is known while σ is unknown. Find a uniformly most powerful test for testing H_0 ($\sigma \leqslant \sigma_0$) against H_1 ($\sigma > \sigma_0$). Compare with Problem 16.1.

16.7. Let $f(\zeta, Q)$ have the same meaning as in Problem 16.4 and let it be presentable in form $(*)$ of the mentioned problem. For testing the hypothesis H_0 ($Q \leqslant Q_1$ or $Q \geqslant Q_2$) against H_1 ($Q_1 < Q < Q_2$), there exists a uniformly most powerful test given by the test function

$$\phi(\zeta) = \begin{cases} 1 & \text{as} & c_1 < T(\zeta) < c_2, \\ \gamma_i & \text{as} & T(\zeta) = c_i \quad (i = 1, 2), \\ 0 & \text{as} & T(\zeta) < c_1 \text{ or } T(\zeta) > c_2, \end{cases}$$

where $\gamma_1, \gamma_2, c_1, c_2$ are determined from the equations

$$P[c_1 < T(\zeta) < c_2 \mid Q = Q_i] + \gamma_1 P[T(\zeta) = c_1 \mid Q = Q_i]$$
$$+ \gamma_2 P[T(\zeta) = c_2 \mid Q = Q_i] = \alpha \quad (i = 1, 2).$$

16.8. Apply the assertion of Problem 16.7 to find a uniformly most powerful test for testing the hypothesis H_0 ($m \leqslant m_1$ or $m \geqslant m_2$) against the hypothesis H_1 ($m_1 < m < m_2$), where m is the unknown mean value of a normal distribution $N(m, 1)$.

16.9. Show that if $\phi(\zeta)$ is a test function for testing a hypothesis about the parameter Q and U is a sufficient estimate of Q, then the conditional expected value $E[\phi(\zeta) \mid u]$ is also a test function and its power is equal to that of $\phi(\zeta)$.

16.10. Let the density $f(\zeta, q)$ depend upon the unknown vector $q = (Q_1, \ldots, Q_r, Q_{r+1})$. Let the composite null hypothesis $H_0 (Q_1 = Q_{10}, \ldots, Q_r = Q_{r0})$ not specify the value of parameter Q_{r+1}. Let $q_0 = (Q_{10}, \ldots, Q_{r0}, Q_{r+1})$, and let ω and Ω denote, respectively, the set of vectors q_0 and the parameter space. We assume that (1) $f(\zeta, q_0)$ is differentiable an arbitrary number of times for almost all values of Q_{r+1}; (2) the relation

$$\frac{\partial^2 \log f(\zeta, Q_0)}{\partial Q_{r+1}^2} = A + B \frac{\partial \log f(\zeta, Q_0)}{\partial Q_{r+1}},$$

where A and B depend upon Q_{r+1} but not upon ζ, holds. Show that for every significance level α, the set w in sample space is similar to the whole sample space with respect to ω if and only if, for $k = 1, 2, \ldots$

$$\int_w \frac{\partial^k f(\zeta, g_0)}{\partial Q_{r+1}^k} \, d\zeta = 0.$$

We introduce the concept of the likelihood ratio test. Let the density $f(\zeta, q)$ of the point ζ in the sample space depend upon the unknown vector $q = (Q_1, \ldots, Q_r, Q_{r+1}, \ldots, Q_{r+s})$. The density $f(\zeta, q)$ is the likelihood function (see Section 13.7.A). Let Ω denote the set of all vectors q, and let ω denote the set of all vectors $q_0 = (Q_{10}, \ldots, Q_{r0}, Q_{r+1}, \ldots, Q_{r+s})$ with $Q_{10}, \ldots, Q_{r0}$ fixed, and with the remaining $Q - s$ arbitrary. We want to test the hypothesis $H_0 (q \in \omega)$.

Denote

$$(*) \qquad\qquad \lambda = \frac{\sup\limits_{\omega} f(\zeta, q)}{\sup\limits_{\Omega} f(\zeta, q)}.$$

The *likelihood ratio test* rejects H_0 when $\lambda \leqslant A$, where A is determined from the relation

$$(**) \qquad\qquad \sup\limits_{\omega} P(\lambda \leqslant A \mid q \in \omega) = \alpha.$$

Notice that relation $(**)$ is weaker than similarity to the whole sample space with respect to ω.

For discrete random variables, the likelihood ratio test can be constructed in an analogous way. The density function $f(\zeta, q)$ should then be replaced by the corresponding probability function.

16.11. We have r normal populations $N(m_i; \sigma_i)$ $(i = 1, \ldots, r)$ and r simple samples drawn from these populations of sizes n_i, respectively. Suppose that $n_1 + n_2 + \ldots + n_r = N$. Let us denote by $\bar{x}_i$ and s_i^2 the sample mean and sample variance of the ith sample, respectively. The null hypothesis $H_0 (m_1 = \ldots = m_r; \ \sigma_1 = \ldots = \sigma_r)$ does not specify the common expected value nor the common standard deviation, while the alternative hypothesis H_1 admits all real values of m_i and all positive σ_i. Show that

$$\lambda = \prod_{i=1}^{r} \left(\frac{s_i^2}{s_0^2} \right)^{n_i/2},$$

where $s_0^2 = \frac{1}{N} \sum_{i=1}^{r} n_i[(\bar{x}_i - \bar{x}_0)^2 + s_i^2]$ and where $\bar{x}_0 = \frac{1}{N} \sum_{i=1}^{r} n_i \bar{x}_i$.

16.12. (Continued). We preserve all notation and assumptions of the preceding problem except that hypothesis H_0 is replaced by the hypothesis H_0' $(\sigma_1 = \ldots = \sigma_r)$. Show that

$$\lambda = \prod_{i=1}^{r}\left(\frac{s_i^2}{s_a^2}\right)^{n_i/2},$$

where $s_a^2 = \dfrac{1}{N}\displaystyle\sum_{i=1}^{r} n_i s_i^2$.

16.13. Prove that if the parameters m and σ of a normal population are unknown and the hypothesis H_0 $(m = m_0)$ does not specify the value of σ, then the likelihood ratio test leads to Student's t-test.

16.14. Prove that if the parameters $m_1, \sigma_1, m_2, \sigma_2$ of two normal populations $N(m_1; \sigma_1)$ and $N(m_2; \sigma_2)$ are unknown then (a) testing the hypothesis H_0 $(\sigma_1 = \sigma_2)$ without any restrictions on m_1 and m_2 by using the likelihood ratio test leads to a test based on Fisher's Z; (b) testing the hypothesis H_0 $(m_1 = m_2)$ (if it is known that $\sigma_1 = \sigma_2$) by using the likelihood ratio test leads to Student's t-test.

16.15. Let x be normally distributed $N(m; \sigma)$ with unknown m and σ. Consider the composite hypothesis H_0 $(\sigma \leqslant \sigma_0)$ against H_1 $(\sigma > \sigma_0)$, while both H_0 and H_1 do not specify the value of m. Show that the test rejecting H_0 when $s^2 \geqslant c$, with c determined from $P[s^2 \geqslant c \mid \sigma = \sigma_0] = \alpha$, is uniformly most powerful. Compare with Problem 16.6(b).

16.16. Let $f(\zeta, Q)$ have the same meaning as in Problem 16.4 and let it be presentable in form (*) of that problem. For testing H_0 $(Q_1 \leqslant Q \leqslant Q_2)$ against H_1 $(Q < Q_1$ or $Q > Q_2)$, there exists a uniformly most powerful unbiased test given by the test function

$$(*) \qquad \phi(\zeta) = \begin{cases} 1 & \text{as} & T(\zeta) < c_1 \quad \text{or} \quad T(\zeta) > c_2, \\ \gamma_i & \text{as} & T(\zeta) = c_i \quad (i = 1, 2), \\ 0 & \text{as} & c_1 < T(\zeta) < c_2, \end{cases}$$

where $c_1, c_2, \gamma_1, \gamma_2$ are determined from

$$E[\phi(\zeta) \mid Q = Q_i] = \alpha \qquad (i = 1, 2).$$

16.17. Let $f(\zeta, Q)$ have the same properties as in the preceding problem. For testing H_0 $(Q = Q_0)$ against H_1 $(Q \neq Q_0)$, there exists a uniformly most powerful unbiased test given by the test function (*) of Problem 16.16, where the constants $c_1, c_2, \gamma_1, \gamma_2$ are determined from the equations

$$E[\phi(\zeta) \mid Q = Q_0] = \alpha,$$
$$E[T(\zeta)\phi(\zeta) \mid Q = Q_0] = \alpha E[T(\zeta) \mid Q = Q_0].$$

16.18. Let x be normally distributed $N(m; \sigma)$ with both m and σ unknown.

(a) Show that for testing H_0 $(\sigma \geqslant \sigma_0)$ against H_1 $(\sigma < \sigma_0)$, with m not specified by both H_0 and H_1, the test rejecting H_0 when $(s^2/\sigma_0^2) \leqslant c$ with c determined from $P(s^2 \leqslant \sigma_0^2 c) = \alpha$, is an unbiased uniformly most powerful test. (Compare with Problem 16.15.)

(b) Show that the uniformly most powerful unbiased tests for testing either H_0 $(m \leqslant m_0)$ against H_1 $(m > m_0)$, or H_0 $(m = m_0)$ against H_1 $(m \neq m_0)$, with σ not specified by the hypotheses H_0 and H_1, are based on Student's t test.

16.19. Let x_1 and x_2 be independent random variables, each having a Poisson distribution, and let the mean value of x_i be λ_i $(i = 1, 2)$. We want to test the composite hypothesis $H_0\,(\lambda_1 = \lambda_2)$ against $H_1\,(\lambda_1 \neq \lambda_2)$. Derive a uniformly most powerful test for testing this hypothesis. (Przyborowski and Wilenski [1]).

16.20. (Notation of Section 16.5). A test for testing the hypothesis $H_0\,(Q = Q_0)$ against the alternative $H_1\,(Q = Q_1)$ is called a *type A unbiased test* at the significance level α if it satisfies relations (16.3.5) and (16.5.5), and

(*)
$$\left[\frac{\partial^2 M(w, Q)}{\partial Q^2}\right]_{Q = Q_0} = \text{maximum.}$$

Show that if the assumptions of theorem 16.5.1 are satisfied and, moreover, the derivative $f_2(\zeta, Q) = \partial^2 f(\zeta, Q)/\partial Q^2$ exists at each interior point $Q \in K$ and, for every set w in sample space, we have

$$\frac{\partial}{\partial Q} \int_w f_1(\zeta, Q)\, d\zeta = \int_w f_2(\zeta, Q)\, d\zeta,$$

then among the type A unbiased tests there exists a most powerful w' which consists of those and only those points ζ for which

(**)
$$f_2(\zeta, Q_0) \geqslant c_1 f_1(\zeta, Q_0) + c_2 f(\zeta, Q_0).$$

The constants c_1 and c_2 are determined in such a way that relations (16.3.5), (16.5.5), and (*) are satisfied (Neyman and Pearson [4]).

16.21. Show that the most powerful unbiased test in example 16.5.2 is of type A.

16.22. The characteristic x has the normal distribution $N(0; \sigma)$ with σ unknown, and we want to test the hypothesis $H_1\,(\sigma = \sigma_0)$ against the alternative $H_1\,(\sigma = \sigma_1 \neq \sigma_0)$. Construct a most powerful type A unbiased test for this hypothesis.

16.23. The random variable $x_i\,(i = 1, 2)$ has a continuous distribution function $F_i(x)$. Let us denote by $x_{1l}\,(l = 1, \ldots, n_1)$ and $x_{2j}\,(j = 1, \ldots, n_2)$ independent observations of the random variables x_1 and x_2 and by $y_m\,(m = 1, \ldots, n_1 + n_2)$ the observations x_{1l}, x_{2j} arranged in increasing order. Let us further denote by $m_1, \ldots, m_{n_2}$ the ranks of those y_m which correspond to x_{2j}. By $M_s\,(s = 1, \ldots, n_2)$ we denote the random variable which assumes the values m_s. Show (Hoeffding [1], Lehmann [3]) that if $F_2(x) = g[F_1(x)]$, where $g(\cdot)$ is a continuous, differentiable, nondecreasing function and $g(0) = 0$, $g(1) = 1$, then

(*) $$P(M_1 = m_1, \ldots, M_{n_2} = m_{n_2}) = \frac{n_1!\, n_2!}{(n_1 + n_2)!}\, E[g'(u_{m_1}) \ldots g'(u_{m_{n_2}})],$$

where $u_{m_s}\,(s = 1, \ldots, n_2)$ is the value of the order statistic $\zeta_{m_s}^{(n_1 + n_2)}$ of a simple sample from a population with the uniform distribution in $[0.1]$.

16.24. (Continued). Show (Lehmann [3]) that if $F_2(x) = [F_1(x)]^k$ (k a positive integer), the right-hand side of (*) has the form

$$\frac{n_1!\, n_2!}{(n_1 + n_2)!}\, k^{n_2} \prod_{s=1}^{n_2} \frac{\Gamma(m_s + ks - s)}{\Gamma(m_{s+1} + ks - s)} \cdot \frac{\Gamma(m_{s+1})}{\Gamma(m_s)}.$$

16.25. (Continued).   By the term *rank test* we mean a test which depends only on the observed ranks $m_1, \ldots, m_{n_2}$.

Construct a most powerful rank test for testing the hypothesis H_0 $[F_1(x) \equiv F_2(x)]$ against the alternative H_1 $\{F_2(x) \equiv g[F_1(x)]\}$, where $g(\ )$ has the properties mentioned in Problem 16.23 (Lehmann [3]).

16.26. (Continued).   Construct a most powerful rank test for testing the hypothesis H_0 $[F_1(x) \equiv F_2(x)]$ against the alternative H_1 $\{F_2(x) = (1 - a)F(x) + a[F(x)]^2\}$, where $0 < a \leqslant 1$ (Lehmann [3], J. R. Savage [1], Uhlmann [1]).

CHAPTER 17

Elements of Sequential Analysis

17.1 PRELIMINARY REMARKS

A In the testing procedure discussed thus far, the sample size was constant. Proceeding in this way, we do not make use of all the information contained in the sample. In fact, the sample size does not depend upon which elements were selected first.

It is easy to give an example in which such a procedure is evidently disadvantageous in the sense that we make unnecessary drawings of elements. Let us consider, for example, a population in which the characteristic x has the normal distribution $N(m; 1)$, with the expected value m unknown. We want to test the hypothesis H_0 ($m = 0$), and to do so we draw a simple sample of size n from this population. Suppose that all the $n - 1$ elements first chosen differ very little from 5. It is obvious that since the standard deviation equals 1, the drawing of the nth element will not change our decision to reject H_0, for if $m = 0$, it is extremely improbable to obtain $n - 1$ successive elements, all of them close to 5. This example is not, of course, typical, but it clearly indicates that the method of hypotheses testing with the sample size fixed in advance may be disadvantageous. The situation is different for sequential testing of hypotheses.

B Such a procedure of testing statistical hypotheses in which the sample size is not fixed in advance, and the value of each element of the sample may cause either the rejection or the acceptance of the hypothesis tested, or the continuation of testing, that is, choosing the next element of the sample, is called a *sequential procedure*. A test using a sequential procedure is called a *sequential test*.

The domain of problems connected with sequential procedures for testing statistical hypotheses is called *sequential analysis*.

Dodge and Romig [1] should be regarded as the precursors of sequential analysis; they constructed a scheme of double sampling in which from the results of the first sample the decision is made whether or not to draw the second sample. In the special case of testing a hypothesis concerning the expected value of a binomial distribution, Bartky [1] considered a scheme close to the sequential probability ratio test, which is discussed in the next section. The foundations of sequential analysis and many fundamental results are due to Wald. The basic ideas of sequential analysis and many of its applications can be found in the works of Wald [2, 4].

Before we start a detailed discussion of sequential tests, we consider the following question: Is it possible that the successive drawing of elements will go on indefinitely, hence will lead to neither the acceptance nor the rejection of the hypothesis tested? The answer is that such procedures are possible, but they have no practical meaning. The sequential tests which we consider have the property that the probability of making a decision after a finite number of observations equals one.

The sequential test is characterized, as in the theory of statistical tests considered up to now, by its power function or by its OC function, and in addition, by the expected size of the sample. The number n of observations necessary for reaching a decision is a random variable; this number may be different in different experiments. The expected value of this random variable[1] characterizes a sequential test; that is, the smaller is this expected value, the better is the test.

17.2   THE SEQUENTIAL PROBABILITY RATIO TEST

Let the distribution of the random variable x depend on the parameter Q; $f(x, Q)$ denotes the density of x if it is of the continuous type, and $f(x, Q)$ denotes the probability function of x if it is of the discrete type. Now we describe a special sequential test, which is called the *sequential probability ratio test*.

Let the parameter Q of the function $f(x, Q)$ be unknown. We wish to test the hypothesis H_0 ($Q = Q_0$) against the alternative hypothesis H_1 ($Q = Q_1$), where Q_1 is a given number different from Q_0. As before, α and β denote, respectively, the probabilities of an error of the first and second kind. For given α and β, we determine two numbers A and B, where $0 < B < 1 < A$ (the method of determining these numbers is discussed later). We select elements for a simple sample. Let us imagine that we have one observation, say x_1. Let

$$(17.2.1) \qquad p_{11} = f(x_1, Q_1), \qquad p_{01} = f(x_1, Q_0).$$

Thus, for a random variable of the discrete type, p_{11} is the probability that $x = x_1$ and for a random variable of the continuous type it is the value of

---

[1] We restrict ourselves to the case when this expected value exists.

the density function at $x = x_1$, provided H_1 is true. Similarly, we can interpret p_{01}. Let us consider the ratio p_{11}/p_{01}.

If this ratio satisfies the inequality

$$\frac{p_{11}}{p_{01}} \geqslant A,$$

we reject H_0 and accept its alternative H_1; if

$$\frac{p_{11}}{p_{01}} \leqslant B,$$

we accept H_0; finally, if

$$B < \frac{p_{12}}{p_{02}} < A,$$

we include another element in the sample, say x_2. Furthermore, let

$$p_{12} = f(x_1, Q_1)f(x_2, Q_1), \qquad p_{02} = f(x_1, Q_0)f(x_2, Q_0).$$

We proceed in a manner similar to that already used, that is, we reject H_0 and accept H_1 if $p_{12}/p_{02} \geqslant A$; we accept H_0 if $p_{12}/p_{02} \leqslant B$; finally, we include a third element in the sample if

$$B < \frac{p_{11}}{p_{01}} < A.$$

In general, if the sample containing $m - 1$ elements does not lead to a decision, to reject or accept H_0, we draw the mth element, and the further procedure depends upon the ratio

$$\frac{p_{1m}}{p_{0m}} = \frac{f(x_1, Q_1)f(x_2, Q_1)\dots f(x_m, Q_1)}{f(x_1, Q_0)f(x_2, Q_0)\dots f(x_m, Q_0)},$$

namely, if

$$\frac{p_{1m}}{p_{0m}} \geqslant A,$$

we reject H_0 and accept H_1; if, however,

$$\frac{p_{1m}}{p_{0m}} \leqslant B,$$

we accept H_0; finally, if

$$B < \frac{p_{1m}}{p_{0m}} < A,$$

we include another element in the sample.

Such a procedure for testing a statistical hypothesis is called the *sequential probability ratio test*.

The procedure just described requires explanation on the following points:

1. Under what conditions will the probability be one that this procedure will lead us to a decision after a finite number of observations?

2. How do we determine the OC function of the sequential probability ratio test?

3. How do we determine the expected number $E(n)$ of observations necessary for making a decision?

4. For given α and β, what numbers A and B should we choose so that the probabilities of an error of the first and second kind will be α and β, respectively, for a random variable of the continuous type, and will not exceed α and β, respectively, for a random variable of the discrete type?

In further considerations, it is more convenient to deal with the logarithms of A and B and with the random variables

$$(17.2.2) \qquad z_i = \log \frac{f(x_i, Q_1)}{f(x_i, Q_0)} \qquad (i = 1, 2, \ldots).$$

The sequential probability ratio test considered may then be characterized as follows: Let us have m elements in the sample ($m \geqslant 1$). Then if

$$\sum_{i=1}^{m} z_i \geqslant \log A,$$

we reject H_0 and accept H_1; if

$$\sum_{i=1}^{m} z_i \leqslant \log B,$$

we accept H_0; finally, if

$$\log B < \sum_{i=1}^{m} z_i < \log A,$$

we draw the next, that is, the $(m + 1)$st element.

The solution of the four problems connected with the sequential probability ratio test just formulated follows from the solution of the same problems for a more general sequential model, namely, when the random variables z_i satisfy some general assumptions, but are not restricted to the particular form (17.2.2).

17.3 AUXILIARY THEOREMS

A   Let us consider a sequence $\{z_i\}$ of independent, identically distributed random variables whose variance exists. Let us denote by z a random variable with the same distribution as that of z_i, and by Z_m the sum of the first m terms of the sequence z_i,

$$(17.3.1) \qquad Z_m = z_1 + z_2 + \ldots + z_m.$$

As before, let the numbers A and B satisfy the inequality $0 < B < 1 < A$. We can write

$$\log B = b < 0, \qquad \log A = a > 0.$$

Let n denote the smallest integer for which Z_n lies outside the open interval (b, a), that is, either $Z_n \leqslant b$ or $Z_n \geqslant a$.

If for every m ($m = 1, 2, \ldots$), we have the inequalities $b < Z_m < a$, we say that $n = \infty$.

Theorem 17.3.1.   *If the variance of z is different from zero, the probability that $n = \infty$ is equal to zero.*

Proof. Let $c = a - b$. Let us split the infinite sequence $z_1, z_2, \ldots$ into segments S_k of r terms, where r is some integer. The first segment S_1 consists of the terms $z_1, z_2, \ldots, z_r$, the second segment S_2 consists of the terms $z_{r+1}, z_{r+2}, \ldots, z_{2r}$. Generally, the kth segment S_k consists of the terms

$$z_{(k-1)r+1}, z_{(k-1)r+2}, \ldots, z_{kr}.$$

Let us denote the sum of the kth segment by ξ_k, that is

$$\xi_k = \sum_{i=(k-1)r+1}^{kr} z_i.$$

Assume that $n = \infty$. Then for an arbitrary integer k we have the inequality

(17.3.2) $\xi_k^2 < c^2.$

In fact, for every m we have, by assumption, the inequalities

$$b < \sum_{i=1}^{m} z_i < a;$$

hence for arbitrary k and r

$$b < \sum_{i=1}^{kr} z_i < a.$$

Therefore,

$$b - \sum_{i=1}^{(k-1)r} z_i < \xi_k = \sum_{i=(k-1)r+1}^{kr} z_i < a - \sum_{i=1}^{(k-1)r} z_i.$$

By the inequality

$$b < \sum_{i=1}^{(k-1)r} z_i < a,$$

we obtain

(17.3.3) $-c = b - a < \xi_k < a - b = c.$

Inequality (17.3.2) follows from (17.3.3).

To show that the probability that n is finite equals one, it is sufficient to show that for a suitable r the probability that inequality (17.3.2) holds for every k equals zero.

Let

$$\pi_k = P(\xi_k^2 < c^2).$$

Since the random variables $z_i\,(i = 1, 2, \ldots)$ are independent and identically distributed, the random variables $\xi_k\,(k = 1, 2, \ldots)$ also are identically distributed and independent. It follows, first, that the π_k are identical for all k (hence we can denote this probability simply by π). In addition, it follows that the probability of the event that inequality (17.3.2) holds for $k = 1, 2, \ldots, j$ is π^j.

To complete the proof, it is enough to show that $\pi < 1$.

By assumption, the variance $D^2(z) \neq 0$. By the independence of the z_i, we have

$$E(\xi_k^2) = E\left[\left(\sum_{i=(k-1)r+1}^{kr} z_i\right)^2\right] = r[E(z^2)] + r(r - 1)[E(z)]^2$$
$$= r\{D^2(z) + r[E(z)]^2\}.$$

Thus, if the number r of elements in a segment is large enough, $E(\xi_k^2)$ will be arbitrarily large; hence $\pi < 1$. (See Problem 17.1)

It follows from theorem 17.3.1 that if the variance of

$$z = \log\frac{f(x, Q_1)}{f(x, Q_0)}$$

exists, then the probability that the sequential probability ratio test just described will lead to a decision after drawing a finite sample equals one.

Thus, we have settled the first of the four problems formulated in Section 17.2 in connection with the sequential probability ratio test.

B   We now make an additional assumption about the distribution of the random variable z and we prove a theorem which is of great importance in sequential analysis.

Theorem 17.3.2. Let z be a random variable satisfying the following assumptions:

I. *The variance $D^2(z)$ exists and is different from zero.* (Then the expected value $E(z)$ also exists.)

II. *There exists a positive number δ such that the inequalities*

$$(17.3.4) \quad P[z < \log(1 - \delta)] > 0, \qquad P[z > \log(1 + \delta)] > 0$$

hold.

III. *For every real h the expected value $g(h) = E(e^{hz})$ exists.*

IV. *The first two derivatives of the function $g(h)$ exist, and they can be obtained by differentiation under the sign E.* (This is to be understood as differentiation under the integral sign if z is of the continuous type and differentiation under the summation sign if z is of the discrete type.) *Thus*

(17.3.5) $g'(h) = E(ze^{hz})$,

(17.3.6) $g''(h) = E(z^2 e^{hz})$.

The assertion of the theorem is the following: *If $E(z) \neq 0$, then there exists one and only one real number $h_0 \neq 0$ such that*

(17.3.7) $E(e^{h_0 z}) = 1$;

if $E(z) = 0$, then equality (17.3.7) holds only for $h_0 = 0$.

Proof. Let $h > 0$ be an arbitrary positive number. Let us consider the random variable $y = e^z$. This random variable can take on only positive values; hence the same applies to the random variable y^h. Hence by theorem 3.3.1, we have

(17.3.8) $g(h) = E(y^h) \geqslant (1 + \delta)^h P[y^h \geqslant (1 + \delta)^h]$
$$= (1 + \delta)^h P(e^z \geqslant 1 + \delta).$$

It follows from assumption II that $P(e^z \geqslant 1 + \delta) > 0$, hence, by the relation

$$\lim_{h \to +\infty} (1 + \delta)^h = +\infty,$$

we obtain from (17.3.8)

(17.3.9) $\lim_{h \to +\infty} g(h) = +\infty$.

Similarly, we can show that for any $h < 0$

$$g(h) \geqslant (1 - \delta)^h P(e^z \leqslant 1 - \delta).$$

From assumption II and the relation

$$\lim_{h \to -\infty} (1 - \delta)^h = +\infty,$$

we obtain

(17.3.10) $\lim_{h \to -\infty} g(h) = +\infty$.

Let us next consider the second derivative of the function $g(h)$ satisfying, by assumption IV, formula (17.3.6). Making use of assumption II in this formula, we obtain

(17.3.11) $g''(h) > 0$.

Formulas (17.3.9) to (17.3.11) show that there exists only one real number h^* for which $g(h)$ takes on its minimum value; hence $g'(h^*) = 0$. If $E(z) \neq 0$, then $g'(0) = E(z) \neq 0$, thus $h^* \neq 0$ and $g(h^*) < g(0) = 1$.

Next, we observe that the function $g(h)$ is strictly decreasing in the interval $(-\infty, h^*)$ and strictly increasing in the interval $(h^*, +\infty)$. Since $g(0) = 1$ and $g(h^*) < 1$, there exists only one $h_0 \neq 0$ such that $g(h_0) = E(e^{h_0 z}) = 1$. If, however, $E(z) = g'(0) = 0$ and all the assumptions I to IV are satisfied, then from the proof it is easy to see that $h = 0$ is the only real number for which $g(h) = 1$. Theorem 17.3.2 is proved.

17.4 THE FUNDAMENTAL IDENTITY

We now prove the Fundamental Identity, which plays an important role in sequential analysis.

Let $\{z_i\}$ be a sequence of independent random variables with the same distribution as that of the random variable z. Let Z_m denote the sum of the first m terms of the sequence $\{z_i\}$. If z satisfies assumptions I to IV of the previous section, then for every h in the interval D, where D denotes that part of the real axis for which

(17.4.1) $$g(h) = E(e^{zh}) \geqslant 1,$$

we have the identity

(17.4.2) $$E\{e^{Z_n h}[g(h)]^{-n}\} = 1,$$

where n is the smallest integer such that Z_n lies outside the open interval (b, a).

Proof. Let us consider the identity

(17.4.3) $$E\{\exp[Z_n h + (Z_N - Z_n)h]\} = E(\exp Z_N h)$$
$$= E\{\exp[(z_1 + z_2 + \ldots + z_N)h]\} = [g(h)]^N,$$

where N is an integer.

Let us denote by P_N the probability that $n \leqslant N$. We denote by $E_N(u)$ the conditional expected value of the random variable u provided $n \leqslant N$, and by $E_N^*(u)$ the conditional expected value of u provided $n > N$.

Using this notation, we can write (17.4.3) in the form

(17.4.4) $$P_N E_N\{\exp[Z_n h + (Z_N - Z_n)h]\}$$
$$+ (1 - P_N)E_N^*(\exp Z_N h) = [g(h)]^N.$$

Since, for every fixed $n \leqslant N$, the random variable $Z_N - Z_n = z_{n+1} + z_{n+2} + \ldots + z_N$ is independent of $Z_n = z_1 + z_2 + \ldots + z_n$, we can write

(17.4.5) $$E_N\{\exp[Z_n h + (Z_N - Z_n)h]\} = E_N\{[\exp(Z_n h)\exp(Z_N - Z_n)h]\}$$
$$= E_N\{[\exp(Z_n h)][g(h)]^{N-n}\}.$$

From formulas (17.4.5) we obtain

$$P_N E_N\{e^{Z_n h}[g(h)]^{N-n}\} + (1 - P_N)E_N^*(e^{Z_N h}) = [g(h)]^N.$$

Dividing both sides by $[g(h)]^N$, we obtain

$$(17.4.6) \qquad P_N E_N\{e^{Z_n h}[g(h)]^{-n}\} + (1 - P_N)\frac{E_N^*(e^{Z_N h})}{[g(h)]^N} = 1.$$

Note that if h_0, determined by formula (17.3.7), is negative, the domain D defined by (17.4.1) consists of the whole h axis excluding the open interval $(h_0, 0)$, and if $h_0 > 0$, the domain D consists of the whole h axis excluding the open interval $(0, h_0)$.

Since $\lim_{N \to \infty} P(n \leqslant N) = 1$, we have $\lim_{N \to \infty} (1 - P_N) = 0$. Observe next that $E_N^*(e^{Z_N h})$ is uniformly bounded with regard to N. In fact, this expression is the expected value of $e^{Z_N h}$ provided $n > N$; hence $b < Z_N < a$. Thus, for $h > 0$

$$e^{bh} < e^{Z_N h} < e^{ah},$$

and for $h < 0$

$$e^{ah} < e^{Z_N h} < e^{bh}.$$

It follows that $E_N^*(e^{Z_N h})$ is uniformly bounded.

Finally, because we consider only the domain D of values of h for which $g(h) \geqslant 1$, we obtain

$$(17.4.7) \qquad \lim_{N \to \infty} \frac{(1 - P_N)E_N^*(e^{Z_N h})}{[g(h)]^N} = 0$$

and

$$(17.4.8) \qquad \lim_{n \to \infty} P_N E_N\{e^{Z_n h}[g(h)]^{-n}\} = E\{e^{Z_n h}[g(h)]^{-n}\}.$$

Using (17.4.7) and (17.4.8), we obtain (17.4.2) from (17.4.6).

Blom [1] proved Wald's Fundamental Identity under weaker assumptions than those used here, and he generalized this identity to random variables z_i that are not supposed to have the same distribution. Blackwell and Girshick [1] generalized the Fundamental Identity to random vectors.

17.5 THE OC FUNCTION OF THE SEQUENTIAL PROBABILITY RATIO TEST

A For brevity we denote by $L(Q)$ the OC function of the sequential probability ratio test.

Let us consider the sequence $\{z_i\}$ of independent random variables defined by (17.2.2) and satisfying assumptions I to IV of theorem 17.3.2. By that theorem, if the random variable

$$z = \log\frac{f(x, Q_1)}{f(x, Q_0)}$$

satisfies assumptions I to IV, then if $E(z) \neq 0$, there exists one and only one $h_0 \neq 0$ which satisfies (17.3.7); if $E(z) = 0$, this condition holds only for $h_0 = 0$.

First we restrict our considerations to $E(z) \neq 0$. Since the distribution of z depends on Q, the number h_0 determined from (17.3.7) also depends on Q. Thus, let us write $h_0 = h_0(Q)$ and let us rewrite formula (17.3.7) as

$$(17.5.1) \qquad g[h_0(Q)] = E\{\exp [z h_0(Q)]\} = 1.$$

If x is a random variable of the continuous type with the density $f(x, Q)$, the last equality may be written in the form

$$(17.5.2) \qquad g[h_0(Q)] = \int_{-\infty}^{+\infty} \left[\frac{f(x, Q_1)}{f(x, Q_0)}\right]^{h_0(Q)} f(x, Q)\, dx = 1.$$

If, however, x is a random variable of the discrete type, equality (17.5.1) takes the form

$$(17.5.3) \qquad g[h_0(Q)] = \sum_j \left[\frac{f(x_j, Q_1)}{f(x_j, Q_0)}\right]^{h_0(Q)} f(x_j, Q) = 1,$$

where x_j ($j = 1, 2, \ldots$) are the jump points of x and $f(x_j, Q)$ is the probability that $x = x_j$.

Now let us write identity (17.4.2) for $h = h_0$. Since $g(h_0) = 1$, the identity takes the form

$$(17.5.4) \qquad E_Q\{\exp [Z_n h_0(Q)]\} = 1.$$

The subscript Q indicates that the left-hand side of (17.5.4) is a function of the parameter Q.

Next denote by $E_Q{}^*$ the conditional expected value of the random variable $\exp [Z_n h_0(Q)]$ provided we accept the hypothesis $H_0(Q = Q_0)$, that is, that $Z_n \leqslant \log B$, and by $E_Q{}^{**}$ the conditional expected value of the random variable $\exp [Z_n h_0(Q)]$ provided we accept the hypothesis $H_1(Q = Q_1)$, that is, $Z_n \geqslant \log A$. By definition of $L(Q)$, we obtain from (17.5.4)

$$L(Q)E_Q{}^* + [1 - L(Q)]E_Q{}^{**} = 1.$$

Hence

$$(17.5.5) \qquad L(Q) = \frac{E_Q{}^{**} - 1}{E_Q{}^{**} - E_Q{}^*}.$$

B We now find an approximate expression for $L(Q)$. To do so, we consider the equalities $Z_n = \log B$ and $Z_n = \log A$ instead of the inequalities

$Z_n \leqslant \log B$ and $Z_n \geqslant \log A$, respectively. Thus, if we put $Z_n = \log B$ in the expression $\exp [Z_n h_0(Q)]$, we obtain

$$E_Q^*\{\exp [Z_n h_0(Q)]\} \cong B^{h_0(Q)},$$

since for $Z_n = \log B$ we accept the hypothesis H_0. Similarly, if we put $Z_n = \log A$, we obtain

$$E_Q^{**}\{\exp [Z_n h_0(Q)]\} \cong A^{h_0(Q)}.$$

Substituting these values in (17.5.5), we obtain the approximate formula for the OC function of the sequential probability ratio test

$$(17.5.6) \qquad L(Q) \cong \frac{A^{h_0(Q)} - 1}{A^{h_0(Q)} - B^{h_0(Q)}}.$$

Formula (17.5.6) has been derived under the assumption that $E(z) \neq 0$; hence $h_0(Q) \neq 0$. If $E(z) = 0$, then $h_0(Q') = 0$, where Q' is that value of parameter Q for which $E(z) = 0$, and then the right-hand side of (17.5.6) has the limit $\lim_{Q \to Q'} L(Q) = L(Q')$ equal to the ratio of the derivatives of the numerator and denominator with respect to Q at Q'. Thus assuming the existence of the derivative of $h_0(Q)$ at Q', we obtain

$$(17.5.7) \qquad L(Q') \cong \frac{\log A}{\log A - \log B}.$$

Formula (17.5.2) or (17.5.3) allows us to express Q as a function of h_0, and formula (17.5.6) expresses $L(Q)$ in terms of $h_0(Q)$. Hence for any real number h_0 we can determine the point in the plane with coordinates $Q, L(Q)$. The locus of these points will be the approximate graph of the OC function.

Let us consider the approximation used in deriving formulas (17.5.6) and (17.5.7) for the OC function.

If we take $Z_n = \log B$ or $Z_n = \log A$, we neglect the possibility of the inequality $Z_n < \log B$ or $Z_n > \log A$ at the moment of terminating the sequential procedure. In other words, we write $Z_n = \log B$ although actually $Z_n < \log B$, or we write $Z_n = \log A$ although actually $Z_n > \log A$.

If z can take on only two values d and $-d$, Z_n can only equal Kd, where K is an integer, positive or negative. If, in addition, $(\log A)/d$ and $(\log B)/d$ are integers, the decision could be made only if $Z_n = \log B$ or $Z_n = \log A$, and then formulas (17.5.6) and (17.5.7) would be exact. Even if the domain of possible values of the random variable does not consist only of two values d and $-d$, the approximate expressions for $L(Q)$ are rather good in practice if $|E_Q(z)|$ and $D^2(z)$ are not large. The assumption that Z_n does not exceed log A or is not less than log B at the moment of terminating the sequential procedure is often made in applications of the sequential probability ratio test.

17.6 THE EXPECTED VALUE $E(n)$

A   As was said in Section 17.1, the expected number of observations $E(n)$ characterizes the sequential test. We find $E(n)$ as a function of the unknown parameter Q when using the sequential probability ratio test for testing the hypothesis $H_0(Q = Q_0)$ against the alternative $H_1(Q = Q_1)$. We make use of the Fundamental Identity.

Let us assume first that $E(z) \neq 0$. Wald [3] showed that if assumptions I to IV from 17.3 are satisfied, the identity (17.4.2) may be differentiated with respect to h under the sign E an arbitrary number of times. Thus, let us differentiate (17.4.2) with respect to h at $h = 0$. We obtain

$$(17.6.1) \qquad E_Q\{Z_n e^{Z_n h}[g(h)]^{-n} - ne^{Z_n h}[g(h)]^{-n-1}g'(h)\}_{h=0} = 0.$$

The subscript Q indicates, as before, that this expected value depends on Q. Since $g(0) = 1$ and $g'(0) = E_Q(z)$, we have

$$E_Q[Z_n - nE_Q(z)] = 0,$$

and since, by assumption, $E(z) \neq 0$, we have

$$(17.6.2) \qquad E_Q(n) = \frac{E_Q(Z_n)}{E_Q(z)}.$$

We now determine $E_Q(Z_n)$. Let $E_Q{}^*(Z_n)$ denote the conditional expected value of the random variable Z_n provided $Z_n \leqslant \log B$, and $E_Q{}^{**}(Z_n)$ the conditional expected value of Z_n provided $Z_n \geqslant \log A$. We have

$$(17.6.3) \qquad E_Q(Z_n) = L(Q)E_Q{}^*(Z_n) + [1 - L(Q)]E_Q{}^{**}(Z_n).$$

Formula (17.6.2) takes the form

$$(17.6.4) \qquad E_Q(n) = \frac{L(Q)E_Q{}^*(Z_n) + [1 - L(Q)]E_Q{}^{**}(Z_n)}{E_Q(z)}.$$

If we can neglect the differences $\log B - Z_n$ and $Z_n - \log A$, that is, if we can assume that when the decision is made $Z_n = \log B$ or $Z_n = \log A$, we can take $\log B$ as $E_Q{}^*(Z_n)$ and $\log A$ as $E_Q{}^{**}(Z_n)$. Then the formula for the expected sample size takes the form

$$(17.6.5) \qquad E_Q(n) \cong \frac{L(Q) \log B + [1 - L(Q)] \log A}{E_Q(z)}.$$

This formula is exact if z can take on only two values d and $-d$, as discussed in the previous section. If the distribution of z is not restricted to these two values, but $|E_Q(z)|$ and $D^2(z)$ are not large, then, as we know, we may neglect the fact that Z_n exceeds $\log A$ or is less than $\log B$. In this case, formula (17.6.5) gives an approximation to $E_Q(n)$.

B   So far, we have assumed that $E(z) \neq 0$. Now let us assume that $E(z) = 0$. After differentiating the Fundamental Identity (17.4.2) twice, we obtain

(17.6.6)

$$E_{Q'} \left[\left\{ \left(Z_n - n \frac{g'(h)}{g(h)} \right)^2 - \frac{ng''(h)g(h) - n[g'(h)]^2}{[g(h)]^2} \right\} e^{Z_n h} [g(h)]^{-n} \right] = 0.$$

Taking the derivative at $h = 0$ and using the fact that $g(0) = 1, g'(0) = E_{Q'}(z) = 0$, and $g''(0) = E_{Q'}(z^2) \neq 0$, we obtain from (17.6.6)

$$E_{Q'} [Z_n^2 - nE_{Q'}(z^2)] = 0.$$

Finally,

(17.6.7)
$$E_{Q'}(n) = \frac{E_{Q'}(Z_n^2)}{E_{Q'}(z^2)}.$$

Let us now determine $E_{Q'}(Z_n^2)$. Let $E_{Q'}^* (Z_n^2)$ and $E_{Q'}^{**}(Z_n^2)$ denote the conditional expected values of Z_n^2 under the condition that $Z_n \leqslant \log B$ and $Z_n \geqslant \log A$, respectively. Then we have

(17.6.8) $E_{Q'}(Z_n^2) = L(Q')E_{Q'}^*(Z_n^2) + [1 - L(Q')]E_{Q'}^{**}(Z_n^2).$

If we can neglect the differences $\log B - Z_n$ and $Z_n - \log A$, then $E_{Q'}^*(Z_n^2)$ and $E_{Q'}^{**}(Z_n^2)$ are equal to $(\log B)^2$ and $(\log A)^2$, respectively. Using formula (17.5.7) for $L(Q')$, which was derived under the assumption $E(z) = 0$, we obtain

(17.6.9)
$$E_{Q'}(Z_n^2) \simeq \frac{\log A}{\log A - \log B} (\log B)^2 + \left(1 - \frac{\log A}{\log A - \log B} \right)(\log A)^2$$

$$= -\log A \log B.$$

From (17.6.7) and (17.6.9) it follows that

(17.6.10)
$$E_{Q'}(n) \simeq - \frac{\log A \log B}{E_{Q'}(z^2)}.$$

The last formula was obtained by Wallis [1] and by Wald.

We direct the reader's attention to the fact that formulas (17.6.2), (17.6.4), and the approximate formulas (17.6.5) and (17.6.10) are satisfied for all random variables satisfying the assumptions of theorem 17.3.2, hence not only for the z_i defined by formula (17.2.2) for the sequential probability ratio test.

Blackwell [1], Blackwell and Girshick [1], and Wolfowitz [2] investigated various generalizations of formula (17.6.2). In the last paper, this formula is generalized to random variables z_i that do not have the same distribution and are dependent (see also Problem 17.2).

17.7   THE DETERMINATION OF A AND B

We still must solve the last of the four problems formulated in connection with the sequential probability ratio test, the problem of determining A and B as functions of the probabilities α and β of errors of the first and second kind. We restrict ourselves to the case in which we can assume that when the decision is made $Z_n = \log B$ or $Z_n = \log A$.

We first observe that from (17.5.2) and (17.5.3) it follows that

$$(17.7.1) \qquad h_0(Q_0) = 1, \qquad h_0(Q_1) = -1.$$

Let us substitute these values in formula (17.5.6). Using the fact that $L(Q_0) = 1 - \alpha$ and $L(Q_1) = 1 - \beta$, we obtain

$$(17.7.2) \qquad L(Q_0) \simeq \frac{A - 1}{A - B} = 1 - \alpha,$$

$$(17.7.2') \qquad L(Q_1) \simeq \frac{1/A - 1}{1/A - 1/B} = \beta.$$

Solving this system for A and B gives

$$(17.7.3) \qquad A \simeq \frac{1 - \beta}{\alpha}, \qquad B \simeq \frac{\beta}{1 - \alpha}.$$

These formulas are, of course, approximate, but they usually are good enough for practical purposes.

17.8   TESTING A HYPOTHESIS CONCERNING THE PARAMETER p OF A ZERO-ONE DISTRIBUTION

A   Let us now consider the application of the sequential probability ratio test just presented to testing a hypothesis concerning the parameter p of a zero-one distribution.

Let the random variable x have a zero-one distribution, that is, $P(x = 1) = p$ and $P(x = 0) = 1 - p$ $(0 < p < 1)$, where the value of p is unknown. For instance, let the population consist of good and defective items. Suppose that we assign the number one to the drawing of a defective item from this population and the number zero to the drawing of a good one. According to the notation used in this chapter, we have

$$(17.8.1) \qquad f(1, p) = p, \qquad f(0, p) = 1 - p.$$

We want to test the hypothesis $H_0(p = p_0)$ against the alternative $H_1(p = p_1)$, where p_0 and p_1 are given values with $0 < p_0 < p_1 < 1$. The probabilities of errors of the first and second kind are, respectively, α and β.

We apply the sequential probability ratio test. We denote

$$(17.8.2) \qquad z = \log \frac{f(x, p_1)}{f(x, p_0)}.$$

The random variable z can take on two values, $z = \log p_1 - \log p_0$ if $x = 1$, and $z = \log (1 - p_1) - \log (1 - p_0)$ if $x = 0$.

Let us check whether z satisfies assumptions I to IV of Section 17.3. Assumption I is satisfied, since the variance of z exists and is different from 0. In fact,

$$(17.8.3)$$
$$E(z) = p \log \frac{p_1}{p_0} + (1 - p) \log \frac{1 - p_1}{1 - p_0} = \log \left[\left(\frac{p_1}{p_0} \right)^p \left(\frac{1 - p_1}{1 - p_0} \right)^{1 - p} \right],$$

where p is the true value of the unknown parameter. Next,

$$E(z^2) = p \left(\log \frac{p_1}{p_0} \right)^2 + (1 - p) \left(\log \frac{1 - p_1}{1 - p_0} \right)^2.$$

After some transformations, we obtain

$$(17.8.4) \qquad D^2(z) = p(1 - p) \left[\log \frac{p_1(1 - p_0)}{p_0(1 - p_1)} \right]^2.$$

Since, by assumption, $p_0 \neq p_1$ and p_0 and p_1 are different from zero and one, $D^2(z)$ is finite and different from zero.

Let us next consider assumption II. We have

$$(17.8.5) \qquad e^z = \begin{cases} \dfrac{p_1}{p_0} & \text{if } x = 1, \\[2mm] \dfrac{1 - p_1}{1 - p_0} & \text{if } x = 0. \end{cases}$$

By assumption, $p_1 > p_0$; hence $1 - p_1 < 1 - p_0$. Let

$$\frac{p_1}{p_0} - 1 = \delta_1, \qquad 1 - \frac{1 - p_1}{1 - p_0} = \delta_2,$$

$$\delta = \tfrac{1}{2} \min (\delta_1, \delta_2);$$

hence δ is the smaller of the numbers $\tfrac{1}{2}\delta_1$ and $\tfrac{1}{2}\delta_2$. Thus we have the relations

(17.8.6) $\quad P(e^z < 1 - \delta) = P\left(e^z = \frac{1 - p_1}{1 - p_0}\right) = 1 - p > 0,$

$$P(e^z > 1 + \delta) = P\left(e^z = \frac{p_1}{p_0}\right) = p > 0,$$

and assumption II is satisfied.

Next, for every real h we have

(17.8.7) $\quad g(h) = E(e^{hz}) = p\left(\frac{p_1}{p_0}\right)^h + (1 - p)\left(\frac{1 - p_1}{1 - p_0}\right)^h.$

From the assumptions about p it follows that $g(h)$ exists for every h. Thus assumption III is also satisfied.

Next, we have

(17.8.9)

$$g'(h) = p\left(\frac{p_1}{p_0}\right)^h \log \frac{p_1}{p_0} + (1 - p)\left(\frac{1 - p_1}{1 - p_0}\right)^h \log \frac{1 - p_1}{1 - p_0} = E(ze^{hz}).$$

Similarly, we can show that

(17.8.10) $\qquad\qquad g''(h) = E(z^2 e^{hz}).$

Thus assumption IV is also satisfied.

B   The sequential procedure will be as follows: Imagine that we already have m observations, $x = 1$ m_1 times, and $x = 0$ $m - m_1$ times. Hence

(17.8.11) $\qquad Z_m = m_1 \log \frac{p_1}{p_0} + (m - m_1) \log \frac{1 - p_1}{1 - p_0}.$

The constants A and B are determined from (17.7.3). Thus, we reject H_0 if

(17.8.12) $\quad m_1 \log \frac{p_1}{p_0} + (m - m_1) \log \frac{1 - p_1}{1 - p_0} \geqslant \log A,$

and we accept H_0 if

(17.8.13) $\qquad m_1 \log \frac{p_1}{p_0} + (m - m_1) \log \frac{1 - p_1}{1 - p_0} \leqslant \log B.$

Finally, we continue sampling if

$$(17.8.14) \quad \log B < m_1 \log \frac{p_1}{p_0} + (m - m_1) \log \frac{1 - p_1}{1 - p_0} < \log A.$$

After transforming formula (17.8.12), we have

$$(17.8.12') \quad m_1 \geqslant \frac{\log A}{\log \dfrac{p_1}{p_0} - \log \dfrac{1 - p_1}{1 - p_0}} - m \frac{\log \dfrac{1 - p_1}{1 - p_0}}{\log \dfrac{p_1}{p_0} - \log \dfrac{1 - p_1}{1 - p_0}}.$$

Similarly, from (17.8.13) and (17.8.14) we obtain

$$(17.8.13') \quad m_1 \leqslant \frac{\log B}{\log \dfrac{p_1}{p_0} - \log \dfrac{1 - p_1}{1 - p_0}} - m \frac{\log \dfrac{1 - p_1}{1 - p_0}}{\log \dfrac{p_1}{p_0} - \log \dfrac{1 - p_1}{1 - p_0}},$$

$$(17.8.14') \quad \frac{\log B}{\log \dfrac{p_1}{p_0} - \log \dfrac{1 - p_1}{1 - p_0}} - m \frac{\log \dfrac{1 - p_1}{1 - p_0}}{\log \dfrac{p_1}{p_0} - \log \dfrac{1 - p_1}{1 - p_0}} < m_1$$

$$< \frac{\log A}{\log \dfrac{p_1}{p_0} - \log \dfrac{1 - p_1}{1 - p_0}} - m \frac{\log \dfrac{1 - p_1}{1 - p_0}}{\log \dfrac{p_1}{p_0} - \log \dfrac{1 - p_1}{1 - p_0}}.$$

Let us denote the right-hand side of inequality (17.8.12') by r_m; the r_m are called *rejection numbers*. The right-hand side of (17.8.13') is denoted by a_m; the a_m are called *acceptance numbers*. This means that if, in a simple sample of m elements, the number of defective items is not smaller than r_m, we reject the hypothesis $H_0 (p = p_0)$ and accept its alternative $H_1 (p = p_1)$. If, however, the number of defective items is not greater than a_m, we accept the hypothesis $H_0 (p = p_0)$.

For given p_0, p_1, α, and β we can compute the r_m and a_m before we start the sequential procedure.

The r_m and a_m are linear functions of the sample size m; we can write them in the form

$$r_m = cm + d_1, \qquad a_m = cm + d_2,$$

respectively.

It is easy to see that for $p_1 > p_0$ we have $c > 0$, $d_1 > 0$, and $d_2 < 0$.

Figure 17.8.1 shows the typical behavior of the graphs of these functions

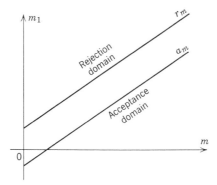

Fig. 17.8.1

in the plane (m, m_1). As we can see from this figure, for points above and on the line r_m, we reject H_0; thus we call this domain the *rejection domain*. Similarly, the domain below and on the line a_m is called the *acceptance domain*. If the point (m, m_1) lies between the lines r_m and a_m, we continue sampling.

Let us find the OC function in this example. Neglecting the value of p for which $E(z) = 0$, by (17.5.3), we obtain for any other value of p

$$(17.8.15) \qquad p\left(\frac{p_1}{p_0}\right)^{h_0(p)} + (1 - p)\left(\frac{1 - p_1}{1 - p_0}\right)^{h_0(p)} = 1,$$

where $h_0(p) \neq 0$.

From this equality we can determine p as a function of h_0,

$$(17.8.16) \qquad p = \frac{1 - \left(\dfrac{1 - p_1}{1 - p_0}\right)^{h_0}}{(p_1/p_0)^{h_0} - [(1 - p_1)/(1 - p_0)]^{h_0}}.$$

Next, from formula (17.5.6) for $L(p)$, using expressions (17.7.3) for A and B, we obtain

$$(17.8.17) \qquad L(p) = \frac{\left(\dfrac{1 - \beta}{\alpha}\right)^{h_0} - 1}{[(1 - \beta)/\alpha]^{h_0} - [\beta/(1 - \alpha)]^{h_0}}.$$

Using formulas (17.8.16), (17.8.17), (17.7.3) and (17.8.3), we compute the expected number of observations from (17.6.5).

The reader can easily find $L(p)$ for a p such that $E(z) = 0$, and the expected number of observations for that p.

In this problem, let us take the following numerical values:

$$p_0 = 0.05, \qquad p_1 = 0.1, \qquad \alpha = \beta = 0.05,$$

and let p be the unknown defectiveness in a lot which contains good and defective items. Then we obtain

$$\log A = \log \frac{1 - \beta}{\alpha} = \log 19 = 1.27875,$$

$$\log B = \log \frac{\beta}{1 - \alpha} = -\log 19 = -1.27875.$$

The random variable z defined by (17.8.2) may take on two values, $z = \log 2 = 0.30103$ if $x = 1$ (that is, if a defective item appears in the sample), and $z = \log 18/19 = -0.02348$ if $x = 0$ (that is, if we draw a good item). Thus the sequential probability ratio test procedure consists in rejecting the hypothesis H_0 ($p = 0.05$) and accepting the alternative H_1 ($p = 0.1$) if after drawing the mth item ($m = 1, 2, \ldots$) it turns out that we have m_1 defective and $m - m_1$ good items in the sample, where

$$0.30103 m_1 - 0.02348(m - m_1) \geqslant 1.27875.$$

On the other hand, if

$$0.30103 m_1 - 0.02348(m - m_1) \leqslant -1.27875,$$

we accept the hypothesis H_0 ($p = 0.05$). Finally, if

$$-1.27875 < 0.30103 m_1 - 0.02348(m - m_1) < 1.27875,$$

we take the next, $(m + 1)$st, element for the sample.

The rejection numbers r_m defined by the right-hand side of (17.8.12') here have the form

$$r_m = \frac{1.27875}{0.30103 + 0.02348} - m \frac{-0.02348}{0.30103 + 0.02348} = 3.94 + 0.072m.$$

Similarly,

$$a_m = \frac{-1.27875}{0.30103 + 0.02348} - m \frac{-0.02348}{0.30103 + 0.02348} = -3.94 + 0.072m.$$

The OC function of this test can be determined from (17.8.16) and (17.8.17), which here have the form

$$p = \frac{1 - (\frac{18}{19})^{h_0}}{2^{h_0} - (\frac{18}{19})^{h_0}}, \qquad L(p) = \frac{19^{h_0} - 1}{19^{h_0} - (\frac{1}{19})^{h_0}}.$$

According to (17.7.2), and (17.7.2'), we have

$$L(0.05) = 0.95, \qquad L(0.10) = 0.05.$$

Let us now compute the expected number of observations for $p = 0.05$ and $p = 0.1$, respectively. From (17.6.5) we have

$$E_{0.05}(n) = \frac{L(0.05)(-1.27875) + [1 - L(0.05)]1.27875}{E_{0.05}(z)}.$$

From (17.8.3) we obtain

$$E_{0.05}(z) = \log 2^{0.05} + \log \left(\tfrac{18}{19}\right)^{0.95} = -0.00725.$$

Finally,

$$E_{0.05}(n) \cong 159.$$

Similarly, we obtain

$$E_{0.1}(n) \cong 128.$$

It might seem that these expected values of n are rather large. However, we shall show that the sequential probability ratio test is more powerful than a most powerful nonsequential test.

To do so, let us test the hypothesis H_0 ($p = 0.05$) against the alternative H_1 ($p = 0.1$), using a nonsequential test, for

$$n = \tfrac{1}{2}[E_{0.05}(n) + E_{0.10}(n)] \cong 144.$$

We again take $\alpha = 0.05$. First, we have to find a number k such that the probability that the number of defective items exceeds k when the hypothesis $H_0(p = p_0 = 0.05)$ is true does not exceed 0.05. This number k may be determined from the relation

$$\sum_{j=k+1}^{144} \binom{144}{j} 0.05^j \cdot 0.95^{144-j} \leqslant 0.05.$$

Since n is large, we use the Poisson distribution with expected value $np_0 = 7.2$ (see Section 5.5). We find $k = 12$.

Let us now compute the probability β of an error of the second kind. Thus we assume that $p = p_1 = 0.1$. We want to compute the value of

$$\beta = 1 - \sum_{j=13}^{144} \binom{144}{j} 0.1^j \cdot 0.9^{144-j}.$$

We find it from the Poisson distribution with expected value $np_1 = 14.4$. We have $\beta = 0.3203$; hence the probability of an error of the second kind is here more than six times greater than for the sequential probability ratio test with almost the same expected number of observations.

To explain thoroughly the advantage of applying the sequential procedure to this problem, we compute what n should be for α and β to be close to 0.05.

Since n will certainly be large (larger than 144), we use the de Moivre-Laplace theorem. Let us denote by y the number of defective items in a sample of size n. This random variable has the asymptotically normal distribution $N(0.05n; \sqrt{0.0475n})$ if the hypothesis H_0 $(p = 0.05)$ is true, and the asymptotically normal distribution $N(0.1n; \sqrt{0.09n})$ if the hypothesis H_1 $(p = 0.1)$ is true.

To test this hypothesis, we take for the critical set w the set of points in the sample space for which

$$P(\bar{y} \geqslant A) = 0.05,$$

where $\bar{y} = y/n$ has the normal distribution $N(0.05; \sqrt{0.0475/n})$. The unknown values of A and n are found from $\alpha = \beta = 0.05$. Using tables of the normal distribution, we get the system of two equations

$$\frac{(A - 0.05)\sqrt{n}}{\sqrt{0.0475}} = 1.65, \qquad \frac{(A - 0.1)\sqrt{n}}{\sqrt{0.09}} = -1.65.$$

From this system we obtain $n \cong 292$, $A \cong 0.071$.

Since A is the critical value of $\bar{y}$, the critical value of y will be $An = 20.7 \cong 21$, that is, we reject the hypothesis H_0 $(p = 0.05)$ and accept the alternative H_0 $(p = 0.1)$ if in a simple sample of size $n = 292$ we find 21 or more defective items. Because of the discrete character of the random variable y and the normal approximation used, the probability of an error of the first or second kind only approximately equals 0.05.

We conclude that we would have to take a sample of $n = 292$ elements in order to obtain a test with the same power; hence almost twice as many as the expected number of observations when applying the sequential probability ratio test.

17.9 TESTING A HYPOTHESIS CONCERNING THE EXPECTED VALUE m OF A NORMAL POPULATION

Let x have the normal distribution $N(Q; 1)$, where Q is the unknown expected value. We want to test the hypothesis H_0 $(Q = Q_0)$ against the alternative H_1 $(Q = Q_1)$, where $Q_1 > Q_0$. We have

$$(17.9.1) \qquad f(x, Q_0) = \frac{1}{\sqrt{2\pi}} \exp\left[-\tfrac{1}{2}(x - Q_0)^2\right],$$

$$(17.9.2) \qquad f(x, Q_1) = \frac{1}{\sqrt{2\pi}} \exp\left[-\tfrac{1}{2}(x - Q_1)^2\right],$$

$$(17.9.3) \qquad z = \log\frac{f(x, Q_1)}{f(x, Q_0)} = \tfrac{1}{2}[(x - Q_0)^2 - (x - Q_1)^2]$$

$$= (Q_1 - Q_0)x + \tfrac{1}{2}(Q_0^2 - Q_1^2).$$

The reader can easily verify that assumptions I to IV of Section 17.3 are here satisfied. We have

$$(17.9.4) \qquad Z_m = \sum_{i=1}^{m} z_i = (Q_1 - Q_0) \sum_{i=1}^{m} x_i + \frac{m}{2}(Q_0^2 - Q_1^2),$$

where the x_i $(i = 1, \ldots, m)$ are independent observations of the random variable x.

We apply the sequential probability ratio test. Let us find the OC function of this test. By (17.5.2) we have

(17.9.5)

$$g[h_0(Q)] = \frac{1}{\sqrt{2\pi}} \int_{-\infty}^{+\infty} \exp\left[-\tfrac{1}{2}(x-Q)^2\right] \left[\frac{\exp\left(-\tfrac{1}{2}(x-Q_1)^2\right)}{\exp\left(-\tfrac{1}{2}(x-Q_0)^2\right)}\right]^{h_0(Q)} dx = 1.$$

After some transformations, we obtain

$$(17.9.6) \quad \frac{1}{\sqrt{2\pi}} \int_{-\infty}^{+\infty} \exp\left\{-\tfrac{1}{2}(x^2 - 2[Q + h_0(Q)(Q_1 - Q_0)]x \right.$$
$$\left. + [Q^2 + h_0(Q)(Q_1^2 - Q_0^2)])\right\} dx = 1.$$

The expression in the braces in the last integral can be written as

$$-\tfrac{1}{2}\{x - [Q + h_0(Q)(Q_1 - Q_0)]\}^2$$
$$- \tfrac{1}{2}h_0(Q)(Q_1 - Q_0)[Q_1 + Q_0 - 2Q - h_0(Q)(Q_1 - Q_0)].$$

Since

$$(17.9.7) \quad \frac{1}{\sqrt{2\pi}} \int_{-\infty}^{+\infty} \exp\left\{-\left(\frac{x - [Q + h_0(Q)(Q_1 - Q_0)]}{2}\right)^2\right\} dx = 1,$$

in order that equation (17.9.6) be satisfied, we must have

$$(17.9.8) \qquad h_0(Q) = \frac{Q_1 + Q_0 - 2Q}{Q_1 - Q_0}.$$

It follows that $h_0(Q)$ defined by (17.9.8) is the solution of equation (17.9.6). Since z satisfies assumptions I to IV, for every Q there exists only one value $h_0(Q)$ satisfying (17.9.5); hence the $h_0(Q)$ given by (17.9.8) is the only solution.

The OC function is obtained from (17.5.6) and (17.9.8). Thus

$$(17.9.9) \qquad L(Q) = \frac{\left(\dfrac{1-\beta}{\alpha}\right)^{h_0(Q)} - 1}{[(1-\beta)/\alpha]^{h_0(Q)} - [\beta/(1-\alpha)]^{h_0(Q)}}.$$

This formula holds for all Q for which $h_0(Q) \neq 0$.

It is easy to see that $h_0(Q) = 0$ for

$$Q' = \tfrac{1}{2}(Q_1 + Q_0).$$

Hence, by (17.5.7), for Q' we have

$$(17.9.10) \qquad L(Q') = \frac{\log \left(\dfrac{1 - \beta}{\alpha}\right)}{\log \left[(1 - \beta)/\alpha\right] - \log \left[\beta/(1 - \alpha)\right]}.$$

Let us now compute $E_Q(n)$, that is, the expected number of observations as a function of the unknown parameter Q.

For $Q \neq Q'$, we have by (17.6.5)

$$(17.9.11) \qquad E_Q(n) = \frac{L(Q) \log B + [1 - L(Q)] \log A}{E_Q(z)}.$$

The OC function, or $L(Q)$, is given by (17.9.9). We still have to find $E_Q(z)$. By (17.9.3), remembering that $E(x) = Q$, we have

$$(17.9.12) \qquad E_Q(z) = (Q_1 - Q_0)Q + \tfrac{1}{2}(Q_0{}^2 - Q_1{}^2).$$

At the point Q' we have $E_{Q'}(z) = 0$, and (17.9.11) cannot be applied. In this case, we find the expected number of observations from the formula

$$(17.9.13) \qquad E_{Q'}(n) = - \frac{\log \left(\dfrac{1 - \beta}{\alpha}\right) \log \left(\dfrac{\beta}{1 - \alpha}\right)}{E_{Q'}(z^2)}.$$

We have to find the denominator. We have

$$E_{Q'}(z^2) = \frac{1}{\sqrt{2\pi}} \int_{-\infty}^{+\infty} \frac{1}{4} [2(Q_1 - Q_0)x + (Q_0{}^2 - Q_1{}^2)]^2$$
$$\times \exp \left\{ -\frac{1}{2}\left[x - \frac{Q_1 + Q_0}{2}\right]^2 \right\} dx.$$

After some simple transformations, we obtain

$$(17.9.14) \qquad E_{Q'}(z^2) = (Q_0 - Q_1)^2.$$

Finally,

$$(17.9.15) \qquad E_{Q'}(n) = - \frac{\log \left(\dfrac{1 - \beta}{\alpha}\right) \log \left(\dfrac{\beta}{1 - \alpha}\right)}{(Q_0 - Q_1)^2}.$$

Let us compute, in particular, $E_{Q_0}(n)$ and $E_{Q_1}(n)$. Since

$$L(Q_0) = 1 - \alpha, \qquad L(Q_1) = \beta,$$

we obtain from (17.9.11) and (17.9.12)

$$(17.9.16) \qquad E_{Q_0}(n) = \frac{(1 - \alpha) \log \dfrac{\beta}{1 - \alpha} + \alpha \log \dfrac{1 - \beta}{\alpha}}{-\frac{1}{2}(Q_1 - Q_0)^2},$$

$$(17.9.17) \qquad E_{Q_1}(n) = \frac{\beta \log \dfrac{\beta}{1 - \alpha} + (1 - \beta) \log \dfrac{1 - \beta}{\alpha}}{\frac{1}{2}(Q_1 - Q_0)^2}.$$

We now express numerically, from the point of view of the expected number of observations, the advantage of applying the sequential probability ratio test to this problem as compared with the most powerful test discussed in Section 16.3.

Suppose that the probabilities α and β are given. We want to test the hypothesis $H_0 (Q = Q_0)$ against the alternative $H_1 (Q = Q_1)$, where Q is the unknown expected value of a population in which x has the normal distribution $N(Q; 1)$ and $Q_1 > Q_0$. Applying a most powerful classical test, we reject H_0 and accept H_1 when the observed value $\bar{x}$ in a simple sample of size n drawn from this population satisfies the relation $\bar{x} \geqslant A$, where A is given by (16.3.18). By (16.3.17), the necessary sample size is

$$n = \frac{(k_\alpha - k_\beta)^2}{(Q_1 - Q_0)^2},$$

where k_α and k_β satisfy the relations

$$P(y \geqslant k_\alpha) = \alpha, \qquad P(y \leqslant k_\beta) = \beta,$$

in which the random variable y has the normal distribution $N(0; 1)$.

Let us denote by λ_0 and λ_1, respectively, the ratios of $E_{Q_0}(n)$ and $E_{Q_1}(n)$ to the sample size n given by (16.3.17). We obtain

$$(17.9.18) \quad \lambda_0 = \frac{E_{Q_0}(n)(Q_1 - Q_0)^2}{(k_\alpha - k_\beta)^2}$$

$$= - \frac{2}{(k_\alpha - k_\beta)^2} \left[(1 - \alpha) \log \frac{\beta}{1 - \alpha} + \alpha \log \frac{1 - \beta}{\alpha} \right],$$

$$(17.9.19) \quad \lambda_1 = \frac{E_{Q_1}(n)(Q_1 - Q_0)^2}{(k_\alpha - k_\beta)^2}$$

$$= \frac{2}{(k_\alpha - k_\beta)^2} \left[\beta \log \frac{\beta}{1 - \alpha} + (1 - \beta) \log \frac{1 - \beta}{\alpha} \right].$$

As we see, λ_0 and λ_1 depend only on α and β, and not on Q_0 and Q_1, where it is only assumed that $Q_1 > Q_0$.

We would have a similar situation if we considered the hypothesis H_0 ($Q = Q_0$) against the alternative H_1 ($Q = Q_1$) for $Q_1 < Q_0$.
Table 17.9.1 gives the values of $100(1 - \lambda_0)$ and Table 17.9.2 the values of $100(1 - \lambda_1)$ for some most frequently applied α and β. Thus these tables give the percentage decrease in sample size when we apply the

| | TABLE 17.9.1 $100(1 - \lambda_0)$ | | | | | | TABLE 17.9.2 $100(1 - \lambda_1)$ | | | | |
| | α | | | | | | α | | | | |
β	0.01	0.02	0.03	0.04	0.05	β	0.01	0.02	0.03	0.04	0.05
0.01	58	54	51	49	47	0.01	58	60	61	62	63
0.02	60	56	53	50	49	0.02	54	56	57	58	59
0.03	61	57	54	51	50	0.03	51	53	54	55	55
0.04	62	58	55	52	50	0.04	49	50	51	52	53
0.05	63	59	55	53	51	0.05	47	49	50	50	51

sequential probability ratio test instead of a classical most powerful test. One can see from these tables that the average decrease is considerable; it amounts to about 50%.

17.10 ADDITIONAL REMARKS

A   In conclusion, we present some information concerning important results obtained in sequential analysis.

Stein [2] showed that for the sequential probability ratio test n has moments of arbitrary order.

Wald and Wolfiwitz [4] proved the following important theorem.

Theorem 17.10.1. *Let S_0 be the sequential probability ratio test for testing the simple hypothesis H_0 against the alternative H_1 and let S_1 be any other test for testing the same hypothesis. Let $\alpha_i(S_j)$ and $E_i{}^j(n)$ $(i, j = 0, 1)$ denote, respectively, the probability of rejecting the hypothesis H_i and the expected sample size, provided hypothesis H_i is true and the test S_j is applied. Suppose $E_i{}^1(n) < \infty$. Then the inequalities*

$$\alpha_i(S_1) \leqslant \alpha_i(S_0) \qquad\qquad (i = 0, 1)$$

imply

$$E_i{}^0(n) \leqslant E_i{}^1(n).$$

From this theorem it follows, in particular, that the average decrease in sample size for the sequential probability ratio test, as compared with tests based on samples of fixed size, established by us in the previous section, also applies to all other sequential tests; hence it is an optimal property which characterizes the sequential probability ratio test.

For the problem treated in Section 17.8, various modifications of the sequential probability ratio test have been designed, aimed at ensuring a similar optimal property, when the true value of Q is contained between Q_0 and Q_1. (See Armitage [2], Kiefer and Weiss [1], and Anderson [2].)

In our discussion of the theory of the sequential probability ratio test, we restricted our considerations to simple hypotheses. The sequential probability ratio test can also be applied to testing composite hypotheses. The reader will find the corresponding information in the book by Wald [4] and in the paper of Girshick [1].

Dvoretzky, Kiefer, and Wolfowitz [2] showed that the theory of the sequential probability ratio test can be extended to apply to observations of a stochastic process with independent and homogeneous increments; hence, in particular, to observations of a Poisson process or of a Wiener process. In this case, the time of observation is a random variable that corresponds to the sample size. For a hypothesis concerning the expected value of a Poisson process, Kiefer and Wolfowitz [1] prepared detailed tables for the application of the sequential probability ratio test.

Sequential methods can also be applied to problems of estimation, as was noticed by Wald [4]. Wolfowitz [2] generalized the Rao-Cramér inequality to the case when the number of observations is determined by a sequential procedure, and found the conditions under which we can replace n by $E(n)$ in (13.5.1) and (13.5.1'). Blackwell and Girshick [2] found the conditions under which one can replace $\geqslant$ by $=$ in the inequalities generalized in this way. Seth [1] continued the investigations of Blackwell and Girshick. The paper of Leimbacher [1] is devoted to the problem of sequential procedure for confidence intervals.

Sequential analysis has not only created possibilities for better solutions of many problems but has also enriched the ideas of mathematical statistics.

B Wald [6] also created a new general theory, called the *theory of decision functions*, which includes as particular cases Neyman-Pearson's classical theory of testing statistical hypotheses and sequential analysis. The main ideas are the following: Denote by $\mathscr{G}$ the set (finite or infinite) of admissible distribution functions $F(x)$ of a certain random variable X, and by g a subset of $\mathscr{G}$. The possible hypotheses are H_g, that the unknown distribution function belongs to g. A hypothesis so formulated may be either nonparametric or parametric. The set of possible decisions, which in Neyman-Pearson's theory contains two elements, the decision to accept the tested hypothesis and the decision to reject it and accept the alternative, and which was extended in sequential analysis, is extended still further in the new theory. This set includes the possible final decisions concerning the validity of one of the hypotheses H_g (there may be an infinite number of them) as well as the decision to continue the investigation. The number of the last type of decisions may be large too, when we perform many experiments at the same time, and we are to make a decision concerning the number of experiments to be continued. Finally, the notions of error of the first and second kind are generalized in this theory by the introduction of functions of the cost of investigation and of the economic loss caused by making a wrong decision.

In the theory of decision functions, new methods for solving problems of testing statistical hypotheses have been created.

Problems and Complements

17.1. Let $\{z_i\}$ ($i = 1, 2, \ldots$) be a sequence of independent, and identically distributed random variables. Then the assertion of theorem 17.3.1 holds if and only if $P(z_i = 0) < 1$. Give a proof of this theorem: (a) using theorem 17.3.1, (b) directly.

17.2. Show that for the validity of formula (17.6.2) it is sufficient to assume that the z_i ($i = 1, 2, 3, \ldots$) are independent, $E(n)$ exists, and that for every Q the expected value $E_Q(z_i)$ is the same for all i (Kolmogorov and Prohorov [1]), (See also Blackwell [1]).
Hint: Use Problem 5.23.

17.3. The defectiveness $p(0 < p < 1)$ of a lot is unknown. We test the hypothesis $H_0 (p = p_0)$ against the alternative $H_1 (p = p_1)$, where $p_0 < p_1$. We apply the following sequential test: Let n_0 be a fixed integer. We independently draw successive elements. If the mth element ($m \leqslant n_0$) is defective, we reject H_0; if all n_0 elements are good, we accept H_0. (a) Find the OC function of this test. (b) Let us denote by n the number of observations which leads to either of the decisions. Find $E(n)$. (c). Put $p_0 = 0.05$, $p_1 = 0.1$, $n_0 = 20$. Compute α, β, and $E(n)$; (d). Compare the power of this test with the power of the sequential probability ratio test in Section 17.8.

17.4. (Continued). Modify the test of Problem 17.3 in the following way: After drawing the mth ($m \leqslant n_0 - 1$) element, continue the investigation if at most one of the drawn elements is defective and reject H_0 if more that one is defective. After drawing n_0 elements, accept H_0 if at most one element is defective and reject H_0 if more than one is defective. Answer all questions of Problem 17.3.

17.5. Represent the sequential probability ratio test procedure in the form of a Markov chain.

17.6. The random variable x has the normal distribution $N(Q; \sigma)$, where Q is known but σ is unknown. We test the hypothesis $H_0 (\sigma = \sigma_0)$ against the alternative $H_1 (\sigma = \sigma_1)$, where $\sigma_0 < \sigma_1$. Let

$$U_m = \sum_{i=1}^{m} (x_i - Q)^2,$$

$$a_m = \left(\frac{1}{\sigma_0^2} - \frac{1}{\sigma_1^2}\right)^{-1} \left(2 \log \frac{\beta}{1 - \alpha} + m \log \frac{\sigma_1^2}{\sigma_0^2}\right),$$

$$r_m = \left(\frac{1}{\sigma_0^2} - \frac{1}{\sigma_1^2}\right)^{-1} \left(2 \log \frac{1 - \beta}{\alpha} + m \log \frac{\sigma_1^2}{\sigma_0^2}\right).$$

(a) Show that if we use the sequential probability ratio test, we should accept H_0, reject H_0, or continue the investigation, according to whether $U_m \leqslant a_m$, $U_m \geqslant r_m$, or $a_m < U_m < r_m$ (Wald [4]).

(b) Find $E(n)$ and the OC function of this test.

17.7. (Continued). Show that if the parameter Q is also unknown, then the sequential test described should be modified by replacing U_m by $\sum_{i=1}^{m} (x_i - \bar{x})^2$, where $\bar{x} = \dfrac{x_1 + \ldots + x_m}{m}$, a_m by a_{m-1}, and r_m by r_{m-1} (Girshick, Stein).

Supplement

The Supplement contains a short outline of certain aspects of measure theory needed for the understanding of some concepts discussed in this book. For the general measure theoretic background needed for the foundations of probability theory, the reader is referred to Kolmogorov [7], Doob [5], Halmos [2] and Loève [4].

Proofs of theorems stated in the Supplement are omitted. They can be found in the quoted monographs.

$$* \quad * \quad *$$

Let E be the basic set (in probability theory, the set of elementary events). By sets we mean subsets of E.

Definition S1. Let A and B be sets. The set

$$A \, \Delta \, B = (A - B) \cup (B - A)$$

is called the *symmetric difference* of A and B.

Definition S2. A *ring* (or *Boolean ring*) of sets is a nonempty class R of sets such that the relations

(S1) $\qquad\qquad A \in R \quad \text{and} \quad B \in R$

imply

(S2) $\qquad (A \cup B) \in R \quad \text{and} \quad (A - B) \in R.$

It follows from definition S1 that if relations (S1) hold, then $(A \, \Delta \, B) \in R$. Since $A \cap B = A \cup B - (A \, \Delta \, B)$, relations (S1) imply

(S3) $\qquad\qquad (A \cap B) \in R.$

It follows by mathematical induction that if R is a ring of sets, for any finite n, the relations

(S1') $\qquad\qquad A_i \in R \quad (i = 1, 2, \ldots, n)$

imply

(S2') $\qquad\qquad \left(\bigcup_{i=1}^{n} A_i \right) \in R.$

Example S1. Take as E the real line and as R the class of all finite unions of bounded semiclosed intervals of the form $[a, b)$. Then R is a ring of sets.

612

Definition S3. A ring R of sets containing the basic set E is called an *algebra* (a *Boolean algebra*).

If R is an algebra of sets, then $A \in R$ implies $\bar{A} \in R$. This follows from $E \in R$ and $\bar{A} = (E - A) \in R$.

Definition S4. A σ-*ring* of sets is a nonempty class of sets such that the relations

(S4) $$A_i \in R \qquad (i = 1, 2, 3, \ldots)$$

imply

(S5) $$\left(\bigcup_{i=1}^{\infty} A_i \right) \in R,$$

(S6) $$(A_{i_1} - A_{i_2}) \in R.$$

Definition S5. A σ-ring of sets containing the basic set E is called a σ-*algebra* (or *Borel field*) of sets.

Definition S5 is equivalent to the following: A σ-*algebra* (*Borel field*) R of sets is a nonempty class of sets such that (S4) implies (S5) and $A \in R$ implies $\bar{A} \in R$.

The reader is referred to Problem 1.6.

Definition S6. Let K be some class of sets. The smallest ring (σ-ring) of sets containing the class K is said to be the *ring* (σ-*ring*) *generated by* K. It is denoted by $g(K)$.

It can be proved that for any class of sets K there exists a unique ring (σ-ring) generated by K.

Example S2. Take as E the real line, as K the class of all bounded, semiclosed intervals of the form $[a, b)$. The ring of sets generated by K is that considered in Example S1. The σ-ring generated by K is the class of all Borel sets on the real line. This σ-ring is at the same time a σ-algebra since it contains the entire line as a union of countably many intervals of the form $[a, b)$.

*   *   *

Definition S7. A *set function* is a function whose domain is a class of sets.

A set function μ defined on a class of sets K is called *finitely additive* if for $n = 2, 3, \ldots$ and for any pairwise disjoint sets $A_i \in K$ $(i = 1, \ldots, n)$

(S7) $$\mu\left(\bigcup_{i=1}^{n} A_i \right) = \sum_{i=1}^{n} \mu(A_i);$$

μ is called *countably* (or *completely*) *additive* if for any sequence of pairwise disjoint sets $A_i \in K$ $(i = 1, 2, \ldots)$

(S8) $$\mu\left(\bigcup_{i=1}^{\infty} A_i \right) = \sum_{i=1}^{\infty} \mu(A_i).$$

Definition S8. A *measure* μ is a real-valued, non-negative, and completely additive set function $\mu(A)$ defined on a Borel field $\mathscr{F}$ of sets and satisfying the equality $\mu(0) = 0$; a measure μ is *finite* if $\mu(E) < \infty$; a measure is a *probability measure* if $\mu(E) = 1$.

A measure μ is called *σ-finite* if there exists a sequence of sets $\{A_n\}$ in $\mathscr{F}$ such that $E \subset \left(\bigcup_n A_n \right)$ and $\mu(A_n) < \infty$ for $n = 1, 2, 3, \ldots$.

The triple $(E, \mathscr{F}, \mu)$ is called a *measure space*. If $\mu(E) < \infty$, the measure space is said to be *finite*; if $\mu(E) = 1$, the triple $(E, \mathscr{F}, \mu)$ is called a *probability space*.

The following important theorem, called the *Extension Theorem*, shows that the assumption that the probability measure $P(A)$ is given for any set A in a Borel field of sets is equivalent to the apparently less restrictive assumption that $P(A)$ is given for any set A in a smaller algebra of sets.

Extension Theorem. *Let $\mu(A)$ be a σ-finite measure defined on an algebra R of subsets of E. Then it is always possible to extend the function $\mu(A)$ in a unique way to all sets of $g(R)$ without losing either of its properties (non-negativeness and complete additivity).*

*   *   *

Let S be a bounded set on the real line. Take $E = (a, b)$, where $(a, b) \supset S$. Consequently, $\bar{S} = (a, b) - S$. Denote by U a finite or countable union of intervals such that

(S8) $$S \subset U \subset (a, b).$$

Represent U as a union of disjoint intervals (see Problem 1.5) and take as measure $L(U)$ the sum of their lengths.

Definition S9. The greatest lower bound of $L(U)$ extended over all U satisfying (S8) is called the *outer measure of S*. We denote it by $\overline{L}(S)$. The expression $\underline{L}(S) = b - a - \overline{L}(\bar{S})$ is called the *inner measure of S*. The set S is called *Lebesgue-measurable* (briefly, *L-measurable*) if $\underline{L}(S) = \overline{L}(S)$. We denote the common value by $L(S)$.

Definition S10. Let S be an unbounded set on the real axis x. S is *L-measurable* if for every $x > 0$ the intersection $[-x, x] \cap S$ is *L*-measurable.

It may be shown that the set function $L(S)$ given by definitions S9 and S10 is a measure defined on the Borel field of *L*-measurable sets on the real line such that if S is a finite interval (a, b), then $L(S) = b - a$.

Every Borel set on the real line is *L*-measurable. The converse is not true; there exist *L*-measurable sets which are not Borel sets. However, if a set S is *L*-measurable without being a Borel set, it can be represented in the form $S = S_1 \cup S_2$, where S_1 is a Borel set and S_2 is *L*-measurable with $L(S_2) = 0$.

Definition S11. A measure μ defined on a Borel field $\mathscr{F}$ is said to be *complete* if the conditions $A \in \mathscr{F}$, $B \subset A$, and $\mu(A) = 0$ imply that $B \in \mathscr{F}$. Lebesgue measure is complete.

*   *   *

The distribution function of a random variable of the discrete (continuous) type will itself be called of the discrete (continuous) type. Any distribution function $F(x)$ can be represented in the form

$$F(x) = a_1 F_1(x) + a_2 F_2(x) + a_3 F_3(x) \qquad (a_i \geqslant 0,\ a_1 + a_2 + a_3 = 1),$$

where $F_1(x)$ and $F_2(x)$ are distribution functions of the continuous and discrete type, respectively, whereas $F_3(x)$ is a singular distribution function, that is, $F_3(x)$ is everywhere continuous and its derivative $F'(x) = 0$ almost everywhere (except at points belonging to a set of L-measure equal to 0). If $a_2 = a_3 = 0$, $F(x)$ is of the continuous type; if $a_1 = a_3 = 0$, $F(x)$ is of the discrete type. We sketch an example of a distribution function with $a_1 = a_2 = 0$.

Example S3. Divide the interval $[0, 1]$ into two parts G_0 and T_0 in the following way:

$$G_0 = \left(\frac{1}{3}, \frac{2}{3}\right) \cup \left[\left(\frac{1}{9}, \frac{2}{9}\right) \cup \left(\frac{7}{9}, \frac{8}{9}\right)\right] \cup \left[\left(\frac{1}{27}, \frac{2}{27}\right) \cup \left(\frac{7}{27}, \frac{8}{27}\right) \cup \left(\frac{19}{27}, \frac{20}{27}\right)\right.$$

$$\left. \cup \left(\frac{25}{27}, \frac{26}{27}\right)\right] \cup \ldots \cup \left[\left(\frac{1}{3^n}, \frac{2}{3^n}\right) \cup \left(\frac{7}{3^n}, \frac{8}{3^n}\right) \cup \ldots \cup \left(\frac{3^n - 6}{3^n}, \frac{3^n - 5}{3^n}\right)\right.$$

$$\left. \cup \left(\frac{3^n - 2}{3^n}, \frac{3^n - 1}{3^n}\right)\right] \cup \ldots$$

$$T_0 = [0, 1] - G_0.$$

The sets T_0 and G_0 are called *Cantor sets*. The set T_0 has Lebesgue measure zero.

Denote by g_{kn} $(n = 1, 2, 3, \ldots;\ k = 1, 2, \ldots, 2^{n-1})$ the kth interval with the denominator 3^n. Define

$$F(x) = \begin{cases} 0 & (x \leqslant 0), \\ \dfrac{2k - 1}{2^n} & (x \in g_{kn}), \\ 1 & (x \geqslant 1). \end{cases}$$

Finally, for $x_0 \in T_0$

$$F(x_0) = \sup_{\substack{x < x_0 \\ x \in G_0}} F(x).$$

The function $F(x)$ thus defined is a continuous distribution function whose derivative equals zero for any $x \in G_0$. Consequently, there does not exist any function satisfying relation (2.3.4) for every real x.

It is worthwhile to be noticed that if the random variables X_1 and X_2 are independent and their distribution functions $F_1(x)$ and $F_2(x)$ are singular, the distribution function $F(x)$ of $X_1 + X_2$ may be absolutely continuous. (See Ranga-Rao and Varadarajan [1].)

* * *

Let $(E, \mathscr{F}, \mu)$ be a finite measure space. Let $f(e)$, $e \in E$, be a real function, measurable (see page 29) with respect to $\mathscr{F}$ and bounded: $-\infty < a \leqslant f(e) \infty b < \infty$. Denote by $\mu(c' \leqslant f(e) < c''$, $e \in A)$, where $A \in \mathscr{F}$, the measure of the intersection of A with the inverse image of (c', c'') under the function f.

Definition S12. If the limit of the expression

(S9) $$\lim_{\varepsilon \to 0} \lim_{n \to \infty} \sum_{i=1}^{n} c_i \mu(c_{i-1} \leqslant f < c_i, e \in A),$$

where $a = c_0 < c_1 < \ldots < c_n = b$ and $\max_{1 \leqslant i \leqslant n} (c_i - c_{i-1}) \leqslant \varepsilon$, exists, it is called the *abstract Lebesgue integral of f over A with respect to μ*. It is denoted by

$$\int_A f \, d\mu.$$

If $f(e)$ is bounded the integral exists for any $A \in \mathscr{F}$ and, in particular, for $A = E$. If $f(e)$ is not bounded, we proceed as follows: Denote for any a and b, where $a < b$,

$$f_{ab}(e) = \begin{cases} a & \text{if } f(e) \leqslant a, \\ f(e) & \text{if } a < f(e) < b, \\ b & \text{if } f(e) \geqslant b. \end{cases}$$

Definition S12'. If the limit

(S9') $$\lim_{\substack{a \to -\infty \\ b \to +\infty}} \int_A f_{ab}(e) \, d\mu$$

exists, it is called the *abstract Lebesgue integral of f over A with respect to μ*. We mention some important properties of the Lebesgue integral.

I. If f is integrable on A and $A = \sum_n A_n (A_n \in \mathscr{F}, n = 1, 2, 3, \ldots)$, where the A_n are pairwise disjoint, then

$$\int_A f \, d\mu = \sum_n \int_{A_n} f \, d\mu.$$

II. If f is integrable, $|f|$ is also integrable.

III. If f_1 and f_2 are integrable and c_1 and c_2 are real constants, $c_1 f_1 + c_2 f_2$ is integrable and

$$\int_A (c_1 f_1 + c_2 f_2)\, d\mu = c_1 \int_A f_1\, d\mu + c_2 \int_A f_2\, d\mu.$$

Now take as μ a probability measure P and let f be the characteristic function of the set $A \in \mathscr{F}$, that is,

$$f(e) = \begin{cases} 1 & \text{if } e \in A, \\ 0 & \text{if } e \notin A. \end{cases}$$

Then

$$\int_A f\, dP = P(A).$$

Now let X be a random variable with distribution function $F(x)$ and let S be a Borel set on the real line. Then

$$P^{(x)}(S) = P^{(x)}(X \in S) = \int_S dF(x) = \int_A dP = P(A),$$

where A is the inverse image of S.

A quite analogous formula holds for random vectors in Euclidian n-dimensional spaces.

If[1] $M(X)$ exists, then

$$M(X) = \int_E x\, dP = \int_{-\infty}^{\infty} x\, dF(x) = \int_{-\infty}^{\infty} x\, dP^{(x)}.$$

If $X = (X_1, X_2, \ldots, X_n)$ is a random vector and $g(X_1, \ldots, X_n)$ is a random variable whose mean value exists, then

$$M[g(X_1, \ldots, X_n)] = \int_E g\, dP = \int_{-\infty}^{\infty} \cdots \int_{-\infty}^{\infty} g\, dP^{(x_1, \ldots, x_n)}.$$

$$* \qquad * \qquad *$$

Before formulating the Fubini theorem, we introduce some preliminary notions.

Definition S13. Let $(E_i, \mathscr{F}_i, \mu_i)$ $(i = 1, 2)$ be two measure spaces. The triple $(E, \mathscr{F}, \mu)$ is called the *Cartesian product measure space* of $(E_1, \mathscr{F}_1, \mu_1)$ and $(E_2, \mathscr{F}_2, \mu_2)$ if

I. $E = E_1 \times E_2$ is the set of all pairs $e = (e_1, e_2)$ with $e_i \in E_i$.

II. $\mathscr{F} = \mathscr{F}_1 \times \mathscr{F}_2$ is the Borel field generated by the "rectangles" $A = A_1 \times A_2$ with $A_i \in \mathscr{F}_i$ (in other words, $e \in A$ means that $e = (e_1, e_2)$ with $e_i \in A_i$).

---

[1] To avoid confusion, we use M in the Supplement as an expectation symbol.

III. for any $A = A_1 \times A_2$ with $A_i \in \mathscr{F}_i$, we have $\mu(A) = \mu_1(A_1)\mu_2(A_2)$.

The measure μ is called the *Cartesian product* of the two measures and is denoted by $\mu_1 \times \mu_2$.

Example S4. Let X and Y be two independent random variables. Take the x axis, y axis, and the plane (x, y) as E_1, E_2, and E and the classes of Borel sets $\mathscr{F}_x$, $\mathscr{F}_y$, and $\mathscr{F}_{(x,y)}$ on E_1, E_2, and E as $\mathscr{F}_1$, $\mathscr{F}_2$, and $\mathscr{F}$, respectively. Finally, take the probability measures $P^{(x)}$, $P^{(y)}$, and $P^{(x,y)}$ of the random variables X, Y, and (X, Y) as μ_1, μ_2, and μ, respectively. Then $(E, \mathscr{F}_{(x,y)}, P^{(x,y)})$ is the Cartesian product measure space of $(E_1, \mathscr{F}_x, P^{(x)})$ and $(E_2, \mathscr{F}_y, P^{(y)})$.

Definition S13′.   Let $A \in \mathscr{F}(A = A_1 \times A_2)$. The *section of A at given* $e_1(e_2)$ is the set of all $e_2(e_1)$ such that $(e_1, e_2) \in A$. We denote it by $A_{e_1}(A_{e_2})$.

Let f be a real function defined on E and measurable with respect to $\mathscr{F}$. The *section* $f_{e_1}(f_{e_2})$ *of* f at given $e_1(e_2)$ is the function defined on $E_2(E_1)$ by the relation $f_{e_1}(e_2) = f(e_1, e_2)$ [or $f_{e_2}(e_1) = f(e_1, e_2)$].

Fubini theorem.   Let $(E_i, \mathscr{F}_i, \mu_i)$ be two finite measure spaces. If the function $f(e_1, e_2)$ defined on $E_1 \times E_2$ is measurable with respect to $\mathscr{F}_1 \times \mathscr{F}_2$ and is either non-negative or integrable on $E = E_1 \times E_2$ with respect to the measure $\mu = \mu_1 \times \mu_2$, then

$$(S10) \qquad \int_E f \, d\mu = \int_{E_1} d\mu_1 \int_{E_2} f_{e_1}(e_2) \, d\mu_2 = \int_{E_2} d\mu_2 \int_{E_1} f_{e_2}(e_1) \, d\mu_1.$$

In the integrable case almost every section (except possibly a set of measure μ equal to zero) is integrable.

The integrals on the right side of (S10) should be computed from the right to the left.

If f is the characteristic function of a set $A \in \mathscr{F}$, relation (S10) takes the form

$$(S11) \qquad \mu(A) = \int_{E_1} \mu_2(A_{e_1}) \, d\mu_1 = \int_{E_2} \mu_1(A_{e_2}) \, d\mu_2.$$

The Fubini theorem allows to give simple proofs of formulas (6.5.6) to (6.5.9). We prove (6.5.6).

Let X and Y be independent random variables and let $Z = f(X, Y)$ be a random variable and $P^{(z)}$ the probability measure of Z. Let S be a Borel set on the z axis and let A be the set in the plane (x, y) such that if the point $(x, y) \in A$ then $f(x, y) \in S$. Formula (S11) gives (see Example S4) then, since $P^{(x,y)}(A) = P^{(z)}(S)$,

$$(S12) \qquad P^{(z)}(S) = \int_{-\infty}^{\infty} P^{(y)}(A_x) \, dP^{(x)} = \int_{-\infty}^{\infty} P^{(x)}(A_y) \, dP^{(y)},$$

where $A_x(A_y)$ is the set of all points on the $y(x)$ axis such that for a given $x(y)$ the relation $f(x, y) \in S$ holds.

Denote by $F_1(x)$, $F_2(y)$, and $F(z)$ the distribution functions of X, Y, and Z, respectively. Taking $S = (-\infty, z)$ and $Z = X + Y$, we have

(S13) $P^{(z)}(S) = F(z)$; $P^{(y)}(A_x) = F_2(z - x)$; $P^{(x)}(A_y) = F_1(z - y)$.

We get immediately from (S12) and (S13) formula (6.5.6.) The other three formulas can be obtained in the same way.

Another use of the Fubini theorem is to prove the following converse of theorem 3.6.2:

Let X and Y be independent random variables satisfying the relations

(S14) $P(X = 0) \neq 1$, $P(Y = 0) \neq 1$,

and let $M(XY)$ exist. Then $M(X)$ and $M(Y)$ exist and

(S15) $M(XY) = M(X)M(Y)$.

For the proof take f in (S10) to be equal XY. The section of f at a given x is xy with constant x. Consequently, taking into account (S14), formula (S10) yields

$$M(XY) = \int_{-\infty}^{\infty}\int_{-\infty}^{\infty} xy \, dP^{(x,y)} = \int_{-\infty}^{\infty} x \int_{-\infty}^{\infty} y \, dP^{(y)} \, dP^{(x)} = M(X)M(Y).$$

$$* \qquad * \qquad *$$

The existence and uniqueness of conditional expected values and conditional probabilities treated in Section 3.6,C follow easily from the Radon-Nikodym theorem.

Let $(E, \mathscr{F}, \mu)$ be a measure space and let $V(A)$ be a countably additive set function defined for $A \in \mathscr{F}$, and let $V(A)$ assume at most one of the values $-\infty$ or $+\infty$.

The set function V is said to be *absolutely continuous with respect to μ* (we denote $V \ll \mu$) if for every $A \in \mathscr{F}$ the equality $\mu(A) = 0$ implies $V(A) = 0$.

If V is finite, non-negative, and $V \ll \mu$, then for every $\varepsilon > 0$ there exists a $\delta > 0$ such that the relation $\mu(A) < \delta$ implies $V(A) < \varepsilon$.

Radon-Nikodym theorem. *Let $(E, \mathscr{F}, P)$ be a probability space and let $Q(A)$ $(A \in \mathscr{F})$ be a completely additive, finite set function. If $Q \ll P$, there exists a random variable $f(e)$ such that for every $A \in \mathscr{F}$*

(S16) $Q(A) = \int_A f(e) \, dP$.

The random variable $f(e)$ is unique in the sense that if, for some other random variable $g(e)$, for every $A \in \mathscr{F}$

$$Q(A) = \int_A g(e) \, dP,$$

then $P[f(e) = g(e)] = 1$.

Let (X, Y) be a random vector and let $M(Y)$ exist.

Definition S14. The random variable $M(Y \mid x)$ satisfying for any Borel set S on the x axis with $P^{(x)}(X \in S) > 0$ the relation

(S17) $$M(Y \mid X \in S) = M[M(Y \mid x) \mid X \in S]$$

is called the *conditional expected value of Y given x*.

We now have to prove the existence and uniqueness (in the sense formulated in the Radon-Nikodym theorem) of $M(Y \mid x)$.

It is convenient now to denote by $\{X \in S\}$ the inverse image of the Borel set S under the random variable X. Multiplying both sides of (S17) by $P(X \in S)$, we obtain

(S18) $$\int_{\{X \in S\}} y \, dP = \int_{\{X \in S\}} M(Y \mid x) \, dP = \int_S M(Y \mid x) \, dP^{(x)}.$$

Let

$$Q(S) = \int_{\{X \in S\}} y \, dP.$$

The function $Q(S)$ is evidently countably additive. In addition, $Q(S)$ is absolutely continuous with respect to the measure $P^{(x)}$, since $P^{(x)}(S) = 0$ is equivalent to $P(X \in S) = 0$, and the last equality implies $Q(S) = 0$.

We see that the assumptions of the Radon-Nikodym theorem, with E the x axis, $\mathscr{F}$ the class of Borel sets on this axis, and $P^{(x)}$ the measure, are satisfied. Consequently the assertions of the Radon-Nikodym theorem are applicable to $M(Y \mid x)$.

References

We use the following abbreviations for the journals most frequently referred to:

AMS   *Annals of Mathematical Statistics*
B   *Biometrika*
CR   *Comptes Rendus de l'Académie des Sciences, Paris*
DAN   *Doklady Akademii Nauk, USSR*
JRSS   *Journal of the Royal Statistical Society*
PCPS   *Proceedings of the Cambridge Philosophical Society*
SA   *Skandinavisk Aktuarietidskrift*
TVP   *Teoria Veroyatnostey i ee Primenenyia*
TAMS   *Transactions of the American Mathematical Society*
UMN   *Uspehi Matematitcheskih Nauk*

A. C. Aitken

[1] *Determinants and matrices*, University Mathematical Texts, 1, Oliver and Boyd, Edinburgh and London, 1944, third edition.

T. W. Anderson

[1] *Introduction to multivariate statistical analysis*, John Wiley, New York–London, 1958.

[2] A modification of the sequential probability ratio test to reduce the sample size, *AMS* **31**, 165 (1960).

[3] Least squares and best unbiased estimates, *AMS* **33**, 266 (1962).

T. W. Anderson, D. A. Darling

[1] Asymptotic theory of certain "goodness of fit" criteria based on stochastic processes, *AMS* **23**, 193 (1952).

T. W. Anderson, L. A. Goodman

[1] Statistical inference about chains, *AMS* **28**, 89 (1957).

F. J. Anscombe

[1] Large sample theory of sequential estimation, *PCPS*, **48**, 600 (1952).

N. Arley

[1] *On the theory of stochastic processes and their application to the theory of cosmic radiation*, John Wiley, New York, 1948.

P. Armitage

[1] A comparison of stratified and unrestricted random sampling from a finite population, *B* **34**, 273 (1947).

[2] Restricted sequential procedures, *B* **44**, 9 (1957).

D. G. Austin

[1] On the existence of the derivative of Markoff transition probability functions, *Proc. Nat. Acad. Sci. USA* **41**, 224 (1955).

[2] Note on differentiating Markoff transition functions with stable terminal states, *Duke Math. J.* **25**, 4, 625 (1958).

V. I. Babkin, P. F. Belayev, Yu I. Maximov

[1] Some remarks to the paper of V. L. Gontcharov "From the domain of combinatorics" (in Russian), *TVP* **4**, 445 (1959).

L. Bachelier

[1] Théorie de la spéculation, *Ann. École Normale Sup. Paris* **17**, 21 (1900).

E. W. Barankin

[1] Extension of a theorem of Blackwell, *AMS* **21**, 280 (1950).

[2] Toward an objectivistic theory of probability, *Proc. Third Berkeley Symp.* **5**, 21 (1956).

E. W. Barankin, M. Katz, Jr.

[1] Sufficient statistics of minimal dimension, *Sankhya* **21**, 217 (1959).

G. A. Barnard

[1] On the Fisher-Behrens test, *B* **37**, 203 (1950).

W. Bartky

[1] Multiple sampling with constant probability, *AMS* **14**, 363 (1943).

M. S. Bartlett

[1] On the theory of statistical regression, *Proc. Royal Soc. Edinburgh* **53**, 260 (1933).

D. E. Barton, F. N. David

[1] Non-randomness in a sequence of two alternatives. II. Run tests, *B* **45**, 253 (1958).

D. E. Barton, F. N. David, C. L. Mallows

[1] Non-randomness in a sequence of two alternatives, I. Wilcoxon's and allied tests statistics, *B* **45**, 166 (1958).

R. Bartoszyński

[1] Some remarks on the convergence of stochastic processes, *Studia Math.* **17**, 313 (1958).

[2] A characterization of the weak convergence of measures, *AMS* **32**, 561 (1961).

D. Basu, R. G. Laha

[1] On some characterizations of the normal distribution, *Sankhya*, **13**, 359 (1954).

G. Baxter, M. D. Donsker

[1] On the distribution of the supremum functional for processes with stationary, independent increments, *TAMS* **85**, 73 (1957).

T. Bażańska

[1] Zastosowanie metod statystyki matematycznej do obliczenia mocy heparyny (unpublished).

Yu. K. Belayev

[1] On non-boundedness of sample functions of a Gaussian process (in Russian), *TVP* **3**, 351 (1958).

[2] Analytic random processes (in Russian), *TVP* **4**, 437 (1959).
[3] Local properties of sample functions of stationary Gaussian processes (in Russian), *TVP* **5**, 128 (1960).

C. B. Bell

[1] On the structure of distribution-free statistics, *AMS* **31**, 703 (1960).

P. O. Berge

[1] A note on a form of Thebycheff's theorem for two variables, *B* **29**, 405 (1957).

S. N. Bernstein

[1] An axiomatic foundation of the theory of probability (in Russian), *Soobschtchenya Khark. Mat. Obschestva* **15** (2), 209 (1917).
[2] Sur les sommes des quantités dépendantes, *Izvestia Acad. Nauk* **20** (6) 1459 (1926).
[3] Sur l'éxtension du théorème limit du calcul des probabilités aux sommes des quantités dépendantes, *Math. Annalen* **97**, 1 (1926).
[4] On some property which characterizes the Gauss' distribution (in Russian), *Trudy Leningr. Polytech. Inst.* **3**, 21 (1941).

A. C. Berry

[1] The accuracy of the Gaussian approximation to the sum of independent variates, *TAMS* **49**, 122 (1941).

H. J. Bhabha, W. Heitler,

[1] The passage of fast electrons and the theory of cosmic showers, *Proc. Royal Soc. A* **159**, 432 (1937).

A. T. Bharucha-Reid

[1] Note on estimation of the number of states in a discrete Markov chain, *Experientia* **12**, 176 (1956).
[2] *Elements of the theory of Markov processes and their applications*, McGraw-Hill, New York–Toronto–London, 1960.

P. Billingsley

[1] The invariance principle for dependent random variables, *TAMS* **83**, 250 (1956).
[2] Statistical methods in Markov chains, *AMS* **32**, 13 (1961).
[3] The Lindeberg-Lévy theorem for martingales, *Proc. Amer. Math. Soc.* **12**, 788 (1961).

G. D. Birkhoff

[1] Proof of the ergodic theorem, *Proc. Nat. Acad. Sci. USA.* **17**, 656 (1931).

A. Birnbaum

[1] A unified theory of estimation, I, *AMS* **32**, 112 (1962).
[2] On the foundations of statistical inference (with discussion), *J. Amer. Stat. Assn.* **57**, 269 (1962).

Z. W. Birnbaum

[1] Numerical tabulation of the distribution of Kolmogorov's statistic for finite sample size, *J. Amer. Stat. Assn.* **47**, 425 (1952).
[2] Distribution-free tests of fit for continuous distribution functions, *AMS* **24**, 1 (1953).
[3] On the power of a one sided test of fit for continuous probability functions, *AMS* **24**, 484 (1953).
[4] On a use of the Mann-Whitney statistic, *Proc. Third Berkeley Symp.* **1**, 13 (1955).

Z. W. Birnbaum, R. A. Hall

[1] Small sample distributions for multi-sample statistics of the Smirnov type, *AMS* **31**, 710 (1960).

Z. W. Birnbaum, A. W. Marshall

[1] Some multivariate Chebyshev inequalities with extension to continuous parameter processes, *AMS* **32**, 687 (1961).

Z. W. Birnbaum, R. C. McCarty

[1] A distribution-free upper confidence bound for $Pr\{Y < X\}$ based on independent samples of X and Y, *AMS* **29**, 558 (1958).

Z. W. Birnbaum, R. Pyke

[1] On some distributions related to the statistic D_n^+, *AMS* **29**, 179 (1958).

Z. W. Birnbaum, H. Rubin

[1] On distribution-free statistics, *AMS* **25**, 593 (1954).

Z. W. Birnbaum, F. H. Tingey

[1] One-sided confidence contours for probability distribution functions, *AMS* **22**, 592 (1951).

D. Blackwell

[1] On an equation of Wald, *AMS* **17**, 84 (1946).

[2] Conditional expectation and unbiased sequential estimation, *AMS* **18**, 105 (1947).

D. Blackwell, M. A. Girshick

[1] On functions of sequences of independent chance vectors with applications to the problem of the "random walk" in k dimensions, *AMS* **17**, 310 (1946).

[2] A lower bound for the variance of some unbiased sequential estimates, *AMS* **18**, 277 (1947).

A. Blanc-Lapierre, R. Fortet

[1] *Théorie des fonctions aléatoires*, Masson, Paris, 1953.

G. Blom

[1] A generalization of Wald's fundamental identity, *AMS* **20**, 439 (1949).

J. R. Blum

[1] On the convergence of empiric distribution functions, *AMS* **26**, 527 (1955).

R. M. Blumenthal

[1] An extended Markov property *TAMS* **85**, 52 (1957).

R. M. Blumenthal, R. K. Getoor

[1] Some theorems on stable processes, *TAMS* **95**, 263 (1960).

[2] A dimension theorem for sample functions of stable processes, *Ill. J. Math.* **4**, 370 (1960).

A. A. Bobrov

[1] Conditions of applicability of the strong law of large numbers, *Duke Math. J.* **12**, 43 (1945).

S. Bochner

[1] *Vorlesungen über Fouriersche Integrale*, Akademische Verlagsgesellschaft, Leipzig, 1932.

[2] Monotone Funktionen, Stieltjessche Integrale und harmonische Analyse, *Math. Annalen* **108**, 378 (1933).

E. Borel

[1] Les probabilités dénombrables et leurs applications arithmétiques, *Rend. Circ. Mat. Palermo*, **27**, 247 (1909).

[2] Sur un probléme de probabilités relatif aux fractions continues, *Math. Annalen* **72**, 578 (1912).

L. Bortkiewicz

[1] *Das Gesetz der kleinen Zahlen*, Teubner, Leipzig, 1898.

[2] *Die Iterationen, Ein Beitrag zur Wahrscheinlichkeitsrechnung*, Springer, Berlin, 1917.

A. H. Bowker

[1] Tolerance limits for normal distributions, *Techniques of Statistical Analysis, Statistical Research Group, Columbia University, New York*, 1947, p. 95.

G. P. Boyev

[1] *Probability theory* (in Russian), GITTL, Moscow, 1950.

L. Breiman

[1] A counterexample to a theorem of Kolmogorov, *AMS* **28**, 811 (1957).

[2] The strong law of large numbers for a class of Markov chains, *AMS* **31**, 801 (1960).

H. D. Brunk

[1] The strong law of great numbers, *Duke Math. J.* **1**, 181 (1948).

F. P. Cantelli

[1] Sulla probabilita come limita della frequenza, *Rend. Accad. Lincei*, **26** (1), 39 (1917).

L. C. Chang

[1] On the ratio of an empirical distribution function to the theoretical distribution function, *Acta Math. Sinica* **5**, 347 (1955).

L. C. Chang, M. Fisz

[1] Asymptotically independent linear functions of empirical distribution functions, *Sci. Record* **1**, 335 (1957).

[2] Exact distributions of the maximal values of some functions of empirical distribution functions, *Sci. Record* **1**, 341 (1957).

D. G. Chapman

[1] Some two-sample tests, *AMS* **21**, 601 (1950).

[2] Estimating the parameters of a truncated gamma distribution, *AMS* **27**, 498 (1956).

[3] A comparative study of several one-sided goodness-of-fit tests, *AMS* **29**, 655 (1958).

P. L. Chebyshev

[1] On mean values (in Russian), *Collected works, Izd Akad. Nauk. USSR*, 2nd Volume Moscow, 1948.

H. Chernoff, E. L. Lehmann

[1] The use of the maximum likelihood estimates in χ^2 test for goodness of fit, *AMS* **25**, 579 (1954).

Y. S. Chow

[1] Martingales in a σ-finite measure space indexed by directed sets, *TAMS* **97**, 254 (1960).

K. L. Chung

[1] The strong law of large numbers, *Proc. Second Berkeley Symp.*, 341 (1951).
[2] Contributions to the theory of Markov chains I, *J. Res. Nat. Bureau Standards* **50**, 203 (1953).
[3] Contributions to the theory of Markov chains, *TAMS* **76**, 397 (1954).
[4] Some new developments in Markov chains, *TAMS* **81**, 195 (1956).
[5] *Markov chains with stationary transition probabilities*, Springer, Berlin–Göttingen–Heidelberg, 1960.

K. L. Chung, W. Feller

[1] On fluctuations of coin tossing, *Proc. Nat. Acad. Sci.* **35**, 605 (1949).

Z. Ciesielski

[1] Hölder conditions for realizations of Gaussian processes, *TAMS* **99**, 403 (1961).

W. G. Cochran

[1] Relative accuracy of systematic and stratified samples for a certain class of populations, *AMS* **17**, 164 (1946).
[2] The χ^2-test of goodness of fit, *AMS* **23**, 315 (1952).

R. Cogburn

[1] Lois limites des termes variationelles des sommes normées, *CR* **246**, 3408 (1958).

R. Courant, D. Hilbert

[1] *Die Methoden der Mathematischen Physik*, Springer, Göttingen, 1930.

H. Cramér

[1] *Random variables and probability distributions*, University Press, Cambridge, 1937.
[2] *Mathematical methods in statistics*, Princeton Univ. Press, Princeton, 1946.
[3] On the composition of elementary errors, *SA* **11**, 13; 141 (1928).
[4] Über eine Eigenschaft der normalen Verteilungsfunktion, *Math. Zeitschrift* **41**, 405 (1936).
[5] On the representation of a function by certain Fourier integrals, *TAMS* **46**, 191 (1939).
[6] A contribution to the theory of stochastic processes, *Proc. Second Berkeley Symp.*, 329 (1951).

H. Cramér, H. Wold

[1] Some theorems on distribution functions, *J. London Math. Soc.* **11**, 290 (1936).

E. Czuber

[1] *Die philosophischen Grundlagen der Wahrscheinlichkeitstheorie*, Teubner, Leipzig–Berlin, 1923.

T. Dalenius

[1] The multivariate sampling problem, *SA*, 92 (1953).
[2] *Sampling in Sweden*, Almquist and Wiksell, Stockholm, 1957.

H. E. Daniels

[1] Some problems of statistical interest in wool research, *Supplement JRSS*, **5**, 89 (1938).
[2] The statistical theory of the strength of bundles of threads, I, *Proc. Royal Society*, ser. *A* **189**, 405 (1945).

D. van Dantzig

[1] On the consistency and the power of Wilcoxon's two sample test, *Proc. Nederl. Akad. Wet. Amsterdam, ser. A* **54,** 1 (1951).

G. B. Dantzig

[1] On the non-existence of tests of "Student's" hypothesis having power functions independent of σ, *AMS* **11,** 186 (1940).

G. B. Dantzig, A. Wald

[1] On the fundamental lemma of Neyman and Pearson, *AMS* **22,** 87 (1951).

D. A. Darling

[1] The Kolmogorov-Smirnov, Cramér-von Mises tests, *AMS* **28,** 823 (1957).

G. Darmois

[1] Sur les lois des probabilités á estimation exhaustive, *CR* **200,** 1265 (1935).
[2] Sur une propriété caractéristique de la loi de probabilité de Laplace, *CR* **232,** 1999 (1951).

F. N. David

[1] *Tables of the correlation coefficient*, London, 1938.

H. T. David

[1] A three-sample Kolmogorov-Smirnov test, *AMS* **29,** 842 (1958).

F. N. David, J. Neyman

[1] Extension of the Markoff theorem on least squares, *Stat. Res. Mem.* **2,** 105 (1938).

O. L. Davies, E. S. Pearson

[1] Methods of estimating from samples the population standard deviation, *Supplement JRSS*, **1,** 76 (1934).

W. E. Deming

[1] *Some theory of sampling*, John Wiley, New York, 1950.

A. P. Dempster

[1] Generalized D_n^+ statistics, *AMS* **30,** 593 (1959).

C. Derman, S. Litauer, H. Solomon

[1] Tightened multi-level continuous sampling plans, *AMS* **28,** 395 (1957).

R. L. Dobrushin

[1] On the regularity conditions of homogeneous in time Markov processes with a countable number of possible states (in Russian), *UMN* **7,** No. 6, 185 (1952).
[2] Conditions of regularity of Markov processes with a finite number of possible states (in Russian), *Mathematicheski Sbornik* **34,** (76), 541 (1954).
[3] A central limit theorem for non-homogeneous Markov chains (in Russian), *DAN* **102,** 5 (1955).
[4] A lemma on the limit of a superposition of random functions (in Russian), *UMN* **10,** No. 2, 157 (1955).
[5] Conditions for continuity of sample functions of a martingale (in Russian), *TVP* **3,** 97 (1958).
[6] The properties of sample functions of a stationary Gaussian process (in Russian), *TVP* **5,** 132 (1960).

628   REFERENCES

H. F. Dodge, H. G. Romig

[1] A method of sampling inspection, *Bell system Tech. J.*, **8**, 613 (1929).

W. Doeblin

[1] Sur les propriétés asymptotiques de mouvments régis par certains types des châines simples, *Bull. Math. Soc. Roum. Sci.*, **39** (2), 3 (1937).

[2] Sur deux problèmes de M. Kolmogoroff concernant les châines dènombrables, *Bull. Math. Soc. Math. France*, **66**, 210 (1938).

[3] Sur certains mouvements aléatoires discontinus, *SA* **22**, 21 (1939).

[4] Éléments d'une théorie générale des châines simples constantes de Markoff, *Ann. École Normale Sup. Paris* (3), **57**, 61 (1940).

M. D. Donsker

[1] An invariance principle for certain probability limit theorems, *Memoirs Amer. Math. Soc.*, **6**, 1 (1951).

[2] Justification and extension of Doob's heuristic approach to the Kolmogorov-Smirnov limit theorems, *AMS* **23**, 277 (1952).

J. L. Doob

[1] Stochastic processes depending on a continuous parameter, *TAMS* **42**, 107 (1937).

[2] Topics in the theory of Markoff chains, *TAMS* **52**, 37 (1942).

[3] Markoff chains-denumerable case, *TAMS* **58**, 455 (1945).

[4] Heuristic approach to the Kolmogorov-Smirnov theorems, *AMS* **20**, 393 (1949).

[5] *Stochastic Processes*, John Wiley, New York–London, 1953.

[6] Probability and statistics, *TAMS* **36**, 759 (1934).

[7] Regularity properties of certain families of chance variables, *TAMS* **47**, 455 (1940).

[8] Application of the theory of martingales, *Le Calcul des Probabilitiés et ses applications, Coll. Intern. Center Nat. Rep. Sci., Paris*, **23** (1949).

[9] Continuous parameter martingales, *Proc. Second Berkeley Symp.* 269 (1951).

[10] Discrete potential theory and boundaries, *J. Math. Mech.* **8**, 433 (1959).

[11] Notes on martingale theory, *Proc. Fourth Berkeley Symp.*, **2**, 95 (1961).

D. Dugué

[1] Application des propriétés de la limite au sens du calcul des probabilités à l'etude des diverses questions d'estimation, *J. École Polytechnique*, 305, (1937).

[2] *Traité de statistique théorique et appliquée, Analyse aléatoire-algebre aléatoire*, Masson, Paris, 1958.

A. Dvoretzky

[1] On the strong stability of a sequence of events, *AMS* **20**, 296 (1949).

A. Dvoretzky, P. Erdös, S. Kakutani

[1] Nonincrease everywhere of the Brownian Motion Process, *Proc. Fourth Berkeley Symp.*, **2**, 103 (1961).

A. Dvoretzky, J. Kiefer, J. Wolfowitz

[1] Sequential decision problems for processes with continuous time parameter. Testing hypotheses, *AMS* **24**, 254 (1953).

[2] Asymptotic minimax character of the sample distribution function and of the classical multinomial estimator, *AMS* **27**, 642 (1956).

A. Dvoretzky, J. Wolfowitz

[1] Sums of random integers reduced modulo *m*, *Duke Math. J.* **18**, 501 (1951).

M. Dwass

[1] The distribution of a generalized D_n^+ statistic, *AMS* **30**, 1024 (1959).

[2] Some k-sample rank order tests, *Contribution to probability and statistics, Essays in honor of Harold Hotelling*, Stanford Univ. Press, 504 (1960).

E. B. Dynkin

[1] Criteria of continuity and absence of second kind discontinuities for trajectories of Markov random processes (in Russian), *Izvestia Acad. Nauk USSR* **16**, 563 (1952).

[2] On some limit theorems for Markov chains (in Russian), *Ukr. Math. J.* **6**, 21 (1954).

[3] *Foundations of the theory of Markov processes* (in Russian), Gos. Izd. Fiz-Mat. Lit., Moscow, 1959.

A. Ehrenfeucht, M. Fisz

[1] A necessary and sufficient condition for the validity of the weak law of large numbers, *Bull. Pol. Acad. Sci.* **8**, 583 (1960).

A. Einstein

[1] Zur Theorie der Brownschen Bewegung, *Ann. Phys.* **IV**, 19, 371 (1906).

P. and T. Ehrenfest

[1] Über zwei bekannte Einwände gagen das Boltzmannsche H-theorem, *Physikalische Zeitschrift* **8**, 311, (1907).

G. Elfwing

[1] The asymptotical distribution of range in samples from a normal population, *B* 35 111 (1947).

P. Erdös

[1] On the law of the Iterated Logarithm, *Annals Math.* 2, **43**, 419 (1942).

P. Erdös, M. Kac

[1] On certain limit theorems of the theory of probability, *Bull. Amer. Math. Soc.* **53**, 292 (1946).

[2] On the number of positive sums of independent random variables, *Bull. Amer. Math. Soc.* **53**, 1011 (1947).

P. Erdös, A. Rényi

[1] On the central limit theorem for samples from a finite population, *Matem. Kutato Intezet Közlem.*, **4**, 49 (1959).

[2] On Cantor's series with convergent $\Sigma \dfrac{1}{q_n}$, *Ann. Univ. Sci. Budapestinensis, Sectio Math.*, **2**, 93 (1959).

A. K. Erlang

[1] *The life and work of*, The Copenhagen Telephone Co., 1948.

C. G. Esseen

[1] Fourier analysis of distribution functions. A mathematical study of the Laplace-Gaussian law, *Acta Math.* **77**, 1 (1945).

[2] A moment inequality with an application to the central limit theorem, *SA* **39**, 160 (1956).

M. Ezekiel

[1] *Methods of correlation analysis*, John Wiley, New York, 1950.

630 REFERENCES

W. Feller

[1] Zur Theorie der Stochastischen Prozesse, *Math. Annalen* **113**, 113 (1936).

[2] Über den zentralen Grenzwertsatz der Wahrscheinlichkeitsrechnung, *Math. Zeitschirift* **40**, 521 (1935); **42**, 301 (1937).

[3] Die Grundlagen der Volterraschen Theorie des Kampfes ums Dasein in wahrschein-lichkeitstheoretischer Behandlung, *Acta Biotheoretica*, **5**, 11 (1939).

[4] On the integro-differential equations of purely discontinuous Markov processes, *TAMS* **48**, 488 (1940).

[5] On the Kolmogorov-Smirnov limit theorems for empirical distributions, *AMS* **19**, 177 (1948).

[6] On the theory of stochastic processes, with particular reference to applications, *Proc. Berkeley Symp.*, 403 (1949).

[7] *An introduction to probability theory and its applications*, John Wiley, New York, 1950.

[8] Über das Gesetz der Grossen Zahlen, *Acta Litt. Acad. Sci. Sect. Sci. Math. Szeged* **8**, 191 (1936–1937).

[9] Note on regions similar to the sample space, *Stat. Res. Memoirs* **2**, 117 (1938).

[10] The general form of the so-called Law of the Iterated Logarithm, *TAMS* **54**, 373 (1943).

[11] On the normal approximation to the binomial distribution, *AMS*, **16**, 319 (1945).

T. Ferguson

[1] A method of generating best asymptotically normal estimates with application to the estimation of bacterial densities, *AMS* **29**, 1046 (1958).

C. D. Ferris, F. E. Grubbs, C. L. Weaver

[1] Operating characteristics for the common statistical tests of significance, *AMS* **17**, 178 (1946)

B. de Finetti

[1] Sulle funzioni a incremento aleatorio, *Rend. Accad. Lincei Cl. Sci. Fis. Mat.* **10**, (6), 163 (1929).

R. A. Fisher

[1] Frequency distributions of the values of the correlation coefficient in samples from an infinitely large population, *B* **10**, 507 (1915).

[2] On the mathematical foundations of theoretical statistics, *Phil. Trans. Royal Soc.* **222**, 309 (1921).

[3] On the "probable error" of a coefficient of correlation deduced from a small sample, *Metron* **1**, No. 4, 1 (1921).

[4] The conditions under which χ^2 measures the discrepancy between observation and hypothesis, *JRSS* **87**, 442 (1924).

[5] Application of Student's distribution, *Metron* **5**, No. 3, 90 (1925).

[6] Theory of statistical estimation, *PCPS* **22**, 700 (1925).

[7] *Statistical methods for research workers*, Oliver and Boyd, Edinburgh–London, 1941.

[8] On an absolute criterion for fitting frequency curves, *Memoir Math* **41**, 155 (1912).

[9] On a property connecting the χ^2 measure of discrepancy with the method of maximum likelihood, *Atti Congr. Intern. Math. Bologna* **6**, 94 (1928).

[10] Two new properties of mathematical likelihood, *Proc. Royal Soc. London, ser. A*, **144**, 285 (1934).

[11] The logic of inductive inference, *JRSS* **98**, 39 (1935).

[12] The fiducial argument in statistical inference, *Ann. Eugenics Cambridge* **6**, 391 (1936).

[13] The distribution of the partial correlation coefficient, *Metron* **3**, 329 (1924).

[14] On a distribution yielding the error functions of several well-known statistics, *Proc. Intern. Math Congr. Toronto*, 805 (1924).

[15] The general sampling distribution of the multiple correlation coefficient, *Proc. Royal Soc. London, ser. A*, **121**, 654 (1928).

R. A. Fisher, F. Yates

[1] *Statistical tables*, Oliver and Boyd, London, 1938.

M. Fisz

[1] The limiting distribution of sums of arbitrary independent and equally distributed r-point random variables, *Studia Math.* **14**, 111 (1953).

[2] Grenzverteilungen der multinomialen Verteilung, *Bericht Tagung Wahrscheinlich-keitsrechung, Berlin*, 51 (1954).

[3] Realizations of some stochastic processes, *Studia Math.* **15**, 359 (1956).

[4] A limit theorem for empirical distribution functions, *Bull. Pol. Acad. Sci.* **5**, 699 (1957).

[5] A limit theorem for non-decreasing random functions, *Bull. Pol. Acad. Sci.* **6**, 485 (1958).

[6] On necessary and sufficient conditions for the validity of the strong law of large numbers expressed in terms of moments, *Bull. Pol. Acad. Sci.* **7**, 229 (1959).

[7] Characterization of sample functions of stochastic processes by some absolute probabilities, *Proc. Fourth Berkeley Symp.* **2**, 143 (1961).

M. Fisz, K. Urbanik

[1] Analytical characterization of some composed non-homogeneous Poisson process, *Studia Math.* **15**, 328 (1956).

E. Fix

[1] Tables of the Noncentral χ^2, *Univ. Calif. Publ. Stat.* **1**, 15 (1949).

E. Fix, J. L. Hodges

[1] Significance probabilities of the Wilcoxon test, *AMS* **26**, 301 (1955).

K. Florek, E. Marczewski, C. Ryll-Nardzewski

[1] Remarks on the Poisson stochastic process, I, *Studia Math.* **13**, 122 (1953).

S. Fogelson

[1] Mediana i jej wyznaczenie, *Kwartalnik Statystyczny*, **7**, 867 (1930).

R. Fortet

[1] Quelques travaux récents sur le mouvement Brownien, *Ann. Inst. H. Poincaré* **11**, 175 (1949).

[2] *Calcul des probabilités*, Centre Nat. Res. Sci., Paris, 1950.

[3] Random functions from a Poisson process, *Proc. Second Berkeley Symp.* 373 (1951).

R. Fortet, E. Mourier,

[1] Convergence de la répartition empirique vers la répartition théorique, *CR* **236**, 1739 (1953).

P. Franklin

[1] *A treatise on advanced calculus*, John Wiley, New York, 1940.

D. A. S. Fraser

[1] *Nonparametric methods in statistics*, John Wiley, New York, 1957.

M. Fréchet

[1] *Recherches théoretiques modernes sur la théorie des probabilités* II, Gauthiers-Villars, Paris, 1937–1938.
[2] Sur la loi de probabilité de l'écart maximum, *Ann. Soc. Polonaise Math.* **6**, 93 (1927).

R. Frisch

[1] Correlation and scatter in statistical variables, *Nord. Stat. Tidskr.* **8**, 36 (1929).
[2] *Statistical confluence analysis by means of complete regression systems*, Oslo, 1934.

M. Fryde

[1] Teoria i rzeczywistość w statystyce, *Ekonomista* **3**, 28 (1923).

A. Fuchs

[1] Sur la continuité stochastique des processus stochastiques réels de Markoff, *CR* **237**, 1329 (1953).
[2] Some limit theorems for non-homogeneous Markov processes, *TAMS* **86**, 511 (1957).

W. H. Furry

[1] On fluctuation phenomena in the passage of high energy electron through lead, *Phys, Rev.* **52**, 569 (1937).

K. R. Gabriel

[1] The distribution of the number of successes in a sequence of dependent trials, *B* **46**, 454 (1959).

C. F. Gauss

[1] *Teoria motus corporum coelestium*, Perthes and Besser, Hamburg, 1809.

R. C. Geary

[1] Distribution of Student's ratio for non-normal samples, *Supplement to JRSS* **3**, 178 (1936).
[2] The estimation of many parameters, *JRSS* **105**, 213 (1942).

H. Gebelein

[1] Das statistische Problem der Korrelation als Variations—und Eigenwertproblem und sein Zusammenhang mit der Ausgleichsrechnung, *Zeitschrift für Angewandte Math. Mech.* **21**, 364 (1941).

S. Geisser, N. Mantel

[1] Pairwise independence of jointly dependent variables, *AMS* **33**, 290 (1962).

R. K. Getoor

[1] The shift operator for non-stationary stochastic processes, *Duke Math. J.* **23**, 175 (1956).
[2] On characteristic functions of Banach space valued random variables, *Pacific J. Math.* **7**, 885 (1957).

P. S. Ghosh

[1] A note on stratified random sampling with multiple characteristics, *Calcutta Stat. Assn. Bulletin*, **8**, 81 (1958)

I. I. Gihman

[1] On the empirical distribution function in the case of grouping of observations (in Russian), *DAN* **82**, No. 6, 837 (1952).
[2] On some limit theorems for conditional distributions and related problems of mathematical statistics (in Russian), *Ukr. Math. J.* **5**, 413 (1953).
[3] Markov processes in problems of mathematical statistics (in Russian), *Ukr. Math. J.* **6**, No. 1, 28 (1954).
[4] On the asymptotic properties of some statistics analogous to the χ^2 statistic (in Russian), *TVP* **1**, 344 (1956).
[5] On a non-parametric criterion for the homogenity of k-samples (in Russian), *TVP* **2**, 380 (1957).

I. I. Gihman, B. W. Gnedenko, N. V. Smirnov

[1] Non-parametric statistical methods (in Russian), *Trudy Tretego Vsesoyuznogo Sezda* **3**, 320 (1958).

W. M. Gilbert

[1] Projections of probability distributions, *Acta Math. Acad. Hung. Sci.* **6**, 195 (1955).

M. A. Girshick

[1] Contributions to the theory of sequential analysis, *AMS* **17**, I, 123, II, III, 282 (1946).

V. I. Glivenko

[1] Sulla determinazione empirica delle leggi di probabilità, *Giorn. Ist. Ital. Attuari* **4**, 92 (1933).

B. W. Gnedenko

[1] On the characteristic functions (in Russian), *Bull. MGU, sec.* A, **1** (5), 17 (1937).
[2] On the convergence of distribution laws of sums of independent terms (in Russian), *DAN* **18**, 231 (1938).
[3] On the theory of limit theorems for sums of independent random variables (in Russian), *Izvestia Acad. Nauk USSR, Math. Series*, 181, 643 (1939).
[4] Limit laws for sums of independent random variables (in Russian), *UMN* **10**, 115 (1944).
[5] On a local limit theorem in probability theory (in Russian), *UMN* **3**, No. 3, 187 (1948).
[6] *Course in probability theory* (in Russian), GITTL, Moscow, 1950.
[7] Testing homogeneity of two independent samples (in Russian), *Math. Nachrichten* **12**, 29 (1954).
[8] Sur la distribution limite du terme maximum d'une série aléatoire, *Annals Math.* **44**, 423 (1943).
[9] Elements of the theory of distribution functions of random vectors (in Russian), *UMN* **10**, 230 (1944).
[10] On a theorem of S.N. Bernstein (in Russian), *Izvestia Acad. Nauk USSR* **12**, 97 (1948).
[11] On Wilcoxon's criterion for comparison of two samples (in Russian), *Bull. Pol. Acad. Sci.* **6**, 611 (1958).

B. W. Gnedenko, A. N. Kolmogorov

[1] *Limit distributions for sums of independent random variables*, Translation from the Russian, Addison-Wesley, Cambridge, 1954.

B. V. Gnedenko, V. S. Koroluk

[1] On the maximum divergence of two empirical distributions (in Russian), *DAN* **80**, No. 4, 525 (1951).

B. V. Gnedenko, V. S. Mihalevitch

[1] On the distribution of the number of excesses of one empirical distribution function over another one (in Russian), *DAN* **82**, 841 (1952).

B. V. Gnedenko, E. L. Rvatcheva

[1] On a problem of comparison of two empirical distribution functions, *DAN* **82**, No. 4, 513 (1952).

I. J. Good

[1] The likelihood ratio test for Markov chains, *B* **42**, 531 (1955); *Corrigenda B* **44**, 301 (1957).

L. A. Goodman

[1] Simplified run tests and likelihood ratio tests for Markov chains, *B* **45**, 181 (1958).
[2] On some statistical test for mth order Markov chains, *AMS* **30**, 154 (1959).

E. L. Grant

[1] *Statistical quality control*, McGraw-Hill, New York, 1956.

M. Greenwood, G. U. Yule

[1] The statistics of anti-typhoid and anti-cholera inoculations and the interpretation of such statistics in general, *Proc. Royal Soc. Medicine* **8**, 113 (1915).
[2] An inquiry into the nature of frequency distributions representative of multiple happenings with particular reference to the occurrence of multiple attacks of disease or of repeated accidents, *JRSS* **83**, 255 (1920).

U. Grenander

[1] On empirical spectral analysis of stochastic processes, *Ark. Mat.* **1**, 503 (1952).

U. Grenander, M. Rosenblatt

[1] *Statistical analysis of stationary time series*, John Wiley, New York, Almquist and and Wiksell, Stockholm, 1957.

F. E. Grubbs, C. L. Weaver

[1] The best unbiased estimate of the population standard deviation based on group ranges, *J. Amer. Stat. Assn.* **42**, 224 (1947).

J. P. Guilford

[1] *Fundamental statistics in psychology and education*, McGraw-Hill, New York Toronto–London 1956.

E. J. Gumbel

[1] The distribution of the range, *AMS* **18**, 384 (1947).
[2] *Statistics of extremes*, Columbia University Press, 1958.
[3] Multivariate distributions with given margins and analytical examples, *Bull. de l'Inst. Intern. de Stat.* **37**, No. 3, 3 (1960).
[4] Bivariate logistic distributions, *J. Amer. Stat. Assoc.* **56**, 335 (1961).

J. Hájek

[1] *Prispevky k teorii statistického odhadu, kandidátská disertace*, Praha, 1955.
[2] Linear estimation of the mean value of a stationary random process with convex correlation function, *Czechoslovak Math. J.* **6**, 81, 94 (1956).

J. Hájek, A. Rényi

[1] Generalization of an inequality of Kolmogorov, *Acta Math. Acad. Sci. Hung.* **6**, 281 (1955).

J. Hajnal

[1] The ergodicity properties of non-homogeneous finite Markov chains, *Proc. Cambridge Phil. Soc.* **52**, 67 (1956).

[2] Weak ergodicity in non-homogeneous Markov chains, *PCPS* **54**, 233 (1958).

A. Hald

[1] *Statistical tables and functions*, John Wiley, New York, 1952

P. Halmos

[1] The theory of unbiased estimation, *AMS* **17**, 34 (1946).

[2] *Measure theory*, Van Nostrand, Toronto–New York–London, 1950.

P. Halmos, L. J. Savage

[1] Application of the Radon-Nikodym theorem to the theory of sufficient statistics, *AMS* **20**, 225 (1949).

H. Hamburger

[1] Über eine Erweiterung des Stieltjesschen Momentenproblems, *Mathemat. Annalen* **81**, 235 (1920); **82**, 120 and 168 (1921).

E. T. Hannan

[1] *Time series analysis*, Methunen's monographs on appl. prob. stat., Methuen, London; John Wiley, New York, 1960.

B. J. Harley

[1] Relation between the distributions of the non-central t and of a transformed correlation coefficient, *B* **44**, 219 (1957).

T. E. Harris

[1] Branching processes, *AMS* **19**, 474 (1948).

H. O. Hartley, E. S. Pearson

[1] The probability integral of the range in samples of n observations from a normal population, *B* **32**, 301 (1942).

F. Haussdorf

[1] Beitrage zur Wahrscheinlichkeitsrechnung, *Berichte Ver. Sachs. Gesell. Wiss. Leipzig Math. Phys. Classe* **53**, 152 (1901).

[2] *Grundzüge der Mengenlehre*, Gruyter, Berlin, 1937.

F. R. Helmert

[1] Über die Wahrscheinlichkeit der Potenzsummen und über einige damit in Zusammenhange stehende Fragen, *Zeitschrift für Math. Phys.* **21**, 192 (1876).

A. Heppes

[1] On the determination of probability distribution of more dimensions by their projections, *Acta Math. Acad. Sci. Hung.* **7**, 403 (1956).

G. Herglotz

[1] Über Potenzreihen mit positivem reelen Teil im Einheitskreis, *Ber. Ver. Sächs. Gesell. Wiss. Leipzig Math. Phys. Classe*, **63**, 501 (1911).

H. O. Hirschfeld

[1] A connection between correlation and contingency, *PCPS* **31**, 520 (1935).

J. L. Hodges, Jr.

[1] The significance probability of the Smirnov two-sample test, *Ark. Math.* **3**, 469 (1957).

J. L. Hodges, Jr., L. LeCam

[1] The Poisson approximation to the Poisson binomial distribution, *AMS* **31**, 737 (1960).

W. Hoeffding

[1] "Optimum" non-parametric tests, *Proc. Second Berkeley Symp.* 83 (1951).

P. G. Hoel

[1] A test for Markov chains, *B* **41**, 430 (1954).

H. B. Horton

[1] A method for obtaining random numbers, *AMS* **19**, 81 (1948).

B. Hostinský

[1] *Méthodes generales du calcul des probabilités*, Gauthiers-Villars, Paris, 1931.

H. Hotelling

[1] The consistency and ultimate distribution of optimum statistics, *TAMS* **32**, 360 (1930).

P. L. Hsu

[1] Contributions to the theory of Student's *t*-test, *Stat. Res. Memoirs* **2**, 1 (1938).
[2] A new proof of the joint product moment distribution, *PCPS* **35**, 336 (1939).
[3] Analysis of variance from the power function standpoint, *B* **32**, 62 (1941).

G. A. Hunt

[1] Markoff processes and potentials, *Ill. J. Math.* **1**, 44 (1957); **2**, 151 (1958).
[2] Markoff processes and Martin boundaries, *Ill. J. Math.* **4**, 313 (1960).

I. A. Ibragimov

[1] Some limit theorems for stationary in the strict sense random processes (in Russian), *DAN* **125**, 711 (1959).

J. O. Irwin

[1] Mathematical theorems involved in the analysis of variance, *JRSS* **94**, 284 (1931).

S. L. Isaacson

[1] On the theory of unbiased tests of simple statistical hypotheses specifying the values of two or more parameters, *AMS* **22**, 217 (1951).

G. Ishii

[1] On the exact probabilities of Rényi's tests, *Ann. Inst. Stat. Math. Tokyo* **11**, 17 (1959).

S. Ito

[1] Brownian motion in a topological group and in its covering group, *Rend. Circ. Mat. Palermo* 2, **1**, 40 (1952).

J. Janko

[1] *Statistické tabulky*, Čes. Slov. Akad. Věd., Praha, 1958.

L. Jánossy, A. Rényi, J. Aczél

[1] On compound Poisson distributions, *Acta Math. Acad. Sci. Hung.* **1**, 209 (1950).

H. Jeffreys

[1] *Theory of probability*, The Clarendon Press, Oxford, 1939.

G. M. Jenkins, M. B. Priestley

[1] The spectral analysis of time series, *JRSS*, ser. B, **19**, 1 (1957).

M. Jiřina

[1] Stochastic branching processes with continuous state space, *Czechoslovak Math. J.* **8**, 83, 292 (1958).

N. L. Johnson, B. L. Welch

[1] Applications of the non-central t-distribution, *B* **31**, 362 (1939).

M. Kac

[1] On a characterization of the normal distribution, *Amer. J. Math.* **61**, 726 (1939).

[2] On distributions of certain Wiener functionals, *TAMS* **65**, 1 (1949).

[3] On deviations between theoretical and empirical distributions, *Proc. Nat. Acad. Sci.* **35**, 252 (1949).

[4] On some connections between probability theory and differential and integral equations, *Proc. Second Berkeley Symp.* 189 (1951).

[5] *Probability and related topics in physical sciences*, Interscience Publishers, London–New York, 1959.

M. Kac, J. F. Siegert

[1] An explicit representation of a stationary Gaussian process, *AMS* **18**, 438 (1947).

M. Kadyrov

[1] *Tables of random numbers* (in Russian), Tashkent, 1936.

K. Karhunen

[1] Über lineare Methoden in der Wahrscheinlichkeitsrechnung, *Ann. Acad. Sci. Fennicae*, ser. A, **I**, *Math-Phys*, No. 37 (1947).

S. Karlin, J. L. McGregor

[1] The differential equations of birth-and-death processes and the Stieltjes moment problem, *TAMS* **85**, 489 (1957).

[2] The classification of birth and death processes, *TAMS* **86**, 366 (1957).

[3] Linear growth birth and death processes, *J. Math. Mech.* **7**, 643 (1958).

[4] Many server queueing processes with Poisson input and exponential service times, *Pacific J. Math.* **8**, 87 (1958).

J. Kaucky

[1] Quelques remarques sur les châines de Markoff, *Spisy Vydávané Přirodovodeckou Fakultou Masarykovy University*, No. 131 (1930).

T. Kawata, H. Sakamoto

[1] On the characterization of the normal population by the independence of the sample mean and the sample variance, *J. Math. Soc. Japan* **1**, 111 (1949).

J. G. Kemeny, J. L. Snell

[1] *Finite Markov chains*, Van Nostrand, Princeton, 1960.

[2] On Markov chain potentials, *AMS* **32**, 709 (1961).

[3] Potentials for denumerable Markov chains, *J. Math. and Appl.* **3**, 196 (1961).

O. Kempthorne

[1] *The design and analysis of experiments*, John Wiley, New York, 1952.

D. G. Kendall

[1] Stochastic processes and population growth, *JRSS*, ser. B, **9**, 250 (1948).
[2] On the generalized "birth and death" process, *AMS* **19**, 1 (1948).

M. G. Kendall

[1] *The advanced theory of statistics*, Griffin, London, 1948.

M. G. Kendall, B. Smith

[1] *Tables of random numbers*, Tracts for computers, No. 24 (1940).

A. Khintchin

[1] Sur la loi des grandes nombres, *CR* **188**, 477 (1929).
[2] *Asymptotische Gesetze der Wahrscheinlichkeitsrechnung*, Springer, Berlin, 1933.
[3] Korrelationstheorie der stationären stochastischen Prozesse, *Math. Annalen* **109**, 604 (1934).
[4] Sul dominio di atrazione della legge di Gauss, *Giorn. Ist. Ital. Attuari* **6**, 371 (1935).
[5] Zur Theorie der unbeschränkt teilbaren Verteilungsgesetze, *Matematicesky Sbornik* **2**, 44, 79 (1937).
[6] *Limit theorems for sums of independent random variables* (in Russian), GONTI, Moscow–Leningrad, 1938.
[7] *Mathematical methods of the theory of mass service* (in Russian), Trudy Mat. Inst. Steklova, Moscow, 1955.
[8] Über einen Satz der Wahrscheinlichkeitsrechnung, *Fundamenta Math.* **6**, 9 (1924).

A. Khintchin, A. N. Kolmogorov

[1] On the convergence of series, *Rec. Math. Soc. Moscow* **32**, 668 (1925).

J. Kiefer

[1] K-sample analogues of the Kolmogorov-Smirnov and Cramér-von Mises tests, *AMS* **30**, 420 (1959).

J. Kiefer, L. Weiss

[1] Some properties of generalized sequential probability ratio tests, *AMS* **28**, 57 (1957).

J. Kiefer, J. Wolfowitz

[1] Sequential tests of hypotheses about the mean occurrence time of a continuous parameter Poisson process, *Naval Research Logistics Quarterly* **3**, No. 3, 205 (1956).
[2] Consistency of the maximum likelihood estimator in the presence of infinitely many incidental parameters, *AMS* **27**, 887 (1956).
[3] The deviation of the empiric distribution function of vector chance variables, *TAMS* **87**, 173 (1958).

J. R. Kinney

[1] Continuity properties of sample functions of Markov processes, *TAMS* **74**, 280 (1953).

A. N. Kolmogorov

[1] Über die Summen durch den Zufall bestimmter unabhängiger Grösse, *Math. Annalen* **99**, 309 (1928)
[2] Sur la loi forte des grandes nombres, *CR* **191**, 910 (1930).

[3] Über die analytischen Methoden der Wahrschienlichkeitsrechnung, *Math. Annalen* **104**, 415 (1931).

[4] Sur le problème d'attente, *Matematicesky Sbornik* **38**, 101 (1931).

[5] Sulla forma generale di un prozesso stocastico omogeneo, *Rend. Accad. Lincei Cl. Sci. Fis. Mat.* **15**, (6), 805 (1932).

[6] Sulla determinazione empirica di una legge di distribuzione, *Giorn. Ist. Ital. Attuari* **4**, 83 (1933).

[7] *Grundbegriffe der Wahrscheinlichkeitsrechnung*, Springer, Berlin, 1933.

[8] La transformation de Laplace dans les espaces linéaires, *CR* **200**, 1717 (1935).

[9] Markov chains with a countable number of possible states (in Russian), *Bull. MGU*, **1**, No. 3, 1 (1937).

[10] Curves in Hilbert space invariant with respect to an one parametric group of movements (in Russian), *DAN* **26**, 6 (1940).

[11] Stationary sequences in Hilbert space (in Russian), *Bull. MGU*, **2**, No. 6, 1 (1941).

[12] Interpolation and extrapolation of stationary sequences (in Russian), *Izvestia Acad. Nauk USSR*, **3** (1941).

[13] A local limit theorem for homogeneous Markov chains (in Russian), *Izvestia Acad. Nauk USSR* **13**, 287 (1949).

[14] Unbiased estimates (in Russian), *Izvestia Acad. Nauk USSR*, **14**, 303 (1950).

[15] On the problem of differentiability of the transition probabilities of time homogeneous Markov processes with a countable number of possible states (in Russian), *Ucenye Zapisky MGU*, ser. 148, *Mathematika* **4**, 53 (1951).

[16] Das Gesetz des iterierten Logarithmus, *Math. Annalen* **101**, 126 (1929).

A. N. Kolmogorov, N. A. Dmitriev

[1] Branching stochastic processes (in Russian), *DAN* **56**, 7 (1947).

A. N. Kolmogorov, Yu V. Prohorov

[1] On the sums of a random number of random terms (in Russian), *UMN* **4**, No. 4, 168 (1949).

S. Kolodziejczyk

[1] On an important class of statistical hypotheses, *B* **27**, 161 (1935).

M. Konêcny

[1] Sur la théorie des châines de Markoff, *Spisy Vydávané Přirodovodeckou Fakultou, Masarykovy University*, No. 147.

B. O. Koopman

[1] On distributions admitting a sufficient statistic, *TAMS* **39**, 399 (1936).

V. S. Koroluk

[1] On the discrepancy of empirical distributions for the case of two independent samples (in Russian), *Izvestia Acad. Nauk USSR* **19**, 81 (1953).

I. Kotlarski

[1] On random variables whose quotient follows the Cauchy law, *Coll. Math.* **7**, 277 (1960).

S. I. Kreczmer

[1] The investigation of micropulsation of the temperature field in atmosphere (in Russian), *DAN* **84**, 55 (1952).

W. H. Kruskal

[1] A nonparametric test for the several sample problem, *AMS* **23**, 525 (1952).

L. Kubik

[1] O rozkładach granicznych sum r-punktowych zmiennych losowych, *Prace Matematyczne* **4**, (1), 111 (1960).

N. H. Kuiper

[1] Alternative proof of a theorem of Birnbaum and Pyke, *AMS* **30**, 251 (1959).
[2] Tests concerning random points on a circle, *Proc. Nederl. Akad. Wet. Amsterdam, Ser. A* **63**, *Indag. Math.* **22**, 38 (1960).

S. Kullback

[1] An application of characteristic functions to the distribution problem of statistics, *AMS* **5**, 263 (1934).

I. D. Kvit

[1] On Smirnov's theorem concerning the comparison of two samples (in Russian), *DAN* **71**, No. 2, 229 (1950).

R. G. Laha

[1] On a characterization of the Gamma distribution, *AMS* **25**, 784 (1954).
[2] On a characterization of the multivariate normal distribution, *Sankhya* **14**, 367 (1955).
[3] On some properties of the normal and gamma distributions, *Proc. Amer. Math. Soc.* **7**, 172 (1956).
[4] On a characterization of the normal distribution from properties of suitable linear statistics, *AMS* **28**, 126 (1957).
[5] An example of a non-normal distribution where the quotient follows the Cauchy law, *Proc. Nat. Acad. Sci. USA* **24**, 222 (1958).
[6] On a class of distribution functions where the quotient follows the Cauchy law. *TAMS* **93**, 205 (1959).

J. Lamperti

[1] An invariance principle in renewal theory, *Technical Report No. 8, Appl. Math. Stat. Lab., Stanford University*, 1961.

O. Lange

[1] Statistical estimation of parameters in Markov processes, *Coll. Math.* **3**, 147 (1955).

P. S. Laplace

[1] *Théorie analytique des probabilités*, Paris, 1820.

A. M. Lapunov

[1] Sur une proposition de la théorie des probabilités, *Bull. Acad. Sci. St. Pétersburg* **13**, 359 (1900).
[2] Nouvelle forme du théorème sur la limite de probabilité, *Mem. Acad. Sci. St. Pétersburg* **12**, No. 5 (1901).

L. LeCam

[1] On the asymptotic theory of estimation and testing hypotheses, *Proc. Third Berkeley Symp.* **1**, 129 (1956).
[2] Convergence in distribution of stochastic processes, *Univ. Calif. Publ. Stat.* **2**, 207 (1957).
[3] Les propriétés asymptotiques des solutions de Bayes, *Publ. Inst. Stat. Univ. Paris* **7** No. 3–4, 17 (1959).

A. M. Legendre

[1] *Nouvelles méthodes pour la determination des orbites des comètes*, Didot, Paris, 1805.

E. L. Lehmann

[1] A general concept of unbiasedness, *AMS* **22**, 587 (1951).

[2] Consistency and unbiasedness of certain non-parametric tests, *AMS* **22**, 165 (1951).

[3] The power of rank tests, *AMS* **24**, 23 (1953).

[4] *Testing statistical hypotheses*, John Wiley, New York-London, 1959.

E. L. Lehmann, H. Scheffé

[1] Completeness, similar regions and unbiased estimation, Part I, *Sankhya* **10**, 305 (1950).

W. R. Leimbacher

[1] On some classes of sequential procedures for obtaining confidence intervals of given length, *Univ. Calif. Pub. Stat.*, **2**, 1 (1953).

H. Levene

[1] On the power function of tests of randomness based on runs up and down, *AMS* **23**, 34 (1952).

P. Lévy

[1] *Calcul des probabilités*, Gauthiers-Villars, Paris, 1925.

[2] Sur les integrals dont les éléments sont des variables aléatoires indépendantes, *Annali R. Sci. Norm. Sup. Pisa*, **3** (2), 337 (1934).

[3] *Théorie de l'addition des variables aléatoires*, Gauthiers-Villars, Paris, 1937.

[4] *Processus stochastiques et mouvement Brownien*, Gauthiers-Villars, Paris, 1948.

[5] Systèms Markoviens et stationaires. Cas denombrable, *Ann. Sci. École Normale Sup. Paris Suppl.* (3), **68**, 327 (1951).

[6] Wiener's random functions, and other Laplacian random functions, *Proc. Second Berkeley Symp.* 471 (1951).

[7] Sur certains processus stochastiques homogénes, *Compositio Math.* **7**, 283 (1939).

[8] Le mouvment Brownien plan, *Am. J. Math.* **62**, 487 (1940).

[9] Complements a l'etude des processus de Markoff, *Ann. Sci. École Normale Sup. Paris* **69**, 203 (1952).

[10] Systèmes Semi-Markoviens à au plus une infinité dénombrable d'etats possibles, *Proc. Int. Congr. Math. Amst.* **2**, 294 (1954); **3**, 416 (1954).

[11] Le mouvement brownien a $n = 2p + 1$ parametres, I, *CR* **239** 1181 (1954), II, *CR* **239**, 1584 (1954); III, *CR* 240, 1043 (1955).

W. Lexis

[1] *Abhandlungen zur Theorie der Bevölkerungs und Moralsstatistik*, Gustav Fischer, Jena, 1903.

G. J. Lieberman, H. Solomon

[1] Multilevel continuous sampling plans, *AMS* **26**, 686 (1955).

J. W. Lindeberg

[1] Eine neue Herleitung des Exponentialgesetzes der Wahrscheilichkeitsrechnung, *Math. Zeitschrift*, **15**, 221 (1922).

Yu. V. Linnik

[1] Linear statistics and the normal distribution (in Russian), *DAN* **83**, 353 (1952).

[2] A problem on characteristic functions of probability distributions (in Russian), *UMN* **10**, (1), 137 (1955).

[3] *The method of least squares and the basic methods of analysis of observations* (translated from the Russian), Pergamon Press, New York, 1961.

642 REFERENCES

M. Loéve

[1] Sur les functions aléatoires stationnaires de second ordre, *Rev. Sci.* **83**, 297 (1945).
[2] Fonctions aléatoires de second ordre, *CR* **220**, 380 (1945), **222**, 469 (1946).
[3] On almost sure convergence, *Proc. Second Berkeley Symp.* 279 (1951).
[4] *Probability theory*, Van Nostrand, New York, 1955.
[5] Ranking limit problems, *Proc. Third Berkeley Symp.* **2**, 177 (1956).
[6] Ètudes asymptotiques des sommes de variables aléatoires, *J. de Math.* **24**, 249 (1945).

A. Lomnicki

[1] Nouveaux fondements du calcul des probabilités, *Fundamenta Math.* **4**, 34 (1923).

Z. A. Lomnicki, S. K. Zaremba

[1] On estimating the spectral density function of a stochastic process, *JRSS, ser.* B, **19**, 13 (1957).

L. Lorch, D. J. Newman

[1] The Lebesgue constants for regular Hausdorf methods, *Canad. J. Math.* **13**, 283 (1961).

E. Lukacs

[1] A characterization of the normal distribution, *AMS* **13**, 91 (1942).
[2] Applications of Faà di Bruno's formula in mathematical statistics, *Amer. Math. Monthly* **62**, 340 (1955).
[3] Characterization of populations by properties of suitable statistics, *Proc. Third Berkeley Symp.* **2**, 215 (1956).
[4] Some extension of a theorem of Marcinkiewicz, *Pacific J. Math.* **8**, 487 (1958).
[5] *Characteristic functions*, No. 5 of Griffins Statistical Monographs and Courses, New York, 1960.

J. Lukaszewicz, W. Sadowski

[1] O porównywaniu kilku populacji z populacją kontrolną, *Zastosowania Mat.* **3**, 204 (1957).

O. Lundberg

[1] *On random processes and their applications to sickness and accidental statistics*, University of Stockholm, thesis, Uppsala, 1940.

W . G. Madow, L. H. Madow

[1] On the theory of systematic sampling, *AMS* **15**, 1 (1944).

P. C. Mahalanobis

[1] On large sample surveys, *Phil. Trans. Royal Soc. London, ser. B,* **231**, 329 (1944).

G. Malécot

[1] Sur un problème de probabilites en chaîne que pose la génétique, *CR* **219**, 379 (1944).

S. Malmquist

[1] On a property of order statistics from a rectangular distribution, *SA* **33**, 214 (1950).
[2] On certain confidence contours for distribution functions, *AMS* **25**, 523 (1954).

G. M. Manija

[1] Generalization of the criterion of A. N. Kolmogorov for an estimate of the distribution law from the empirical data (in Russian), *DAN* **69**, No. 4, 495 (1949).

H. B. Mann

[1] *Analysis and design of experiments*, Dover, New York, 1949.

H. B. Mann, A. Wald

[1] On the choice of the number of class intervals in the application of the chi-square test, *AMS* **13**, 306 (1942).

H. B. Mann, D. R. Whitney

[1] On a test whether one of two random variables is stochastically larger than the other, *AMS* **18**, 50 (1947) .

K. Marbe

[1] *Die Gleichförmigkeit in der Welt*, Munich, 1916.

J. Marcinkiewicz

[1] Sur une propriété de la loi de Gauss, *Math. Zeitschrift* **44**, 612 (1939).

J. Marcinkiewicz, A. Zygmund

[1] Quelques théorèmes sur les fonctions independantes, *Studia Math.* **7**, 104 (1938).

E. Marczewski

[1] Remarks on the Poisson stochastic process, II, *Studia Math.* **13**, 130 (1953).

A. A. Markov

[1] Extension of the law of large numbers to dependent random variables (in Russian), *Izv. Mat. Fiz. Ob. pri Kazanskom Univ. ser.* 2, **15**, 135 (1906).

[2] Investigation of an important case of dependent trials (in Russian). *Izvestia Acad. Nauk SPB*, **VI**, *ser.* I, 61 (1907).

[3] Extension of limit theorems of probability theory to a sum of variables connected in a chain (in Russian), *Memoires de l'Academie Sci. St. Petersburg, ser.* 8, 22 (1908).

[4] An example of statistical investigation of the poem "Eugene Onegin" illustrating the connection of events in a chain (in Russian), *Izvestia Acad. Nauk* **7** (6) (1913).

[5] *Calculus of probability* (in Russian), GIZ, Moscow, 1924.

P. Masani

[1] The prediction theory of multivariate stochastic processes, III, *Acta Math.* **140**, 1–2, 141 (1960).

F. J. Massey

[1] A note on the power of a nonparametric test, *AMS* **21**, 440 (1950).

[2] The distribution of the maximum deviation between two sample cumulative step functions, *AMS* **22**, 125 (1951).

J. G. Mauldon

[1] Characterizing properties of statistical distributions, *Quarterly J. Math. Oxford.*, *ser.* 2, **7**, 155 (1956).

S. Mazurkiewicz

[1] *Podstawy rechunku prawdopodobienstwa*, PWN, Warsaw, 1956.

H. P. McKean

[1] Sample functions of stable processes, *Annals of Math. 2nd Ser.* **61**, 564 (1955).

D. G. Mejzler

[1] On the limit distribution of the maximal term of a variational series (in Ukrainian), *Dopovidi Akad. Nauk Ukr. SSR* **1**, 3 (1950).

644 REFERENCES

D. G. Mejzler, O. S. Parasiuk, E. L. Rvatcheva

[1] On a multidimensional local limit theorem of probability theory (in Russian), *Ukr. Math. J.* **1**, 9 (1949).

M. Merrington, E. S. Pearson

[1] An approximation to the distribution of noncentral t, *B* **45**, 484 (1958).

V. S. Mihalevitch

[1] On the mutual position of two empirical distribution functions (in Russian), *DAN* **85**, 485 (1952).

R. v. Mises

[1] Grundlagen der Wahrscheinlichkeitsrechnung, *Math. Zeitschrift* **4**, 1 (1919).
[2] *Wahrscheinlichkeit, Statistik und Wahrheit*, Springer, Vienna, 1936.
[3] La distribution de la plus grande de n valeurs, *Revue Math. de l'Union Interbalkanique*, **1**, 1, 141 (1939).
[4] On the correct use of Bayes' formula, *AMS* **13**, 156 (1942).
[5] Über die Wahrscheinlichkeit seltener Ereignisse, *Zeitschrift Angewandter Math. Mech.* **1**, 121 (1921).
[6] *Wahrscheinlichkeitsrechnung und ihre Anwendung in der Statistik, Fehlertheorie und in der theoretischen Physik*, Franz Deuticke, Vienna-Leipzig, 1931.

A. M. Mood

[1] The distribution theory of runs, *AMS* **11**, 367 (1940).
[2] *Introduction to the theory of statistics*, McGraw-Hill, New York, 1950.

F. Mosteller

[1] A k-sample slippage test for an extreme population, *AMS* **19**, 58 (1948).

E. Mourier

[1] *Eléments aléatoires dans un espace de Banach, Thése*, Paris, 1954.

J. E. Moyal

[1] Multiplicative population chains, *Technical Report No. 6, Appl. Math. Stat. Lab., Stanford University* (1961).
[2] Multiplicative population processes, *Technical Report No. 7, Appl. Math. Stat. Lab., Stanford University*, (1961).

S. V. Nagayev

[1] Some limit theorems for homogeneous Markov processes (in Russian), *TVP* **2**, 389 (1957).

S. W. Nash

[1] An extension of the Borel-Cantelli lemma, *AMS* **25**, 165 (1954).

J. Neyman

[1] *Zarys teorii i praktyki badania struktury ludności metoda reprezentacyjna*, GUS, Warsaw, 1933.
[2] On the problem of confidence intervals, *AMS* **6**, 111 (1935).
[3] Su un teorema concernente le cosidette statistiche sufficienti, *Giorn. Ist. Ital. Attuari* **6**, 320 (1935).
[4] Outline of a theory of statistical estimation based on the classical theory of probability, *Phil. Trans. Royal Soc. London, ser. A* **236**, 333 (1937).

[5] L'estimation statistique traitée comme un problème classique de probabilité, *Act. Scient. Ind.* **739**, (1938).

[6] *Lectures and conferences on mathematical statistics*, U.S. Department of Agriculture, Washington, 1938.

[7] Basic ideas and some recent results of the theory of testing statistical hypotheses, *JRSS* **105**, 292 (1944).

[8] Contributions to theory of the χ^2 test, *Proc. Berkeley Symp.* 239 (1949).

J. Neyman, E. S. Pearson

[1] On the use and interpretation of certain test criteria for purposes of statistical inference, *B* **20A**, 175, 263 (1928).

[2] On the problem of two samples, *Bull. Int. Acad. Cracovie* **A**, 73 (1930).

[3] On the problem of the most efficient tests of statistical hypotheses, *Phil. Trans. Royal Soc. London ser.* A **231**, 289 (1933).

[4] Contributions to the theory of testing statistical hypotheses, *Stat. Res. Memoir* **1**, 1 (1936); **2**, 25 (1938).

[5] Sufficient statistics and uniformly most powerful tests of statistical hypotheses, *Stat. Res. Memoir* **1**, 113 (1936).

J. Neyman, B. Tokarska

[1] Errors of the second kind in testing Student's hypothesis, *J. Amer. Stat. Assn.*, **31**, 318 (1936)

W. L. Nicholson

[1] A computing formula for the power of the analysis of variance test, *AMS* **25**, 607 (1954).

I. Olkin, J. W. Pratt

[1] A multivariate Tchebysheff inequality, *AMS* **29**, 226 (1958).

O. Onicescu, G. Mihoc

[1] Sur l'application de la notion de fonction caractéristique dans la théorie des chaînes de Markoff, Premier memoire (cas discontinu), *Matematica Cluj*, 1938.

[2] Sur les sommes de variables enchaînées, Second memoire, *Bull. Math. Soc. Roum.* **41**, 99 (1939).

V. Ozols

[1] Gnedenko-Koroluka teoremas visparinâjums uz tris izlasem pie divãm vienpuzigãn robezãm, *Latvijas PRS Zinatum Akademijas Vestis*, **10**, 141 (1956)

C. Palm

[1] Intensitätsschwankungen im Fernsprechverkehr, *Ericson Technics*, Stockholm **44**, 1 (1943).

E. Parzen

[1] On consistent estimates of the spectrum of a stationary time series, *AMS* **28**, 329 (1952).

[2] *Modern probability theory and its applications*, John Wiley, New York-London, 1960.

E. Paulson

[1] An optimum solution to the k-sample slippage problem for the normal distribution, *AMS* **23**, 610 (1952).

K. Pearson

[1] *Tables of the incomplete Γ-function*, Cambridge University Press, London, 1922.
[2] *Tables for statisticians and biometricians*, Cambridge University Press, Cambridge, vol I (1924); vol. II (1931).
[3] Researches on the mode of distribution of the constants of samples taken at random from a bivariate normal population, *Proc. Royal Soc.*, A **112**, 1 (1926).
[4] *Tables of the incomplete B-function*, Cambridge University Press, London, 1934.
[5] On the criterion that a given system of deviations from the probable in the case of a correlated system of variables is such that it can be reasonably supposed to have arisen from random sampling, *Phil. Magazine*, V, 50, 157 (1900).
[6] On the probable errors of the frequency constants, *B* **73**, 113 (1920).
[7] Contributions to the mathematical theory of evolution, *Phil. Trans. Royal Soc. London, ser.* A, **185**, 71 (1894).
[8] Contributions to the mathematical theory of evolution II. Skew variation in homogeneous material, *Phil. Trans. Royal Soc. London, ser.* A, **186**, 343 (1895).
[9] On the systematic fitting of curves to observations and measurements, *B* **1**, 265 (1902), *B* **2**, (1902).

J. Pfanzagl

[1] Ein kombiniertes Test- und Klassifikations-Problem, *Metrika* **2**, 1, 11 (1959).

E. J. G. Pitman

[1] The closest estimation of statistical parameters, *PCPS* **33**, 212 (1937).
[2] On the derivatives of a characteristic function at the origin, *AMS* **27**, 1156 (1956).

H. Poincaré

[1] *Calcul des probabilités*, Gauthier-Villars, Paris, 1912.

S. D. Poisson

[1] *Recherches sur la probabilité des jugements en matière criminelle et en matière civile*, Paris, 1837.

G. Pólya

[1] Sur quelques points de la théorie des probabilités, *Ann. Inst. H. Poincaré* **1**, 117 (1930).
[2] Herleitung des Gausschen Gesetzes aus einer Funktionalgleichung, *Math. Zeitschrift* **18**, 96 (1923).
[3] Über den zentralen Grenzwertsatz der Wahrscheinlichkeitsrechnung und das Momentenproblem, *Math. Zeitschrift* **8**, 171 (1920)

A. Prékopa

[1] On composed Poisson distributions, IV, *Acta Math. Acad. Sci. Hung.* **3**, 317 (1952).

Yu. V. Prohorov

[1] On the strong law of large numbers (in Russian), *DAN* **69**, 607 (1949).
[2] Probability distributions in functional spaces (in Russian), *UMN* **8**, No. 3, 165 (1953).
[3] Convergence of random processes and limit theorems in probability theory (in Russian), *TVP* **1**, 177 (1956).
[4] Strong stability of sums and infinitely divisible distributions (in Russian), *TVP* **3**, 152 (1958).
[5] Some remarks on the strong law of large numbers (in Russian), *TVP* **4**, 215 (1959).

Yu V. Prohorov, M. Fisz

[1] A characteristic property of the normal distribution in Hilbert space, *TVP* **2**, 475 (1957).

R. Pyke

[1] The supremum and infimum of the Poisson process, *AMS* **30**, 568 (1959).

[2] Markov renewal properties: definitions and preliminary properties, *AMS* **32**, 1231 (1961).

[3] Markov renewal processes with finitely many states, *AMS* **32**, 1243 (1961).

J. Przyborowski, H. Wilenski

[1] Homogeneity of results in testing samples from Poisson series, *B* **31**, 313 (1939).

D. Raikov

[1] On the decomposition of Gauss' and Poisson's laws (in Russian), *Izvestia Acad. Nauk USSR*, **2**, 91 (1938).

A. Rajchman

[1] Zaostrzone prawo wielkich liczb, *Mathesis Polska* **6**, 145 (1932).

R. Ranga-Rao

[1] Relations between weak and uniform convergence of measures with applications, *AMS* **33**, 659 (1962).

R. Ranga-Rao, V. S. Varadarajan

[1] On the decomposition of Haar measure in compact groups, *Fundamenta Math.* **49**, 119 (1961).

C. R. Rao

[1] Information and accuracy obtainable in an estimation of a statistical parameter, *Bull. Calcutta Math. Soc.* **37**, 81 (1945) .

O. Reiersöl

[1] Confluence analysis by means of instrumental sets of variables, *Ark. Math.* **32A**, No. 4 (1945).

A. Rényi

[1] On some problems concerning Poisson processes, *Publicationes Math.* **2**, 66 (1957).

[2] On composed Poisson distributions II, *Acta Math. Acad. Sci. Hung.* **2**, 83 (1951).

[3] On the theory of order statistics, *Acta Math. Acad. Sci. Hung.* **4**, 191 (1953).

[4] On a new axiomatic theory of probability, *Acta Math. Acad. Sci. Hung.* **6**, 285 (1955).

[5] On projections of probability distributions, *Acta Math. Acad. Sci. Hung.* **3**, 131 (1952).

[6] A characterization of Poisson process, *Publ. Math. Inst. Hung. Acad. Sci.* **1**, 526 (1956).

[7] On the asymptotic distribution of the sum of a random number of independent random variables, *Acta Math. Acad. Sci. Hung.* **8**, 193 (1957).

[8] New version of the probabilistic generalization of the large sieve, *Acta Math. Acad. Sci. Hung.* **10**, 217 (1959).

[9] On measures of dependence, *Acta Math. Acad. Sci. Hung.* **10**, 441 (1959).

[10] Neues Kriterion zum Vergleich zweier Stichproben, *Magyar Tud. Akad. Mat. Inst.* **2**, 243 (1953).

G. J. Resnikoff, G. J. Lieberman

[1] *Tables of the noncentral t-distribution*, Appl. Math, Stat. Lab., Stanford University, California, 1957.

G. E. H. Reuter, W. Ledermann

[1] On the differential equations for the transition probabilities of Markov processes with enumerably many states, *PCPS* **49**, 247 (1953).

W. Richter

[1] Local limit theorems for large deviations (in Russian), *TVP* **1**, 155 (1956).
[2] Multidimensional local limit theorems for large deviations (in Russian), *TVP* **3**, 107 (1958).

P. J. Rijkoort

[1] A generalization of Wilcoxon's test, *Proc. Royal Inst. Acad. Sci. ser.* A, **55**, 394 (1952).

J. Riordan

[1] *Stochastic service systems*, John Wiley, New York, 1962.

H. Robbins

[1] The asymptotic distribution of the sum of a random number of random variables, *Bull. Amer. Math. Soc.* **54**, 1151 (1948).
[2] Convergence of distributions, *AMS* **19**, 72 (1948).
[3] On the equidistribution of sums of independent random variables, *Proc. Amer. Math. Soc.* **4**, 786 (1953).
[4] An empirical Bayes' Approach to statistics, *Proc. Third Berkeley Symp.*, **1**, 157 (1956).

V. I. Romanowsky

[1] On the moments of standard deviations and correlation coefficient in samples from normal populations, *Metron* **5**, No. 4, 3 (1925).
[2] On the distribution of the regression coefficient in samples from normal populations, *Izvestia Acad. Nauk USSR*, **20**, 643 (1929).
[3] *Application of mathematical statistics to experimentation* (in Russian), 1948.
[4] *Discrete Markov chains* (in Russian), Moscow-Leningrad, 1949.

S. Rosenbaum

[1] Tables for a nonparametric test of dispersion, *AMS* **24**, 663 (1953).

M. Rosenblatt

[1] Some regression problems in time series, *Proc. Third Berkeley Symp.* **1**, 165 (1956).
[2] A multidimensional prediction problem, *Ark. Math.* **3**, 407 (1958).

Yu. A. Rozanov

[1] Spectral theory of multi-dimensional random processes with discrete time parameter (in Russian), *Izvestia Acad. Nauk USSR* **13**, 2, 93 (1958).

H. Rubin, H. G. Tucker

[1] Estimating the parameters of a differential process, *AMS* **30**, 641 (1959).

E. L. Rvatcheva

[1] On the maximum discrepancy between two empirical distributions (in Russian), *Ukr. Mat. J.* **4**, No. 4, 373 (1952).

Cz. Ryll-Nardzewski

[1] On the non-homogeneous Poisson process, I, *Studia Math.* **14**, 124 (1953).

[2] Remarks on the Poisson stochastic process, III, *Studia Math.* **14**, 314 (1953).

W. Sadowski, et al.

[1] *Tablice statystyczne*, PWN, Warsaw, 1957.

[2] O nieparametrycznym tescie na porównywanie rozsiewów, *Zastosowania Mat.* **2**, 161 (1955).

S. Saks, A. Zygmund

[1] *Analytic functions*, Polish Math. Society, Warsaw, 1952.

T. A. Sarymsakov

[1] On the ergodicity of nonhomogeneous Markov chains (in Russian), *DAN* **90**, 25 (1953).

J. R. Savage

[1] Contribution to the theory of Rank Order Statistics—the two-sample case, *AMS* **27**, 590 (1956).

L. J. Savage

[1] The foundations of statistics reconsidered, *Proc. Fourth Berkeley Symp.* **1**, 575 (1961).

H. Scheffé

[1] *The analysis of variance*, John Wiley, New York, 1959.

L. Schmetterer

[1] *Einführung in die mathematische Statistik*, Springer, Vienna, 1956.

[2] Bemerkungen zur Theorie der erwantungstreuen Schätzfunkzionen, *Mitteilungsblatt für Math. Statistik Anwendungsgebiete*, **9**, 2, 147 (1957).

[3] Über nichtparametrische Methoden in der Mathematischen Statistik, *Jahresbericht Deutschen Math. Vereinigung* **61**, 3, 104 (1959).

P. Schmid

[1] On the Kolmogorov and Smirnov limit theorems for discontinuous distribution functions, *AMS* **29**, 1011 (1958).

G. R. Seth

[1] On the variance of estimates, *AMS* **20**, 1 (1949).

A. Sevastianov

[1] On the theory of branching random processes (in Russian), *UMN* **6**, No. 6, 47 (1951).

R. Sikorski

[1] *Real functions*, PWN and Pergamon Press, Warsaw-London.

S. H. Sirazhdinov

[1] Sharpening of limit theorems for Markov chains (in Russian), *DAN* **84**, 1143 (1952).

J. Sittig, H. Freudenthal

[1] *De juiste maat*, Leiden, 1951.

V. P. Skitovitch

[1] Linear forms of independent random variables and the normal distribution (in Russian), *Izvestia Acad. Nauk USSR* **18**, 185 (1954).

A. V. Skorohod

[1] On passing from a sequence of independent random variables to a homogeneous random process with independent increments (in Russian), *DAN* 104, 364 (1955).
[2] Limit theorems for Markov processes (in Russian), *TVP* 3, 217 (1958).

E. E. Slutsky

[1] Über stochastische Asymptoten und Grenzwerte, *Metron* 5 (3), 3 (1925).
[2] Alcuni proposizioni sulla theoria degli funzioni aleatorie, *Giorn. Ist. Ital. Attuari*. 8, 183 (1937).
[3] Sur les fonctions aléatoires presque periodique et sur la décomposition des fonctions aléatoires stationnaires en composantes, *Act. Sci. Ind*. 738, 33 (1938).

N. V. Smirnov

[1] On the estimation of the discrepancy between empirical curves of distributions for two independent samples (in Russian), *Bull. MGU* 2, No. 2, 3(1939).
[2] Approximate laws of distribution of random variables from empirical data (in Russian), *UMN* X, 179 (1944).
[3] *Limit distributions for the terms of a variational series* (in Russian), Trudy Mat. Inst. Steklova, Moscow, 1941.

W. L. Smith

[1] Regenerative stochastic processes, *Proc. Royal Soc. London, ser.* A, **232**, 6 (1955).

M. Smoluchowski

[1] Über Brownsche Molekularbewegung, *Ann. Phys*. 48, 110, 3 (1915).

J. L. Snell

[1] Application of martingale system theorems, *TAMS* 73, 293 (1952).

F. Spitzer

[1] A combinatorial lemma and its application to probability theory, *TAMS* **82**, 323 (1956).

A. B. L. Srivastava

[1] Effect of non-normality on the power function of *t*-test, *B* 45, 421 (1958).

C. Stein

[1] A two-sample test for a linear hypothesis whose power is independent of the variance, *AMS* 16, 243 (1945).
[2] A note on cumulative sums, *AMS* 17, 498 (1946).
[3] Unbiased estimates with minimum variance, *AMS* 21, 406 (1950).

H. Steinhaus

[1] Les probabilités dénombrables et leur raport a la théorie de la measure, *Fundamenta Math*. 4, 286 (1923).
[2] Quality control by sampling (A plea for Bayes rule), *Coll. Math*. 2, 98 (1950).
[3] *Tablice liczb przetasowanych*, Rozprawy Matematyczne, VI, Math. Inst. Pol. Acad. Sci. Warsaw, 1954.

V. E. Stepanov

[1] Some statistical tests for Markov chains (in Russian), *TVP* 2, 143 (1957).

W. L. Stevens

[1] Distribution of groups in a sequence of alternatives, *Annal. Eugenics* 9, 10 (1939).

J. Stieltjes

[1] Recherches sur les fractions continues, *Oeuvres complètes*, Groningen, Amsterdam, 1918.

Student

[1] The probable error of a mean, *B* **6,** 1 (1908).

P. V. Sukhatme

[1] Contribution to the theory of the representative method, *JRSS, Suppl.* **2,** 253 (1935).
[2] A contribution to the problem of two samples, *Proc. Ind. Acad. Sci. A* **2,** 584 (1936).
[3] *Sampling theory of surveys with applications*, State College Press, Ames, Iowa, 1954.

E. Sverdrup

[1] The limit distribution of a continuous function of random variables, *Memorandum fra Universitete Socialekonomiske Inst.*, Oslo, 1951.

F. S. Swed, C. Eisenhart

[1] Tables for testing randomness of grouping in a sequence of alternatives, *AMS* **14,** 66 (1943).

R. Syski

[1] *Introduction to congestion theory in telephone systems*, Oliver and Boyd, Edinburgh, 1960.

L. Takács

[1] Occurrence and coincidence phenomena in case of happenings with arbitrary distribution law of duration, *Acta Math. Acad. Sci. Hung.* **2,** 275 (1951).
[2] On secondary processes generated by a Poisson process and their applications in physics, *Acta Math. Acad. Sci. Hung.* **5,** 203 (1954).
[3] On processes of happenings generated by means of a Poisson process, *Acta Math. Acad. Sci. Hung.* **6,** 81 (1955).
[4] On stochastic processes connected with certain physical recording apparatuses, *Acta Math. Acad. Sci. Hung.* **6,** 363 (1955).
[5] On a sojourn time problem, *TVP* **3,** 61 (1958).
[6] *Introduction to the theory of queues*, Oxford University Press, 1962.

P. C. Tang

[1] The power function of the analysis of variance tests with tables and illustrations of their use, *Stat. Res. Memoir* **2,** 126 (1938).

H. Teicher

[1] On the convergence of projected distributions, *Ann. Inst. Stat., Math.* **9,** 79 (1958).

W. R. Thompson

[1] On confidence ranges for the median and other expectation distributions for populations of unknown distribution form, *AMS* **7,** 122 (1936).

G. Tintner

[1] Multiple regression for systems of equations, *Econometrica*, **14,** 5 (1946).

L. H. C. Tippet

[1] On the extreme individuals and the range of samples taken from a normal population, *B* **17,** 364 (1925).
[2] *Random sampling numbers*, Tracts for computers, No. 15, 1927.

K. D. Tocher

[1] Extension of the Neyman-Pearson theory of tests to discontinuous variates, *B* **37**, 130 (1950).

C. K. Tsao

[1] An extension of Massey's distribution of the maximum deviation between two-sample cumulative step functions, *AMS* **25**, 587 (1954).

N. N. Tschentsov

[1] Weak convergence of random processes whose trajectories have no discontinuities of the second kind (in Russian), *TVP* **1**, 155 (1956).

A. A. Tschuprov

[1] On the mathematical expectation of the moments of the frequency distributions in the case of correlated observations, *Metron* **2**, 4,646 (1923).

H. G. Tucker

[1] A generalization of the Glivenko-Cantelli theorem, *AMS* **30**, 1267 (1959).

S. H. Tumanian

[1] Asymptotic distribution of the χ^2 criterion in the case when the number of groups increases simultaneously with the number of observations (in Russian), *TVP* **1**, 131 (1956).

W. Uhlmann

[1] Zu einem nichtparametrischen Test von E. L. Lehmann, *Metrika* **2**, 3, 169 (1959).

K. Urbanik

[1] Limit properties of homogeneous Markoff processes with a denumerable set of states, *Bull. Pol. Acad. Sci.*, **2**, 371 (1954).

[2] On a problem of the theory of birth and death processes (in Russian), *Acta. Math. Acad. Sci. Hung.* **7**, 99 (1956).

J. V. Uspensky

[1] *Introduction to mathematical probability*, McGraw-Hill, New York, 1937.

H. R. van der Vaart

[1] Some remarks on the power function of Wilcoxon's test for the problem of two samples, *Proc. Royal Inst. Acad. Sci. ser.* A, **I, 53**, 494 (1950); **II, 53**, 507 (1950).

[2] On the characteristic functions of absolutely continuous distribution functions, *AMS* **33**, 824 (1962).

V. S. Varadarajan

[1] Weak convergence of measures on separable metric spaces, *Sankhya* **19**, 15 (1958).

[2] On the convergence of sample probability distributions, *Sankhya* **19**, 23 (1958).

J. Venn

[1] *The logic of chance*, Recent edition, Chelsea, New York, 1962.

A. Vessereau

[1] Sur les conditions d'application de criterion χ^2 de Pearson, *Rev. Stat. Appl.* **6**, No. 2, 83 (1958).

E. Vielrose

[1] *Tablice liczb losowych*, GUS, Warsaw, 1951.

I. Vincze

[1] Einige zweidimensionale Verteilungs—und Grenzverteilungsgesetze in der Theorie der geordneten Stichproben, *Publ. Math. Inst. Hung. Acad. Sci.* **2**, 183 (1957).

[2] On some joint distributions and joint limiting distributions in the theory of order statistics, *Publ. Math. Inst. Hung. Acad. Sci.* **4**, 29 (1959).

B. L. van der Waerden

[1] Testing a distribution function, *Proc. Royal Inst. Acad. Sci.*, ser. A, **56**, 201 (1953).

[2] Order tests for the two-sample problem, *Proc. Royal Inst. Sci.*, *ser.* A, **I**, **55**, 453 (1952); **II**, **56**, 303 (1953); **III**, **56**, 311 (1953).

A. Wald

[1] An extension of Wilks' method for setting tolerance limits, *AMS* **14**, 45 (1943).

[2] On cumulative sums of random variables, *AMS* **15**, 283 (1944).

[3] Differentiation under the expectation sign in the fundamental identity of sequential analysis, *AMS* **17**, 493 (1946).

[4] *Sequential analysis*, John Wiley, New York, and Chapman and Hall, London, 1947.

[5] Note on the consistency of the maximum likelihood estimate, *AMS* **20**, 595 (1949).

[6] *Statistical decision functions*, John Wiley, New York, 1950.

A. Wald, J. Wolfowitz

[1] Confidence limits for continuous distribution functions, *AMS* **10**, 105 (1939).

[2] On a test whether two samples are from the same population, *AMS* **11**, 147 (1940).

[3] Tolerance limits for a normal distribution, *AMS* **17**, 208 (1946).

[4] Optimum character of the sequential probability ratio test, *AMS* **19**, 326 (1948).

W. A. Wallis

[1] Techniques of statistical analysis, *Statistical Research Group, Columbia University, New York*, 1946, Chapter 17.

[2] Tolerance intervals for linear regression, *Proc. Second Berkeley Symp.* 43 (1951).

S. J. Wang

[1] On the limiting distribution of the ratio of two empirical distributions, *Acta Math. Sinica* **5**, 253 (1955).

G. S. Watson

[1] The χ^2 goodness-of-fit test for normal distributions, *B* **44**, 336 (1957).

[2] On chi-square goodness-of-fit tests for continuous distributions, *JRSS B* **20**, 44 (1958).

B. L. Welch

[1] The significance of the difference between two means when the population variances are unequal, *B* **29**, 350 (1938).

N. Wiener

[1] Differential space, *J. Math. Phys. Math. Inst. Techn.* **2**, 131 (1923).

[2] Generalized harmonic analysis, *Acta Math.* **55**, 117 (1930).

[3] *Extrapolation, interpolation and smoothing of stationary time series, with engineering applications.* Cambridge-New York, 1949.

N. Wiener, P. Masani

[1] The prediction theory of multivariate stochastic processes, I, *Acta Math.* **98**, 111 (1957); II, *Acta Math.* **99**, 93 (1958).

F. Wilcoxon

[1] Individual comparisons by ranking methods, *Biometrics Bull.* **1**, 80 (1945).

S. S. Wilks

[1] Determination of sample size for setting tolerance limits, *AMS* **12**, 91 (1941).

[2] Certain generalisations in the analysis of variance, *B* **24**, 471 (1932).

C. A. Williams

[1] On the choice of the number and width of classes for the chi-square test of goodness of fit, *J. Amer. Stat. Assn.* **45**, 77 (1950).

J. Wishart

[1] The generalized product-moment distribution in samples from a normal multivariate population, *B* **20** A, 32 (1928).

J. Wishart, H. O. Hirschfeld

[1] A theorem concerning the distribution of joins between line segments, *J. London Math. Soc.* **11**, 227 (1936).

W. Wiszniewski

[1] Pogoda naszych zim, *Gazeta Obserwatora PIHM* **1**, No. 3 (1948).

H. Wold

[1] *A study in the analysis of stationary time series*, Almquist and Wiksell, Uppsala, 1938.

J. Wolfowitz

[1] Additive partition functions and a class of statistical hypotheses, *AMS* **13**, 247 (1942).

[2] The efficiency of sequential estimates and Wald's equation for sequential processes, *AMS* **18**, 215 (1947).

[3] On Wald's proof of the consistency of the maximum likelihood estimate, *AMS* **20**, 602 (1949).

[4] Nonparametric statistical inference, *Proc. Berkeley Symp.*, 93 (1949).

[5] Generalization of a theorem of Glivenko-Cantelli, *AMS* **25**, 131 (1954).

[6] Estimation by the minimum distance method in nonparametric stochastic difference equations, *AMS* **25**, 203 (1954).

[7] The minimum distance method, *AMS* **28**, 75 (1957).

[8] Convergence of the empiric distribution function on half spaces, *Contribution to Probability and Statistics, Essays in Honor of Harold Hotelling*, Stanford University Press, 1960, p. 198.

W. N. Woods, A. T. Bharucha-Reid

[1] Age dependent branching stochastic processes in cascade theory, II, *Il Nuovo Cimento, ser* 10, **10**, 569 (1958).

A. M. Yaglom

[1] Introduction to the theory of stationary random functions (in Russian), *UMN* **7**, No. 5, 3 (1952).

G. U. Yule

[1] On the association of attributes in statistics, *Phil. Trans.* A **194**, 257 (1900).

[2] On the methods of measuring the association between two attributes, *JRSS* **75**, 579 (1912).

[3] A mathematical theory of evolution, based on conclusions of Dr. J. C. Willis, FRS, *Phil. Trans. Royal Soc. London, ser.* B, **213**, 21 (1924).

A. A. Yushkevitch

[1] On the differentiability of transition probabilities of homogeneous Markov processes with a countable set of states (in Russian), *Ucenye Zapisky MGU* **9**, 186, 141 (1959).

A. A. Zinger

[1] On independent samples from a normal population (in Russian), *UMN* **6**, 172 (1951).

A. A. Zinger, Yu. V. Linnik

[1] On an analytic extension of a theorem of Cramér and its applications (in Russian), *Vestnik Leningr. Univ.* **10**, 51 (1855).

S. Zubrzycki

[1] Remarks on random, stratified and systematic sampling in a plane, *Coll. Math.* **6**, 251 (1958).

A. Zygmund

[1] A remark on characteristic functions, *AMS* **18**, 272 (1947).
[2] A remark on characteristic functions, *Proc. Second Berkeley Symp.* 369 (1951).

Statistical Tables

STATISTICAL TABLES

TABLE I

POISSON DISTRIBUTION

The probability function of a random variable X with a Poisson distribution.

$$P(X = r) = \frac{\lambda^r}{r!}\, e^{-\lambda}$$

r	λ 0.1	0.2	0.3	0.4	0.5	0.6	0.7	0.8
0	0.904837	0.818731	0.740818	0.670320	0.606531	0.548812	0.496585	0.449329
1	0.090484	0.163746	0.222245	0.268128	0.303265	0.329287	0.347610	0.359463
2	0.004524	0.016375	0.033337	0.053626	0.075816	0.098786	0.121663	0.143785
3	0.000151	0.001092	0.003334	0.007150	0.012636	0.019757	0.028388	0.038343
4	0.000004	0.000055	0.000250	0.000715	0.001580	0.002964	0.004968	0.007669
5	—	0.000002	0.000015	0.000057	0.000158	0.000356	0.000696	0.001227
6	—	—	0.000001	0.000004	0.000013	0.000036	0.000081	0.000164
7	—	—	—	—	0.000001	0.000003	0.000008	0.000019
8	—	—	—	—	—	—	0.000001	0.000002

r	λ 0.9	1.0	1.5	2.0	2.5	3.0	3.5	4.0
0	0.406570	0.367879	0.223130	0.135335	0.082085	0.049787	0.030197	0.018316
1	0.365913	0.367879	0.334695	0.270671	0.205212	0.149361	0.105691	0.073263
2	0.164661	0.183940	0.251021	0.270671	0.256516	0.224042	0.184959	0.146525
3	0.049398	0.061313	0.125510	0.180447	0.213763	0.224042	0.215785	0.195367
4	0.011115	0.015328	0.047067	0.090224	0.133602	0.168031	0.188812	0.195367
5	0.002001	0.003066	0.014120	0.036089	0.066801	0.100819	0.132169	0.156293
6	0.000300	0.000511	0.003530	0.012030	0.027834	0.050409	0.077098	0.104196
7	0.000039	0.000073	0.000756	0.003437	0.009941	0.021604	0.038549	0.059540
8	0.000004	0.000009	0.000142	0.000859	0.003106	0.008102	0.016865	0.029770
9	—	0.000001	0.000024	0.000191	0.000863	0.002701	0.006559	0.013231
10	—	—	0.000004	0.000038	0.000216	0.000810	0.002296	0.005292
11	—	—	—	0.000007	0.000049	0.000221	0.000730	0.001925
12	—	—	—	0.000001	0.000010	0.000055	0.000213	0.000642
13	—	—	—	—	0.000002	0.000013	0.000057	0.000197
14	—	—	—	—	—	0.000003	0.000014	0.000056
15	—	—	—	—	—	0.000001	0.000003	0.000015
16	—	—	—	—	—	—	0.000001	0.000004
17	—	—	—	—	—	—	—	0.000001

r	λ 4.5	5.0	6.0	7.0	8.0	9.0	10.0
0	0.011109	0.006738	0.002479	0.000912	0.000335	0.000123	0.000045
1	0.049990	0.033690	0.014873	0.006383	0.002684	0.001111	0.000454
2	0.112479	0.084224	0.044618	0.022341	0.010735	0.004998	0.002270
3	0.168718	0.140374	0.089235	0.052129	0.028626	0.014994	0.007567
4	0.189808	0.175467	0.133853	0.091226	0.057252	0.033737	0.018917
5	0.170827	0.175467	0.160623	0.127717	0.091604	0.060727	0.037833
6	0.128120	0.146223	0.160623	0.149003	0.122138	0.091090	0.063055
7	0.082363	0.104445	0.137677	0.149003	0.139587	0.117116	0.090079
8	0.046329	0.065278	0.103258	0.130377	0.139587	0.131756	0.112599
9	0.023165	0.036266	0.068838	0.101405	0.124077	0.131756	0.125110
10	0.010424	0.018133	0.041303	0.070983	0.099262	0.118580	0.125110
11	0.004264	0.008242	0.022529	0.045171	0.072190	0.097020	0.113736
12	0.001599	0.003434	0.011264	0.026350	0.048127	0.072765	0.094780
13	0.000554	0.001321	0.005199	0.014188	0.029616	0.050376	0.072908
14	0.000178	0.000472	0.002228	0.007094	0.016924	0.032384	0.052077
15	0.000053	0.000157	0.000891	0.003311	0.009026	0.019431	0.034718
16	0.000015	0.000049	0.000334	0.001448	0.004513	0.010930	0.021699
17	0.000004	0.000014	0.000118	0.000596	0.002124	0.005786	0.012764
18	0.000001	0.000004	0.000039	0.000232	0.000944	0.002893	0.007091
19	—	0.000001	0.000012	0.000085	0.000397	0.001370	0.003732
20	—	—	0.000004	0.000030	0.000159	0.000617	0.001866
21	—	—	0.000001	0.000010	0.000061	0.000264	0.000889
22	—	—	—	0.000003	0.000022	0.000108	0.000404
23	—	—	—	0.000001	0.000008	0.000042	0.000176
24	—	—	—	—	0.000003	0.000016	0.000073
25	—	—	—	—	0.000001	0.000006	0.000029
26	—	—	—	—	—	0.000002	0.000011
27	—	—	—	—	—	0.000001	0.000004
28	—	—	—	—	—	—	0.000001
29	—	—	—	—	—	—	0.000001

TABLE II

THE NORMAL DISTRIBUTION

The density of a random variable X with the normal distribution $N(0; 1)$.

$$\phi(x) = \frac{1}{\sqrt{2\pi}} e^{-x^2/2}$$

x	$\phi(x)$	x	$\phi(x)$	x	$\phi(x)$	x	$\phi(x)$	x	$\phi(x)$	x	$\phi(x)$
0.00	0.3989	0.50	0.3521	1.00	0.2420	1.50	0.1295	2.00	0.0540	2.50	0.0175
0.05	0.3984	0.55	0.3429	1.05	0.2299	1.55	0.1200	2.05	0.0488	2.55	0.0154
0.10	0.3970	0.60	0.3332	1.10	0.2179	1.60	0.1109	2.10	0.0440	2.60	0.0136
0.15	0.3945	0.65	0.3230	1.15	0.2059	1.65	0.1023	2.15	0.0396	2.65	0.0119
0.20	0.3910	0.70	0.3123	1.20	0.1942	1.70	0.0940	2.20	0.0355	2.70	0.0104
0.25	0.3867	0.75	0.3011	1.25	0.1826	1.75	0.0863	2.25	0.0317	·2.75	0.0091
0.30	0.3814	0.80	0.2897	1.30	0.1714	1.80	0.0790	2.30	0.0283	2.80	0.0079
0.35	0.3752	0.85	0.2780	1.35	0.1604	1.85	0.0721	2.35	0.0252	2.85	0.0069
0.40	0.3683	0.90	0.2661	1.40	0.1497	1.90	0.0656	2.40	0.0224	2.90	0.0060
0.45	0.3605	0.95	0.2541	1.45	0.1394	1.95	0.0596	2.45	0.0198	2.95	0.0051
										3.00	0.0044

TABLE III

THE NORMAL DISTRIBUTION

The distribution function of a random variable X with the normal distribution $N(0; 1)$.

$$\Phi(x) = \frac{1}{\sqrt{2\pi}} \int_{-\infty}^{x} e^{-t^2/2} \, dt$$

x	$\Phi(x)$	x	$\Phi(x)$	x	$\Phi(x)$	x	$\Phi(x)$	x	$\Phi(x)$	x	$\Phi(x)$
0.00	0.500000	0.50	0.691463	1.00	0.841345	1.50	0.933193	2.00	0.977250	2.50	0.993790
0.05	0.519939	0.55	0.708840	1.05	0.853141	1.55	0.939429	2.05	0.979818	2.55	0.994614
0.10	0.539828	0.60	0.725747	1.10	0.864334	1.60	0.945201	2.10	0.982136	2.60	0.995339
0.15	0.559618	0.65	0.742154	1.15	0.874928	1.65	0.950528	2.15	0.984222	2.65	0.995975
0.20	0.579260	0.70	0.758036	1.20	0.884930	1.70	0.955434	2.20	0.986097	2.70	0.996533
0.25	0.589706	0.75	0.773373	1.25	0.894350	1.75	0.959941	2.25	0.987776	2.75	0.997020
0.30	0.617911	0.80	0.788145	1.30	0.903200	1.80	0.964070	2.30	0.989276	2.80	0.997445
0.35	0.636831	0.85	0.802338	1.35	0.911492	1.85	0.967843	7.35	0.990613	2.85	0.997814
0.40	0.655422	0.90	0.815940	1.40	0.919243	1.90	0.971283	2.40	0.991802	2.90	0.998134
0.45	0.673645	0.95	0.828944	1.45	0.926471	1.95	0.974412	2.45	0.992857	2.95	0.998411
										3.00	0.998650

TABLE IV

THE χ^2 DISTRIBUTION

The table gives the values of χ_α^2 for some values of α, where the probability that a random variable χ^2 having the χ^2-distribution with n degrees of freedom is not smaller than χ_α^2 is equal to α.

$$P(\chi^2 \geqslant \chi_\alpha^2) = \frac{1}{2^{n/2}\Gamma(\tfrac{1}{2}n)} \int_{\chi_\alpha^2}^{\infty} e^{-u^2} u^{n/2-1} \, du = \alpha$$

n	α								
	0.80	0.70	0.50	0.30	0.20	0.10	0.05	0.02	0.01
1	0.064	0.148	0.455	1.074	1.642	2.706	3.841	5.412	6.635
2	0.446	0.713	1.386	2.408	3.219	4.605	5.991	7.824	9.210
3	1.005	1.424	2.366	3.665	4.642	6.251	7.815	9.837	11.345
4	1.649	2.195	3.357	4.878	5.989	7.779	9.488	11.668	13.277
5	2.343	3.000	4.351	6.064	7.289	9.236	11.070	13.388	15.086
6	3.070	3.828	5.348	7.231	8.558	10.645	12.592	15.033	16.812
7	3.822	4.671	6.346	8.383	9.803	12.017	14.067	16.622	18.475
8	4.594	5.527	7.344	9.524	11.030	13.362	15.507	18.168	20.090
9	5.380	6.393	8.343	10.656	12.242	14.684	16.919	19.679	21.666
10	6.179	7.267	9.342	11.781	13.442	15.987	18.307	21.161	23.209
11	6.989	8.148	10.341	12.899	14.631	17.275	19.675	22.618	24.725
12	7.807	9.034	11.340	14.011	15.812	18.549	21.026	24.054	26.217
13	8.634	9.926	12.340	15.119	16.985	19.812	22.362	25.472	27.688
14	9.467	10.821	13.339	16.222	18.151	21.064	23.685	26.873	29.141
15	10.307	11.721	14.339	17.322	19.311	22.307	24.996	28.259	30.578
16	11.152	12.624	15.338	18.418	20.465	23.542	26.296	29.633	32.000
17	12.002	13.531	16.338	19.511	21.615	24.769	27.587	30.995	33.409
18	12.857	14.440	17.338	20.601	22.760	25.989	28.869	32.346	34.805
19	13.716	15.352	18.338	21.689	23.900	27.204	30.144	33.687	36.191
20	14.578	16.266	19.337	22.775	25.038	28.412	31.410	35.020	37.566
21	15.445	17.182	20.337	23.858	26.171	29.615	32.671	36.343	38.932
22	16.314	18.101	21.337	24.939	27.301	30.813	33.924	37.659	40.289
23	17.187	19.021	22.337	26.018	28.429	32.007	35.172	38.968	41.638
24	18.062	19.943	23.337	27.096	29.553	33.196	36.415	40.270	42.980
25	18.940	20.867	24.337	28.172	30.675	34.382	37.652	41.566	44.314
26	19.820	21.792	25.336	29.246	31.795	35.563	38.885	42.856	45.642
27	20.703	22.719	26.336	30.319	32.912	36.741	40.113	44.140	46.963
28	21.588	23.647	27.336	31.391	34.027	37.916	41.337	45.419	48.278
29	22.475	24.577	28.336	32.461	35.139	39.087	42.557	46.693	49.588
30	23.364	25.508	29.336	33.530	36.250	40.256	43.773	47.962	50.892

TABLE V

STUDENT'S t-DISTRIBUTION

The table gives the values of t_α for some values of α, where the probability that a random variable t having Student's t-distribution with n degrees of freedom is not smaller than t_α in absolute value is equal to α.

$$P(|t| \geqslant t_\alpha) = \frac{2}{\sqrt{n}\, B(\frac{1}{2}, \frac{1}{2}n)} \int_{t_\alpha}^{\infty} \frac{1}{(1 + t^2/n)^{(n+1)/2}}\, dt = \alpha$$

n	α							
	0.80	0.60	0.40	0.20	0.10	0.05	0.02	0.01
1	0.325	0.727	1.376	3.078	6.314	12.706	31.821	63.657
2	0.289	0.617	1.061	1.886	2.920	4.303	6.965	9.925
3	0.277	0.584	0.978	1.638	2.353	3.182	4.541	5.841
4	0.271	0.569	0.941	1.533	2.132	2.776	3.747	4.604
5	0.267	0.559	0.920	1.476	2.015	2.571	3.365	4.032
6	0.265	0.553	0.906	1.440	1.943	2.447	3.143	3.707
7	0.263	0.549	0.896	1.415	1.895	2.365	2.998	3.499
8	0.262	0.546	0.889	1.397	1.860	2.306	2.896	3.355
9	0.261	0.543	0.883	1.383	1.833	2.262	2.821	3.250
10	0.260	0.542	0.879	1.372	1.812	2.228	2.764	3.169
11	0.260	0.540	0.876	1.363	1.796	2.201	2.718	3.106
12	0.259	0.539	0.873	1.356	1.782	2.179	2.681	3.055
13	0.259	0.538	0.870	1.350	1.771	2.160	2.650	3.012
14	0.258	0.537	0.868	1.345	1.761	2.145	2.624	2.977
15	0.258	0.536	0.866	1.341	1.753	2.131	2.602	2.947
16	0.258	0.535	0.865	1.337	1.746	2.120	2.583	2.921
17	0.257	0.534	0.863	1.333	1.740	2.110	2.567	2.898
18	0.257	0.534	0.862	1.330	1.734	2.101	2.552	2.878
19	0.257	0.533	0.861	1.328	1.729	2.093	2.539	1.861
20	0.257	0.533	0.860	1.325	1.725	2.086	2.528	2.845
21	0.257	0 532	0.859	1.323	1.721	2.080	2.518	2.831
22	0.256	0.532	0.858	1.321	1.717	2.074	2.508	2.819
23	0.256	0.532	0.858	1.319	1.714	2.069	2.500	2.807
24	0.256	0.531	0.857	1.318	1.711	2.064	2.492	2.797
25	0.256	0.531	0.856	1.316	1.708	2.060	2.485	2.787
26	0.256	0.531	0.856	1.315	1.706	2.056	2.479	2.779
27	0.256	0.531	0.855	1.314	1.703	2.052	2.473	2.771
28	0.256	0.530	0.855	1.313	1.701	2.048	2.467	2.763
29	0.256	0.530	0.854	1.311	1.699	2.045	2.462	2.756
30	0.256	0.530	0.854	1.310	1.697	2.042	2.457	2.750
40	0.255	0.529	0.851	1.303	1.684	2.021	2.423	2.704
60	0.254	0.527	0.848	1.296	1.671	2.000	2.390	2.660
120	0.254	0.526	0.845	1.289	1.658	1.980	2.358	2.617
∞	0.253	0.524	0.842	1.282	1.645	1.960	2.326	2.576

TABLE VI

FISHER'S Z-DISTRIBUTION

The table gives the values of z_0 such that the probability that a random variable with Fisher's Z-distribution with (r_1, r_2) degrees of freedom is not smaller than z_0 equals 0.05;

$$P(Z \geqslant z_0) = \int_{z_0}^{\infty} f(z)\, dz = 0.05,$$

where $f(z)$ is given by (9.7.5).

r_2	r_1									
	1	2	3	4	5	6	8	12	24	∞
1	2.5421	2.6479	2.6870	2.7071	2.7194	2.7276	2.7380	2.7494	2.7588	2.7693
2	1.4592	1.4722	1.4765	1.4787	1.4800	1.4808	1.4819	1.4830	1.4840	1.4851
3	1.1577	1.1284	1.1137	1.1051	1.0994	1.0953	1.0899	1.0842	1.0781	1.0716
4	1.0212	0.9690	0.9429	0.9272	0.9168	0.9093	0.8993	0.8885	0.8767	0 8639
5	0.9441	0.8777	0.8441	0.8236	0.8097	0.7997	0.7862	0.7714	0.7550	0.7368
6	0.8948	0.8188	0.7798	0.7558	0.7394	0.7274	0.7112	0.6931	0.6729	0.6499
7	0.8606	0.7777	0.7347	0.7080	0.6896	0.6761	0.6576	0.6369	0.6134	0.5862
8	0.8355	0.7475	0.7014	0.6725	0.6525	0.6378	0.6175	0.5945	0.5682	0.5371
9	0.8163	0.7242	0.6757	0.6450	0.6238	0.6080	0.5862	0.5613	0.5324	0.4979
10	0.8012	0.7058	0.6553	0.6232	0.6009	0.5843	0.5611	0.5346	0.5035	0.4657
11	0.7889	0.6909	0.6387	0.6055	0.5822	0.5648	0.5406	0.5126	0.4795	0.4387
12	0.7788	0.6786	0.6250	0.5907	0.5666	0.5487	0.5234	0.4941	0.4592	0.4156
13	0.7703	0.6682	0.6134	0.5783	0.5535	0.5350	0.5089	0.4785	0.4419	0.3957
14	0.7630	0.6594	0.6036	0.5677	0.5423	0.5233	0.4964	0.4649	0.4269	0.3782
15	0.7568	0.6518	0.5950	0.5585	0.5326	0.5131	0.4855	0.4532	0.4138	0.3628
16	0.7514	0.6451	0.5876	0.5505	0.5241	0.5042	0.4760	0.4428	0.4022	0.3490
17	0.7466	0.6393	0.5811	0.5434	0.5166	0.4964	0.4676	0.4337	0.3919	0.3366
18	0.7424	0.6341	0.5753	0.5371	0.5099	0.4894	0.4602	0.4255	0.3827	0.3253
19	0.7386	0.6295	0.5701	0.5315	0.5040	0.4832	0.4535	0.4182	0.3743	0.3151
20	0.7352	0.6254	0.5654	0.5265	0.4986	0.4776	0.4474	0.4116	0.3668	0.3057
21	0.7322	0.6216	0.5612	0.5219	0.4938	0.4725	0.4420	0.4055	0.3599	0.2971
22	0.7294	0.6182	0 5574	0.5178	0.4894	0.4679	0.4370	0.4001	0.3536	0.2892
23	0.7269	0.6151	0.5540	0.5140	0.4854	0.4636	0.4325	0.3950	0.3478	0.2818
24	0.7246	0.6123	0.5508	0.5106	0.4817	0.4598	0.4283	0.3904	0.3425	0.2749
25	0.7225	0.6097	0.5478	0.5074	0.4783	0.4562	0.4244	0.3862	0.3376	0.2685
26	0.7205	0.6073	0.5451	0.5045	0.4752	0.4529	0.4209	0.3823	0.3330	0.2625
27	0.7187	0.6051	0.5427	0.5017	0.4723	0.4499	0.4176	0.3786	0.3287	0.2569
28	0.7171	0.6030	0.5403	0.4992	0.4696	0.4471	0.4146	0.3752	0.3248	0.2516
29	0.7155	0.6011	0.5382	0.4969	0.4671	0.4444	0.4117	0.3720	0.3211	0.2466
30	0.7141	0.5994	0.5362	0.4947	0.4648	0.4420	0.4090	0.3691	0.3176	0.2419
40	0.7037	0.5866	0.5217	0.4789	0.4479	0.4242	0.3897	0.3475	0.2920	0.2057
60	0.6933	0.5738	0.5073	0.4632	0.4311	0.4064	0.3702	0.3255	0.2654	0.1644
20	0.6830	0.5611	0.4930	0.4475	0.4143	0.3885	0.3506	0.3032	0.2376	0.1131
∞	0.6729	0.5486	0.4787	0.4319	0.3974	0.3706	0.3309	0.2804	0.2085	0.0000

TABLE VII

FISHER'S Z-DISTRIBUTION

The table gives the values of z_0 such that the probability that a random variable with Fisher's Z-distribution with (r_1, r_2) degrees of freedom is not smaller than z_0 equals 0.01;

$$P(Z \geq z_0) = \int_{z_0}^{\infty} f(z)\, dz = 0.01,$$

where $f(z)$ is given by (9.7.5).

r_2	r_1									
	1	2	3	4	5	6	8	12	24	∞
1	4.1535	4.2585	4.2974	4.3175	4.3297	4.3379	4.3482	4.3585	4.3689	4.3794
2	2.2950	2.2976	2.2984	2.2988	2.2991	2.2992	2.2994	2.2997	2.2999	2.3001
3	1.7649	1.7140	1.6915	1.6786	1.6703	1.6645	1.6569	1.6489	1.6404	1.6314
4	1.5270	1.4452	1.4075	1.3856	1.3711	1.3609	1.3473	1.3327	1.3170	1.3000
5	1.3943	1.2929	1.2449	1.2164	1.1974	1.1838	1.1656	1.1457	1.1239	1.0997
6	1.3103	1.1955	1.1401	1.1068	1.0843	1.0680	1.0460	1.0218	0.9948	0.9643
7	1.2526	1.1281	1.0682	1.0300	1.0048	0.9864	0.9614	0.9335	0.9020	0.8658
8	1.2106	1.0787	1.0135	0.9734	0.9459	0.9250	0.8983	0.8673	0.8319	0.7904
9	1.1786	1.0411	0.9724	0.9299	0.9006	0.8791	0.8494	0.8157	0.7769	0.7305
10	1.1535	1.0114	0.9399	0.8954	0.8646	0.8419	0.8104	0.7744	0.7324	0.6816
11	1.1333	0.9874	0.9136	0.8674	0.8354	0.8116	0.7785	0.7405	0.6958	0.6408
12	1.1166	0.9677	0.8919	0.8443	0.8111	0.7864	0.7520	0.7122	0.6649	0.6061
13	1.1027	0.9511	0.8737	0.8248	0.7907	0.7652	0.7295	0.6882	0.6386	0.5761
14	1.0909	0.9370	0.8581	0.8082	0.7732	0.7471	0.7103	0.6675	0.6159	0.5500
15	1.0807	0.9249	0.8448	0.7939	0.7582	0.7314	0.6937	0.6496	0.5961	0.5269
16	1.0719	0.9144	0.8331	0.7814	0.7450	0.7177	0.6791	0.6339	0.5786	0.5064
17	1.0641	0.9051	0.8229	0.7705	0.7335	0.7057	0.6663	0.6199	0.5630	0.4879
18	1.0572	0.8970	0.8138	0.7607	0.7232	0.6950	0.6549	0.6075	0.5491	0.4712
19	1.0511	0.8897	0.8057	0.7521	0.7140	0.6854	0.6447	0.5964	0.5366	0.4560
20	1.0457	0.8831	0.7985	0.7443	0.7058	0.6768	0.6355	0.5864	0.5253	0.4421
21	1.0408	0.8772	0.7920	0.7372	0.6984	0.6690	0.6272	0.5773	0.5150	0.4294
22	1.0363	0.8719	0.7860	0.7309	0.6916	0.6620	0.6196	0.5691	0.5056	0.4176
23	1.0322	0.8670	0.7806	0.7251	0.6855	0.6555	0.6127	0.5615	0.4969	0.4068
24	1.0285	0.8626	0.7757	0.7197	0.6799	0.6496	0.6064	0.5545	0.4890	0.3967
25	1.0251	0.8585	0.7712	0.7148	0.6747	0.6442	0.6006	0.5481	0.4816	0.3872
26	1.0220	0.8548	0.7670	0.7103	0.6699	0.6392	0.5952	0.5422	0.4748	0.3784
27	1.0191	0.8513	0.7631	0.7062	0.6655	0.6346	0.5902	0.5367	0.4685	0.3701
29	1.0164	0.8481	0.7595	0.7023	0.6614	0.6303	0.5856	0.5316	0.4626	0.3624
29	1.0139	0.8451	0.7562	0.6987	0.6576	0.6263	0.5813	0.5269	0.4570	0.3550
30	1.0116	0.8423	0.7531	0.6954	0.6540	0.6226	0.5773	0.5224	0.4519	0.3481
40	0.9949	0.8223	0.7307	0.6712	0.6283	0.5956	0.5481	0.4901	0.4138	0.2952
60	0.9784	0.8025	0.7086	0.6472	0.6028	0.5687	0.5189	0.4574	0.3746	0.2352
120	0.9622	0.7829	0.6867	0.6234	0.5774	0.5419	0.4897	0.4243	0.3339	0.1612
∞	0.9462	0.7636	0.6651	0.5999	0.5522	0.5152	0.4604	0.3908	0.2913	0.0000

TABLE VIII

THE KOLMOGOROV-SMIRNOV λ-DISTRIBUTION

The table gives the values of the function $Q(\lambda)$ defined by the formula

$$Q(\lambda) = \sum_{k=-\infty}^{\infty} (-1)^k e^{-2k^2\lambda^2}$$

λ	$Q(\lambda)$	λ	$Q(\lambda)$	λ	$Q(\lambda)$	λ	$Q(\lambda)$	λ	$Q(\lambda)$	λ	$Q(\lambda)$
0.32	0.0000	0.66	0.2236	1.00	0.7300	1.34	0.9449	1.68	0.9929	2.00	0.9993
0.33	0.0001	0.67	0.2396	1.01	0.7406	1.35	0.9478	1.69	0.9934	2.01	0.9994
0.34	0.0002	0.68	0.2558	1.02	0.7508	1.36	0.9505	1.70	0.9938	2.02	0.9994
0.35	0.0003	0.69	0.2722	1.03	0.7608	1.37	0.9531	1.71	0.9942	2.03	0.9995
0.36	0.0005	0.70	0.2888	1.04	0.7704	1.38	0.9556	1.72	0.9946	2.04	0.9995
0.37	0.0008	0.71	0.3055	1.05	0.7798	1.39	0.9580	1.73	0.9950	2.05	0.9996
0.38	0.0013	0.72	0.3223	1.06	0.7889	1.40	0.9603	1.74	0.9953	2.06	0.9996
0.39	0.0019	0.73	0 3391	1.07	0.7976	1.41	0.9625	1.75	0.9956	2.07	0.9996
0.40	0.0028	0.74	0.3560	1.08	0.8061	1.42	0.9646	1.76	0.9959	2.08	0.9996
0.41	0.0040	0.75	0.3728	1.09	0.8143	1.43	0.9665	1.77	0.9962	2.09	0.9997
0.42	0.0055	0.76	0.3896	1.10	0.8223	1.44	0.9684	1.78	0.9965	2.10	0.9997
0.43	0.0074	0.77	0.4064	1.11	0.8299	1.45	0.9702	1.79	0.9967	2.11	0.9997
0.44	0.0097	0.78	0.4230	1.12	0.8374	1.46	0.9718	1.80	0.9969	2.12	0.9997
0.45	0.0126	0.79	0.4395	1.13	0.8445	1.47	0.9734	1.81	0.9971	2.13	0.9998
0.46	0.0160	0.80	0.4559	1.14	0.8514	1.48	0.9750	1.82	0.9973	2.14	0.9998
0.47	0.0200	0.81	0.4720	1.15	0.8580	1.49	0.9764	1.83	0.9975	2.15	0.9998
0.48	0.0247	0.82	0.4880	1.16	0.8644	1.50	0.9778	1.84	0.9977	2.16	0.9998
0.49	0.0300	0.83	0.5038	1.17	0.8706	1.51	0.9791	1.85	0.9979	2.17	0.9998
0.50	0.0361	0.84	0.5194	1.18	0.8765	1.52	0.9803	1.86	0.9980	2.18	0.9999
0.51	0.0428	0.85	0.5347	1.19	0.8823	1.53	0.9815	1.87	0 9981	2 19	0.9999
0.52	0.0503	0.86	0.5497	1.20	0.8877	1.54	0.9826	1.88	0.9983	2.20	0.9999
0.53	0.0585	0.87	0.5645	1.21	0.8930	1.55	0.9836	1.89	0.9984	2.21	0.9999
0.54	0.0675	0.88	0.5791	1.22	0.8981	1.56	0.9846	1.90	0.9985	2.22	0.9999
0.55	0.0772	0.89	0.5933	1.23	0.9030	1.57	0.9855	1.91	0.9986	2.23	0.9999
0.56	0 0876	0.90	0.6073	1.24	0.9076	1.58	0.9864	1.92	0.9987	2.24	0.9999
0.57	0.0987	0.91	0.6209	1.25	0.9121	1.59	0.9873	1.93	0.9988	2.25	0.9999
0.58	0.1104	0.92	0.6343	1.26	0.9164	1.60	0.9880	1.94	0.9989	2.26	0.9999
0.59	0.1228	0.93	0.6473	1.27	0.9206	1.61	0.9888	1.95	0.9990	2.27	0.9999
0.60	0.1357	0.94	0.6601	1.28	0.9245	1.62	0.9895	1.96	0.9991	2.28	0.9999
0.61	0.1492	0.95	0.6725	1.29	0.9283	1.63	0.9902	1.97	0.9991	2.29	0.9999
0.62	0.1632	0.96	0.6846	1.30	0.9319	1.64	0.9908	1.98	0.9992	2.30	0.9999
0.63	0.1778	0.97	0.6964	1.31	0.9354	1.65	0.9914	1.99	0.9993	2.31	1.0000
0.64	0.1927	0.98	0.7079	1.32	0.9387	1.66	0.9919				
0.65	0.2080	0.99	0.7191	1.33	0.9418	1.67	0.9924				

Author Index

665

Subject Index

Errata

Page and Line	For	Substitute
13, l. 12 from top	and e_k	and (e_k)
16, l. 4 from top	Is	It
130, l. 12 from bottom	so that	since here
214, l. 10 from top	6.6.1b	6.6.1a
241, l. 16, 17 from bottom	Show that	Check whether
244, l. 15 from top	tends	tends uniformly
246, l. 18 from top	(a) and (b)	(b)
280, l. 13 from top	$j < i$	$j > i$
281, l. 4-5 from bottom	the process . . . and the	X_t is a homogeneous Markov process and satisfies a
316, l. 2 from top	$R(\tau)$	$\overline{R(\tau)}$
322, l. 6 from bottom	$\dfrac{1}{2}\pi$	$\dfrac{1}{2\pi}$
327, l. 15 from top	$\underline{W_i(t)}$	$\underline{W_1(t)}$
341, l. 2 from top	for	with $\sigma = 1$ for
347, l. 17 from bottom	this sample	these samples
347, l. 13 from bottom	sample	samples
356, l. 18 from top	exceed	exceed or equal
385, l. 10 from top	$h_{kn}(x)$	$h_{kn}(z)$
388, l. 5 from bottom	dy	dy_1
394, l. 7 from bottom	$\displaystyle\sum_{k=-\infty}^{n}$	$\displaystyle\sum_{k=-\infty}^{\infty}$
404, l. 3 from top	exceed	exceed or equal
404, l. 5 from top	greater	not less
412, l. 13 from top	Prove	Prove (Malmquist [1])
412, l. 15 from top	$V_{j(j=1,\dots,n-1)}$	$V_j(j = 1, \dots, n - 1)$
426, l. 11 from top	more extreme	as extreme or more
464, l. 17 from top	$E(X)$	$E(X) = m$
465, l. 8 from bottom	$u, y_1, \dots, y_{n-1}$	$(u, y_1, \dots, y_{n-1})$
478, l. 13 from bottom	singular	degenerate
514, l. 4 from bottom	$\dfrac{1}{n}$	$= \dfrac{1}{n}$

Page and Line	For	Substitute
542, l. 9 from top	exceeding	not less than
553, l. 4 from top	$\geqslant$	$\leqslant$
567, l. 6, 8, 13 from bottom; 568, l. 3, 4 from top	$D_n{}^+$	$D_n{}^-$
567, l. 8 from bottom; 568, l. 8, 11 from top	$d_n{}^+$	$d_n{}^-$
579, l. 9 from top	$=$	$\leqslant$
582, l. 12 from bottom	observations	observations of
582, l. 11 from bottom	observations of	observations
582, l. 7 from bottom	decreasing	nondecreasing